PROCEEDINGS OF THE FIRST
MARCEL GROSSMANN MEETING ON GENERAL RELATIVITY

M A R C E L   G R O S S M A N N

9 April 1878 Budapest 7 September 1936 Zurich

Marcel Grossmann studied mathematics at the Zurich Polytechnikum and earned his doctorate in 1912. He was appointed professor of descriptive geometry at the Eidgenössische Technische Hochschule in 1907; he was a teacher of outstanding ability and gave to many mathematicians their training in geometry.

Marcel Grossmann was Albert Einstein's classmate. When Einstein sought to formulate mathematically his ideas on the general theory of relativity he turned to Grossmann for assistance. Grossmann introduced Einstein to the differential calculus, started by Elwin Bruno Christoffel (1864) and fully developed at the University of Padova by Gregorio Ricci Curbastro and Tullio Levi Civita (1901). The collaboration between Einstein and Grossmann is significantly documented in their article "Entwurf einer verallgemeinerten Relativitätstheorie und einer Theorie der Gravitation" in Zeit. für Mathem. und Phys. 62, 3 (1913).

By allowing the encounter of the mathematical achievements of the Italian geometers and the profound physical insight of Einstein, Marcel Grossmann facilitated the unique synthesis of mathematical and theoretical physics reached by Albert Einstein in the most elegant and powerful field theory of physics: The General Theory of Relativity.

PROCEEDINGS OF THE FIRST

MARCEL GROSSMANN MEETING ON GENERAL RELATIVITY

organized and held at the

INTERNATIONAL CENTRE FOR THEORETICAL PHYSICS, TRIESTE

and

ISTITUTO DI FISICA TEORICA, UNIVERSITY OF TRIESTE
7-12 July 1975

Edited by

REMO RUFFINI

Department of Physics
University of Rome

N·H
P C

1977

NORTH-HOLLAND PUBLISHING COMPANY
AMSTERDAM · NEW YORK · OXFORD

North-Holland ISBN: 0 7204 0707 9

Published by:
NORTH-HOLLAND PUBLISHING COMPANY – AMSTERDAM • NEW YORK • OXFORD

Distributors for the U.S.A. and Canada:
Elsevier North-Holland, Inc.
52 Vanderbilt Avenue
New York, N.Y. 10017

Library of Congress Cataloging in Publication Data

Marcel Grossmann Meeting on General Relativity, 1st, Trieste, 1975.
Proceedings of the First Marcel Grossmann Meeting on General Relativity.

Bibliography: p.
Includes index.
1. General relativity (Physics)--Congresses.
I. Ruffini, Remo. II. International Centre for Theoretical Physics. III. Triesto. Università. Istituto di fisica teorica.
QC173.6.M37 1975 530.1'1 76-57709
ISBN 0-7204-0707-9

PRINTED IN GREAT BRITAIN

PREFACE

The Marcel Grossmann meeting was conceived with the scope of reviewing recent advances in gravitation and general relativity with major emphasis on their mathematical foundations and their physical predictions. The main objective was to elicit contributions, deepening our understanding of space-time structure, as well as review the status of experiments verifying Einstein's theory of gravitation. The volume is a record of this meeting.

The contributions range from alternative approaches to the quantization of gravity, to the theoretical significance of gravitational experiments; from recent advances in vacuum polarization in strong gravitational fields, to alternative classical microphysical structure of space-time.

The first day of the meeting was dedicated to the quantization of gravitation. Four different treatments of this problem are presented in these proceedings: the covariant approach (P. van Nieuwenhuizen), the super-space, or rather the total of configuration space approach (D. Christodoulou), the string model approach (T. Regge and C. Teitelboim) and finally the geometric of symplectic formalism approach (J.M. Souriau). More on these topics is presented in the communications section: the possible advantage of using null co-ordinates formalism (C. Aragone), a review of the traditional canonical quantization programme (M. Pilati and C. Teitelboim) and more details on the symplectic structure approach (W. Szczyrba). Related communications deal with possible new unified approach to electromagnetism and gravitation (L. Halpern), on strong gravity (A. Inomata) and Segal's cosmology (J. Tarski).

From the wealth of the approaches presented, it may be concluded that this profound and fundamental problem of quantization of gravitation is still very much at the frontier of the theoretical understanding. No consensus has yet been reached either on the approach or the techniques to be followed. Each one of the contributions presented contains some desirable feature either in respect of the direct understanding of the physics involved, or in respect of elegance and simplicity of the formalism.

The second day and part of the third day of the meeting were devoted to the progress made in our understanding of general properties of solutions of Einstein's field equations and of space-time structure. The reports cover, once again, a vast variety of topics: general properties of singularities, conjectures and their entropy and on their connection to the arrow of time (R. Penrose), the description of the angular momentum, spin and gyromagnetic ratio of a relativistic system in the manifold of complex asymptotically shear free light cones (E. Newman), a new formulation of the asymptotic conditions of an isolated system (B. Schmidt), general properties of maximal surfaces in closed and open space times (D. Brill), an analysis of the field equations and singularity properties of Bianchi universes (A. Taub), a review of the advances towards the proof of the vacuum black hole uniqueness theorem (B. Carter) and finally, in a somewhat different line, a conceivable role for torsion in general relativity (F. Hehl).

More on these topics* is presented in the communications section: modification of black holes horizons due to external fields (R. Hajicek), background field formulation of general relativity (A. Quale), electromagnetic fields in stationary geometries (B. Lukács and Z. Perjes), radiation reaction in a freely falling charge

---

* Not contained in the proceedings are the reports presented at the meeting by E. Lifshitz, H. Sato, E. Schücking, A. Tomimatsu and Abdus Salam, since the material presented, though highly relevant to the work of the sessions concerned, had already appeared elsewhere in print.

D. Wilkins) and finally, contributions on possible effects of torsion and spin on general relativity (C. Lopez, E. Mielke, N. Hari Dass and V. Radhakrishnan, and A. Prasanna).

In the afternoon of the third day of the meeting, a round table was held on the advances in perturbation techniques in a given stationary background metric. The following reports of this round table are contained in these proceedings: electromagnetic perturbations of a rotating black hole in the Chandrasekhar formalism (S. Detweiler), electromagnetic perturbations in Reissner-Nordstrøm spacetimes in the Newman-Penrose formalism (M. Johnston), perturbations in any analytic background using the Hadamard's elementary solution (M. Peterson and R. Ruffini), forced perturbations of a Reissner-Nordstrøm geometry in the Dirac-Arnowitt-Deser-Misner formalism (J. Weinstein).

A scan of these contributions will convince the reader that in all the topics covered, very conspicuous progress has been made in recent years. One may safely forecast that future work on singularities and their properties and on the complexification of spacetime will lead to a new understanding of entropy in cosmological models and to a simpler and deeper explanation of spinning relativistic systems and spacetime structures. In a different direction, work on the uniqueness and perturbation techniques of black holes are giving a tool of unprecedented power, by providing a complete set of eigen functions, designed to describe and predict physical processes occurring in the field of gravitationally collapsed stars. These advances could lead to the direct identification and observation of some of the novel non-linear effects of general relativity in the limit of very strong gravitational fields. Finally, the work on torsion and on the possible modifications of general relativity in microphysics is likely to promote a revival of the classical Riemann programme presented as far back as 1857 on the possibility that "the metric relations of space in the infinitely small do not conform to the hypotheses of geometry". This message could very likely play a major role in future developments of physics and general relativity, especially with regards to the quantization problems.

The last three days of the meeting were devoted to topics which have already become, or are expected to become, of paramount importance for the understanding of experimental observations in physics and astrophysics. The basic theory of relativistic magnetohydrodynamics (A. Lichnerowicz) and the extensive theoretical and numerical work on magnetohydrodynamics in the fields of black holes (J. Wilson and communications by R. Hanni and W. Kundt) have by now become the basic tools for building up models able to explain the observed x-ray fluxes from close binary systems (R. Ruffini). Similarly, the coupling between electromagnetic and gravitational radiation (Y. Choquet-Bruhat) and the analysis of processes of vacuum polarization in strong gravitational fields (T. Damour, N. Deruelle and A. Starobinsky and communications by L. Davis and by W. Zaumen) are likely to be subjects of direct experimental verification within the forthcoming years (R. Ruffini). One of the topics of great interest in the meeting was the analysis of black holes thermodynamics along the lines suggested by Christodoulou, Bekenstein and Hawking. In particular the Hawking process of black holes evaporation was discussed from varying approaches, some of which are contained in these proceedings (G. Gibbons, T. Damour, P. Davies, U. Gerlach, H. Rumpf and W. Unruh). It may be thought unlikely that the Hawking process can actually be observed in the real physical world for some time to come. The analysis of the process, even viewed as a Gedanken experiment, however, is surely likely to lead to a new, deep and unifying view of thermodynamics and relativistic field theories in a curved background.

In a final session K. Nordtvedt reviewed the theoretical significance of present day gravity experiments, F. Everitt gave a critical historical review of experiments on gravitation leading to an analysis of the most sophisticated experiments currently being performed in the solar system, while B. Partridge gave an experimentalist's review of the recent status of observational cosmology.

If one may summarize the feelings of those attending the meeting, one had the overwhelming impression that general relativity has moved from being an extremely elegant mathematical theory, all the way to having acquired the maturity of a modern field theory at the very centre of physics and astrophysics. On the one hand, nearly the totality of the recent discoveries in astrophysics seem to have found their basic explanations in the framework of general relativity. On the other, the processes of vacuum polarization in highly curved spacetimes give an occasion to formulate (in a fully relativistic language) some of the basic notion of quantum field theories (the vacuum, the concept and definition of positive and negative frequencies, the particle number operator, functional integral methods and the like). General relativity offers an important example of a complex non-linear field theory, presenting a challenge and demanding a consistent formulation as a quantized field theory, while at the same time arrogating to itself the deepest notions on the very basic structure of space and time.

It is our hope that these impressions, so sharply conveyed during the meeting, are reflected in these proceedings. Clearly, what we have not been able to reproduce is the lively and beautiful presentation of the material by the lecturers themselves. If one may select one set of lectures for mention, perhaps the most outstanding were the lectures on cosmology by E. Lifschitz: all the participants, we are sure, will keep in their memory the three superb lectures he delivered, and even more, the passionate discussions following each lecture. This was a splendid example of strong interaction between speaker and audience, leading to flashes of unsuspected insights.

We would like now to thank the people who have made this meeting possible. Besides the sponsors of the International Centre for Theoretical Physics, these include the Consorzio per l'Incremento degli Studi e delle Ricerche degli Instituti di Fisica dell'Università di Trieste, the University of Trieste and the Alfred P. Sloan Foundation.

We would also like to thank the speakers and participants who came from all parts of the world. Most of the success of the meeting was due to the efficacy of the chairmen of the sessions - B. Bertotti, K. Bleuler, S. Chandrasekhar, J. Ehlers, W. Fairbank, H. Sato and F. Zerilli - who kept an excellent equilibrium between the presentations, interventions and the general discussions, and gave, with their scientific knowledge, due perspective to important new contributions.

Our deep gratitude goes to the local organizers of the meeting in Trieste; first of all Gallieno Denardo and then to the entire staff of the International Centre for Theoretical Physics, who worked devotedly and tirelessly for its success.

Finally, the publication of these proceedings would never have been possible without massive help from T. Damour and the generous and careful secretarial help and typing work of Mrs. D. Battle and Mrs. C. Kappes.

Trieste, 18 November 1976

Remo Ruffini
Abdus Salam

CONTENTS

List of participants

P.C. Aichelburg - Inst. für Theoretische Physik, Wien, Austria.

S. Ames - Los Alamos Scientific Laboratory, Los Alamos, U.S.A.

C. Aragone - Dep. de Fisica, Universidad S. Bolivar, Caracas, Venezuela.

A. Barnes - School of Mathematics, Dublin, Ireland.

R. Beig - Inst. für Theoretische Physik, Wien, Austria.

B. Bertotti - Istituto di Fisica Teorica, Università, Pavia, Italy.

N.T. Bishop - Dep. of Mathematics, University, Southampton, England.

K. Bleuler - Inst. für Theoretische Kernphysik, Bonn, W. Germany.

S. Bonanos - N.R.C. "Demokritos", Attiki, Greece.

D. Brill - University of Maryland, College Park, U.S.A./W. Germany.

M.R. Brown - Dep. of Astrophysics, Oxford, England.

H.A. Buchdahl - Dep. of Theoretical Physics, Aust. Nat. University, Canberra, Australia.

C. Callias - Dep. of Physics, Princeton University, U.S.A./Greece.

M. Calvani - Istituto di Astronomia, Padova, Italy.

P. Campbell - Dep. of Mathematics, University of Lancaster, Bailrigg, England.

B. Carter - D.A.F. Observatoire de Paris, Meudon, Paris, France/England.

R. Catenacci - Istituto di Fisica Teorica, Università, Pavia, Italy.

A. Chamorro - Dep. de Fisica, Universidad de Bilbao, Spain.

S. Chandrasekhar - The Enrico Fermi Institute, University of Chicago, Illinois, U.S.A.

J. Chela-Flores - Inst. Venezolano Investigaciones Cientificas, Caracas, Venezuela.

Y. Choquet-Bruhat - Dep. de Mécanique, Université de Paris, France.

D. Christodoulou - International Centre for Theoret. Physics, Trieste, Italy/Greece.

F. Cooperstock - Dep. of Physics, University of Victoria, Canada.

T. Damour - Dep. of Physics, Princeton University, U.S.A./France.

P. Davies - Dep. of Mathematics, King's College, London, England.

L.R. Davis - Dep. of Physics, Princeton University, U.S.A.

F. De Felice - Istituto di Fisica "G. Galilei", Padova, Italy.

J.C. Dell - Physics Dep., University of Maryland, College Park, U.S.A.

L. De Luca - Istituto di Macchine, Università, Milano, Italy.

G. Denardo - Istituto di Fisica Teorica, Università, Trieste, Italy.

W. Deppert - Philosophisches Seminar der Universität Kiel, W. Germany.

N. Deruelle - Ecole Normale Superieure de Jeunes Filles, Paris, France.

S. Detweiler - Dep. of Physics, University of Maryland, College Park, U.S.A.

W. Dietz - Physikalisches Institut der Universität, Würzburg, W. Germany.

C.T.J. Dodson - Dep. of Mathematics, University of Lancaster, Bailrigg, England.

P. Dolan - Dep. of Mathematics, Imperial College, London, England.

J.S. Dowker - Dep. of Theoretical Physics, University of Manchester, England.

C. Duval - Dep. de Physique, Université de Marseille, France.

R. Ebert - Physikalisches Institut der Universität, Würzburg, W. Germany.

J. Ehlers - Max-Planck Institut für Physik, Munich, W. Germany.

J. Elhadad - U.E.R. Mathématiques, Université de Provence, Marseille, France/Marocco.

H.G. Ellis - Dep. of Mathematics, University of Colorado, Boulder, U.S.A.

C.W.F. Everitt - W.W. Hansen Laboratories, Stanford University, Cal., U.S.A./England.

W.M. Fairbank - Physics Dep., Stanford University, Cal., U.S.A.

E. Fliche - U.E.R. Mathématiques, Université de Provence, Marseille, France.

H. Friedrich - Max-Planck Institut für Physik, Munich, W. Germany.

W. Garczynski - Institute of Theoretical Physics, University of Wrocław, Poland.

J. Geheniau - Université Libre de Bruxelles, Belgium.

U. Gerlach - Dep. of Mathematics, Ohio State University, Columbus, U.S.A.

G.W. Gibbons - D.A.M.T.P., University of Cambridge, England.

L. Girardello - Istituto di Fisica, Università di Milano, Italy.

A. Goddard - Mathematical Institute, Oxford, England.

R. Goldoni - Scuola Normale Superiore, Pisa, Italy.

W. Graf - Fachbereich Physik der Universität, Konstanz, W. Germany.

R. Grassini - Istituto di Matematica "Renato Caccioppoli", Napoli, Italy.

F. Grilli - Via Montanari 29, Mirandola, Modena, Italy.

P. Hajicek - Institute for Theoret. Phys., Univ. of Berne, Switzerland/Czechoslovakia.

R. Hakim - Section d'Astrophysique, Observatoire de Paris, Meudon, France.

L. Halpern - Dep. of Physics, Florida State University, Tallahassee, U.S.A./Austria.

R. Hanni - Physics Dep., Stanford University, Cal., U.S.A.

N.D. Hari Dass - Max-Planck Institut, Munich, W. Germany.

F.W. Hehl - Institut für Theoretische Physik, Clausthal-Zellerfeld, W. Germany.

J. Hoek - Instituut voor Theoretische Fysica, Amsterdam, The Netherlands.

A. Inomata - Dep. of Physics, University of New York, Albany, U.S.A./Japan.

M.D. Johnston - Massachusetts Institute of Technology, Cambridge, U.S.A.

C.-G. Källman - NORDITA, Copenhagen, Denmark/Finland.

I. Khan - Dep. of Physics, University of Khartoum, Sudan/Pakistan.

F.S. Klotz - Dep. of Applied Mathematics, Dublin, Ireland/U.S.A.

E. Krotscheck - Institut für Theoretische Physik, Hamburg, W. Germany/Austria.

W. Kundt - Institut für Theoretische Physik, Hamburg, W. Germany.

J. Ławrynowicz - Polish Academy of Sciences, Łodź Branch, Poland.

C.W. Lee - Dep. of Mathematics, University of Lancaster, Bailrigg, England.

A. Lichnerowicz - College de France, Paris, France.

E.M. Lifshitz - Institute for Physical Problems, Moscow, U.S.S.R.

C.A. Lopez - Dep. de Fisica, Universidad de Chile, Santiago, Chile.

R. Mansouri - Institut für Theoretische Physik der Universität Wien, Austria/Iran.

M. Martellini - Istituto di Scienze Fisiche, Università di Milano, Italy.

M.A. Mashkour - Dep. of Physics, University of Mosul, Iraq.

P. Mathieu - Université Catholique de Louvain, Belgium.

E.W. Mielke - Institut für Reine und Angewandte Kernphysik, Universität Kiel, W. Germany.

H. Müller Zum Hagen - Hochschule der Bundeswehr Hamburg, W. Germany.

A. Nduka - Dep. of Physics, University of Ife, Nigeria.

E.T. Newman - Dep. of Physics, University of Pittsburgh, U.S.A.

K. Nordtvedt - Dep. of Physics, Montana State University, Bozeman, U.S.A.

E. Nowotny - Fachbereich Physik, Universität Konstanz, W. Germany.

M.A. Obeid - College of Engineering, Riyadh, Saudi Arabia.

F. Occhionero - Consiglio Nazionale delle Ricerche, Roma, Italy.

H.C. Ohanian - Rensselaer Polytechnic Institute, Troy, New York, U.S.A.

J. O'Hanlon - Dep. of Applied Mathematics, University of Dublin, Ireland.

H. Okamura - Dep. of Physics, Kogakuin University, Tokyo, Japan.

D. Papadopoulos - Astronomy Dep., University of Thessaloniki, Greece.

L. Papaloucas - Mathematics Institute, University of Athens, Greece.

G. Papini - Dep. of Physics, University of Regina, Canada.

R.B. Partridge - Astronomy Dep., Haverford College, U.S.A.

R. Penrose - Mathematical Institute, Oxford, England.

Z. Perjes - Central Research Institute for Physics, Budapest, Hungary.

M.J. Perry - University of Cambridge, England.

S. Persides - Astronomy Dep., University of Thessaloniki, Greece.

M. Peterson - Physics Dep., Amherst College, U.S.A.

G. Platania - Osservatorio Astronomico di Capodimonte, Napoli, Italy.

L. Platania - Dep. of Mathematics, Università di Cosenza, Italy.

A.R. Prasanna - Institut für Theoretische Kernphysik, Universität Bonn, W. Germany/India.

A. Qadir - Mathematics Dep., Islamabad University, Pakistan.

A. Quale - Institute of Physics, Oslo University, Norway.

H. Quintana - Dep. of Astronomy, University of Chile, Santiago, Chile.

T. Regge - Institute for Advanced Study, Princeton, U.S.A./Italy.

L. Rodrigues - Faculté des Sciences Mirande, Dijon, France/Brazil.

W. Roos - Institut für Theoretische Physik, Universität Hamburg, W. Germany.

M. Rosenbaum - Universidad Nacional Autonoma de Mexico, Mexico.

F. Rosso - Mathematical Institute, Università di Napoli, Italy.

R. Ruffini - Dep. of Physics, Princeton, U.S.A./Italy.

H. Rumpf - Institut für Theoretische Physik der Universität Wien, Austria.

M.P. Ryan - Universidad Nacional Autonoma de Mexico, Mexico/U.S.A.

F. Sacchetti - Comitato Nazionale per l'Energia Nucleare, Roma, Italy.

J.L. Safko - Dep. of Physics, University of South Carolina, Columbia, U.S.A.

Abdus Salam - International Centre for Theoretical Physics, Trieste, Italy/Pakistan.

N.G. Sanchez - Institute of Astronomy, Buenos Aires, Argentina.

H. Sato - Res. Institute for Fundamental Physics, Yukawa Hall, Kyoto, Japan.

B. Schmidt - Max-Planck Institut für Physik, Munich, W. Germany.

E.L. Schucking - Dep. of Physics, New York University, U.S.A./W. Germany.

H.-J. Seifert - Hochschule der Bundeswehr Hamburg, W. Germany.

A. Selloni - Istituto di Fisica "G. Marconi", Roma, Italy.

M. Soffel - Johann W. Goether University, Frankfurt/Main, W. Germany.

J.M. Souriau - Centre de Physique Théorique, Marseille, France.

P. Spindel - Faculté des Sciences, Université de l'Etat à Mons, Belgium.

E. Streeruwitz - Institut für Theoretische Physik der Universität Wien, Austria.

G. Strini - Istituto di Fisica, Università di Milano, Italy.

W. Szczyrba - Polish Academy of Sciences, Warsaw, Poland.

J. Tafel - Uniwersytet Warszawski, Instytut Fizyki Teoretycznej, Warsaw, Poland.

A.H. Taub - Mathematics Dep., University of California, Berkeley, U.S.A.

C. Teitelboim - Dep. of Physics, Princeton University, U.S.A./Chile.

L.R. Thomlinson - King's College, Cambridge, England.

H. Tilgner - Mathematische Physik, Freie Universität Berlin, W. Germany.

A. Tomimatsu - Res. Institute for Theoretical Physics, Takehara, Hiroshima, Japan.

D.A. Tranah - Dep. of Mathematics, University of Southampton, England.

A. Treves - Università di Milano, Italy.

D. Trevese - Osservatorio Astronomico di Roma, Italy.

M.J. Trinkala - Physics Dep., State University of New York, Albany, U.S.A.

W.G. Unruh - McMaster University, Hamilton, Ontario, Canada/U.S.A.

P. Urban - Institut für Theoretische Physik der Universität Graz, Austria.

H. Urbantke - Institüt fur Theoretische Physik der Universität Wien, Austria.

F. Vagnetti - Laboratorio Astrofisica Spaziale, Frascati, Italy.

E. Van Der Spuy - Atomic Energy Board, Pretoria, South Africa.

P. Van Nieuwenhuizen - Brandeis University, Waltham, Mass., U.S.A./The Netherlands.

L. Verhoustraete - Institut de Physique Théorique, Louvain, Belgium.

J. Weinstein - Dep. of Physics, Princeton University, U.S.A.

D.C. Wilkins - University of California, Santa Barbara, Cal., U.S.A.

R.M. Williams - Girton College, Cambridge, England.

J.R. Wilson - Lawrence Livermore Laboratory, Livermore, Cal., U.S.A.

W.T. Zaumen - Lockheed Palo Alto Res. Laboratory, Cal., U.S.A.

F. Zerilli - Dep. of Physics, University of Washington, Seattle, U.S.A.

D. Zivanovic - Institute of Physics, Beograd, Yugoslavia.

A. Zupancic - Istituto di Fisica Teorica, Università di Trieste, Italy.

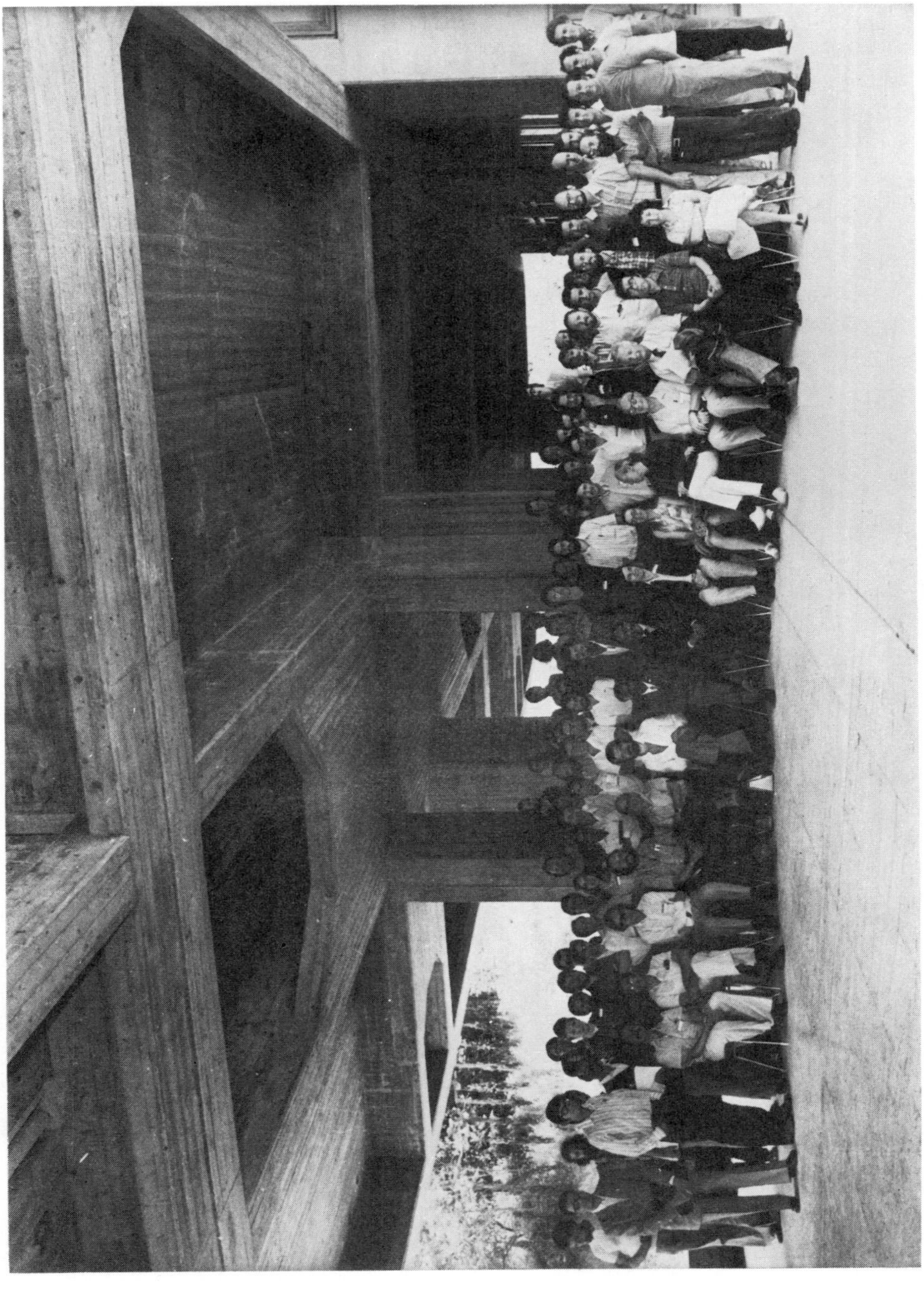

AN INTRODUCTION TO COVARIANT QUANTIZATION OF GRAVITATION*

P. van Nieuwenhuizen

Institute for Theoretical Physics
State University of New York at
Stony Brook, L.I., New York 11794

CONTENTS

1. INTRODUCTION

Gravitation is the gauge field theory describing massless self-interacting spin 2^+ bosons, just as Yang-Mills fields describe massless self-interacting spin 1^- bosons and quantumelectrodynamics massless non-self-interacting spin 1^- bosons. During the past five years important developments have been made in the general theory of covariant quantization of gauge fields, the main results obtained being of course the renormalization of field theories which unify the weak and electromagnetic interactions. Also for gravitation, however, our advancement in understanding quantized gauge theories has had significant consequences. In this review we will discuss some of the applications of these new ideas about covariant quantization to gravitation as a theory of particle physics.†

Covariant quantization of a gauge theory consists of deriving Feynman rules for the propagators and vertices of this gauge theory. Once the propagators and vertices are given, Feynman diagrams can be constructed in terms of which the S-matrix can be expressed. One may then investigate whether the theory is renormalizable, i.e. whether one can eliminate the unavoidable infinities of loop diagrams in such a way that the theory gets some predictive power. Covariant quantization is not the only way to quantize gauge field theories; one can also use canonical quantization (in which case one determines the p's and the q's of the theory). When one wants to determine S-matrix elements, covariant quantization is by far the simpler scheme. For Yang-Mills theory (and electrodynamics) both quantization schemes have been shown to be equivalent [1] but for gravitation

*Supported in part by the NSF Grant #MPS-74-13208 A01.

†In particle physics a distinction is sometimes made between relativistic particle theory and relativistic field theory. In this review we always talk about relativistic field theory.

a similar proof is still absent in the literature. Most physicists believe that this is only a matter of time.

According to the particle physics approach, gravitons are treated on exactly the same basis as other particles such as photons and electrons. In particular, particles (including gravitons) are always in flat Minkowski space and move as if they followed their geodesics in curved spacetime because of the dynamics of multiple graviton exchange.† This particle physics approach is entirely equivalent to the usual geometrical approach. Pure relativists often become somewhat uneasy at this point because of the following two aspects entirely peculiar to gravitation:

1. In canonical quantization one must decide before quantization which points are spacelike separated and which are timelike separated, in order to define the basic commutation relations. However, it is only after quantization that the fully quantized metric field can tell us this spacetime structure. It follows that the concept of space- or time-like separation has to be preserved under quantization, and it is not clear whether this is the case. (One might wonder whether the causal structure of space-time need be the same in covariant quantization as in canonical quantization.)
2. Suppose one wanted to quantize the fluctuations (for example of a scalar field, or even of the gravitational field itself) about a given curved classical background instead of about flat Minkowski spacetime. In order to write the field operators corresponding to these fluctuations in second-quantized form, one needs positive and negative frequency (annihilation and creation) solutions. In non-stationary spaces it is not clear whether one can define such solutions. (It may help to think of non-stationary space-time as giving rise to a time-dependent Hamiltonian.)

The strategy of particle physics has been to ignore these two problems for the time being, in the hope that they will ultimately be resolved in the final theory. Consequently we will not discuss them here any further.

Historically, there have been many (and often conflicting) opinions on whether and how to quantize gravitation. Heisenberg [2] already pointed at that feature which we know nowadays to be the chief problem of quantum gravity: the dimension of an inverse mass of the gravitational coupling constant κ. He

---

†As suggested by this picture, only asymptotically flat spaces are considered. For any finite number of gravitons the corresponding amplitudes are singularity free. It is only by summing the whole perturbation series that one can expect to produce such singularities as for example are present in the Schwarzschild solution. Most of the results we will discuss are to first order in quantum effects only (one-loop diagrams). Consequently we expect that these results are only adequate approximations to solutions which are singularity free.

predicted that this would force gravitation to be nonrenormalizable. Pauli, Salam and others [3,4] on the other hand hoped that gravitation, when coupled to matter, would quench all infinities, including its own. Wheeler [5] suggested that one should quantize the geometry of spacetime itself rather than the fields in it ("no fixed arena"). Penrose [6] goes even further; if I interpret his work correctly, he wants to quantize the points themselves. (Since in his approach points are intersections of two quantized twistors, they are no longer sharply defined but subject to quantum oscillations.) Some people believe that gravitation is something like van der Waals forces: present where matter is, but not propagating in free radiation modes (which might then explain the limited success in detecting gravitational radiation). Others, like Zumino [7], have speculated that general relativity is a low-energy phenomenological reflection of an unknown deeper-lying renormalizable theory. Weinberg [8] considers the possibility that the graviton is a bound state, with the extra positive-metric zero helicity part of the gravitational field serving to cancel a negative-metric Goldstone boson generated by spontaneous symmetry breakdown of scale invariance. (The generation of a scalar boson, the dilation, is known in the literature [9], but a tensor boson has not yet been produced.)

This review was written for pure relativists who are not familiar with particle physics techniques. Most results are therefore stated without derivations and the physical aspects are stressed. Time and space being finite, such interesting topics as nonpolynomial Lagrangians [4,10,11] (Delbourgo, Isham, Salam and Strathdee and others), f-gravity [12], super-symmetry in gravitation (Arnowitt, Nath and Zumino) and particle creation in curved spacetime (Hawking, DeWitt, Boulware, and others) have been omitted with reluctance. This seemed to us necessary in order to exhibit in a clear way the fundamental problems. In particular all results will be presented in the language of normal field theory, although many results were first obtained in the background [13-23] field formalism of DeWitt. Although this elegant formalism greatly reduces the horrendous algebra of quantum gravity, most physicists are still more familiar with normal field theory.

The following conventions are used in this review: $\delta_{\mu\nu} = (+1,+1,+1,+1)$ and $p^1 \cdot p^2 = \vec{p}^1 \cdot \vec{p}^2 + p^1{}_4 p^2{}_4$. The scalar curvature R is related to the Ricci tensor $R_{\mu\nu}$ by $R = g^{\mu\nu} R_{\mu\nu}$ with $R_{\mu\nu} = R^\alpha{}_{\mu\nu\alpha}$ and the Riemann-Christoffel tensor is defined by

$$R^\alpha{}_{\mu\nu\beta} = (\partial_\nu \Gamma^\alpha_{\mu\beta} + \Gamma^\alpha_{\nu\lambda} \Gamma^\lambda_{\mu\beta}) - (\nu \leftrightarrow \beta)$$

The Christoffel symbols are defined by

$$\Gamma^\sigma_{\alpha\beta} = \tfrac{1}{2} g^{\sigma\rho} [\partial_\alpha g_{\beta\rho} + \partial_\beta g_{\alpha\rho} - \partial_\rho g_{\alpha\beta}] .$$

The Einstein field equations read $G_{\mu\nu} \equiv R_{\mu\nu} - \tfrac{1}{2} g_{\mu\nu} R = - \frac{\kappa^2}{4} T_{\mu\nu}$ where $T_{\mu\nu}$ is the generally covariant matter energy momentum tensor and $\kappa^2 = 32\pi G$ with G = Newton's constant. Quantum gravitational fields $h_{\mu\nu}$ are defined by $g_{\mu\nu} = \delta_{\mu\nu} + \kappa h_{\mu\nu}$ and have the dimension of a mass, as any boson field should.

2. COVARIANT QUANTIZATION OF EINSTEIN GRAVITATION

We begin by reviewing briefly how covariant quantization of a gauge theory works. Consider first canonical quantization. In this method one looks for the dynamical p's and q's and expresses the other variables in terms of these. Although the theory is still Lorentz and gauge invariant, manifest Lorentz and gauge invariance is lost. It turns out that for calculations of cross sections it is easier to keep manifest Lorentz invariance at all stages, at the price of dealing with more variables than is strictly necessary. This is the virtue of covariant quantization. Manifest gauge invariance is, however, lost when one applies the rules of covariant quantization to a gauge theory within the framework of normal field theory. Only in the background field [13-23] method does one also retain some form of manifest gauge invariance. We restrict ourselves here to normal field theory.

In covariant quantization one does not determine the true dynamical variables and one does not eliminate the other variables, but instead one works with all variables in the action on a par. This is achieved by considering instead of the classical action another closely related action which no longer describes a gauge theory so that all its variables are dynamical. In physical quantities such as S-matrices, however, only the true dynamical variables contribute and give the same result as is obtained by canonical quantization.

The recipe for the construction of this non-gauge-invariant action from the original classical gauge invariant action consists always of two steps [24-27]:

1. Add a so-called gauge fixing term $\mathcal{L}^B$ to the classical Langrangian $\mathcal{L}^C$. Physically this means that one chooses a particular gauge. One can show that the S-matrix (but not the Green's functions) is independent of the gauge chosen, i.e. it is independent of $\mathcal{L}^B$. This gauge fixing term must be gauge dependent itself (in order to specify the gauge in which one works) and chosen such that it makes the kinetic matrix of the fields in the action regular. If there are N gauge parameters $\Lambda_1(x),\ldots,\Lambda_N(x)$ in the theory, then this term must be of the form of N squares

$$\mathcal{L}^B = -\tfrac{1}{2}\sum_{j=1}^{N} (C_j)^2 \ . \qquad (1)$$

where the functions C_j depend on the gauge fields and may contain derivatives. This procedure was invented by Fermi [28] who added to the Maxwell action $-\frac{1}{4}F^2_{\mu\nu}$ the gauge fixing term $-\frac{1}{2}(\partial_\mu A_\mu)^2$.

2. Add to this sum a second term, the so-called ghost term $\mathcal{L}^G$. Its physical role is to restore unitarity at the quantum level (see the end of this section for an explanation). It is derived from the gauge fixing term in the following way

$$G = \chi^*_j \; (\delta C_j \; / \; \delta\Lambda_k)_{\Lambda=0} \;\; \chi_k \; . \tag{2}$$

The complex fields χ_k are the Feynman-DeWitt-Faddeev-Popov ghost fields [24,28]. They are called ghosts because they appear only inside Feynman diagrams and never come out. Although they are boson fields, they satisfy Fermi-statistics: for each closed ghost-loop an extra minus sign is added to the amplitude.

Expanding the quantum action $\int d^4x\,(\mathcal{L}^C + \mathcal{L}^B + \mathcal{L}^G)$ into ghost and non-ghost quantum fields, the terms independent of and linear in quantum fields vanish in the cases we consider below. The terms quadratic in quantum fields define the kinetic matrix. Since the kinetic matrix of the sum of these three terms is non-singular (for the classical fields by construction, and also, as it turns out, for the ghost fields) one can invert it to obtain well-defined propagators. The terms with three or more quantum fields define the vertices. Below we will apply this general procedure to the Einstein action, but first we continue with some general theory.

One can build two kinds of Feynman diagrams: tree diagrams and loop diagrams. The tree diagrams reproduce the classical theory [29,30]. For example, the gravitational field $g_{\mu\nu}(x)$ at the point x outside a sphere with mass M can either be calculated from the Einstein field equations, yielding the Schwarzschild solution, or it can equally well be determined by summing the infinite series of tree diagrams in Figure 1*. The loop diagrams on the other hand introduce genuine quantum effects. However, they also introduce infinities and it is therefore necessary to find a method of regularization which makes the loop integrals finite and well-defined.

It is very convenient for the study of renormalizability of a gauge theory to use a regularization scheme which respects the hidden symmetry of the theory.** Although one has broken the local gauge invariance of the action by adding a gauge fixing term,† the symmetry of the original classical action is still at work and can be made explicit by certain formal relations between the unregularized (and

---

*Quantum corrections of the Schwarzschild solution have been discussed by M. J. Duff [31].

**For a calculation using a different regularization scheme, see M. R. Brown [32].

†Global gauge invariance (Lorenta invariance) may be broken by $\mathcal{L}^B$ as well.

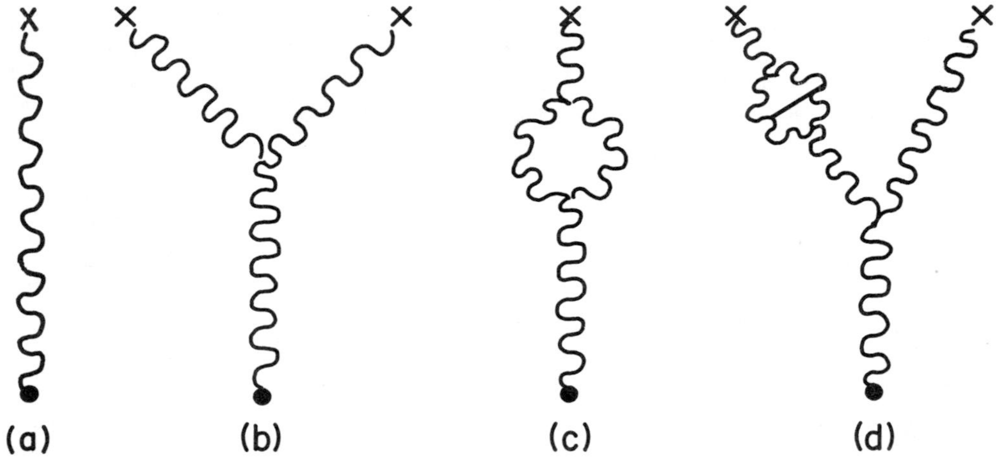

Figure 1. The classical gravitational field outside a massive sphere (denoted by a cross) at a point x (denoted by the dot) is the sum of the tree diagrams. Diagram "a" reproduces the Newtonian limit, diagram "b" the post-Newtonian limit. Quantum corrections are due to loop diagrams. Diagram "c" gives the lowest quantum correction, diagram "d" a higher order quantum correction.

therefore divergent and undefined) Feynman diagrams, so-called Ward-identities.* (Capper, Duff, Halper, Leibbrandt and Medrano [33-34] have derived in a very elegant way Ward identities in quantum gravity and verified them in perturbation theory. For a general discussion see A. A. Slavnov [35]. The actual form of these Ward identities depends on $\mathcal{L}^B$.) It is now very useful although not necessary to employ such a regularization scheme that after regularization the (now well-defined) Feynman diagrams still satisfy the Ward-identities. One of the oldest regularization schemes is the Pauli-Villars method. It turns out that this scheme respects the Ward identities of such Abelian theories as quantum electrodynamics but not of non-Abelian theories such as Yang-Mills fields and gravitation.

Fortunately there also exist regularization schemes which respect the Ward identities of non-Abelian gauge theories. The best known is dimensional regularization [36-39]. In this scheme one reformulates the theory in n-dimensions instead of in four dimensions. Four dimensional integrals $\int d^4k$ become n-dimensional integrals $\int d^nk$, where n is a positive integer, and after integrating over the (n-1) angles one is left with a final radial integration depending on the discrete positive integer parameter n. By continuing n into the complex plane, this last integral becomes finite and after evaluating it one returns to the physical point n=4. The original logarithmic ultraviolet divergences show up as poles at n=4.

*A familiar example in electromagnetism is $k_\mu \Pi_{\mu\nu} = 0$, where $\Pi_{\mu\nu}$ is the photon self-energy.

One has therefore a well-defined theory: propagators and vertices are given by the quantum action $\mathcal{L}^C + \mathcal{L}^B + \mathcal{L}^G$ while loop integrals are regularized by dimensional regularization. In the next sections we will discuss whether all ultraviolet poles at n=4 can be eliminated from the amplitudes by renormalization.

Let us now consider Einstein gravitation and apply the ideas of covariant quantization to it. The classical action for Einstein gravitation is given by

$$I^E(g) = \int [-2\kappa^{-2} g^{\frac{1}{2}} R(g)] \, d^4 x \tag{3}$$

where R and κ were defined in the introduction and $g = \det(g_{\mu\nu})$. This action is gauge-invariant; if one defines

$$g_{\mu\nu}'(x) = g_{\mu\nu}(x) + \frac{\partial\eta^\alpha(x)}{\partial x^\mu} g_{\alpha\nu}(x) + \frac{\partial\eta^\alpha(x)}{\partial x^\nu} g_{\mu\alpha}(x) + \eta^\alpha(x)\partial_\alpha g_{\mu\nu}(x) \tag{4}$$

then $I^E(g) = I^E(g')$. In this equation $\kappa^{-1}\eta^\alpha$ are the four infinitesimal dimensionless gravitational gauge parameters. It is easily verified that the following (deDonder-or harmonic-type) gauge-fixing term

$$\mathcal{L}^B = -\frac{1}{2\gamma}(\partial_\mu h_{\mu\nu} - \tfrac{1}{2}\partial_\nu h_{\mu\mu})^2 \tag{5}$$

is not invariant under the transformation in Equation (4) for any η^α and any γ. (Summation over repeated indices is always understood.) This equation looks and is not gauge invariant on purpose, since its purpose is to choose a gauge. The kinetic matrix of the classical action in Equation (3) is obtained by collecting terms quadratic in $h_{\mu\nu}$ (terms linear in $h_{\mu\nu}$ and without $h_{\mu\nu}$ vanish in Equation (3), and we recall the definition $g_{\mu\nu} = \delta_{\mu\nu} + \kappa h_{\mu\nu}$). It is given by the unique [40] Pauli-Fierz [41] free field Lagrangian for a massless spin 2 particle

$$\mathcal{L}^0(\text{spin } 2) = -\frac{1}{2} h_{\alpha\beta,\mu}^2 + h_\mu^2 - h_\mu h_{,\mu} + \frac{1}{2} h_{,\mu}^2 \tag{6}$$

where we have defined $h_\mu \equiv h_{\mu\nu,\nu}$, $h \equiv h_{\mu\mu}$ and $h_{\alpha\beta,\mu}^2$ stands for $h_{\alpha\beta,\mu} h_{\alpha\beta,\mu}$ with summation over repeated indices. Adding Equations (6) and (5) one finds that the total kinetic matrix for the quantum gravitational field $h_{\mu\nu}$ is non-singular. Choosing for convenience $\gamma = 1/2$, this total kinetic matrix, of the form hMh, becomes

$$\mathcal{L}^0(\text{spin } 2) + \mathcal{L}^B(\gamma = \tfrac{1}{2}) = -\tfrac{1}{2} h_{\alpha\beta,\mu}^2 + \tfrac{1}{4} h_{,\mu}^2 \tag{7}$$

from which the Feynman [42] propagator for gravitons is obtained by inverting the kinetic matrix M

$$P_{\mu\nu,\rho\sigma} = \frac{-i}{2}(\delta_{\mu\rho}\delta_{\nu\sigma} + \delta_{\mu\sigma}\delta_{\nu\rho} - \delta_{\mu\nu}\delta_{\rho\nu})\, k^{-2} \tag{8}$$

The ghost term is obtained from Equation (5) by the prescription in Equation (2), C_j being in this case $C_j = \sqrt{2}(\partial_\mu h_{\mu j} - 1/2\, \partial_j h_{\mu\mu})$ with j = 1,4 and Λ_k being the $\kappa^{-1}\eta^\alpha$ in Equation (4). One finds

$$G = \chi^*_\mu \quad \chi_\mu + \kappa\chi^*_\mu \begin{bmatrix} h_\beta\partial_\mu + h_{\mu\beta,\alpha}\partial_\alpha + h_{\mu\beta} + h_{\mu\alpha,\beta}\partial_\alpha \\ + h_{\mu,\beta} - h_{\alpha\beta,\mu}\partial_\alpha + \tfrac{1}{2} h,_\beta \partial_\mu - \tfrac{1}{2} h,_{\beta\mu} \end{bmatrix} \chi_\beta \tag{9}$$

Since the fields χ^* and χ are independent, one may always replace χ^* by $c\chi^*$ where c is a constant while leaving χ unaltered. We have used this freedom to cast the kinetic term in Equation (9) in canonical form. We have written this formula out in full to underline the algebraic complexity of quantum gravity. The vertices without ghost fields, obtained by expanding the Einstein action in three or more fields $h_{\mu\nu}$, are even worse. The ghost propagator is obtained from the kinetic term in Equation (9) as

$$P_{\mu\nu} = -i\delta_{\mu\nu}\, k^{-2} \tag{10}$$

while the other terms in Equation (9), proportional to κ, represent the ghost-ghost-graviton vertices which are needed to restore unitarity in loop diagrams.

We conclude with a few remarks. The non-linear structure of Einstein gravitation is particularly clear by writing the action in first-order (Palatini) form. One finds schematically

$$E = \int [(\partial\Gamma - \partial\Gamma + \delta^{\mu\nu}\Gamma\Gamma) + (\kappa\, h^{\mu\nu}\Gamma\Gamma)]\, d^4x$$

where the first terms produce the Pauli-Fierz action of Equation (6), while the last term, clearly on order h^3 and higher, contains all the horrendous non-linearity [42]. One can, in fact, derive full non-linear Einstein theory by coupling the linear gravitational field $h_{\mu\nu}$ described by Equation (4) consistently to its own energy tensor. (Gravitation is infinitely non-linear because $\sqrt{-g}$ and $g^{\mu\nu}$ are so)

A free graviton has two helicities, just as a free photon. This we can understand as follows: there are ten gravitational fields $h_{\mu\nu}$ and four gravitational degrees of gauge freedom, hence eight (twice as many) fields can be gauged away. For the photon this is, of course, "4-2x1=2." Each degree of gauge freedom can be used twice; for example, in electromagnatism once to obtain the Lorentz condition $\partial_\mu A_\mu = 0$, and once more because one can still gauge $A_\mu \to A_\mu + \partial_\mu \Lambda$ while keeping the Lorentz condition, provided $\Box\Lambda = 0$. (For details in the gravitational case, see the book by Landau and Lifshitz.)

The Pauli-Fierz action is the unique linear theory of gravitation. The reasoning goes as follows: suppose that the force between two objects is due to the exchange of a particle with spin S. The long-range nature of gravitation implies that the graviton mass is zero. Weinberg (see his book) has shown that massless fields with spin s $\geq$ 2 are inconsistent. Exchange of a spin 1 particle leads to repulsion instead of attraction (think of a photon between two equal charges). This still allows spin 0 and spin 2 for the graviton, but the only coupling of a spin 0 particle to a relativistic generalization of the mass is coupling to the trace of the energy tensor. From experiment we know that light deflects at the sun; on the other hand $T^{\mu}_{\mu} = 0$ for light. Hence the graviton has spin 2 and mass zero. Finally, the only linear massless Lagrangian field theory containing only spin 2 is the Pauli-Fierz Lagrangian [40].

Actually, one can even show that the mass of the graviton is strictly zero. This is thus different from the case of electromagnetism where the best one can do is to give a (very small) upper limit on the mass of the photon. The reason for this discontinuity is that a massive graviton has a zero-helicity component which does not decouple from matter. Both for massive and massless gravitons, the helicity ±1 components always decouple from matter as a consequence of energy conservation. In this respect the situation is as in electromagnetism, where the helicity zero photon decouples as a consequence of electrical current conservation. However, the helicity zero graviton couples to the trace of the energy tensor $T_{\mu}{}^{\mu}$ which does not vanish, consequently there is a finite discrete difference between the predictions of the massless and the massive theory. For example: if in both theories Newton's law holds, then the angle of light bending at the sun differs by a factor 4/3. Consequently, from a measurement with an inaccuracy of, say, 15% we can rigorously conclude that the graviton has strictly zero mass.

The two constants in the Einstein action, the κ in $g_{\mu\nu} = \delta_{\mu\nu} + \kappa h_{\mu\nu}$ and the k in the factor $2\kappa^{-2}$ in front of R in Equation (3), are equal and defined by the two requirements that the kinetic energy of the fields $h_{\mu\nu}$ have the canonical form and that Newton's law be obtained from one-graviton exchange. For γ tending to zero in Equation (5), this gauge fixing term becomes a delta function in the path integral formalism (which we won't discuss here) so that in this limit one is really working in the deDonder or harmonic gauge $\partial_{\mu} h_{\mu\nu} - \frac{1}{2}\partial_{\nu} h_{\mu\mu} = 0$. (For $\gamma \neq 0$ one speaks of a deDonder-type gauge.)

The occurrence of a complex vector ghost field can be understood physically by considering the unitarity relation

$$\text{Im}\, T = \sum_{n} T^{\dagger} |n\rangle \langle n| T . \tag{11}$$

According to this relation the imaginary part of a one-loop diagram should be equal to the product of two tree graphs (the Born amplitudes). In the loop, all ten helicity components of $h_{\mu\nu}$ contribute, but in the intermediate states $|n\rangle$ only the two physical polarizations are present. For non-spinning or massive particles, Equation (11) is always satisfied; it is only the nonphysical helicities of massless particles in loop diagrams which cause trouble. It is now clear that a complex vector field has precisely the number of degrees of freedom needed to cancel the eight unphysical helicities in the loop diagram,* and it is also

*The complex vector field in Equation (9) is not a gauge field hence it really has eight degrees of freedom. For physical particles such a Langrangian should be rejected since it has no lower bound on the energy (one should use in

clear that ghost fields should only appear inside loop diagrams. It can indeed be shown that ghost fields and unphysical helicities cancel each other to any number of closed loops. (It turns out that the ghost fields are even needed to make the S-matrix gauge independent, i.e. independent of the choice of $\mathcal{L}^B$, although they were primarily introduced to restore only unitarity.) In the same way the reader may understand why in quantumelectrodynamics the ghost is a complex scalar field. In flat space, the electromagnetic ghost term is given by $\chi^* \Box \chi$ hence in this case the electromagnetic ghost particle is decoupled from all other particles and can be neglected, but in curved space the electromagnetic ghost couples to gravitons.

3. FINITENESS OF THE ONE-LOOP CORRECTION TO THE S-MATRIX OF PURELY GRAVITATIONAL PROCESSES

In 1973 't Hooft and Veltman [43] published a fundamental paper in which they demonstrated that the S-matrix for processes involving only gravitons but no matter particles, is finite when one considers all diagrams for a given process with at most one-loop. Then they went on and considered what can be said about the system consisting of quantized scalars interacting with quantized gravitons. Although these authors used the background field formalism of DeWitt, we will present their results in the language of normal field theory. The background field method is equivalent to normal field theory and simplifies algebraic calculations [22,23] considerably, but most physicists are perhaps more familiar with normal field theory.

Let us start by considering the following simple diagram

It contains two propagators, two vertices and two external lines. For large momenta p, each propagator behaves as p^{-2} (see Equation (8)), while any N-graviton vertex contains exactly two momentum factors p. (This is easily seen from the fact that all terms obtained by expanding R in terms of $h_{\mu\nu}$ contain two derivatives, see the definition of R in the introduction.) The diagram therefore behaves (if no cancellations occur) as

---

in this case the complex Maxwell action).

$$\int d^4p \, \frac{p^2 p^2}{p^2 p^2} = \text{quartically divergent} \qquad (12)$$

Gauge invariance usually reduces the degrees of divergence of (the sum of) certain diagrams somewhat. For example, in quantum electrodynamics, power counting tells us that the photon propagator correction might be quadratically divergent, whereas it actually is logarithmically divergent.* In gravitation it is precisely such cancellations which lead to the 't Hooft-Veltman result.

In most field theories, the convergence of a one-loop diagram improves if more external lines are added to a loop. Only ("one particle irreducible," see later) diagrams with a few external lines are divergent. This is not the case for gravitation: for each extra propagator (factor p^{-2}) an extra vertex is created (factor p^2). So we see that any one-loop diagram with arbitrarily many external gravitons is quartically divergent (the same holds for diagrams consisting of a ghost loop with arbitrarily many external gravitons, see Equation (9)). Expanding the numerator in leading powers of loop momentum, these diagrams usually also contain logarithmic subdivergences, hence by calculating these diagrams by means of dimensional regularization, poles at $n = 4$ occur. The dependence of the vertices on p^2 is due to the dimensional character of κ. Consider for example a 3- graviton vertex. It is of the generic form $\kappa\partial\partial hhh$ and the two derivatives are needed to give this expression the dimension of a $(\text{mass})^4$ (being a Lagrangian density). If κ were dimensionless, only one momentum factor would be needed; in that case the theory would be very much like Yang-Mills theory and probably would be renormalizable. Thus the root of the nonrenormalizability of quantum Einstein gravity is the dimensional character of the gravitational coupling constant.

In order to investigate whether one can renormalize a theory, one must first determine the divergences of the so-called one-particle irreducible Green's functions. For one-loop diagrams these are the divergences of diagrams which consist of one loop onto which are attached any number of external gravitons but no trees. A vertex with N fields $h_{\mu\nu}$ is proportional to κ^{N-2} because $g_{\mu\nu} = \delta_{\mu\nu} + \kappa h_{\mu\nu}$ while there is a factor κ^{-2} in front of the action in Equation (3). A vertex with N fields $h_{\mu\nu}$ in a one-loop diagram contributes (N-2) external lines and 2 internal lines. It follows that in a one-loop diagram there are as many factors κ as there are external lines.† Hence a one-loop diagram with M external $h_{\mu\nu}$ lines is proportional to κ^M (the reader may check this for the case of the simple diagram in the beginning of this section). Hence: terms in the <u>action</u> with M fields $h_{\mu\nu}$ are proportional to κ^{M-2}, but <u>divergences</u> depending on M external fields are proportional to κ^M. Since both types of terms must be of dimension 4 (being a Lagrangian density), it follows that terms in the divergences

---

*If one uses a regularization scheme which respects the symmetry of the theory such as dimensional regularization.

†This is also true for ghost loops, see Eq. (9).

have two more derivatives than terms in the action. In other words, these two types of terms are very different in functional form: the action is $\sim\kappa^{-2}R$ while (as we shall see) the sum of all one-loop divergences is $\sim aR_{\mu\nu}R^{\mu\nu} + bR^2$.

A theory is renormalizable if one can absorb the divergences of the one-particle irreducible Green's functions in to the action ($\mathcal{L}^E + \mathcal{L}^B + \mathcal{L}^G$) by rescaling of the physical parameters (the scales of the gravitational and ghost fields, the coupling constant κ and possibly some parameters in the gauge-fixing term). Due to the difference in functional form mentioned before, this is not possible. It could be that the coefficients of these divergences would vanish by accident but the explicit calculation of 't Hooft and Veltman [43]* shows this not to be the case. In fact, infinitely many diagrams with infinitely many external gravitons are equally divergent.

However, a peculiar phenomenon occurs in the S-matrix. When one calculates the one-loop radiative corrections to the S-matrix for a given process, the sum of all residues of the poles at n = 4 cancels. The one-loop radiative corrections to the S-matrix are determined by all one-loop diagrams into which are inserted arbitrarily many external tree structures at the end of which are in-fields on their mass-shell ($p^2 = 0$) with physical polarizations (transverse and traceless $\varepsilon_{\mu\nu}$). (In general no cancellation occurs between subclasses of diagrams, because only the sum of all diagrams is gauge invariant, unlike in quantum-electrodynamics where one usually can find gauge invariant subclasses due to the Abelian character of that gauge theory.) The reason for this cancellation is the following: sums of trees determine a field $g^C_{\mu\nu}(x)$ which satisfies the classical field equations [29] (see Section 2). Gauge invariance of the S-matrix requires that all one-loop divergences sum up to a local, generally covariant function of $g^C_{\mu\nu}(x)$, with dimension 4 (being a Lagrangian density). The proof that in normal field theory the sum of all divergences of the S-matrix, when expressed as a function of $g^C_{\mu\nu}$ (which represents the sum of all tree diagrams) is generally covariant, is most easily proved in the background formalism, but its result is also valid in normal field theory because the S-matrix is equal in both cases. We saw before that if each external line $h_{\mu\nu}$ in a one-loop divergence picks up one factor κ, then no κ's are left. The same holds if the external line $h_{\mu\nu}$ is replaced by a sum of trees $h^C_{\mu\nu}$. It follows that, as a function $g^C_{\mu\nu} = \delta_{\mu\nu} + \kappa\, h^C_{\mu\nu}$, the one-loop

*Actually these authors calculated the divergences of one-particular irreducible Green's functions in the background field formalism; a similar calculation in normal field theory is not available. In the literature most results concern the S-matrix for which background and normal field theory give the same results [23]. Green's functions have been considered much less, because different definitions of the quantum gravitational field (for example [41] the linear part of $g_{\mu\nu} - \eta_{\mu\nu}$ or of $g^{\frac{1}{2}}g^{\mu\nu} - \eta_{\mu\nu}$) yield different Green's functions, both in background and in normal field theory. The S-matrix is also independent of this definition.

divergences do not depend explicitly on κ. The sum of all one-loop divergences of the purely gravitational S-matrix has therefore the following properties (i) it is a generally covariant function of $g^C_{\mu\nu}$, (ii) it is independent of κ and (iii) it has canonical dimension 4, being a Lagrange density. Its most general form is therefore

$$S^{div}(g^C_{\mu\nu}) = \frac{1}{n-4}\int d^4 x\ g^{\frac{1}{2}}\ [aR_{\mu\nu\rho\sigma}\ R^{\mu\nu\rho\sigma} + bR_{\mu\nu}R^{\mu\nu} + cR^2] \qquad (13)$$

Now comes a surprise. The Gauss-Bonnet theorem in four dimensions leads to the following integral relation [44] for any field $g_{\mu\nu}$ (it is thus an identity)

$$\int g^{\frac{1}{2}} \left(R_{\mu\nu\alpha\beta}\ R^{\mu\nu\alpha\beta} - 4\ R_{\mu\nu}R^{\mu\nu} + R^2\right) = 0. \qquad (14)$$

On the other hand, since $g^C_{\mu\nu}$ satisfy the classical field equations, $R_{\mu\nu}$ and R in Equation (13) vanish.[††] Hence $S^{div} = 0$! In other words: <u>the sum of all one-loop radiative corrections to the total S-matrix is finite</u>! The same is true for the S-matrix of a given physical process since one can expand the total S-matrix in κ and in the number of external lines. (Some people find this as dull as the vanishing of the Dirac action on shell. Others believe that it is a glimpse of the power of symmetry in Einstein theory, which heralds good news for the higher order quantum corrections of pure gravitation.)

Summarizing we note that the one-loop divergences of the purely gravitational Green's functions cannot be renormalized away, due to the dimensional character of the gravitational constant, but that all one-loop divergences in the S-matrix cancel due to a miracle. At the two-loop level the divergences in the S-matrix will be of the form $\sim\kappa^2\ [R_{\mu\nu\alpha\beta}\ (g^C_{\kappa\lambda})]^3$. No integral relations similar to Equation (14) seem to exist for products of three curvature tensors, and it requires strong optimism to hope that the miracle will repeat itself. Grisaru et al. [45] , have concluded that if helicity is conserved in graviton-graviton scattering, then such a miracle will happen at the two-loop level but nothing can be said about the n-loop level. It is not known whether helicity is conserved in pure gravitation.

4. NONRENORMALIZABILITY OF INTERNAL GRAVITATIONAL INTERACTION

Processes involving both quantized matter and quantized gravitational fields are nonrenormalizable when gravitons appear inside loops (internal gravitons[†]). This was first shown for the system consisting of a quantized scalar and the quantized gravitational field by 't Hooft and Veltman [43]. A number of authors have extended their results to other matter systems: charged scalars [46], photons [47], spin 1/2 fermions [48], Yang-Mills bosons [49]. Strictly speaking, one only knows that in all these examples the theory is unrenormalizable,

---

[†] Gravitons which appear inside tree diagrams will not be called internal.

[††] Because the source-free Einstein equations read $R_{\mu\nu} = 0$.

but arguments based on unitarity [48] seem to indicate that any matter system, when coupled to gravitation, produces nonrenormalizable divergences. One case has been investigated where there is more than one matter constituent: quantum electrodynamics coupled to gravitation is nonrenormalizable [57].

Consider as an example scalar-scalar scattering through graviton exchange. In the preceding section we saw that according to power counting any purely gravitational one-loop diagram is quartically divergent. In particular Diagram A below is quartically divergent, as is its ghost companion, Diagram B. The action for the

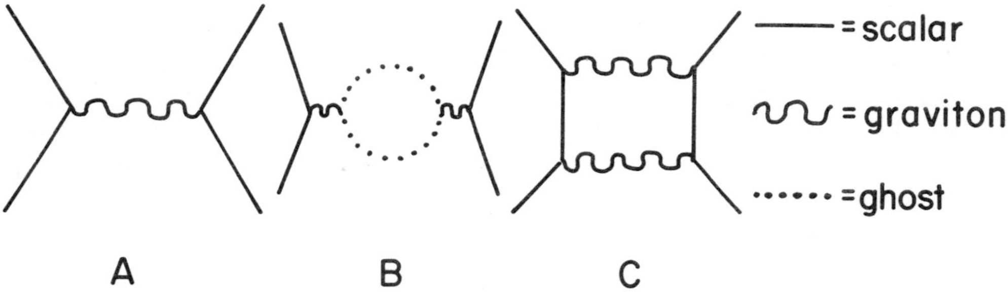

scalar graviton system is the sum of the purely gravitational action and the generally covariant scalar action [see Equations (5) and (9)]

$$I = \int d^4 x \, g^{\frac{1}{2}} \left[-2\kappa^{-2} R - \mathcal{L}^{B} - \mathcal{L}^{G} - \frac{1}{2}\partial_\mu \phi \partial_\nu \phi \, g^{\mu\nu} \right] \tag{15}$$

from which we deduce two things: (i) vertices, containing scalar fields, contain exactly two scalar fields, no more and no less, and arbitrarily many gravitational fields, and (ii) each scalar line carries one momentum factor p. Let us now calculate the degree of divergence of diagram C. At each vertex only one scalar field is inside the loop integral, hence each vertex contributes one factor p to the loop integral. Since all propagators (of scalars as well as of gravitons) tend to p^{-2} for large momenta, diagram C diverges in this limit as

$$\int d^4p \, p^4 \, (p^{-2})^4 = \text{logarithmically divergent.} \tag{16}$$

The reader will have no difficulty in deriving the following theorem along the same lines:

<u>THEOREM</u>: one-particle irreducible one-loop diagrams with arbitrarily many external gravitons and with

0		quartically	
2		quadratically	divergent
4	external scalars are	logarithmically	
6		convergent	

The same theorem holds when one replaces "scalars" by "photons." For fermions, which behave as if they were a square root of bosons, diagrams with up to eight external fermion lines are divergent!

We will first show that the scalar-gravity system is nonrenormalizable because the one-loop divergences of the one-particle irreducible Green's functions cannot be absorbed into the action by rescaling of the physical parameters. Then we will investigate whether a similar miracle as occurred in pure gravitation will make the S-matrix finite.

Consider all diagrams consisting of a single loop with four external scalar lines with arbitrary momenta and no external gravitons nor tree structures attached (for example diagram C above but not diagrams A and B). These diagrams define the one-particle irreducible Green's functions with four external ϕ-fields. Renormalizability requires that it must be possible to absorb the divergent terms of these functions into the original action by rescaling of the physical parameters (the scales of the gravitational, the ghost and the scalar fields, the coupling constant κ and the parameters in the gauge fixing term). A fortiori this must be possible when the external scalars are on-shell ($p^2 = 0$). An explicit calculation of the divergences of all one-particle irreducible one-loop diagrams with exactly four external scalar fields which are moreover on-shell ($p^2 = 0$) by Grisaru et al [23] has yielded the following results

$$\frac{5}{16}\kappa^2 (s^2 + t^2 + u^2) \frac{-i}{8\pi^2(n-4)} \tag{17}$$

From this result we see that the functional form of the ϕ-dependent terms in $\Delta\mathcal{L}$ and in the action is quite different: in the action there are only terms with at most two ϕ-fields, while the above-mentioned calculation has established the non-vanishing of the coefficient of the one-particle irreducible one-loop divergences with four ϕ-fields. No rescaling can ever absorb terms with four ϕ-fields into an action containing terms with at most two ϕ-fields. It follows that the Einstein-scalar system is nonrenormalizable.

The next result these authors obtained was the sum of all one-loop divergences in the S-matrix for scalar-scalar scattering (we recall: these are the diagrams where trees are also inserted in the loop, while the four scalars are on their mass shell). They [23] found that the S-matrix is still divergent (the sum of the residues of the poles at $n = 4$ does not vanish). It follows that the miracle, which made the one-loop radiative corrections to the S-matrix of pure gravitational processes finite, does not work in the system consisting of scalars and gravitons.

One might believe that photons are more fundamental than scalars. However, an explicit calculation by Deser et al. along the same lines as 't Hooft and Veltman's analysis of the scalars revealed that the Einstein-Maxwell system is also one-loop nonrenormalizable. Then these authors turned to spin 1/2 fermions [48] since these particles are the fundamental units of nature as far as angular momentum is concerned. (In this case a small technical complication is that one must quantize the 16 vierbein instead [50,7] of the 10 metric fields. This was done and it turned out that the six antisymmetric vierbein components did not propagate. It was also found that the vierbein and metric approach for bosons are in the quantum domain equivalent, just as they are equivalent in the classical domain.) The sum of the divergences of all one-particle irreducible diagrams with 8 external fermions and no gravitons was found to be nonzero and proportional to $(n-4)^{-1}$ $\kappa^8(\bar{\psi}\,\gamma^a\gamma^5\psi)$ (since real]Majorama] spinors can still couple to gravitation [but not to electromagnetism], only the invariants constructed from $\bar{\psi}\gamma_5\psi, \bar{\psi}\gamma_\mu\gamma_5\psi$ and $\bar{\psi}\psi$ could have been found because $\bar{\psi}\gamma_\mu\psi$ and $\bar{\psi}\sigma_{\mu\nu}\psi$ vanish for Majorana spinors.) Since the functional form of this expression is again drastically different from the terms present in the Dirac-Einstein action, nonrenormalizability was again concluded. Finally, with H. S. Tsao, the Einstein-Yang-Mills [49] system was investigated in the hope that the free Yang-Mills self-coupling constant g could be adjusted such that the S-matrix would be finite. It turned out that for no value of g this was possible; the closest one could come was the case of Maxwell theory. In order to be definite we quote here the result. The sum of the one-loop divergences in the sum of all S-matrices for graviton-Yang-Mills-boson processes is given by

$$S^{div}\left(g^C_{\mu\nu},\, W^{a,C}_\mu\right) = \int \frac{d^4x\, g}{8\pi^2(n-4)} \{[\frac{137}{60} + \frac{(r-1)}{60}]R_{\mu\nu}R^{\mu\nu} - \frac{11}{12} g^2 C G_{\mu\nu}{}^a G^{\mu\nu,a}\} \,,\quad rC = f_{abc}f^{abc} \tag{18}$$

where r is the number of Yang-Mills bosons and f_{abc} the structure constants. The last term is renormalizable; it is proportional to the Yang-Mills action but the first term is not. Its coefficient is positive and smallest when r = 1, i.e. in the case of Maxwell theory. Several other a priori possible nonrenormalizable divergences, such as $(G^a_{\mu\nu})^4$, are actually absent. This cancellation can be traced back to the invariance of the Einstein system under duality rotations [79]

$(\delta F_{\mu\nu} = \delta\lambda\, {}^*F_{\mu\nu}$ and [80] $\delta A_\mu(x) = \int d^4y\, f^\nu(y) F_{\mu\nu}(x-y)$ with $\delta_\nu f^\nu(y) = \partial_\nu f^\nu(y) = \delta^4(y)$; f_ν fixes the gauge) and by relating that system to the Yang-Mills system.

Corrections to the graviton propagator due to virtual photons and neutrinos have been calculated by Capper and Duff and Halpern [31]. Since the matter

Lagrangians are scale-invariant, the counter terms must be so and indeed are found to be part of the only scale-invariant gravitational counter term $R^2_{\mu\nu}$ - 1/3 R^2.

Taylor and Nouri-Mogadam [46] have investigated whether modifying the action by adding certain nonminimal interactions leads to finite results at the one-loop level. They find that these extra terms lead to new one-loop divergences with increasing powers of derivatives which are even more nonrenormalizable, so that also this approach seems of no help.

Summarizing we can say that processes involving quantized gravitons and quantized matter particles are nonrenormalizable because the divergences look drastically different from the terms in the action. This is due to the dimensional character of the gravitational coupling constant κ.

5. RENORMALIZABILITY OF EXTERNAL GRAVITATIONAL INTERACTIONS

D. Z. Freedman, I. J. Muzinich, E. J. Weinberg and So-Young-Pi [51,53] have investigated what happens with a flat-space renormalizable field theory of conventional particle physics during an earthquake. Let us explain what they meant by this statement. They considered arbitrary diagrams (technically, the Green's functions) of such renormalizable field theories as $\lambda\phi^4$ and spontaneously broken non-Abelian gauge theories and attached arbitrarily many external gravitons to these diagrams (the earthquake). They found that these diagrams (which are finite in flat-space because loop-divergences can be renormalized there) stay finite no matter how many external gravitons are attached and no matter what momenta these gravitons carry (off- or on-shell)--provided one applies standard renormalization of the flat-space theory and renormalizes in the case of $\lambda\phi^4$ theory the coefficient of the improvement term† of Callan, Coleman and Jackiw [54].

This is a very strong and positive result. It tells us that when we allow external gravitons, physics stays finite, but that only when gravitons are internal, divergences are produced which we cannot handle (as we saw in Section 3). The role of the improvement term is curious. Originally introduced to define a locally scale invariant theory for scalar particles, it actually corresponds to a nonminimal interaction between scalars and gravitons. Once this term is added to the usual scalar-graviton system (thereby defining a locally scale invariant action) this system turns out to be renormalizable, provided the gravitons stay external.

These general results are confirmed in explicit calculations. R. Delbourgo and P. Phocas-Cosmetatos [55,56] considered the radiative corrections to the electron-graviton and to the photon-graviton vertex. The graviton stays external

†This term is given by 1/12 $R\phi^2$ where ϕ is the scalar field and is to be added to Eq. (15).

in these examples. The electron-graviton vertex receives radiative corrections from a virtual photon, while the photon-graviton vertex is corrected by taking into consideration virtual electron and pion loop contributions. The corresponding flat-space theories (quantum-electrodynamics for fermions or scalars) are renormalizable and the corrections are indeed finite. Obviously it is of interest to see whether the effects given by these radiative corrections can be tested. The first place these authors considered was the frequency dependence of light bending at the sun; however, the effects are much too small to be seen. Also applications in high-energy physics evade any experimental test.

The radiative corrections (due to virtual photons) to the charged-scalar-graviton vertex have also been considered [57]. It was found that they were finite when the improvement term mentioned above was included. At the one-loop level one need not yet renormalize the coefficient of this term.

For zero-momentum of the external graviton, these vertices all reduce to their Born approximation, because (due to simple Ward identities relating different loop-diagrams) the radiative corrections vanish at this kinematical point. Physically this means that a low-energy theorem holds: a particle with inertial mass M weighs the same, whether or not that particle is charged. So this is the quantum version of the equivalence principle.

Delbourgo and Salam [58] have determined the gravitational counterpart of the Adler anomaly. Considering one-fermion-loops with two external gravitons, they found a finite gravitational correction (A_μ = axial current, P = pseudoscalar)

$$\partial_\mu A_\mu = 2\,m\,P + \frac{e^2}{16\pi^2}\,\varepsilon_{\mu\nu\rho\sigma}F^{\mu\nu}F^{\rho\sigma} + \frac{1}{768\pi^2}\,\varepsilon_{\mu\nu\rho\sigma}R^{\mu\nu}{}_{\kappa\lambda}R^{\kappa\lambda\rho\sigma} \tag{19}$$

again in agreement with our earlier general remarks.

6. FINITENESS OF THE CORRECTIONS TO g-2 DUE TO INTERNAL GRAVITONS

Berends and Gastmans [59], arguing that quantumelectrodynamics describes the interactions between electrons and photons so well, felt that there was perhaps a chance that inclusion of internal gravitons would not lead immediately to meaningless results in this particular case. They calculated the gravitational corrections to the electron and muon anamalous magnetic moment and found a finite result

$$\mu = \frac{eh}{2mc}\ \left[1 + \frac{\alpha}{2\pi} + \frac{7}{4}\frac{Gm^2}{\pi} + \ldots.\right] \tag{20}$$

In this equation the factor $(\frac{eh}{2mc})$ is the value Dirac theory predicts, the term $(\alpha/2\pi)$ is the famous purely electrodynamical one-loop radiative correction of Schwinger, and the last term is the one-loop gravitational radiative correction found by these authors. (G = Newton's constant and m is the lepton mass.) The diagrams which contribute are sketched below. This is the first example of an observable quantity whose gravitational radiative corrections are finite. Unfortunately, the value of these corrections is much too small to be observed. (As a

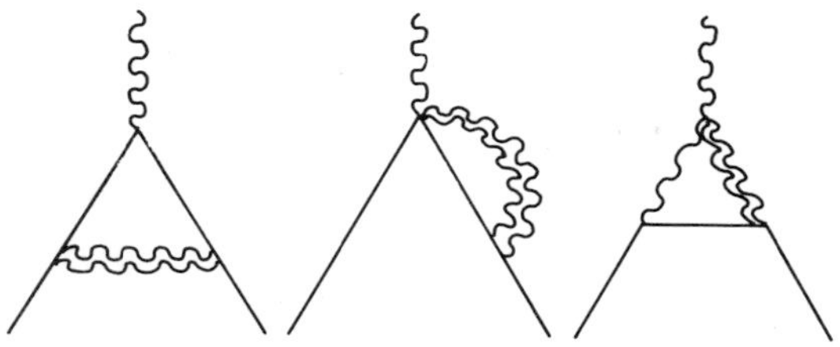

Diagrams contributing to the lepton anomaly.
—— lepton, ∿∿ = photon and ≈≈ = graviton.

comparison we mention that one can nowadays measure the three-loop corrections of pure quantumelectrodynamics to the electron anomalous magnetic moment [60] which are of the order of $(\alpha/\pi)^3 \approx 10^{-8}$ while for an electron $G\,m^2 \approx 10^{-45}$).

Grisaru et al. [57] found that Q.E.D. interacting with gravitation is non renormalizable.They calculated the radiative corrections of the photon-electron vertex due to virtual graviton exchange. They found that the magnetic form factor stays ultraviolet and infrared finite at all values of the momentum transfer q^2, not only at $q^2 = 0$ where it is equal to the anomalous magnetic moment. The electric form factor, however, is both ultraviolet and infrared divergent at q^2 different from zero. At $q^2 = 0$ a low energy theorem holds: the electric form factor equals unity at this point. This means physically that whether or not there are gravitational corrections inside an electron, its charge stays fixed and equal to e. Naive power counting reveals that the separate contributions to the gravitational corrections of the electric form factor diverge quadratically. The low energy theorem which we mentioned can reduce this degree of divergence to logarithmic but no further. This explains why the electric form factor diverges in the ultraviolet region.

7. BORN DIAGRAMS, RENORMALIZABILITY AND THE CLASSICAL THEORY

Most of the features of the renormalizable gauge theories for the weak interactions can be rederived by requiring that Born amplitudes do not grow with energies for fixed angles. It thus seems worthwhile to apply the same arguments to gravitation and to see whether this can shed some light on the renormalization problem.

The argument goes as follows: The imaginary part of a one-loop diagram is equal to the product of two Born graphs, as we discussed in Section 2. If the Born graphs do not grow too fast with energy, the imaginary part also behaves well for high energies and one needs no subtraction constant when determining the real part from the imaginary part by means of a dispersion integral. The argument is completed by stating that subtraction terms have something to do with nonrenormalizable counter terms.

Born amplitudes for graviton-graviton scattering (and for other gravitational processes) were calculated by DeWitt and Cooke [13] and recalculated by

Berends and Gastmans [61], who used Veltman's algebraic manipulation program Schoonschip† (which means "clean up the mess" in Dutch). After very tedious calculations the following results were obtained

$$F_{22;22} = \kappa^2 s^3/(4tu) \qquad F_{2,-2;-22} = \kappa^2 t^3/(4su)$$

$$F_{2-2;2-2} = \kappa^2 u^3/(4st) \tag{21}$$

while the other amplitudes are related by parity or are zero. The variables s,t and u are the Mandelstam variables (s+t+u = 0) and the indices 2 refer to the helicities of the four gravitons. These simple results suggest that it should be possible to derive them in a much easier way. Grisaru et al. [62] showed that these amplitudes are completely determined, up to an overall factor, by kinematical singularities and crossing symmetry. If one assumes (i) Lorentz invariance, (ii) that propagators have the usual form $\sim p^{-2}$ and (iii) that the amplitudes are proportional to κ^2 (i.e. no non-minimal couplings are considered) then one can write down the graviton-graviton amplitudes at once; an overall scale factor is also easily determined by choosing a convenient kinematical point.

These results tell us that the amplitudes are proportional to the energy-squared for fixed angles, which follows already from dimensional arguments, and that some of them are nonzero. This should be a sign of nonrenormalizability. However, we know from Section 3 that pure gravity is one-loop finite. Berends and Gastmans [61] conclude that the connection between subtraction terms and nonrenormalizability is not tight; moreover, infrared divergences complicate the explicit calculation of the imaginary part of the one-loop corrections to graviton-graviton scattering.

In the context of the Feynman-Weinberg program, Boulware and Deser [63], showed that the low-frequency limit of the pure gravitational and matter-gravity vertices is determined by Lorentz invariance alone. The equivalence of the long-wavelength limit of quantum gravity and Einstein theory does not exclude the presence of nonminimal couplings (which are absent in Einstein theory) because these contain higher derivatives, hence vanish at low frequencies. Since these authors do not consider quantum effects, they do not introduce ghost particles. It thus follows that any quantum theory of gravitation must limit to Einstein theory for low frequencies, but the high frequency behavior is unrestricted.

8. Alternative Theories for Gravitation

The road of covariant quantization of gravitation has turned into a cul-de-sac. The rules of covariant quantization seem well established. Also the

---

†As in quantumelectrodynamics, also in gravitation algebraic manipulation programs are becoming indispensable for complicated analytic calculations. Besides "Schoonschip," there is also "Reduce" by A. C. Hearn and "Ashmedai" by J. Levine.

Einstein action is unique in its low frequencies. Nevertheless, when we apply the rules of covariant quantization to Einstein gravitation, the resulting theory is nonrenormalizable. If we insist that we should be able to reconcile quantum theory and general relativity, then either the rules of covariant quantization must be changed (as they were when Feynman and DeWitt introduced the unitarity restoring ghosts) or the models for gravitation must be modified. We will discuss this latter point and enumerate some alternatives.

(i) The chief problem of quantum gravity is its dimensional coupling constant κ. Weinberg [8] has suggested to consider Brans-Dicke [64] theory instead of Einstein theory, because in this theory the role of κ is taken over by a scalar field ϕ and no dimensional coupling constant is present. Unfortunately, it has been shown [47] that pure Brans-Dicke theory is equivalent to Einstein theory + a scalar particle, and this system is nonrenormalizable as we saw in Section 3.

(ii) Nonminimal couplings can be added. For example, in the scalar-gravity system, addition of the nonminimal coupling $R\phi^2$ makes the system better (but not well enough) behaved [43] in its ultraviolet region. Clearly, there is a large number of possibilities here.

(iii) If propagators would tend to p^{-4} instead of p^{-2} for large momenta, then the momentum dependence of the vertices might be compensated. A possibility is to replace Einstein theory by Weyl models with Lagrangians $\alpha R_{\mu\nu}R^{\mu\nu} + \beta R^2 + \gamma R$ [65]. A nice feature is that such Lagrangians are quadratic in the Christoffel symbols $\partial\Gamma$, in the same way as Maxwell theory is quadratic in the field tensor $F_{\mu\nu}$. It turns out that there are both terms proportional to $\kappa^2 p^{-4}$ and to p^{-2} in the propagator [66]. The unwanted terms proportional to p^{-2} do not contribute on the tree level due to gauge invariance but in loops this is not clear. There is another serious problem: theories with higher derivatives [67] lead to negative energy ghosts and/or to tachyons (particle with imaginary masses). There is no known method to get rid of these complications.

(iv) Strictly speaking, we only know that Einstein theory contains non-renormalizable divergences at the one-loop level when we couple to either scalars, or photons, or fermions, or Yang-Mills bosons, or to quantum electrodynamics.

In principle, complicated matter multiplets might give rise to extra cancellations between the divergences. Such cancellations usually happen in particle physics when a theory contains an extra symmetry. One might therefore turn to the matter multiplets of supersymmetry. Since Prof. Salam will discuss supersymmetry, I will not say more about this new symmetry between fermions and bosons. (A few remarks on the restrictions of unitarity to the divergence of arbitrary matter multiplets can be found in Reference 48.

(v) Massless particle theories have the unwelcome property that the threshholds of N-particle production coincide for all N. Perhaps one should turn this disadvantage into an advantage and apply nonperturbative methods to quantum gravity. Instead of using a basis with well-defined numbers of gravitons, one might turn to a description in terms of collective modes. One such framework is the theory of solitons, which is at the moment in the center of interest. It might even be that quantum gravity is not Borel-summable and that perturbation calculations necessarily lead to meaningless results.

(vi) As already remarked in the introduction, it could be that gravitation is a reflection of a deeper-lying, more fundamental, renormalizable theory. S. Weinberg has drawn a parallel with chiral theories: if instead of the renormalizable $\lambda\phi^4$ theory for pions somebody had started with one of the non-linear chiral Lagrangian theories for pions, then we might be stuck nowadays with the nonrenormalizable theory of "general pion dynamics." What can the deeper-lying theory for gravitation be? One possibility is that the graviton is a zero-mass spin 2 bound state of, say, two scalars. Lorentz invariance then leads to Einstein theory at low frequencies [68]. Another possibility is provided by dual models. When one requires unitarity at the one-loop level for systems of particles with spin 0, ½ and 1, this criterion selects the spontaneously broken gauge theories which unify the weak and electromagnetic interactions. When spin 2 particles are included, one-loop unitarity cannot be implemented unless an infinite number of high-spin particles is also added. The only known examples of relativistic quantum theories with an infinite spectrum of high-spin states and good asymptotic behavior are dual models. Scherk and Schwarz [69] have shown that one obtains Einstein gravitation by taking the zero-slope limit of a particular dual model. Since dual models are renormalizable, it is conceivable that the field-theoretic analogue of dual models (not only in the zero slope limit but for finite non-zero slope) provides a renormalizable model for quantum gravity.

(vii) Scale invariance might be useful as a symmetry which prohibits certain divergences. R. Kallosh [70] has investigated this possibility within the context of the background formalism. One may start from such scale invariant actions as Brans-Dicke theory or the square of the Weyl tensor or Yang's new theory [71]. Masses of matter constituents break local scale invariance and should be generated by spontaneous symmetry breaking [9]. However, it is known that local scale invariance is broken at the quantum level because anomalies are present [78, 79]. Although very interesting, these attempts are still in an exploratory stage.

9. Infrared Divergences

In the preceding sections we have discussed the ultraviolet divergences of quantum gravity. But there are two kinds of divergences in field theories: those due to the large momenta of virtual particles (the ultraviolet divergences) and those which occur when momenta of zero-mass virtual or real particles vanish (the infrared divergences). In this section we devote a few words to the infrared divergences in gravitation; they present no difficulties at all.

Physically there is little difference between a virtual or a real graviton when its momentum is small. The rather unphysical break-up of amplitudes into amplitudes with virtual gravitons and amplitudes with real gravitons (Bremsstrahlung amplitudes) makes both amplitudes infrared divergent. One expects, however, that adding in the cross section the squares of the virtual and real graviton amplitudes will again eliminate these unnatural infrared divergences. This is exactly what happens.

The square of the real graviton polarization tensor, summed over the two physical helicity states, is given by [73]

$$\sum_{\lambda=\pm} \varepsilon^{\lambda}_{\mu\nu}\varepsilon^{\lambda}_{\rho\sigma} = \tfrac{1}{2}(\bar{\delta}_{\mu\sigma}\bar{\delta}_{\nu\sigma} + \bar{\delta}_{\mu\sigma}\bar{\delta}_{\nu\rho} - \bar{\delta}_{\mu\nu}\bar{\delta}_{\rho\sigma}) \tag{22}$$

where $\bar{\delta}_{\mu\nu}(k) = \delta_{\mu\nu} - (k_\mu \bar{k}_\nu + k_\nu \bar{k}_\mu)/(k\cdot\bar{k})$ with $\bar{k}_\mu$ the parity reversed of k_μ and defined by $k = (\underset{\sim}{k}, ik_0)$ and $\bar{k} = (-\underset{\sim}{k}, ik_0)$. The k-dependent terms drop out of the amplitude due to gravitational gauge invariance and the remainder is the numerator of the graviton propagator in Eq. (8). The numerators of the squares of the virtual and real graviton contributions are thus equal, and the cancellation of infrared divergences proceeds from this point on, <u>both</u> for pure gravitation <u>and</u> for gravity-matter interactions as in quantum electrodynamics [74].

There is, however, one difference between gravitation and quantumelectrodynamics with respect to the infrared divergences: an emitted Bremsstrahlung graviton with energy E can emit in its turn a second soft graviton, and so on. As shown by S. Weinberg [75], the effective coupling constant for emission of a very soft graviton from a soft graviton (or photon) line with energy E is proportional to E and no infrared divergences are generated by such cascades of soft gravitons. (In Yang-Mills theories the vertices are not proportional to E and this theory is incurably(?) infrared divergent.) It follows that in pure gravitation diagrams with graviton-trees grafted on a Bremstrahlung gravitons are individually infrared convergent, while the cancellation of the remaining infrared divergences (those one also finds in quantum electrodynamics) proceeds as in quantum electrodynamics.

Capper [76] has verified in an example that no infrared divergences are created in Green's functions involving massive scalars and gravitons when the mass of the scalar particle tends to zero. It is satisfactory to know that the results of dimensional regularization for massless theories can also be obtained from the better defined massive theories by taking the limit of vanishing mass.

Finally we want to mention an interesting technical point concerning infrared divergences and dimensional regularization [77]. Both the untraviolet and the infrared divergences show up as poles at n=4. After one has parameterized a given Feynman diagram by means of auxiliary variables, the ultraviolet divergences show up during the loop-integration $\int d^n p$, while the infrared divergences show up during the auxiliary variable integration (for example in

$\int dx\, x^{n-5}$). A given integral can either be ultraviolet divergent or infrared divergent, but not both. Since the cross sections of quantum gravity are free from infrared divergences but contain ultraviolet divergences, a momentary hope that both kinds of divergences might cancel each other in quantum gravity is dashed.

10. Conclusions and Outlook

One can straightforwardly apply the ideas of covariant quantization of general gauge theories to gravitation. In the case of Einstein gravitation the results are

(1) The one-loop radiative corrections to the S-matrix of any purely gravitational process are finite. Two-loop properties depend on helicity conservation.

(2) All investigated matter-gravity systems are nonrenormalizable when both matter and gravity is quantized. This is due to the derivative couplings of internal gravitons, which in its turn is due to the dimension of G. In particular, Q.E.D. coupled to Einstein gravitation is unrenormalizable.

(3) Matter systems which are renormalizable in flat space remain renormalizable when only external gravitons are admitted. Particle physics in a curved classical background stays therefore finite.

An exceptional case is the gravitational radiative corrections to the lepton anomalous magnetic moment; these are finite.

As to future developments, many subjects will be investigated. Pure gravitation with a Weyl action (proportional to $R_{\mu\nu}^2$) might be freed from its negative energy ghosts. Nonminimal interactions might improve the high frequency behavior of quantized matter-gravity system. Yang's new theory of gravitation will certainly be investigated. Nonperturbative treatments (solitons) may give radically different results, as they correspond to a sum of divergent perturbation diagrams. Extra symmetries such as local scale invariance or supersymmetry might restrict the possible counterterms enough to obtain a finite S-matrix.

The failure to combine the particle physics version of quantum theory and general relativity pose a fundamental problem, since gravitation undeniably exists as a force in nature. Either our quantum theory must be modified, or other gravitational models should be considered, or we must leave gravitation unquantized which might be inconsistent according to a Bohr argument. Perhaps a total new idea is needed. When computations are made order by order, quantum gravity is seriously ill so far as conventional renormalizability is concerned. It will be interesting to hear next year what progress has been made.

REFERENCES

[1] R. Arnowitt and S. Fickler, Phys. Rev. 127, 1821 (1962).
[2] W. Heisenberg, Solvay Ber Kap. III, IV (1939).

[3] C. J. Isham, A. Salam and J. Strathdee, Anns. of Phys. 62, 98 (1971).
[4] R. Delbourgo, A. Salam and J. Strathdee, Lett. Nuov, Cim 2, 354 (1969); C. J. Isham, A. Salam and J. Strathdee, Phys. Rev. D3, 1805 (1971) and Phys. Rev. D5, 2548 (1972); R. Delbourgo and A. Hunt, Lett. Nuov. Cim. 4, 1010 (1970); J. Ashmore and R. Delbourgo, J. Math. Phys. 14, 176 (1973).
[5] J. A. Wheeler in "Geometrodynamics," Academic Press (N.Y. 1962).
[6] R. Penrose in "Magic Without Magic," W. H. Freeman and Co. (1972).
[7] B. Zumino in Brandeis Lectures in Theoretical Physics 1970 MIT Press (1970).
[8] S. Weinberg, private communication.
[9] F. Englert, E. Gunzig, C. C. Truffin and P. Winday, Phys. Lett. 57B, 73 (1975).
[10] A. Salam and J. Strathdee, Lett. Nuov. Cim 4, 101 (1970).
[11] C. J. Isham, A. Salam and J. Strathdee, Phys. Rev. D3, 867 (1971).
[12] C. J. Isham, A. Salam and J. Strathdee, Phys. Rev. D3, 867 (1971).
[13] B. S. DeWitt, Phys. Rev. 160, 1113 (1967) and 162, 1195 (1967) and 162, 1239 (1967).
[14] B. S. DeWitt in "Dynamical Theory of Groups and Fields," Les Houches 1963.
[15] B. S. DeWitt, "Magic Without Magic," W. H. Freeman & Co., San Francisco (1972).
[16] B. S. DeWitt, "Quantum Field Theory in Curved Spacetime," Physics Reports
[17] J. Honerkamp, Nucl. Phys. B48, 269 (1972).
[18] R. Kallosh, Nucl. Phys. B78, 293 (1974).
[19] S. Sarkar and H. Strubbe, Nucl. Phys. B90, 45 (1975).
[20] H. Kluberg-Stern and J.B. Zuker, Saclay preprint D-Ph-T 74-56, 74-83 and 75-28; I. Ya. Arefieva, L.D. Faddeev and A.A. Slavnov, Theor. and math. phys. (USSR)Dec(1973).
[21] M. Brown and M.J. Duff, Oxford preprint 1975.
[22] G. 't Hooft, Nucl. Phys. B62, 444 (1973).
[23] M. T. Grisaru, P. van Nieuwenhuizen and C.C. Wu, Phys. Rev. D. 12, 3203 (1975).
[24] L. D. Faddeev and V. N. Popov, Phys. Lett. 25B, 29 (1967).
[25] E. S. Fradkin and I.V. Tyutin, Phys. Rev. D2, 2841 (1970); C.J. Isham in "Quantum Gravity," ICTP preprint 72-8.
[26] R. P. Feynman, Act. Phys. Polon. 26, 697 (1973) and in "Magic Without Magic," W. H. Freeman and Co., San Francisco (1972); L. D. Faddeev and V. N. Popov, Sov. Phys Uspekhi 16, 777 (1975).
[27] S. Mandelstam, Phys. Rev. 175, 158 and 1604 (1968).
[28] E. Fermi, Rev. Mod. Phys. 4, 87 (1932).
[29] D. M. Capper and M. J. Duff, Nucl. Phys. B82, 147 (1974); D. G. Boulware and L. S. Brown, Phys. Rev. 172, 1628 (1968).
[30] M. J. Duff, Phys. Rev. D7, 2317 (1973); Sardelis (unpublished).
[31] M. J. Duff, Phys. Rev. D9, 1837 (1974).
[32] M. R. Brown, Nucl. Phys. B56, 194 (1973).
[33] D. M. Capper, G. Leibbrandt and M. Ramon Medrano, Phys. Rev. D8, 4320 (1973); D. M. Capper and M. Ramon Medrano. Phys. Rev. D9, 1641 (1974).
[34] D. M. Capper, M. J. Duff and L. Halpern, Phys. Rev. D10, 461 (1974).
[35] A. A. Slavnov, Theor. and Math. Phys. 10, 99 (1973).
[36] G. 't Hooft and M. Veltman, Nucl. Phys. B44, 189 (1972).
[37] C. G. Bollini and J. J. Giambiagi, Nuov, Cim, 12B, 20 (1972); J. F. Ashmore, Lett. Nuov. Cim. 4, 289 (1972).
[38] G. Leibbrandt, Introduction to dim. regularization, Rev. Mod. Phys. (Oct. 1975).
[39] D. M. Capper and G. Leibbrandt, Lett. Nuov. Cim. 6, 117 (1973); N. Nouri-Moghadam. and J. G. Taylor, Journal of Physics A, 8, 334 (1975).
[40] P. van Nieuwenhuizen, Nucl. Phys. B60, 478 (1973).
[41] M. Fierz and W. Pauli, Proc. Roy. Soc. 173A, 211 (1939).
[42] S. Deser, JGRG 1, 1 (1970).
[43] G. 't Hooft and M. Veltman, Ann. Inst. H. Poincare, XX, 1, 69 (1974).
[44] R. Bach, Math. Z. 9, 140 (1921); C. Lanczos. Ann. Math. 39, 842 (1938).
[45] M. T. Grisaru, P. van Nieuwenhuizen and C.C. Wu, Phys. Rev. D. 12, 1563 (1975).

[46] J. G. Taylor and M. Nouri-Moghadam, Proc. Roy. Soc. 344, of (1975) and "Non Rev. of Einstein - Spont. Symm. breaking vector and scalar meson int." King's College preprint March 1975.
[47] S. Deser and P. van Nieuwenhuizen, Phys. Rev. Letters 32, 245 (1974) and Phys. Rev. D10, 401 (1974).
[48] S. Deser and P. van Nieuwenhuizen, Lett. Nuov. Cim. 11, 218 (1974) and Phys. Rev. D10, 411 (1974).
[49] S. Deser, H-S Tsao, and P. van Nieuwenhuizen, Phys. Lett. 50B, 491 (1974), and Phys. Rev. D10, 3337 (1974).
[50] H. Weyl, Phys. Rev. 77, 699 (1950); T.W.B. Kibble, J. Math. Phys. 4, 1433 (1963).
[51] D. Z. Freedman, I. J. Muzinich and E. Weinberg, Ann. of Phys. (N.Y.) 87, 959 (1974).
[52] D. Z. Freedman and E. Weinberg, Ann. of Phys. (N.Y.) 87, 354 (1974).
[53] D. Z. Freedman and So-Young Pi, Ann. of Phys. (N.Y.)
[54] C. G. Callan, S. Coleman and R. Jackiw, Ann. of Phys. (N.Y.) 59, 42 (1970).
[55] R. Delbourgo and P. Phocas-Cosmetatos, Phys. Lett. 41B, 533 (1972).
[56] R. Delbourgo and P. Phocas-Cosmetatos, Lett. Nuov. Cim. 5, 420 (1972).
[57] M. T. Grisaru, P. van Nieuwenhuizen and C.C. Wu, Phys. Rev. D. 12, 1813 (1975).
[58] R. Delbourgo and A. Salam, Phys. Lett. 40B, 38 (1972).
[59] F. A. Berends and R. Gastmans, Phys. Lett. 55B, 311 (1975).
[60] R. Z. Roskies, Pittsburgh preprints 132, 137.
[61] F. A. Berends and R. Gastmans, Nucl. Phys. B88, 99 (1975).
[62] M. T. Grisaru, P. van Nieuwenhuizen and C. C. Wu, Phys. Rev. D. 12, 397 (1975).
[63] D. G. Boulware and S. Deser, Ann. of Phys. 89, 193 (1975).
[64] C. Brans and R. H. Dicke, Phys. Rev. 124, 925 (1961).
[65] A. Gavielides, T. K. Kuo and S. Y. Lee, to be published in Phys. Rev. D; H. Weyl in "Space, Time and Matter" (Dover, N.Y. 1950); P.A.M. Dirac, Proc. Roy. Soc. A 333, 403 (1973).
[66] M. Nouri-Moghadam and J. G. Taylor, King's College preprints, Feb. and March, 1975; K. Stelle, private communication; J. G. Taylor, Problems in quantum gravity, King's College preprint, June 1974.
[67] C. Bernard and A. Duncan, Phys. Rev. D11, 848 (1975).
[68] S. Weinberg, Phys. Rev. 138, B988 (1965).
[69] J. Scherk and J. H. Schwarz, Phys. Lett. 52B, 347 (1974).
[70] R. Kallosh, Phys. Lett. 55B, 321 (1975).
[71] C. N. Yang, Phys. Rev. Lett. 33, 445 (1974).
[72] D. M. Capper and M. J. Duff, Nuov. Cim. 23A, 173 (1974).
[73] H. van Dam and M. Veltman, Nucl. Phys. B22, 397 (1970).
[74] D. Yennie, S. Frautschi and H. Suura, Ann. Phys. (N.Y.) 13, 379 (1961).
[75] S. Weinberg, Phys. Rev. 140B, 516 (1965).
[76] D. M. Capper, Lett. Nuov. Cim. 10, 413 (1974).
[77] R. Gastmans and R. Meuldermans, Nucl. Phys. B63, 277 (1973).
[78] D. M. Capper and M. J. Duff, Phys. Lett. (to be published).
[79] S. Deser, M. T. Grisaru, P. van Nieuwenhuizen and C. C. Wu, Phys. Lett. 58 B, 355 (1975).
[80] J. Schwinger, UCLA preprint, August 1975.

THE CHRONOS PRINCIPLE AND THE INTERACTIONS OF FIELDS OF SPIN 0 AND 1

D. Christodoulou*

International Centre for Theoretical Physics, Trieste, Italy

ABSTRACT

The definition of time for a physical system consisting of the geometry of space in which there is any number of scalar and vector fields, is established as a fundamental physical principle. From this principle, which we call "chronos principle," in combination with the relativity principle, the Yang-Mills type of field interaction is uniquely deduced, without making any assumption about the form of the gauge transformation of the fields which leaves the action invariant.

I. INTRODUCTION AND SUMMARY

A new approach to the problem of quantization of general relativity theory was followed in Ref. [1], based on the idea that the intrinsic geometry of 3-dimensional space is the only directly measurable entity in general relativity theory if matter fields are absent. The initial value equations of the theory were understood to give the definition of time (cf. also Ref. [2]), once a sequence of 3-dimensionalgeometries is given. The quantum laws were then directly deduced from Feyman's formulation of the quantum principle, as applied to geometrodynamics.

In Ref. [3] the definition of time for a purely geometrical system was established in its abstract form as a fundamental principle. From this principle, which we called "chronos principle" in combination with the relativity principle, the geometrical Lagrangian of general relativity theory was uniquely deduced.

In Ref. [4], the work of Ref. [1] was extended to the case where non-interacting scalar and vector fields are present. It was shown that the definition of time was generalized in a natural way, the spatial configurations of the fields being included in the directly measurable entities.

In the present paper we consider the case of a physical system composed of the geometry of space in which there is any number of scalar fields and any number of vector fields. In Sec. II the definition of distance in the total configuration space is investigated, the quantum principle is reformulated and the generalized chronos principle is introduced. In Sec. III it is shown that the most general field Lagrangian which is consistent with the generalized chronos and relativity principles is the Yang-Mills Lagrangian [5-7], in which the vacuum expectation values of the scalar fields are translated by constants, thus giving mass to the vector fields [8-10]. Finally, the quantum theory is derived in Sec. IV and the consistency with the canonical approach to quantization is demonstrated.

II. THE GENERALIZED CHRONOS PRINCIPLE

The fundamental object in what we are about to consider is the space $\mathcal{C}_{TOT}$, defined by

$$\mathcal{C}_{TOT}\,[\mathcal{g},\ \phi_1\ldots\phi_n,\ W^1\ldots W^N] =$$

$$= \mathcal{C}[\mathcal{g}] \times \mathcal{C}[\phi_1] \times\ldots\times \mathcal{C}[\phi_n] \times \mathcal{C}[W^1] \times\ldots\times \mathcal{C}[W^N] \qquad (2.1)$$

where $\mathcal{C}[\mathcal{g}]$ is superspace [2,11], the configuration space of the 3-dimensional geometry of space [12], while $\mathcal{C}[\phi_a]$ is the configuration space of the scalar field ϕ_a and $\mathcal{C}[W^A]$ is the configuration space of the vector field W^A [13]. A

*On leave of absence from Department of Physics, University of Athens, Greece.

A particular element of $\mathcal{C}_{TOT}$ is a specific $[\mathcal{g}, \phi_1 \ldots \phi_n, W^1 \ldots W^N]$, namely a specific space geometry $\mathcal{g}$ endowed with the particular scalar field configurations $\phi_1 \ldots \phi_n$ and the particular vector field configurations $W^1 \ldots W^N$.

Any acceptable definition of the distance between two elements of $\mathcal{C}_{TOT}$ must be consistent with the requirement that it be a measure of the difference of the elements in question.

In order that the concept of distance of two elements of $\mathcal{C}_{TOT}$ is at all definable, it is necessary that there exists a one-to-one correspondence between the points of the two space geometries which the two elements contain. This correspondence must be a continuous mapping whose inverse is also continuous. As a consequence, the distance between two elements of $\mathcal{C}_{TOT}$ is definable only when the corresponding space geometries belong to the same topology. It follows that $\mathcal{C}_{TOT}$ is composed of disjoint connected subspaces each encompassing elements containing space geometries of a distinct topology.

We introduce in the usual way tangent vectors $[d\mathcal{g}, d\phi_1 \ldots d\phi_n, dW^1 \ldots dW^N]$ associated with differentiable lines passing through an element $[\mathcal{g}, \phi_1 \ldots \phi_n, W^1 \ldots W^N]$ of $\mathcal{C}_{TOT}$. The concept of orthogonality of two such vectors $[d\mathcal{g}, d\phi_1 \ldots d\phi_n, dW^1 \ldots dW^N]$ and $[dy', d\phi_1' \ldots d\phi_n', dW^{1'} \ldots dW^{N'}]$ is defined by a relation of the form

$$K[\mathcal{g}, \phi_1 \ldots, W^1 \ldots; d\mathcal{g}, d\phi_1 \ldots, dW^1 \ldots; d\mathcal{g}', d\phi_1' \ldots, dW^{1'} \ldots] = 0 \qquad (2.2)$$

We now demand that there be a one-to-one correspondence between the elements of the set of mutually orthogonal vectors defined at a certain element of $_{TOT}$ and those of the set of linearly independent vectors defined at that element. It follows that K must be linear in both vectors.

The length $d\ell_{TOT}$ of the vector $[d\mathcal{g}, d\phi_1 \ldots d\phi_n, dW^1 \ldots dW^N]$, as well as the distance of the elements $[\mathcal{g}, \phi_1 \ldots \phi_n, W^1 \ldots W^N]$ and $[\mathcal{g} + d\mathcal{g}, \phi_1 + d\phi_1, \ldots, \phi_n + d\phi_n, W^1 + dW^1, \ldots, W^N + dW^N]$ of $\mathcal{C}_{TOT}$ the infinitesimal difference of which this vector represents, is then accordingly defined as

$$d\ell^2_{TOT} = K[\mathcal{g}, \phi_1 \ldots, W^1 \ldots; d\mathcal{g}, d\phi_1 \ldots, dW^1 \ldots; d\mathcal{g}, d\phi_1 \ldots, dW^1 \ldots] \qquad (2.3)$$

If now there is a third vector $[d\mathcal{g}'', d\phi_1'' \ldots d\phi_n'', dW^{1''} \ldots dW^{N''}]$ which is the sum of the two aforementioned orthogonal vectors, namely $d\mathcal{g}'' = d\mathcal{g} + d\mathcal{g}'$, $d\phi_1'' = d\phi_1 + d\phi_1', \ldots, d\phi_n'' = d\phi_n + d\phi_n'$, $dW^{1''} = dW^1 + dW^{1'}, \ldots, dW^{N''} = dW^N + dW^{N'}$, it follows from the definition of orthogonality that

$$d\ell''^2_{TOT} = d\ell^2_{TOT} + d\ell'^2_{TOT} \qquad (2.4)$$

Once a one-to-one correspondence has been established between the points of the space geometries $\mathcal{g}$ and $\mathcal{g} + d\mathcal{g}$ belonging to the two aforementioned infinitesimally differing elements of $\mathcal{C}_{TOT}$, the components of the vector $[d\mathcal{g}, d\phi_1 \ldots d\phi_n, dW^1 \ldots dW^N]$ representing their difference can be expressed as $[dg_{ij}(^3x), d\phi_1(^3x) \ldots d\phi_n(^3x), dW^1_i(^3x) \ldots dW^N_i(^3x)]$, namely as the metric and field function differences at the corresponded points.

Now let us consider a special vector $[d\mathcal{g}_1, d\phi_{11} \ldots d\phi_{n1}, dW^1_1 \ldots dW^N_1]$ whose components $[dg^1_{ij}(^3x), d\phi_{11}(^3x) \ldots d\phi_{n1}(^3x), dW^1_{i1}(^3x) \ldots dW^N_{i1}(^3x)]$ are different from zero only in an infinitesimal neighbourhood of the point $(^3x_1)$ of volume d^3x. Such a vector represents an infinitesimal change in the geometry and field configuration of only a localized neighbourhood of space. We postulate that the length of this local vector is independent of the geometry and field configuration of all other neighbourhoods. It then follows from the definition of orthogonality that the square of its length $d\ell^2_{TOT}(^3x_1)\, d^3x$ can only be of the form:

$$d\ell^2_{TOT}(^3x_1)d^3 = (G^{ijmn}dg_{ij}dg_{mn} + 2G^{ija}dg_{ij}d\phi_a + 2G^{jAm}dg_{ij}dW^A_m + G^{ab}d\phi_a d\phi_b + G^{AmBn}dW^A_m dW^B_n + 2G^{aAm}d\phi_a dW^A_m)(^3z_1)d^3x \tag{2.5}$$

where the coefficients G^{ijmn}, G^{ija}, G^{ijAm}, G^{ab}, G^{AmBn} and G^{aAm} are functions of the metric and field functions and their derivatives at the point $(^3x_1)$.

Consider now another such local vector $[d\mathcal{g}_2, d\phi_{12} \ldots d\phi_{n2}, dW^1_2 \ldots dW^N_2]$, with components $[dg^2_{ij}(^3x), d\phi_{12}(^3x) \ldots d\phi_{n2}(^3x), dW^1_{i2}(^3x) \ldots dW^N_{i2}(^3x)]$ different from zero only in the infinitesimal neighbourhood of the point $(^3x_2)$, which is disjoint from that of the point $(^3x_1)$. We postulate that these two local vectors are orthogonal, since they represent independent geometrical and field changes. It then follows (cf. Ref. [3], Sec. 2) that the length $d\ell_{TOT}$ of a general vector $[d\mathcal{g}, d\phi_1 \ldots d\phi_n, dW^1 \ldots dW^N]$ is given by

$$d\ell^2_{TOT} = \int d\ell^2_{TOT}(^3x)\, d^3_x \tag{2.6}$$

As has been mentioned above, the components

$[dg_{ij}(^3x), d\phi_1(^3x) \ldots d\phi_n(^3x), dW^1_i(^3x) \ldots dW^N_i(^3x)]$ of the vector representing the difference of the elements $[\mathcal{g}, \phi_1 \ldots \phi_n, W^1 \ldots W^N]$ and $[\mathcal{g} + d\mathcal{g}, \phi_1 + d\phi_1 \ldots \phi_n + d\phi_n, W^1 + dW^1 \ldots W^N + dW^N]$ of $\mathcal{C}_{TOT}$ are defined only after a certain one-to-one correspondence has been established between the points of the geometries $\mathcal{G}$ and $\mathcal{G} + d\mathcal{G}$. Now this correspondence can a priori be done in an infinity of ways. We may think of the two elements in question as being the points corresponding to the values σ and σ + dσ of a path in $\mathcal{C}_{TOT}$, the parametric equations of which are

$$\mathcal{G} = \mathcal{G}(\sigma),\ \phi_1 = \phi_1(\sigma)\ldots\phi_n = \phi_n(\sigma),\ W^1 = W^1(\sigma)\ldots W^N = W^N(\sigma) \tag{2.7}$$

A change of the way of doing the correspondence will then be a transformation in the co-ordinate systems in which the geometries and field configurations are described

$$x^i = f^i({}^3x^*, \sigma) \tag{2.8}$$

the transformation depending on σ and hence being different for σ and σ + dσ. The correspondence change is thus manifested by the infinitesimal vectors

$$\xi^i = \frac{\partial f^i}{\partial \sigma} d\sigma \tag{2.9}$$

and the new metric and field function differences are given in terms of the original ones by:

$$dg^*_{mn} = \frac{\partial x^i}{\partial x^{*m}} \frac{\partial x^j}{\partial x^{*n}} (dg_{ij} + \xi_{j;j} + \xi_{i;i}) \tag{2.10}$$

$$d\phi^*_a = d\phi_a + \phi_a{}^{,i}\xi_i \tag{2.11}$$

$$dW^{A*}_m = \frac{\partial x^i}{\partial x^{*m}} (dW^A_i + (\xi \times \mathrm{rot}\ W^A)_i + (W^A_k \xi^k)_{,i}) \tag{2.12}$$

A different value of $d\ell^2_{TOT}$ will thus result from changing the correspondence between the two geometries belonging to the two given elements of $\mathcal{C}_{TOT}$. It is necessary, however, to demand the distance of two infinitesimally different elements of $\mathcal{C}_{TOT}$ to be unique. Therefore, a particular way of doing the correspondence should be preferred. In addition we must demand the distance between identical elements to be zero. It can be seen that it follows from these demands that we must accept only the correspondence which extremizes $d\ell^2_{TOT}$ and only the extremal $d\ell^2_{TOT}$ must we call true distance of the two elements of $\mathcal{C}_{TOT}$ in question.

Summarizing, we first make an arbitrary correspondence between the points of the two space geometries belonging to the two given elements of $\mathcal{C}_{TOT}$. We then change this correspondence through vectors ξ^i, and we vary those vectors ξ^i in order to extremize $d\ell^2_{TOT}$. The extremization conditions are

$$\frac{1}{2}\ \frac{\delta d\ell^2_{TOT}}{\delta\xi^i} = -2(G^{*ijmn}dg^*_{mn})_{;*j} + \phi^{*,*i}_a G^{ab} d\phi^*_b$$

$$- g^{*ip}\delta_{pjk}G^{A*\ell B*j}dW^{*B}_\ell(\mathrm{rot}\ W^{*A})^k - W^{*Ai}(G^{A*kB*j}dW^{*B}_j)_{;*k}$$

$$- 2(G^{*ija}d\phi^*_a)_{;*j} + \phi^{*,*i}_a G^{*kja}dg^*_{kj} \qquad (2.13)$$

$$- 2(G^{*ijA*m}dW^{*A}_m)_{;*j} - g^{*ip}\delta_{pmn}G^{*k\ell A*m}dg^*_{k\ell}\ (\mathrm{rot}\ W^{*A})^n$$

$$- W^{*Ai}(G^{*k\ell A*j}dg^*_{k\ell})_{;*j} + \phi^{*,*i}_a G^{aA*m}dW^{*A}_m$$

$$- g^{*ip}\delta_{pmn}\ G^{aA*m}d\phi^*_a\ (\mathrm{rot}\ W^{*A})^n - W^{*Ai}(G^{aA*m}d\phi^*_a)_{;*m} = 0$$

The above equations (through Eqs. (2.10)-(2.12)) are to be considered as equations determining the vectors ξ^i, namely the necessary correction to the way of doing the correspondence that must be made so that a consistent definition of distance in $\mathcal{C}_{TOT}$ may be obtained [14].

The correspondence of points implied by Eq. (2.13) involves the global vector $[d\mathcal{g}, d\phi_1 \ldots d\phi_n, dW^1 \ldots dW^N]$. It will thus be called "global correspondence" to distinguish it from the local one we shall speak of later.

We now assert that the quantum-mechanical history of the physical system under consideration is given by a continuous (1-parameter) sequence of functionals $\varphi[\mathcal{g}, \phi_1 \ldots \phi_n, W^1 \ldots W^N; \sigma]$. For any specific set of its arguments the functional is interpreted physically as the "probability amplitude" for the universe to arrive at the elements $[\mathcal{g}, \phi_1 \ldots \phi_n, W^1 \ldots W^N]$ of $\mathcal{C}_{TOT}$ with the value σ of the sequential parameter.

We further introduce the assumption that there exists a unique connecting rule in the aforementioned sequence, namely a rule by means of which from any given member of the sequence one can obtain any other member, and do so uniquely. It follows that there is a linear connection between the probability amplitude at a certain value of the parameter σ and that at a value infinitesimally differing from the first. Such a connection in general has the form

$$\varphi[\mathcal{g}', \phi_1' \ldots, W^{1'} \ldots; \sigma+\delta\sigma] = r \int_{\mathcal{C}_{TOT}} f[\mathcal{g}'(\sigma+\delta\sigma), \phi_1'(\sigma+\delta\sigma) \ldots, W^{1'}(\sigma+\delta\sigma) \ldots;$$

$$\mathcal{g}(\sigma), \phi_1(\sigma) \ldots, W^1(\sigma) \ldots] \, \varphi \, [\mathcal{g}, \phi_1 \ldots, W^1 \ldots; \sigma] \, dV_{C_{TOT}} \tag{2.14}$$

where r is a normalization constant and the integration is taken over all $\mathcal{C}_{TOT}$ for the set of arguments $[\mathcal{g}, \phi_1 \ldots \phi_n, W^1 \ldots W^N]$, $dV_{\mathcal{C}_{TOT}}$ being the volume element of $\mathcal{C}_{TOT}$, deduced in the usual way from its metric. It is noted that the connecting factor f depends on the elements of an infinitesimal path or tangent vector in $\mathcal{C}_{TOT}$. The most natural physical interpretation of f which is consistent with the given interpretation of the functional φ is that it represents the probability amplitude for the physical system to follow, during the parameter interval $d\sigma$, this infinitesimal path in $\mathcal{C}_{TOT}$ the end points of which is the pair of sets of arguments of f --provided that it is known to be at the end point $[\mathcal{g}, \phi_1 \ldots \phi_n, W^1 \ldots W^N]$ at σ.

By repeated application of the infinitesimal connecting rule (2.14), it can be shown that the connection between the probability amplitude at σ and that at σ', the interval between σ and σ' being finite, is given by

$$\varphi[\mathcal{g}', \phi_1' \ldots, W^{1'} \ldots; \sigma'] = R \int_{\mathcal{C}_{TOT}} \sum_{\text{all paths}} f[\mathcal{g}(\sigma), \phi_1(\sigma) \ldots, W^1(\sigma) \ldots]_\sigma^{\sigma'} \varphi[\mathcal{g}, \phi_1 \ldots, W^1 \ldots; \sigma] \, dV_{\mathcal{C}_{TOT}} \tag{2.15}$$

where R is an overall normalization constant and the summation is over all paths in $\mathcal{C}_{TOT}$ that have as end points the elements $[\mathcal{g}, \phi_1 \ldots \phi_n, W^1 \ldots W^N]$ at σ and $[\mathcal{g}', \phi_1' \ldots \phi_n', W^{1'} \ldots W^{N'}]$ at σ'. The connecting factor f in the above equation is the probability amplitude of one such path,

$$\mathcal{g} = \mathcal{g}(\sigma), \ \phi_1 = \phi_1(\sigma) \ldots \phi_n = \phi_n(\sigma), W^1 = W^1(\sigma) \ldots W^N = W^N(\sigma),$$

defined to be the product of the probability amplitudes of the infinitesimal paths which constitute it:

$$f[\mathcal{g}(\sigma), \phi_1(\sigma) \ldots, W^1(\sigma) \ldots]_\sigma^{\sigma'} = f[\mathcal{g}(\sigma'), \phi_1(\sigma') \ldots W^1(\sigma') \ldots; \mathcal{g}(\sigma'-\delta\sigma),$$

$$\phi_1(\sigma'-\delta\sigma), W^1(\sigma'-\delta\sigma) \ldots] \tag{2.16}$$

$$\times \ldots \times f[\mathcal{g}(\sigma+\delta\sigma), \phi_1(\sigma+\delta\sigma)\ldots, W^1(\sigma+\delta\sigma)\ldots; \mathcal{g}(\sigma), \phi_1(\sigma)\ldots, W^1(\sigma)\ldots]$$

Let us now specify the general form of the infinitesimal connecting factor which enters Eq. (2.14). First, it is evident from our physical interpretation of this factor that continuity of the probability amplitude distribution requires it to be only infinitesimally different from unity. We then postulate that this difference is proportional to the length $d\ell_{TOT}$ of the tangent vector $[\mathcal{g}' - \mathcal{g}, \phi_1' - \phi_1 \ldots, W^{1'} - W^1 \ldots]$ to which the connecting factor refers

$$f[\mathcal{g}'(\sigma+\delta\sigma), \phi_1'(\sigma+\delta\sigma)\ldots, W^{1'}(\sigma+\delta\sigma)\ldots; \mathcal{g}(\sigma), \phi_1(\sigma)\ldots, W^1(\sigma)\ldots] = 1 + iA_{TOT}d\ell_{TOT} \tag{2.17}$$

where A_{TOT} is an, as yet, arbitrary, complex in general, functional of $[\mathcal{g}, \phi_1 \ldots \phi_n, W^1 \ldots W^N]$. We can then deduce from Eq. (2.16) [3] the form of the finite connecting factor:

$$f[\mathcal{g}(\sigma), \phi_1(\sigma)\ldots, W^1(\sigma)\ldots]_{\sigma}^{\sigma'} = \exp\{ i\int_{\sigma}^{\sigma'} A_{TOT}d\ell_{TOT} \} \tag{2.18}$$

If the physical interpretation of the functional φ is to hold true, it is necessary that the law of conservation of probability

$$\frac{d}{d\sigma}\left\{\int_{\mathcal{C}_{TOT}} |\varphi[\mathcal{g}, \phi_1 \ldots \phi_n, W^1 \ldots W^N; \sigma]|^2 \, dV_{\mathcal{C}_{TOT}}\right\} = 0 \tag{2.19}$$

should be satisfied. It can be shown from Eqs. (2.18) and (2.15) that in order that this is so we have to set $\operatorname{Im} A_{TOT} = 0$.

We are therefore led to the following statement:

$$\varphi[\mathcal{g}', \phi_1' \ldots, W^{1'} \ldots; \sigma'] = R \int_{\mathcal{C}_{TOT}} \sum_{\text{all paths}} \left(\exp\{ i\int_{\sigma}^{\sigma'} A_{TOT}d\ell_{TOT}\}\right) \varphi[\mathcal{g}, \phi_1 \ldots, W^1 \ldots; \sigma] dV_{\mathcal{C}_{TOT}} \tag{2.20}$$

which constitutes the quantum principle for the physical system under consideration.

Finally, we do not consider as physically meaningful the question "what is the probability amplitude for the physical system to arrive at the configuration $[\mathcal{g}, \phi_1 \ldots \phi_n, W^1 \ldots W^N]$ with the value σ of the parameter?" Viewing the parameter as non-observable, we rather seek the probability amplitude for assuming this configuration independently of the value of the parameter. We thus define the physical probability amplitude $\psi[\mathcal{g}, \phi_1 \dots \phi_n, W^1 \dots W^N]$ as follows:

$$\psi[\mathcal{g}, \phi_1 \dots \phi_n, W^1 \dots W^N] = \int_{\text{all } \sigma} \varphi[\mathcal{g}, \phi_1 \dots \phi_n, W^1 \dots W^N; \sigma] d\sigma \tag{2.21}$$

The above requirement ensures the consistency of the quantum principle with the chronos principle to be stated next.

Consider now a continuous and piecewise differentiable (1-parameter) sequence $[\mathcal{g}, \phi_1 \dots \phi_n, W^1 \dots W^N; \sigma]$. Such a sequence represents a line in $_{TOT}$ possessing length. It is in relation to such sequences that we introduce the concept of time, defined through the concept of distance in $\mathcal{C}_{TOT}$ by

$$d\tau^2 = \frac{d\ell^2_{TOT}}{B_{TOT}} \quad , \tag{2.22}$$

where B_{TOT} us an as yet arbitrary functional of $[\mathcal{g}, \phi_1 \dots \phi_n, W^1 \dots W^N]$. Further, we assume that B_{TOT} has the form

$$B_{TOT} = \int b_{TOT} \sqrt{g}\, d^3x, \tag{2.23}$$

where b_{TOT} is an as yet arbitrary ordinary function of the metric and field functions and their derivatives.

It should be noted that the integrals in Eqs. (2.6) and (2.23) extend over the entire 3-dimensional space manifold. Correspondingly, the time defined by Eq. (2.22) refers to the change in configuration of the entire physical system. This time will therefore be called "global" time. On the other hand, since both $d\ell^2_{TOT}$ and B_{TOT} are simple integrals, we can restrict them both to a region U of space, defining

$$d\ell^2_{TOT_U} = \int_U d\ell^2_{TOT}(^3x)\, d^3x \tag{2.24}$$

and

$$B_{TOT_U} = \int_U b_{TOT} \sqrt{g}\, d^3x \tag{2.25}$$

We can then define the concept of "local" time $d\tau_U$ for the region U by:

$$d\tau_U^2 = \frac{d\ell^2_{TOT_U}}{B_{TOT_U}}$$

refering to the change in configuration in the region U alone. Finally, in the limit of an infinitesimal neighbourhood of the point $({}^3x)$,

$$d\tau({}^3x)^2 = \frac{d\ell^2_{TOT}({}^3x)}{b_{TOT}\sqrt{g}({}^3x)} \tag{2.26}$$

will be the local time at the point $({}^3x)$. We must point out however that the correspondence of points which is required in order to calculate $d\ell^2_{TOT}({}^3x)$ in the above equation is not the global one we have already discussed, through which $d\ell^2_{TOT}$ in Eq. (2.22) is calculated, but rather the local correspondence to which we shall come later.

Eqs. (2.22) and (2.26) constitute the chronos principle. They contain the statement that time is not a separate physical entity in which the changing of the physical system takes place. It is the measure of the changing of the physical system itself that is time.

Let us be given a line $[\mathcal{G} = \mathcal{G}(\sigma),\ \phi_1 = \phi_1(\sigma)\ldots,\ W^1 = W^1(\sigma)\ldots]$ in $\mathcal{C}_{TOT}$, and let us establish one-to-one correspondences [15] between the points of the space geometries which nearby elements of $\mathcal{C}_{TOT}$ on the line contain. Let us then define through Eq. (2.26) the local time intervals between these elements at every space point. The projection $\mathcal{G} = \mathcal{G}(\sigma)$ of the line in superspace constitutes a 4-dimensional riemannian space-time manifold if we define:

1) the 3-dimensional space geometries to be the intrinsic geometries of hypersurfaces in spacetime,

2) the infinitesimal normal vectors to these hypersurfaces to join the corresponded points of nearby 3-geometries,

3) the squares of the lengths of these normal vectors to be equal in absolute value to the squares of the local time intervals at the same space points.

It follows from the above definitions that the resulting space-time geometry is given by the following expression for the square of the 4-dimensional element of distance

$$^{(4)}ds^2 = \pm\, d\tau({}^3x)^2 + g^*_{mn}({}^3x^*,\sigma)\,dx^{*m}dx^{*n} \tag{2.27}$$

Wishing, further, to differentiate between spacelike and timelike distances, we choose the minus sign for the coefficient of $d\tau({}^3x)^2$ in the above equation.

If we identify the sequential parameter σ with the 4th co-ordinate x^4 and define

$$N(^4x) = d\tau(^3x)/dx^4 \tag{2.28}$$

we can write Eq. (2.27) in the form

$$^{(4)}ds^2 = -N^2(dx^4)^2 + g_{ij}(^4x)(dx^i - N^i dx^4)(dx^j - N^j dx^4) \tag{2.29}$$

where we have expressed the spatial part of $^{(4)}ds^2$ in the co-ordinate system ^3x (cf. Eq. (2.8)) and we have made the identification $\xi^i = N^i\, dx^4$. The components of the normal vectors joining the hypersurfaces x^4 = const. and $x^4 + dx^4$ are given in the co-ordinate system ^3x by

$$N^\mu = (N^i, 1) \tag{2.30}$$

since

$$g_{\mu i}\, N^\mu = 0 \tag{2.31}$$

while the length of those vectors, which is the normal separation of the hypersurfaces, is given by

$$g_{\mu\nu}\, N^\mu N^\nu = -N^2 \tag{2.32}$$

It should be stressed that Eq. (2.29) gives the space-time geometry in a specific manner: by giving the intrinsic and extrinsic geometries of a specific sequence of hypersurfaces. Space-time, however, can be sliced into spacelike hypersurfaces (hypersurfaces the intrinsic geometries of which possess a positive definite metric) in an arbitrary manner, and an infinity of such sequences may be obtained. Each sequence offers through Eq. (2.29) a particular "representation" of the space-time geometry. Since every sequence of spacelike hypersurfaces represents a line in superspace, it follows that, although to each such line corresponds a single space-time geometry, the reverse is not true. The 4-dimensional geometry of space-time must thus be envisaged as the equivalence class of all lines in superspace that are related by a change in the way of slicing:

$$x'^4 = h(^3x, x^4) \tag{2.33}$$

The statement that the laws of physics are independent of the way of slicing space-time into spacelike hypersurfaces, depending only on the 4-dimensional space-time geometry and not on its particular representation, constitutes the relativity principle. Since a change in the way of slicing as given by Eq. (2.33) may be combined by a spatial co-ordinate transformation as that of Eq. (2.8), namely

$$x'^i = f^i(^3x, x^4)$$

to get a general 4-dimensional co-ordinate transformation

$$x'^\mu = f^\mu(^4x) \tag{2.34}$$

the relativity principle demands the laws of physics to be of a form which is invariant under general 4-dimensional co-ordinate transformations.

Let us now return to the quantum principle expressed by Eq. (2.20), and let us call "total action" S_{TOT} the phase of the exponential

$$S_{TOT} = \int_{path} A_{TOT} d\ell_{TOT} \tag{2.35}$$

As we have shown above, the projection in superspace of a line in $\mathcal{C}_{TOT}$ represents a space-time geometry if the chronos principle is taken into account. We may thus express S_{TOT}, which is an as yet arbitrary path integral in $\mathcal{C}_{TOT}$, as a space-time integral, by employing the definition of global time [Eq. (2.22)]. In doing so, we must take into account the following two restrictions motivated by the necessity that the global correspondence conditions [Eq. (2.13)] and the definition of global time be part of the laws of the physical theory:

1) The extremization of S_{TOT} with respect to the correspondence of space points should give back the global intercorrespondence conditions.

2) The extremization of S_{TOT} with respect to the global time interval $d\tau$ should give back the definition of global time.

It can be seen that restriction 1) tells us that the vector $[d\mathcal{g}, d\phi_1 \ldots d\phi_n, dW^1 \ldots dW^N]$ enters S_{TOT} only in terms of its length $d\ell_{TOT}$. The most general expansion of S_{TOT} from a path integral in $\mathcal{C}_{TOT}$ to a space-time integral that conforms with this restriction is the following:

$$S_{TOT} = \int_{path} A_{TOT} d\ell_{TOT} = \frac{1}{\sum_{n=0}^{\infty} c_n} \int \left\{ c_0 B_{TOT}^{1/2} + c_1 \frac{d\ell_{TOT}}{d\tau} + \frac{c_2}{B_{TOT}^{1/2}} \left(\frac{d\ell_{TOT}}{d\tau}\right)^2 + \ldots \right\} A_{TOT} d\tau \tag{2.36}$$

where the coefficients c_n are constants. Restriction 2) then requires us to set

$$c_0 = c_2 + 2c_3 + 3c_4 + \dots \tag{2.37}$$

We now take into account the relativity principle and we require S_{TOT}, in which through the quantum principle the physical laws are contained, to be invariant under 4-dimensional co-ordinate transformations. Since in Eq. (2.36) only a single integration over τ occurs, it is clear that S_{TOT} must be a simple invariant integral over the 4-dimensional space-time which is represented by the projection in superspace of the given path in $\mathcal{C}_{TOT}$. Such an integral generally has the form

$$S_{TOT} = \int l_{TOT} \sqrt{-{}^{(4)}g}\ d^4x \tag{2.38}$$

It then follows from the fact that B_{TOT} and $d\ell^2_{TOT}$ are simple space integrals that

$$\begin{aligned} c_1 &= 0 \\ c_3 &= c_4 = \dots = 0 \end{aligned} \tag{2.39}$$

and

$$A_{TOT} = c' B^{1/2}_{TOT} \tag{2.40}$$

c' being a constant. Hence S_{TOT} can only be of the form [cf. Eq. (2.37)]

$$S_{TOT} = \frac{c'}{2} \int \left\{ \left(\frac{d\ell_{TOT}}{d}\right)^2 + B_{TOT} \right\} d\tau \tag{2.41}$$

It was shown in Ref. [3] that in the case where the fields are absent the only relativistically invariant functional consistent with the above equation is

$$S_G = k \int ({}^{(4)}R + \lambda) \sqrt{-{}^{(4)}g}\ d^4x \tag{2.42}$$

The constant k can be set equal to 1/2 by an appropriate choice of the unit of distance (choosing the latter to be equal to the Plank length $\sqrt{8\pi G\hbar/c^3}$). Then we can choose c' to be equal to 1 without loss of generality. If space-time is sliced in such a way that the normal separation of nearby hypersurfaces is constant over the hypersurfaces, then [cf. Eq. (2.29)] N is independent of ${}^3x$ and all local time intervals are equal to the global time interval

$$d\tau({}^3x) = d\tau \tag{2.43}$$

In this frame, the "geometrical action" S_G assumes the form given by Eq. (2.41) after the necessary integrations by parts have been performed:

$$S_G = \frac{1}{2}\iint \left\{ \frac{1}{4}\left(\frac{g^{*im}g^{*jn} + g^{*in}g^{*jm}}{2} - g^{*ij}g^{*mn} \right) \frac{dg^*_{ij}}{d\tau}\frac{dg^*_{mn}}{d\tau} + R^* + \lambda \right\} \sqrt{g^*}\, d^3x^*\, d\tau \tag{2.44}$$

By comparing the above expression with that of Eq. (2.41) we conclude that the geometric parts of $d\ell^2_{TOT}$ and B_{TOT} are given by

$$d\ell^2_G = \int \frac{1}{4}\left(\frac{g^{*im}g^{*jn} + g^{*in}g^{*jm}}{2} - g^{*ij}g^{*mn} \right) dg^*_{ij}\, dg^*_{mn} \sqrt{g^*}\, d^3x^* \tag{2.45}$$

$$B_G = \int (R^* + \lambda) \sqrt{g^*}\, d^3x^* \tag{2.46}$$

In a general frame S_G is given by

$$S_G = \frac{1}{2}\int d\tau(^3x) \int \left\{ \left(\frac{d\ell_G(^3x)}{d\tau(^3x)} \right)^2 + b_G \sqrt{g^*} \right\} d^3x^* \tag{2.47}$$

which results directly from Eq. (2.44) by replacing the global time interval $d\tau$ by the local time interval $d\tau(^3x)$. Correspondingly, S_{TOT} is expressed in a general frame by:

$$S_{TOT} = \frac{1}{2}\int d\tau(^3x) \int \left\{ \left(\frac{d\ell_{TOT}(^3x)}{d\tau(^3x)} \right)^2 + b_{TOT} \sqrt{g^*} \right\} d^3x^* \tag{2.48}$$

By extremizing the above expression with respect to the correspondence of space points, or, equivalently, with respect to N^i, we obtain the local correspondence conditions, which result directly from the global ones given by Eq. (2.13) by replacing

$$dg^*_{ij} \text{ by } \frac{dg^*_{ij}}{d\tau(^3x)}\, d\tau, \quad d\phi^*_a \text{ by } \frac{d\phi^*_a}{d\tau(^3x)}\, d\tau, \quad dW^{A*}_i \text{ by } \frac{dW^{A*}_i}{d\tau(^3x)}\, d\tau$$

In every local region in which $d\tau(^3x)$ = const. the local correspondence conditions will naturally be of a form which is identical to that of the global ones. The extremization of S_{TOT}, as given by Eq. (2.48), with respect to $d\tau(^3x)$, or, equivalently, with respect to N, will reproduce the definition of local time, Eq. (2.26).

We now draw attention to the fact that relativistic invariance requires that W_i^A to be the 3-dimensional components of 4-vectors W_μ^A. On the other hand, the W_4^{A*} do not enter S_{TOT} as given by Eq. (2.48). The only way in which this can be consistent with the relativity principle is if S_{TOT} is invariant under gauge transformations of the fields, involving N arbitrary functions χ^A. Then Eq. (2.48) must be said to express S_{TOT} in the gauge

$$W_4^{A*} = 0 \tag{2.49}$$

or [written in the co-ordinate system $({}^3x)$],

$$N^\mu W_\mu^A = 0 \tag{2.50}$$

No derivatives of W_4^A with respect to x^4 should be contained in S_{TOT} expressed in a general gauge, if the chronos principle is to be consistent with the relativity principle. It is clear that in S_{TOT} as given by Eq. (2.48) no such derivatives of N^i and N exist. The same should also be true in a general gauge. In the language of the canonical formalism we shall say that N^i, N and W_4^A are not dynamical variables.

Let us now consider two vectors $[d\mathcal{g}, 0...0, 0...0]$ and $[0, d\phi_1...d\phi_n, dW^1...dW^N]$ in $\mathcal{C}_{TOT}$, one of which is purely geometrical and the other purely field-like. We introduce the assumption that these vectors are orthogonal motivated from the desire that they refer to independent changes. We therefore set

$$G^{ija} = 0 \ , \ G^{ijAm} = 0 \tag{2.51}$$

In addition we assume that the length of a purely geometrical vector is independent of the field configuration. It follows that this length is the same as in the case when the fields are absent. Hence,

$$G^{ijmn} = \frac{\sqrt{g}}{4}\left(\frac{g^{im}g^{jn}+g^{in}g^{jm}}{2} - g^{ij}g^{mn}\right) \tag{2.52}$$

It follows from Eqs. (2.51) and (2.52) that if we write $d\ell^2_{TOT}$ as

$$d\ell^2_{TOT} = d\ell^2_G + d\ell^2_F \tag{2.53}$$

then the entity $d\ell^2_F$ thus defined, is the square of the element of distance in the configuration space of the fields $\mathcal{C}[F]$ defined by

$$\mathcal{C}[F] = \mathcal{C}[\phi_1] \times \ldots \times \mathcal{C}[\phi_n] \times \mathcal{C}[W^1] \times \ldots \times \mathcal{C}[W^N] \tag{2.54}$$

It follows from Eq. (2.53) that the volume element of $_{TOT}$ is given by

$$dV_{\mathcal{C}_{TOT}} = dV_{\mathcal{C}[g]} \, dV_{\mathcal{C}[F]} \tag{2.55}$$

where $dV_{\mathcal{C}[g]}$ is the volume element of superspace

$$dV_{\mathcal{C}[g]} = \prod_{\text{all } ^3x} G^{1/2}(^3x) \prod_{i \geq j} dg_{ij}(^3x) \tag{2.56}$$

while $dV_{[F]}$ is the volume element of the configuration space of the fields.

If we then define B_F by

$$B_{TOT} = B_G + B_F \tag{2.57}$$

and, finally, the "field action" S_F by

$$S_{TOT} = S_G + S_F \tag{2.58}$$

then S_F, which is expressed in a general frame by

$$S_F = \frac{1}{2} \int d\tau(^3x) \int \left\{ \left(\frac{d\ell_F(^3x)}{d\tau(^3x)} \right)^2 + b_F \sqrt{g^*} \right\} d^3x^* \; , \tag{2.59}$$

does not contain time derivatives of g_{ij}. It can easily be proved that the only way in which this can hold in consistency with the relativity principle, is if space derivatives are also absent, indeed if derivatives of the 4-dimensional metric tensor $g_{\mu\nu}$ with respect to all four space-time co-ordinates are absent from S_F.

We now introduce the additional assumption that $\mathcal{C}[F]$ is euclidean. We can then, without loss of generality, set

$$G^{ab} = \delta^{ab} \sqrt{g}, \; G^{AmBn} = \delta^{AB} g^{mn} \sqrt{g} \; ; \; G^{aAm} = 0 \tag{2.60}$$

Hence

$$d\ell_F^2 = \sum_{a=1}^{n} d\ell_{\phi_a}^2 + \sum_{A=1}^{N} d\ell_{W^A}^2 \tag{2.61}$$

where

$$d\ell_{\phi_a}^2 = \int (d\phi_a)^2 \sqrt{g} \, d^3x \tag{2.62}$$

and

$$d\ell^2_W{}^A = \int g^{mn} dW^A_m dW^A_n \sqrt{g}\ d^3x \tag{2.63}$$

Then the volume element of $\mathcal{C}[F]$ is given by

$$dV_{\mathcal{C}[F]} = dV_{\mathcal{C}[\phi_1]} \cdots dV_{\mathcal{C}[\phi_n]} dV_{\mathcal{C}[W^1]} \cdots dV_{\mathcal{C}[W^N]} \tag{2.64}$$

where

$$dV_{\mathcal{C}[\phi_a]} = \prod_{\text{all } ^3x} g^{1/4}(^3x)\, d\phi_a(^3x) \tag{2.65}$$

and

$$dV_{\mathcal{C}[W^A]} = \prod_{\text{all } ^3x} g^{1/4}(^3x)\, d^3W^A(^3x) \tag{2.66}$$

In the classical limit, i.e., the limit of large S_{TOT}, only the paths in $\mathcal{C}_{TOT}$ for which S_{TOT} is extremal contribute in the sum of Eq. (2.20). These extremization conditions are

$$\frac{\delta S_{TOT}}{\delta g_{ij}} = 0, \quad \frac{\delta S_{TOT}}{\delta \phi_a} = \frac{\delta S_F}{\delta \phi_a} = 0, \quad \frac{\delta S_{TOT}}{\delta W^A_i} = \frac{\delta S_F}{\delta W^A_i} = 0 \tag{2.67}$$

The above equations are supplemented by the correspondence conditions and the definition of time which, as we saw, are expressed by the equations obtained by extremizing S_{TOT} with respect to N^i and N, or, equivalently, by

$$\frac{\delta S_{TOT}}{\delta g_{4i}} = 0, \quad \frac{\delta S_{TOT}}{\delta g_{44}} = 0 \tag{2.68}$$

In addition we must extremize S_{TOT}, expressed in a general gauge, with respect to W^A_4, setting

$$\frac{\delta S_{TOT}}{\delta W^A_4} = \frac{\delta S_F}{\delta W^A_4} = 0, \tag{2.69}$$

to obtain the full set of classical equations

$$\frac{\delta S_{TOT}}{\delta g_{\mu\nu}} = 0, \quad \frac{\delta S_{TOT}}{\delta \phi_a} = 0, \quad \frac{\delta S_{TOT}}{\delta W^A_\mu} = 0 \tag{2.70}$$

generally covariant in 4-dimensions as required by the relativity principle.

It can be seen from the fact that g_{4i}, g_{44} and W^A_4 are non-dynamical variables, and from the form of S_{TOT} given by Eq. (2.48) that Eqs. (2.68) and (2.69) contain only first derivatives with respect to x^4, while Eqs. (2.67) contain also second, and those second derivatives only linearly. It follows that Eqs. (2.68) and (2.69) impose constraints on the initial value data necessary in order to integrate Eqs. (2.67). Now it has been proved [16] that it follows from the relativistic invariance of S_{TOT} that given that Eqs. (2.68) are satisfied initially and that all the field equations ($\delta S_F/\delta\phi_a = 0$, $\delta S_F/\delta W^A_\mu = 0$) are satisfied at all x^4, the dynamical geometrical equations

$$\frac{\delta S_{TOT}}{\delta g_{ij}} = 0$$

guarantee that Eqs. (2.68) are always satisfied. We shall prove in the next section that it similarly follows from the gauge invariance of S_F that Eq. (2.69) hold always provided only that they are initially satisfied and that the dynamical field equations

$$\frac{\delta S_F}{\delta\phi_a} = 0, \quad \frac{\delta S_F}{\delta W^A_i} = 0$$

hold at all x^4. It will thus turn out that Eqs. (2.68) and (2.69) are truly initial value equations.

III. DEDUCTION OF THE FIELD LAGRANGIAN

We now define the field Lagrangian function I_F by

$$S_F = \int I_F \sqrt{-{}^{(4)}g}\, d^4x \tag{3.1}$$

From Eqs. (2.59), (2.61) and (2.11), (2.12) and (2.28) we obtain the following expression for I_F in the gauge (2.49) or, equivalently, (2.50):

$$I_F = \frac{1}{2}\frac{d\phi^*_a}{d\tau(^3x)}\frac{d\phi^*_a}{d\tau(^3x)} + \frac{1}{2}g^{*ij}\frac{dW^{A*}_i}{d\tau(^3x)}\frac{dW^{A*}_j}{d\tau(^3x)} + \frac{b^*_F}{2}\,(g^*_{ij};\ \phi^*_1\ldots\phi^*_n,\ W^{1*}\ldots W^{N*} \text{ and their spatial derivatives})$$

$$= \frac{1}{2N^2}\left(\frac{\partial\phi_a}{\partial x^4} + N^i\phi_{a,i}\right)\left(\frac{\partial\phi_a}{\partial x^4} + N^j\phi_{a,j}\right)$$
$$+ \frac{g^{ij}}{2N^2}\left(\frac{\partial W^A_i}{\partial x^4} + (N\mathrm{xrot}\, W^A)_i + (N^kW^A_k)_{,i}\right)\left(\frac{\partial W^A_j}{\partial x^4} + (N\mathrm{xrot}\, W^A)_j + (N^\ell W^A_\ell)_{,j}\right)$$
$$+ \frac{b_F}{2}\,(g_{ij};\ \phi_1\ldots\phi_n,\ W^1\ldots W^N \text{ and their spatial derivatives}) \tag{3.2}$$

To this Lagrangian we shall, in this section, apply the requirements of relativistic and gauge invariance.

Let us first consider the case of a single scalar field ϕ. In this case I_F is given by

$$I_F = \frac{1}{2N^2} \left(\frac{\partial \phi}{\partial x^4} + N^i \phi_{,i} \right) \left(\frac{\partial \phi}{\partial x^4} + N^i \phi_{,j} \right) + \frac{b_F}{2} \quad (g_{ij}; \ \phi \text{ and its spatial derivatives})$$

(3.3)

Since there are no vector fields present, there are no gauge functions, and the only requirement we need to impose on I_F is relativistic invariance. Observing the fact that the components $g^{\mu\nu}$ of the contravariant 4-dimensional metric tensor are given by

$$g^{44} = -\frac{1}{N^2}, \quad g^{4i} = -\frac{N^i}{N^2}, \quad {}^{(4)}g^{ij} = g^{ij} - \frac{N^i N^j}{N^2}$$

(3.4)

we conclude that the first term on the right-hand side of Eq. (3.3) can be made relativistically invariant if a part

$$-\frac{1}{2} g^{ij} \phi_{,i} \phi_{,j}$$

is added to it from the second term $\frac{b_F}{2}$. Then the first term becomes $-\frac{1}{2} g^{\mu\nu} \phi_{,\mu} \phi_{,\nu}$, the so-called "kinetic" part of the Lagrangian. In order to maintain relativistic invariance, no more spatial derivatives should be contained in b_F. Hence the most general Lagrangian for a single scalar field which is consistent with our hypotheses is

$$I_F = -\frac{1}{2} \phi_{,\mu} \phi^{,\mu} + f(\phi)$$

(3.5)

devoid of derivative self-coupling.

In the case of many scalar fields ϕ_a the same argument shows that the Lagrangian should be of the form:

$$I_F = -\frac{1}{2} \phi_{a,\mu} \phi_a^{,\mu} + f.(\phi_1 \ldots \phi_n)$$

(3.6)

again without derivative couplings.

If we have many scalar fields ϕ_a which are complex rather than real as we have hitherto assumed, we may decompose each of them in two real scalar fields:

$$\phi_a = \text{Re}\phi_a + i\,\text{Im}\phi_a$$

(3.7)

Then, from (3.6), the Lagrangian is given by

$$l_F = -\frac{1}{2}(\mathrm{Re}\phi_a)_{,\mu}(\mathrm{Re}\phi_a)^{,\mu} - \frac{1}{2}(\mathrm{Im}\phi_a)_{,\mu}(\mathrm{Im}\phi_a)^{,\mu}$$

$$+ f(\mathrm{Re}\phi_1 \ldots \mathrm{Re}\phi_n;\ \mathrm{Im}\phi_1 \ldots \mathrm{Im}\phi_n)$$

$$= -\frac{1}{2}\phi_{a,\mu}\phi_a^{*,\mu} + f(\phi_1 \ldots \phi_n;\ \phi_1^* \ldots \phi_n^*) \tag{3.8}$$

where the function f is real.

Let us now turn to the case of a single vector field W and no scalar fields. Observing the fact that Eq. (3.2) refers to the gauge (2.50) where $W_i N^i = -W_4$, we can write the relevant Lagrangian in the form

$$l_F = \frac{g^{ij}}{2N^2}\left(\frac{\partial W_i}{\partial x^4} + (N\times\mathrm{rot}\ W)_{,i} - W_{4,i}\right)\left(\frac{\partial W_i}{\partial x^4} + (N\times\mathrm{rot}\ W)_{,j} - W_{4,j}\right)$$

$$+ \frac{b_F}{2}(g_{ij};\ W \text{ and its spatial derivatives}). \tag{3.9}$$

Again, taking into account Eqs. (3.4) we conclude that the first term on the right-hand side of the above equations can be made relativistically invariant if a part

$$-\frac{1}{4}g^{im}g^{jn}f_{ij}f_{mn}, \text{ where } f_{ij} = W_{i,j} - W_{j,i}$$

is added to it from $\frac{b_F}{2}$. The kinetic part of the Lagrangian then becomes

$$-\frac{1}{4}g^{\mu\alpha}g^{\nu\beta}f_{\mu\nu}f_{\alpha\beta}, \text{ where } f_{\mu\nu} = W_{\mu,\nu} - W_{\nu,\mu}$$

Again, no more spatial derivatives of W_i should be contained in b_F, and the Lagrangian in a general gauge is given by

$$l_F = -\frac{1}{4}f_{\mu\nu}f^{\mu\nu} + g(g_{\mu\nu}, W_\mu) \tag{3.10}$$

To the action resulting from this Lagrangian we must now apply the requirement of invariance under gauge transformations involving one arbitrary function χ. Taking into account the fact that the field Lagrangian does not contain 2nd or higher order derivatives of the fields, we see that the infinitesimal form of this transformation should, most generally, be:

$$W'_\mu = W_\mu + g_\mu\chi + h\chi_{,\mu} \tag{3.11}$$

(not involving derivatives of the gauge function χ beyond those of the first order) where

$$g_\mu = g_\mu(g_{\alpha\beta}, W_\alpha, f_{\alpha\beta}), \quad h = h(g_{\alpha\beta}, W_\alpha, f_{\alpha\beta}), \tag{3.12}$$

keeping in mind the fact that no metric derivatives enter the field Lagrangian. We must then demand the change in S_F resulting from this transformation to vanish:

$$\delta S_F = \int \frac{\delta S_F}{\delta W_\mu} \delta W_\mu d^4x = \int \frac{\delta S_F}{\delta W_\mu} (g_\mu \chi + h\chi_{,\mu}) d^4x$$

$$= \int \left\{ - \left(h \frac{\delta S_F}{\delta W_\mu} \right)_{;\mu} + g_\mu \frac{\delta S_F}{\delta W_\mu} \right\} \chi d^4x = 0 \tag{3.13}$$

Since the above should be true for an arbitrary infinitesimal gauge function χ we conclude that we must have

$$- \left(h \frac{\delta S_F}{\delta W_\mu} \right)_{;\mu} + g_\mu \frac{\delta S_F}{\delta W_\mu} = 0 \tag{3.14}$$

for any vector field W_μ. The above equation is a functional differential constraint on S_F. Inserting the Lagrangian of Eq. (3.10) and taking into account the identity

$$f^{\mu\nu}{}_{;\nu\mu} = 0 \tag{3.15}$$

we obtain

$$(g_\mu - h_{,\mu}) f^{\mu\nu}{}_{;\nu} - \left(h \frac{\partial g}{\partial W_\mu} \right)_{;\mu} + g_\mu \frac{\partial g}{\partial W_\mu} = 0 \tag{3.16}$$

The first thing we can deduce from the above equation is that h does not depend on $f_{\alpha\beta}$, since if it did, the part $-h_{,\mu} f^{\mu\nu}{}_{;\nu}$ of the first term would be the only one in Eq. (3.16) quadratic in the second derivatives of W_μ, and it would have to vanish separately. It then follows that the term

$$(g_\mu - h_{,\mu}) f^{\mu\nu}{}_{;\nu}$$

is the only one in Eq. (3.16) containing second derivatives of W_μ. Hence the coefficient of those derivatives should vanish

$$g_\mu - h_{,\mu} = 0 \tag{3.17}$$

Since, however, any invariant h other than a simple constant should contain the metric tensor $g_{\alpha\beta}$, there is no g_μ which is independent of metric derivatives that satisfies Eq. (3.17). Hence $g_\mu = 0$ and h = const., which can be taken equal to one without loss of generality. It then follows from Eq. (3.16) that g = const., absorbed in the constant λ of S_G [cf. Eq. (2.42)]. We conclude that the most general Lagrangian for a single vector field W which is consistent with our hypotheses is

$$l_F = -\frac{1}{4} f_{\mu\nu} f^{\mu\nu} \tag{3.18}$$

consisting only of the kinetic term and invariant under the gauge transformation

$$W'_\mu = W_\mu + \chi_{,\mu} \tag{3.19}$$

We should note that Eq. (3.19) expresses not only the infinitesimal but also the finite form of the gauge transformation. It follows from Eq. (3.18) that:

1) a vector field can have no self-interaction;

2) a single vector field can have no mass.

Going back to Eq. (3.14), it follows from the gauge invariance of S_F under the gauge transformation (3.19) that the classical field equations

$$\frac{\delta S_F}{\delta W_\mu} = 0$$

satisfy the identities

$$\left(\frac{\delta S_F}{\delta W_\mu}\right)_{;\mu} = 0 \tag{3.20}$$

It is evident from these identities that if the equation $\delta S_F/\delta W_4 = 0$ is satisfied on one spacelike hypersurface, the dynamical field equations $\delta S_F/\delta W_i = 0$ guarantee that it is satisfied over all space-time.

Let us now consider the case of many vector fields W^A $(A = 1\ldots N)$. We write the Lagrangian as

$$l_F = l_{F_0} + l_{F_1} \quad \text{and correspondingly } S_F = S_{F_0} + S_{F_1} \tag{3.21}$$

where l_{F_0} is the "free" Lagrangian, the sum of the Lagrangians of the single vector fields

$$l_{F_0} = -\frac{1}{4} f^A_{\mu\nu} f^{A\mu\nu}, \text{ where } f^A_{\mu\nu} = W^A_{\mu,\nu} - W^A_{\nu,\mu} \tag{3.22}$$

It then follows from Eq. (3.2) that the interaction Lagrangian l_{F_1} is of the form

$$l_{F_1} = l_{F_1}(g_{ij} \ ; W^1\ldots W^N \text{ and their spatial derivatives}), \tag{3.23}$$

expressed in the gauge (2.50). Here, however, contrary to the case of a single vector field, it does not follow from the requirement of relativistic invariance of l_{F_1} that it should be free of any derivatives of the W^A's. Indeed if such derivatives enter only through expressions

$$f^{ABC} = f^{A}_{\mu\nu} W^{B\mu} W^{C\nu} \ , \ f^{ABC} = - f^{ACB}$$

then in the gauge (2.50) no derivatives of W^A_i with respect to x^4 occur. We conclude that the interaction Lagrangian must be of the form

$$I_{F_1} = I_{F_1}(g_{\mu\nu}, W^A_\mu, f^{ABC}) \tag{3.24}$$

We shall now apply the criterion of invariance of the resulting field action under gauge transformations involving N arbitrary functions χ_A. Similarly to the case of a single vector field, the infinitesimal form of this transformation should be

$$W^{A'}_\mu = W^A_\mu + g_{\mu AB}\, \chi^B + h_{AB}\, \chi^B_{,\mu} \tag{3.25}$$

where

$$g_{\mu AB} = g_{\mu AB}(g_{\alpha\beta}, W^P_\alpha, f^P_{\alpha\beta}), \ h_{AB} = h_{AB}(g_{\alpha\beta}, W^P_\alpha, f^P_{\alpha\beta}) \tag{3.26}$$

The demand that the change in S_F as a result of the above transformation vanishes implies that

$$- \left(h_{BA} \frac{\delta S_F}{\delta W^B_\mu} \right)_{;\mu} + g_{\mu BA} \frac{\delta S_F}{\delta W^B_\mu} =$$

$$- \left(h_{BA} \frac{\delta S_{F_1}}{\delta W^B_\mu} \right)_{;\mu} + g_{\mu BA} \frac{\delta S_{F_1}}{\delta W^B_\mu} + (g_{\mu BA} - h_{BA,\mu})\, f^{B\mu\nu}{}_{;\nu} = 0 \tag{3.27}$$

Taking into account Eq. (3.24), we express $\delta S_{F_1}/\delta W^A_\mu$ as follows:

$$\frac{\delta S_{F_1}}{\delta W^A_\mu} = \frac{\partial I_{F_1}}{\partial W^A_\mu} + \frac{\partial I_{F_1}}{\partial f^{PQR}} \frac{\partial f^{PQR}}{\partial W^A_\mu} - 2\left(\frac{\partial I_{F_1}}{\partial f^{PQR}} \frac{\partial f^{PQR}}{\partial f^A_{\mu\nu}} \right)_{;\nu} \tag{3.28}$$

Then, observing the facts that

$$\frac{\partial f^{PQR}}{\partial W^A_\mu} = f^{P\mu\nu} (\delta^Q_A W^R_\nu - \delta^R_A \delta^Q_\nu) \tag{3.29}$$

and

$$\frac{\delta f^{PQR}}{\delta f^A_{\mu\nu}} = \frac{1}{2} \delta^P_A (W^{Q\mu} W^{R\nu} - W^{Q\nu} W^{R\mu}) \tag{3.30}$$

as well as that the operator $;\mu$, as applied to functions having the same arguments as I_{F_1}, is given by

$$;\mu = W^A_{\alpha;\mu} \frac{\partial}{\partial W^A_\alpha} + f^{ABC}{}_{,\mu} \frac{\partial}{\partial f^{ABC}} , \tag{3.31}$$

we expand Eq. (3.27) as follows:

$$h_{BA} \frac{\partial^2 I_{F_1}}{\partial f^{BPR} \partial W^Q_\lambda} R^\alpha{}_{\lambda\mu\nu} W^Q_\alpha W^{P\mu} W^{R\nu}$$

$$+ \left[h_{BA} \left(\frac{\partial^2 I_{F_1}}{\partial W^B_\nu \partial f^{QMN}} + 2 \frac{\partial^2 I_{F_1}}{\partial f^{PBR} \partial f^{QMN}} f^{P\nu\mu} W^R_\mu \right) \right.$$

$$\left. + 2 (g_{\mu BA} - h_{BA,\mu}) \frac{\partial^2 I_{F_1}}{\partial f^{BPR} \partial f^{QMN}} W^{P\mu} W^{R\nu} \right] f^{QMN}{}_{,\nu}$$

$$- \left(2 h_{BA} \frac{\partial I_{F_1}}{\partial f^{PBR}} W^R_\mu - g_{\mu PA} + h_{PA,\mu} \right) f^{P\mu\nu}{}_{;\nu}$$

$$- h_{BA} \left(\frac{\partial I_{F_1}}{\partial f^{PBR}} + \frac{\partial I_{F_1}}{\partial f^{RBP}} \right) f^{P\mu\nu} f^R{}_{\mu\nu} - (g_{\mu BA} - h_{BA,\mu}) \frac{\partial I_{F_1}}{\partial W^B_\mu}$$

$$+ h_{BA} \frac{\partial^2 I_{F_1}}{\partial W^B_\mu \partial W^Q_\lambda} W^Q_{\lambda;\mu} - 2 (g_{\mu BA} - h_{BA,\mu}) \frac{\partial I_{F_1}}{\partial f^{PBR}} f^{P\mu\nu} W^R_\nu$$

$$+ 2 (g_{\mu BA} - h_{BA,\mu}) \frac{\partial I_{F_1}}{\partial f^{BPR}} (W^{P\mu} W^{R\nu})_{;\nu}$$

$$+ 2 \left[h_{BA} \frac{\partial^2 I_{F_1}}{\partial f^{PBR} \partial W^Q_\lambda} f^{P\nu\mu} W^R_\mu + (g_{\mu BA} - h_{BA,\mu}) \frac{\partial^2 I_{F_1}}{\partial f^{BPR} \partial W^Q_\lambda} W^{P\mu} W^{R\nu} \right] W^Q_{\lambda;\nu} = 0 \tag{3.32}$$

The first thing we can say about the above equation is that the first term is the only one containing components of the curvature tensor. Hence its coefficient must vanish, giving

$$\frac{\partial^2 I_{F_1}}{\partial f^{BPR} \partial W^Q_\lambda} = 0 \tag{3.33}$$

It follows that

$$I_{F_1} = g(g^{ABC}) + h(g_{\mu\nu}, W^A_\mu) \tag{3.34}$$

and that the last term in Eq. (3.32) vanishes.

If h_{AB} depends of $f^P_{\alpha\beta}$, then the part $h_{PA,\mu} f^{P\mu\nu}{}_{;\nu}$ of the third term in Eq. (3.32) would be the only one containing the second derivatives of W^A_μ quadratically. Hence, as in the case of a single vector field, h_{AB} should be independent of $f^{P}_{\alpha\beta}$. It then follows that the only terms containing second derivatives of W^{A}_{μ} in Eq. (3.32) are the second and the third, in $f^{QMN}{}_{,\nu}$ and $f^{B\mu\nu}{}_{;\nu}$, respectively. Since these expressions cannot cancel each other, their coefficients must vanish separately. Taking into account Eq. (3.33), it is easy to see that the vanishing of the coefficient of $f^{QMN}{}_{,\nu}$ implies

$$\frac{\partial^2 I_{F_1}}{\partial f^{BPR} \partial f^{QMN}} = 0 \tag{3.35}$$

Hence

$$g = g_{ABC} f^{ABC}, \text{ where } g_{ABC} \text{ are (real) constants.} \tag{3.36}$$

It follows from the antisymmetry of f^{ABC}, with respect to the last two indices, that also

$$g_{ABC} = - g_{ACB}. \tag{3.37}$$

The vanishing of the coefficient of $f^{P\mu\nu}{}_{;\nu}$ then gives

$$2h_{BA} g_{PBR} W^{R}_{\mu} - g_{\mu PA} + h_{PA,\mu} = 0 \tag{3.38}$$

Any invariant h_{PA} ($g_{\mu\nu}$, W^{A}_{μ}) other than a simple constant should contain the metric tensor $g_{\mu\nu}$. Therefore, the third term in the above equation necessarily contains derivatives of $g_{\mu\nu}$. However, such derivatives are not contained in the first two terms. It follows that, as the case of a single vector field, the h_{AB}'s can only be constants. Then by a suitable re-definition of the gauge functions χ^{A} we may set

$$h_{AB} = \delta_{AB} . \tag{3.39}$$

It then follows from Eq. (3.38) that

$$g_{\mu AB} = 2\, g_{ABC}\, W^{C}_{\mu} \tag{3.40}$$

which fixes the form of the infinitesimal gauge transformation (3.25).

Taking into account Eqs. (3.34), (3.36), (3.39) and (3.40), the remaining terms in Eq. (3.32) become

$$- (g_{PAR} + g_{RAP})\, f^{P\mu\nu} f^{R}_{\mu\nu}$$

$$+ \frac{\partial^2 h}{\partial W^{A}_{\mu} \partial W^{Q}_{\lambda}} W^{Q}_{\lambda;\mu} - 4\, g_{BAC} g_{PBR} f^{PCR} + 4\, g_{BAC} g_{BPR} (W^{P\mu} W^{R\nu})_{;\nu} W^{C}_{\mu}$$

$$- 2 g_{BAC} \frac{\partial h}{\partial W^B_\mu} W^C_\mu \qquad = 0 \tag{3.41}$$

In the above equation, the first term is the only one containing the first derivatives of W^A_μ quadratically. Its vanishing implies that

$$g_{ABC} = - g_{CBA} \ . \tag{3.42}$$

Considering then Eq. (3.37) we conclude that g_{ABC} is fully antisymmetric.

From the remaining terms, the second, third and fourth terms are linear in the first derivatives of W^A_μ, while the last term does not contain any derivatives of W^A_μ. The vanishing of the coefficient of the first derivative of W^A_μ gives the following equation for h:

$$\frac{\partial^2 h}{\partial W^A_\mu \partial W^B_\nu} = - 4 \ (2\lambda_{AQBP} - \lambda_{APBQ}) \ W^{P\mu} W^{Q\nu}$$

$$+ 2 \ (\lambda_{APBQ} + \lambda_{AQBP}) \ W^P_\rho W^{Q\rho} g^{\mu\nu} \tag{3.43}$$

where we have defined

$$\lambda_{ABCD} = g_{KAB} g_{KCD} \tag{3.44}$$

The symmetries of λ_{ABCD} are the following:

$$\lambda_{ABCD} = - \lambda_{BACD} = - \lambda_{ABDC}$$

$$\lambda_{ABCD} = \lambda_{CDAB} \tag{3.45}$$

The integrability conditions of Eq. (3.43) are

$$\frac{\partial^3 h}{\partial W^A_\mu \partial W^B_\nu \partial W^C_\lambda} = \frac{\partial^3 h}{\partial W^C_\lambda \partial W^B_\nu \partial W^A_\mu} \tag{3.46}$$

Considering that, from Eq. (3.43),

$$\frac{\partial^3 h}{\partial W^A_\mu \partial W^B_\nu \partial W^C_\lambda} = - 4 \ (2\lambda_{AQBC} - \lambda_{ACBQ}) \ g^{\mu\lambda} W^{Q\nu}$$

$$- 4 \ (2\lambda_{ACBQ} - \lambda_{AQBC}) \ g^{\nu\lambda} W^{Q\mu}$$

$$+ 4 \ (\lambda_{ACBQ} + \lambda_{AQBC}) \ W^{Q\lambda} g^{\mu\nu} \tag{3.47}$$

the integrability conditions (3.46) are satisfied if and only if λ_{ABCD} possesses, in addition to those given by Eq. (3.45), the symmetry

$$\lambda_{ABCD} + \lambda_{BCAD} + \lambda_{CABD} = 0 \tag{3.48}$$

which, through Eq. (3.44), is a condition on the constants g_{ABC}. It should be noted that this is the necessary and sufficient condition in order that the g_{ABC} be the structure constants of a Lie group.

Taking into account Eq. (3.48), the factor $(2\lambda_{AQBP} - \lambda_{APBQ})$ in the first term on the right-hand side of Eq. (3.43) becomes $(\lambda_{ABQP} + \lambda_{PBQA})$ and Eq. (3.43) integrates to·

$$h = -\frac{1}{2} (\lambda_{ABQP} + \lambda_{PBQA}) \, W^{A\rho} W^{B\sigma} W^{P}_{\rho} W^{Q}_{\sigma} + \text{const.} \tag{3.49}$$

where the additive constant can be absorbed in the constant λ of S_G.

We now turn to the final term in Eq. (3.41). Its vanishing implies that [15]

$$\xi_{A(BC)(DE)} + \xi_{A(DE)(BC)} = 0 \tag{3.50}$$

where we have defined

$$\xi_{ABCDE} = g_{KAB} g_{LCD} g_{LEK} \tag{3.51}$$

It can be proved that Eq. (3.50) is an identity, once condition (3.48) is satisfied.

We conclude that in the case of many vector fields W^A $(A = 1...N)$ the most general interaction Lagrangian which is consistent with our hypotheses is

$$l_{F_1} = g_{ABC} f^{A}_{\mu\nu} W^{B\mu} W^{C\nu} + g_{KAB} g_{KPQ} W^{A\rho} W^{B\sigma} W^{P}_{\rho} W^{Q}_{\sigma} \tag{3.52}$$

where the constants g_{ABC} are the structure constants of a Lie group in the representation in which they are fully antisymmetric. Defining

$$F^{A}_{\mu\nu} = f^{A}_{\mu\nu} + 2 \, g_{ABC} W^{B}_{\mu} W^{C}_{\nu} \tag{3.53}$$

the total field Lagrangian l_F is simply

$$l_{F_1} = -\frac{1}{4} F^{A}_{\mu\nu} F^{A\mu\nu}, \tag{3.54}$$

invariant under the infinitesimal gauge transformation

$$W^{A'}_{\mu} = W^{A}_{\mu} + 2 \, g_{ABC} \chi^{B} W^{C}_{\mu} + \chi^{A}_{,\mu} \tag{3.55}$$

It follows from Eq. (3.54) that even in the case of many vector fields, the vector fields can have no bare mass.

Going back to Eq. (3.27) and taking into account Eqs. (3.39) and (3.40), and (3.37) and (3.42), we conclude that the classical field equations

$$\frac{\delta S_F}{\delta W^A_\mu} = 0$$

satisfy the identities

$$\left(\frac{\delta S_V}{\delta W^A_\mu}\right)_{;\mu} + g_{ABC}\frac{\delta S_F}{\delta W^B_\mu} W^C_\mu = 0 \qquad (3.56)$$

It follows from these identities that in the gauge (2.50) if the equations $\delta S_F/\delta W^A_4 = 0$ are satisfied on one spacelike hypersurface, the dynamical field equations $\delta S_F/\delta W^A_i = 0$ guarantee that it is satisfied over all space-time.

Let us finally turn to the general case in which many vector fields W^A (A = 1...N) as well as many complex scalar fields ϕ_a (a = 1...n) are present. We write the Lagrangian as

$$l_F = l_{F_0} + l_{F_1}, \text{ and correspondingly } S_F = S_{F_0} + S_{F_1}. \qquad (3.57)$$

Here l_{F_0} is the sum of the Lagrangian (3.54) of the vector fields and the Lagrangian (3.8) of the complex scalar fields if we subtract the function f of Eq. (3.8) which includes the scalar interaction:

$$l_{F_0} = -\frac{1}{2}\phi_{a,\mu}\phi^{*,}_a - \frac{1}{4}F^A_{\mu\nu}F^{A\mu\nu} \qquad (3.58)$$

The interaction between the vector and the scalar fields as well as the scalar interaction is contained in the Lagrangian l_{F_1}, which Eq. (3.2) requires to be of the form:

$$l_{F_1} = l_{F_1}(g_{ij}, W^1 \ldots W^N; \phi_1 \ldots \phi_n, \phi^*_1 \ldots \phi^*_n \text{ and their spatial derivatives}) \qquad (3.59)$$

expressed in the gauge (2.50). Relativistic invariance of l_{F_1} requires the derivatives of ϕ_a to enter l_{F_1} only through the expressions

$$\beta^A_a = \phi_{a,\mu}W^{A\mu}$$

Then, in the gauge (2.50) no derivatives of ϕ_a with respect to x^4 occur. We conclude that the interaction Lagrangian l_{F_1} is given by

$$l_{F_1} = l_{F_1}(g_{\mu\nu}, W^A_\mu, \phi_a, \phi^*_a, \beta^A_A, \beta^{*A}_a) \tag{3.60}$$

We shall now demand the resulting field action to be invariant under gauge transformations in which the vector fields transform as given by Eq. (3.55), while the scalar fields transform as (taking again into account the fact that the field Lagrangian is independent of field derivatives of order higher than the first)

$$\phi'_a = \phi_a + \xi^A_a \chi^A + \eta^{A\mu}_a \chi^A_{,\mu} \tag{3.61}$$

where

$$\xi^A_a = \xi^A_a (\phi_a, \phi_{a,\mu}, \phi^*_a, \phi^*_{a,\mu}, W^A_\mu, g_{\mu\nu})$$

$$\eta^{A\mu}_a = \eta^{A\mu}_a (\phi_a, \phi_{a,\mu}, \phi^*_a, \phi^*_{a,\mu}, W^A_\mu, g_{\mu\nu}) \tag{3.62}$$

keeping in mind the fact that no derivatives of W^A_μ enter the interaction Lagrangian l_{F_1} gauge invariance then implies that

$$-\left(\frac{\delta S_F}{\delta W^A_\mu}\right)_{;\mu} - 2\, g_{ABC} \frac{\delta S_F}{\delta W^B_\mu} W^C_\mu + \xi^A_a \frac{\delta S_F}{\delta \phi_a} + \xi^{*A}_a \frac{\delta S_F}{\delta \phi^*_a}$$

$$- \left(\eta^{A\mu}_a \frac{\delta S_F}{\delta \phi_a}\right)_{;\mu} - \left(\eta^{*A\mu}_a \frac{\delta S_F}{\delta \phi^*_a}\right)_{;\mu} =$$

$$= -\left(\frac{\delta S_{F_1}}{\delta W^A_\mu}\right)_{;\mu} - 2\, g_{ABC} \frac{\delta S_{F_1}}{\delta W^B_\mu} W^C_\mu + \xi^A_a \frac{\delta S_{F_1}}{\delta \phi_a} + \xi^{*A}_a \frac{\delta S_{F_1}}{\delta \phi^*_a}$$

$$+ \frac{1}{2} (\xi^A_a \phi^{*,\mu}_a{}_{;\mu} + \xi^{*A}_a \phi^{,\mu}{}_{;\mu}) - \left(\eta^{A\mu}_a \frac{\delta S_{F_1}}{\delta \phi_a}\right)_{;\mu} - \left(\eta^{*A\mu}_a \frac{\delta S_{F_1}}{\delta \phi^*_a}\right)_{;\mu} \tag{3.63}$$

We shall temporarily assume that $\eta^{A\mu}_a = \eta^{*A\mu}_a = 0$. Then, at the end of the calculation, we shall demonstrate that this is necessarily so.

Taking into account Eq. (3.60), we express $\delta S_{F_1}/\delta W^A_\mu$, $\delta S_{F_1}/\delta\phi_a$ and $\delta S_{F_1}/\delta\phi^*_a$ as follows:

$$\frac{\delta S_{F_1}}{\delta W^A_\mu} = \frac{\partial l_{F_1}}{\partial W^A_\mu} + \frac{\partial l_{F_1}}{\partial \beta^A_a} \phi_a{}^{,\mu} + \frac{\partial l_{F_1}}{\partial \beta^{*A}_a} \phi^{*,\mu}_a \tag{3.64}$$

$$\frac{\delta S_{F_1}}{\delta \phi_a} = \frac{\partial I_{F_1}}{\partial \phi_a} - \left(\frac{\partial I_{F_1}}{\partial \beta^A_a} W^{A\mu}\right)_{;\mu} \tag{3.65}$$

$$\frac{\delta S_{F_1}}{\delta \phi^*_a} = \frac{\partial I_{F_1}}{\partial \phi^*_a} - \left(\frac{\partial I_{F_1}}{\partial \beta^{*A}_a} W^{A\mu}\right)_{;\mu} \tag{3.66}$$

Then, keeping in mind that the operator ;μ, as applied to functions with the same arguments as I_{F_1} of Eq. (3.60), is given by

$$;\mu = W^A_{\nu;\mu} \frac{\partial}{\partial W^A_\nu} + \phi_{a,\mu} \frac{\partial}{\partial \phi_a} + \phi^*_{a,\mu} \frac{\partial}{\partial \phi^*_a} + \beta^A_{a,\mu} \frac{\partial}{\partial \beta^A_a} + \beta^{*A}_{a,\mu} \frac{\partial}{\partial \beta^{*A}_a} \tag{3.67}$$

we expand Eq. (3.63) as follows (assuming that $\eta^{A\mu}_a$ vanishes):

$$\left(\frac{\partial^2 I_{F_1}}{\partial W^A_\mu \partial W^B_\nu} + \frac{\partial^2 I_{F_1}}{\partial \beta^A_a \partial W^B_\nu} \phi_a{}^{,\mu} + \frac{\partial^2 I_{F_1}}{\partial \beta^{*A}_a \partial W^B_\nu} \phi^{*,\mu}_a + \xi^A_a \frac{\partial I_{F_1}}{\partial \beta^B_a} g^{\mu\nu} + \xi^{*A}_a \frac{\partial I_{F_1}}{\partial \beta^{*B}_a} g^{\mu\nu}\right.$$

$$\left. + \xi^A_a \frac{\partial^2 I_{F_1}}{\partial \beta^C_a \partial W^B_\nu} W^{C\mu} + \xi^{*A}_a \frac{\partial^2 I_{F_1}}{\partial \beta^{*C}_a \partial W^B_\nu} W^{C\mu}\right) W^B_{\nu;\mu}$$

$$+\left(\frac{\partial I_{F_1}}{\partial \beta^A_a} - \frac{1}{2}\xi^{*A}_a\right) \phi_a{}^{,\mu}{}_{;\mu} + \left(\frac{\partial I_{F_1}}{\partial \beta^{*A}_a} - \frac{1}{2}\xi^A_a\right) \phi^{*,\mu}_a{}_{;\mu}$$

$$+\left(\frac{\partial^2 I_{F_1}}{\partial W^A_\mu \partial \beta^B_b} + \frac{\partial^2 I_{F_1}}{\partial \beta^A_a \partial \beta^B_b} \phi^{,\mu}_a + \frac{\partial^2 I_{F_1}}{\partial \beta^{*A}_a \partial \beta^B_b} \phi^{*,\mu}_a\right.$$

$$\left. + \xi^A_a \frac{\partial^2 I_{F_1}}{\partial \beta^C_a \partial \beta^B_b} W^C + \xi^{*A}_a \frac{\partial^2 I_{F_1}}{\partial \beta^{*A}_a \beta^B_b} W^{C\mu}\right) \beta^B_{b,\mu}$$

$$+\left(\frac{\partial^2 I_{F_1}}{\partial W^A_\mu \partial \beta^{*B}_b} + \frac{\partial^2 I_{F_1}}{\partial \beta^A_a \partial \beta^{*B}_b} \phi_a{}^{,\mu} + \frac{\partial^2 I_{F_1}}{\partial \beta^{*A}_a \partial \beta^{*B}_b} \phi^{*,\mu}_a\right.$$

$$\left. + \xi^A_a \frac{\partial^2 I_{F_1}}{\partial \beta^C_a \partial \beta^{*B}_b} W^{C\mu} + \xi^{*A}_a \frac{\partial^2 I_{F_1}}{\partial \beta^{*C}_a \partial \beta^{*B}_b} W^{C\mu}\right) \beta^{*B}_{b,\mu}$$

$$+ \frac{\partial^2 I_{F_1}}{\partial\beta^A_a \partial\phi_b} \phi_{b,\mu}\phi_a{}^{,\mu} + \frac{\partial^2 I_{F_1}}{\partial\beta^{*A}_a \partial\phi^*_b} \phi^*_{b,\mu}\phi^{*,\mu}_a$$

$$+ \left(\frac{\partial^2 I_{F_1}}{\partial\beta^A_a \partial\phi^*_b} + \frac{\partial^2 I_{F_1}}{\partial\beta^{*A}_b \partial\phi_a} \right) \phi^*_{b,\mu}\phi_a{}^{,\mu}$$

$$+ \frac{\partial^2 I_{F_1}}{\partial W^A_\mu \partial\phi_a} \phi_{a,\mu} + \frac{\partial^2 I_{F_1}}{\partial W^A_\mu \partial\phi^*_a} \phi^*_{a,\mu}$$

$$+ \left(-2\, g_{ABC} \frac{\partial I_{F_1}}{\partial\beta^C_b} + \xi^A_a \frac{\partial^2 I_{F_1}}{\partial\beta^B_a \partial\phi_b} + \xi^{*A}_a \frac{\partial^2 I_{F_1}}{\partial\beta^{*B}_a \partial\phi_b} \right) \beta^B_b$$

$$+ \left(-2\, g_{ABC} \frac{\partial I_{F_1}}{\partial\beta^{*C}_b} + \xi^A_a \frac{\partial^2 I_{F_1}}{\partial\beta^B_a \partial\phi^*_b} + \xi^{*A}_a \frac{\partial^2 I_{F_1}}{\partial\beta^{*B}_a \partial\phi^*_b} \right) \beta^{*B}_b$$

$$+ 2\, g_{ABC} \frac{\partial I_{F_1}}{\partial W^B_\mu} W^C_\mu - \xi^A_a \frac{\partial I_{F_1}}{\partial\phi_a} - \xi^{*A}_a \frac{\partial I_{F_1}}{\partial\phi^*_a} = 0 \qquad (3.68)$$

The first thing we can say about the above equation is that since $W^B_{\nu;\mu}$ is not one of the arguments of I_{F_1} its coefficient in the above equation must vanish. Considering that

$$\beta^B_{b,\mu} = \phi_b{}^{,\nu} W^B_{\nu;\mu} + \phi_b{}^{,\nu}{}_{;\mu} W^B_\nu \, , \qquad (3.69)$$

the vanishing of this coefficient is given by

$$\frac{\partial^2 I_{F_1}}{\partial\beta^A_a \partial W^B_\nu} \phi_a{}^{,\mu} + \frac{\partial^2 I_{F_1}}{\partial W^A_\mu \partial\beta^B_b} \phi_b{}^{,\nu} + \frac{\partial^2 I_{F_1}}{\partial\beta^{*A}_a \partial W^B_\nu} \phi^{*,\mu}_a + \frac{\partial^2 I_{F_1}}{\partial W^A_\mu \partial\beta^{*B}_b} \phi^{*,\nu}_b$$

$$+ \frac{\partial^2 I_{F_1}}{\partial\beta^A_a \partial\beta^B_b} \phi_a{}^{,\mu}\phi_b{}^{,\nu} + \frac{\partial^2 I_{F_1}}{\partial\beta^{*A}_a \partial\beta^{*B}_b} \phi^{*,\mu}_a \phi^{*,\nu}_b$$

$$+ \frac{\partial^2 I_{F_1}}{\partial\beta^A_a \partial\beta^{*B}_b} \phi_a{}^{,\mu}\phi^{*,\nu}_b + \frac{\partial^2 I_{F_1}}{\partial\beta^{*A}_a \partial\beta^B_b} \phi^{*,\mu}_a \phi_b{}^{,\nu}$$

$$+ \left(\xi_a^A \frac{\partial^2 I_{F_1}}{\partial \beta_a^C \partial \beta_b^B} + \xi_a^{*A} \frac{\partial^2 I_{F_1}}{\partial \beta_a^{*C} \partial \beta_b^B} \right) \phi_b{}^{,\nu} W^{C\mu}$$

$$+ \left(\xi_a^A \frac{\partial^2 I_{F_1}}{\partial \beta_a^C \partial \beta_b^{*B}} + \xi_a^{*A} \frac{\partial^2 I_{F_1}}{\partial \beta_a^{*C} \partial \beta_b^{*B}} \right) \phi_b^{*,\nu} W^{C\mu}$$

$$+ \frac{\partial^2 I_{F_1}}{\partial W_\mu^A \partial W_\nu^B} + \xi_a^A \frac{\partial I_{F_1}}{\partial \beta_a^B} g^{\mu\nu} + \xi_a^{*A} \frac{\partial I_{F_1}}{\partial \beta_a^{*B}} g^{\mu\nu}$$

$$+ \xi_a^A \frac{\partial^2 I_{F_1}}{\partial \beta_a^C \partial W_\nu^B} W^{C\mu} + \xi_a^{*A} \frac{\partial^2 I_{F_1}}{\partial \beta_a^{*C} \partial W_\nu^B} W^{C\mu} = 0 \qquad (3.70)$$

Since now $\phi_a{}^{,\mu}\ \phi_b{}^{,\nu}$, $\phi_a^{*,\mu}\ \phi_b^{*,\nu}$ and $\phi_a{}^{,\mu}\ \phi_b^{*,\nu}$, $\phi_a^{*,\mu}\ \phi_b{}^{,\nu}$ are not arguments of I_{F_1}, their coefficients in the above equation should vanish giving

$$\frac{\partial^2 I_{F_1}}{\partial \beta_a^A \partial \beta_b^B} = \frac{\partial^2 I_{F_1}}{\partial \beta_a^{*A} \mu \beta_b^{*B}} = 0 \qquad (3.71)$$

and

$$\frac{\partial^2 I_{F_1}}{\partial \beta_a^A \partial \beta_b^{*B}} = 0 \qquad (3.72)$$

respectively. Then the demand that the coefficients of $\phi_a{}^{,\mu}$, $\phi_b{}^{,\nu}$ and $\phi_a^{*,\mu}$, $\phi_b^{*,\nu}$ should also vanish gives

$$\frac{\partial^2 I_{F_1}}{\partial W_\mu^A \partial \beta_b^B} = \frac{\partial^2 I_{F_1}}{\partial W_\mu^A \partial \beta_b^{*B}} = 0 \qquad (3.73)$$

It follows from Eqs. (3.71), (3.72) and (3.73) that

$$I_{F_1} = f_A^a(\phi_a,\phi_a^*)\ \beta_a^A + f_A^{*a}\ (\phi_a,\phi_a^*)\ \beta_a^{*A}$$

$$+ g(g_{\mu\nu}, W_\mu^A, \phi_a, \phi_a^*), \qquad (3.74)$$

Keeping in mind the requirement that I_{F_1} be real. Then Eq. (3.70) becomes

$$\frac{\partial^2 g}{\partial W_\mu^A \partial W_\nu^B} + (\xi_a^A f_B^a + \xi_a^{*A} f_B^{*a})\ g^{\mu\nu} = 0 \qquad (3.75)$$

The general solution of the above equation is

$$g = -\frac{1}{2}\left(\xi_a^A f_B^a + \xi_a^{*A} f_B^{*a}\right) W^{A\mu} W_\mu^B + e \tag{3.76}$$

where e is a real function of the ϕ_a's and ϕ_a^*'s only.

We now return to Eq. (3.68). The coefficients of $\phi_a{}^{,\mu}{}_{;\mu}$ and $\phi_a^{*,\mu}{}_{;\mu}$ should vanish, since I_{F_1} does not depend on these expressions. Hence,

$$\xi_a^A = 2f_A^{*a}\ ,\quad \xi_a^{*A} = 2f_A^a \tag{3.77}$$

Taking into account Eqs. (3.74), (3.76) and (3.77), Eq. (3.68) reduces to:

$$\begin{aligned}
&\frac{\partial f_A^a}{\partial \phi_b}\phi_{b,\mu}\phi_a{}^{,\mu} + \frac{\partial f_A^{*a}}{\partial \phi_b^*}\phi_{b,\mu}^*\phi_a^{*,\mu} \\
&+\left(\frac{\partial f_A^a}{\partial \phi_b^*} + \frac{\partial f_A^{*b}}{\partial \phi_a}\right)\phi_{b,\mu}^*\phi_a{}^{,\mu} \\
&-2\left(f_B^a\frac{\partial f_A^{*a}}{\partial \phi_b} + f_B^{*a}\frac{\partial f_A^a}{\partial \phi_b} + f_A^{*a}\frac{\partial f_B^b}{\partial \phi_a} + f_A^a\frac{\partial f_B^b}{\partial \phi_a^*} - g_{ABC}f_C^b\right)\beta_b^B \\
&-2\left(f_A^{*a}\frac{\partial f_A^a}{\partial \phi_b^*} + f_B^a\frac{\partial f_A^{*a}}{\partial \phi_b^*} + f_A^a\frac{\partial f_B^{*b}}{\partial \phi_a^*} + f_A^{*a}\frac{\partial f_B^{*b}}{\partial \phi_a} - g_{ABC}f_C^{*b}\right)\beta_b^{*B} \\
&+2\left(f_A^{*a}\frac{\partial}{\partial \phi_a}\left(f_P^{*b}f_Q^b + f_P^b f_Q^{*b}\right) + f_A^a\frac{\partial}{\partial \phi_a^*}\left(f_P^{*b}f_Q^b + f_P^b f_Q^{*b}\right)\right. \\
&\left.- g_{ABQ}\left(f_P^{*a}f_B^a + f_P^a f_B^{*a}\right) - g_{ABP}\left(f_Q^{*a}f_B^a + f_Q^a f_B^{*a}\right)\right)W^{P\mu}W_\mu^Q \\
&+ f_A^{*a}\frac{\partial e}{\partial \phi_a} + f_A^a\frac{\partial e}{\partial \phi_a^*} = 0
\end{aligned} \tag{3.78}$$

In the above equation, the unknown functions f_A^a and e are functions of the ϕ_a's and ϕ_a^*'s only. It follows that the coefficients of the expressions

$$\phi_{b,\mu}\phi_a{}^{,\mu},\ \phi_{b,\mu}^*\phi_a^{*,\mu},\ \phi_{b,\mu}^*\phi_a{}^{,\mu},\ \beta_b^B,\ \beta_b^{*B} \text{ and } W^{P\mu}W_\mu^Q$$

should all vanish separately. The vanishing of the coefficients of $\phi_{b,\mu}\phi_a{}^{,\mu}$ and $\phi_{b,\mu}^*\phi_a^{*,\mu}$ gives

$$\frac{\partial f^a_A}{\partial \phi_b} = \frac{\partial f^{*a}_A}{\partial \phi^*_b} = 0 \tag{3.79}$$

Hence

$$f^a_A = f^a_A (\phi^*_a) \ , \ f^{*a}_A = f^{*a}_A (\phi_a) \tag{3.80}$$

The vanishing of the coefficient of $\phi^*_{b,\mu}\phi^{,\mu}_a$ gives

$$\frac{\partial f^a_A}{\partial \phi^*_b} + \frac{\partial f^{*b}_A}{\partial \phi_a} = 0 \tag{3.81}$$

Since, from Eq. (3.80), the first term in the above equation is a function of the ϕ^*_a's only; while the second term is a function of the ϕ_a's only, Eq. (3.81) can hold if and only if

$$\frac{\partial f^{*b}_A}{\partial \phi_a} = \frac{1}{2} \ i \ T^A_{ba} \qquad \text{(constants)} \tag{3.82}$$

It then follows from Eq. (3.81) that

$$T^{*A}_{ab} = T^A_{ba} \ . \tag{3.83}$$

In other words, the matrix T^A is hermitian. Eq. (3.82) integrates to

$$f^{*a}_A = \frac{1}{2} \ i \ T^A_{ab}\phi_b + \frac{1}{2} \ i \ P^A_a \ , \tag{3.84}$$

where the P^A_a's are constants.

Taking into account Eqs. (3.84) and (3.83), the vanishing of the coefficient of β^{*B}_b in Eq. (3.78) gives

$$\begin{aligned}(T^A_{ba} \ T^B_{ac} - T^B_{bc} \ T^A_{ac} - 2ig_{ABC}T^C_{bc}) \ \phi_c \\ + T^A_{ba}P^B_a - T^B_{ba}P^A_a - 2ig_{ABC}P^C_b = 0\end{aligned} \tag{3.85}$$

Clearly, both the coefficient of ϕ_c and the remaining constant terms in the above equation should vanish. The vanishing of the coefficient of ϕ_c gives

$$[T^A, T^B] = 2ig_{ABC}T^C \tag{3.86}$$

Thus the g_{ABC}'s are the structure constants of the group whose N generators are represented in n dimensions by the matrices T^A. Since these matrices are hermitian, the group is a unitary one. Condition (3.48) is nothing other than the

Jacobi identity:

$$[T^A,[T^B,T^C]] + [T^B,[T^C,T^A]] + [T^C,[T^A,T^B]] = 0 \tag{3.87}$$

The vanishing of the remaining terms in Eq. (3.85) follows from Eq. (3.86) if we set

$$P^A_a = T^A_{ab} P_b \tag{3.88}$$

The demand that the coefficient of β^B_b in Eq. (3.78) vanishes gives just the complex conjugates of Eqs. (3.86) and (3.88). [It should be noted here that the reality of the structure constants which was originally required so that the Lagrangian of the vector fields be real, follows also from the fact that the group is unitary. Taking into account Eq. (3.83), the complex conjugate of Eq. (3.86) is

$$[\tilde{T}^A,\tilde{T}^B] = - 2ig^*_{ABC}\ \tilde{T}^C \tag{3.89}$$

where $\tilde{T}^A$ is the matrix which is the transpose of T^A. On the other hand, the transpose of Eq. (3.86) gives

$$[\tilde{T}^A, \tilde{T}^B] = - 2ig_{ABC}\tilde{T}^C \tag{3.90}$$

Comparing Eqs. (3.89) and (3.90), we conclude that the constants g_{ABC} can only be real.]

The demand that the coefficient of $W^{P\mu}\ W^Q_\mu$ in Eq. (3.78) vanishes gives (taking account of Eqs. (3.84) and (3.83)):

$$\begin{aligned} &T^P T^Q T^A - T^A T^Q T^P + T^Q T^P T^A - T^A T^P T^Q = \\ &= 2ig_{QAB}(T^B T^P + T^P T^B) + 2ig_{PAB}(T^B T^Q + T^Q T^B) \end{aligned} \tag{3.91}$$

which is seen to follow directly from Eq. (3.86) if we add onto the left-hand side the vanishing expression

$$- T^P T^A T^Q + T^Q T^A T^P - T^Q T^A T^P + T^P T^A T^Q$$

We now turn to the final constraint imposed by the vanishing of the remaining terms in Eq. (3.78). Taking account of Eqs. (3.84) and (3.88), this gives the following equations for the function e:

$$(\phi_b + P_b)\ \frac{\partial e}{\partial \phi_a} + (\phi^*_a + P^*_a)\ \frac{\partial e}{\partial \phi^*_b} = 0 \tag{3.92}$$

The general solution of the above equations is

$$e = e((\phi_a + P_a)(\phi^*_a + P^*_a)) \tag{3.93}$$

which completes the solution of Eq. (3.68).

We now return to Eq. (3.63) in order to demonstrate that $\eta^{A\mu}_a$ vanishes. If the contrary were true, then derivatives of the ϕ_a's and ϕ^*_a's of the third order would appear in Eq. (3.68). These derivatives are all contained in the terms

$$-\frac{1}{2}\eta^{*A\mu}_a \phi_a{}^{,\nu}{}_{;\nu\mu} - \frac{1}{2}\eta^{A\mu}_a \phi^*_a{}^{,\nu}{}_{;\nu\mu}$$

$$+ \left(\eta^{A\mu}_a \frac{\partial^2 l_{F_1}}{\partial\beta^A_a \partial\beta^B_b} + \eta^{*A\mu}_a \frac{\partial^2 l_{F_1}}{\partial\beta^{*A}_a \partial\beta^B_b}\right) W^{A\nu}\,{}^B_{b;\nu\mu}$$

$$+ \left(\eta^{A\mu}_a \frac{\partial^2 l_{F_1}}{\partial\beta^A_a \partial\beta^{*B}_b} + \eta^{*A\mu}_a \frac{\partial^2 l_{F_1}}{\partial\beta^{*A}_a \partial\beta^{*B}_b}\right) W^{A\nu}\beta^{*B}_{b;\nu\mu}$$

The expressions $\phi_a{}^{,\nu}{}_{;\nu\mu}$ and $\beta^A_{a;\mu\nu}$ cannot cancel each other, and neither can, of course, their complex conjugates. It follows that their coefficients should vanish separately. The coefficients of $\phi_a{}^{,\nu}{}_{;\nu\mu}$ and $\phi^{*,\nu}_a{}_{;\nu\mu}$ then give

$$\eta^{A\mu}_a = \eta^{*A\mu}_a = 0 \tag{3.94}$$

We conclude that the interaction Lagrangian l_{F_1} is given by

$$l_{F_1} = \frac{1}{2}[-i\, T^A_{ba}(\phi^*_b + P^*_b)\phi_{a,\mu} + i\, T^A_{ab}(\phi_b + P_b)\phi^*_{a,\mu}]\, W^{A\mu}$$

$$-\frac{1}{2} T^A_{ak} T^B_{kb}(\phi^*_a + P^*_a)(\phi_b + P_b)\, W^{A\mu} W^B_\mu$$

$$+ e((\phi^*_a + P^*_a)(\phi_a + P_a)) \tag{3.95}$$

and the infinitesimal form of the gauge transformation of the scalar fields is given by

$$\phi'_a = \phi_a + i\, T^A_{ab}(\phi_b + P_b)\chi^A \tag{3.96}$$

Adding the constant P_a to both sides of the above equation, it is seen that the T^A generate unitary transformations on the

$$\tilde{\phi}_a = \phi_a + P_a \tag{3.97}$$

Since the addition of constants to the scalar fields does not alter the kinetic Lagrangian of the scalar fields, we can write the total field Lagrangian in the form

$$l_F = l_{\tilde\phi} + l_W , \tag{3.98}$$

where

$$l_{\tilde\phi} = -\frac{1}{2}(\tilde\phi_{a,\mu} - iW^A_\mu T^A_{ab}\tilde\phi_b)(\tilde\phi^{*,\mu}_a + iW^{B\mu} T^B_{ca}\tilde\phi^*_c)$$

$$+ e(\tilde\phi_a\tilde\phi^*_a) \tag{3.99}$$

and

$$l_W = -\frac{1}{4} F^A_{\mu\nu} F^{A\mu\nu} \tag{3.100}$$

In conclusion, the only Lagrangian for many complex scalar and many vector fields in interaction which is consistent with our hypotheses is the generalized Yang-Mills Lagrangian [5-7].

The finite form of the gauge transformation of the translated scalar fields $\tilde\phi_a$ is the following:

$$\tilde\phi'_a = \left(e^{i\,T^A\chi^A}\right)_{ab} \tilde\phi_b \tag{3.101}$$

Going back to Eq. (3.63) and taking into account Eqs. (3.77), (3.83), (3.84) and (3.94), we conclude that the classical field equations

$$\frac{\delta S_F}{\delta W^A_\mu} = 0 \ , \ \frac{\delta S_F}{\delta\tilde\phi_a} = \frac{\delta S_F}{\delta\tilde\phi^*_a} = 0$$

satisfy the identities

$$\left(\frac{\delta S_F}{\delta W^A_\mu}\right)_{;\mu} + 2\, g_{ABC}\frac{\delta S_F}{\delta W^B_\mu} W^C_\mu - i\frac{\delta S_F}{\delta\tilde\phi_a} T^A_{ab}\tilde\phi_b + i\frac{\delta S_F}{\delta\tilde\phi^*_a} T^A_{ba}\tilde\phi^*_b = 0 \tag{3.102}$$

It follows from these identities that in the gauge (2.50), if the equations $\delta S_F/\delta W^A_4 = 0$ are satisfied on one spacelike hypersurface, the dynamical field equations $\delta S_F/\delta W^A_i = 0$, $\delta S_F/\delta\tilde\phi_a = \delta S_F/\delta\tilde\phi^*_a = 0$ guarantee that they are satisfied over all space-time.

Returning to Eq. (3.95), we see that the term

$$-\frac{1}{2} T^A_{ak}T^B_{kb}P^*_a P_b W^{A\mu}W^B_\mu$$

depends only on the vector fields. Therefore it should be added to the Lagrangian of the vector fields. Thus the total field Lagrangian is alternatively written as

$$l_F = l_\phi + l_W + l_{\phi W} , \tag{3.103}$$

where l_ϕ is the Lagrangian of the scalar fields

$$l_\phi = \phi_{a,\mu}\phi_a^{*,\mu} + e((\phi_a + P_a)(\phi_a^* + P_a^*)) \tag{3.104}$$

l_W is the Lagrangian of the vector fields

$$l_W = - \frac{1}{4} F^A_{\mu\nu} F^{A\mu\nu} - \frac{1}{2} (M^2)_{AB} W^{A\mu} W^B_\mu \tag{3.105}$$

and $l_{\phi W}$ is the interaction Lagrangian

$$l_{\phi W} = \frac{1}{2} [- iT^A_{ba}(\phi^*_b + P^*_b)\phi_{a,\mu} + iT^A_{ab}(\phi_b + P_b)\phi^*_{a,\mu}] W^{A\mu}$$

$$- \frac{1}{2} T^A_{ak} T^B_{kb} (\phi^*_a\phi_b + P^*_a\phi_b + \phi^*_a P_b) W^{A\mu} W^B_\mu \tag{3.106}$$

In Eq. (3.104) we have defined the matrix

$$(M^2)_{AB} = T^A_{ak} T^B_{kb} P^*_a P_b = P^{*A}_k P^B_k \tag{3.107}$$

which is clearly both hermitian and positive definite. This matrix is the bare mass matrix of the vector fields, if the function e has a minimum at the values P_a of the scalars $\tilde{\phi}_a$ [8-10]. In this case the scalar fields ϕ_a have vanishing vacuum expectation values.

If we perform an orthogonal transformation

$$W^A_\mu = K_{AB}\overline{W}^B_\mu \quad , \quad K_{AB}K_{AC} = \delta_{BC} \tag{3.108}$$

on the vector fields, the K_{AB}'s being real constants, and we define transformed generators

$$\overline{T}^A = K_{BA} T^B$$

and transformed structure constants

$$\overline{g}_{PQR} = K_{AP}K_{BQ}K_{CR}g_{ABC}$$

which are, evidently, still fully antisymmetric, the commutation relations (3.86) and the first term in Eq. (3.105) remain unaltered in form, while the mass matrix transforms into

$$(\overline{M}^2)_{AB} = K_{CA}(M^2)_{CD}K_{DB} \tag{3.109}$$

By performing a suitable orthogonal transformation, we may thus diagonalize the mass matrix, deducing the vector fields possessing definite masses.

IV. QUANTIZATION

We now go back to Eq. (2.48) and, considering the results of the previous section, we conclude that, in the case of a physical system consisting of the geometry of space containing n complex scalar fields $\tilde{\phi}_a$ (the vacuum expectation values of which are constants p_a) and N vector fields W^A,

$$d\ell^2_{TOT}(^3x) = d\ell^2_G(^3x) + d\ell^2_{\tilde{\phi}}(^3x) + d\ell^2_W(^3x) \tag{4.1}$$

where

$$d\ell^2_{\tilde{\phi}}(^3x) = \sqrt{g}\ (dRe\tilde{\phi}_a dRe\tilde{\phi}_a + dIm\tilde{\phi}_a dIm\tilde{\phi}_a) = g\ d\tilde{\phi}_a d\tilde{\phi}^*_a \ , \tag{4.2}$$

$$d\ell^2_W(^3x) = \sqrt{g}\ \ g^{ij}dW^A_i dW^A_j \ , \tag{4.3}$$

and

$$b_{TOT} = b_G + b_{\tilde{\phi}} + b_W \tag{4.4}$$

where

$$b_{\tilde{\phi}} = -\ (\tilde{\phi}_{a,m} - i\ T^A_{ab}W^A_m\tilde{\phi}_b)(\tilde{\phi}^{*,m}_a + i\ T^B_{ca}W^{Bm}\tilde{\phi}^*_c\ ) \ , \tag{4.5}$$

$$b_W = -\ \frac{1}{2}\ F^A_{mn}F^{Amn} \tag{4.6}$$

Eq. (2.48) refers of course to the gauge (2.49), or equivalently (2.50). If space-time is sliced in such a way that the normal separation of nearby hypersurfaces is constant over the hypersurfaces, then Eq. (2.43) holds and S_{TOT} assumes the form given by Eq. (2.41). Substituting this form of S_{TOT} in the statement of the quantum principle Eq. (2.20) and following an approach completely analogous to that employed in Ref. [1] we are led to the following functional differential equation for the probability amplitude functional $\varphi[\mathcal{g}, \tilde{\phi}_1 \ldots \tilde{\phi}^*_n, \tilde{\phi}^*_1 \ldots \tilde{\phi}^*_n, W^1 \ldots W^N; \tau]$, the role of the sequential parameter σ taken here by τ:

$$\mathcal{H}_{TOT}\ \varphi = i\ \frac{\partial}{\partial\tau}\ \varphi \tag{4.7}$$

The Hamiltonian operator $\mathcal{H}_{TOT}$ is given by

$$\mathcal{H}_{TOT} = \mathcal{H}_G + \mathcal{H}_{\tilde{\phi}} + \mathcal{H}_W \tag{4.8}$$

where

$$\mathcal{H}_G = -\frac{1}{2}\left(\frac{d^2}{d\mathcal{G}^2} + B_G\right) \tag{4.9}$$

and

$$\mathcal{H}_{\tilde{\phi}} = -\frac{1}{2}\left(\frac{d^2}{d\tilde{\phi}^2} + B_{\tilde{\phi}}\right) \quad , \quad \mathcal{H}_W = -\frac{1}{2}\left(\frac{d^2}{dW^2} + B_W\right) \tag{4.10}$$

Here [16]

$$\frac{d^2}{d\mathcal{G}^2} = \int d^3x \, \frac{1}{G^{1/2}} \frac{\delta}{\delta g_{ij}} \, G^{1/2} \, G^{-1}_{ijmn} \frac{\delta}{\delta g_{mn}} \tag{4.11}$$

$$\frac{d^2}{d\tilde{\phi}^2} = \int d^3x \, \frac{1}{\sqrt{g}} \left(\frac{\delta^2}{\delta \mathrm{Re}\tilde{\phi}_a \delta \mathrm{Re}\tilde{\phi}_a} + \frac{\delta^2}{\delta \mathrm{Im}\tilde{\phi}_a \delta \mathrm{Im}\tilde{\phi}_a}\right) = \int d^3x \, \frac{4}{\sqrt{g}} \frac{\delta^2}{\delta\tilde{\phi}_a \delta\tilde{\phi}^*_a} \tag{4.12}$$

$$\frac{d^2}{dW^2} = \quad d^3x \, \frac{g_{ij}}{\sqrt{g}} \frac{\delta^2}{\delta W^A_i \delta W^A_j} \tag{4.13}$$

Eq. (4.7) is invariant under co-ordinate transformations in $\mathcal{C}_{TOT}$, namely under changes of the way of describing the geometry and the fields. It is then obviously invariant under co-ordinate transformations in ordinary space, since such transformations are nothing but linear and strictly local transformations in $\mathcal{C}_{TOT}$.

It can be shown that the operator $\mathcal{H}_{TOT}$ is hermitian. As a consequence, the condition expressed by Eq. (2.19) is indeed satisfied. Since this operator does not depend explicitly on τ, Eq. (4.7) may be solved by separation of the variables $[\mathcal{G}, \tilde{\phi}_1 \ldots \tilde{\phi}_n, \tilde{\phi}^*_1 \ldots \tilde{\phi}^*_n, W^1 \ldots W^N]$ and τ. The general solution is given by

$$\varphi[\mathcal{G}, \tilde{\phi}_1 \ldots, \tilde{\phi}^*_1 \ldots, W^1 \ldots; \tau] = \frac{1}{2\pi} \int_{\mathrm{all}\,\lambda} e^{-i\lambda\tau} \, \psi_\lambda[\mathcal{G}, \tilde{\phi}_1 \ldots, \tilde{\phi}^*_1 \ldots, W^1 \ldots] d\lambda \tag{4.14}$$

where the ψ_λ's are eigenfunctions of $\mathcal{H}_{TOT}$ with eigenvalues λ. It follows from the above equation and the definition of the physical probability amplitude $[\mathcal{G}, \tilde{\phi}_1 \ldots \tilde{\phi}_n, \tilde{\phi}^*_1 \ldots \tilde{\phi}^*_n, W^1 \ldots W^N]$ given by Eq. (2.21), that the latter is the eigenfunction of $\mathcal{H}_{TOT}$ with eigenvalue zero. Thus the physical probability amplitude satisfies the equation:

$$\mathcal{H}_{TOT} \, \psi = 0. \tag{4.15}$$

It will be shown in the following that the above equation, for a co-ordinate and gauge-invariant functional ψ, contains all the laws of the quantum theory of the physical system under consideration. In order to do this we shall turn our attention away from the path integral formulation of the quantum principle which we have hitherto followed, and we shall formulate that principle in terms of the canonical formalism. We shall then show that the complete set of quantum laws derived in the canonical way is contained in Eq. (4.15).

Defining the integral Lagrangian L_{TOT} by

$$S_{TOT} = \int L_{TOT}\, dx^4, \tag{4.16}$$

we have

$$L_{TOT} = L_G + L_{\tilde{\phi}} + L_W \tag{4.17}$$

where, in a general frame and a general gauge,

$$L_G = \frac{1}{2}\int N\left\{\frac{G^{ijmn}}{N^2}\left(\frac{\partial g_{ij}}{\partial x^4} + N_{i;j} + N_{j;i}\right)\left(\frac{\partial g_{mn}}{\partial x^4} + N_{m;n} + N_{n;m}\right) + R\sqrt{g}\right\} d^3x, \tag{4.18}$$

$$L_{\tilde{\phi}} = \frac{1}{2}\int N\sqrt{g}\, d^3x\left\{\frac{1}{N^2}\left(\frac{\partial\tilde{\phi}_a}{\partial x^4} + N^m\tilde{\phi}_{a,m} - iN^\mu W^A_\mu T^A_{ab}\tilde{\phi}_b\right)\left(\frac{\partial\tilde{\phi}^*_a}{\partial x^4} + N^n\tilde{\phi}^*_{a,n} + iN^\nu W^B_\nu T^B_{ca}\tilde{\phi}^*_c\right) + b_{\tilde{\phi}}\right\} \tag{4.19}$$

$$L_W = \frac{1}{2}\int N\sqrt{g}\, d^3x\left\{\frac{g^{mn}}{N^2}\left(\frac{\partial W^A_m}{\partial x^4} + (N\times\mathrm{rot}W^A)_m - W^A_{4,m} + 2g_{ABC}W^B_m N^\mu W^C_\mu\right)\left(\frac{\partial W^A_n}{\partial x^4} + (N\times\mathrm{rot}W^A)_n - W^A_{4,n} + 2g_{APQ}W^P_n N^\nu W^Q_\nu\right) + b_W\right\} \tag{4.20}$$

We then define the dynamical variables canonically conjugate to g_{ij}, $\tilde{\phi}_a, \tilde{\phi}^*_a, W^A_i$, namely the canonical momenta

$$\pi^{ij} = \frac{\delta L_{TOT}}{\delta(\partial g_{ij}/\partial x^4)} = \frac{\delta L_G}{\delta(\partial g_{ij}/\partial x^4)} = \frac{G^{ijmn}}{N}\left(\frac{\partial g_{mn}}{\partial x^4} + N_{m;n} + N_{n;m}\right) \tag{4.21}$$

$$\tilde{\sigma}_a = \frac{\delta L_{TOT}}{\delta(\partial\tilde{\phi}_a/\partial x^4)} = \frac{\delta L_{\tilde{\phi}}}{\delta(\partial\tilde{\phi}_a/\partial x^4)} = \frac{\sqrt{g}}{2N}\left(\frac{\partial\tilde{\phi}^*_a}{\partial x^4} + N^m\tilde{\phi}^*_{a,m} + iN^\mu W^A_\mu T^A_{ba}\tilde{\phi}^*_b\right) \tag{4.22}$$

$$\tilde{\sigma}^*_a = \frac{\delta L_{TOT}}{\delta(\partial\tilde{\phi}^*_a/\partial x^4)} = \frac{\delta L_{\tilde{\phi}}}{\delta(\partial\tilde{\phi}^*_a/\partial x^4)} = \frac{\sqrt{g}}{2N}\left(\frac{\partial\tilde{\phi}_a}{\partial x^4} + N^m\tilde{\phi}_{a,m} - iN^\mu W^A_\mu T^A_{ab}\tilde{\phi}_b\right) \tag{4.23}$$

$$\omega^{Ai} = \frac{\delta L_{TOT}}{\delta(\partial W_i^A/\partial x^4)} = \frac{\partial L_W}{\delta(\partial W_i^A/\partial x^4)} = \frac{\sqrt{g}\, g^{ij}}{N}\left(\frac{\partial W_j^A}{\partial x^4} + (N\times \mathrm{rot} W^A)_j - W^A_{4,j} + 2\, g_{ABC} W_j^B N^\mu W_\mu^C\right) \tag{4.24}$$

Finally we define H_{TOT}, the generator of translations in x^4, by

$$H_{TOT} = \int \left(\pi^{ij} \frac{\partial g_{ij}}{\partial x^4} + \tilde{\sigma}_a \frac{\partial \tilde{\phi}_a}{\partial x^4} + \tilde{\sigma}_a^* \frac{\partial \tilde{\phi}_a^*}{\partial x^4} + \omega^{Ai} \frac{\partial W_i^A}{\partial x^4}\right) d^3x - L_{TOT}$$

$$= \int \left\{ N \mathcal{H}_{TOT}(^3x) + N^i P_i + W_4^A \theta^A \right\} d^3x \tag{4.25}$$

where

$$\theta^A = \omega^{Am}{}_{;m} + 2\, g_{ABC}\, \omega^{Bm} W_m^C - i\tilde{\sigma}_a T^A_{ab} \tilde{\phi}_b + i\tilde{\sigma}_a^* T^A_{ba} \tilde{\phi}_b^* \tag{4.26}$$

$$- P^m = - 2\pi^{mn}{}_{;n} + \tilde{\sigma}_a \tilde{\phi}_a{}^{,m} + \tilde{\sigma}_a^* \tilde{\phi}_a^{*,m} - (\omega^A \times \mathrm{rot} W^A)^m$$

$$- i\tilde{\sigma}_a W^{Am} T^A_{ab} \tilde{\phi}_b + i\tilde{\sigma}_a^* W^{Am} T^A_{ba} \tilde{\phi}_b^* + 2\, g_{ABC}\, \omega^{Ai} W_i^B W^{Cm} \tag{4.27}$$

and $\mathcal{H}_{TOT}(^3x)$ is the Hamiltonian density

$$\mathcal{H}_{TOT}(^3x) = \frac{1}{2}\left\{ G^{-1}_{ijmn} \pi^{ij} \pi^{mn} + 4\, (\sqrt{g})^{-1} \tilde{\sigma}_a \tilde{\sigma}_a^* + (\sqrt{g})^{-1} g_{ij} \omega^{Ai} \omega^{Aj} - (R + b_{\tilde{\phi}} + b_W)\sqrt{g}\right\} \tag{4.28}$$

Among Hamilton's equations, the equations [17]

$$\frac{\partial g_{ij}}{\partial x^4} = \left\{ g_{ij}, H_{TOT}\right\}, \quad \frac{\partial \tilde{\phi}_a}{\partial x^4} = \left\{\tilde{\phi}_a, H_{TOT}\right\}, \quad \frac{\partial W_i^A}{\partial x^4} = \left\{ W_i^A, H_{TOT}\right\} \tag{4.29}$$

reproduce the definitions of the canonical momenta Eqs. (4.21) to (4.24), while the equations

$$\frac{\partial \pi^{ij}}{\partial x^4} = \left\{\pi^{ij}, H_{TOT}\right\}, \quad \frac{\partial \tilde{\sigma}_a}{\partial x^4} = \left\{\tilde{\sigma}_a, H_{TOT}\right\}, \quad \frac{\partial \omega^{Ai}}{\partial x^4} = \left\{\omega^{Ai}, H_{TOT}\right\} \tag{4.30}$$

reproduce the classical dynamical equations

$$\frac{\delta S_{TOT}}{\delta g_{ij}} = 0, \quad \frac{\delta S_{TOT}}{\delta \tilde{\phi}_a} = 0, \quad \frac{\delta S_{TOT}}{\delta W_i^A} = 0$$

respectively. The remaining classical equations, namely the initial value equations

$$\frac{\delta S_{TOT}}{\delta W_4^A} = 0 \ , \ \frac{\delta S_{TOT}}{\delta g_{4i}} = 0 \ , \ \frac{\delta S_{TOT}}{\delta g_{44}} = 0$$

give the constraints

$$\theta^A = 0 \ , \ p^m = 0 \ , \text{ and } \mathcal{H}_{TOT}(^3x) = 0 \qquad (4.31)$$

respectively.

The canonical formulation of the quantum principle is that Eqs. (4.29), (4.30) and (4.31) are to be re-interpreted as operator equations, replacing Poisson brackets by commutators multiplied by (1/i). Thus, we postulate the following commutation relations among the dynamical variables:

$$[g_{ij}(^3x), \pi^{mn}(^3x')] = \frac{i}{2} (\delta_i^m\delta_j^n + \delta_j^m\delta_i^n) \ \delta^3(x-x') \qquad (4.32)$$

$$[\tilde{\phi}_a(^3x), \tilde{\sigma}_b(^3x')] = [\tilde{\phi}_a^*(^3x), \tilde{\sigma}_b^*(^3x')] = i\delta_{ab}\delta^3(x-x') \qquad (4.33)$$

$$[\tilde{\phi}_a^*(^3x), \tilde{\sigma}_b(^3x')] = [\tilde{\phi}_a(^3x), \tilde{\sigma}_b^*(^3x')] = 0 \qquad (4.34)$$

$$[W_i^A(^3x), \omega^{Bj}(^3x')] = i\delta_{AB}\delta_i^j\delta^3(x-x') \qquad (4.35)$$

It follows from the above relations that if the operators $g_{ij}(^3x)$, $\tilde{\phi}_a(^3x)$, $\tilde{\phi}_a^*(^3x)$, $W_i^A(^3x)$ are represented by multiplication by $g_{ij}(^3x)$, $\tilde{\phi}_a(^3x)$, $\tilde{\phi}_a^*(^3x)$, $W_i^A(^3x)$ respectively, then the operators $\pi^{ij}(^3x)$, $\tilde{\sigma}_a(^3x)$, $\tilde{\sigma}_a^*(^3x)$ and $\omega^{Ai}(^3x)$ must be represented by:

$$\pi^{ij}(^3x) = \frac{1}{i}\frac{\delta}{\delta g_{ij}(^3x)} \ , \ \tilde{\sigma}_a(^3x) = \frac{1}{i}\frac{\delta}{\delta\tilde{\phi}_a(^3x)} \ , \ \tilde{\sigma}_a^*(^3x) = \frac{1}{i}\frac{\delta}{\delta\tilde{\phi}_a^*(^3x)} \ ,$$

$$\omega^{Ai}(^3x) = \frac{1}{i}\frac{\delta}{\delta W_i^A(^3x)} \qquad (4.36)$$

The quantum analogues of Eqs. (4.29) and (4.30) are included in the following equation for the probability amplitude functional

$$\psi[\mathcal{g}, \tilde{\phi}_1 \ldots \tilde{\phi}_n, \tilde{\phi}_1^* \ldots \tilde{\phi}_n^*, W^1 \ldots W^N; x^4]$$

$$H_{TOT}\psi = i\frac{\partial\psi}{\partial x^4} \qquad (4.37)$$

and the quantum analogue of Eqs. (4.31) are the following constraints on ψ:

$$\theta^A \psi = 0 \ , \ p^m \psi = 0 \ , \mathcal{H}_{TOT}(^3x)\psi = 0 \tag{4.38}$$

It is clear from Eq. (4.25) that, provided that the first two of the above constraints are satisfied, the third contains Eq. (4.37), if we take ψ independent of x^4. Thus the constraints (4.38) represent the full set of laws of the quantum theory of the physical system under consideration.

Taking into account Eqs. (4.36), the first of the constraints (4.38) is expressed by

$$\left(\frac{\delta\psi}{\delta W^A_m}\right)_{;m} + 2\, g_{ABC} \frac{\delta\psi}{\delta W^B_m} W^{Cm} - i \frac{\delta\psi}{\delta\tilde{\phi}_a} T^A_{ab}\tilde{\phi}_b + i \frac{\delta\psi}{\delta\tilde{\phi}^*_a} T^A_{ba}\tilde{\phi}^*_b = 0 \tag{4.39}$$

independently of the choice of ordering of factors. It is evident from Eq. (3.102) that the above equation is the condition that ψ be invariant under spatial gauge transformations. Consequently, it is an identity for a gauge invariant functional ψ.

The second of the constraints (4.38) is expressed by

$$-2\left(\frac{\delta\psi}{\delta g_{mn}}\right)_{;n} + \tilde{\phi}_a{}^{,m} \frac{\delta\psi}{\delta\tilde{\phi}_a} + \tilde{\phi}^{*,m}_a \frac{\delta\psi}{\delta\tilde{\phi}^*_a} - \left(\frac{\delta\psi}{\delta W^A} \times \text{rot } W^A\right)^m$$

$$- iW^{Am} \frac{\delta\psi}{\delta\tilde{\phi}_a} T^A_{ab}\tilde{\phi}_b + iW^{Am} \frac{\delta\psi}{\delta\tilde{\phi}^*_a} T^A_{ba}\tilde{\phi}^*_b + 2\, g_{ABC} \frac{\delta\psi}{\delta W^A_i} W^B_i W^{Cm} \tag{4.40}$$

again independently of factor ordering. Let us perform an infinitesimal spatial co-ordinate transformation,

$$x^m = x'^m + \xi^m \tag{4.41}$$

The resulting changes in the metric and field functions are given by:

$$\delta g_{mn} = \xi_{m;n} + \xi_{n;m} \tag{4.42}$$

$$\delta\tilde{\phi}_a = \tilde{\phi}_a{}^{,m}\xi_m \tag{4.43}$$

$$\delta W^A_i = (\xi \text{xrot } W^A)_m + (\xi^n W^A_n)_{,m} \tag{4.44}$$

Since ψ should be independent of the co-ordinate system used to describe the geometry and the fields, the change in ψ as a result of this transformation should vanish:

$$\delta\psi = \int \left\{ \frac{\delta\psi}{\delta g_{mn}} (\xi_{m;n} + \xi_{n;m}) + \frac{\delta\psi}{\delta\phi_a} \phi_a{}^{,m} + \frac{\delta\psi}{\delta\phi_a^*} \phi_a^{*,m} \right.$$

$$\left. + \frac{\delta\psi}{\delta W_m^A} [(\xi \mathrm{x} \operatorname{rot} W^A)_m + (\xi^n W_n^A)_{,m}] \right\} d^3x$$

$$= \int \left\{ -2\left(\frac{\delta\psi}{\delta g_{mn}}\right)_{;n} + \phi_a{}^{,m} \frac{\delta\psi}{\delta\phi_a} + \phi_a^{*,m} \frac{\delta\psi}{\delta\phi_a^*} \right.$$

$$\left. - \left(\frac{\delta\psi}{\delta W^A} \mathrm{x} \operatorname{rot} W^A\right)^m - W^{Am} \left(\frac{\delta\psi}{\delta W_n^A}\right)_{;n} \right\} \xi_m \, d^3x = 0 \qquad (4.45)$$

Since the above should be true for an arbitrary infinitesimal vector ξ_m, we conclude that we must have

$$-2\left(\frac{\delta\psi}{\delta g_{mn}}\right)_{;n} + \phi_a^{,m} \frac{\delta\psi}{\delta\phi_a} + \phi_a^{*,m} \frac{\delta\psi}{\delta\phi_a^*} - \left(\frac{\delta\psi}{\delta W^A} \mathrm{x} \operatorname{rot} W^A\right)^m - W^{Am}\left(\frac{\delta\psi}{\delta W_n^A}\right)_{;n} = 0 \qquad (4.46)$$

Replacing the last term in the above equation by its equal from Eq. (4.39), we recover Eq. (4.40). We conclude that Eq. (4.40) is an identity for a co-ordinate and gauge-invariant functional ψ.

We finally turn our attention to the last of the constraints (4.38). Here substitution of the representations (4.36) of the canonical momenta does not lead to a unique equation because of the arbitrariness in the choice of factor ordering of the first term in the expression given by Eq. (4.28). However, since Eq. (4.37) should be identical to Eq. (4.7) (replacing φ by ψ) in the frame N = 1 to which Eq. (4.7) refers, and since in that frame

$$H_{TOT}\, \psi = \int \mathcal{H}_{TOT}(^3x)\, \psi \, d^3x$$

if Eqs. (4.39) and (4.40) are satisfied, we conclude that the Hamiltonian density operator $\mathcal{H}_{TOT}(^3x)$ should simply be the integrant of the Hamiltonian operator $\mathcal{H}_{TOT}$ of Eq. (4.7):

$$\mathcal{H}_{TOT} = \int d^3x\, \mathcal{H}_{TOT}(^3x) \qquad (4.47)$$

Hence

$$\mathcal{H}_{TOT}(^3x) = \mathcal{H}_G(^3x) + \mathcal{H}_{\tilde{\phi}}(^3x) + \mathcal{H}_W(^3x) \tag{4.48}$$

where

$$\mathcal{H}_G(^3x) = -\frac{1}{2}\left(\frac{d^2}{d\mathcal{g}(^3x)^2} + b_G\sqrt{g}\right) \tag{4.49}$$

$$\mathcal{H}_{\tilde{\phi}}(^3x) = -\frac{1}{2}\left(\frac{d^2}{d\tilde{\phi}(^3x)^2} + b_{\tilde{\phi}}\sqrt{g}\right), \quad \mathcal{H}_W(^3x) = -\frac{1}{2}\left(\frac{d^2}{dW(^3x)^2} + b_W\sqrt{g}\right) \tag{4.50}$$

and

$$\frac{d^2}{d\mathcal{g}(^3x)^2} = \frac{1}{G^{1/2}}\frac{\delta}{\delta g_{ij}} G^{1/2} G^{-1}_{ijmn}\frac{\delta}{\delta g_{mn}} \tag{4.51}$$

$$\frac{d^2}{d\tilde{\phi}(^3x)^2} = \frac{4}{\sqrt{g}}\frac{\delta^2}{\delta\tilde{\phi}_a\delta\tilde{\phi}^*_a}, \quad \frac{d^2}{dW(^3x)^2} = \frac{g_{ij}}{\sqrt{g}}\frac{\delta^2}{\delta W^A_i\delta W^A_j} \tag{4.52}$$

We see that the factor ordering problem we encounter in the canonical formulation has been solved by the path integral approach.

We shall show in the following that a co-ordinate and gauge-invariant set of solutions of Eq. (4.15) may be chosen to satisfy identically the last of the constraints (4.31) at all space points (^3x).

First we observe that, classically,

$$\left\{\mathcal{H}_{TOT}(^3x), \mathcal{H}_{TOT}\right\} = -p^m{}_{;m} \tag{4.53}$$

which vanishes if the second of the classical constraints (4.31) is satisfied. Similarly it can be shown that, quantum mechanically,

$$[\mathcal{H}_{TOT}(^3x), \mathcal{H}_{TOT}]\,\psi =$$

$$= \left\{-2\left(\frac{\delta\psi}{\delta g_{mn}}\right)_{;n} + \tilde{\phi}_a{}^{,m}\frac{\delta\psi}{\delta\tilde{\phi}_a} + \tilde{\phi}^{*,m}_a\frac{\delta\psi}{\sigma\tilde{\phi}^*_a} - \left(\frac{\delta\psi}{\delta W^A}\times \mathrm{rot}\, W^A\right)^m \right.$$

$$\left. - iW^{Am}\frac{\delta\psi}{\delta\tilde{\phi}_a}T^A_{ab}\tilde{\phi}_b + iW^{Am}\frac{\delta\psi}{\delta\tilde{\phi}^*_a}T^A_{ba}\tilde{\phi}^*_b + 2\,g_{ABC}\frac{\delta\psi}{\delta W^A_i}W^{Bi}W^{Cm}\right\}_{;m} \tag{4.54}$$

since Eq. (4.40) is identically satisfied if ψ is a co-ordinate and gauge-invariant functional. Here we note that as a consequence of the facts that $d\ell^2_G$ and B_G do not depend on the fields and $d\ell^2_F$ and B_F do not depend on the derivatives of the metric tensor,

$$[\mathcal{H}_G({}^3x), \mathcal{H}_{\tilde{\phi}}] + [\mathcal{H}_{\tilde{\phi}}({}^3x), \mathcal{H}_G] = 0 \qquad (4.55)$$

and

$$[\mathcal{H}_G({}^3x), \mathcal{H}_W] + [\mathcal{H}_W({}^3x), \mathcal{H}_G] = 0 \qquad (4.56)$$

Also, as a consequence of the facts that $d\ell_W^2$ and B_W do not depend on the scalar fields and $d\ell_{\tilde{\phi}}^2$ and $B_{\tilde{\phi}}$ do not depend on the derivatives of the vector fields.

$$[\mathcal{H}_{\tilde{\phi}}({}^3x), \mathcal{H}_W] + [\mathcal{H}_W({}^3x), \mathcal{H}_{\tilde{\phi}}] = 0 \qquad (4.57)$$

Since $\mathcal{H}_{TOT}({}^3x)$ and $\mathcal{H}_{TOT}$ commute, it follows that they possess the same set of eigenfunctions. Thus, a suitable set of linearly independent solutions of the equation

$$\mathcal{H}_{TOT}\,\psi = \lambda\psi \qquad (4.58)$$

satisfies also equations of the type

$$\mathcal{H}_{TOT}({}^3x)\;\psi = \rho({}^3x)\psi \qquad (4.59)$$

where

$$\int \rho({}^3x)\; d^3x = \lambda \qquad (4.60)$$

We now introduce the entity T_ε defined by

$$T_\varepsilon = \int d^3x\; g_{ij}\varepsilon^i\;(P^j + W^{Aj}\theta^A) \qquad (4.61)$$

where ε^i is an infinitesimal vector whose (contravariant) components are, in a certain co-ordinate system, the same at all $({}^3x)$. This entity generates infinitesimal spatial translations by the vector ε: For a classical variable $F[g_{ij}({}^3x), \tilde{\phi}_a({}^3x), \tilde{\phi}_a^*({}^3x), W_i^A({}^3x), \pi^{ij}({}^3x), \tilde{\sigma}_a({}^3x), \tilde{\sigma}_a^*({}^3x), \omega^{Ai}({}^3x)]$, T_ε itself considered as a variable

$$\left\{T_\varepsilon,\; F({}^3x)\right\} = F_{,i}\varepsilon^i \qquad (4.62)$$

while for a quantum operator $F[g_{ij}({}^3x), \tilde{\phi}_a({}^3x), \tilde{\phi}_a^*({}^3x), W_i^A({}^3x), \delta/\delta g_{ij}({}^3x), \delta/\delta\tilde{\phi}_a({}^3x), \delta/\delta\tilde{\phi}_a^*({}^3x), \delta/\delta W_i^A({}^3x)]$, T_ε itself considered as an operator

$$\frac{1}{i}\,[T_\varepsilon, F({}^3x)] = F_{,i}\varepsilon^i \qquad (4.63)$$

We can thus express the Hamiltonian density operator at the point $({}^3x + {}^3\varepsilon)$ in terms of the same operator at the point ${}^3x$:

$$\mathcal{H}_{TOT}({}^3x + {}^3\varepsilon)\psi = (\mathcal{H}_{TOT}({}^3x) + \frac{1}{i}\,[T_\varepsilon, \mathcal{H}_{TOT}({}^3x)])\psi \qquad (4.64)$$

Then, taking into account Eq. (4.59), we have

$$\rho(^3x + {}^3\varepsilon)\psi = \rho(^3x)\psi + i(\mathcal{H}_{TOT}(^3x) - \rho(^3x))T_\varepsilon\psi \tag{4.65}$$

However, it is obvious from Eq. (4.46), that for a co-ordinate invariant functional ψ,

$$T_\varepsilon\psi = 0 \tag{4.66}$$

identically. It follows that $\rho(^3x + {}^3\varepsilon) = \rho(^3x)$ or ρ = const., in the co-ordinate system in which the vector ε^i is constant. Then, Eq. (4.60) gives ρ.(total volume of space) = λ. Hence for co-ordinate and gauge-invariant solutions of Eqs. (4.58) and (4.59), with $\lambda = 0$, implies $\rho(^3x) = 0$. The latter being true in one co-ordinate system it is true in all, since ρ is a scalar density. This gives the result we had sought to prove [20].

We conclude that all the laws of the quantum theory of the physical system under consideration are included in Eq. (4.15) for a co-ordinate and gauge-invariant functional ψ. Since no special choice in slicing of the space-time and no special gauge condition was used in the canonical approach, as was done in the path integral approach, Eq. (4.15) is entirely independent of such choices.

We note that since the third of the constraints (4.38) represents the quantum analogue of the definition of local time, Eq. (4.15) must represent the quantum analogue of the definition of global time.

The physical interpretation of the functional ψ is the following: Let measurements of the geometry and of the fields be made at every region of space. The quantity

$$P_V = \int_V |\psi[\ ,\phi_1 \ldots \phi_n, \phi_1^* \ldots \phi_n^*, W^1 \ldots W^N]|^2 \, dV_{\mathcal{C}_{TOT}} \tag{4.67}$$

where

$$dV_{\mathcal{C}_{TOT}} = dV_{\mathcal{C}[g]}\, dV_{\mathcal{C}[\phi]}\, dV_{\mathcal{C}[W]} \tag{4.68}$$

and

$$dV_{\mathcal{C}[\phi]} = \prod_{\text{all } ^3x} g^{n/2}(^3x) \prod_{a=1}^{n} d\mathrm{Re}\phi_a(^3x)\, d\mathrm{Im}\phi_a(^3x) \tag{4.69}$$

$$dV_{[W]} = \prod_{\text{all } ^3x} g^{N/4}(^3x) \prod_{A=1}^{N} d^3W^A(^3x) \tag{4.70}$$

is the probability that the synthesis of the outcomes gives a geometry and field configuration of the whole which belongs to the region V of $\mathcal{G}_{TOT}$. If two such sets of measurements are performed, giving outcomes differing by $[d\mathcal{g}(^3x), d\tilde{\phi}_1(^3x)\ldots, d\tilde{\phi}_1^*(^3x)\ldots, dW^1(^3x)\ldots]$, then, provided that the corresponding $d\ell^2_{TOT}\ (^3x)/b_{TOT}\sqrt{g}(^3x)$ are all either positive or zero, there exist classical time intervals $d\tau(^3x)$ between the measurements at every local region, defined by

$$d\tau(^3x) = \frac{d\ell_{TOT}(^3x)}{(b_{TOT}\sqrt{g}(^3x))^{1/2}} \tag{4.71}$$

where we have assumed that the measurements of the vector fields are registered in the gauge (2.49).

We would conclude that the functional ψ corresponds classically not to the configuration of the physical system on a particular spacelike hypersurface but rather to the 4-dimensional configuration of the system, namely the 4-geometry of space-time endowed with the 4-dimensional configuration of the fields.

ACKNOWLEDGMENTS

The author wishes to thank Professor Abdus Salam, the International Atomic Energy Agency and UNESCO for hospitality at the International Centre for Theoretical Physics, Trieste, Italy.

REFERENCES

[1] D. Christodoulou, Proceedings of the Academy of Athens, 47 (20 Jan. 1972).

[2] J. A. Wheeler, "Geometrodynamics and the issue of the final state," chapter in Relativity, Groups and Topology, Les Houches (1963).

[3] D. Christodoulou, Nuovo Cimento 26 B, 56 (1975).

[4] D. Christodoulou, Nuovo Cimento 26 B, 335 (1975).

[5] C. N. Yang and R. S. Mills, Phys. Rev. 96, 191 (1954).

[6] R. Utiyama, ibid. 101, 1597 (1956).

[7] M. Gell-Mann and S. Glashow, Ann. Phys. (N.Y.) 15, 437 (1961).

[8] P. W. Higgs, Phys. Rev. Letters 12, 132 (1964); 13, 508 (1964); Phys. Rev. 145, 1156 (1966).

[9] T. W. Kimble, Phys. Rev. 155, 1554 (1967).

[10] S. Weinberg, Phys. Rev. D7, 1068 (1973).

[11] A. Fisher in Relativity, Carmeli, Fickler and Witten, eds., Plenum Press (1970).

[12] Here "space" geometry means a riemannian geometry with a positive definite metric.

[13] Both scalar and vector fields here are taken to be real.

[14] For one of the earliest uses of the concept of distance in superspace see: B. DeWitt, Phys. Rev. 160, 1113 (1967).

[15] The proper way of doing these correspondences is such as to satisfy the local correspondence conditions.

[16] A. Lichnerowicz, J. Math. Pure Appl. 23, 37 (1944).

[17] The symbol ( ) denotes symmetrization with respect to the indices it contains.

[18] G^{-1}_{ijmn} are the components of the matrix inverse to that with components $G^{mnk\ell}$, i.e. $G^{-1}_{ijmn}\, G^{mnk\ell} = \frac{1}{2}(\delta^k_i \delta^\ell_j + \delta^k_j \delta^\ell_i)$. Further G is the determinant of the matrix G^{ijmn}.

[19] The Poisson bracket A,B of two functionals of the dynamical variables of the physical system under consideration is defined as follows:

$$\left\{A,B\right\} = \int d^3x \left(\frac{\delta A}{\delta g_{ij}} \frac{\delta B}{\delta \pi^{ij}} + \frac{\delta A}{\delta \tilde{\phi}_a} \frac{\delta B}{\delta \tilde{\sigma}_a} + \frac{\delta A}{\delta \tilde{\phi}^*_a} \frac{\delta B}{\delta \tilde{\sigma}^*_a} + \frac{\delta A}{\delta W^A_i} \frac{\delta B}{\delta \omega^{Ai}} - \frac{\delta B}{\delta g_{ij}} \frac{\delta A}{\delta \pi^{ij}} - \frac{\delta B}{\delta \tilde{\phi}_a} \frac{\delta A}{\delta \tilde{\sigma}_a} - \frac{\delta B}{\delta \tilde{\phi}^*_a} \frac{\delta A}{\delta \tilde{\sigma}^*_a} - \frac{\delta B}{\delta W^A_i} \frac{\delta A}{\delta \omega^{Ai}} \right)$$

[20] V. Moncrief, Phys. Rev. D5, 277 (1972). In this paper it was shown that, in the case of pure geometry, if a coördinate-independent functional satisfies the Hamiltonian density constraint for some particular space point, then that functional will necessarily satisfy this constraint for all space points.

GENERAL RELATIVITY A LA STRING: A PROGRESS REPORT*

Tullio Regge

Institute for Advanced Study, Princeton, New Jersey 08540

and

Claudio Teitelboim

Joseph Henry Laboratories,
Princeton University, Princeton, New Jersey 08540

ABSTRACT

Preliminary results on a canonical formulation of general relativity based on an analogy with the string model of elementary particles are presented. Rather than the metric components, the basic fields of the formalism are taken to be the functions describing the embedding of four dimensional spacetime in a ten or possibly higher dimensional manifold. So far, the main drawback of the formalism is that the generator of normal deformations ("fourth constraint") cannot be written down in closed form. The present approach is compared and contrasted with the usual one and with the canonical description of the relativistic string.

It is our intention in this note to analyze some formal analogies existing in different relativistic systems and to assess some of the current difficulties in the canonical formalism for general relativity. In particular we shall compare two different approaches to general relativity (the conventional one (A) [1,2] and a new one based on the notion of external variables (B)) with the string model of elementary particles [3].

A. General Relativity: Conventional Formalism

In the usual canonical formalism for Einstein's theory of gravitation developed by Dirac [1] and Arnowitt Deser and Misner (ADM) [2] the starting point is the Hilbert action

$$S = \int (-{}^4g)^{1/2}\ {}^{(4)}R\, d^4x \tag{1}$$

which is regarded as a functional of the metric tensor $g_{\mu\nu}(x), x \in R^4$.

It is an important feature of the Hilbert action that by adding a suitable divergence to the integrand in (1) one can switch to an alternate action density - the Dirac - ADM - lagrangian density

*Research sponsored by the National Science Foundation under Grants No. GP-30799X to Princeton University and GP-40768X to the Institute for Advanced Study.

$$\mathscr{L} = (-{}^{(4)}g)^{1/2}\,{}^{(4)}R + \frac{\partial Y^\alpha}{\partial x^\alpha} \tag{2}$$

which contains no second time derivatives of the $g_{\mu\nu}$ and furthermore contains no first time derivatives of the $g_{o\mu}$.

The $g_{o\mu}$ have then vanishing conjugate momenta and enter the theory as arbitrary functions. At this stage the remaining degrees of freedom are thus those represented by the spatial metric components g_{ij} and their conjugates π^{ij}. The fields g_{ij}, π^{ij} are however not independent, but they are restricted by the constraint(*) equations

$$\mathscr{H}_\perp = g^{-1/2}(\pi_{ij}\pi^{ij} - \tfrac{1}{2}(\pi^i_i)^2) - g^{1/2}R \approx 0 \tag{3a}$$

$$\mathscr{H}_i = -2\pi_i{}^j{}_{|j} \approx 0 \quad . \tag{3b}$$

Geometrically speaking the $\mathscr{H}_i$ generate arbitrary reparametrizations of the spacelike hypersurface on which the state is defined whereas $\mathscr{H}_\perp$ generates deformations which change the location of the hypersurface in the ambient spacetime. The fact that the hypersurfaces are embedded in a common spacetime is expressed through the closure relations [4,5]

$$[\mathscr{H}_\perp(x), \mathscr{H}_\perp(x')] = (g^{rs}(x)\mathscr{H}_s(x) + g^{rs}(x')\mathscr{H}_s(x'))\,\delta,_s(x,x'), \tag{4a}$$

$$[\mathscr{H}_r(x), \mathscr{H}_\perp(x')] = \mathscr{H}_\perp(x)\ \delta,_r(x,x') \quad , \tag{4b}$$

$$[\mathscr{H}_r(x), \mathscr{H}_s(x')] = \mathscr{H}_r(x')\,\delta_{,s}(x,x') + \mathscr{H}_s(x)\,\delta_{,r}(x,x') \quad . \tag{4c}$$

In this theory it is in principle possible to fix the gauges by imposing particular coordinate conditions on the surface and also by fixing the time slicing. The fixation of the spacetime coordinates amounts therefore to bring in four extra constraints besides (3). After this is done one is left with only two independent pairs of canonical variables per space point. These degrees of freedom appear in the weak field approximation as the two polarization states

---

*The "weak equality" symbol is used to emphasize that $\mathscr{H}_\perp$ and $\mathscr{H}_i$ have non-vanishing Poisson brackets with the canonical variables of the theory. The vertical slash denotes covariant differentiation in the spatial metric g_{ij}. Spacetime covariant derivatives are indicated by a semicolon. The letter R denotes the spatial curvature and g is the determinant of the spatial metric. To avoid confusion some spacetime quantities carry an upper left index (4) as in (1).

per wave vector $\tilde{k}$ of a massless spin two graviton propagating on a flat background.

The practical implementation of the coordinate fixing is unfortunately frought with difficulties which have prevented so far the construction of an actual canonical quantum theory of gravity. In the first place it is not a simple matter to fix the gauge freedom in such a manner as to ensure a proper parametrization of spacetime through coordinates, although some of the proposed choices look reasonable [2,6,7]. A second difficulty is that the reduced Hamiltonian associated to the coordinate conditions proposed so far cannot be written down in closed form and usually appears as a highly non-local expression in the canonical fields. This brings virtually to a halt the construction of the quantum theory because of the formidable problems of ordering which must be solved ex-novo at each order of perturbation theory in the expression for the Hamiltonian.

Yet another difficulty arises in the so-called maximal slicing ($\pi^i{}_i = 0$), which appears to be the gauge condition most exhaustively investigated from the point of view of ensuring a proper parametrization of spacetime [7]. The difficulty in question is that ordering problems appear here already at the level of interpreting the Poisson brackets of the basic fields as commutators, because q-numbers appear nontrivially on the right hand side of the commutation relations. Such difficulties do not arise however for the ADM variables [2,8], but unfortunately there is not much evidence that the ADM gauge defines a good system of spacetime coordinates.

The difficulties mentioned above are by no means exclusive to the gravitational field and they also appear, for example, in the string model which bears in many respects a striking analogy with Einstein's theory of gravitation. In the case of the string, because of the simple geometrical nature of the model, it is possible to circumvent the ordering problem by means of the DDF variables [9] as suggested by the interpretation of the theory in the framework of the dual models of hadrons.

In what follows we would like to examine the possibility of extending some of the useful concepts of the string model into general relativity. Although we have not been successful in this attempt we feel that the comparative discussion of the two systems is interesting by itself and leads to useful critical remarks.

B. The String Model

Here we consider n + 1 fields $y^A(x,t)$, $x \in R$. The functions y^A parametrize a two dimensional surface V_2 embedded in an N + 1 dimensional Minkowski space of metric

$$ds^2 = d\tilde{y}\cdot d\tilde{y} = \eta_{AB}dy^A dy^B = -(dy^o)^2 + \sum_1^N (dy^A)^2 \quad . \tag{5}$$

The two dimensional surface is spanned by the motion of the (one dimensional) string in the N + 1 dimensional space.

The action for the system is taken to be

$$S = \int (-{}^{(2)}g)^{1/2} dx\, dt \tag{6}$$

where $(-{}^{(2)}g)^{1/2}\, dx\, dt$ is the area element on V_2. The string is assumed to have a finite length and one has to impose Poincaré invariant boundary conditions at its ends in order to obtain a relativistic theory. The boundary conditions imply that the endpoints move transversally with the speed of light. The canonical formalism based on (6) leads to a vanishing canonical Hamiltonian (due to the time reparametrization invariance of (6)) and to constraints of the form

$$\mathcal{H}_1 = \tilde{\pi}\cdot \frac{\partial \tilde{y}}{\partial x} \approx 0 \quad , \tag{7a}$$

$$\mathcal{H}_\perp = \frac{1}{2} \left| \frac{\partial \tilde{y}}{\partial x} \right|^{-1} \left(\tilde{\pi}^2 + \left(\frac{\partial \tilde{y}}{\partial x}\right)^2\right) \approx 0 \quad . \tag{7b}$$

The functions (7) admit again the geometrical interpretation of generating tangential and normal deformations of the string. They satisfy closure relations analogous to (4), namely [8]

$$[\mathcal{H}_\perp(x), \mathcal{H}_\perp(x')] = \left(\left|\frac{\partial \tilde{y}}{\partial x}\right|^{-2}(x)\,\mathcal{H}_1(x) + \left|\frac{\partial \tilde{y}}{\partial x}\right|^{-2}(x')\,\mathcal{H}_1(x')\right)\delta'(x,x')$$

$$+2\left(\left|\frac{\partial y}{\partial x}\right|^{-3}(x)\,\mathcal{H}_\perp(x)\,\mathcal{H}_1(x) + \left|\frac{\partial y}{\partial x}\right|^{-3}(x')\,\mathcal{H}_1(x')\right)\delta'(x,x') , \tag{8a}$$

$$[\mathcal{H}_1(x), \mathcal{H}_\perp(x')] = \mathcal{H}_\perp(x)\,\delta'(x,x') \quad , \tag{8b}$$

$$[\mathcal{H}_1(x), \mathcal{H}_1(x')] = [\mathcal{H}_1(x) + \mathcal{H}_1(x')]\,\delta'(x,x') \quad . \tag{8c}$$

We note that the only difference between (8) and (4) is the presence of the term quadratic in the constraints on the right hand side of (8a). This term has however weakly vanishing brackets with everything, which means that (7) still ensures that all the strings are embedded in a common two dimensional Riemannian surface.

In GGRT [3] the problem of accounting for the constraints and fixing the coordinate system on the surface spanned by the string is solved by introducing a system of null surfaces $y^0 - y^1 = t$ in R^{N+1} which reduces the problem to dealing with N-1 independent modes per point on the string*. It is also possible to introduce a more conventional spacelike gauge [12] $y^0 = t$. In the latter case the Dirac brackets of the basic fields are given typically by expressions of the form

$$[\alpha_m^A, \alpha_n^B] = m\delta_{m,-m}\delta^{AB} + \sum_{M\neq 0} \frac{mn}{M} (p^0)^{-2} \alpha_{m-M}^A \alpha_{n+M}^B \tag{9a}$$

where

$$y^A(x,t) = q^A + p^A t + i \sum_{n\neq 0} \frac{1}{n} \alpha_n^A \cos nx e^{-int} \quad . \tag{9b}$$

Equation (9) shows that the fields y^A, π_A are related to the fundamental canonical variables of the theory by a non-elementary expression. It is in fact extremely hard to approach the quantization procedure by considering the y^A as operators and (9a) as a commutation relation because of the ordering problem. A better approach is to consider the DDF operators which appear in the integral form

$$D_n^A = \frac{1}{2} \int_{\pi}^{2\pi} \frac{dy^A(0,t)}{dt} \exp[n(\tilde{k}\cdot\tilde{p})^{-1} \tilde{k}\cdot\tilde{y}(0,t)]dt \quad , \tag{10}$$

(here $\tilde{k}$ is an arbitrary null vector) and which obey a simple algebra. The whole string model can be built upon a systematic exploitation of this algebra.

The underlying pseudo Euclidean structure of R^{N+1} is necessary for the use of the DDF operators in that from it follows that there are orthonormal coordinates x,t such that the equations of motion can be explicitly solved in the form

$$y^A(x,t) = f^A(t-x) + f^A(t+x) \quad , \tag{11}$$

an equation which is crucial in defining the Fourier transform used by DDF.

A solution similar to (11) is of course not available in general relativity but it is nevertheless of interest to investigate what happens if one tries to cast general relativity in a string-like

*Null surfaces have been introduced to analyze the dynamics of gravity by Aragone and Gambini [10] and Kaku [11]. There is however no analog in the discussion given by those authors of an ambient flat space which is heavily relied upon in the string model.

form, which we pass to do now.

C. General Relativity a la String

By analogy with the string model we postulate here that ordinary curved spacetime V_4 is embedded in some Minkowski space R^{N+1} with a sufficiently high dimensionality $N \geq 9$ so as to be able to accommodate locally a generic four-dimensional pseudo Riemannian manifold. We thus consider the spacetime V_4 as a "trajectory" swept by a three dimensional string in R^{N+1}.

The key difference between the present formalism and the usual approach described in Section (A) above, is that the metric components $g_{\mu\nu}(x)$ are no longer the basic variables but, rather, they are regarded now as derived objects constructed from the functions $y^A(x^0,x^1,x^2,x^3)$ determining the (time dependent) embedding of V_3 in R^{N+1}.

The metric tensor is thus given by

$$g_{\mu\nu}(x) = \tilde{y}_{,\mu} \cdot \tilde{y}_{,\nu} = \eta_{AB} \frac{\partial y^A}{\partial x^\mu} \frac{\partial y^B}{\bar{\partial} x^\nu} \tag{12}$$

with $\eta_{AB} = \text{diag}\ (-1,1,\ldots\ldots,1)$ $\quad A,B\ ,\ 0,\ldots\ldots,N$.
We shall use the same action as in (A), namely

$$S[y] = \int \mathscr{L} d^4x \quad , \tag{13}$$

where $\mathscr{L}$ is the Dirac-ADM lagrangian density appearing in (2), regarded this time as a functional of the y^A through (12).

The fact that $\mathscr{L}$ contains no time derivatives of the $g_{o\mu}$ implies that only first time derivatives of the y^A enter into the action (13). (Eq. 12) shows that $y^A_{,o}$ can enter $\mathscr{L}$ through $g_{o\alpha}$ only). As solely first time derivatives of the y^A appear in the action we see that we are still dealing with a system that can be put in canonical form by standard methods. We have already paid however, a stiff price by introducing the external variables y^A, namely, we have to retain all the fields instead of being able to eliminate four of them (the g_o) at an early stage as was done in (A).

A worse feature is that requiring the action (13) to be stationary under arbitrary variations of the y^A does not reproduce the equations of motion of general relativity

$$(\text{Einstein tensor})^{\alpha\beta} = G^{\alpha\beta} = 0 \quad , \tag{14}$$

but gives rather the weaker set

$$G^{\alpha\beta}\,\tilde{y}_{;\alpha\beta} = 0 \quad . \tag{15}$$

Equations (15) are the analog of the string equations

$$g^{\alpha\beta}\,\tilde{y}_{;\alpha\beta} = 0 \quad \text{(String)} \tag{16}$$

in which case α and β refer to the two dimensional spanned by the string.

Equations (14) do not imply $G^{\alpha\beta} = 0$ due to the identities

$$\tilde{y}_{;\alpha\beta} \cdot \tilde{y}_{,\gamma} = 0 \tag{17}$$

which show that in the generic case only six among the N + 1 equations are independent. We note in passing that the identities (17) avoid the paradoxical implication $g^{\alpha\beta} = 0$ in (16). The difficulty of having only six independent equations in (15) instead of the full Einstein set is not unsurmountable and could be circumvented by imposing in an ad-hoc fashion the additional constraints

$$G_{\perp\alpha} = 0 \tag{18}$$

where the symbol $\perp$ refers to the unit normal to V_3 lying in V_4 and $\alpha = \perp, 1, 2, 3$. Examination of the canonical formalism for the external variables shows in fact that one may expect (18) not to be an entirely unreasonable addition to the equations (15).

Canonical Formalism for External Variables

We start from the Dirac-ADM Lagrangian density (13), which written down in detail reads

$$\mathcal{L} = g^{1/2} N (R + K_{ab}K^{ab} - (K^a_a)^2) , \tag{19}$$

where R is the curvature scalar of V_3 and where the extrinsic curvature K_{ab} of V_3 with respect to V_4 is given by

$$K_{ab} = (2N)^{-1} (-\dot{g}_{ab} + N_{a|b} + N_{b|a}) \quad . \tag{20}$$

The symbols N and N_a stand for the lapse and shift functions

$$N = (-{}^4g^{oo})^{1/2} \quad , \quad N_a = g_{oa} \quad . \tag{21}$$

A dot denotes differentiation with respect to x^o. The Lagrangian density (19) is expressed as a functional of the y^A by means of

(12),(20) and (21).

The canonical momenta are defined by

$$\tilde{\pi}(x) = \frac{\delta}{\delta \dot{\tilde{y}}(x)} \int d^3x' \, \mathscr{L}(x') \tag{22}$$

which gives, after some calculation,

$$\tilde{\pi} = g^{1/2}\{-2G_{\perp\perp}\tilde{n} + 2(K^{ab}-K^m_m g^{ab})\ \tilde{y}_{|ab}\} \quad . \tag{23}$$

Here $\tilde{n}$ denotes the unit normal to V_3 lying in V_4:

$$\tilde{n} = N^{-1}[\dot{\tilde{y}} - (\dot{\tilde{y}} \cdot \tilde{y}^{|i})\tilde{y}_{,i}] \tag{24}$$

and $G_{\perp\perp}$ is the double projection of the Einstein tensor along $\tilde{n}$:

$$-2G_{\perp\perp} = K_{ab}K^{ab} - (K_m)^2 - R \quad . \tag{25}$$

The normal (24) satisfies the normalization condition,

$$\tilde{n} \cdot \tilde{n} = -1 \quad , \tag{26}$$

and the extrinsic curvature is related to $\tilde{n}$ by

$$K_{ab} = \tilde{n} \cdot \tilde{y}_{|ab} \tag{27}$$

Now we note that the six vectors $\tilde{y}_{|ab}$ are perpendicular to V_3 (this is just the V_3 version of the identities (17)). Also the normal $\tilde{n}$ is perpendicular to V_3. It thus follows that the three components of $\tilde{\pi}$ on V_3 vanish. We then get the three constraints

$$\mathscr{H}_i = \tilde{\pi} \cdot \tilde{y}_{,i} \approx 0 \quad , \tag{28}$$

which are the analog of (7a) for the string. The $\mathscr{H}_i$ defined by (28) generate reparametrizations on V_3 and they satisfy consequently the closure relations (4c). It follows from (28), for example that y^A and π_A transform as scalars and scalar densities respectively under changes of coordinates in V_3, which was of course to be expected.

The fourth constraint (analog to (7b) for the string) is obtained in principle by solving (23) as a system of nonlinear algebraic equations for n^A as a function of π^A and y^A and imposing afterwards the normalization condition (26). In the case of the string this procedure yields (7b). In fact the string analog of (23) reads simply

$$\tilde{\pi} = \left|\frac{\partial \tilde{y}}{\partial x}\right| \tilde{n} \quad \text{(string)}, \tag{29}$$

which upon squaring and using (26), gives (7b). The solution of (23) is however considerably harder and there seems to be no way of obtaining a simple closed form for

$$\tilde{n} = \tilde{n}(\tilde{y}, \tilde{\pi}) \quad . \tag{30}$$

This problem might be circumvented to some extent with the help of the additional constraints (18) which we pass to discuss now.

As we mentioned before, even if we imagine having the solution (30), the formalism does not reproduce Einstein's theory. In fact, if we count the number of independent degrees of freedom of the theory we find: 2(N+1)-4 (first class constraints) -4 (gauge conditions) = 2(N-3). That is we have N-3 degrees of freedom per point, which recalling that $N \geq 9$ is at least an excess of four over the required number of two for general relativity.

We see therefore that even if we bring N down to its minimum value of 9, as we shall do tentatively from now on, we need four additional first class constraints besides the $\mathcal{H}_\mu$. It is quite reasonable to take these new constraints to be (18). In fact the $G_{\perp\mu}$ are constructed from the g_{ab} and K_{ab} only, which means that they can in principle be expressed, via (30), as functions of the canonical variables $\tilde{y}, \tilde{\pi}$.

Now, if we are going to impose the constraints (18), we need to solve (23) for $\tilde{n}$ only when $G_{\perp\perp} = 0$. This will result in changing the constraints by linear combinations of themselves and will therefore not change the dynamics of the system.

When $G_{\perp\perp} = 0$ (23) can be written as

$$\pi^A \approx W^A_B n^B \quad , \tag{31}$$

with

$$W^A_B = 2g^{1/2}(g^{ad}g^{bc} - g^{ab}g^{cd})y^A{}_{|ab}y_{B|cd} \quad .$$

The matrix W defined by (32) regarded as a mapping of R^{10} onto R^{10} does not have an inverse because it maps to zero the three vectors $\tilde{y}_{,i}$ (i=1 2 3). However when restricted to the sub-space orthogonal to the $\tilde{y}_{,i}$, W will have an inverse in the generic case. Let us denote that inverse by M. The matrix M is therefore defined as giving that solution of (31),

$$n^B = M^B_A \, {}^A \tag{33}$$

which satisfies

$$\tilde{n} \cdot \tilde{y}_{,i} = 0 \quad . \tag{34}$$

It follows from (32) that M is constructed from the y^A and their derivatives and that $M_{AB} \equiv \eta_{AC} M^C_B$ is symmetric. The eight constraints of the theory can then be expressed in terms of M as follows

$$-2G_{\perp\perp} = K_{ab}K^{ab} - (K^m_m)^2 - R \approx \frac{1}{2} g^{-1/2} M_{AB} \pi^A \pi^B - R \approx 0 \quad , \tag{35a}$$

$$-G_{\perp i} = (K^k_i - K^m_m \delta^k_i)_{|k} \approx (M_{AB}\pi^B)_{,i} y^{A|m}{}_{|m} - (M_{AB}\pi^B)_{,m} y^{A|m}{}_{|i} \approx 0 \quad , \tag{35b}$$

$$\mathcal{H}_\perp = g^{1/2}(\tilde{n}^2 + 1) \approx g^{1/2}((M^2)_{AB}\pi^A\pi^B + 1) \approx 0 \quad , \tag{35c}$$

$$\mathcal{H}_i = \tilde{\pi} \cdot \tilde{y}_{,i} = 0 \quad . \tag{35d}$$

The essential problem at this point is to prove that the eight constraints (34) are first class. It does not seem possible to do this without knowing more about the form of M_{AB}. We plan to investigate this matter in the future.

If the constraints (35) are indeed first class their compatibility is ensured and the theory is consistent. Furthermore we are then sure that we are dealing exactly with Einstein's equations because the only way in which $G_{\perp\mu}$ can vanish on every three dimensional space like hypersurface of V_4 is that all ten equations $G_{\alpha\beta}=0$ hold. On the other hand, if the system (35) is not first class we would be merely selecting by means of (35a,b) special coordinates on V_4 (i.e., fixing the gauge) instead of reducing the number of physical degrees of freedom of the theory. The formalism would not reproduce Einstein's theory in that case.

Final Remarks

The theory as we have presented it here is not complete but we feel it deserves further investigation. It is quite possible that the actual value of N is not relevant in a final, as yet hypothetical, complete form. We must keep in mind in this connection that the possibility of embedding a four dimensional manifold in R^{10} holds only in a very local sense and that non-trivial problems are already encountered in trying to embed globally a smooth two dimensional manifold [13] in R^3. However the existence of the variables y^A gives us more freedom to construct field variables which do not exist in the conventional theory and which could possibly lead to a canonical formulation of general relativity different from the conventional one. In this sense it could be interesting to try to find the analog of the DDF operators for the string model. Unfortunately we have not been able as yet to obtain any definite result along this direction.

Acknowledgments

The authors are indebted to Professors Abdus Salam and Gallieno Denardo for their kind hospitality at Trieste. One of us (C.T.) would also like to thank John Wheeler for much encouragement.

REFERENCES

[1] P.A.M. Dirac, Proc. Roy. Soc. (London) A246, (1958) 333.
[2] R. Arnowitt, S. Deser and C.W. Misner in Gravitation: An Introduction to Current Research [L. Witten, Ed.], Wiley, New York (1962).
[3] P. Goddard, J. Goldstone, C. Rebbi and C.B. Thorne, Nucl. Phys. B56, (1973) 109.
[4] P.A.M. Dirac, Lectures on Quantum Mechanics, Belfer Graduate School of Science, Yeshiva University, New York (1964).
[5] C. Teitelboim, Ann. Phys. (N.Y.) 79 (1973) 542.
[6] P.A.M. Dirac, Phys. Rev. 114 (1958) 924.
[7] J.W. York, Phys. Rev. Lett. 28 (1972) 1082.
[8] See, for example, A.J. Hanson, T. Regge and C. Teitelboim, Constrained Hamiltonian Systems, Accademia Nazionale dei Lincei, Rome (1975).
[9] E. Del Giudice, P. DiVecchia and S. Fubini, Ann. Phys. (N.Y.) 70 (1972) 378.
[10] C. Aragone and R. Gambini, Nuovo Cimento 18B (1973) 311.
[11] M. Kaku, to be published.
[12] P. Goddard, A.J. Hanson and G. Ponzano, to be published.
[13] J. Nash, Annals of Mathematics (2), 60 (1954) 383.

GEOMETRIC QUANTIZATION AND GENERAL RELATIVITY

Jean-Marie Souriau

Universite de Provence and Centre de Physique Theorique
13274 Marseille Cedex 2, France

INTRODUCTION

The purpose of geometric quantization is to give a rigorous mathematical content to the "correspondence principle" between classical and quantum mechanics. The main tools are borrowed on one hand from differential geometry and topology (differential manifolds, differential forms, fiber bundles, homology and cohomology, homotopy), on the other hand from analysis (functions of positive type, infinite dimensional group representations, pseudo-differential operators). Some satisfactory results have been obtained in the study of dynamical systems, but some fundamental questions are still waiting for an answer. The "geometric quantization of fields," where some further well known difficulties arise, is still in a preliminary stage. In particular, the geometric quantization of the gravitational field is still a mere project. The situation is even more uncertain due to the fact that there is no experimental evidence of any quantum gravitational effect which could give us a hint towards what we are supposed to look for. The first level of both Quantum Theory and General Relativity describes passive matter: influenced by the field without being a source of it (first quantization and equivalence principle respectively). In both cases this is only an approximation (matter is always a source). But this approximation turns out to be the least uncertain part of the description, because on one hand the first quantization avoids the problems of renormalization and on the other hand the equivalence principle does not imply any choice of field equations (it is known that one can modify Einstein equations at short distances without changing their geometrical properties).

One hopes to go beyond that first level: that is, to quantize field equations, to build a synthetic theory, which would contain the other two at least as approximations. It is difficult to be a physicist--to believe in the possibility of understanding nature--without being obliged to bet that such a theory can exist, can be mathematically consistent. Maybe the stumbling block is not only technical but also epistemological. Maybe some a prioris must be rejected, and some new concepts found in order to make any progress. In this methodological perspective, the study of the first level may still be useful, both because it can prevent arbitrary choices and because the symmetries of the first level can give us hints concerning the symmetries of the synthetical theory.

I. HYPERSPACE AND CONDENSED STATES

In order to avoid controversies about the meaning of Einstein "general relativity principle," it is necessary to endow it with a precise mathematical content. Let V_4 be the space-time manifold (see Figure 1)

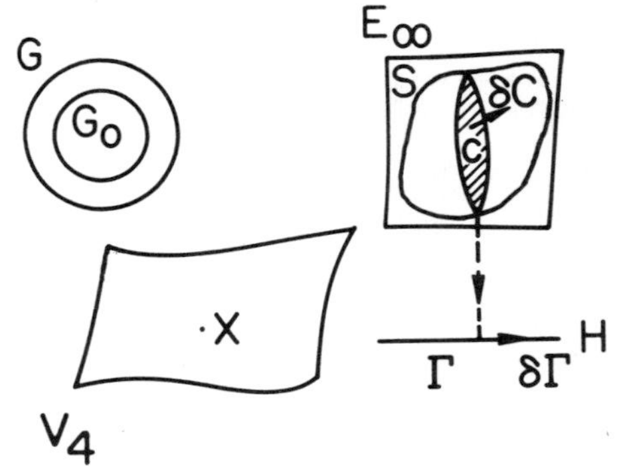

Let E_∞ denote the set of symmetric covariant differentiable tensor fields of second degree on V_4. E_∞ is an infinite dimensional vector space. The space-time metric $X \to g$ is a point C in E_∞; it belongs to the set S of tensor fields whose signature is (+---) for every $x \in V_4$. The set G of the diffeomorphisms of V_4 (C^∞ one to one map with nonvanishing jacobian) is an infinite parameter group. The differentiable geometric objects (vectors, forms, tensors, connexions, densities, etc.) share the property of being acted upon by G. The principle of General Relativity says, on one hand, that physical objects are differentiable objects; on the other hand that a simultaneous transformation by the same diffeomorphism is unobservable. In other words, V_4, X, g, etc. are merely abstract auxiliary constructions and the concrete arena of the physical world is the quotient of such a mathematical structure by the action of G. This is, then, the choice of a structural symmetry. In particular the action of G on the tensor g defines a linear representation of G on E_∞, leaving S invariant: S is a union of G-orbits (the dashed area in Figure 1 denotes the orbit of C). The quotient space H is called hyperspace (this is the space of admissible geometries of V_4). At this stage a technical difficulty arises: H is not free of singularities, since the stability subgroup of a point C in S is not always of the same structure. The singular points of H are the geometries which possess a nontrivial isometry group. This difficulty can be cured by replacing G by its invariant subgroup G_0 of diffeomorphisms with compact support (diffeomorphisms which reduce to the identity outside of a compact subset of V_4). If V_4 is not compact, G_0 does not contain any nontrivial isometry for any given geometry, hence H is free of singularities.

It is possible to extend this structure by adding the electromagnetic potentials A_ρ to the gravitational ones $g_{\mu\nu}$, and by adding the group of electromagnetic gauge transformations $A_\rho \to A_\rho + \partial_\rho \alpha$ (which are, in principle, unobservable too) to the group of diffeomorphisms of V_4. More precisely diffeomorphisms and gauge transformations generate a group G which is a semi-direct product.* The elements in G with compact support still constitute an invariant subgroup G_0. G and G_0 act linearly on the new space E_∞ of potentials C ($X \to (g,A)$). The stability subgroups in G_0 are trivial. In the following we shall consider $H = S/G_0$ as the hyperspace.

It is possible to endow H with a differential structure. Let us try, for instance, to define a differential one-form μ at a point Γ (projection of C on H). This must be a linear form on the variations $\delta\Gamma$, which are the projections on H of the field variations δC (see Fig. 1). This will allow to lift μ to a linear form on δC, i.e. to a tensor-distribution. Moreover, this tensor-distri-

---

*This is a consequence of a natural action of a diffeomorphism on A_ρ and α. This group arises most naturally in the Kaluza — Klein formalism, see [18].

bution must vanish when $\delta\Gamma = 0$, that is when δC is vertical. This occurs when δC is given by the infinitesimal action of an element of the Lie algebra of G_0, characterized by a vector field δX and a scalar field $\delta\alpha$ with compact supports. The theory of Lie derivatives then leads to the following theorem (see demonstration in [21]).

Let us suppose that μ is lifted to a completely continuous measure:

$$\mu(\delta\Gamma) = \int_{V_4} [\frac{1}{2} T^{\mu\nu}\delta g_{\mu\nu} + \mathcal{J}^{\rho}\delta A_{\rho}]\, dv(X) \tag{1}$$

where $T^{\mu\nu}$ is symmetric, v denotes the Riemannian measure associated to g, and δ is an arbitrary field variation. It follows that the fields $X \to T$ and $X \to$ defined by Equation (1) must verify the equations:

$$\hat{\partial}_{\rho}\mathcal{J}^{\rho} = 0 \qquad \hat{\partial}_{\mu}T^{\mu}_{\nu} + F_{\nu\rho}\mathcal{J}^{\rho} = 0 \tag{2}$$

($\hat{\partial}$ = covariant derivative and $F_{\nu\rho} = \partial_{\nu}A_{\rho} - \partial_{\rho}A_{\nu}$). One recognizes the principles of the electrodynamics of continuous media, $\mathcal{J}$ being the 4-current density and T being the energy-momentum tensor. Therefore we have introduced a duality between the laws of motion and the geometric structure of the theory. What we learn from this duality is that the universality of Equation (2) is correlated to the choice of the group G_0. Any modification of one of them requires a corresponding modification of the other one.*

This duality has also a technical interest: it can be directly adapted to the description of condensed states of matter. It is sufficient to assume that μ is no longer completely continuous in δC, but is a distribution whose support is a submanifold of V_4. The simplest case is that of a particle: choosing as support of μ a worldline Λ, and assuming that μ a distribution of order one, one obtains the following result [21]:

(a) At each point X of Λ, there exist a number q, a vector P^{μ} and two skew-symmetric tensors $S^{\mu\nu}$ and $M^{\mu\nu}$ such that:

$$\mu(\delta\Gamma) = \int \left\{ \begin{array}{l} \frac{1}{2}\,(P^{\mu}\frac{dX^{\nu}}{ds} + M^{\mu\lambda}F^{\nu}_{\lambda})\,\delta g_{\mu\nu} + q\,\frac{dX^{\rho}}{ds}\,\delta A_{\rho} \\ \frac{1}{2}\,S^{\lambda\mu}\frac{dX^{\nu}}{ds}\,\hat{\partial}_{\lambda}\delta g_{\mu\nu} + M^{\lambda\rho}\,\hat{\partial}_{\lambda}\delta A_{\rho} \end{array} \right\} ds \tag{3}$$

(b) These quantities satisfy the equations

$$dq/ds = 0$$

---

*The elementary character of the laws (2) is therefore an argument for unifying diffeomorphisms and electromagnetic gauge transformations for which there are no a priori reasons. If one adopts a 5-dimensional viewpoint, in order to give a geometrical meaning to G and G_0, one is lead to modify the equations of motions.

$$\hat{d}\, P_\sigma/ds = qF_{\sigma\rho}dX^\rho/ds + \frac{1}{2} M^{\mu\nu}\hat{\partial}_\sigma F_{\mu\nu} - \frac{1}{2} R_{\mu\nu,\rho\sigma}\, S^{\mu\nu}dX^\rho/ds \qquad (4)$$

$$\hat{d}\, S^{\mu\nu}/ds = P^\mu dX^\nu/ds - P^\nu dX^\mu/ds - M^{\mu\rho}F^\nu_\rho + M^{\nu\rho}F^\mu_\rho$$

($R_{\mu\nu,\rho\sigma}$ = Riemann Christoffel tensor)

These are the universal equations describing the motion of a test particle in an external electromagnetic and gravitational field, provided that quadrupole moments are neglected (see [16] [19]). q is interpreted as the electric charge, P as the linear momentum, S as the spin-tensor, and M as the electromagnetic dipole moment. The role of these four quantities is to generate a hyperspace one-form through Eq. (3). Eq. (4) are then merely the necessary and sufficient consistency relations of (3). This role can be checked by a statistical argument. By exploiting the bilinearity of the application $(\mu, \delta\Gamma) \to \mu(\delta\Gamma)$, one can interpret a statistical ensemble of particles with spin by means of an average functional of the kind of Eq. (1). For instance by describing a magnet as a macroscopic assembly of particles with spin, one uncovers not only the equivalence of the magnet with a solenoid but also, and by necessity, the gyromagnetic effect and the magnetostriction [21]. The completeness of such a description raises serious doubts about the legitimacy of some concepts, in particular the one of a macroscopic density of spin which, a priori, could be considered as a source for the gravitational field.

On the other hand, the geometric structure defined by Eq. (1) or (3) implies the following: to every symmetry of the field corresponds a conserved quantity--a well-known result in the case of continuous media.

Let us examine the case of a single particle. We noticed before that the stability subgroup of C in G_0 is necessarily reduced to the identity. Let us suppose that C has a nontrivial stability subgroup in G. This will be, by necessity, a Lie group of finite dimensions n ($n \leq 11$). The n conserved quantities associated to this group are all contained in the formula:

$$P_\mu V^\mu + \frac{1}{2} S^{\mu\nu}\, \hat{\partial}_\mu V_\nu + q[u + A_\rho V^\rho] \qquad (5)$$

$\delta X^\mu = V^\mu$, $\delta A_\rho = \partial_\rho u$ being the infinitesimal action on the points and the potentials of an arbitrary element in the Lie algebra of the stability subgroup. A direct geometrical proof can be found in [21]. A straightforward calculation checks that Eq. (5) gives a first integral of the differential system (4). This is a remarkable result because these 11 equations do not constitute a deterministic system for the 21 variables X, q, P, S, M. Therefore, this is not covered by Noether's theorem. A particle of a given kind will be characterized by some equations of state restricting these universal variables (and possibly other internal variables). In this way we can hope to obtain a deterministic model.

For instance one can try the following hypotheses:

$$S^{\mu\nu}P_\nu = 0 \qquad M^{\mu\nu} = \lambda S^{\mu\nu} \tag{6}$$

(monolocality, vanishing electric dipole moment)

Consistency with the universal Eq. (4) implies that the variables

$$P^2 = P_\mu P^\mu \quad \text{and} \quad \alpha = S^{\mu\nu}F_{\mu\nu} \tag{7}$$

are functionally related along the world line. Whence the introduction of a last equation of state:

$$P^2 = f(\alpha) \tag{8}$$

which gives the contribution to the energy of the particle due to the component of the magnetic field along the spin. Now the system (4), (6), (8) is deterministic. In the experimental situations corresponding to the precise measurements of the magnetic moments of the electron and the muon (negligible gravitational effects, constant magnetic field) one recovers the equations of Bargmann-Michel-Telegdi [17] which proves at least for this case their consistency with experiment.

II. GEOMETRIC QUANTIZATION OF DYNAMICAL SYSTEMS

The fundamental law of the classical dynamics of a system of n mass points can be stated in the form of the principle of virtual work (or d'Alembert principle)

$$\sum_j \langle m_j d\vec{v}_j/dt - \vec{F}_j, \delta\vec{r}_j \rangle = 0 \tag{9}$$

where the brackets $\langle , \rangle$ denote the scalar product in R^3. Each point, with mass m_j, position $\vec{r}_j$, and velocity

$$\vec{v}_j = d\vec{r}_j/dt \tag{10}$$

is acted upon by a force $\vec{F}_j$, which is, a priori, function of

$$y = (\vec{r}_i,\ldots,\vec{r}_n, \vec{v}_i,\ldots,\vec{v}_n, t) \tag{11}$$

$\delta\vec{r}_j$ denotes an arbitrary variation of the position $\vec{r}_j$. In the case of a constrained system, it is understood that $\delta\vec{r}_j$ is compatible with the constraints at time t. It is possible to enlarge this principle by considering positions and velocities as independent variables. One considers the expression

$$\sigma(dy,\delta y) = \sum_j \langle m_j d\vec{v}_j - \vec{F}_j dt, \delta\vec{r}_j - \vec{v}_j\delta t\rangle - \langle m_j\delta\vec{v}_j - \vec{F}_j\delta t, d\vec{r}_j - \vec{v}\,dt\rangle \tag{12}$$

which reduces to Eq. (9) (within a factor dt) when $\delta t = 0$, $\delta\vec{v}_j = 0$. Let us notice that Eq. (12) defines a tensor σ in the evolution space V parametrized by the variable y (11). The minus sign in Eq. (12) corresponds to a seemingly arbitrary choice made by Lagrange as early as 1808 [1], it makes σ antisymmetric. Its interest lies in giving an "invariant intégral absolu" [2] for the equations of motion , a result explicitly stated by Lagrange, and rediscovered by Elie Cartan, a century later [2]. Let us note that the equations of motion (including Eq. (10)) now read:

$$\sigma(dy,\delta y) = 0 \quad \forall \delta y \tag{13}$$

each motion (a curve in V) is a one-dimensional leaf of the fibering defined in V by Eq. (13) (the dimension of V is 6n + 1; the rank of σ, which must be even by a well known algebra theorem, is here equal to 6n). There exists a quotient manifold U, of dimension 6n, each point of which represents a motion (see Fig. 2).

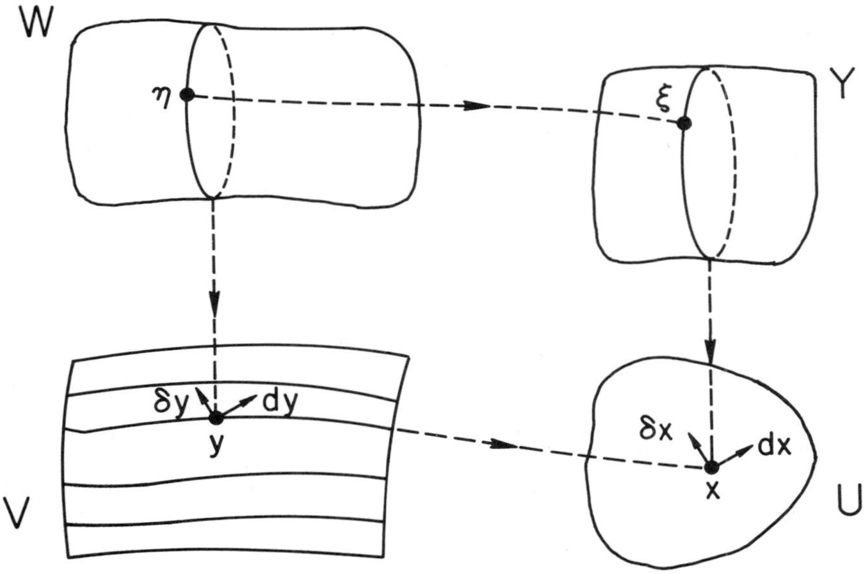

If the forces $\vec{F}_j$ are derivable from a potential (even function of time) the form σ passes to the quotient: there exists a 2-form σ_u defined by

$$\sigma(dy,\delta y) = \sigma_u(dx,\delta x) \tag{14}$$

dx and δx being the variations of the motion x generated by arbitrary variations dy and δy of the initial condition y. σ_u is a non-singular flat two-form (there exist coordinates in U such that the components of σ_u are constant). A manifold U endowed with such a form is called a symplectic manifold. It is always of even dimension. The symmetry of the theory is defined by the group Symp (U) of those diffeomorphisms of U which leave σ_u invariant ("symplectomorphisms"). It is an infinite dimensional group which will contain the manifest symmetries of the system. To every Lie group of symplectomorphisms corresponds a system of conserved quantities (the "moment"), with values in the dual of the Lie algebra of the group see [8]. Together with this construction, one can define a certain cohomology for Lie groups ("symplectic cohomology"). To each symplectic action of a group is associated a symplectic cohomology class.

For instance an isolated classical dynamic system defines a symplectic action of the Galilean group. The cohomology of this group is one dimensional, hence every system is characterized by a measurable quantity m: m is its mass now endowed with a geometrical meaning.

Symplectic structure is not limited to the realm of classical mechanics: for instance, a relativistic free particle defines a symplectic action of the Poincare group. The elementary character of the particle corresponds to the fact that this action is transitive. Since the group cohomology vanishes, it can be shown that this is an action in the sense of Kirillov [10], i.e. U is isomorphic to an orbit of Poincare group in the dual representation of the adjoint representation. This fact provides an a priori classification of particle models. Noetherian quantities associated to each model (linear momentum, energy, position, intrinsic angular momentum (spin), helicity, parity) distinguish among these candidates those which can describe real particles (spin or spinless particles, massless particles [8]). This universal symplectic structure plays an important role also in statistical mechanics (see [8]); but it appears to be even more fundamental in the quantization process.

The first stage --"prequantization" in Kostant's terminology--requires the construction of a circle bundle Y over U (see Fig. 2) with a one-form $\bar{\omega}$ such that:

- the exterior derivative of $\bar{\omega}$ coincides with the pullback of σ_U by the projection $\xi \to x$;
- $\bar{\omega}$ is everywhere non-vanishing along the circular fibres; the loop integral along the fibres $\oint \bar{\omega}\,(d\xi/ds)ds$ equals Planck's constant $2\pi\hbar$. (15)

In order that this construction be possible, σ_U must verify certain homological conditions; it is then possible to solve Dirac's problem, that is to realize the Lie algebra of dynamical variables (under Poisson bracket) on a Hilbert space, namely on a space of functions defined on Y. In case U is the space of motions of a free particle, one can go further: the method of polarization (Kostant, Sourian) allows to derive and to interpret the classical wave equations (Schroedinger, Klein-Gordon, Dirac, Maxwell, etc.) (see [8]). The case of particles subject to an external field poses a much more delicate problem, for which a solution is foreseen, thanks to the technique of pairing (Blattner, Kostant, Sternberg) and the one of Maslov index (Maslov, Arnold, Leray) (see [12], [12], [14]; [5], [7], [12], [14]). The problem amounts to construct certain unitary representations of the group of quantomorphisms of Y (diffeomorphisms which leave $\overline{\omega}$ invariant), or, more precisely, of a covering of this group. For instance, in the case of a linear oscillator (of any dimension), the quantomorphisms of Y whose projection on U are affine transformations define a Lie group (Weyl-Heisenberg group); using the pairing and the Maslov index, one can construct a representation space H of the double covering of this group (metaplectic group); the representation is isomorphic to the Shale-Weil representation. To each point of H there corresponds a solution of Schroedinger equation, integrated by an explicit formula which continues the Feynman resolvent kernel (Souriau [14]).

III. COMPLEMENTARITY

The two approaches that we have been sketching are conceptually very different although they can use similar mathematical tools: the notion of space-time is fundamental in the former and "forgotten" in the latter. The intervening groups (diffeomorphisms-gauge on one hand, quantomorphisms on the other hand) admit interpretations nearly foreign to each other. General Relativity leads automatically to a definition of the dynamical quantities P, S, etc. (see Eq. (3)) and to a prediction for the action of the field (Eq. (4)). The symplectic approach, on the contrary, gives a privileged role to the case of a free particle, i.e., of the maximum symmetry interpreted as the absence of any field. It gives the definition of P and S (noetherian first integrals) only in this case. It ignores not only the action of the field but even the permanent character of the quantities P and S when the symmetry is broken. At the same time, Relativity gives only partial information on the motion and non-deterministic equations. It ignores, a priori, the internal degrees of freedom. On the contrary the symplectic approach demands not only a deterministic but a determined description. Since its conceptual basis is the space of motions U whose definition requires the preliminary integration of the equations of motion.

The positive aspect of these differences is the complementarity of the two theories. The two approaches must be consistent when dealing with a well defined physical object. Let us study the case of a spinning particle. As we said before, the universal equations (4) and the equations of state (6), (8) yield a deterministic system for a given field. One can define an initial condition y for this system by means of a point X in space-time, a vector I and a tensor Ω in this point, with the following constraints:

$$\Omega^{\mu\nu} + \Omega^{\nu\mu} = 0 \qquad \Omega^{\mu\nu} I_\nu = 0$$
$$\Omega_{\mu\nu}\Omega^{\mu\nu} = 2 \qquad I_\mu I^\mu = 1 \tag{16}$$

this define an evolution space V with dimension 9. The other variables are given by:

$$S^{\mu\nu} = s_0 \Omega^{\mu\nu} \quad (s_0 \text{ is a first integral-spin-whose value is supposed to be given}) \tag{17}$$

$$\alpha = S_0 \Omega^{\mu\nu} F_{\mu\nu} \tag{17}$$

$$P^\mu = I^\mu f(\alpha) \quad , \quad M^{\mu\nu} = S_0 (f'(\alpha)/f(\alpha))(P_\lambda dX^\lambda/ds)\Omega^{\mu\nu}$$

Then the equations of motion are A = 0, B = 0 and C = 0 with:

$$A_\sigma = \frac{\hat{d}P_\sigma}{ds} - qF_{\sigma\rho}\frac{dX^\rho}{ds} - \frac{1}{2}\frac{f'(\alpha)}{f(\alpha)} P_\lambda \frac{dX^\lambda}{ds} S^{\mu\nu}\hat{\partial}_\sigma F_{\mu\nu} + \frac{1}{2} R_{\mu\nu,\rho\sigma} S^{\mu\nu}\frac{dX^\rho}{ds}$$

$$B^{\mu\nu} = \frac{\hat{d}}{ds} S^{\mu\nu} - P^\mu \frac{dX^\nu}{ds} + P^\nu \frac{dX^\mu}{ds} + \frac{f'(\alpha)}{f(\alpha)} P_\lambda \frac{dX^\lambda}{ds} [S^{\mu\rho}F^\nu_\rho - S^{\nu\rho}F^\mu_\rho] \tag{18}$$

$$C^\sigma = \frac{dX^\sigma}{ds} - \frac{1}{f(\alpha)} P^\sigma P_\lambda \frac{dX^\lambda}{ds} + \frac{1}{S_0^2} S^\sigma_\mu S^\mu_\nu \frac{dX^\nu}{ds}$$

As in classical mechanics, these equations can be deduced from a principle of virtual work of the type (9) or better (12) by considering the expression:

$$\sigma\left(\frac{dy}{ds}, \delta y\right) = C^\sigma \hat{\delta} P_\sigma - A_\sigma \delta X^\sigma - S_0^{-2} B^{\mu\nu} S_{\nu\rho} \hat{\delta} S^\rho_\mu \tag{19}$$

After some algebra one can write (19) as:

$$\sigma\left(\frac{dy}{ds}, \delta y\right) = \frac{dX^\sigma}{ds}\hat{\delta}P_\sigma - \delta X^\sigma \frac{\hat{d}P_\sigma}{ds} + qF_{\sigma\rho}\frac{dX^\rho}{ds}\delta X^\sigma - \frac{1}{2} R_{\mu\nu,\rho\sigma} S^{\mu\nu}\frac{dX^\rho}{ds}\delta X^\sigma$$

$$- S_0^{-2} \frac{\hat{d}S^\mu{}_\nu}{ds} S^\nu_\rho \hat{\delta} S^\rho_\mu$$

where the antisymmetry in d/ds and δ is apparent. The following is analog to the case of classical mechanics. The equations of motion are $\sigma(dy, \delta y) = 0 \ \forall\delta$ (see Eq. (13)) and define a one-dimensional fibering of V as well as a quotient manifold U of dimension 8 (Fig. 2). Moreover the exterior differential of σ vanishes (the computation uses Eq. (16), their differentiation, the appearance of the curvature tensor in second order differentiations and Bianchi identities). The theorem of Darboux allows to show that σ, passes to the quotient U by a formula of the kind (14), and, endows U with a symplectic structure. In the particular case of vanishing field ($A_\mu \equiv 0$, V_4 minkowskian), U is nothing else but a coadjoint orbit of the Poincare group, already met in II as a model for a spinning particle.

This coincidence implies not only the existence of a global symplectomorphism, but also the equality between the noetherian quantities and the quantities (5) associated respectively by each theory to the Poincare symmetry. This equality is maintained when the dimension of the symmetry group of the field decreases.

Let us come back to a general field and try to quantize. One must first solve the cohomological problem allowing the construction of the fiber bundle Y over U. One shows that a necessary condition is that the spin S_0 (17) is an integer multiple of $\hbar/2$. Let us study the case $S_0 = \hbar/2$.

One must first assume the existence of a spinor structure associated to the metric of V_4 (therefore to C). This means the existence of a global differentiable Lorentz-frame field (vierbein field). Then let us write:

$$\eta = (X,\theta) \qquad (21)$$

X being a point in V_4 and θ a spinor at this point fulfilling the constraints:

$$\bar{\theta}\theta = 1 \qquad \bar{\theta}\gamma_5\theta = 0 \qquad (22)$$

η defines a ten dimensional manifold W which can be projected on the evolution space V (Fig. 2) by the formulae:

$$\bar{\theta}\gamma_\mu\theta = I_\mu \qquad \mathrm{Re}(i^{-1}\bar{\theta}\gamma_\mu\gamma_\nu\theta) = \Omega_{\mu\nu} \qquad (23)$$

which imply the constraints (16) without fixing the phase of θ.

One defines on W a one form $\bar{\omega}_W$ by the formula

$$\bar{\omega}_W(\delta\eta) = \hbar\,\mathrm{Re}(i^{-1}\bar{\theta}\hat{\delta}\theta) - [P_\mu + qA_\mu]\,\delta X^\mu$$

where $\hat{\delta}\theta$ is the covariant derivative of the spinor θ for an arbitrary variation δ of η. The exterior derivative of $\bar{\omega}_W$ coincides with the pull back of σ par the projection $\eta \to y$. By doing the quotient of W by the characteristic fibering of $\bar{\omega}_W$, one obtains a 9-dimensional manifold Y to which the form $\bar{\omega}$ passes, and, a commutative diagram of projections between the manifolds W, V, Y, U (Fig. 2). Y yields the prequantization of U in the sense (15). Using the homotopy of U one can show that this prequantization is essentially unique.

In the case of vanishing field, one gets back the prequantization indicated in II which leads to Dirac's equations. In the case of a general field, the quantization problem is not completely solved because one must use complex polarizations (Kostant) which complicates the pairing technique (see Blattner [15]). However, it seems probable that the Dirac equation with minimal coupling or with a Pauli term (abnormal moment) corresponds to the case where the function f defined in (8) is affine [22].

CONCLUSION

We have just seen, on this example, that it is possible to connect the two theories thanks to a whole sequence of coincidences. One must now interpret this possibility.

REFERENCES

On symplectic structure and geometric quantization:

[1] J. L. Lagrange, Mécanique Analytique, 1788-1808 (réédition A. Blanchard, Paris, 1965).

[2] E. Cartan, Lecons sur les invariants intégraux (Paris, 1920).

[3] J. M. Souriau, Alger Mathématiques, I, 2 (1954), pp. 240-265.

[4] I. Segal, Trans. Amer. Math. Soc. 81 (1956), pp. 106-134.

[5] A. Maslov, Théorie des perturbations et méthodes asymptotiques, Moscou (1965) (French translation: Dunod, Paris, 1972).

[6] J. M. Souriau, Comm. Math. Phys., I (1966), pp. 374-398.

[7] V. I. Arnold, Journal d'Analyse fonctionnelle (in Russian) I (1967), pp. 1-14 (French translation in [5]).

[8] J. M. Souriau, Structure des systèmes dynamiques (Paris, Dunod, 1969).

[9] B. Kostant, Orbits and quantization theory, Intern. Congr. of Math., Nice (1972).

[10] A. Kirillov, Eléments de la théorie des représentations (in French, Moscow, 1975).

[11] V. Guillemin, S. Sternberg, Geometrical Asymptotics (to be published).

A review and a detailed bibliography in the following proceedings:

[12] Geometria simplecttica e fisica matematica, Rome, janvier 1973 (Symposia Mathematica XIV, Ac. Press, London, 1974).

[13] Géométrie symplectique et physique mathématique, Aix-en-Provence, juil, 1974 (Coll. internat. C.N.R.S. 237, Paris, 1976).

[14] Group theoretical Methods in Physics, Nijmegen, juin 1975 (to be published).

[15] Differential geometrical methods in mathematical physics, Bonn, juil, 1975 (to be published in Springer Lecture Notes in Math).

Some references about particles models:

[16] A. Papapetrou, Proc. Roy. Soc. A 209 (1951), pp. 248-258.

[17] V. Bargmann, L. Michel, V. L. Telegdi, Phys. Rev. Lett. 2 (1959), p. 435.

[18] J. M. Souriau, Géométrie et relativité (Hermann, Paris, 1964).

[19] W. G. Dixon, Nuovo Cim. 38, 4 (1965), p. 1616.

[20] H. P. Kunzle, J. Math. Phys. 13 (1972), pp. 729-744.

[21] J. M. Souriau, Ann. Inst. H. Poincaré, XX, 4 (1974), pp. 315-364.

[22] Ch. Duval, preprint 75 P 767, Centre Phys. Th., Marseille (1975).

GRAVITY IN THE RAY GAUGE

C. Aragone*

International Centre for Theoretical Physics
Trieste, Italy

ABSTRACT: 1. Linearized gravity in the null-plane co-ordinates. 2. Pure gravity in the ray gauge. 3. Gravity coupled to matter. 4. Gravitational fields with vanishing null-energy density and the spin coefficient approach.

I. LINEARIZED GRAVITY IN THE NULL-PLANE CO-ORDINATES

The analysis of the linearized Einstein theory of gravity shows some interesting features when, diversely to the usual 3 + 1 approach, it is performed along a light-like co-ordinate u taken as "time" [1].

Let us assume from now on that the four co-ordinates labelling an event of the flat Minkowski space are x_i, u, v, i = 1, 2, where

$$u \equiv 2^{-\frac{1}{2}} x^0 - 2^{-\frac{1}{2}} x^3, \quad v \equiv 2^{-\frac{1}{2}} x^0 + 2^{-\frac{1}{2}} x^3. \tag{1}$$

In terms of their null-plane components, the inner product of two vectors X^α and Y^β becomes: $X^\alpha Y_\alpha = X^i Y^i - X^u Y^v - X^v Y^u \equiv \eta_{\alpha\beta} X^\alpha Y^\beta$, where the flat metric $\eta_{\alpha\beta}$ is diagonal in the pure two-dimensional space $\eta_{ij} = \delta_{ij}$, and antidiagonal in the two-dimensional u-v plane, $\eta_{uv} = -1$. The four-dimensional volume element also has a simple aspect, $d^4\tau = d^4x = d^2x\, du\, dv$; therefore we can write down the first-order linearized version of the Palatini action:

$$dA(h_{\mu\nu}, \Gamma^\alpha_{\mu\nu}) = \{(\tfrac{1}{2} h^\alpha_\alpha \eta^{\mu\nu} - h^{\mu\nu})(\partial_\alpha \Gamma^\alpha_{\mu\nu} - \partial_\mu \Gamma_{\nu\alpha}{}^{\alpha}) + \\ + \eta^{\mu\nu} \Gamma^\alpha_{\mu\nu} \Gamma^\beta_{\alpha\beta} - \Gamma^\alpha_{\mu\beta} \Gamma^\beta_{\nu\alpha})\}\, d^2x\, dudv\ . \tag{2}$$

We already learned from previous analysis of the linearized gravity [2] that some of the 10 (for $h_{\nu\mu} = h_{\nu\mu}$) plus 40 (for $\Gamma^\alpha_{\mu\nu} = \Gamma^\alpha_{\nu\mu}$) = 50 independent variables are not dynamically relevant and can be eliminated using the algebraic constraints. We also know that besides the dynamical variables the theory contains four gauge functions which functionally parametrize it.

*Permanent address: Departamento de Fisica, Universidad Simon Bolivar, Apartado 80659, Caracas 108, Venezuela.

In the null-plane formulation there are also four differential constraints C^u_μ associated with the four Lagrangian multipliers $h_{u\mu}$. They can be solved using the T+L decomposition technique for two-dimensional tensors and vectors. A two-vector V_i, for instance, can be split up into its transverse part V^T_i plus its longitudinal part V^L_i,

$$V_i \equiv \varepsilon_{i\ell}\, \rho_\ell\, V^T + \rho_i\, V^L \equiv V^T_i + V^L_i ,\quad V^T_{i}{}_{,i} = 0 , \tag{3a}$$

where ρ_i is the Riesz operator:

$$\rho_i(\cdot) = (-\Delta_2)^{-\frac{1}{2}}\, \partial_i(\cdot) . \tag{3b}$$

In a similar way the symmetric tensor h_{ij} can be written as the sum of its transverse part h^T_{ij} (containing only one degree of freedom) plus a gage-dependent part:

$$h_{ij} \equiv h^T_{ij} + h_{i,j} + h_{j,i} \equiv \varepsilon_{i\ell}\, \varepsilon_{jm}\, \rho_\ell\, \rho_m\, h^T + h_{i,j} + h_{j,i} ,$$

$$h^T_{ij,j} = 0 \tag{3c}$$

It was found that the two gauge-independent, dynamically significant field variables were the transverse component of h_{ij}, h^T and the transverse part Γ^T of the two-vector $\Gamma_{iu}{}^u \equiv \varepsilon_{i\ell}\, \rho_\ell\, \Gamma^L$.

In terms of these two variables, after performing the reduction process, the unconstrained form of the initial action (2) was obtained:

$$D\, dA^{**} = \{h^{\cdot T}\, h^{T'} + 4\dot{\Gamma}^T \Gamma^{T'} - [1/2\, h^T_{,i}\, h^T_{,i} + 2\, \Gamma^T_{,i}\, \Gamma^T_{,i}]\}\, d^2x\, du dv. \tag{4}$$

where the dot means $(\cdot)^{\bullet} \equiv \partial_u(\cdot)$ and the prime is a short notation for the partial v derivative $(\cdot)' \equiv \partial_v(\cdot)$.

Of course, it is immediately seen from Eq.(4) that independent variations of h^T, Γ^T indicate that both two-dynamical variables of our massless neutral helicity-2 free field obey the wave equation $\Box h^T = 2\dot{h}'^T - \Delta_2 h^T = 0 = \Box\Gamma^T$, as is expected on physical grounds.

II. PURE GRAVITY IN THE RAY GAUGE

In the exact Einstein theory of gravity [3] one can take as the "null-plane" co-ordinate u a co-ordinate labelling characteristic hypersurfaces u = constant of the metric $g_{\mu\nu}$, which in general are not going to be flat. This condition implies ${}_4g^{uu} = 0$. Moreover, the associated quasi-null co-ordinate v can be chosen such that the lines x_i = constant, u = constant, v = variable, are light rays with v the affine parameter along them. These geometrical conditions imply that ${}_4g^{ui} = 0$, ${}_4g^{uv} = -1$. More generally, we shall take the conditions $(\det g_{ij})^{\ell}\, g^{u\mu} = -\delta^{\mu}_{v}$ as the definition of the ray gauge. They have been used (with $\ell = 0$) by Robinson and Trautman [4] and Newman and Penrose [5] in their analysis of the asymptotic behaviour of uniformly smooth gravitational fields. Recently, Kaku [6] introduced the possibility of $\ell \neq 0$ and obtained a ghost-free formulation of quantum gravity which, as he has shown, becomes specially simple for $\ell = 0$ and $\ell = \frac{1}{4}$. In the following we shall assume, unless otherwise stated, $\ell = 0$.

We postulate the exact Palatini action to develop the full theory. In terms of the metric tensor ${}_4g_{\mu\nu} = {}_4g_{\nu\mu}$ and the torsionless affinity $\Gamma^{\alpha}_{\mu\nu}$ it is [7] (${}_4g^{\mu\nu}$: ${}_4g^{\mu\alpha}\,{}_4g_{\alpha\sigma} = \delta^{\mu}_{\sigma}$):

$$dA({}_4g_{\mu\nu},\ \Gamma^{\alpha}_{\mu\nu}) = {}_4g^{\mu\nu}\ R_{\mu\nu}(\Gamma)\ (-{}_4g)^{\frac{1}{2}}\ d^2x\ dudv\ . \tag{5a}$$

Actually, instead of $g_{\mu\nu}$ we shall adopt another set of 10 independent variables, which have a much closer geometrical meaning [3] when one is working with null co-ordinates. They are $g_{ij} \equiv {}_4g_{ij}$, N^i_u, N^i_v, n_u, n_v, and m, where ${}_4g_{ij}\, N^j_u = {}_4g_{iu}$, ${}_4g_{ij}\, N^j_v = {}_4g_{iv}$, ${}_4g_{uu} = -2n_v m^{-2}$, ${}_4g_{vv} = -2n_u\, m^{-2}$, ${}_4g_{uv} = -(1 + n_u n_v)m^{-2}$. Defining g^{ij}, the two-dimensional inverse of g_{ij}, as is usually done: $g^{ij}\, g_{j\ell} = \delta^i_{\ell}$, and introducing these new metric components into Eq. (5a) we obtain the action:

$$dA(g_{ij},\ n_u,\ n_v,\ N^i_u,\ N^i_v,\ m,\ \Gamma^{\alpha}_{\mu\nu}) = \{m^2\, g^{ij}\, R_{ij} - 2n_v\, N^i_u\, N^j_u\, R_{ij} -$$

$$- 2n_u\, N^i_v\, N^j_v\, R_{ij} - 2(1+n_u\, n_v)\, N^i_u\, N^j_v + 4n_v\, N^i_u\, R_{iu} + 4n_u\, N^i_v\, R_{iv}$$

$$+ 2(1 + n_u\, n_v)N^i_v\, R_{iu} + 2(1 + n_u\, n_v)\, N^i_u\, R_{iv} - 2n_v\, R_{uu} -$$

$$- 2n_u R_{vv} - 2(1 + n_u n_v) R_{uv}\} \; {}_2g^{\frac{1}{2}} \, d^2x \, dudv \; , \qquad {}_2g = \det g_{ij} . \tag{5b}$$

Independent variations of g_{ij}, N^i_u, N^i_v, n_u, n_v and m yield a set of equations algebraically equivalent to the vacuum Einstein equation:

$$G_{ij} = G_{iu} = G_{iv} = G_{uu} = G_{vv} = G_{uv} = 0 \; , \tag{6a}$$

while variations of $\Gamma^{\alpha}_{\mu\nu}$ give a set of equations equivalent to stating that the Γ's are Riemannian, i.e.

$$\Gamma^{\alpha}_{\mu\nu} = \{^{\alpha}_{\mu\nu}\} \, (g_{\sigma\rho}) \; , \tag{6b}$$

where $g_{\mu\nu} = g_{\mu\nu} \, (g_{ij}, n_u, n_v, N^i_u, N^i_v, m)$. We now pick up a special set of co-ordinates: the ray gauge. That means making $n_v = 0$, $N^i_v = 0$, $m = 1$ and thereafter introducing them into the action (5b) and also in the field equations (6). The action becomes simpler ($n_u = n$, $N^i_u = N$):

$$dA(g_{ij}, n, N^i, \Gamma^{\alpha}_{\mu\nu}) = \{g^{ij} \, {}_4R_{ij} + 2 \, N^i \, {}_4R_{iv} - 2n \, {}_4R_{vv} -$$

$$- 2 \, R_{uv}\} \; {}_2g^{\frac{1}{2}} \, d^2x \, dvdu. \tag{7}$$

The set of field equations is still the set (6) where the values of the gauge functions $n_v = 0$, $m = 1$, $N^i_v = 0$ has been introduced throug-throughout.

At this state we can eliminate the algebraic constraints appearing in the set (6). This gives rise to a new action A_*, depending only upon $3 + 2 + 1 = 6$ variables: g_{ij}, N^i, n:

$$dA*(g_{ij}, n, N^i) = \{\tfrac{1}{2} \, {}_2g^{\frac{1}{2}} \, (g^{i\ell} g^{jm} - g^{ij} g^{\ell m}) \, g'_{\ell m} \, g_{ij} +$$

$$+ g^{\frac{1}{2}} \, N^i \, D_j \, [g'^{\,j}_{i} - \delta_i^{\,j} \, g'^{\,\ell}_{\ell}] + nC_v + \tag{8}$$

$$+ \tfrac{1}{2} \, g_{ij} \, N^{i'} N^{j'} \, g^{\frac{1}{2}} + D^{\ell}_{\;,\ell} + D^{v'}\} \, d^2x \, dudv \quad ,$$

where C_v is:

$$C_v \equiv g^{ij} g''_{ij} - \frac{1}{2} g'^{i}_{j} g'^{j}_{i} \quad , \tag{9}$$

$D_j(\cdot)$ means the covariant derivative with respect to the two-dimensional affinity $_2\Gamma^{\ell}_{ij} \equiv \{^{\ell}_{ij}\} ({}_2g_{pq})$, $g'^{i}_{j} \equiv g^{i\ell}(g_{\ell j})'$ and $\partial_\ell D^\ell + D'$ is an exact three-divergence and will therefore be discarded.

Variations of A^* with respect to N^i provide two differential constraints:

$$\delta A^*/\delta N^i = 0 \rightleftarrows [g^{\frac{1}{2}} g_{ij} N^{j'}]' = \tag{10a}$$
$$= g^{\frac{1}{2}} Dj [g'^{j}_{i} - \delta^j_i g'^{\ell}_{\ell}] \equiv g^{\frac{1}{2}} Ci \quad ,$$

which have a very different aspect, compared with the 3 + 1 constraints of the canonical formulation of Arnowitt, Deser and Misner [8]. This is one of the advantages of the null co-ordinates: the differential constraints of any theory which, in a canonical 3 + 1 formulation, are strongly coupled partial differential equations on three-dimensional spaces, also exist in the null-co-ordinate frames, but they are found to be just ordinary differential equations in the quasi-null co-ordinate v.

From Eq. (10a) one can obtain $N^i(g_{\ell m})$:

$$N^i = \frac{1}{\partial_v} [g^{ij} g^{-\frac{1}{2}} \frac{1}{\partial_v} g^{\frac{1}{2}} (D_\ell g'^{\ell}_{j} - D_j g'^{\ell}_{\ell})] \tag{10b}$$

and substitute it into A^*. That gives us:

$$dA'^* = \{\frac{1}{2} g^{\frac{1}{2}}(g^{i\ell} g^{jm} - g^{ij} g^{\ell m}) g'_{\ell m} \dot{g}_{ij} -$$
$$- \tfrac{1}{2} g_{ij} g^{\frac{1}{2}} N^{i'} N^{j'} + n C_v g^{\frac{1}{2}}\} d^2x \, du dv \quad , \tag{11}$$

where it is evident that n is the Lagrangian multiplier associated with the highly non-linear differential constraint $C_v = 0$, and $N^i = N^i(g_{\ell m})$ as given by Eq. (10b):

$$\delta A'_*/\delta n = 0 \rightleftarrows C_v = 0 \ . \tag{12a}$$

So, if one wants to reach an unconstrained formulation of gravity, one still has to solve the second-order differential equation $C_v = 0$, which in terms of the Ricci tensor is nothing but $G^u_u = 0$ (as the vector constraint (10a) is exactly $G^u_i = 0$). This has been done by Kaku [6] taking the new variables $h \equiv g^{\frac{1}{2}}$ and $e_{ij} \equiv h^{-1} g_{ij}$ which decouple the two-volume from the pure deformation e_{ij} tensor ($\det e_{ij} = 1$), rewriting Eq. (12a) in terms of them and thereafter performing a functional integration. The new version of C_v turns out to be:

$$(\ell n h)'' - 2a(\ell n h)'^2 + \tfrac{1}{4} e^{ij} e''_{ij} = 0 \ , \quad (a = -\tfrac{1}{4}), \tag{12b}$$

whose functional solution is

$$h = \left[\frac{1}{2} \sum_{n=0}^{\infty} a^n u_n\right]_{a=-\frac{1}{4}} , \quad u_n = \frac{1}{\partial_v} \left[\sum_{\ell=0}^{n-1} u_\ell \, u_{n-\ell}\right] , \quad n \geq 1 ,$$

$$u_o = \frac{1}{\partial_v} (-\tfrac{1}{2} e^{ij} e''_{ij}) \ . \tag{12c}$$

Now, if we introduce $h = g^{\frac{1}{2}}$, as given by Eq. (12c), and $N^i(h)(e_{ij}),\ e_{ij})$, as determined by Eq. (10b), into A'_*, we end up with the unconstrained action A_{**}, a functional of only two independent variables, α, β, parametrizing the unimodular matrix e_{ij},

$$e_{ij} \equiv \begin{pmatrix} e^{2\alpha} \cosh 2\beta & \sinh 2\beta \\ \sinh 2\beta & e^{-2\alpha} \cosh 2\beta \end{pmatrix} , \tag{13}$$

$$A^{**}(\alpha,\beta) = \int \left[\frac{1}{2} g^{\frac{1}{2}} (g^{i\ell} g^{jm} - g^{ij} g^{\ell m}) \, g'_{ij} \, \dot{g}_{\ell m} - \tfrac{1}{2} g^{\frac{1}{2}} g_{ij} N^{i'} N^{j'} \right] d^2x \, du dv \ . \tag{14}$$

Remarkable is the final dynamical structure of A_{**}; it has dynamical term, some kind of distorted $\sim q'\dot{q}$ term minus an explicitly non-negative definite quantity $J^u \equiv \frac{1}{2} g_{ij} \, g^{\frac{1}{2}} N^{i'} N^{j'}$ which, being the generator of the u evolution of the field, can be called the null-energy density. The structure (14) is the curved analogue of the flat model

$$\sim dA_{flat} = d^2x \, dudv \left\{ \sum_a q'_a \, \dot{q}_a - \frac{1}{2} q^2_{a,i} - \frac{1}{2} m^2_a \, q^2_a \right\} .$$

III. GRAVITATING MATTER

In order to understand how a source can modify the structure of gravity, we studied [9] the model constituted by an Abelian massive vector field coupled to gravity. We chose A_μ, a covariant vector, and $F^{\mu\nu}$, an antisymmetric contravariant density, to represent the massive vector. When it is minimally coupled to gravity, its first-order action can be written:

$$dA^M \equiv K^2 \Big\{ A_{\mu,\nu} F^{\mu\nu} + \frac{1}{4} F^{\mu\nu} F_{\mu\nu} (-g)^{-\frac{1}{2}} - \qquad (15)$$
$$- \frac{1}{2} m^2 A_\mu A^\mu (-g)^{\frac{1}{2}} \Big\} d^2x \, dudv \quad ,$$

where $K^2 \equiv K_g^2/K_m^2$ is the ratio of the two respective characteristic strengths. The whole gravitating system has its evolution determined by the total action A:

$$dA = dA^G + dA^M \quad , \qquad (16)$$

where dA^G is given by Eq.(5).

After spoiling (16) of all the algebraic constraints it carries, we are left with the reduced action:

$$dA^* = \Big\{ \frac{1}{2} (g^{i\ell} g^{jm} - g^{ij} g^{\ell m}) g'_{ij} \dot{g}_{\ell m} + K^2 g^{ij} F_{vi} \dot{A}_j + K^2 E\dot{A}_v +$$
$$+ g^{\frac{1}{2}} N^i (C_i + K^2 C_i^M) + \frac{1}{2} g^{\frac{1}{2}} g_{ij} N^{i'} N^{j'} - \frac{1}{2} K^2 g^{\frac{1}{2}} B^2 -$$
$$- \frac{1}{2} K^2 g^{-\frac{1}{2}} E^2 - \frac{1}{2} K^2 m^2 g^{ij} A_i A_j + K^2 A_u C_Q +$$
$$+ ng^{\frac{1}{2}} [C_v + K^2 C_v^M] + g^{\frac{1}{2}} N^i [C_i + K^2 C_i^M] \Big\} d^2x \, dudv \, , \qquad (17)$$

where we use the natural variables $F_{vi}(A_i, A_v)$, E, $B(A_i)$:

$$F_{vi} \equiv g^{\frac{1}{2}}(A_i' - A_{v,i}), \quad E \equiv F^{vu}, \quad B \equiv g^{-\frac{1}{2}} \varepsilon^{ij} \partial_i A_j \quad . \qquad (18)$$

Moreover, the quantities C_Q, C_v^M and C_i^M have the explicit values:

$$C_Q \equiv E' + (g^{ij} F_{jv}(A_i, A_v))_{,i} + m^2 g^{\frac{1}{2}} A_v, \tag{19a}$$

$$C_v^M \equiv g^{-1} F_{vi} F_{vj} g^{ij} + m^2 A_v^2 = T_{vv}^M , \tag{19b}$$

and
$$C_i^M \equiv g^{-1} F_{vi} E + F_{vj} B\varepsilon_{ji} = - T_{iv}^M . \tag{19c}$$

Let us recall, for the sake of completeness, that $T_{uv}^M = \frac{1}{2} g^{\frac{1}{2}} B^2 + \frac{1}{2} g^{-\frac{1}{2}} E^2 + \frac{1}{2} m^2 A_i A^i \equiv J_u^M$, is the null-energy density of the matter field. Looking at Eq. (17), we see that A_u is the Lagrange multiplier associated with the Coulomb constraint $C_Q = 0$,

$$C_Q \equiv E' + (g^{ij} F_{jv})_{,i} + m^2 g^{\frac{1}{2}} A_v = 0 , \tag{20a}$$

which can be solved immediately for $E = E(A_i, A_v)$ with the result:

$$E = \frac{1}{\partial_v} [(g^{\frac{1}{2}} g^{ij} A_{v,j})_{,i} - (g^{\frac{1}{2}} g^{ij} A_j')_{,i} + m^2 g^{\frac{1}{2}} A_v] . \tag{20b}$$

Variations of N^i in A_* provide the generalization of the vectorial constraint (10a) when we have matter acting as a source of the gravitational field:

$$\delta_{N^i} A^* = 0 \rightleftarrows [g^{\frac{1}{2}} g_{ij} N^{j\prime}]' = g^{\frac{1}{2}} C_i + K^2 g^{\frac{1}{2}} C_i^M , \tag{21a}$$

which, after one integration, turns out to be:

$$N^i = \frac{1}{\partial_v} \{g^{-\frac{1}{2}} g^{ij} \frac{1}{\partial_v} [g^{\frac{1}{2}} C_i + K^2 g^{\frac{1}{2}} C_i^M]\} \tag{21b}$$

Introduction of this value of N^i into A_* allows us to go one step forward with the unconstrained formulation, even in the presence of matter. We find:

$$dA_*' = \Big\{ \frac{1}{2} (g^{i\ell} g^{jm} - g^{ij} g^{\ell m}) g_{ij}' \dot{g}_{\ell m} + K^2 g^{ij} F_{vi} \dot{A}_j + K^2 E\dot{A}_v - [j_u^G + K^2 j_u^M] + ng^{\frac{1}{2}} [C_v + K^2 T_{vv}^M] \Big\} d^2x \, du dv . \tag{22}$$

The last step is to realize that again in this case n has the character of a multiplier associated with the second-order ordinary differential constraint $C_v + K^2 T_{vv}^M = 0$. In terms of h, e_{ij}, the differential equation for h, now has the aspect:

$$\delta A'^{*}/\delta n = 0 \rightleftarrows C_v + K^2 T^M_{vv} \equiv (\ell n h)'' + \frac{1}{2} (\ell n h)'^2 + \tfrac{1}{4} e^{ij} e''_{ij}$$

$$+ K^2 h^{-1} (A'_i - A_{v,i}) (A'_j - A_{v,j}) e^{ij} \quad . \tag{23}$$

Assuming that one can solve Eq.(23), obtaining $h = h\ (e_{ij}, A_i, A_v)$, the final unconstrained action can be written:

$$A^{**}(\alpha,\beta,A_i,A_v) = \int\left\{\frac{1}{2} (g^{i\ell} g^{jm} - g^{ij} g^{\ell m}) g^{\frac{1}{2}} g'_{ij} \dot{g}_{\ell m} + \right.$$

$$\left. + K^2 g^{ij} F_{vi} \dot{A}_j + K^2 E\dot{A}_v - [j^G_u + K^2 j^M_u] \right\} d^2x\ du dv \ , \tag{24}$$

where e_{ij} is parametrized according to Eqs. (13).

IV. GRAVITATIONAL FIELDS WITH VANISHING NULL-ENERGY DENSITY AND THE SPIN COEFFICIENT APPROACH

Regarding the physical significance of the new generator J^G_{uu}, it is worth investigating (at least in the absence of matter) what kind of fields have $J^G_u = 0$. Of course $J^G_u = 0$ implies $N^{i\,\prime} = 0$; thus N^i must have the structure:

$$N^i = N^i(\vec{x},u) \quad . \tag{25}$$

If at this point one adds some extra assumption regarding the symmetry of the solutions, for instance that $g_{ij} = h\delta_{ij}$, one is led either to the spherically-shear-free Robinson-Trautman solution or to the Kundt [10] non-expanding plane front waves. This does not imply that $J^G_u = 0$ necessarily leads to vanishing shear.

It can be shown that there exist vanishing null-energy solutions which do not have vanishing shear, for instance one can exhibit the exact metric [11] (θ is a real parameter),

$$ds^2 = v^{1 + \sin\theta} [v^{\cos\theta} dx^2 + v^{-\cos\theta} dy^2] - 2\ du dv, \tag{26}$$

whose shear has the value: $\sigma(\partial_v) = \frac{1}{2} \cos\theta\ v^{-2}$, and which is non-flat for any value of θ different from zero.

A broad understanding of this situation, what the physical meaning of J^G_u is, what the connection is between this quantity and the shear of the congruence of the light rays $\{v = \text{variable},\ u = u_0,\ x^i = x^i_0\}$, is gained when one finds the connection between the present dynamical formulation and the spin coefficient approach of Newman and Penrose [5], [12].

We have found [13] that by taking as the basic null tetrad

$$\ell \equiv \partial_v \ , \ n \equiv \partial_u + n\partial_v - N^\ell \partial_\ell$$

$$m \equiv m^i \partial_i = \frac{1}{2} h^{-\frac{1}{2}} e^{-\alpha}(e^\beta + ie^{-\beta})\partial_1 + \frac{1}{2} h^{-\frac{1}{2}} e^{\alpha}(-e^\beta + ie^{-\beta})\partial_2 \ , \tag{27}$$

it turns out that the null-energy density J^u_G can be expressed exactly in terms of any of the two spin coefficients π, τ:

$$\pi = \bar{\tau}$$

and

$$J^u_G = 4g^{\frac{1}{2}} \ |\tau|^2 = 4g^{\frac{1}{2}} \ |\pi|^2 \ , \tag{28}$$

where $\tau \equiv \ell_{\mu;\nu} \ m^\mu n^\nu$ and $\pi \equiv -n_{\mu;\nu} \ \bar{m}^\mu \ell^\nu$.

It is worth recalling that τ describes precisely how the direction of the light-like vector ℓ changes as we move along n, the other null vector associated to ℓ, while π has exactly the converse meaning.

Moreover, it can be shown that a mass formula can be given in terms of g_{ij} and n:

$$8\pi M(u) = \lim_{v\to\infty} \int_{u = \text{const} \cap v = \text{const}} v \ n'' \ g^{\frac{1}{2}} \ d^2x. \ . \tag{29}$$

ACKNOWLEDGEMENTS

The author wishes to thank Professor Abdus Salam, the International Atomic Energy Agency and UNESCO for hospitality at the International Centre for Theoretical Physics, Trieste.

REFERENCES

[1] C. Aragone and R. Gambini, Nuovo Cimento 18B (1973) 311.
[2] R. Arnowitt and S. Deser, Phys. Rev. 113 (1959) 745.
[3] C. Aragone and J. Chela-Flores, Nuovo Cimento 25B (1975) 225.

[4] I. Robinson and A. Trautman, Proc. Roy. Soc. (London) A265 (1962) 463.

[5] E. Newman and R. Penrose, J. Math. Phys. 3 (1962) 566.

[6] M. Kaku, Nucl. Phys. B91 (1975) 99 .

[7] We are using the signature $R_{\mu\nu}(\Gamma) \sim \partial_\alpha \Gamma_{\mu\nu}^{\alpha} - \partial_\mu \Gamma_{\nu\alpha}^{\alpha}$.

[8] R. Arnowitt, S. Deser and C. W. Misner, Phys. Rev. 116 (1959) 1322.

[9] C. Aragone and A. Restuccia, Universidad Simon Bolivar preprint SB. FM. F/8 (1974) (to be published in Phys. Rev. D).

[10] W. Kundt, Z. Physik 163 (1961) 77.

[11] H. Bondi, F. A. Pirani and I. Robinson, Proc. Roy. Soc. (London) A251 (1959) 519.

[12] E. Newman and T. Unti, J. Math. Phys. 3 (1962) 891.

[13] C. Aragone, R. Gambini and A. Restuccia, ICTP, Trieste, preprint IC/75/94 (1975).

AN APPROACH TO A UNIFIED TREATMENT OF ELECTROMAGNETIC AND GRAVITATIONAL THEORY EMERGING FROM THE COVARIANT DIRAC EQUATION

Leopold Halpern*
Department of Physics, The Florida State University
Tallahassee, Florida 32306 U.S.A.

I INTRODUCTION: Dirac's equation has become a foundation of modern physics far beyond the description of relativistic spinning particle waves which it achieves. O. Klein and independently A. Zakharov have shown that even General Relativity and some of its quantum modifications can be derived from the covariant formulation of Dirac's equation [1].

The aim of the present communication is to make full use of the invariance properties of Dirac's equation - not only of its covariance w.r.t. coordinate transformations and its invariance w.r.t. electromagnetic gauge transformations but also of its invariance w.r.t. the group of similarity transformations of which the group of gauge transformations is an invariant sub-group. The result is a gauge theory which unifies the electromagnetic and gravitational fields. The gravitational part of the Lagrangian of the gauge fields is not $\sqrt{-g}R$ of general relativity but a nonlinear Lagrangian which gives rise to field equations with fourth derivatives. These field equations yield, however, the Schwarzschild solution of the vacuum. In the most general case there appear superpositions of the non-linear Lagrangians and the Einstein Lagrangian which are definitely not in contradiction with experiments.

The formalism required to investigate Klein's suggestion has been admirably worked out by D. Laurent [2]. Laurent investigated also the possibilities of independent variation of γ_μ and Γ_μ which occur in the covariant form of the Dirac equation. Expressing the Lagrangian of general relativity in terms of these two entities he arrives in the case of independent variation at a modified theory which differs only in higher orders of the gravitational constant from Einstein's theory. Electromagnetism is not considered in this paper. We are in the following making use of Klein's and Laurent's formalism and notation and of Bargmann's formulation of the covariant Dirac equation [3]. Our approach is, however, fundamentally different from that of the previous authors.

II Outline of the Formalism of the covariant Dirac Equation

We deal with entities which are four dimensional spin matrices and have a definite character w.r.t. coordinate transformations. They are called correspondingly scalar, vector- and tensor operators. The vector operators γ^μ fulfill the modified Dirac algebra:

$$\{\gamma^\mu, \gamma^\lambda\} = 2g^{\mu\lambda}(x) \tag{1}$$

There exists a vector operator Γ'_μ so that the extended covariant derivative (e.c.d.) of the γ^λ vanishes:

$$\gamma^\mu_{|'\lambda} \equiv \gamma^\mu{}_{;\lambda} + [\gamma^\mu, \Gamma'_\lambda] = 0 \qquad \gamma^\mu{}_{;\lambda} \equiv \gamma^\mu{}_{,\lambda} + \Gamma^\mu_{\lambda\alpha}\gamma^\alpha \tag{2}$$

*This research was supported in part by U.S.E.R.D.A. under number AT-(40-1)-3509.

This entity is determined up to its trace by eq. (2). Γ'_ν is expressible in terms of the γ-matrices and their covariant derivatives as follows:

$$\Gamma'_\nu = -\frac{1}{8}\{\frac{1}{3}\mathrm{Tr}(s_{\alpha\rho}\gamma^\rho{}_{;\nu})\gamma^\alpha + \mathrm{Tr}(\gamma_\beta\gamma_{\alpha;\nu})s^{\alpha\beta} +$$

$$+ (g^{-1})\mathrm{Tr}(\gamma_5\gamma_{\alpha;\nu})\gamma_5\gamma^\alpha - 1/4(g)^{-1}\ (\gamma_5\gamma_\rho\gamma^\rho{}_{;\nu})\gamma_5\} + A_\nu 1 \qquad (2a)$$

where $g \equiv -\det(g_{\mu\nu})$ and A_ν is an arbitrary vector field. We shall later consider more general e.c.d. with vector operators Γ_μ (prime ommitted) for which $\gamma^\mu{}_{|\lambda} \neq 0$. The Dirac spinor Ψ transforms as scalar w.r.t. coordinate transformations. One needs a further scalar operator α, the hermitizing matrix with the properties:

$$\alpha^+ = \alpha\ ,\quad -\gamma^{\mu\,+} = \alpha\gamma^\mu\alpha^{-1} \qquad (3)$$

The e.c.d. $\alpha_{|\lambda} \equiv \alpha_{;\lambda} + \alpha\Gamma_\lambda + \Gamma^+_\lambda\alpha$ (4)

is supposed to vanish for $\Gamma_\lambda = \Gamma'_\lambda$, for which also $\gamma^\mu{}_{|'\lambda}$ vanishes. α is determined by equs. (3) up to a real factor. One defines: $\bar{\Psi} \equiv \Psi^+\alpha$. It is useful to introduce also: $\xi \equiv (g)^{1/4}\Psi$ and $\bar{\xi} \equiv \xi^+\alpha \equiv (g)^{1/4}\bar{\Psi}$.

Similarity transformations with nonsingular matrix S transform the above entities as follows:

$$\Psi \to S^{-1}\Psi\ ,\quad \gamma^\mu \to S^{-1}\gamma^\mu S\ ,\quad \alpha \to S^+\alpha S\ ,\quad \Gamma_\mu \to S^{-1}\Gamma_\mu S - S^{-1}S_{,\mu} \qquad (5)$$

We use also $\gamma_\alpha = \gamma^\mu g_{\mu\alpha}$. The Leibniz rule applies to products for the e.c.d. and $g_{\mu\alpha;\lambda} \equiv g_{\mu\alpha,\lambda} - g_{\zeta\alpha}\Gamma^\zeta_{\mu\lambda} - g_{\mu\zeta}\Gamma^\zeta_{\mu\alpha} \equiv 0$

The field: $\Phi_{\mu\nu} = \Gamma_{\nu,\mu} - \Gamma_{\mu,\nu} - [\Gamma_\mu, \Gamma_\nu]$ (6)

transforms homogenously: $\Phi_{\mu\nu} \to S^{-1}\Phi_{\mu\nu}S$ according to the adjoint reresentation. If the e.c.d. of the γ^μ and of α vanish, one finds the relation:

$$\Phi'_{\lambda\mu} = 1/4 R_{\mu\lambda\alpha\beta}\ s^{\alpha\beta} + (A_{\mu,\lambda} - A_{\lambda,\mu})\cdot lie \qquad (7)$$

where $s^{\alpha\beta} = 1/2[\gamma^\alpha, \gamma^\beta]$, $A_\mu(x)$ is arbitr. real vectorfield, e = real constant, i = imaginary unit.

We note the relations:

$$R = R^{\mu\lambda}{}_{\lambda\mu} = -1/2\ \mathrm{Tr}(s^{\mu\lambda}\Phi'_{\mu\lambda}),\quad R_{\mu\lambda\alpha\gamma} = 1/2\ \mathrm{Tr}\ (s_{\mu\lambda}\Phi'_{\alpha\gamma}) \qquad (8)$$

Therefore:

$$\mathrm{Tr}(\Phi'_{\mu\nu}\Phi'^{\mu\nu}) = -1/2\, R_{\mu\lambda\alpha\gamma}R^{\mu\lambda\alpha\gamma} - 4\, e^2\, f_{\mu\lambda}f^{\mu\lambda} \tag{9}$$

with $f_{\mu\lambda} = A_{\lambda,\mu} - A_{\mu,\lambda}$.

The e.c.d. of a spinor is defined as:

$$\Psi_{|\mu} \equiv \Psi_{;\mu} - \Gamma_\mu \Psi \quad \text{and} \quad \bar{\Psi}_{|\mu} \equiv \bar{\Psi}_{;\mu} + \bar{\Psi}\Gamma_\mu \tag{10}$$

We write the Dirac equation for $\Gamma_\mu = \Gamma'_\mu$:

$$\gamma^\mu \Psi_{|'\mu} + \mu\Psi = 0\ , \quad \bar{\Psi}_{|'\mu}\gamma^\mu - \mu\Psi = 0 \qquad \mu = \frac{mc}{\hbar} \tag{11}$$

The spinor Ψ in (11) may be replaced by ξ because the covariant derivative of g vanishes. The action of the Dirac field is then:

$$\mathscr{L}_D = 1/2(\bar{\xi}\gamma^\alpha \xi_{|'\alpha} - \bar{\xi}_{|'\alpha}\gamma^\alpha \xi) + \mu\bar{\xi}\xi \tag{11a}$$

Γ'_μ can be transformed away at any point by suitable coordinate transformations and S-transformations. The most general form for Γ_μ is thus: $\Gamma_\mu = \Gamma'_\mu + M_\mu$ where M_μ is a vector operator which transforms according to the adjoint representation $M_\mu \to S^{-1}M_\mu S$.

The e.c.d. become:

$$\gamma^\mu_{|\lambda} \equiv \gamma^\mu_{;\lambda} + [\gamma^\mu, \Gamma_\lambda] = [\gamma^\mu, M_\lambda] \tag{2'}$$

and $\alpha_{|\lambda} \equiv \alpha_{;\lambda} + \alpha\Gamma_\lambda + \Gamma^+_\lambda \alpha = \alpha M_\lambda + M^+_\lambda \alpha$ (3')

The hermitian Lagrangian would assume in this general case an additional term:

$$1/2\xi^+(M^+_\mu \alpha + \alpha M_\mu)\ \ \gamma^\mu \xi \tag{11a'}$$

and the Dirac equation becomes:

$$\gamma^\mu \psi_{|\mu} + 1/2[\gamma^\mu, M\mu]\psi \ + 1/2(M^+_\mu \alpha + \alpha M\mu)\gamma^\mu \psi - \mu\psi = 0$$

respect.

$$\bar{\psi}_{|\mu}\gamma^\mu + 1/2\bar{\psi}[\gamma^\mu, M_\mu] - 1/2\ \psi^+(\alpha M_\mu + M^+_\mu \alpha)\gamma^\mu + \bar{\psi}\mu = 0 \tag{11'}$$

We shall deal in general only with Γ_μ for which $\alpha_{|\mu}=0$.

III The Group of Similarity Transformations and its Gauge Field

The equations (11) and (11a) are not only covariant under general coordinate transformations but also w.r.t. similarity transformations with all nonsingular matrices S. (see equ. 5). The electromagnetic gauge transformations are special transformations of this type which form an invariant subgroup. The matrix X is here of the special form: $S = e^{i\chi}.1$. These transformations give rise to a gauge potential $A_\mu = \chi_{,\mu}/e$. In units where $\hbar = c = 1$, e is a dimensionless constant which is determined empirically. The electromagnetic gauge potentials occur only in the diagonal terms. We have to introduce a gauge potential for each of the generators of the full group of similarity transformations to generalize the procedure. We may choose as generators G_{ab} of the matrix group the matrix with only the one element in the ath line and bth column differing from zero. To adjust ourselves to the electromagnetic case we choose the group coordinates such that these elements assume all the value (-e). Writing the gauge potentials as a matrix $\mathcal{A}_{\mu(x)}$ we can replace formally Γ_μ by $e\mathcal{A}_\mu$. The transformation according to equ. (5) for $\mathcal{A}_\mu$ transforms the gauge potential correctly [4]. The electromagnetic potentials are equal to $-i/4\ \mathrm{Tr}(\mathcal{A}_\mu) = A_\mu$. Gauge fields can be formed out of the potentials. We denote the matrix of these fields by $F_{\mu\lambda}$:

$$F_{\mu\lambda} = \mathcal{A}_{\lambda,\mu} - \mathcal{A}_{\mu,\lambda} - e\ [\mathcal{A}_\mu, \mathcal{A}_\lambda] \tag{12}$$

The Lagrangians of the electromagnetic field and the other terms of the gauge field can be united in the form:

$$\mathcal{L}_G = C\mathrm{Tr}(F'_{\mu\nu}F^{\mu\nu}) + \mathrm{c.c.} \tag{13}$$

The Lagrangian presented is the most suggestive and symmetric rather than the most general one. One may add also to the Lagrangian (11a) of the Dirac field the Lagrangian which corresponds to the Einstein Lagrangian $K^{-1}\sqrt{-g}R = -eK^{-1}\sqrt{-g}\mathrm{Tr}(s^{\mu\lambda}F_{\mu\lambda})$ and a linear combination of the nonlinear Lagrangians:

$$\sqrt{-g}\mathrm{Tr}(F_{\mu\lambda}F^{\mu\lambda}),\ \sqrt{-g}\mathrm{Tr}(F_{\alpha\lambda}s^{\alpha\mu}s^{\lambda\gamma}F_{\mu\gamma}) \text{ and } \sqrt{-g}\mathrm{Tr}(F_{\mu\alpha}s^{\alpha\lambda}F^{\mu}{}_{\lambda})$$

The first two of the latter three contain also the electromagnetic field and the sum of their coefficients is determined by the latter. K is the Einstein constant $8\pi G_{\mathrm{Newton}}/c^4$ so that the first term, if present, may dominate and assure agreement with the Newtonian limit. The symmetry becomes more apparent in dimensionless units by choosing the remaining units of length for example, such that $K = e$. The general combined Lagrangian obtained remarkably is of the form predicted by quantum field theory and the coefficients are of the same magnitude [5].

We have seen that if Γ_μ obeys equ. (2), it can be expressed in terms of the γ^μ and their derivatives and therefore in terms of the metric tensor according to equ. (1). One may vary the total Lagrangian formed out of equs. (11a) and (13) w.r.t. $\xi, \bar{\xi}, (\mathcal{A}_\mu)^{ab}$ and the $(\gamma^\mu)^{ab}$ taking account of the relations between the γ^μ and the $\mathcal{A}_\mu$. The result in this case is the Dirac and Maxwell equations in a gravitational field resulting from the nonlinear Lagrangian (see equ. (9)). Independent variation of the $\mathcal{A}_\mu$ and the γ^α is expected to give rise to a different theory to be investigated in another paper. (See also [6].)

ACKNOWLEDGEMENTS

I thank Prof. P. A. M. Dirac and Prof. J. E. Lannutti for the possibility to work in their Institute.

REFERENCES

[1] O. Klein, Nuclear Physics B21, (1970) 253.
A. Zakharov, Doklady Akad. Nauk. SSR 177 (1967) 70-71.

[2] B. Laurent, Arkiv F. Fysik, Vol. 16, Nr. 25 (1959) 263.

[3] V. Bargmann, Berl. Ber. (1932) 346.

[4] B. DeWitt, Les Houches Lecture (1963).

[5] L. Halpern, Ark. f. Fysik, Vol. 34, Nr. 43 (1967) 539.

[6] L. Halpern, Lecture at the Symposium on Differential Geometry in Mathematical Physics, Bonn (1975). To appear in Springer Tracts on Modern Physics.

ON STRONG-GRAVITY*

Akira Inomata**
Sektion Physik der Universitat Munchen
8 Munchen 2, Germany

INTRODUCTION: What I would like to talk about here is not a story of strong normal gravity but a story of anomalous strong-gravity whose coupling is 10^{38} times stronger in the order of magnitude than the Newtonian coupling. Developing the theory of general relativity, Einstein suggested its possible application in two directions: cosmology and elementary particle physics. As is indicated by enthusiasm in this conference, general relativity appears to be widely accepted as a physics in the astronomical scale. On the other hand, there has been no definite sign that general relativity plays an essential role in particle physics or otherwise. To Einstein [1], the problem of elementary particles primarily meant the atomic model for which the geometrical approach was of no practical importance. To us, elementary particles are much smaller in scale and more complicated in structure than atoms, which are totally open for the general-relativistic consideration.

Earlier in this conference, Professor Salam talked about a possible mechanism of suppressing ultraviolet divergences in electrodynamics by means of the gravitational effect [2]. Not one or two, but counting the effect of an infinite number of gravitons, he could regulate the divergences. The contribution of a finite number of gravitons is negligible in the present perturbative framework of field theory. The correction to, say, the self-energy of an electron by an infinite number of gravitons would imply taking some geometrical structure of the electron into account. This attempt seems to be an important step in understanding the role of gravity in the microscale.

The electromagnetic self-mass of an electron so obtained is $\delta m \simeq (3/2\pi)\alpha m \log(m_o/m)$ where α is the fine structure constant. The effective cutoff occurs at the value of the Planck mass $m_o \sim 10^{-5}$g. In order to expect an effective cutoff for the proton at a few GeV by a similar suppression mechanism [3], the coupling k of gravitons to the nuclear matter must be 10^{38} times greater in the order of magnitude than the usual Newtonian value $k_o = 8\pi G$. Such a speculative gravity is called <u>strong-gravity</u>. Notice that the ratio of the nuclear force to the gravitational force between two protons is of the order of 10^{38}. Strong-gravity means a gravity which might possibly be associated with strong interactions. Since strongly interacting particles are called <u>hadrons</u>, a less confusing, though no better, word for strong-gravity may be <u>hadrogravity</u>. The purpose of this report is to show how far one can go with the idea that the nuclear force is a kind of gravity.

STRONG-GRAVITY DOMINANCE

To begin with we characterize strong-gravity or hadrogravity to be (i) geometrized (tensor gravity), (ii) short-ranged ($\sim 10^{-13}$cm), (iii) of strong coupling ($\sim 10^{38} k_o$). Then we ask ourselves. How can it be reconciled with the Einstein gravity? How can it be dominant at short distances?

In the two-tensor theory of Isham, Salam and Strathdee [4], strong-gravity is mixed with the normal gravity in analogy with the ρ-γ mixing theory in the electromagnetic processes of hadrons; the strong tensor field $f_{\mu\nu}$, carrying a mass μ, dominates at short distances while the usual massless tensor field $g_{\mu\nu}$ remains effective at far distances. It was a conjecture that the mixed field equation

*Supported in part by the Research Foundation of State University of New York.

**Permanent address: Physics Department, State University of New York at Albany, Albany, New York 12222.

when solved for a static case would yield the metric of the form [3]

$$g_{oo} = 1 - \frac{M}{4\pi}\{\frac{k_o}{r} + \frac{k \exp[-\mu r]}{r}\} \quad . \tag{1}$$

Here the Yukawa potential is superimposed on the usual Schwarzschild potential due to a hadron mass M. However, the presence of two metric tensors at every space-time point causes the question of bi-causality; the motion of a particle, for instance, must be recorded by two clocks and two metersticks. Furthermore, the mass of the f-field, which is necessarily non-zero for the short-range behavior, breaks the general covariance of the theory and makes the geometrical interpretation of strong-gravity difficult. It is as yet unclear whether the mixed solution (1) is at all possible in the two-tensor theory.

To circumvent these problems, we consider a scalar-tensor theory of the Brans-Dicke type. In contrast to the Brans-Dicke case, we assume that the coupling constant is a function of a scalar field of non-zero mass [5]. As usual, we take the action integral of the form

$$I = \int \{f(\varphi)\ R + L_m\}\ \sqrt{-g}\ \ d^4x \tag{2}$$

where $f(\varphi)$ is a function of the scalar field φ of mass μ_o and L_m is the Lagrangian density for matter. From the action (2) follows the field equation

$$G_{\mu\nu} = -\ (2f)^{-1}\ \theta_{\mu\nu} \tag{3}$$

where the source tensor is the so-called improved stress-energy tensor [6],

$$\theta_{\mu\nu} = T_{\mu\nu}\ (\text{matter}) + 2(\nabla_\mu\nabla_\nu - g_{\mu\nu}\Box)f\ , \tag{4}$$

with the usual stress-energy tensor $T_{\mu\nu}$ for the matter fields. We make a specific choice of the function $f(\varphi)$:

$$f(\varphi) = (2\ k_o)^{-1}\ \exp[\sqrt{k}\varphi]\ . \tag{5}$$

The coupling function $(2f)^{-1}$ is now characterized by the two coupling constants, k_o normal and k strong. Substitution of (5) into (3) yields

$$G_{\mu\nu} = -\ k_o\ \exp\ [-\sqrt{k}\varphi]T_{\mu\nu} - k\ \tilde{\theta}_{\mu\nu} \tag{6}$$

where

$$\tilde{\theta}_{\mu\nu} = (g_{\mu\nu}\ g_{\rho\sigma} - g_{\mu\rho}\ g_{\nu\sigma})(\partial^\rho\varphi\partial^\sigma\varphi - \frac{1}{\sqrt{k}}\nabla^\rho\nabla^\sigma\varphi). \tag{7}$$

Now strong-gravity dominance can be exhibited in terms of the sources strongly coupled with geometry. As φ diminishes, Eq. (6) reduces to Einstein's equation in the standard form and matter interacts with gravity in the ordinary manner. In the domain where φ is significant, the second term unique to the varying coupling theory is to dominate because of the extremely large value of k in comparison with k_o. Clearly, geometry in a certain domain of space-time can dominantly be determined by the sources with the strong coupling constant k.

Whether strong-gravity dominance can occur at short distances or not has to be seen by solving the field equation (6). The action (2) with the Lagrangian involving only the scalar field leads also to the field equation for φ :

$$\Box \varphi + \mu_o{}^2 \varphi - R \frac{df}{d\varphi} = 0 \;, \tag{8}$$

where $R = -g^{\mu\nu} G_{\mu\nu}$. Thus, for the geometry in strong-gravity dominance, we have

$$G_{\mu\nu} = - k \tilde{\theta}_{\mu\nu} \tag{9}$$

where

$$\tilde{\theta}_{\mu\nu} = \{\frac{3}{2} \mu^2 \exp [- \sqrt{k}\varphi] \varphi^2 g_{\mu\nu} + \tilde{\theta}_{\mu\nu}\} \tag{10}$$

with $\mu = \mu_o\{2 k_o/3 k\}^{\frac{1}{2}}$. Solving (8) and (9) asymptotically for the static and spherically symmetric case, we find the Yukawa behavior of the scalar field φ:

$$\varphi \simeq A \frac{\exp [-\mu r]}{r} \tag{11}$$

and the metric component in which, as expected in (1), the Yukawa potential is superimposed on the Schwarzschild potential:

$$g_{oo} \simeq 1 - \frac{M}{4\pi} \{\frac{k_o}{r} + \frac{k \exp [- \mu r]}{r}\} \quad . \tag{12}$$

In fact, for $\mu \sim 1$ GeV, the Yukawa potential damps off rapidly in the range of $10^{-14} \sim 10^{-13}$cm.

BROKEN SYMMETRY AND THE de SITTER STRUCTURE

Recent developments of gauge theories with spontaneously broken symmetry seem to suggest a general principle of unifying all kinds of fundamental interactions [7]. An important ingredient in renormalizable gauge theories of the Weinberg-Salam type is the Higgs field whose vacuum expectation value does not vanish because of spontaneous symmetry breakdown. The idea that the Higgs field may have a cosmological character has been also proposed [8]. If the cosmological term arises from the nonzero vacuum expectation value of the stress-energy tensor $T_{\mu\nu}$ via the Higgs mechanism, then the mass of the Higgs field must be, to be consistent with the experimental upper limit of the cosmological constant, as small as 10^{-27} times the electron mass. However such a small mass limit has been strongly questioned [9].

In strong-gravity dominance, the role of the Higgs field could be more serious. Here, without referring to any specific mechanism, we simply assume that the vacuum expectation value of the scalar field φ is not zero: $\langle\varphi\rangle \neq 0$. In other words, a certain symmetry built in our starting Lagrangian is assumed to be spontaneously broken. As a result, the vacuum expectation value of $\theta_{\mu\nu}$ in (9) becomes proportional to the metric tensor $g_{\mu\nu}$. Thus we expect that the short-distant geometry corresponding to the lowest state of the hadronic matter is given by

$$G_{\mu\nu} = - k\langle \tilde{\tilde{\theta}}_{\mu\nu} \rangle = - k\rho \, g_{\mu\nu} \quad . \tag{13}$$

Here the porportionality constant ρ is to be understood as the energy density of the hadronic matter.

The geometry (13) describes a de Sitter space-time of constant curvature:

$$R_{\mu\nu\rho\sigma} = \lambda^2 (g_{\mu\rho} \, g_{\nu\sigma} - g_{\mu\sigma} \, g_{\nu\rho}) \ , \tag{14}$$

where the radius of curvature λ^{-1} is related to the energy density ρ as

$$3 \lambda^2 = k\rho \quad . \tag{15}$$

In earlier papers [10], we have considered the de Sitter geometry (13) as a consequence of the breaking of the conformal symmetry SO(4,2) into SO(4,1). Although we do not get into the details of the symmetry breaking mechanism, I would like to point out that the de Sitter geometry is the simplest model for the possible curved structure of hadrons in the lowest state.

What one can immediately read off from the relation (15) is that the nuclear density $\rho \sim 10^{15}$g and the hadronic size $\lambda^{-1} \sim 10^{-14}$cm are consistently related [11], so far as the order of magnitude is concerned, by the strong-gravity constant $k \sim 10^{38} \, k_o$. Determination of a more detailed value of the constant k will be discussed later.

The mass of the proton may be estimated by

$$M = \frac{4\pi}{3\lambda^3} \rho \quad . \tag{16}$$

Combining (15) and (16), we find

$$\lambda^{-1} = \frac{kM}{4\pi} \tag{17}$$

which has the value of the Schwarzschild radius for the proton in strong gravity. It is interesting to note that the proton in the ground state looks like a black hole in strong-gravity. This could be a mechanism for the quark confinement. This is also consistent with the picture of the growing proton. The fragmentation of the proton at high energies would occur only in excited states for which the de Sitter geometry is not necessarily a good approximation.

DETERMINATION OF STRONG-GRAVITY CONSTANT

Next we attempt to determine the strong-gravity constant k. As is well-known, the group of motion allowed in the de Sitter space-time is SO(4,1). This de Sitter group is locally isomorphic to the five-dimensional rotation group and its ten generators J_{ab}(a,b = 1,2,3,0,5) satisfy the following commutation relations:

$$[J_{ab}, J_{cd}] = i(\eta_{ad} J_{bc} - \eta_{bd} J_{ac} + \eta_{bc} J_{ad} - \eta_{ac} J_{bd}), \tag{18}$$

where $\eta_{11} = \eta_{22} = \eta_{23} = -\eta_{oo} = \eta_{55} = 1$ and $\eta_{ab} = 0$ for $a \neq b$. Let $J_{\mu\nu}(\mu, \nu = 1,2,3,0)$ be the generators of the Lorentz group SO(3,1) which is a subgroup of SO(4,1) and π_μ designate $\lambda J_{5\mu}$. Then, in the limit $\lambda \to 0$, the algebra for $J_{\mu\nu}$ and π_μ reduces to that of the Poincare group; π_μ assumes the role of the generators of translation in the Minkowskian space-time [12].

For the quantum consideration, we make use of this symmetry. Namely, we look upon the de Sitter hadron state as an eigenstate of the de Sitter rotator:

$$\tfrac{1}{2} \lambda^2 J_{ab} J^{ab} |\psi\rangle = M_o^2 |\psi\rangle \quad . \tag{19}$$

Furthermore, in order to see the spin-dependence of the mass, we characterize the state $|\psi\rangle$ by the spin quantum number J on the basis of the Lorentz group;

$$J_{\mu\nu} J^{\mu\nu} \; | J \rangle = 2 \{ K^2 - J(J+1) \} \; | J \rangle \tag{20}$$

where K is a constant. The eigenstates $| J \rangle$ form the base of an irreducible representation of the Lorentz group. Decomposing the Casimir invariant $\lambda^2 J_{ab} J^{ab}$ into the form

$$2 \; \pi_\mu \pi^\mu + \lambda^2 J_{\mu\nu} J^{\mu\nu} \tag{21}$$

and identifying $\pi_\mu \pi^\mu$ with the mass operator in analogy with $P_\mu P^\mu = M^2$, one can find a mass spectrum [13]

$$M^2 = \lambda^2 J(J+1) + C \tag{22}$$

where $C = M_o^2 - K^2\lambda^2$. Notice that the spectrum is completely degenerate in the Minkowskian background (i.e., $M^2 \to M_o^2$ as $\lambda \to 0$).

Suppose we have an ensemble of the de Sitter rotators. Every individual rotator is to be specified by the values of M, J and λ. The symmetry predicts the mass-spin relation (22) for a group of rotators with a common value of λ, but gives no restriction on their geometrical dimension. On the other hand, the geometric equation (13) applied to a single hadron decides the size λ^{-1} when the mass is known. Thus, combining (15) and (22), we obtain a self-consistent mass formula [14]

$$M^4 \simeq \left(\frac{4\pi}{k}\right) J(J+1) + C\, M^2. \tag{23}$$

Here the iso- and unitary-spin dependence is suppressed in the constant C. It is interesting to compare this formula with the Kerr-Newman inequality

$$M^4 > \left(\frac{8\pi}{k_o}\right)^2 J^2 + \left(\frac{8\pi}{k_o}\right) Q^2 M^2 \quad . \tag{24}$$

By regarding hadrons as nearly collapsing objects in strong-gravity, Salam has in fact suggested a similar mass formula [3].

For $J \gtrsim \frac{1}{2}$ and C small, we find from (23) a linearly rising trajectory [14]

$$J \simeq \alpha' M^2 + \alpha(0) \tag{25}$$

with $\alpha' = k/4\pi$, and $\alpha(0)$ = constant. Comparison of the slope of this trajectory with the universal slope $\alpha' \simeq 0.95\ \mathrm{GeV}^{-2}$ of the Chew-Frautschi plot gives us $k \simeq 0.75 \times 10^{38}\ k_o$ ($k_o = 8\pi G$; G the Newtonian gravitational constant).

RELATION TO THE DUAL RESONANCE MODEL

Now I would like to describe briefly a recipe of reducing the geometric model to the dual resonance model [15]. The duality is an indispensable concept in formalizing the bootstrap feature of hadron dynamics. The scattering amplitude for a hadron process, if properly chosen, should exhibit the resonance behavior in the low energy limit and the Reggeon exchange behavior in the high energy limit. The first example for such an amplitude is that of Veneziano. Although here are many kinds of variation, for the sake of simplicity, we confine ourselves to the conventional model [16]. It is already well-known that quantization of a relativistic string leads to the dual resonance model [17]. Certainly the string is the simplest relativistic extended object. Even from the dimensional consistency of the action integral with the linear trajectory, the simple string model appears more plausible than the models of higher dimensional relativistic objects such as membranes and bags. What I would like to stress here is not the advantage of the de Sitter model over the string model but the possibility of understanding the concept of duality within the framework of the geometrical approach.

In deriving the mass spectrum (23), the spin states have been taken as the base of the representation. Here, we look at the same de Sitter rotator from a different aspect. Let us decompse $J_{ab}\ J^{ab}$ in the form $J_{5\nu}\ J^{5\nu} + (J_{\mu\nu}\ J^{\mu\nu} + J_{\mu 5}\ J^{\mu 5})$ and express the second part in terms of the five-dimensional Cartesian variables $q_a (a = 1,2,3,0,5)$ and their conjugate momenta p_a. If the commutation relations,

$$[q_a, p_b] = i\ \eta_{ab}, \tag{26}$$

and the constraint,

$$(q_a\ p^a + p_a q^a)\ |\psi\rangle = 0\ , \tag{27}$$

are imposed, then the wave equation (19) for the rotator can be written as

$$\{\pi^2 + (4\pi/k) \sum_n (R_n^{\ 2}\ p_n^{\ 2} + m_n^{\ 2}\ q_n^{\ 2})\}\ |\psi\rangle = 2M_o^{\ 2}\ |\psi\rangle \tag{28}$$

Here $\pi^2 = \sum (4\pi R_n^{\ 2}/k)\pi^2\ |n\rangle\langle n|$, $q_n^{\ 2} = q_\mu\ q^\mu\ |n\rangle\langle n|$ and $p_n^{\ 2} = p_\mu p^\mu\ |n\rangle\langle n|$ with the states $|\ n\ \rangle$ $(n = 1,2,3,\ldots\ldots)$ which diagonalize $q_a\ q^a$ and $p_a p^a$ simultaneously; i.e., $q_a q^a\ |\ n\ \rangle$ and $p_a p^a\ |\ n\ \rangle = m_n^{\ 2}\ |\ n\ \rangle$. The second term on the left hand side of (28), acting upon $|\psi\rangle$, represents a collection of harmonic oscillators with frequency

$$\omega_n = 4\pi\ m_n R_n/k. \tag{29}$$

Introducing the operators $a_{n\mu}$ and $a_{n\mu}^{\ +}$ by

$$a_{n\mu} = \sqrt{R_n/(2m_n)}\ p_{n\mu} - i\ \sqrt{m_n/(2R_n)}\ q_{n\mu} \tag{30a}$$

$$a_{n\mu}^{\ +} = \sqrt{R_n/(2m_n)}\ p_{n\mu} + i\ \sqrt{m_n/(2R_n)}\ q_{n\mu} \tag{30b}$$

we rewrite (28) as

$$\{\pi^2 + 2 \sum_n \omega_n\ a_{n\mu}^{\ +}\ a_n^{\ \mu} - C'\}\ |\psi\rangle = 0 \tag{31}$$

where $C' = 2M_o^{\ 2} - 4\ \Sigma\omega_n$. This is similar in form to that suggested by Nambu [18]. If the geometrical condition (17) and the linear relation (25) are both valid for each m_n, the frequency ω_n of (29) will become an integral multiple of the ground frequency ω_o:

$$\omega_n = n\,\omega_o \,, \tag{32}$$

where $\omega_o = 4\pi/k$. This frequency condition is essential to the factorization of (31).

Then we introduce a local Yukawa interaction of an external scalar field $V(x)$ into the wave equation (31) which is prepared with the Minkowskian variables:

$$\{P_\mu P^\mu + \frac{1}{2} i\omega_o \sum_n \sqrt{\frac{n}{2}}(a_n{}^\mu - a_n{}^{\mu +})P_\mu + 2\omega_o \sum_n n\, a_n{}^{\mu +} a_{n\mu} - C' + V(x)\}|\psi\rangle = 0 \,, \tag{33}$$

where P_μ is a modified displacement operator which is expressed in the Minkowskian background as

$$P_\mu = -i\,\partial_\mu - \tfrac{1}{2}\omega_o(1 - \tfrac{1}{4}\omega_o\, x_\sigma x^\sigma)^{-1}\, x^\nu\, J_{\mu\nu} \,.$$

An appropriate unitary transformation reduces (33) to the form

$$\{\tilde{P}_\mu \tilde{P}^\mu + 2\,\omega_o \sum_n n\, a_{n\mu}^{+}\, a_n{}^\mu - C' + V(x - \phi)\}\,|\tilde{\psi}\rangle = 0 \tag{34}$$

with

$$\phi^\mu = \sum_n \sqrt{2/n}\,(a_n{}^\mu + a_n{}^{\mu +}) \,.$$

Note that the interaction becomes non-local. Following the standard method [19] we can derive from (34) the Veneziano amplitude.

Recently it has been pointed out by Scherk and Schwarz [20] that the zero slope limit $\alpha' \to 0(\omega_o \to 0)$ of a dual model leads to a Lagrangian theory in which space-time has a non-Riemannian geometry. Although their approach is completely different from ours, this indicates also a possible connection between the general-relativistic view and the dual resonance theory.

CONCLUSION

As we have observed, the idea that elementary particles can be general-relativistic objects does not seem too wild. The effect of a few gravitons is negligible in any elementary particle process because of their weak coupling, but the collective effect of a curved structure, if any, could be vital at short distances. For hadrons to be geometrical objects, the usual Einsteinian effect seems still too small; the concept of strong-gravity has to be introduced. In fact, Einstein's equation in an extended form can describe a geometry in strong-gravity dominance and the geometry due to strong-gravity is seen somehow compatible with the structure of hadrons. It is questionable that general relativity provides an effective tool to calculate physical quantities involved in particle interactions.

Nevertheless, it would be important to see if the basic principles of general relativity are at all functioning in the small scale.

More practically, we would get some insight for quantization of gravity. If Einstein's equation is regarded as a field equation analogous to Maxwell's equations, the method in quantum electrodynamics should be a good guide for quantum gravidynamics. However, the nonlinear aspect of gravity, which would be more essential for certain problems, cannot be properly counted. If, in contrast, Einstein's equation either in the original form or in a modified form is considered to be an equation describing geometry of a compositie system, its possible quantum characters may be more appropriately compared with those of infinite component theory or dual resonance models. For instance, an infinite component wave equation may be formulated with the global symmetry of a curved geometry interpreted as the dynamical symmetry of the composite. The quantum features of a single composite are to be realized on the base of the c-number fields. Second-quantization would be needed only when interactions of two or more composites are involved. The system described by Einstein's equation, insofar as its global geometry is important, would have to be quantized as a composite.

Finally I wish to thank Professor Salecker and Professor Barut for the hospitality at the University of Munich where part of the work was done. I would also like to acknowledge gratefully Dr. Friedl's assistance and contribution to the work.

REFERENCES

[1]. A. Einstein, Sitz. Preuss. Akad. Wiss. (1919) reprinted in Das Relativitatsprinzip (Teubner, 1922) and The Principle of Relativity (Dover, 1923).
[2]. A. Salam and J. Strathdee, Lett. Nuovo Cimento 4 (1970) 101; C. J. Isham, A. Salam and J. Strathdee, Phys. Rev. D3, (1971) 1805; D5, (1972) 2548.
[3]. A. Salam, in Nonpolynomial Lagrangians, Renormalization and Gravity, 1971 Coral Gables Conference on Fundamental Interactions at High Energies, vol.1 (New York, 1971) p.3.
[4]. C. J. Isham, A. Salam, and J. Strathdee, Phys. Rev. D3 (1971) 867.
[5]. J. Friedl and A. Inomata, Lett. Nuovo Cimento 9 (1974) 83.
[6]. C. G. Callan, S. Coleman and R. Jackiw, Ann. Phys. 59 (1970) 42.
[7[. See, e.g., S. Weinberg, Rev. Mod. Phys. 46 (1974) 255.
[8]. J. Dreitlein, Phys. Rev. Lett. 33 (1974) 1243.
[9]. M. Veltman, Phys. Rev. Lett. 34 (1974)777.
[10]. A. Inomata, J. Friedl and A. Bakesigaki, Lett. Nuovo Cimento 1 (1971) 1013; A. Inomata, in Lectures in Theoretical Physics, vol 13, edited by A.O.Barut and W. E. Brittin (Boulder, Col., 1971) p.377.
[11]. A. Inomata and D. Peak, Progr. Theor. Phys. 42 (1969) 134.
[12]. See, e.g., F. Gürsey, in Group Theoretical Concepts and Methods in Elementary Particle Physics, (New York, 1962).
[13]. A. O. Barut and A. Böhm, Phys. Rev. 139, B1107 (1965); P. Roman, in Noncompact Groups in Particle Physics, edited by Y. Chow (New York, 1966)p.89; A. Böhm, Phys. Rev. 175 (1968) 1767.
[14]. A. Bakesigaki and A. Inomata, Lett. Nuovo Cimento 2 (1971) 697.
[15]. J. Friedl and A. Inomata, to be published.
[16]. G. Veneziano, Nuovo Cimento 57A (1968) 190.
[17]. See, e.g., J. Scherk, Rev. Mod. Phys. 47 (1975) 123.
[18]. Y. Nambu, in Symmetries and Quark Models, (Wayne State Univ., 1965).
[19]. L. Susskind, Nuovo Cimento 69 (1970) 457.
[20]. J. Scherk and J. H. Schwarz, Phys. Lett. 52B, (1974) 347.

REMARKS ON THE MANY-TIME CANONICAL QUANTIZATION OF GRAVITATION*

Martin Pilati† and Claudio Teitelboim
Joseph Henry Laboratories, Princeton University,
Princeton, New Jersey 08540, U.S.A.

We discuss some aspects of the construction of a many-time canonical quantum theory of gravity. It is shown that, if the time variable is constructed from the trace of the extrinsic curvature, the following can be done: (i.) A first order, many-time wave equation can be (implicitly) written down; (ii.) An inner product that is conserved under arbitrary hypersurface deformations can be defined; (iii.) The spacelike character of the surface on which the state is defined is automatically taken care of, leading to no additional restriction in the quantum theory; (iv.) For the time evolution to be continuous, the state functional must obey a well defined set of boundary conditions that require zero probability for singular manifolds. The analysis is developed separately for open and closed spaces. In the latter case some points remain to be clarified.

I. INTRODUCTION

An arbitrary deformation of a three-dimensional spacelike surface embedded in a four-dimensional spacetime changes both its intrinsic and extrinsic geometries. The Hamiltonian formulation of general relativity[1] treats this change as a dynamical process. The intrinsic metric g_{ij} and a linear combination π^{ij} of the extrinsic curvature components enter the theory as canonically conjugate quantities. The components $g_{o\mu}$ of the spacetime metric play the role of arbitrary functions (the evolution of which is not restricted by Hamilton's equations) associated with the freedom of making arbitrary deformations of the surface on which the state is defined.

The general procedure for quantizing constrained Hamiltonian systems[2,3] allows one to follow two routes to a canonical quantum theory. One may either eliminate the freedom of making arbitrary surface deformations by first imposing coordinate conditions at the classical level and then quantizing ("single-time theory") or one may quantize leaving the freedom in question open ("many-time theory"). While the two approaches are in principle physically equivalent, it is more attractive to pursue the second. There are at least two reasons for saying

---

*Work supported in part by the National Science Foundation under Grant No. GP30799X to Princeton University.

†N.S.F. predoctoral fellow.

this. First of all, the freedom of making arbitrary surface deformations, and the related commutation rules for the generators of the deformations, may be regarded as the basic structure of the classical theory.[4] Indeed, the form of the Hamiltonian may be deduced from such properties.[5] Thus one would also like to permit arbitrary surface deformations in the quantum theory. Secondly, on a more technical ground, the available coordinate conditions are usually only good in a region of spacetime and not globally. A many-time theory avoids this problem by allowing one to smoothly adapt the coordinates to suit a particular region of spacetime.

There are three basic problems that must be solved before one can even begin to worry about the detailed structure of the theory, namely:

(A) There must exist a wave equation that gives the change of the state functional (probability amplitude) under an arbitrary deformation of the spacelike hypersurface on which the state is defined. This equation must be of first order in the parameters describing the surface in order to comply with the basic principles of quantum mechanics.[6]

(B) There must exist an inner product that is conserved under arbitrary deformations of the hypersurface.

(C) The spacelike character of the surface must be accounted for.

Problem (C) is important because, in order to solve (A), it is necessary to transform from the variables g_{ij},π^{ij} to another set. The condition for a surface to be spacelike in the variables g_{ij},π^{ij} is expressed by the requirement that g_{ij} be positive definite. This condition reexpressed in the new variables could restrict the range of the variables, and this indeed happens in certain models (see e.g., ref. 6). This restriction must be taken over to the quantum theory, which causes no problems in the model because the condition only involves the parameters that describe the location of the surface; there is no reason for this to be true in general. If the condition had non-trivial dependence on the independent canonical variables then serious difficulties would arise with the inner product in which one integrates over all values of the independent variables.[7] There are no such problems with the variables discussed here.

A definitive solution of problems (A)-(C) requires solution of the factor ordering problem. However, one can reach quite far in the analysis by relying exclusively on the classical theory in Hamiltonian form. We shall thus ignore the ordering problem. With this cautionary remark in mind, our conclusion will be that problems (A)-(C) can be satisfactorily dealt with.

II. SUMMARY OF THE CANONICAL FORMULATION

In the Hamiltonian formulation of general relativity[1] the change F of an arbitrary functional F of the canonical variables induced by a deformation $N^{\mu}\delta t$ that connects two neighboring surfaces of coordinate time t and t+δt (Fig. 1) is given by

$$\delta F = \int d^3x\{F, \mathcal{H}_{\mu}(x)\}N^{\mu}\delta t \quad . \tag{2.1}$$

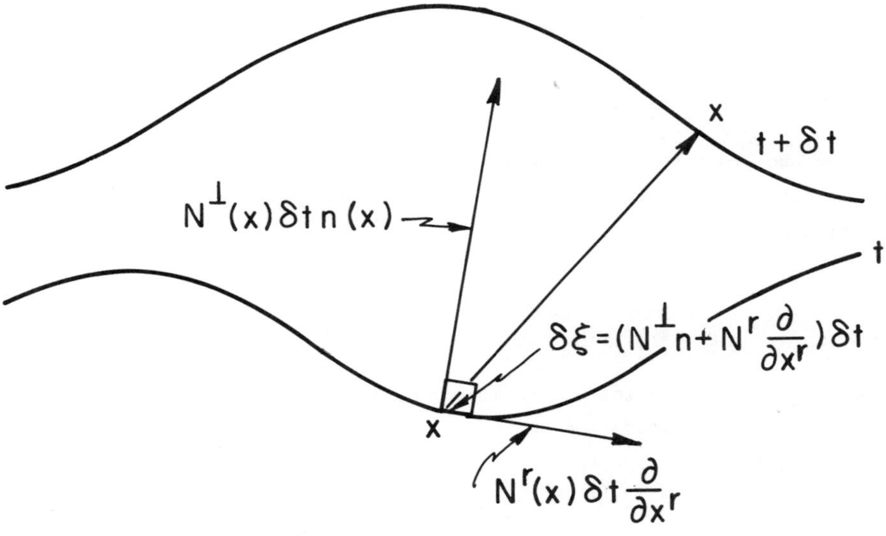

Figure 1. Deformation of a coordinatized surface. Starting from a given surface t with coordinate (x) one goes to a neighboring surface t+δt by means of a deformation $\delta\xi(x)=[N^{\perp}(x)\ n(x) + N^{r}(x)(\partial/\partial x^{r})]\delta t$. Note that the defomation defines the coordinate system on t+δt by giving the same coordinate value to points at the tail and at the tip of the deformation. The orientation of the normal ("future direction") is arbitrary. $N^{\perp}(x)$ and $N^{r}(x)$ are the lapse and shift functions of Arnowitt, Deser, and Misner.[1]

The generators $\mathcal{H}_\mu$ ($\mu = \perp,1,2,3$) are

$$\mathcal{H}_\perp = \frac{1}{2}g^{-1/2}(g_{ik}g_{jl}+g_{il}g_{jk}-g_{ij}g_{kl})\pi^{ij}\pi^{kl}-g^{1/2}R \ , \tag{2.2}$$

$$\mathcal{H}_i = -2\pi_i{}^j{}_{|j} = -2g_{ik}\pi^{kj}{}_{,j}-(2g_{ki,j}-g_{kj,i})\pi^{kj} \ , \tag{2.3}$$

where a vertical slash denotes covariant differentiation in the metric g_{ij}. The $\mathcal{H}_\mu$ are constrained to be zero. Matter fields are not included in (2.2) or (2.3), but their inclusion, at least for non-derivative couplings to gravity, would not change any of our conclusions. For asymptotically flat spaces certain surface terms have to be added[8] to the integral $\int d^3x\, N^\mu \mathcal{H}_\mu$ which plays the role of a Hamiltonian in (2.1).

The canonical variables g_{ij}, π^{ij} satisfy the Poisson bracket relations

$$\{g_{ij}(x), \pi^{kl}_{(x')}\} = \delta^{kl}_{ij}\delta(x,x') \tag{2.4}$$

where

$$\delta^{kl}_{ij} = \frac{1}{2}(\delta^k_i\ \delta^l_j + \delta^l_i\ \delta^k_j) \ . \tag{2.5}$$

If we want to pass to the quantum theory without fixing the coordinates, Dirac's general prescription[2,3] is to impose the first-class constraints $\mathcal{H}_\mu = 0$ as restrictions on the state functional Ψ:

$$\mathcal{H}_\perp\Psi = 0 \tag{2.6}$$

and

$$\mathcal{H}_i\Psi = 0 \ . \tag{2.7}$$

Equation (2.7) is equivalent to the statement that the state functional is invariant under reparametrizations of the three dimensional space.

If the quantization prescription (2.6) is followed naively in the metric representation ($\Psi = \Psi[g_{ij}]$, $\pi^{ij} \rightarrow \frac{h}{i}\frac{\delta}{\delta g_{ij}}$) one arrives at the so-called Wheeler-DeWitt[9,10] equation. This equation is of second order in all functional derivatives and therefore does not allow for the interpretation of Ψ as a probability amplitude. In order to obtain a first order many-time wave equation we need to consider a different set of fields as our primary canonical variables. Following Dirac[11] and York[12] we take our basic variables to be

$$\tau \equiv \frac{2}{3}g^{-\frac{1}{2}}g_{ij}\,\pi^{ij} \ , \tag{2.8}$$

$$p \equiv -g^{\frac{1}{2}} \ , \tag{2.9}$$

$$\tilde{g}_{ij} \equiv g^{-1/3}\,g_{ij} \ , \tag{2.10}$$

$$\tilde{\pi}^{ij} \equiv g^{1/3}\left(\pi^{ij} - \frac{1}{3}g^{ij}\,g_{mn}\,\pi^{mn}\right) \ . \tag{2.11}$$

(For asymptotically flat space one further transformation of τ and p will be necessary later.)

The variable $\tilde{g}_{ij}$ is not changed if the original g_{ij} is multiplied by a conformal factor. In this sense $\tilde{g}_{ij}$ describes the conformal intrinsic geometry of three-space.

The only non-vanishing Poisson brackets among variables (2.8)-(2.11) are

$$\{\tau(x),p(x')\} = \delta(x,x') \quad , \tag{2.12}$$

$$\{\tilde{g}_{ij}(x),\tilde{\pi}^{kl}(x')\} = \tilde{\delta}^{kl}_{ij}\,\delta(x,x') \quad , \tag{2.13}$$

$$\{\tilde{\pi}^{ij}(x),\tilde{\pi}^{kl}(x')\} = \frac{1}{3}\left(\tilde{g}^{kl}\tilde{\pi}^{ij}-\tilde{g}^{ij}\tilde{\pi}^{kl}\right)\delta(x,x') \quad , \tag{2.14}$$

where

$$\tilde{\delta}^{kl}_{ij} = \frac{1}{2}\left(\delta^{k}_{i}\delta^{j}_{l}+\delta^{l}_{i}\delta^{k}_{j}\right) - \frac{1}{3}\tilde{g}_{ij}\tilde{g}^{kl} \quad . \tag{2.15}$$

The presence of fields on the right side of (2.13) and (2.14) is the source of the severe factor ordering problems encountered elsewhere.[13] The quantity $\tilde{g}^{ij}$ appearing in (2.14) and (2.15) is the inverse of the conformal metric $\tilde{g}_{ij}$ $(\tilde{g}^{lm}\tilde{g}_{ms} = \delta^{l}_{s})$ and is related to the full metric by $\tilde{g}^{ij} = g^{1/3}g^{ij}$. The "conformal Kronecker delta" defined by (2.15) has the properties

$$\tilde{\delta}^{ij}_{ij} = 5 \tag{2.16}$$

$$\tilde{\delta}^{ij}_{kl}\tilde{\delta}^{kl}_{mn} = \tilde{\delta}^{ij}_{mn} \tag{2.17}$$

$$\tilde{\delta}^{kl}_{ij}\tilde{g}_{kl} = \tilde{\delta}^{kl}_{ij}\,g^{ij} = 0 \quad , \tag{2.18}$$

$$\tilde{\delta}^{kl}_{ij}\tilde{\pi}^{ij} = \tilde{\pi}^{kl} \tag{2.19}$$

The meaning of (2.16) is that there are five independent canonical variables per space point in accordance with the conditions

$$\tilde{g} \equiv \det||g_{ij}|| = 1 \quad , \tag{2.20}$$

$$\tilde{\pi} \equiv g_{ij}\,\tilde{\pi}^{ij} = 0 \quad . \tag{2.21}$$

It should be pointed out that $\tilde{g}$ and $\tilde{\pi}$ have zero Poisson brackets with everything, consequently they can be set strongly equal to unity and zero respectively. The symmetry in all indices can thus be preserved even though one is dealing with more variables than are needed (it is not necessary to solve for one of the variables in terms of the others).

III. ASYMPTOTICALLY FLAT SPACE

From now on we restrict ourselves to the case of open, asymptotically flat three-space. Relatively few changes have to be made to deal with closed spaces, and these will be discussed in a separate section.

Our assumption of asymptotic flatness is not only applied to the intrinsic geometry, but also to the extrinsic geometry of the surface.* This assumption severely restricts the permissible surface deformations at infinity, but allows for an arbitrary deformation everywhere else (subject to the requirement of continuity).

The freedom that is left at infinity still allows for asymptotic Poincaré transformations and some angle dependent terms.[8] For our present purpose it is sufficient to consider only "pure time translations" at infinity. This means that the functions N^{μ} behave asymptotically as

$$N^{\mu} \rightarrow (N^{\perp}(\infty),0,0,0) \tag{3.1}$$

where $N^{\perp}(\infty)$ is a constant. Assuming that the parameter t is a good coordinate at infinity, we may, without loss of generality, choose it to coincide with the proper time at infinity. We then have $N^{\perp}(\infty) = 1$. However, it is also useful sometimes to deal with two neighboring surfaces that, although not identical everywhere, coincide asymptotically. In that case one has $N^{\perp}(\infty) = 0$.

The boundary conditions on the canonical variables will be taken as[9]

$$\tau \sim r^{-(3+\varepsilon)} \tag{3.2}$$

$$p + 1 \sim r^{-1} \tag{3.3}$$

$$\tilde{g}_{ij} - \delta_{ij} \sim r^{-1} \tag{3.4}$$

$$\tilde{\pi}^{ij} \sim r^{-2} \quad . \tag{3.5}$$

Equations (3.3)-(3.5) ensure, roughly speaking, that the total energy-momentum is finite. Condition (3.2) is needed for the inverse Laplacian used later to be well defined.

IV. MANY TIME WAVE EQUATION

We now show that a many-time wave equation with the desired properties exists. Our analysis is an application to this particular set of variables of the more general discussion given in reference 7.

---

* One might think that it is enough to just restrict the enveloping spacetime to be asymptotically flat, allowing the surfaces to be arbitrary at infinity (neither intrinsically nor extrinsically flat) provided that they respect the embedding conditions. It appears, however, that the Hamiltonian formalism needs to have surfaces that are intrinsically and extrinsically flat separately; for example, without this requirement one is not assured of a finite action. (For a related discussion see ref. 8.)

We begin by showing that the Hamiltonian constraint

$$\mathcal{H}_{\perp} = 0 \tag{4.1}$$

can be reexpressed in the form

$$p + K[\tilde{g}_{ij}, \tilde{\pi}^{ij}, \tau] = 0 \tag{4.2}$$

The condition for the infinite system of equations (4.1) (one equation for each x^1, x^2, x^3) to be solvable in the form (4.2) is that the infinite matrix

$$\mathcal{M}^{x}{}_{x'} = \frac{\delta \mathcal{H}_{\perp}(x')}{\delta p(x)} \tag{4.3}$$

be invertible. (The functional differentiation in (4.3) is carried out at constant τ, $\tilde{g}_{ij}$, $\tilde{\pi}^{ij}$.) More precisely we require that

$$\int d^3x' \frac{\delta \mathcal{H}_{\perp}(x')}{\delta p(x)} N^{\perp}(x') = 0 \tag{4.4}$$

implies

$$N^{\perp}(x) = 0 \tag{4.5}$$

if

$$N^{\perp}(\infty) = 0 \tag{4.6}$$

The boundary condition (4.6) is imposed because only testing functions which vanish at infinity are necessary to completely characterize the asymptotically vanishing function $\mathcal{H}_{\perp}(x)$, i.e. if $\int d^3x N^{\perp}(x)\, \mathcal{H}_{\perp}(x)$ is known for every $N^{\perp}$ then we know $\mathcal{H}_{\perp}$.

Now the left hand side of (4.4) is nothing other than the Poisson bracket

$$\dot{\tau} \equiv \int d^3x' \{\tau(x), \mathcal{H}_{\perp}(x')\} N^{\perp}(x') \tag{4.7}$$

which gives the change of $\tau(x)$ under a purely normal deformation $N^{\perp}$. We can thus summarize the condition for (4.2) to be possible by the statement

$$\dot{\tau}(x) = 0 \text{ for all } x \Longrightarrow N^{\perp}(x) = 0 \text{ for all } x\ . \tag{4.8}$$

To prove (4.8) we write the Poisson bracket (4.7) in detail obtaining

$$\dot{\tau} = \frac{4}{3} \{\Delta N^{\perp} - (g^{-1} \tilde{\pi}_{ij} \tilde{\pi}^{ij} + \frac{3}{16} \tau^2) N^{\perp}\} \tag{4.9}$$

where Δ is the Laplacian in the metric g_{ij}. If we now demand $\dot{\tau} \equiv 0$ we find an equation of the form

$$\Delta N^{\perp} - V(x) N^{\perp} = 0 \quad , \quad V(x) \geqslant 0 \quad . \tag{4.10}$$

The only solution to (4.10) with the boundary condition $N^{\perp}(\infty) = 0$ is $N^{\perp} \equiv 0$.

One might think that, having established the existence of (4.2), we have achieved our goal. In fact, upon quantization (4.2) becomes a restriction on the state:

$$\frac{\hbar}{i}\frac{\delta\psi}{\delta\tau(x)} + K[\tilde{g}_{ij},\tilde{\pi}^{ij},\tau]\psi = 0 \tag{4.11}$$

which is a first order many time equation in the variable $\tau(x)$. The problem here, as emphasized by DeWitt[9] (see also 6,8,14), is that the state functional depends not only on $\tilde{g}_{ij}$, $\tilde{\pi}^{ij}$ and τ, but also on the asymptotic location of the surface, which is described by the parameter t. This follows from the fact that when non-vanishing asymptotic deformations are allowed in the classical theory the Hamiltonian does not vanish, but is rather equal to a field dependent surface integral. Now (4.11) only determines the change in ψ induced by a change of τ, and does not provide any information about the response of ψ to a change in t. Using classical language, we might say that (4.11) determines the change in ψ under a deformation that vanishes at infinity.

It is not possible to supplement equation (4.11) with a separate equation describing the change in ψ induced by a change in t keeping τ constant. The reason for this is best understood by going back to the classical theory and observing that, for continuity reasons, it is not possible to deform a surface at infinity (changing t) without deforming anywhere else (changing τ). The converse statement, deforming "inside" but not at infinity, is perfectly reasonable, hence (4.11) is well defined.[4,8]

To obtain an equation which describes the behavior of ψ under a deformation not vanishing at infinity, we have to find a variable T(x) the change of which can account for both the asymptotic and non-asymptotic parts of a deformation. Such a variable will have to be dependent on t. The clue for defining T(x) is provided by the surface term in the Hamiltonian[8]:

$$H = \int d^3x\, N^\mu \mathcal{H}_\mu + \frac{4}{3}\oint d^2S_k\, P_{,k} \quad . \tag{4.12}$$

The Hamiltonian (4.12) does not vanish when the constraints hold, but rather becomes equal to the surface term. This is just a reflection of the fact that the canonical variables $\tilde{g}_{ij}$, $\tilde{\pi}^{ij}$, τ, p do not include information about the location of the surface at infinity. As one says, the theory is not "fully parametrized." <u>In order to obtain a fully parametrized theory we must change the variables so that the new Hamiltonian is constrained to be zero</u>. To achieve this result it is most convenient to write the action as

$$S = \int dt \int d^3x \left(p\frac{\partial\tau}{\partial t} + \tilde{\delta}^{ij}_{kl}\,\tilde{\pi}^{kl}\,\frac{\partial\tilde{g}_{ij}}{\partial t} - N^\mu\mathcal{H}_\mu - \frac{4}{3}\Delta_\flat p\right) \tag{4.13}$$

where we have rewritten the surface term as a volume integral.

Next we define

$$T = -\frac{3}{4}\Delta_\flat^{-1}\tau + t \tag{4.14}$$

$$\Pi_T = -\frac{4}{3}\Delta_\flat p \tag{4.15}$$

where $\Delta_\flat$ and $\Delta_\flat^{-1}$ are respectively the flat space Laplacian and its inverse. By definition the action of $\Delta_\flat^{-1}$ always yields a function that vanishes at infinity. Equations (4.14)-(4.15) define a time dependent canonical transformation. When rewritten in terms of the new variables the action (4.13) reads

$$S = \int dt \int d^3x \left(\pi_T \frac{\partial T}{\partial t} + \tilde{\delta}^{ij}_{kl}\, \tilde{\pi}^{kl} \frac{\partial \tilde{g}_{ij}}{\partial t} - N^\mu \mathcal{H}_\mu\right) . \tag{4.16}$$

The new Hamiltonian is just $\int N^\mu \mathcal{H}_\mu$, without the surface term, and it vanishes weakly.

Equation (4.16) shows that

$$\{T(x), \pi_T(x')\} = \delta(x,x') \tag{4.17}$$

so that T and π_T are canonically conjugate. The reason for using the flat space Laplacian in (4.14)-(4.15) instead of the Laplacian $\tilde{\Delta}$ of the metric $\tilde{g}_{ij}$ is that $\tilde{\Delta}$ involves $\tilde{g}_{ij}$; so we would have obtained a time variable and a conjugate that had non-zero brackets with $\tilde{\pi}^{ij}$ making the identification $\pi_T \to \frac{h}{i}\frac{\delta}{\delta T}$ impossible.

It is important to emphasize that the variable T defined by (4.14) behaves asymptotically as

$$T \xrightarrow[r\to\infty]{} t \tag{4.18}$$

and thus contains information about the asymptotic location of the surface, a property that none of the original variables (2.8)-(2.11) possessed. However, not every variable satisfying (4.18) is a good T variable. For example, had we defined $T = \tau + t$ we would not have been able to eliminate the surface term, and consequently not all of the information on the dynamics would be contained in the equation $\mathcal{H}_\perp \psi = 0$.

The constraint (4.2) can now be written as

$$\pi_T + K_T[\tilde{g}_{ij}, \tilde{\pi}^{ij}, T] = 0 \tag{4.19}$$

with

$$K_T[\tilde{g}_{ij}, \tilde{\pi}^{ij}, T] \equiv -\frac{4}{3}\Delta_\flat K[\tilde{g}_{ij}, \tilde{\pi}^{ij}, -\frac{4}{3}\Delta_\flat T] .$$

The many-time Schroedinger equation then reads

$$\frac{\hbar}{i}\frac{\delta\psi}{\delta T} + K_T\psi = 0 \tag{4.20}$$

where* $\psi = \psi[\tilde{g}_{ij}, T]$. This equation gives the evolution of Ψ both at infinity and in the interior, and thus contains all of the dynamics of the system.

Although we are ignoring the factor ordering problem, it is appropriate at

---

* We use the metric representation $\Psi[\tilde{g}_{ij}] = \langle\tilde{g}_{ij}|\Psi\rangle$. This is a good choice because the $\tilde{g}_{ij}$ commute with each other. The "momentum representation" does not exist because the $\tilde{\pi}^{ij}$ do not commute.

this point to ask what the operator expression for $\tilde{\pi}^{ij}$ is. One can verify that the definition

$$\tilde{\pi}^{ij}_{op} = \frac{\hbar}{i}\,\tilde{\delta}^{ij}_{kl}\,\frac{\delta}{\delta\tilde{g}_{kl}} = \frac{\hbar}{i}\left(\frac{\delta}{\delta\tilde{g}_{ij}} - \frac{1}{3}\tilde{g}^{ij}\,\tilde{g}_{kl}\,\frac{\delta}{\delta\tilde{g}_{kl}}\right) , \tag{4.21}$$

turns the Poisson bracket relations (2.13), (2.14) into commutation relations by the usual rule: (Poisson bracket) → $i\hbar$(commutator). In order to evaluate the action of $\tilde{\pi}^{ij}_{op}$ on a functional Ψ defined on the subspace $\tilde{g} = 1$, one does not need to go outside of that subspace. This follows from the property

$$\tilde{g}_{ij}\,\tilde{\pi}^{ij}_{op} \equiv 0 \quad . \tag{4.22}$$

Thus Ψ may be regarded as undefined outside the subspace $\tilde{g} = 1$, although, for practical purposes, it may be continued off that subspace in any convenient manner.

Lastly a word about the tangential constraints

$$\mathcal{H}_i = -2\pi_i{}^j{}_{|j} = 0 \quad . \tag{4.23}$$

The generator of reparametrizations $\mathcal{H}_i$ can be written as

$$\mathcal{H}_r = (\Delta_{\flat} T_{,r})\Delta_{\flat}^{-1}\pi_T + \tilde{g}_{ab,r}\tilde{\pi}^{ab} - 2\left(\tilde{g}_{rb}\tilde{\pi}^{ab}\right)_{,a} = 0 \quad . \tag{4.24}$$

This becomes upon quantization the restriction on the wave functional

$$(\Delta_{\flat} T_{,r})\Delta_{\flat}^{-1}\frac{\delta\psi}{\delta T} + \tilde{g}_{ab,r}\frac{\delta\psi}{\delta\tilde{g}_{ab}} - 2\left[\left(\tilde{g}_{rb}\frac{\delta\psi}{\delta\tilde{g}_{ab}}\right)_{,a} - \frac{1}{3}\left(\tilde{g}_{ab}\frac{\delta\psi}{\delta\tilde{g}_{ab}}\right)_{,r}\right] = 0 \quad , \tag{4.25}$$

which is the statement of the invariance of ψ under reparametrizations which become the identity at infinity.

V. PROPERTIES OF THE WAVE EQUATION

The functional K appearing in (4.19) is not known in closed form, but, at the classical level, is the solution of a non-linear differential equation obtained by writing $\mathcal{H}_\perp = 0$ in terms of the variables $\tilde{g}_{ij}, \tilde{\pi}^{ij}, p, \tau$, namely[12,15]

$$\tilde{\Delta}\phi + \frac{1}{8}\tilde{\pi}_{ij}\tilde{\pi}^{ij}\,\phi^{-7} - \frac{1}{8}\tilde{R}\phi - \frac{3}{64}\tau^2\phi^5 = 0 \tag{5.1}$$

where $\phi = (-p)^{1/6} = g^{1/12}$, $\tilde{\pi}_{ij} = \tilde{g}_{il}\tilde{g}_{lj}\tilde{\pi}^{lk}$ and $\tilde{R}, \tilde{\Delta}$ are the curvature scalar and Laplacian for the metric $\tilde{g}_{ij}$. The existence and the uniqueness of solutions to (5.1) (with $\phi > 0$ everywhere and $\phi \to 1$ at infinity) have been studied.[16,17] It is found that solutions exist and are unique (for open spacelike sections) if and only if the following conditions hold:

(a) $\tilde{g}_{ij}$ is conformally related to a metric $\tilde{g}'_{ij}$ with the property that $\tilde{R}'<0$ at all points where $\tilde{\pi}^{ij} = 0$ but $\tau \neq 0$.

(b) $\tilde{g}_{ij}$ is conformally related to a metric $\tilde{g}'_{ij}$ with the property that $\tilde{R}'>0$ at all points where $\tau = 0$ but $\tilde{\pi}^{ij} \neq 0$.

(c) $\tilde{g}_{ij}$ is conformally related to a metric $\tilde{g}'_{ij}$ such that $\tilde{R}' = 0$ at points where $\tilde{\pi}^{ij} = 0$ and $\tau = 0$.

If $\tilde{\pi}^{ij}(x) \neq 0$ and $\tau(x) \neq 0$ for all x then a unique solution always exists.

When $\tilde{g}_{ij}$ does not satisfy any one of conditions (a), (b), or (c) the constraint $\mathcal{H}_\perp = 0$ cannot be solved classically. This means that the functional K_T appearing in the constraint

$$\pi_T + K_T[\tilde{g}_{ij},\tilde{\pi}^{ij},T] = 0 \tag{5.2}$$

does not exist for those $\tilde{g}_{ij}$. This fact will have consequences for the quantum equations. To see how this comes about let us assume that the state ψ is such that

$$\tilde{\pi}^{ij}_{op}(x_o)\psi = 0 \tag{5.3}$$

for some point x_o. If we look at the wave equation obtained from (5.2), namely

$$\frac{\hbar}{i}\frac{\delta\psi}{\delta T(x)} + K_T(x)[\tilde{g}_{ij},\tilde{\pi}^{ij}_{op},T]\psi = 0 \quad . \tag{5.4}$$

We see that K_T has no definite limit when the function variable $\tilde{g}_{ij}(x)$ approaches a function that does not fulfill either condition (a) or (c) at x_o. For the many-time rate of change $(\delta\psi/\delta T(x))$ to be regular at all points x the state must satisfy the boundary condition

$$\psi[\tilde{g}_{ij},T] = 0 \tag{5.5}$$

for all g_{ij} not satisfying (a) or (c) at x_o whenever (5.3) holds. Similarly when $\tau(x_o) = 0$, i.e.

$$\Delta_b T\big|_{x_o} = 0 \tag{5.6}$$

we obtain the boundary condition

$$\psi[\tilde{g}_{ij},T] = 0 \tag{5.7}$$

whenever $\tilde{g}_{ij}$ does not satisfy either of (b) or (c) above.

Work of Brill[18] (see appendix A) on the solutions to (5.1) for the case of an axially symmetric surface of time symmetry suggests a possible physical interpretation for (5.5) and (5.7). Brill showed that, for this special case, a $\tilde{g}_{ij}$ for which K_T does not exist corresponds to a manifold that is either closed or has a singular scalar curvature or both. We interpret (5.5) and (5.7) in this case as meaning that the probability for a closed or singular manifold is zero. There appears to be no possibility for a quantum transition from an open topology to a closed one.

It is conceivable that the precise statement of this interpretation may have to be modified in the general case depending on the details of the singular behavior of K_T. However, we would not expect a drastic change in this interpretation to be necessary.

VI. MANY TIME INNER PRODUCT

We have seen that the problem of finding a first order many-time wave equation can be satisfactorily dealt with, at least in rough terms. We now turn to the examination of the inner product that goes with this equation.

The interpretation of the functional $\psi[\tilde{g}_{ij}, T(x)]$ is that of a probability amplitude for measuring the conformal three geometry $\tilde{g}_{ij}$ on the hypersurface* $T(x)$. We define the scalar product

$$(\Psi,\phi)\Big|_{T(x)} = \int \Psi^*[\tilde{g}_{ij},T]\,\Phi[\tilde{g}_{ij},T]\,\mathcal{D}\tilde{g}_{ij} \tag{6.1}$$

where the integration runs over all $\tilde{g}_{ij}$ for fixed $T(x)$, and $\mathcal{D}\tilde{g}_{ij}$ symbolizes some appropriate integration measure.

The problem that arises now is that $T(x)$ in (6.1) is not just any function of three coordinates (at least a priori), but it has to correspond to the quantum analog of a spacelike hypersurface. In fact, in all of our previous classical analysis we have assumed that the surface $T(x)$ was spacelike. This assumption is built into the canonical formalism at a fundamental level and cannot simply be dropped in passing to the quantum theory.

If we go back to the classical theory, there is no reason to expect that the hypersurface will be spacelike for any smooth function $T(x)$. To establish this property, if at all true, one needs to examine in detail the properties of T (or, more precisely, of $\tau = -\frac{4}{3}\Delta_b T$). In fact, in specific models it appears that an extra condition must be imposed on the function playing the role of many-time variable. Due to the high symmetry of these models the condition for the surface to be spacelike involves only the many time variable itself and is independent of the other canonical variables ($\tilde{g}_{ij}, \tilde{\pi}^{ij}$ in our case). There is no reason to expect that when the symmetry conditions are relaxed the spacelike condition would not also involve the dynamical modes. If this were the case we would run into

---

*The use of the word hypersurface here is actually bad terminology. In fact in the quantum theory one cannot speak any longer about a sharply defined spacetime. Consequently the various $\tilde{g}_{ij}$ cannot be regarded as describing the conformal structure of a generic three dimensional cut through a four dimensional spacetime which solves Einstein's equations. We will indulge however in the common practice of calling quantum objects by the name of their classical analogs, hoping that not too much confusion will arise.

serious difficulty with the inner product (6.1) because the integration range for $\tilde{g}_{ij}$ would depend on T(x) and, worse than that, possibly on the state ψ itself (through $\tilde{\pi}^{ij} \rightarrow \tilde{\pi}^{ij}_{op}\,\psi$), a feature that would make (6.1) of no use as an inner product.

Fortunately a remarkable property of the extrinsic curvature scalar τ is that the surface is spacelike everywhere, provided that τ is bounded. This is consequence of the fact that the solution ϕ of (5.1) never goes through zero if conditions (a)-(c) hold[16,17]. A simple geometrical argument which sheds light on this point is given in appendix B. Therefore the only restriction on the arguments of the functionals ψ,ϕ in (6.1) is that*

$$\Delta_{\flat} T(x) < \infty \tag{6.2}$$

and the integration over $\tilde{g}_{ij}$ can be carried out unrestricted.

The inner product (6.1) will be conserved under arbitrary deformations provided that we can find an integration measure that makes the Hamiltonian density K_T Hermitian. A definite answer to this problem of course requires a solution to the factor ordering problem (which in turn has to be reconsidered to every order of approximation in the perturbation expansion for K_T used to solve (5.1)). We will not address ourselves to this difficult problem here; so we merely assume that a solution exists. Nevertheless it is appropriate to mention that the measure $\mathfrak{D}\tilde{g}_{ij}$ should restrict the integration to the subspace $\tilde{g} = 1$ for ψ's restricted to that subspace. Alternatively, one may extend the domain of definition of ψ outside the space $\tilde{g} = 1$ by taking ψ to be the same for any two conformally related $\tilde{g}_{ij}$. It is then permissible to integrate over all $\tilde{g}_{ij}$, without the restriction $\tilde{g} = 1$, provided one uses a conformally invariant measure.

VII. CLOSED SPACES

We summarize here some preliminary results that partially establish for closed spaces conclusions analogous to those presented above for open spaces.

For closed spaces (spacelike sections are compact, without boundary) we choose our basic variables to be (2.8)-(2.11). There are no surface terms to worry about, no deformations at infinity to describe; so we do not need variables such as T and π_T. The variable $\tau(x)$ itself is used to describe the global location of spacelike surfaces in the ambient spacetime. This means that if we can solve the constraint $\mathcal{H}_\perp = 0$ in the form

*We write $\Delta_{\flat}T<\infty$ instead of $T<\infty$ because there is at least one important case of interest where T diverges at infinity (but $\Delta_{\flat}T$ remains bounded), namely an asymptotic boost where $T \xrightarrow[r\to\infty]{} \underset{\sim}{\beta}\cdot\underset{\sim}{x}$. This is, of course, excluded by our condition $N^{\perp} \rightarrow 1$ for large r, but our discussion can be easily generalized to admit any asymptotic Poincare transformation.

$$p + K[\tilde{g}_{ij}, \tilde{\pi}^{ij}, \tau] = 0 \quad , \tag{7.1}$$

the associated many-time wave equation

$$\frac{h}{i}\frac{\delta\psi}{\delta\tau(x)} + K\psi = 0 \tag{7.2}$$

will contain all the dynamics.

To derive (7.1) it must be true that, under a deformation, the original surface and the deformed surface coincide if and only if $\delta\tau \equiv 0$. This is equivalent to the statement that the lapse $N^{\perp}(x)$ is identically zero if and only if $\dot{\tau}(x) \equiv 0$ (our treatment here follows the arguments given for the open case). As before the lapse satisfies the equation

$$\Delta N^{\perp} = (g^{-1}\,\tilde{\pi}^{ij}\tilde{\pi}_{ij} + \frac{3}{16}\tau^2)N^{\perp} - \frac{3}{4}\dot{\tau}^2 \ . \tag{7.3}$$

With $\dot{\tau} \equiv 0$ this becomes

$$\Delta N^{\perp} = (g^{-1}\tilde{\pi}^{ij}\,\tilde{\pi}_{ij} + \frac{3}{16}\tau^2)N^{\perp} \ . \tag{7.4}$$

If we do not have both $\tilde{\pi}^{ij} = 0$ and $\tau^2 = 0$ everywhere, the unique solution to (7.4) is $N^{\perp} \equiv 0$ and the desired result follows. If $\tilde{\pi}^{ij} \equiv 0$ and $\tau^2 \equiv 0$ (i.e. the surface is time symmetric) the solution to (7.4) is not unique, any $N^{\perp} \equiv$ constant solves the equation. This is a sign that the functional K is multivalued at $\tilde{\pi}^{ij} \equiv 0$ and $\tau \equiv 0$. Indeed, (5.1) determines ϕ in this case only up to a constant multiplicative factor.[17] This multivaluedness could be removed by requiring K to be continuous in the limit $\tilde{\pi}^{ij} \to 0$, $\tau \to 0$ with $\tilde{g}_{ij}$ held constant. Such a procedure would be viable only if the limit exists (i.e. if the scale equation (5.1) does not have an isolated solution for $\tilde{\pi}^{ij} \equiv 0$, $\tau \equiv 0$). Further investigation is needed to elucidate this point. From now on we will assume that the surfaces under consideration are not time symmetric.

Detailed examination of the scale equation (5.1) shows[16,17] that a solution ϕ, that never goes through zero, exists provided the same conditions (a)-(c) of section V. hold. Therefore the boundary conditions (5.5), (5.7) on the state functional also apply to the case of a closed space. They are to be interpreted as assigning zero probability amplitude to a singular space.

The inner product associated with (7.2) is naively taken to be the same as the one for the open case ((6.1), with τ replacing T). However, as is well known, difficulties arise here in relation to reparametrization invariance because a closed manifold cannot in general be covered by a single coordinate patch. The spacelike character of the surface follows from the fact that the solution of the scale equation never goes through zero. Again, the arguments of Appendix B concerning the boundedness of τ provide additional insight.

Summarizing, all the results obtained for an open space carry over to the closed case with the exception that the many-time Hamiltonian density K may be ill-defined on a surface of time symmetry.

Acknowledgements. It is a pleasure to thank Tullio Regge for very many discussions and John Wheeler for much encouragement. The kind hospitality of Professors Abdus Salam and Gallieno Denardo at Trieste is also gratefully acknowledged.

APPENDIX A

INTERPRETATION OF NON-ALLOWED CONFORMAL GEOMETRIES

We give here the basis of our interpretation of conditions (5.5), (5.7). The argument is for the special case of an axially symmetric surface of time symmetry. Our treatment is an essentially verbatim (although abbreviated) reproduction of Brill's treatment.[18]

For a time symmetric surface eq.(5.1) becomes

$$\tilde{\Delta}\psi - \frac{1}{8}\tilde{R}\psi = 0 \tag{A.1}$$

(we are using ψ where we used ϕ before so as to cause no confusion with the coordinate ϕ). An axially symmetric metric g_{ij} has the form

$$ds^2 = e^{2q}(d\rho^2+dz^2)+\rho^2 d\phi^2 \tag{A.2}$$

where ρ, z, and ϕ are curved space cylindrical coordinates. The metric (A.2) does not have $\tilde{g} = 1$. This is all right because both equation (A.1) and conditions (i)-(iii) below are conformally invariant (provided the conformal factor goes to unity at infinity); so we can always conformally transform (A.2) to a metric with $\tilde{g} = 1$ without affecting our results.

The function q satisfies the following boundary conditions (determined by axial symmetry and condition (3.4))

$$q = 0 \quad \text{at } \rho = 0 \quad , \tag{A.3}$$

$$q_{,\rho} = 0 \quad \text{at } \rho = 0 \quad , \tag{A.4}$$

$$q_{,\phi} = 0 \quad \text{everywhere} \quad , \tag{A.5}$$

$$q \to 0 \quad \text{as } r = \sqrt{\rho^2+z^2} \to \infty \quad . \tag{A.6}$$

ψ satisfies the conditions

$$\psi_{,\rho} = 0 \quad \text{at } \rho = 0 \text{ ("no cusp")} \quad , \tag{A.7}$$

$$\psi \to 1 \quad \text{as } r \to \infty \quad . \tag{A.8}$$

The scalar curvature is computed to be

$$\tilde{R} = -2e^{-2q}(q_{,\rho\rho} + q_{,zz}) \tag{A.9}$$

and equation (A.1) becomes

$$(\nabla^2 + \tfrac{1}{4}(q_{,\rho\rho} + q_{,zz}))\psi = 0 \tag{A.10}$$

where

$$\nabla^2\psi \equiv \psi_{,\rho\rho}+\psi_{,\phi\phi}+\psi_{,zz} \quad . \tag{A.11}$$

We define a "potential" by

$$V = -\frac{1}{4}(q_{,\rho\rho}+q_{,zz}) \quad . \tag{A.12}$$

Equation (A.10) is now

$$(\nabla^2 - V)\psi = 0 \quad . \tag{A.13}$$

(A.12) is the 2-dimensional Poisson equation with source -4V and (A.13) is the Schroedinger equation with potential V. Since q satisfies the greatest number of boundary conditions, we choose to specify it and solve (A.12) and (A.13) for V and ψ. Varying q corresponds to choosing different $\tilde{g}_{ij}$.

Solutions to (A.13) (ignoring the asymptotic boundary condition on ψ) are of three types:

(i) $\psi \to 1$ at infinity and is nowhere zero. When the choice of q gives a solution of this type, $\tilde{g}_{ij}$ is allowed in the sense of conditions (a)-(c) in section V.

(ii) $\psi \to 0$ at infinity and is nowhere zero.

(iii) ψ is zero for finite r and either $\psi \to 1$ or 0 at infinity.

Types (ii) and (iii) correspond to non-allowed $\tilde{g}_{ij}$, i.e. for those $\tilde{g}_{ij}$ we cannot find a solution to (5.1) meeting the conditions $\psi > 0$ and $\psi \to 1$ as $r \to \infty$. We get solutions of type (i) when the "potential" is weak; types (ii) and (iii) result from stronger potentials.

Type (ii) corresponds to a universe that is closed. The metric is given by $g_{ij} = \psi^4\tilde{g}_{ij}$, and, since $\psi \to 0$ at infinity, all points at infinity are separated by zero distance (remember that we are on a spacelike surface and the metric has signature (+++)). By introducing the variable $r' = {}^1/_r$ one can check that the universe closes smoothly in all directions.

Type (iii), with $\psi = 0$ at some finite distance, corresponds to a singularity in the scalar curvature. The argument we give is due to Bargmann and Wheeler as quoted by Brill.[18] Let p_o be a point such that $\psi(p_o) = 0$ and take a sphere S of radius ε about p_o. If we assume that ψ is analytic in r then $\psi \propto r$ for ε small enough (r and ε are proper distances in the metric $\tilde{g}_{ij}$). The radius r_g of the sphere in the metric $g_{ij} = \psi^4\tilde{g}_{ij}$ has the property

$$r_g \sim \int_o^\varepsilon \psi^2(\text{slowly varying function})\, dr \propto \int_o^\varepsilon r^2 dr \sim \varepsilon^3 \quad . \tag{A.14}$$

Let $d\tilde{A}$ and dA be the surface elements on the sphere in the metrics $\tilde{g}_{ij}$ and g_{ij} respectively. The area of the sphere is given by

$$A_g = \int_s h^{\frac{1}{2}} dA \sim \int_s \psi^4_{(\varepsilon)}\, d\tilde{A} \sim \psi^4_{(\varepsilon)} \int_s d\tilde{A} \sim \varepsilon^4 \tag{A.15}$$

where h is the determinant of the metric h_{ij} induced on the sphere by g_{ij} (h_{ij} is 2×2, $h_{ij} = \psi^4 \tilde{h}_{ij}$). Now take the ratio of the area of the sphere to the radius squared, and the limit as the radius goes to zero.

$$\lim_{r_g \to 0} \left(A_g / r_g^2\right) \propto \lim_{r_g \to 0} \left(r_g^{4/3}\right) / r_g^2 \to \infty \quad ; \tag{A.16}$$

so the curvature is infinite at p_o.

APPENDIX B

BEHAVIOR OF τ NEAR THE LIGHT CONE

We show by a direct geometrical argument that τ blows up when the surface becomes null.

We have a spacelike surface S and would like to compute the quantity $\tau = \frac{2}{3} g^{-\frac{1}{2}}$ at the point p_o of S' in a Riemann normal coordinate system. In this coordinate system the metric is the flat metric $\eta_{\mu\nu}$ and the Christoffel symbols $\Gamma^{\alpha}_{\beta\gamma}$ are zero at the point p_o.

The equation for the surface is (in a neighborhood about p_o)

$$t = S(x^i) \tag{B.1}$$

where t is the time coordinate and x^i the space coordinates of this coordinate system. The (non-unit) normal is

$$\overset{*}{n}_\mu = (-1, S_{,j}) \; ; \; n^{*\mu} = (1, S_{,j}) \quad . \tag{B.2}$$

For the surface to be spacelike at p_o we must have

$$\eta_{\mu\nu} n^{*\mu} n^{*\nu} < 0 \text{ (i.e. } \overset{*}{n}_\mu \text{ is timelike)}$$

or

$$1 - S_{,i} S_{,i} > 0 \tag{B.3}$$

where there is a summation over repeated indices. The metric on the surface at p_o is

$$g_{ij} = \delta_{ij} - S_{,i} S_{,j} \tag{B.4}$$

and

$$g \equiv \det g_{ij} = 1 - S_{,i} S_{,i} > 0 \quad . \tag{B.5}$$

The unit normal is

$$n_\mu = g^{-\frac{1}{2}}(-1, S_{,j}) \; ; \; n^\mu = g^{\frac{1}{2}}(1, S_{,j}) \tag{B.6}$$

with

$$\eta_{\mu\nu} n^\mu n^\nu = -1 \quad .$$

The extrinsic curvature is defined by

$$K_{\alpha\beta} = P^\mu_\alpha P^\nu_\beta n_{\mu;\nu} \tag{B.7}$$

where

$$P^{\alpha}_{\beta} = \delta^{\alpha}_{\beta} + n^{\alpha} n_{\beta} \quad . \tag{B.8}$$

Computing $K_{\alpha\beta}$ in our coordinate system we find

$$\begin{aligned} K_{oo}(p_o) &= g^{-5/2} S_{,i}S_{,j}S_{,ij} \\ K_{ok}(p_o) &= -g^{-3/2} S_{,i}S_{,ik} - g^{-5/2} S_{,i}S_{,j} S_{,ij} S_{,k} \\ K_{mn}(p_o) &= g^{-\frac{1}{2}}S_{,mn} + g^{-3/2} S_{,i} S_{,im} S_{,n} \\ &\quad + g^{-3/2} S_{,i}S_{,in}S_{,m} + g^{-5/2} S_{,i}S_{,j}S_{,ij} S_{,m}S_{,n} \ . \end{aligned} \tag{B.9}$$

To find τ we need to know the trace of $K_{\alpha\beta}$. It is

$$K(p_o) = S_{,mm}g^{-\frac{1}{2}} + S_{,i}S_{,j} S_{,ij} g^{-3/2} \ . \tag{B.10}$$

K is related to π by $\pi = 2g^{\frac{1}{2}}K$; so

$$\tau(p_o) = \frac{4}{3} K(p_o) = \frac{4}{3}(S_{,mm} g^{-\frac{1}{2}} + S_{,i}S_{,j}S_{,ij} g^{-3/2}) \ . \tag{B.11}$$

When the surface becomes null at p_o, $g \to 0$ and $\tau \to \infty$. If $\tau(p_o)$ is infinite and the surface is smooth (i.e. $S_{,ij}(p_o)$ is finite) then $g(p_o) = 0$ and the surface is null at p_o . We conclude that as long as τ remains bounded the surface cannot become null. The argument fails if $S_{,mn}(p_o) = 0$, namely when the surface is locally time symmetric. In that case one has to rely on the positivity of the solution of the scale equation to guarantee that the surface is spacelike.

REFERENCES

[1.] P.A.M. Dirac, Proc. Roy. Soc., A246, 333 (1958).
R. Arnowitt, S. Deser, and C. W. Misner, in Gravitation: An Introduction to Current Research, edited by L. Witten (Wiley, New York, 1962).

[2.] P.A.M. Dirac, Lectures on Quantum Mechanics, Belfer Graduate School of Science (Yeshiva University, New York, 1964).

[3.] A. Hanson, T. Regge, and C. Teitelboim, Constrained Hamiltonian Systems (Accademia Nazionale Lincei, Rome, 1976).
C. Teitelboim, mimeographed lecture notes (1976).

[4.] C. Teitelboim, Ann. Phys. (N.Y.), 79, 542 (1973).
C. Teitelboim, The Hamiltonian Structure of Spacetime, Ph.D. Thesis, Princeton University, 1973 (unpublished).

[5.] S. Hojman, K. Kuchař, and C. Teitelboim, Nature Phys. Sci., 245, 97 (1973).
S. Hojman, K. Kuchař, and C. Teitelboim, Ann. Phys. (N.Y.), 96, 88 (1976).

[6]. K. Kuchař, Phys. Rev. D, 4, 955 (1971).

[7.] C. Teitelboim, Phys. Lett., 56B, 376 (1975).
[8.] T. Regge, and C. Teitelboim, Phys. Lett., 53B, 101 (1974).
T. Regge, and C. Teitelboim, Ann. Phys. (N.Y.), 88, 286 (1974).
[9.] B. S. DeWitt, Phys. Rev., 160, 1113 (1967).
[10.] J. A. Wheeler, "Superspace," in Analytic Methods in Mathematical Physics, edited by R. D. Gilbert and R. Newton (Gordon and Breach, New York, 1970).
[11.] P.A.M. Dirac, Phys. Rev., 114, 924 (1958).
[12.] J. W. York, Phys. Rev. Lett., 28, 1082 (1972).
[13.] T. Regge, and C. Teitelboim, these Proceedings.
[14.] K. Kuchař, J. Math. Phys., 11, 3322 (1970).
[15.] A. Lichnerowicz, J. Math. Pure Appl., 23, 37 (1944).
[16.] N. O'Murchadha, and J. W. York, J. Math. Phys., 14, 1551 (1973).
[17.] N. O'Murchadha, Existence and Uniqueness of Solutions to the Hamiltonian Constraint of General Relativity, Ph.D. Thesis, Princeton University, 1973 (unpublished).
[18.] D. Brill, Ann. Phys. (N.Y.), 7, 466 (1959).
D. Brill, Time Symmetric Solutions of the Einstein Equations: Initial Value Problem and Positive Definite Mass, Ph.D. Thesis, Princeton University, 1959 (unpublished).

A GEOMETRIC CONSTRUCTION OF THE SYMPLECTIC FORM IN GENERAL RELATIVITY

Wiktor Szczyrba*

I. Introduction

The concept of the canonical quantization plays an important role in passing from classical to quantum systems. This procedure, well known in mechanics is based on the Hamilton (canonical) formulation of physical systems with finite degrees of freedom [1], [3.1]. In recent fifteen years an elegant geometric approach to the Hamilton formalism has been found [1], [3 1]. In this formulation we consider a 2n-dimensional manifold $\mathcal{P}_{2n}$ - the phase space of the system and a closed non-degenerate differential 2-form Ω on $\mathcal{P}_{2n}$. The form Ω defines a Lie algebra structure in the set $\mathcal{F}$ of all smooth functions on $\mathcal{P}_{2n}$. In many examples of this theory the manifold $\mathcal{P}_{2n}$ is simply the co-tangent bundle of an n-dimensional manifold V (the configuration space). In this case Ω is the canonical 2-form on T*V (of. [1], [8], [3 1]) and if (q^i) are local coordinates in V, (p_i, q^i) are local coordinates in $\mathcal{P}_{2n}$ = T*V then

$$\Omega = \sum_{i=1}^{n} dp_i \wedge dq^i \tag{1.1}$$

The form Ω defines an isomorphism between the bundles $T(\mathcal{P}_{2n})$ and $T^*(\mathcal{P}_{2n})$. If $X \in T_p(\mathcal{P}_{2n})$ then this isomorphism is given by:

$$T_p(\mathcal{P}_{2n}) \ni X \rightarrow X^b \in T_p^*(\mathcal{P}_{2n}), \text{ where } \langle Y, X^b \rangle = -\Omega(X,Y), \; Y \in T_p(\mathcal{P}_{2n}) \tag{1.2}$$

We denote the inverse mapping by # .

$$\text{For } f_1, f_2 \in \mathcal{F} = C^\infty(\mathcal{P}_{2n}) \; \{f_1, f_2\} = \Omega((df_1)^\#, (df_2)^\#) \tag{1.3}$$

In local coordinates we have the known classical formula:

$$\{f_1, f_2\} = \sum_{i=1}^{n} \left(\frac{\partial f_1}{\partial p_i} \frac{\partial f_2}{\partial q_i} - \frac{\partial f_2}{\partial p_i} \frac{\partial f_1}{\partial q^i}\right) \tag{1.4}$$

The above formulation of mechanics is not convenient for a generalization for the case of field theory. A more suitable is the homogeneous approach. We consider an 2n+1 dimensional submanifold $\mathcal{P}_{hom}$ of the bundle $T^*(V^*\mathbb{R})$ given by the constraint equation

$$H = H(p_i, q^j, t) \tag{1.5}$$

*Permanent Address: Institute of Mathematics, Polish Academy of Sciences 00-950 Warsaw, Poland.

where t is the global coordinate in $\mathbb{R}$ and -H is the conjugate coordinate in $T^*(V\times\mathbb{R})$. The canonical 2-form on $T^*(V\times\mathbb{R})$ generates a closed differential 2-form ω_{hom} on $\mathcal{P}_{hom}$. In local coordinates

$$\omega_{hom} = \sum_{i=1}^{n} dp_i \wedge dq^i - dH \wedge dt \tag{1.6}$$

The form ω_{hom} is degenerate and therefore we can distinguish a family $\mathcal{H}$ of 1-dimensional trajectories (submanifolds) of $\mathcal{P}_{hom}$ such that for $C \in \mathcal{H}$ and any vector field X tangent to $\mathcal{P}_{hom}$ and defined on C we have

$$(X \rfloor \omega_{hom}) | C = 0 \tag{1.7}$$

(| denotes the pull-back of the 1-form $X \rfloor \omega_{hom}$ onto manifold C). It is easy to know that elements of $\mathcal{H}$ are in one to one corerspondence with trajectories in V. If C is parametrized by the coordinate t

$C = \{(p_i(t), q^i(t), t) \in \mathcal{P}_{hom} : t \in \mathbb{R}\}$ then

$$\frac{dq^i}{dt} = \frac{\partial H}{\partial p_i}, \quad \frac{dp_i}{dt} = -\frac{\partial H}{\partial q^i} \quad i = 1, \ldots, n \tag{1.8}$$

The space $\mathcal{H}$ is 2n-dimensional because it can be parametrized by initial values of coordinates and momenta and it is diffeomorphic to $\mathcal{P}_{2n}$ (different instants of time give different diffeomorphisms). Therefore the space $\mathcal{H}$ carries a natural symplectic structure induced by the 2-form Ω in $\mathcal{P}_{2n}$. This structure is independent of a choice of an initial instant of time (motion is a symplectic diffeomorphism).

The homogeneous formulation of mechanics can be generalized for multidimensional cases i.e. for field theories. This generalization is based on the geometric theory of the calculus of variations (cf. [8], [20], [32]). It turns out that for any variational problem with a fixed boundary in the space-time M there exists a bundle $\mathcal{P}$ over M and a closed 5-form $\overset{5}{\omega}$ on $\mathcal{P}$.

The equation

$$(X \rfloor \overset{5}{\omega}) | C = 0 \tag{1.9}$$

where C is a 4-dimensional submanifold of $\mathcal{P}$ (the image of a section of $\mathcal{P}$) and X is an arbitrary vector field on C (tangent to $\mathcal{P}$), gives a family $\mathcal{H}$ of submanifolds in $\mathcal{P}$. The elements of $\mathcal{H}$ are in one to one correspondence with the set of solutions of field equations of our variational problem. The multisymplectic approach to the field theory has been investigated in papers [19], [22], [32], where it was pointed out that the multisymplectic formulation is more useful then the lagrangian one. In the present paper we construct the multisymplectic structure for General Relativity despite of non-existence of a covariant lagrangian density depending on first derivatives of the metric tensor.

Having the multisymplectic structure of the given field theory i.e. the manifold $\mathcal{P}$ and the closed 5-form $\overset{5}{\omega}$ one can define a presymplectic structure in the space of states $\mathcal{H}$. Such a construction has been recently done by J. Kijowski and the author [23]. It means that we endow the space $\mathcal{H}$ with a pseudo-differential structure and define on $\mathcal{H}$ a closed differential 2-form Ω. In theories with a gauge the form Ω is degenerate. Its degeneracy distribution W is involutive and therefore we can try to construct the space $\tilde{\mathcal{H}}$ such that $T(\tilde{\mathcal{H}}) = T(\mathcal{H})/W$. Projecting Ω onto $\tilde{\mathcal{H}}$ we obtain a non-degenerate closed 2-form $\tilde{\Omega}$. Such a structure enables to define a Lie algebra structure in the set of smooth functionals (physical quantities) on $\tilde{\mathcal{H}}$. But the degeneracy of Ω implies restrictions on the set of functionals to be considered. We can consider only gauge independent functionals i.e. functionals which can be projected on $\tilde{\mathcal{H}}$. For instance in electrodynamics the degeneration of Ω corresponds to an invariance of the Maxwell equations with respect to the gradient gauge $A_\mu \to A_\mu + \partial_\mu \psi$ and the potentials A_μ do not define physical quantities but the field strengths B and E do (cf. [23]).

In the present paper we give the construction of the form Ω in gravidynamics. First at all, we construct a multisymplectic manifold $(\mathcal{P}, \overset{5}{\omega})$ such that sections of the bundle $\tau : \mathcal{P} \to M$ satisfying (1.9) are in one to one correspondence with the set of solutions of the Einstein equations:

$$\Gamma^{\lambda}_{\mu\nu} = \frac{1}{2} g^{\lambda\alpha} (\partial_\mu g_{\nu\alpha} + \partial_\nu g_{\mu\alpha} - \partial_\alpha g_{\mu\nu}) \tag{1.10}$$

$$R_{\mu\nu} - \frac{1}{2} g_{\mu\nu} R + \lambda g_{\mu\nu} = 0 \tag{1.11}$$

Having this structure we define the presymplectic form Ω according to the general formula given in [23]. It turns out that the diagonalization of the form Ω distinguish the ADMW coordinate system (cf. [3], [33]). If $\sigma \subset M$ is a space-like surface and g_{ij}, K_{ij} are its first and second fundamental forms then the form Ω has a diagonal expression in terms of the infinitesimal translations δg_{ij}, $\delta\pi^{ij}$. The quantities π^{ij} are components of a tensor density on σ and are defined by:

$$\pi^{ij} = -\sqrt{g}\,(K_{pq} - g_{pq}\,\mathrm{tr}K)\,\bar{g}^{ip}\bar{g}^{jq} \tag{1.12}$$

Using the diagonal expression for Ω we prove that its degeneration distribution W is determined by an action of the diffeomorphism group of M in the space $\mathcal{H}$. If we divide $\mathcal{H}$ by this action we obtain the superphase space $\tilde{\mathcal{H}}$ for General Relativity.

The presented approach elucidates some classical problems of General Relativity. In particular, the geometric character of the ADMW coordinates is proved. We discuss also the problem of degrees of freedom for gravitational field and show that their numbers is equal to four (in the phase space).

The considered here problems are classical and were investigated earlier by Bergmann, Dirac, Wheeler, Arnowitt-Deser-Misner, De Witt, Fadeev, Fischer and Marsden. cf. [5], [33], [3], [9], [12], [13], [15], [18]. The detailed comparison between theirs and ours results will be given elsewhere.

This paper summarizes only our main results. The complete exposi tion with proofs will appear in Communications in Mathematical Physics.

2. The Multisymplectic structure of General Relativity

Let M be a 4-dimensional smooth manifold - the space-time and λ be a real number - the cosmological constant. Let $S^2_*T^*M$ be the bundle of symmetric 2-covariant non-degenerate tensors (metrics) on M with the negative determinant $g = \det g_{\mu\nu}$. We construct over $S^2_*T^*M$ the bundle of coefficients of the affine connection Aff. If $(x^\lambda, g_{\lambda\mu\nu})$ are local coordinates in $S^2_*T^*M$ then we have local coordinates $(x^\lambda, g_{\mu\nu}, \Gamma^\lambda_{\mu\nu})$ $(\Gamma^\lambda_{\mu\nu} = \Gamma^\lambda_{\nu\mu})$ in Aff with known transformation properties:

$$g_{\mu'\nu'} = \frac{\partial x^\mu}{\partial x^{\mu'}} \frac{\partial x^\nu}{\partial x^{\nu'}} g_{\mu\nu}$$

$$\Gamma^{\lambda'}_{\mu'\nu'} = \frac{\partial x^{\lambda'}}{\partial x^\lambda} \frac{\partial x^\mu}{\partial x^{\mu'}} \frac{\partial x^\nu}{\partial x^{\nu'}} \Gamma^\lambda_{\mu\nu} + \frac{\partial^2 x^\sigma}{\partial x^{\mu'} \partial x^{\nu'}} \frac{\partial x^{\lambda'}}{\partial x^\sigma} \qquad (2.1)$$

We put $\mathcal{P}$ = Aff and

$$\overset{4}{\theta} = g^{\alpha\beta} \sqrt{-g}\, dx^0 \wedge \cdots \wedge \overbrace{d\Gamma^\tau_{\alpha\beta}}^{\tau} \wedge \cdots \wedge dx^3 - g^{\alpha\tau} \sqrt{-g}\, dx^0 \wedge \cdots \wedge \overbrace{d\Gamma^\beta_{\beta\alpha}}^{\tau} \wedge \cdots \wedge dx^3 +$$
$$- (g^{\mu\rho}(\Gamma^\mu_{\tau\nu} \Gamma^\tau_{\mu\rho} - \Gamma^\mu_{\mu\tau} \Gamma^\tau_{\nu\rho}) + 2\lambda) \sqrt{-g}\, dx^0 \wedge \cdots \wedge dx^3 \qquad (2.2)$$

Proposition 1

The formula (2.2) is covariant with respect to the coordinate transformations (2.1) and therefore it defines a differential 4-form on $\mathcal{P}$. We define $\overset{5}{\omega} = d\overset{4}{\theta}$ (2.3)

The couple $(\mathcal{P}, \overset{5}{\omega})$ determines a multisymplectic structure of General Relativity. Let $f: M \to \mathcal{P}$ be a section of the bundle $\tau: \mathcal{P} \to M$ and X be a τ-vertical vector field on $C = f(M)$. We look for such sections of τ that for every X $(X \lrcorner \overset{5}{\omega})| C = 0$ (2.4)

($|$ denotes the pull-back of the differential 4-form $X \lrcorner \overset{5}{\omega}$ onto C).

In local coordinates we have $f(x^\lambda) = \{(x^\lambda, g_{\mu\nu}(x^\lambda), \Gamma^\tau_{\mu\nu}(x^\lambda))$ (2.5)
$X = \sum_{\mu\leqslant\nu} Q_{\mu\nu} \frac{\partial}{\partial g_{\mu\nu}} + \sum_{\mu\leqslant\nu} P^\tau_{\mu\nu} \frac{\partial}{\partial \Gamma^\tau_{\mu\nu}}$. It is easy to see that (2.4) implies

$$\Gamma^\lambda_{\mu\nu} = \frac{1}{2} g^{\lambda\alpha} (\partial_\mu g_{\nu\alpha} + \partial_\nu g_{\mu\alpha} - \partial_\alpha g_{\mu\nu}) \tag{2.6}$$

$$G_{\mu\nu} = R_{\mu\nu} - \frac{1}{2} g_{\mu\nu} R + \lambda g_{\mu\nu} = 0 \tag{2.7}$$

$$\text{Where } R^\beta_{\mu\alpha\nu} = \partial_\alpha \Gamma^\beta_{\mu\nu} + \Gamma^\beta_{\tau\alpha} \Gamma^\tau_{\mu\nu} - \Gamma^\beta_{\tau\nu} \Gamma^\tau_{\mu\alpha} - \partial_\nu \Gamma^\beta_{\mu\alpha}$$

$$R_{\mu\nu} = R^\alpha_{\mu\alpha\nu} \ ; \ R = g^{\mu\nu} R_{\mu\nu} \tag{2.8}$$

We have obtained

Proposition 2

Sections of $\tau : \mathcal{P} \to M$ satisfying (2.4) are in one to one correspondence with the set of solutions of the Einstein equations (2.6) and (2.7).

We can consider the equation (2.4) as a geometric formulation of the Hamilton equations. It is known (cf. [19], [20], [22], [32]) that for a field theory with a Lagrangian function it is always possible to construct a multisymplectic manifold by means of the Legendre transformation. In General Relativity we have no Lagrangian depending on first derivatives of a metric tensor $g_{\mu\nu}$ and therefore the multisymplectic approach have to be done axiomatically. We see that the Hamilton formulation is more appropriate for General Relativity than the Lagrangian formulation.

However, it is possible to give a non-covariant construction of the multisymplectic 5-form $\overset{5}{\omega}$. Let $\mathcal{L}(g_{\mu\nu}, \Gamma^\tau_{\mu\nu}) =$

$$\sqrt{-g}\ (g^{\nu\rho}(\Gamma^\mu_{\tau\nu} \Gamma^\tau_{\mu\rho} - \Gamma^\mu_{\mu\tau} \Gamma^\tau_{\nu\rho}) + 2\lambda) \tag{2.9}$$

be a non-covariant Lagrangian density for gravidynamics (cf. [2], [29]). Using the Dedecker formula (cf. [32])

$$\psi = \sum_{\mu\leqslant\nu} \frac{\partial\mathcal{L}}{\partial g_{\mu\nu,\lambda}} dx^0 \wedge \cdots \wedge \underbrace{dg_{\mu\nu}}_{\lambda} \wedge \cdots \wedge dx^3 - \Big(\sum_{\mu\leqslant\nu} \frac{\partial\mathcal{L}}{\partial g_{\mu\nu\ \lambda}} g_{\mu\nu\ \lambda} - \mathcal{L}\Big)\, dx^0 \wedge \cdots \wedge dx^3 \tag{2.10}$$

we obtain:

$$\psi = (g^{\alpha\tau} g^{\beta\zeta} \Gamma^\rho_{\tau\zeta} - \tfrac{1}{2}(g^{\alpha\tau} g^{\beta\rho} \Gamma^\zeta_{\zeta\tau} + g^{\alpha\rho} g^{\beta\tau} \Gamma^\zeta_{\zeta\tau}) - \tfrac{1}{2} g^{\alpha\beta} g^{\tau\zeta} \Gamma^\rho_{\tau\zeta} + \tfrac{1}{2} g^{\alpha\beta} g^{\tau\rho} \Gamma^\zeta_{\zeta\tau}) \sqrt{-g}\ dx^0 \wedge \cdots \wedge \underbrace{dg_{\alpha\beta}}_{\rho} \cdots \wedge dx^3 +$$

$$- (g^{\nu\rho}(\Gamma^\mu_{\tau\nu} \Gamma^\tau_{\mu\rho} - \Gamma^\mu_{\mu\tau} \Gamma^\tau_{\nu\rho}) + 2\lambda) \sqrt{-g}\ dx^0 \wedge \cdots \wedge dx^3 \tag{2.11}$$

It turns out, that the formal exterior differentiation of the non-covariant expression (2.11) gives the 5-form $\overset{5}{\omega}$ i.e. $d\psi = \overset{5}{\omega}$ (2.12)

The formula (2.12) together with proposition 2 give the connection between the classical formulation of General Relativity and the mulitsympletic approach presented above. It is interesting that the equation (2.4) can be obatined from the Palatini-variational principle (cf. [29]) for the action intergral

$$S(f) = \int_{f(M)} \overset{4}{\theta} = \int_{M} f^{*}\overset{4}{\theta} \qquad (2.13)$$

where $f: M \to \mathcal{P}$ is a section of $\mathcal{T}$.
(Note that $f^{*}\theta = (R - 2\lambda)\sqrt{-g}\, dx^{o}\wedge\cdots\wedge dx^{3}$) . The detailed discussion of this fact will be given elsewhere.

Having the multisympletic manifold $(\mathcal{P}, \overset{5}{\omega})$ we need an additional structure- a set $\mathcal{C}$ of 3-dimensional submanifolds in $\mathcal{P}$ which determine initial data for the field equations (2.6) and (2.7). We call elements of $\mathcal{C}$ admissible initial surfaces in $\mathcal{P}$ (cf. [22]) . They are lifts of space-like surfaces in M (for a given metric satisfying the Einstein equations) to the bundle $\mathcal{P}$. The description of the set $\mathcal{C}$ in a special coordinate system is given in the section 3.

The set $\mathcal{H}$ of 4-dimensional submanifolds of $\mathcal{P}$ which satisfy (2.4) is called the space of the states in gravidynamics or the pre-phase space.

3. The Cauchy problem and ADMW coordinates in General Relativity.

The Einstein equations form a system of second order differential equations for comonents of the metric tensor $g_{\mu\nu}$ on the manifold M. These equations have the hyperbolic character and we can consider the Cauchy problem for them. The discussion of the problem will be given the ADMW coordinate system (cf. [3], [29], [33]) . Let $g_{\mu\nu}$ be a metric tensor on M with a signature (-1,+3) and σ be a 3-dimensional space-like submanifold of M. We choose such a coordinate system in M that $\sigma = \{x \in M : x^{o} = \text{const.}\}$. The metric $g_{\mu\nu}$ on M generates on σ a positively defined metric tensor g_{ij} . We denote by $\bar{g}^{ij}$ the inverse matrix to g_{ij}. The inequality $g = \det g_{\mu\nu} < 0$ implies $g^{00} < o$. Therefore we put:

$$N = (-g^{oo})^{-\frac{1}{2}}$$

$$N_k = g_{ok}$$

$$N^k = \bar{g}^{kj} N_j$$

$$\bar{g} = \det g_{ij} > 0 \qquad (3.1)$$

It is known that N is a scalar function on σ (a lapse function) and N^k are components of a vector field on σ (a shift vector). We have also formulas:

$$g^{ok} = N^k/N^2$$

$$g_{oo} = -N^2 + N^k N_k$$

$$g^{sp} = \bar{g}^{sp} - N^s N^p/N^2$$

$$\sqrt{-g} = N\sqrt{\bar{g}} \qquad (3.2)$$

The normal unit vector to σ is given by

$$n^\mu = \left(\frac{1}{N}, -\frac{N^k}{N}\right) \qquad (3.3)$$

and the second fundamental form is defined by (cf. [24])

$$K_{ij} = -g_{j\mu} \nabla_i n^\mu \qquad (3.4)$$

In our special coordinate system

$$K_{ij} = -N \Gamma^o_{ij} \qquad (3.5)$$

The metric tensor g_{ij} on σ defines the Christoffel symbols $\bar{\Gamma}^k_{ij}$ and the covariant derivative $\bar{\nabla}_k$ on σ. We have

$$\partial_o g_{ij} = \bar{\nabla}_i N_j + \bar{\nabla}_j N_i + 2N^2 \Gamma^o_{ij} \qquad (3.6)$$

In our further considerations an important role is played by the ADMW tensor density on σ defined by:

$$\pi^{ij} = -\sqrt{\bar{g}}\,(K_{pq} - g_{pq} K_{rs} \bar{g}^{rs}) \bar{g}^{ip} \bar{g}^{jq} \qquad (3.7)$$

It is known (cf. [2], [29]) that the system of the Einstein equation (2.7) is equivalent to the system:

$$R_{ks} = \lambda g_{ks} \quad \text{on } M \tag{3.8a}$$

$$G^{o}_{\mu} = 0 \qquad \text{on } \sigma \tag{3.8b}$$

We consider a neighbourhood of σ in M which is diffeomorphic to $\sigma \times \mathbb{R}$ and take the system of coordinates $(N, N_k, M_\mu, M_{\mu k}, g_{ij}, \pi^{ij}, \bar{\Gamma}^{k}_{ij})$ (where $M_\mu = \partial_\mu N$, $M_{\mu k} = \partial_\mu N_k$) in the corresponding subset of $\mathcal{P}$. The equations (3.8) read:

$$\partial_o \pi^{ij} = - N\sqrt{g}\ (\bar{R}^{ij} - \bar{g}^{ij}\bar{R}) - \frac{2N}{\sqrt{g}}(\pi^{i}_{q}\ \pi^{qj} - \frac{1}{2}\ \mathrm{tr}\pi\ \pi^{ij}) +$$
$$+ \sqrt{g}\ (\bar{\nabla}^{i}\bar{\nabla}^{j}N - \bar{g}^{ij}\bar{\nabla}^{s}\bar{\nabla}_{s}N) + \bar{\nabla}_{u}(N^{u}\pi^{ij}) +$$
$$- \bar{\nabla}_{s}N^{i}\pi^{s}{}_{j} - \bar{\nabla}_{s}N^{j}\pi^{si} - 2\lambda N\sqrt{g}\ \bar{g}^{ij} \tag{3.9a}$$

$$\bar{\nabla}_{i}\ \pi^{ij} = 0 \quad \text{on } \sigma \tag{3.9b'}$$

$$\bar{R} - 2\lambda - \frac{1}{\bar{g}}(\pi_{pq}\ \pi^{pq} - \frac{1}{2}(\mathrm{tr}\pi)^{2}) = 0 \text{ on } \sigma \tag{3.9b''}$$

The equations (3.9b) are simply the constraints imposed on initial data g_{ij}, π^{ij} on σ. These equations do not involve N and N_k therefore we can assign N and N_k in an arbitrary way on σ. The dynamical equations (3.9a) do not contain the time-derivatives of N and N_k. Therefore having a positive defined metric g_{ij} on σ and a tensor density π^{ij} on σ which satisfy the constraints equations (3.9b) we must choose N and N_k in an neighbourhood of σ in M. In this case we have a unique solution for g_{ij} and π^{ij} solving the equations (3.6) and (3.9a). Technical details concerning this problem can be found in [6], [7], [15], [16], [21], [28].

We see that the surface σ in M together with quantities g_{ij}, π^{ij} N, N_k on it determine an admissible initial surface in $\mathcal{P}$. Such a surface generates a family of states i.e. a family of solutions of the Einstein equations. In the following we shall consider two cases: (a) σ-is a compact manifold without boundary; (b) M is asymptotically flat and in spatially distant points σ is a surface x^{o} = const. in the cartosian coordinate system $(x^{o}, x^{1}, x^{2}, x^{3})$. (It imples that $N_k = 0$, $N = 1$ in spatially remote points).

4. A Symplectic Structure of the Set of Einstein Metrics

We consider the space $\mathcal{H}$ of all 4-dimensional submanifolds of satisfying (2.4). This set is simply the set of all Einstein metrics in M (for the given λ). According to the general approach to infinite dimensional manifolds (df. [11], [23],) the tangent space

$T_C(\mathcal{H})$ to $\mathcal{H}$ at $C\varepsilon\mathcal{H}$ can be defined by means of curves in $\mathcal{H}$. Every such a curve generates a τ-vertical vector field X on C which satisfies the linearized system of Einstein equations. If X is given by

$$X = \sum_{\mu\leq\nu} \delta g_{\mu\nu} \frac{\partial}{\partial g_{\mu\nu}} + \sum_{\mu\leq\nu} \delta\Gamma^{\tau}_{\mu\nu} \frac{\partial}{\partial\Gamma^{\tau}_{\mu\nu}} \qquad (4.1)$$

then the components $\delta g_{\mu\nu}$ and $\delta\Gamma^{\tau}_{\mu\nu}$ satisfy:

$$\delta\Gamma^{\tau}_{\mu\nu} = \frac{1}{2} g^{\tau\alpha}(\nabla_{\mu}\delta g_{\nu\alpha} + \nabla_{\nu}\delta g_{\mu\alpha} - \nabla_{\alpha}\delta g_{\mu\nu}) \qquad (4.2a)$$

$$\delta(R_{\mu\nu} - \lambda g_{\mu\nu}) = \sum_{\alpha\leq\beta} \frac{\partial R_{\mu\nu}}{\partial g_{\alpha\beta}} \delta g_{\alpha\beta} + \sum_{\alpha\leq\beta} \frac{\partial R_{\mu\nu}}{\partial\Gamma^{\tau}_{\alpha\beta}} \delta\Gamma^{\tau}_{\alpha\beta} - \lambda\delta g_{\alpha\beta} = 0 \qquad (4.2b)$$

The set of vector fields on C of the type (4.1) satisfying (4.2) is denoted by $\tilde{T}_C(\mathcal{H})$. Of course, $T_C(\mathcal{H}) \subset \tilde{T}_C(\mathcal{H})$ and the problem of equality of these spaces is the problem of the linearization stability of the Einstein equations (cf. [17]). We do not consider it here Let $C\varepsilon\mathcal{H}$, $\sigma\subset M$ be an arbitrary space-like surface and $c<C$ be the admissible initial surface determined by σ and C. Let $\hat{X}_1, \hat{X}_2 \varepsilon \tilde{T}_C(\mathcal{H})$ and X_1, X_2 are vector fields of the type (4.1) representing $\hat{X}_1$, $\hat{X}_2$. We define a skew-symmetric mapping:

$$\tilde{T}_C(\mathcal{H}) \times \tilde{T}_C(\mathcal{H}) \ni (\hat{X}_1,\hat{X}_2) \to \Omega(\hat{X}_1,\hat{X}_2) = \int_c (X_1 \wedge X_2) \lrcorner \overset{5}{\omega} \qquad (4.3)$$

Of course, to provide the convergence of the integral in (4.3) we have to assume that appropriate boundary conditions in the spatial infinity are fulfilled (e.g. one of X_1, X_2 has a compact support on c).

The above construction for an arbitrary hyperbolic field theory has been discussed in [23] to which we refer for details.

Proposition 3 [23]
The integral (4.3) does not depend on a choice of a space-like surface $\sigma \subset M$ (of a choice of an admissible initial surface $c \subset C$).
It has been shown in the paper [23] that the space $\mathcal{H}$ can be endowed with a pseudo-differential structure and the formula (4.3) defines a differential 2-form Ω on $\mathcal{H}$.
Proposition 4 [23]

The form Ω is closed i.e. $d\Omega = 0$.
It can be also proved that Ω is exact. This result will be published elsewhere.

In the ADMW coordinate system a vector $\hat{X}$ at C is represented by a vector field X of the type:

$$X = \delta N \frac{\partial}{\partial N} + \delta N_k \frac{\partial}{\partial N_k} + \sum_{i\leq j} \delta g_{ij} \frac{\partial}{\partial g_{ij}} + \sum_{i\leq j} \delta\pi^{ij} \frac{\partial}{\partial \pi^{ij}} + \delta M_\mu \frac{\partial}{\partial M_\mu} +$$

$$+ \delta M_{\mu k} \frac{\partial}{\partial M_{\mu k}} + \sum_{i\leq j} \delta \bar{\Gamma}^k_{ij} \frac{\partial}{\partial \Gamma^k_{ij}} \quad . \tag{4.4}$$

where the following equations hold:

$$\delta M_\mu = \partial_\mu \delta N \ ; \ \delta M_{\mu k} = \partial_\mu \delta N_k \tag{4.5}$$

$$\delta \bar{\Gamma}^j_{ks} = \frac{1}{2} \bar{g}^{ja}(\bar{\nabla}_k \delta g_{sa} + \bar{\nabla}_s \delta g_{ka} - \bar{\nabla}_a \delta g_{ks}) \tag{4.6}$$

$$\partial_o \delta g_{ij} = \delta(\bar{\nabla}_i N_j + \bar{\nabla}_j N_i + \frac{2N}{\sqrt{\bar{g}}} (g_{ip} g_{jq} \pi^{pq} - \frac{1}{2} g_{ij} \operatorname{tr}\pi)) \tag{4.7}$$

$$\partial_o \delta\pi^{ij} = \delta(- N \sqrt{\bar{g}} (\bar{R}^{ij} - \bar{g}^{ij}\bar{R}) - \frac{2N}{\sqrt{\bar{g}}}(\pi^i{}_q \pi^{qj} - \frac{1}{2} \operatorname{tr} \pi\pi^{ij})) +$$

$$+ \delta(\sqrt{\bar{g}} (\bar{\nabla}^i \bar{\nabla}^j N - \bar{g}^{ij} \bar{\nabla}^s \bar{\nabla}_s N) + \bar{\nabla}_u (N^u \pi^{ij})) +$$

$$+ \delta(-\bar{\nabla}_s N^i \pi^{sj} - \bar{\nabla}_s N^j \pi^{si} - 2\lambda N \sqrt{\bar{g}}\, \bar{g}^{ij}) \tag{4.8a}$$

$$\delta(\bar{\nabla}_j \pi^{ij}) = \bar{\nabla}_j \delta\pi^{ij} + \delta\bar{\Gamma}^i_{ks} \pi^{ks} = 0 \text{ on } \sigma \tag{4.8b'}$$

$$\delta(\bar{R} - 2\lambda - \frac{1}{\bar{g}} (\pi_{pq} \pi^{pq} - \frac{1}{2}(\operatorname{tr} \pi)^2) = 0 \text{ on } \sigma \tag{4.8b''}$$

Therefore a vector $\hat{X} \varepsilon \check{T}_C(\mathcal{H})$ determine 12 quantities $(\delta\pi^{ij}, \delta g_{ij})$ on $c \subset C$ (on $\sigma \subset M$) which satisfy the constraints equations (4.8b) and four arbitrary quantities δN, δN_k given in a neighbourhood of c in C (in a neighbourhood of σ in M). On the other hand, if we have 12 quantities $(\delta\pi^{ij}, \delta g_{ij})$ on c satisfying the constraints equation (4.8b) and four quantities δN, δN_k in a neighbourhood of c in C we can determine the vector field X on C solving the equations (4.7) and (4.8a).

It turns out that in the ADMW coordinates the form Ω is diagonal:

Theorem 1

Let $\hat{X}_1$, $\hat{X}_2$ ε $\tilde{T}_C(\mathcal{H})$ be represented by vector fields X_1, X_2 of the type (4.4) then

$$\Omega (\hat{X}_1, \hat{X}_2) = \int_c (\underset{1}{\delta}\pi^{ij} \underset{2}{\delta} g_{ij} - \underset{2}{\delta}\pi^{ij} \underset{1}{\delta} g_{ij})\, dx^1 \wedge dx^2 \wedge dx^3 \tag{4.9}$$

This theorem shows that the components of the ADMW density π^{ij} can b be treated as the conjugate variables to the spatial components of the metric tensor $g_{\mu\nu}$. But these quantities are not independent, they satisfy the constraints equations (3.9b). These four equations give rise to the degeneration of the form Ω. We investigate this problem in the next section.

5. The Gauge Distribution, An Action fo the Diffeomorphism Group and the Superphase Space for General Relativity

We know that in mechanics the symplectic form (1.1) provides an isomorphism between the tangent and the cotangent bundle of the phase space $\mathcal{P}_{2n}$. The form Ω in General Relativity is degenerate and therefore the correspondence between tangent and cotangent vectors in $\mathcal{H}$ is not one to one. We define the gauge distribution of the form

$$W_C = \{\hat{Y}\varepsilon\tilde{T}_C(\mathcal{H}) : \Omega(\hat{X},\hat{Y}) = 0, \quad \hat{X}\varepsilon\tilde{T}_C(\mathcal{H})\} \quad , C \varepsilon \mathcal{H} \tag{5.1}$$

and two subspaces of $\tilde{T}_C(\mathcal{H})$:

Definition: $\overset{c}{\tilde{T}}_C(\mathcal{H}, c)$ is a linear subspace of $\tilde{T}_C(\mathcal{H})$ consisting of these $\hat{Y}\varepsilon\tilde{T}_C(\mathcal{H})$ which are represented by vector fields Y on C of the type (4.4) such that:

1^0 δN, δN_k, δM_0, δM_{ok} are arbitrary on c

2^0 $\delta M_k = \delta_k\, \delta N$, $\delta M_{sk} = \delta_s\, \delta N_k$ on c

3^0 $\delta\pi^{ij} = 0$, $\delta g_{ij} = 0$ on c (5.2)

Definition: $\overset{1}{\tilde{T}}_C(\mathcal{H}, c)$ is the linear subspace of $\tilde{T}_C(\mathcal{H})$ consisting of vectors which are represented by vector fields on C of the type (4.4) such that:

$$\delta N = 0, \quad \delta N_k = 0 \quad \text{on } C \tag{5.3}$$

Proposition 5

$$\tilde{T}_C(\mathcal{H}) = \overset{c}{\tilde{T}}_C(\mathcal{H}, c) \oplus \overset{1}{\tilde{T}}_C(\mathcal{H}, c) \quad \text{(a direct sum)}$$

Propos ition 6

$$\overset{c}{\tilde{T}}_C(\mathcal{H}, c) \subset W_C .$$

By vitrue of propositions 5 and 6 we can consider only the subspace $\overset{1}{\tilde{T}}_C(\mathcal{H},c)$. Let $C_\sigma = C^\infty(\mathrm{den}\, S^2T(\sigma) \oplus S^2T^*(\sigma))$ be a subspace consisting of couples $\mathcal{X} = (\delta\pi^{ij}, \delta g_{ij})$, where δg_{ij} is a 2-covariant symmetric tensor field on σ and $\delta\pi^{ij}$ is a 2-contravariant symmetric tensor density on σ. This space has the natural scalar product:

$$g(\sigma, C)(\mathcal{X}_1, \mathcal{X}_2) = \int_\sigma \left(\frac{1}{\sqrt{\bar g}} \underset{1}{\delta\pi}^{ij} \, g_{ip} g_{jq} \underset{2}{\delta\pi}^{pq} + \sqrt{\bar g} \, \underset{1}{\delta g}_{ij} \bar g^{ip} \bar g^{jq} \underset{2}{\delta g}_{pq} \right) dx^1 \wedge dx^2 \wedge dx^3 \tag{5.4}$$

Let J be the operator $C_\sigma \to C_\sigma$; $J(\delta\pi^{ij}, \delta g_{ij}) =$

$$= \left(\sqrt{\bar g}\, \delta g_{pq} \, \bar g^{pi} \bar g^{qj}, \; -\frac{1}{\sqrt{\bar g}} g_{ip} g_{jq} \, \delta\pi^{pq}\right) \tag{5.5}$$

The operator J has the property $J^2 = -\,\mathrm{id}$.

Let $g_1(\sigma, C)$ be the scalar product in the space $C^\infty(T(\sigma) \oplus \mathbb{R})$ consisting of couples $U = (u^j, \chi)$, where u^j are components of a vector field on σ and χ is a scalar function on σ .

$$g_1(\sigma, C)(U_1, U_2) = \int_\sigma \left(\underset{1}{u}_j \underset{2}{u}^j \sqrt{\bar g} + \underset{1}{\chi} \underset{2}{\chi} \sqrt{\bar g} \right) dx^1 \wedge dx^2 \wedge dx^3 \tag{5.6}$$

The constraints equations (4.8b) generate a differential operator A: $C^\infty(\mathrm{den}\, S^2 T(\sigma) \oplus S^2 T^*(\sigma)) \to C^\infty(T(\sigma) \oplus \mathbb{R})$. For $\chi = (\delta\pi^{ij}, \delta g_{ij})$ we have:

$$A\mathcal{X} = \left(\frac{1}{\sqrt{\bar g}} (\bar\nabla_j \, \delta\pi^{ij} + \delta\bar\Gamma^i_{ks} \, \delta\pi^{ks}), \; \delta\bar R + \frac{1}{\sqrt{\bar g}} (\bar R - 2\lambda)\delta\bar g - \frac{1}{\sqrt{\bar g}} \delta(\pi^{ks}\pi_{ks} - \frac{1}{2}(\mathrm{tr}\pi)^2)\right) \tag{5.7}$$

By means of the scalar products (5.4) and (5.6) we define the adjoint operator

$$A^*: C\,(T(\sigma) \oplus \mathbb{R}) \to C\,(\mathrm{dem} S^2 T(\sigma) \oplus S^2 T^*(\sigma))$$

$A^*(u^j, \chi) = (\delta\pi^{ij}, \delta g_{ij})$, where

$$\delta\pi^{ij} = -\frac{1}{2}(\bar\nabla^i u^j + \bar\nabla^j u^i)\sqrt{\bar g} - 2(\pi^{ij} - \frac{1}{2}\mathrm{tr}\pi \, \bar g^{ij})\chi$$

$$\delta g_{ij} = -\frac{1}{2\sqrt{\bar g}} (\pi_{ai} \bar\nabla^a u_j + \pi_{aj} \bar\nabla^a u_i - \bar\nabla_a(\pi_{ij} u^a)) + - \frac{2}{g}(\pi_{is}\pi^s{}_j - \frac{1}{2}\mathrm{tr}\,\pi\pi_{ij})\chi + \bar\nabla_i \bar\nabla_j \chi - g_{ij} \bar\nabla^k \bar\nabla_k \chi - \bar R_{ij}\chi + + g_{ij}(\bar R - 2\lambda)\chi \tag{5.8}$$

Proposition 7
im $JA^* \subset$ ker A.

The above proposition enables to construct vectors belonging to $\tilde{T}^1_C(\mathcal{H}, c)$.

Definition: $\overset{1}{W}_C(c)$ is a linear subspace of $\overset{1}{\tilde{T}}_C(\mathcal{H}, c)$ consisting of vectors which are represented by vector fields on C generated by im JA*.

Proposition 8

$\overset{1}{W}_C(c) \subset W_C$

Definition: $\overset{2}{W}_C(c) = \overset{1}{W}_C(c) \oplus \overset{0}{\tilde{T}}_C(\mathcal{H}, c)$ (a direct sum).

If σ is a compact manifold without boundary then

1^0 ker A = (ker A $\cap$ ker AJ) $\oplus$ imJA* (an orthogonal sum)

2^0 $W_C = \overset{2}{W}_C(C)$

3^0 ${}_1\tilde{T}_C(\mathcal{H}) = F_C(c) \oplus W_C$ (a direct sum), where $F_C(c)$ is a subspace of $\tilde{T}_C(\mathcal{H}, c)$ generated by ker A $\cap$ ker AJ.

The proof of this theorem is based on the theory of differential operators with injective symbols (the operator A* has an injective symbol) (cf. [4], [30]). Probably this theorem can be also proved in a non-compact case but it is necessary to impose appropriate boundary conditions at the spatial infinity.

We shall explain now the relation between the gauge distribution W and an action of the diffeomorphism group of M in the space $\mathcal{H}$. The group Diff M acts on the right in the set of all metrics on M having the signature (-1,+3)

$$\text{Diff } M \times C^\infty(S^2_*T^*M) \ni (\gamma, \underline{g}) \to R_\gamma \underline{g} = \gamma^*\underline{g} \in C^\infty(S^2_*T^*M) \qquad (5.9)$$

The action (5.9) generates a left action in the bundle $\mathcal{P}$ and the forms $\overset{4}{\theta}$ and $\overset{5}{\omega}$ are invariant with respect to it. Therefore we have the right action in the space $\mathcal{H}$:

$$\text{Diff } M \times \mathcal{H} \ni (\gamma, C) \to R_\gamma(C) = \gamma^*C \in \mathcal{H} \qquad (5.10)$$

The Lie algebra of Diff M can be identified with the Lie algebra of smooth vector fields on M (cf. [11]) with the commutatior as the Lie bracket. The action (5.10) generates an action of the Lie algebra

$$C^\infty(TM) \ni v \to d\hat{R}_{id}(C)v \in \tilde{T}_C(\mathcal{H}) \qquad (5.11)$$

where $d\hat{R}_{id}(C)$ is the derivative of (5.10) with respect to the first variable at the point (id,C) $\in$ Diff M $\times$ $\mathcal{H}$.

Proposition 9

$\text{im } d\hat{R}_{id}(C) = \overset{2}{W}_C(c)$

Theorem 3

If the manifold M has compact spatial sections i.e. if σ is compact then $\mathrm{im}\ d\hat{R}_{id}(C) = W_C$.

We see that the action (5.10) determines the gauge distribution of Ω.

It has been proved in [23] that the gauge distribution W is involutive, that means the commutator of two vector fields with values in W is also in W. This fact is simply the Frobenius integrability condition and therefore we can look for a space $\tilde{\mathcal{H}}$ for which $T(\mathcal{H})/W$. The theorem 3 suggests to divide the space $\mathcal{H}$ by the action of Diff M and take $\tilde{\mathcal{H}} = \mathcal{H}/\mathrm{Diff}\ M$. (5.12)

The other possibility is to define $\tilde{\mathcal{H}}$ axiomatically in the following way: let $(\delta\pi^{ij}, \delta g_{ij}) \in C_\sigma$ (for a given $\sigma \subset M$) satisfy the constraint equations (4.8b) and four equations obtained from (4.8b) by the transformation

$$\delta\pi^{ij} \to \sqrt{\bar{g}}\ \bar{g}^{ip}\bar{g}^{jq}\ \delta g_{pq},\quad g_{ij} \to -\frac{1}{\sqrt{\bar{g}}}\ g_{ip}g_{jq}\ \delta\pi^{ij} \qquad (5.13)$$

In this case the system $(\delta\pi^{ij}, \delta g_{ij})$ determines a vector belonging to $F_C(c)$. These eight equations for $(\delta\pi^{ij}, \delta g_{ij})$ reduce the number of independent variables to four but these four variables are given in an implicit way. To carry these consideration on the space $\mathcal{H}$ we take on account 12 quantities (π^{ij}, g_{ij}) satisfying (3.9b) on σ. Solving the dynamical equations (3.6) and (3.9a)(for $N = 1$, $N_k = 0$) we obtain an Einstein metric on M. By means this metric we define an action of Diff M in the systems (π^{ij}, g_{ij}). If we denote by Cd (Cauchy data) the set of couples (π^{ij}, g_{ij}) which satisfy (3.9b) we have $\tilde{\mathcal{H}} = \mathrm{Cd}/\mathrm{Diff}\ M$ (5.14)

Remarks:

1^0 In our considerations we have put $N = 1$, $N_k = 0$. If we take another lapse function and another shift vector on M we obtain another action of Diff M in Cd. It can be proved that the corresponding actions of the Lie algebra in the tangent space are isomorphic but it is difficult to prove it for the space Cd.

2^0 It can be proved that the tangent spaces of two possible choices of ((5.12) and (5.14)) are isomorphic. The question arises whether these spaces are isomorphic?

An object of the type (5.14) has been recently proposed by A. Fischer and J. Marsden [15],[18] as a possible choice of a superphase space for the Einstein dynamics. These authors have pointed out that the superphase space is a very complicated object (cf. also [14]) and maybe therefore the problem of the quantization of gravitation is so difficult.

References

[1] Abraham, R., Foundations of Mechanics, (1967) New York: Benjamin.
[2] Adler, R., Bazin, M., Schiffer, M., Introduction to General Relativity. New York: McGraw Hill (1965).
[3] Arnowitt, R., Deser, S., Misner, C.W., The Dynamics of General Relativity. In: Witten, L. (ed), Gravitation - an Introduction to Current Research, New York: John Wiley (1962).
[4] Berger, M., Ebin, D., Some decompositions of the space of symmetric tensors on a Riemannian manifold, Journ. of Differential Geometry 3 (1969) 379-392.
[5] Bergmann, P. G., Komar, A.B., Status report on the quantization of the gravitational field. In: Recent Developments in General Relativity. London-Warsaw Pergamon Press-PWN: London-Warsaw 196 1962.
[6] Choquet-Bruhat, Y., The Cauchy problem. In: Witten, L. (ed) Gravitation - an introduction to current research. New York: J. Wiley 1962.
[7] Choquet-Bruhat, Y., Geroch, R., Global aspects of the Cauchy problem in G. R. Comm. Math. Phys. 14 (1969) 329-335.
[8] Dedecker, P., Calcul des variations, formes differentieles et champs geodesiques. In: Coll. Intern. Geometrie Differ. Strasbourg 1953.
[9] DeWitt, B., Quantum Theory of Gravity I. The canonical theory. Phys. Rev. 160 (1967) 1113-1148.
[10] Dirac, P. A. M., Generalized Hamiltonian Dynamics, Proc. Roy. Soc. (London) A246 (1958) 326-332; The theory of Gravitation in Hamiltonian Form, Proc. Roy. Soc. A246 (1958) 333-346.
[11] Ebin, D. Marsden, J., Group of diffeomorphisms and the motion of an incompressible fluid. Ann. of Math. 92 (1970) 102-163.
[12]. Fadeev, L., Symplectic structure and Quantization of the Einstein Gravitation Theory. In: Actes du Congres Intern. A. Math Nice (1970) 34-39. (vol. 3).
[13] Fadeev, L., Popov, V., A Covariant Quantization of the New Gravitational Field, Uspechi fiz. Nauk 111 (1973) 427-450 (in Russian).
[14] Fischer, A., The theory of Superspaces. In: Carmel, M., Fickler S., Witten, L. (ed) Relativity. New York: Plenum Press 1970.
[15] Fischer, A., Marsde n, J., The Einstein Equations of Evolution. A Geometrie approach, Journ. Math. Phys. 13 (1972) 546-568.
[16] Fischer, A., Marsden, J., The Einstein Evolution Equations as a Quasi-linear First Order A Symmetrie Hyperbolic System I. Comm. Math. Phys. 28 (1972) 1-38.
[17] Fischer, A., Marsden, J., Linearization Stability of the Einstein Equations, Bull. Am. Math. Soc. 79 (1973) 997-1003.
[18] Fischer, A., Marsden, J., General Relativity as a Hamiltonian System. Symposia Mathematica XIV(published by Istituto di Alta Matematica Roma) London-New York Academic Press 1974; 193-205.
[19] Gawedzki, K., On the geometrization of the Canonical Formalism in the Classical Field Theory, Reports on Math. Phys. 3 (1972) 307-326.
[20] Goldschmidt, H., Sternberg, S., The Hamilton-Cartan Formalism in The Calculus of Variations, Ann. Inst. Fourier 23 (1975) 203-267.
[21] Hawking,S. W., Ellis, G- F. R., The Large Scale Structure of Space-time, Cambridge University Press 1973.
[22] Kijowski, J. Szczyrba, W. A., A Canonical Structure of Classical Field Theory, Comm. Math. Phys. 30 (1973) 99-128.

[23] Kijowski, J., Szczyrba, W. A., A Canonical Structure of Classical field Theories (to appear in Comm. Maht. Phys.).
[24] Kobayashi, S., Noizu, K., Foundations of Differential Geometry, New York: Interscience Publ. Vol. 1, 1963, Vol. 2, 1969.
[25] Kostant, B., Quantization and Unitary Representations, in: Lecture Notes in Mathematics 170, Berlin: Springer-Verlag 1970. Symplectic Spinors, in Symposia Mathematica Vol. XIV London New York: Academic Press 1974.
[26] Kuchar, K. A Bubble-time Canonical Formalism for Geometrodynamics, Journ. Math. Phys. 13 (1972) 768-781.
[27] Kundt, W., Canonical Quantization of Gauge Invariant Field Theories, Springer Tracts in Modern Physics 40. Berlin-Heidelberg-New York: Springer-Verlag 1966.
[28] Lichnerowicz, A., Relativistic Hydrodynamics and Magnetohydrodynamics, New York: Benjamin 1967.
[29] Misner, Ch. W., Thorne, K. S., Wheeler, J. A., Gravitation, San Francisco: W. H. Freeman and Co. 1973.
[30] Narasimhan, R., Analysis on Real and Complex Manifolds, Paris: Masson ans Cie 1968.
[31] Souriau, J.M., Structure Des Systemes Dynamiques, Paris: Dunod 1969.
[32] Szczyrba, Lagrangian Formalism in the Classical Field Theory, Ann., Pol. Math. 32 (1975) (to appear).
[33] Wheeler, J. A., Geometrodynamics and the Issue of the Final State. In: DeWitt, B. De Witt, C. (ed), Relativity, Groups and Topology, New York: Gordon and Breach 1964.

INTERACTIONS ON THE EXTENDED CONFORMAL SPACE-TIME

Jan Tarski*

International Centre for Theoretical Physics, Trieste

ABSTRACT

The mechanism suggested by Segal for redshift is described in terms of metrics and wave equations on the space $\tilde{M}$. In addition, some comments on dynamical groups are included.

INTRODUCTION

Some time ago Segal pointed out that certain features of the cosmos might be conveniently described with the help of the space-time, or (for brevity) space $\tilde{M}$ [1]. This is, up to a scale, the covering space of the conformal space M_4^c of Veblen [2], and the latter is a completion of the Minkowski space M_4. In recent years, both of those spaces were investigated and described in detail for also other reasons: to provide an improved foundation for the study of conformal transformations in physics [3,4] and for the study of the asymptotic region in gravitation theory [5].

The preliminary geometrical questions about these spaces having been clarified, the next problem is to describe interactions on $\tilde{M}$. Segal already outlined a simple physical mechanism which might account for the redshift on $\tilde{M}$ (as an alternative to the expansion hypothesis). In this note we analyze this mechanism in greater detail, in terms of wave equations and metrics on $\tilde{M}$. These metrics are induced by the tangent Minkowski spaces at different points.

We also consider briefly the dynamical groups for elementary particles on $\tilde{M}$. This new way of looking at the dynamical groups seems to be of interest for primarily pedagogical reasons. There are, on the other hand, several related topics that we do not discuss: massless electrodynamics [6], which in principle should allow an immediate extension to $\tilde{M}$, and Weyl manifolds [7], which may lead to an adaptation of the ideas under consideration to gravitation theory. Moreover, the massive free fields on $\tilde{M}$, even though they appear to be relevant to this paper, are bypassed.

In the title we spoke of $\tilde{M}$ as the extended conformal space-time, since it is obtained from M_4 by a conformal completion and a covering. However, $\tilde{M}$ is also Einstein's static universe if it is equipped with the metric $\tilde{g}$ (as below) [5].

We next summarize the basic geometrical features of the spaces M_4^c and $\tilde{M}$. Consider R^6 with the metric +----+. The rays of the null cone then yield the space M_4^c (see e.g. [3]). This space is topologically equivalent to $(S^1 \times S^3)/Z_2$, and can be covered by local charts, each having the structure of the usual Minkowski space M_4. Four such local charts suffice. The space $\tilde{M}$ is the simply connected covering space of M_4^c, has a global causal structure, and is topologically equivalent to $R^1 \times S^3$ (see [1,4,5]).

Now, Segal's proposal is to take $R^1 \times S^3(\rho)$ as a model for the cosmos, where $S^3(\rho)$ is the sphere of radius ρ, and ρ is to be thought of as the radius of the universe. Thus, $\tilde{M}$ has a natural metric $\tilde{g}$ with the signature +---. This metric is also induced by imbedding $\tilde{M}$ in the Minkowski space M_5, for which we use the coordinates $(t,u^1,\ldots,u^4)$. The metric $\tilde{g}$ is compatible with the causal structure on $\tilde{M}$, and has the isometry group $R^1 \times SO_4$ [4].

The metric $\tilde{g}$ yields a volume element for $\tilde{M}$, namely $d^4v|\det \tilde{g}(v)|^{\frac{1}{2}}$ in arbitrary coordinates, or $dt d\mu(u)$ where $d\mu(u)$ is the natural volume element of $S^3(\rho)$. We associate with this volume element the δ-function which satisfies

$$I = \int_V d^4v|\det \tilde{g}(v)|^{\frac{1}{2}}\, \delta(v,w) = 1 \text{ if } w\epsilon V, \tag{1}$$

V being an open set; and $I = 0$ if $w \notin V$.

The space $\tilde{M}$ is locally indistinguishable from M_4^c and, like the latter, can be covered by local coordinate patches, each having the appearance of M_4, as follows. For each point $v\epsilon\tilde{M}$, the tangent space $M_4(v)$ yields such a patch through the (onto) map

$$\sigma_v : M_4(v) \to M'(v) \subset \tilde{M}. \tag{2}$$

This map has the property that in $M'(v)$ the causal structure induced by $M_4(v)$ agrees with that of $\tilde{M}$. We will call a patch like $M'(v)$ an M_4-patch. If

$$v_0 = (0,0,0,0,\rho)\epsilon\tilde{M}\subset M_5, \tag{3}$$

then the points of $M'(v_0)$ satisfy the following conditions (among others):

$$-\tfrac{1}{2}\pi\rho < t < \tfrac{1}{2}\pi\rho, \quad u^4 > -\rho. \tag{4}$$

The natural metric on $M_4(v)$ is of course $\tilde{g}(v)$. However, then the map σ_v induces on $M'(v)$ a metric g_v, which differs from $\tilde{g}$. Since the mapping preserves the causal structure, the two metrics must be proportional [8]. Thus, for $w \in M'(v)$ we set

$$g_v(w) = \lambda_v(w)\tilde{g}(w). \tag{5}$$

The factor λ can be found from [5], eqs. (8.4)-(8.9). Let $w_v \in M_4(v)$ correspond to w, i.e.

$$\sigma_v w_v = w. \tag{6}$$

Then

$$\lambda_v(w) = [1 + (w_v^0 - |\vec{w}_v|)^2 / \rho^2][1 + (w_v^0 + |\vec{w}_v|)^2/\rho^2]. \tag{7}$$

In the sequel we largely limit our attention to the case where all points lie on the same M_4-patch. In particular, we will assume $v,w \in M'(x)$ and $v \in M'(w)$. [Note that $v_1 \in M'(v_2)$ iff $v_2 \in M'(v_1)$.]

In closing this introduction, the author would like to gratefully acknowledge discussions with several members of the International Centre for Theoretical Physics, in particular with Professor H. G. Ellis. He thanks Professor Abdus Salam, the International Atomic Energy Agency, and UNESCO for hospitality at the ICTP.

WAVE EQUATIONS AND THE REDSHIFT

We recall the analysis of field equations on the space M_4^c in [3]. One finds there in particular the construction of the operator $P_\square$, which is the extension of the wave operator $\square$ on M_4. Thus the equation $\square\phi = 0$ has a natural analogue on M_4^c. The electromagnetic field can be similarly treated [3], but for simplicity we will restrict ourselves largely to the scalar massless field.

Since $P_\square$ is a local object in M_4^c, its definition can be directly adapted to $\tilde{M}$. However, the resulting operator will depend on a chosen reference point w. In fact, it will be the image of $\square$ on the tangent space $M_4(w)$, under a map determined by σ_w. Homogeneity of $\tilde{M}$ now requires that all such operators for different w's be treated on an equal footing. We will write $P_{(w,\square)}$ to indicate the reference point, and also $P_{(w,\square)}(x)$ to indicate the variable $x \in \tilde{M}$ for differentiation.

We now wish to relate two such operators $P_{(w,\square)}(x)$ and $P_{(v,\square)}(x)$. Let s be an element of $R^1 \times SO_4$ (the isometry group) such that sw = v. It follows that

$$P_{(w,\Box)}\phi = 0 \text{ iff } P_{(v,\Box)}\phi(s\cdot) = 0. \qquad (8a)$$

On the other hand, s is a conformal transformation, and the arguments of [3] can be adapted to yield

$$P_{(v,\Box)}\phi = 0 \text{ iff } P_{(v,\Box)}\phi(s\cdot) = 0. \qquad (8b)$$

We thus conclude

$$P_{(w,\Box)}\phi = 0 \text{ iff } P_{(v,\Box)}\phi = 0. \qquad (9)$$

It is not necessary in these equations that ϕ be a global solution.

Let us now turn to Green's functions, which satisfy equations like

$$P_{(w,\Box)}(x)\ G(x,w) = -\ \delta(x,w). \qquad (10)$$

The causal solution can be expressed in terms of the usual Feynman propagator,

$$G_c(x,w) = D_c(x_w), \qquad (11)$$

where x_w is as in eq. (6).

Before continuing, we should like to comment on the wave equation $\partial d\psi = 0$. This equation is not equivalent to $P_{(w,\Box)}\psi = 0$, and one way to see the non-equivalence is to compare the Green's functions. The retarded or the advanced solution of (10) has its support on the light cone, while the corresponding Green's functions for the operator ∂d have also "tails" inside the light cone [9]. The operator ∂d does not seem to have any particular conformal-theoretic features.

If in eq. (10) we were to replace $P_{(w,\Box)}$ by $P_{(v,\Box)}$ with $v \neq w$, we would not get an equivalence such as in (9), and this brings us to the core of the redshift. For convenience we will speak of emission and of absorption of photons, also in the classical theory.

We now assume the following rule for electromagnetic interactions, which is in agreement with the premises of Segal [1]. Let a photon be emitted at a point $w \in \tilde{M}$ and absorbed at v. Then the emission and the absorption should be described with the help of the metric $\tilde{g}$ at the respective points, while for the free propagation we have equations such as in (9), which do not involve $\tilde{g}$ directly.

One expects of course that $\tilde{g}$ should be basic for describing interactions, and an example in the next section illustrates this. Further examples bearing on related geometrical questions can be found in [10].

One way of exploiting this rule is through comparison of e.g. the causal solution of (10), G_c, with the causal function satisfying

$$P_{(v,\Box)}(x)\ G_c^{(1)}(x,w) = -\ \delta(x,w). \tag{12}$$

Note that for x = w, the two differential operators are proportional,

$$P_{(w,\Box)}(x)|_{x=w} = \lambda_v(w)\ P_{(v,\Box)}(w). \tag{13}$$

To prove this relation, observe first that in coordinates on M'(v) which are induced by Cartesian coordinates on $M_4(v)$ one has

$$\begin{aligned} P_{(v,\Box)}(w) &= -\ (g_v(w))^{\mu\nu}\partial_\mu\partial_\nu \\ &= -\ (\lambda_v(w))^{-1}\ \tilde{g}(w)^{\mu\nu}\partial_\mu\partial_\nu. \end{aligned} \tag{14}$$

Moreover, the resulting coordinates at w are compatible with a set of Cartesian coordinates on $M_4(w)$, so that

$$P_{(w,\Box)}(x)|_{x=w} = -\ \tilde{g}(w)^{\mu\nu}\partial_\mu\partial_\nu\ . \tag{15}$$

We used here $g_w(w) = \tilde{g}(w)$. Equation (13) follows.

Next, for $x \neq w$, eqs. (10) and (12) are equivalent. This fact and eq. (13) suggest that

$$G_c^{(1)}(x,w) = G_c(x,w)\lambda_v(w). \tag{16}$$

We will accept this equation, even though the foregoing is not a rigorous deduction.

One can now proceed to obtain the redshift by making various approximate calculations, e.g. such as those in geometrical optics. We limit ourselves to the following simple estimate.

A photon emitted by a δ-source at w is described by the function G (having suitable boundary conditions). The absorption of this photon at v, however, will be analyzed in terms of the function $G^{(1)}$. Thus, as the photon approaches v, one should no longer write G, but instead, $G^{(1)}/\lambda_v$. Furthermore,

$$G^{(1)}(y) \sim \int d^4k\ e^{iky}/k^2, \tag{17}$$

so that the reduction of $G^{(1)}$, i.e. $G^{(1)} \to G^{(1)}/\lambda_v$, is equivalent to $k^\mu \to k^\mu/\lambda_v^{\frac{1}{2}}$. For propagation along $y^0 = |\vec{y}|$ we then get [cf. (7)]

$$k^\mu \to k^\mu[1+4(y^0/\rho)^2]^{-\frac{1}{2}}. \tag{18}$$

For small y^0/ρ this yields approximately

$$Z = (k^0 - k^0\lambda_v^{-\frac{1}{2}})/k^0\lambda_v^{-\frac{1}{2}} \approx 4(y^0/\rho)^2, \qquad (19)$$

in qualitative agreement with [1].

The existence of the redshift leads to the next problem, namely that of energy conservation. Segal's suggestion is that any energy loss due to redshift should be interpreted as a contribution to the microwave background [1]. However, there appears to be no need for suggesting a new physical process here, since there is necessarily some radiative energy loss in an electro-magnetic interaction.

This completes the main part of our discussion. We should exphasize that the preceding theory is not only very sketchily presented, but is also conceptually incomplete. In particular, we did not include the case where the points of emission w and of absorption v do not satisfy $v \in M'(w)$, and we did not discuss processes involving the microwave background in the initial configuration.

Nevertheless, it might not be premature to think about expressing the foregoing ideas in the language of quantized fields. We should like to note in particular the following problem that arises in the quantized theory: An adaptation of quantum electrodynamics to $\tilde{M}$ must predict the redshift in lowest order. However, in the lowest order processes on the Minkowski space, there is no radiative energy loss. What mechanism gives rise to this loss, when the adaptation to $\tilde{M}$ is made?

COMMENTS ON DYNAMICAL GROUPS

In this section we try to relate the mathematical structures of $\tilde{M}$ and of the dynamical groups. The two developments have indeed one important feature in common, namely both lead to discrete mass spectra by exploiting the conformal group (cf. [11,12]). However, as we already indicated, the subsequent material should not be considered at this time as having a deeper significance for the problem of mass spectrum.

The connection between dynamical groups and the conformal group was described in a recent article by Barut and Bornzin [13]. These authors consider the action of the conformal Lie algebra (i.e. Lie algebra of the conformal group) on a two-particle system, and much of their discussion can be adapted also to the case where the underlying space is $\tilde{M}$. We give a few details.

Suppose that $\psi(x_1,x_2)$ is the wave function for a two-particle system, where each $x_j \varepsilon \tilde{M}$. The support of ψ in each variable is to be restricted to a suitable three-dimensional submanifold of $\tilde{M}$, which can be mapped into a single M_4-patch. Now, in case of a wave function on $M_4 \times M_4$ with a central potential, one would introduce the coordinates

$$\underline{X} = \beta_1 x_1 + \beta_2 x_2, \quad x = x_1 - x_2, \tag{20a,b}$$

where the β_j depend on the masses, and

$$0 < \beta_1, \beta_2, \qquad \beta_1 + \beta_2 = 1. \tag{21}$$

The relations (20) are meaningless on $\tilde{M}$, since this space is not linear. However, as shown in [13], if $x_j \varepsilon M_4$ and $x^2 = 0$, then one can represent not only the x_j but also $\underline{X}$ on the null cone in R^6. Then these can become well-defined points on M_4^c, and also on $\tilde{M}$. The condition for this, $x^2 = 0$, remains meaningful on $\tilde{M}$: The points x_1 and x_2 must be lightlike separated. Then $\underline{X}$ will be on the same light ray, intermediate between x_1 and x_2.

The analogues of the vectors x can be found not directly on $\tilde{M}$, but rather on the tangent bundle to $\tilde{M}$, as follows. Given $x_1, x_2 \varepsilon \tilde{M}$ lightlike separated, determine $\underline{X}$ by the condition that $(x_j)_{\underline{X}}$, $\underline{X}_{\underline{X}} \varepsilon M_4(\underline{X})$ satisfy (20a) [cf. eq. (6)], and ψ then becomes a function $\psi_0(\underline{X};x)$, where now $x = (x_1)_{\underline{X}} - (x_2)_{\underline{X}}$. Moreover, instead of ψ_0, one can take at each $\underline{X}$ the Fourier transform to define the p-space $\hat{M}_4(\underline{X})$ and $\hat{\psi}(\underline{X};p)$.

On each $\hat{M}_4(\underline{X})$ there is a natural action of the conformal Lie algebra, and functions $\Phi(\underline{X};x)$ or $\hat{\Phi}(\underline{X};p)$ with support on the cone $x^2 = 0$ form an invariant set. On this set, the conformal Lie algebra reduces to one of the kind encountered in dynamical groups.

Furthermore, let us suppose that the $\underline{X}$'s lie in the $y^0 = 0$ space associated with a certain M_4-patch. Then the x^0 and the p_0 directions are singled out in all the $M_4(\underline{X})$ and $\hat{M}_4(\underline{X})$, respectively, and $\vec{r}^2$, etc. are well defined, for all $\underline{X}$. Such invariants yield several elements of the dynamical group algebras.

Let us, finally, consider the Schrödinger equation for e.g. the H atom:

$$[(2m)^{-1}|\vec{r}|\vec{p}^2 - \alpha - E|\vec{r}|]\hat{\psi} = 0. \tag{22}$$

We note the following points:

i) This equation applies to functions $\psi(\underline{\overline{X}};p)$, but does not involve $\underline{\overline{X}}$ directly.

ii) The levels of the energy E can be found by e.g. dynamical group techniques [12]. These levels agree exactly with the usual theory, and there are no corrections due to the curvature of $\tilde{M}$.

iii) The metric $\tilde{g}$ enters into (22) in particular in the definition of $|\vec{r}|$ and of $\vec{p}^2$. Thus, the electromagnetic interactions in this case would indeed be described with the help of this metric, in agreement with the discussion of Sec. 2.

iv) The foregoing conclusions also apply to a system that can be described by a dynamical group, without reference to a wave equation.

REFERENCES

[1] I. E. Segal, Astron. Astrophysics 18, 143 (1972) and Proc. Nat. Acad. Sci. USA 71, 765 (1974).

[2] O. Veblen, Proc. Nat. Acad. Sci. USA 19, 503 (1933).

[3] T. H. Go and D. H. Mayer, Reports on Math. Phys. 5, 187 (1974); see also D. H. Mayer, J. Math. Phys. 16, 884 (1975).

[4] T. H. Go, Commun. Math. Phys. 41, 157 (1975).

[5] R. Penrose, in Battelle rencontres, 1967, edited by C. M. DeWitt and J. A. Wheeler (W. A. Benjamin Inc., New York, 1968), especially p. 171ff.

[6] S. L. Adler, Phys. Rev. D6, 3445 (1972).

[7] D. K. Sen and J. R. Vanstone, J. Math. Phys. 13, 990 (1972).

[8] E. C. Zeeman, J. Math. Phys. 5, 490 (1964).

[9] B. S. DeWitt and R. W. Brehme, Ann. Phys. (New York) 9, 220 (1960).

[10] H. D. Doebner and J. Tolar, J. Math. Phys. 16, 975 (1975).

[11] I. Segal, Proc. Nat. Acad. Sci. USA 57, 194 (1967).

[12] A. O. Barut, Dynamical groups and generalized symmetries in quantum theory (University of Canterbury Publications, Christchurch, New Zealand, 1972).

[13] A. O. Barut and G. L. Bornzin, J. Math. Phys. 15, 1000 (1974).

*Author's address during 1975-76: Fakultät für Physik, Universität Bidefeld.

SPACE-TIME SINGULARITIES

Roger Penrose

University of Oxford, England

ABSTRACT

The "singularity theorems" of classical general relativity imply the existence of singularities in space-time (i.e. places where the normal classical space-time descriptions must break down) whenever too large concentrations of mass are present, provided that reasonable requirements of causality and positivity of energy are satisfied. Such singularities may be of the initial exploding big-bang type or the result of final gravitational collapse inside a black hole--or they might possibly be naked singularities resulting from a collapse. Most relativists apparently believe that naked singularities will not occur in realistic collapse situations according to classical general relativity, but a proof is not yet forthcoming.

A convenient way to represent singularities is by the construction of "ideal points" ("TIP's" and "TIF's"). Initial, final and naked singularities may be characterized using this construction and their structure may be examined in detail.

A crucial difference between the geometry of initial and of final singularities seems probable, whereby the conformal curvature vanishes initially, but dominates over Ricci curvature finally. This may plausibly be related to the statistical time's arrow, and, more speculatively, to the minute time-asymmetry of local physical laws.

The existence of singularities in the structure of space-time is strongly indicated from both theory and observation. By such a singularity is meant a region of the universe at which our normal physical space-time picture breaks down. So to describe the nature of such a region in detail a new physical theory would be needed. According to a commonly expressed view, in the neighbourhood of a singularity, the radii of space-time curvature might be approaching 10^{-33} cm - the Planck length, at which quantum fluctuations in the metric ought in any case to render the normal Riemannian manifold picture of space-time inappropriate. Accordingly, the required new theory would be some satisfactory fully quantized version of Einstein's general relativity. Alternatively, it might turn out that when the radii of space-time curvature approach only the much more modest value of, say, 10^{-13} cm, a new theory would already be needed. Indeed, one could not expect much of present-day elementary particle theory to survive at a regime in which the approximation of a flat space-time ceases to be appropriate and the Poincaré group accordingly has no local relevance, various characteristic lengths involved in elementary particles being already of the general order of 10^{-13} cm. Yet again, it might be that space-time curvatures do not approach at all closely either of these values. But if not, this would require a more radical rethinking still of basic physical concepts, since the new physical theory would be needed at an even larger scale--a scale at which there has been as yet no hint from any other direction that such a new theory is necessary for the performing of calculations.

One reason that such a new theory would be needed at some stage, however, is that the existing theory of space-time structure, namely classical general relativity, leads to the conclusion, via the "singularity theorems" (cf. [1],[2]), that under suitable qualitative circumstances singularities in space-time are inevitable. These qualitative circumstances would arise in (at least) two main situations, namely inside black holes and at the "big bang" origin of the universe. It is noteworthy that there is some observational evidence in support of both types of situation occurring in the actual universe. The observations do not directly imply the existence of singularities, but they do suggest that the aforementioned qualitative circumstances actually occur.

I shall not discuss the singularity theorems in detail here, as these have been amply described elsewhere. However, it is worth remarking on the main physical assumption involved (beyond the validity of general relativity itself) namely the energy condition (cf. [1],[2]). This states, in effect, that at each point the energy density component (to a comoving observer) of the energy momentum tensor plus the sum of the principal pressures is non-negative, and that the energy density plus each principal pressure is individually non-negative. If only the second of these assumptions is to hold, this is the weak energy condition. The weak energy condition is a consequence of the assumption that to every observer (not merely those comoving with the matter) the energy density is non-negative. The existence of space-time singularities under appropriate circumstances follows merely from the weak energy condition (cf. [3]), although only if a suitable Cauchy surface is assumed to exist.

So it is clear that the positivity of energy density is closely related, via Einstein's equations, to the existence of space-time singularities. It would seem that on the classical scale, such positivity is a necessary consequence of present-day physical ideas (such as that on some smaller scale quantum field theory is appropriate). However, as soon as the matter needs to be treated quantum mechanically on the scale under consideration, such positivity (or effective positivity) can no longer be assumed. Indeed, as the Hawking "black-hole explosion" phenomenon indicates (as we shall see later), there are reasons for believing that this positivity does not in fact hold at the quantum level. The question involved here is not one of quantum gravity, but of the nature of quantum fields on an effective classical curved space-time background. Even at this level there are fundamental questions of principle which remain unsolved.

Since the physics "at" a space-time singularity is not something that present-day theory can cope with, the precise mathematical discussion of singularities must concern the non-singular portions of the space-time only. This discussion may consist of an investigation of the way in which Einstein's equations describe the evolution of space-time structure as a singularity is approached (cf. the work of the Soviet school [4],[5]) or it may be of a general character, in which the singular points are viewed, rigorously, as boundary points to the non-singular space-time manifold. It is this second approach that I shall adopt--although I shall be biassed here also, in that only one of the several different possible approaches ([6],[7],[8],[9],[10]) to attaching boundaries to space-times will be considered.

Suppose that the space-time $\mathcal{M}$ satisfies an appropriate causality condition (such as strong causality or, at least, future and past-distinguishability [10],[2]). Then each point p of $\mathcal{M}$ may be uniquely associated with the set of points $I^-(p)$ lying to the past of p in $\mathcal{M}$ (i.e. which can be connected to p by a future-timelike curve). If we know $I^-(p)$ then we know p and vice-versa. Such a set is called a PIP ('proper indecomposable past-set'). The PIP $I^-(p)$ may also be described as the set of points lying to the past of any time-like curve γ with future end-point at p. That is, $I^-(p) = I^-[\gamma]$. We say that γ <u>generates</u> the PIP $I^-[\gamma]$. Now the virtue of considering the set $I^-[\gamma]$ as a way of representing the point p is that it generalizes (see [10]). If γ is <u>any</u> timelike curve (not just one with a finite future end-point) then the set $I^-[\gamma]$ is called an IP ('indecomposable past-set'). If $I^-[\gamma]$ is not of the form $I^-(p)$ for $p \in \mathcal{M}$, then $I^-[\gamma]$ is a TIP ('terminal indecomposable past-set'). Thus, timelike curves which go to infinity in the future generate TIP's. So do timelike curves which go to a singularity in the future. We may say that two timelike curves γ and γ' go into the future to the <u>same</u> point at infinity or to the <u>same</u> singular point if they generate the same TIP ($I^-[\gamma] = I^-[\gamma']$). With this criterion we may adjoin new "ideal" points to the space-time manifold, these being associated with the various TIP's. <u>Any</u> timelike curve now has a future "end-point". This may be an ordinary point of $\mathcal{M}$ (when $I^-[\gamma]$ is a PIP) or a future ideal point (when $I^-[\gamma]$ is a TIP). The future ideal points may be further classified as points at infinity or singular points according as the corresponding TIP's can, or cannot, be generated by timelike curves of infinite proper length into the future. (Alternative criteria can also be used.)

In an exactly similar way we can define sets $I^+[\gamma]$, called IF's ('indecomposable future-sets'), which are futures of timelike curves, these being either PIF's (if $I^+[\gamma] = I^+(p)$ for some $p \in \mathcal{M}$) or TIF's (otherwise). The TIF's define the past ideal points for $\mathcal{M}$. (Some ideal points may be considered to be both past and future ideal points at the same time, but I shall not bother with giving a definition for this here). These past ideal points are either points at infinity or singular points, according as the TIF's can or cannot be generated by timelike curves of infinite proper length into the past. In this way, any timelike curve in $\mathcal{M}$ acquires a past "end-point" which is either an actual point of $\mathcal{M}$ or an ideal point at infinity or a singular ideal point.

The ideal point structure of various space-times may be examined in detail. In the case when M is Minkowski space, for example, the different TIP's may be associated uniquely with the different points of future conformal infinity II^+, together with the one point i^+ representing future timelike infinity. Similarly, the different TIF's are uniquely associated with the points of past conformal infinity II^- and i^-. In the case of the Friedmann models we obtain singular ideal points as well as (for some models) points at infinity. The singularities turn out to be three-dimensional sets. In fact, the various past ideal points, representing points of the initial big bang, correspond to the various particle horizons of the model (the sets $\partial I^+[\gamma]$). The singularity structure of the Schwarzschild solution may also be examined. For a black hole this singularity consists of future ideal points (TIP's), whereas for a white hole, it would consist of past ideal points (TIF's). In each case we again have a three-dimensional singular region. Also, as in the Friedmann models, this singular region is, in a well-defined sense, space-like. This can be seen from the fact that no singular TIP contains another singular TIP (for the future singularity) and no singular TIF contains another singular TIF (for the past singularity).

The question of the existence of timelike singularities in space-time models is closely related to what is known as the cosmic censorship hypothesis. A timelike singularity would (at least according to some definitions, cf. [11] be an example of a naked singularity--namely a singularity which lies to the future of some point on an observer's world line and to the past of another point on the same world line. (This is a local definition, which does not require reference to observers at infinity. It means, more technically, that a singular TIP is contained entirely in the past of some point of M, or, equivalently, that a singular TIF is contained entirely in the future of some point of M, cf. [11]. With this definition, the presence of a naked singularity implies the nonexistence of a Cauchy hypersurface for the space-time. Other definitions are also used [2],[12]. Naked singularities are generally considered to be highly undesirable physically, since if they occur, an essential indeterminacy would be introduced into classical theory (over and above that already involved in quantum mechanics). The hypothesis of cosmic censorship asserts that naked singularities should not arise in any physically realistic classical situation.

There have been various attempts at obtaining a contradiction with the cosmic censorship hypothesis [12],[13], but so far none of these can be considered to have done so. Perhaps the failure of some of these attempts lends some small support to the validity of the hypothesis. On the other hand, I know of no argument which could be construed as a serious attempt at a proof of the hypothesis. So cosmic censorship remains a very open question--perhaps the most important unsolved question in general relativity theory.

If we exclude naked singularities, we are still left with the two (spacelike?) types of singularity, namely those given by the past ideal points (big bang) and those given by the future ideal points (in black holes). Superficially, at least, there is a strong resemblance between these two types of singularity. It seems that apart from the difference in the scale involved, the past and future singularities might be essentially time-reverses of one another. On the other hand there is likely to be, in my view, at least one very important additional difference between these two types of singularity.

We have become accustomed to the fact that all the microscopic laws of physics, which have importance in governing the behaviour of matter as we know it (i.e. excluding K^0-decay), are symmetrical in time--whereas there is the gross statistical time asymmetry involved on a macroscopic scale associated with the phenomenon of entropy increase. The "blame" for the statistical asymmetry is to be placed on the boundary conditions. Thus the initial conditions of the universe must be such that the entropy is at a comparatively low value, while the final state would be one of much higher entropy. But to investigate the initial conditions of the universe we must examine the big bang singularity itself. If the universe is closed and recollapses to an all-embracing final singularity, then the entropy of the final singularity must be much greater than that of the initial singularity. (This is assuming that the statistical time-sense of the universe will not somehow reverse itself at the moment of maximum expansion. This possibility has been suggested by some authors, but I prefer to exclude it here. I find it very hard to reconcile with other aspects of our understanding of the physical world.) Accordingly, the nature of this final singularity ought to be very different from that of the initial singularity. The same would apply inside a black hole. To an observer falling into a black hole, the singularity inside the hole would appear all-embracing, like the singularity of a recollapsing closed universe (assuming naked singularities of the type referred to above are excluded). This all-embracing singularity would form his final boundary condition, so if his statistical time-sense is not to reverse as he falls into the hole (and I am excluding this possibility from consideration), this should be a high-entropy singularity, unlike that of the initial big bang.

To emphasize that the time-asymmetry ought to be present in the structure of the space-time singularity itself, not just in the nature of the matter in the neighbourhood of the singularity, it may be recalled that discussions of the nature of matter in the early stages of the universe tend to suppose the matter to be completely thermalized. Thus, as far as the matter itself is concerned, the entropy is as high as possible. If it were not for the degrees of freedom in the geometry of the space-time itself, there would be no room for this entropy to increase further. In some sense it must be that the entropy of the geometry starts off low--and ends up high.

Initially it is the expansion of the universe which allows room for the entropy of the matter in it to increase. The entropy of the matter lags behind the value it would need to have for complete thermalization. And as the matter cools, local gravitational effects begin to play a role. Now the entropy can increase by local agglomerations of matter clumping together. In the way which is peculiar to gravity, higher and higher entropy values can be obtained as the matter clumps more and more. We obtain galaxies, normal stars, then white dwarfs or neutron stars. Finally, the greatest entropy of all is achieved when the matter collapses into black holes.

It appears that it is these gravitational degrees of freedom which are not used at all (or, at least, very little used) in the initial big bang. As far as can be inferred from observation (principally from the isotropy of the 2.7°K black-body radiation), the initial singularity was remarkably free of irregularity, the universe apparently resembling a Friedmann model to a high degree of approximation. This is quite unlike what seems reasonable to infer about the nature of a final singularity, whether it be that inside a black hole or one resulting from the recollapse of the universe as a whole. Any deviations from spherical symmetry in a collapsing body would be expected to get magnified as that body shrinks. (This is assuming, again, that the statistical time-sense does not reverse as the final singularity is approached!) The structure of the resulting singularity must be expected to be very irregular. Indeed, the analysis of Belinskii, Lifshitz and Khalatnikov [5] suggests a particularly complicated geometry for such singularities. The nature of the matter near the singularity, however, becomes quite unimportant. The geometry becomes more and more like that of a vacuum space-time, with the Weyl (conformal) part of the curvature ultimately completely dominating the Ricci part.

The following hypothesis thus suggests itself: the set of past ideal points has a particularly simple structure, the Weyl tensor vanishing in its immediate neighbourhood--on the other hand, at the future singular ideal points the Weyl tensor is likely to dominate. It seems that in some way the Weyl tensor gives a measure of the entropy in the space-time geometry. The initial curvature singularity would then be one with large Ricci tensor and vanishing Weyl tensor (zero entropy in the geometry); the final curvature singularity would have Weyl tensor much larger than Ricci tensor (large entropy in the geometry). These ideas are indeed somewhat imprecise, but it would seem worth while to pursue them further. The vanishing of Weyl tensor at the initial singularity, for example, would of itself seem to be effectively sufficient to ensure that the space-time closely resembles a Friedmann model, at least in its early stages. Of course, some way of introducing irregularities must also be allowed.

One consequence of the above hypothesis is that white holes are excluded. Even if exact spherical symmetry is assumed, the singularity inside the hole, though consisting of past ideal points, is quite different from that of the

Friedmann models. The Weyl tensor diverges to infinity rather than becoming zero. In any case, white holes must be thought of as anti-thermodynamic objects. If they are treated according to a physics subject to the normal statistical time directivity, then it seems that they are unstable [14],[15]. Furthermore, the area of the horizon of a white hole is subject to a law of surface area decrease. The Bekenstein [16] formula which states that the entropy of a black hole is proportional to the area of its horizon leads, if applied to a white hole, to the conclusion that the entropy of a white hole should decrease rather than increase. So it would seem satisfactory that a hypothesis governing the entropy of the universe as a whole should also directly exclude white holes.

At this point, mention should be made of some interesting ideas due to Hawking [17]. In accordance with the ideas of Bekenstein, a black hole ought also to possess a temperature proportional to its surface gravity (although this temperature would be absurdly low for any black hole which could arise by stellar collapse). Hawking predicts that owing to quantum mechanical effects, a black hole ought indeed to radiate as a black body of that temperature. If the temperature of the black hole exceeded that of its surroundings, it would radiate away some of its mass. This would mean that the area of its horizon should decrease, in contradiction with the classical area principle (which states that the area of the absolute event horizon of a black hole should be non-decreasing with time [2]). The occurrence of this phenomenon indicates that the weak energy principle is effectively being violated--because quantum rather than classical fields are involved. This decrease in surface area is now consistent with thermodynamics. The external temperature being lower than that of the black hole, the total entropy of black hole plus surroundings still increases. As the black hole gets smaller it also gets hotter. As it gets hotter it radiates faster, finally exploding with the emission of many baryons and anti-baryons. At its final disappearance, the black hole would momentarily become a naked singularity (not violating the classical cosmic censorship principle since this arises from an essentially quantum effect). If it disappears instantaneously, this naked singularity seems not to cause any real difficulty.

For a solar mass black hole this process would be absurdly slow, taking something of the order of 10^{55} times the present age of the universe for completion. Though the effect could only be of astronomical significance if very small black holes (of radius of the general order of 10^{-13} cm) had been created in the big bang--a possibility which seems at variance with the hypothesis suggested here--it nevertheless is of considerable theoretical interest.

Hawking argued [18] that the thermal equilibrium state of a system enclosed in a box with perfectly reflecting walls should (if the size of the box is suitable, given the total energy) be a black hole in stable equilibrium with black body radiation of the same temperature. He further suggests that being a state of thermal equilibrium, such a configuration ought to be time-symmetric--one reason arising from an appeal to CPT invariance. Hence, a black hole ought to be physically indistinguishable from a white hole! Accordingly, the Hawking (quantum mechanical) emission process ought to be physically indistinguishable from the time-reverse of the normal (classical) accretion process of a black hole. He also argues that although the classical geometry of a black hole differs from that of a white hole (even outside their respective horizons) this does not invalidate their identification, since the geometry cannot be treated strictly classically (so it is claimed). in a situation such as this.

Such an identification would, however, be at variance with the hypothesis put forward here, since white holes are to be excluded but not black holes. I feel that the viewpoint I am presenting receives some support from consideration of the deviations from equilibrium which would occasionally occur with the configuration of a black hole in a perfectly reflecting box. By an initial fluctuation, the hole might emit too much radiation to remain in stable equilibrium with its surroundings. It would then heat up and finally disappear in a burst of baryons and anti-baryons. Eventually it would reform out of the background energy to achieve the original equilibrium maximum entropy state once more. However, it is exceedingly hard to believe that its probable mode of formation would be the time-reverse of this method of disappearance. Baryons and anti-baryons would have to be specially created out of the radiation and aimed simultaneously at a tiny point, as the initial step in the creation of the hole. Only subsequently would ordinary radiation be falling in.

So it seems that this equilibrium is not in fact time-symmetric. And if it is indeed true that white holes are to be excluded, we have the possibility to contemplate an essentially time-asymmetric physics--which might be going on <u>virtually</u> at a sub-microscopic level. If so, we would have small violations not only of T-invariance (recall K°-decay!), but also of CPT-invariance. One might speculate on the closing of a logical loop. If T and CPT non-invariant interactions occur in microphysics, then though unimportant to phenomena <u>now</u>, they might nevertheless be important in the regimes neighbouring space-time singularities. The hypothesis of initially vanishing Weyl tensor could conceivably then have a basis in microscopic physical laws.

REFERENCES

[1] S. W. Hawking & R. Penrose, Proc. Roy. Soc. (Lond) <u>A314</u> (1970) 529.

[2] S. W. Hawking & G. F. R. Ellis, <u>The Large Scale Structure of Space-Time</u> (Univ. Press., Cambridge, 1973).

[3] R. Penrose, Phys. Rev. Lett. <u>14</u> (1965) 57.

[4] V. A. Belinskii, I. M. Khalatnikov & E. M. Lifshitz, Adv. Phys. <u>19</u> (1970) 523.

[5] V. A. Belinskii, I. M. Khalatnikov & E. M. Lifshitz, Soviet Phys. J.E.T.P. 62 (1972) 1606.

[6] H.-J. Seifert, Gen. Rel. & Grav. 1 (1971) 247.

[7] B. Schmidt, Gen. Rel. & Grav. 1 (1971) 269.

[8] R. Geroch, J. Math. Phys. 9 (1968) 450.

[9] B. Schmidt, Comm. Math. Phys. 36 (1974) 73.

[10] R. Geroch, E. H. Kronheimer & R. Penrose, Proc. Roy. Soc. (Lond) A327 (1972) 545.

[11] R. Penrose, in Confrontation of Cosmological Theories with Observational Data (Ed. M. S. Longair, I.A.U., 1974) 263.

[12] R. Penrose, Ann. N.Y. Acad. Sci. 224 (1973) 125.

[13] H. Müller zum Hagen, H.-J. Seifert & P. Yodzis, Comm. Math. Phys. 34 (1973) 135.

[14] Ya. Zel'dovich, in Gravitational Radiation and Gravitational Collapse (ed. C. DeWitt-Morette I.A.U. 1974) 82.

[15] D. Eardley, Death of White Holes in the Early Universe (Caltech Preprint 1974).

[16] J. D. Bekenstein, Phys. Rev. D.7 (1973) 2333, 9 (1974) 3292.

[17] S. W. Hawking, in Quantum Gravity (eds. C. J. Isham, R. Penrose & D. W. Sciama, Clarendon Press, Oxford, 1975) 219.

[18] S. W. Hawking (Caltech Preprint 1974).

H Space - The Manifold of Complex Asymptotically Shear Free Light Cones*

E. Newman

Physics Department

University of Pittsburgh

Pittsburgh, Pennsylvania 15260

ABSTRACT

We describe a manifold (four complex dimensional) constructed from the (complex) null surfaces in the analytic extension of the vacuum Einstein or Maxwell-Einstein geometries which are asymptotically flat. Those null surfaces whose asymptotic shear vanishes are the points of our manifold, H. H inherits complex quadratic line element from the physical space which automatically satisfies the vacuum Einstein equations. Further properties of H are described.

In several earlier [1, 2, 3] paper we pointed out that there were some curious consequences obtainable by extending our physical Minkowski space-time into a four complex dimensional space-time and then considering on this new manifold, analytically continued physical fields (e.g. the Maxwell field) and sources moving on complex world-lines. As an example [1], one can interpret the spin angular momentum of a real particle as the particle's orbital angular momentum but with the particle's center of mass world-line being complex, the spin vector being essentially the displacement of the line away from the real space-time. In a similar fashion the magnetic moment of a charged particle can be interpreted as arising from a complexcenter of charge world-line. If the two lines, happen to coincide one obtains as a purely kinematic result that the real particle must have the Dirac value of the gyromagnetic ratio, i.e. g = 2.

*This is a revised version of a talk previously given at the conference""The Riddle of Gravity" in honor of the 60th birthday of Peter G. Bergmann.

Similar [2] results are obtainable from the complexified linear Einstein and Maxwell equations when the two world-lines act as sources.

It is the purpose of this note to point out that the ideas concerning the complexification of Minkowski space can be greatly extended and generalized to include the asymptotically flat solutions of the vacuum Einstein equations (or more precisely the asymptotic regions of these solutions) and to describe some of the rather surprising results which follow.

We begin with some observations about null surfaces in Minkowski space. The null cone about an arbitrary point $\overset{\circ}{x}{}^a$ can be described by a function $f(x^a, \overset{\circ}{x}{}^a) = \eta_{ab}(x^a - \overset{\circ}{x}{}^a)(x^b - \overset{\circ}{x}{}^b)$. It is however more convenient for our purposes to use instead of the coordinates $\overset{\circ}{x}{}^a$, the null coordinates such that the metric takes the form

$$ds^2 = 2du^2 + 2\,dudr - \frac{1}{2}r^2\, d\zeta d\bar{\zeta}/p_o^2, \quad p_0 = \frac{1}{2}(1 + \zeta\bar{\zeta}) \tag{1}$$

where $u = \frac{1}{\sqrt{2}}(t-r)$ is the retarded time and r is $1/\sqrt{2}$ of the usual r and ζ and $\bar{\zeta}$ are the (complex) stereographic coordinates on the sphere. The null cone about $\overset{\circ}{x}{}^a$ can then be described by

$$u = u(\overset{\circ}{x}{}^a, \zeta, \bar{\zeta}, r) \tag{2}$$

which in the limit as $r \to \infty$ becomes

$$u = Z(\overset{\circ}{x}{}^a, \zeta, \bar{\zeta}) = \overset{\circ}{x}{}^a \ell_a(\zeta, \bar{\zeta}) = \sum_{\ell=0}^{1} \overset{\circ}{x}_{\ell m}\, Y_{\ell m}(\zeta, \bar{\zeta}) \tag{3}$$

with ℓ_a having as space-time components:

$$\ell_a = \frac{1}{2}\sqrt{2}(1 + \zeta\bar{\zeta})^{-1}\,[(1 + \zeta\bar{\zeta}), -(\zeta+\bar{\zeta}), -i(\zeta-\bar{\zeta}), -(\zeta\bar{\zeta}-1)] \tag{4}$$

and the $\overset{\circ}{x}_{\ell m}$ being simply linear combinations of the $\overset{\circ}{x}{}^a$.

Notice that a point in Minkowski space can be characterized by the asymptotic behavior of its light cone. Conversely it can be proven that if one has a null surface which in the asymptotic region $(r\to\infty)$ has a vanishing asymptotic shear $(\sigma = 0)$ then (in Minkowski space) this surface is the light cone of a point $\overset{\circ}{x}{}^a$. We thus conclude that the <u>space of asymptotically shear-free cones</u> is one to one with Minkowski space.

If we had begun with complex Minkowski space $(x^a \to z^a = x^a + iy^a, x^a$ and y^a real, with $ds^2 = \eta_{ab}dz^a dz^b)$ then we could have considered complex null cones from complex points $\overset{\circ}{z}{}^a$. Eq. (3) then becomes

$$u = Z(\overset{\circ}{z}{}^a, \zeta, \tilde{\zeta}) = \overset{\circ}{z}{}^a \ell_a(\zeta \tilde{\zeta}) \qquad (5)$$

where u is complex and $\tilde{\zeta}$ is freed from $\bar{\zeta}$. Thus the space of complex asymptotically shear free null cones (or for short "good" cones) is one to one with complex Minkowski space.

If one takes a second cone (about the point $\overset{\circ}{z}{}^a + dz^a$) then we define the distance dS between the two neighboring cones by

$$dS^2 = [\frac{1}{8\pi i} \oint \frac{d\Omega}{(dZ)^2}]^{-1} \qquad (6)$$

where

$$dZ = Z(\overset{\circ}{z}{}^a + dz^a, \zeta, \tilde{\zeta}) - Z(\overset{\circ}{z}{}^a, \zeta, \tilde{\zeta}) = dz^a \ell_a(\zeta, \tilde{\zeta}) \qquad (7)$$

and $d\Omega = d\zeta \Lambda d\tilde{\zeta}/(1 + \zeta\tilde{\zeta})^2$. The integration is a double line integral in the two complex dimensional space of ζ and $\tilde{\zeta}$; a line integral from $\zeta = 0$ to ∞ in the ζ space, then a contour integral anound the remaining pole in the $\tilde{\zeta}$ space are the paths. On performing this integration we obtain the result that

$$dS^2 = ds^2 = \eta_{ab}\, dz^a\, dz^b, \qquad (8)$$

i.e. the space of (complex) good cones is metrically equivalent to (complex) Minkowski space.

Considering now asymptotically flat solutions to the vacuum Einstein (or Einstein-Maxwell) equations and using a Bondi coordinate system $(u, r, \zeta, \bar{\zeta})$ then the equation describing a null surface could be written as

$$u = u(\zeta, \bar{\zeta}, r) \qquad (9)$$

and in the limit $r \to \infty$

$$u = Z(\zeta, \bar{\zeta}) \ . \qquad (10)$$

(Knowing $Z(\zeta, \bar{\zeta})$, $u(\zeta, \bar{\zeta}, r)$ can be determined).

In general the null surface determined by the choice of Z will have a non-vanishing asymptotic shear. In fact one can show that the condition on the function Z so that the surface has a vanishing asymptotic shear is

$$\eth^2 Z = \sigma^0(Z, \zeta, \bar{\zeta}) \qquad (11)$$

where $\eth$ is an angular differential operator [4] and $\sigma^o(u,\zeta,\bar{\zeta})$ is the asymptotic shear of the Bondi cones, u = constant. From (11) it follows that in general there will be no asymptotically shear free null surfaces for real asymptotically falt space-time.

If however we analytically extend the physical space time in the neighborhood of $r = \infty$ (assuming that the relevant function $\sigma^o(u,\zeta,\bar{\zeta})$ is analytic in its three arguments) then it is possible to find solutions to (11) (with $\bar{\zeta}\to\tilde{\zeta}$ and $u = Z(\zeta,\tilde{\zeta})$) and thus find complex asymptotically shear free null surfaces. The set of these surfaces form a four complex dimensional manifold H (or heaven, the space of good cones). We can thus write the solutions to (11) as

$$u = Z(z^a,\zeta,\tilde{\zeta}) \qquad (12)$$

the z^a labeling points of H or good cones.

A distance dS between neighboring points of H is defined, analogous to Eq. (6), by

$$(dS)^2 = [\frac{1}{8\pi i} \oint \frac{d\Omega}{(dZ)^2}]^{-1} \qquad (13)$$

with

$$dZ(z^a,\ dz^a,\ \zeta,\tilde{\zeta}) = Z_{,a}\ dz^a \qquad (14)$$

and the path of integration being the same as for (6).

After considerable effort one can prove our first major result, namely that dS^2 is quadratic in the dz^a, i.e.

$$dS^2 = g_{ab}(z^c)\ dz^a\ dz^b\ , \qquad (15)$$

and that H is thus a complex Riemannian space. (We wish to emphasize that this result if far from obvious). One can further show that the curvature tensor has the properties that

$$R_{ab} = 0 \qquad (16)$$

and half the Weyl tensor vanishes, i.e., if

$$C_{abcd} = \Psi_{ABCD}\ \varepsilon_{A'B'}\ \varepsilon_{C'D'} + \tilde{\Psi}_{A'B'C'D'}\ \varepsilon_{AB}\ \varepsilon_{CD} \qquad (17)$$

then

$$\Psi_{ABCD} = 0 \qquad (18)$$

(It appears almost certain that the $\tilde{\Psi}_{A'B'C'D'}$ is determined by the second and third derivatives of the shear, i.e. by $\ddot{\sigma}^0$ and $\dddot{\sigma}^0$.) We thus conclude that H satisfies the vacuum Einstein equations, but that one of the two parts (self dual and anti self dual) of the Weyl tensor vanishes and thus there cannot be any real four dimensional submanifold of H that has a real metric on it induced from (15) except in the case of H being flat. Thus H cannot in any obvious sense be considered as the analytic extension of physical space-time.

Though all the results described until now have been purely mathematical, there are several indications of possible physical content in the properties of H. We mention here a few of these.

a) There is a version of the generalized Kerr theorem which states that for each complex world line in H there exists in the real physical space-time an asymptotically shear free but twisting null geodesic congruence. Further, it appears certain that there exists in H a unique world line which can be called the complex center of mass line (obtained from the physical space center of mass-angular momentum aspect). The twist of the associated shearfree congruence can be considered as a measure of the spin angular momentum of the source.

b) If one is dealing with the Einstein-Maxwell equations, again it appears certain that there exists in H a second world line (obtained from the electric and magnetic dipole aspects) which can be called the complex center of charge. If the complex center of charge line coincides with the complex center of mass line it appears likely that the real physical source must have a g = 2 as in our flat space case. (This has been proven [5] for stationary Einstein-Maxwell fields.)

c) If the original physical space time is stationary (or simply non radiative) then H is flat, i.e. it is complex Minkowski space. In the stationary case one can show that the physical space source distribution generates a world line in the real section of the Minkowski space H which satisfies the equations of motion

$$\dot{P}^a = 0, \quad \dot{M}^{ab} + 2\, P^{[a} v^{b]} = 0\,, \quad M^{ab}\, P_b = 0 \tag{19}$$

where P^a and M^{ab} are the momentum 4 vector and angular momentum tensor defined from surface integrals of the Weyl tensor at infinity.

A similar, but far more complicated equation is derivable in the radiation case.

d) Almost the entire Penrose theory of flat space twistors can be re-expressed as geometric objects on H yielding a theory of asymptotic twistors. In fact one can begin with the theory of asymptotic twistors and from it develop the theory of H.

These clues (and several others too technical and speculative to discuss now) make it appear that H space and theclosely related asymptotic twistor space offers a large and potentially valuable source of mathematical and physical problems and even, hopefully, are related to fundamental physical questions.

The details and proof of the material presented here will be given in a paper being prepared with R. Penrose.

REFERENCES

[1] E. T. Newman, and J. Winicour, J. Math. Phys. 15 (1974) 1113.
[2] E. T. Newman, J. Math. Phys. 14 (1973) 102.
[3] R. W. Lind and E. T. Newman, J. Math. Phys. 15 (1974) 1103.
[4] E. T. Newman and R. Penrose, J. Math. Phys. 7 (1966) 863.
[5] B. Branson, Private communication.

CONFORMAL STRUCTURE - SPACELIKE INFINITY

B. Schmidt

Max Planck Institute für Physik und Astrophysik
Munchen, West Germany

ABSTRACT

A new method to formulate asymptotic conditions for the gravitational field of isolated systems is presented. It is based on a boundary attached to spacetime, which is determined by the conformal structure intrinsically.

Application to asymptotically simple spacetimes shows that the conformal boundary contains $\mathcal{J}$ and moreover any generator of $\mathcal{J}$ has a future and past endpoint in the conformal boundary.

Hence using this construction I can define spacelike infinity in a natural and intrinsic way.

If the past endpoints of the generators of $\mathcal{J}^+$ are identified one gets an action of the Poincare group on $\mathcal{J}^+$. In this way a reduction of the BMS group to the Poincare group as an asymptotic symmetry group is achieved. Necessary conditions for this to be the case are given.

Further issues, where the described techniques might be useful, will be discussed.

The idea of isolated systems is intimately related to a concept of asymptotic flatness. Bounded sources should determine a space-time which becomes more and more like Minkowski space far away from the sources.

Essentially two notions of asymptotic flatness are used in General Relativity.

In the context of gravitational radiation Sachs and Bondi formulated asymptotic conditions along outgoing null hypersurfaces. These conditions were further developed and recast by Penrose [1] into the definition of "future null infinity" called $\mathcal{J}$. Space-times which possess $\mathcal{J}$ are called "weakly asymptotically simple" and behave in a precisely defined sense like Minkowski space along null geodesics which terminate at $\mathcal{J}$.

For static and stationary solutions a quite different asymptotic flatness condition has proved to be useful. One demands that a spacelike hypersurface behaves more and more like a 3-plane in Minkowski space. Geroch [2] developed this further in defining spacelike infinity also for non-stationary space-times using similar technics as Penrose.

Up to now no relation between the two approaches is known. In a recent paper [3] I developed a generalization of the b-boundary construction applicable to the conformal and projective structure of space-time. The conformal boundary one gets seems to be very useful to formulate asymptotic flatness conditions in a intrinsic and natural way.

First the construction of the conformal boundary will be described briefly then some new results will be reported and finally some important problems are mentioned, which can be tackled from a new point of view.

A conformal structure can be defined as the reduction P of the frame bundle consisting of all frames which are orthonormal in any metric in the conformal class. The collection of connections, defined by the metrics in the conformal class, define a further reduction P^1 of the frame bundle of P. On the bundle P^1 there exists a natural parallelisation determined by the conformal structure, which is used to define a positive definite metric on P^1. The parallelisation is determined as follows: any connection of a metric in the conformal class defines a section in P^1. Sections passing through the same point with different tangent directions have different Ricci tensors. Under all subspaces of the tangent space one gets this way, there is a unique one determined by the condition that the Ricci tensor vanishes. This complement to the tangent space of the fibre defines the parallelisation.

Constructing the Cauchy completion one defines a boundary of P^1 and via the projection one gets a boundary of space-time, intrinsically defined by its conformal structure. Boundary points can be characterised in the following way: Take a curve $x(\lambda)$ inextensible in V^4. Determine a connection in the conformal class whose Ricci tensor vanishes along $x(\lambda)$. (This is always possible.) If the generalised affine length given by this connection is finite, then the curve defines a point in the boundary.

For Minkowski space the conformal boundary turns out to be $\mathcal{J} = \mathcal{J}^+ \cup \mathcal{J}^-$ together with the three points I^-, I^0, I^+. Hence one gets precisely the boundary attached to Minkowski space by conformal imbedding into the Einstein universe.

The projective structure of space-time defines a boundary $\partial_p V^4$ in a quite analogous way as described for the conformal structure above.

For Minkowski space the boundary coincides with the one which is obtained by the natural projective imbedding of Minkowski space into a 4-sphere. This agrees with the definition of "future projective infinity" defined by Eardley and Sachs recently [5]. The interesting point is that timelike geodesics which in the conformal boundary all terminate at one point I^0, terminate in the projective case on a hypersurface.

Therefore one might conjecture that generally the projective boundary will be useful to describe the behaviour of matter in the distant future.

Constructing the conformal boundary $\partial_c V^4$ of a weakly asymptotically simple space-time one finds that $\mathcal{J}$ is contained in $\partial_c V^4$. More interesting, however, is that any generator of $\mathcal{J}$ gets a future and past endpoint in $\partial_c V^4$! These sets of boundary points are denoted by I^-, $I^0(\mathcal{J}^+)$, $I^0(\mathcal{J}^-)$, I^+. It is, however, not true that the endpoints are identified always to form just three points as for Minkowski space. Therefore weakly asymptotically simple space-times can be naturally classified according to their structure of I^0.

In general I have so far only been able to find sufficient conditions for the structure of $\mathcal{I}^+$ which imply that $I^0(\mathcal{I}^+)$ is a point. The conditions are essentially that the news function tends to zero there. This indicates a relation between the structure of $I^0(\mathcal{I}^+)$ and the amount of radiation produced by the source in the infinite past, which is physically quite plausible. Necessary and sufficient conditions for $I^0(\mathcal{I}^+)$ to be a point and their relation to the outgoing radiation field have to be found by further investigations.

Suppose that $I^0(\mathcal{I}^+)$ is a point. Then one can show that there is a uniquely defined action of the Poincaré group on $\mathcal{I}^+$!

In the case of Minkowski space one can find the Poincaré group in the Bondi Metzner Sachs group as the subgroup of those transformations which are regular at the point I^0. In the general case, as long as $I^0(\mathcal{I}^+)$ is a point, even a singular one, there remains sufficient regularity along $\mathcal{I}^+$ to single the Poincaré group out of the BMS group. This result is of major importance, because it implies the possibility to define energy-momentum and angular momentum. The action of the Poincaré group on $\mathcal{I}^+$ defines a collection of canonical slices of $\mathcal{I}^+$ uniquely up to Poincaré transformations. Using the expressions of Tamburino and Winicour and the canonical slices one can define energy-momentum, angular momentum and calculate the change of these quantities in the radiation process. Because of the supertranslation freedom in the Bondi Metzner Sachs group it was up to now not possible to define angular momentum.

The canonical slices also define a preferred class of coordinate systems near $\mathcal{I}^+$ which are uniquely defined by the slicing of $\mathcal{I}^+$. These coordinate systems can be used to linearise Einsteins equation near $\mathcal{I}^+$. This way one gets a much smaller gauge group as usual.

These are the results obtained so far. Let us now turn to further problems which can be dealt with using the conformal boundary.

Cauchy data on a spacelike hypersurface determine uniquely a space-time. Hence the data specify also the conformal boundary. Its a formidable task to find conditions on the data which imply the existence of $\mathcal{I}$ and a certain structure of I^0. A simpler question is to ask for conditions on the asymptotic behaviour of the data which guarantee the existence of a "piece of $\mathcal{I}$" and a certain structure of I^0. The conformal boundary will certainly be a useful tool in this context and hopefully its application will bring some insight into the relation between null and spacelike asymptotic flatness.

Related to this is the problem of incoming radiation on $\mathcal{I}^-$. For a truly isolated system the radiation field on $\mathcal{I}^-$ should vanish. It is, however, completely unclear whether non-stationary solution satisfying this condition exists at all!

The structure of I^0 which relates in some sense $\mathcal{J}^+$ and $\mathcal{J}^-$ may give first indications. Somehow it seems puzzling that a change in the sign of the second fundamental form of the initial surface should shift the radiation field from $\mathcal{J}^+$ to $\mathcal{J}^-$.

The results obtained so far and the whole range of problems which can be reconsidered from a new point of view, show that the conformal boundary is a useful concept in General Relativity. Hopefully it will lead to further insight into the structure of Einstein's theory of gravitation.

Acknowledgement: This essay would not have been written in this form without many discussions with Martin Walker.

REFERENCES

[1] Penrose, R. Phys. Rev. Letts. 10 (1963).

[2] Geroch, R. J. Math. Phys. 13 (1972).

[3] Schmidt, B. G. Commun. Math. Phys. 36 (1974).

MAXIMAL SURFACES IN CLOSED AND OPEN SPACETIMES*

D. Brill**
Max-Planck-Institut für Physik und Astrophysik
München, West Germany

I. INTRODUCTION

In this paper we study spacelike hypersurfaces in general relativity with vanishing mean extrinsic curvature (trace of second fundamental form). Such surfaces have a direct geometrical meaning as surfaces of stationarity of the "area" (3-volume) functional[1]. Typically they are minimal surfaces in a Riemannian space, and maximal spacelike surfaces in a Lorentz space; we shall use the abbreviation M.S. for any surface of stationary area.

The importance of M.S. in general relativity has been pointed out by a number of authors[2-4]. For example, the Cauchy problem takes a particularly simple form on M.S., and the "kinetic" part of the gravitational energy is positive definite on such surfaces. Thus, for a practicable solution procedure of the Cauchy problem[3], and for the problem of positivity of total energy in general relativity, it is essential to know about the existence and general properties of M.S. In addition, Schücking[4] has emphasized that the description of gravitational fields on M.S. constitutes a particularly appropriate (time) coordinate condition for physical interpretation. A simple example is given in section III.

In general relativistic spacetimes with closed spacelike sections (i.e. cosmological solutions) M.S. have another immediate physical meaning: if a maximal Cauchy surface exists, then any set of observers will find that their space sections, as a function of their proper time, cannot expand forever; whereas if no such surface exists, some family of observers will find an ever-expanding 3-volume, without any recontraction phase. Thus in cosmology the existence of M.S. distinguishes universes that recollapse from those that expand forever.

Closely related to closed M.S. are spacelike surfaces of constant average extrinsic curvature (H-surfaces). York[3] has pointed out that the solution formalism and most of the physical interpretations for M.S. in asymptotically flat spaces can be carried over to H-surfaces in cosmological solutions.

II. M.S. IN ASYMPTOTICALLY FLAT SPACES

Considerable detail is known about M.S. in Euclidean space[1]. In the corresponding general relativity problem, complete M.S. are of particular interest for coordinate surfaces. Schücking[4] has pointed out that if an analog to the Bernstein theorem[5] exists in general relativity, M.S. would be sufficiently unique to make them particularly appropriate coordinate surfaces from the physical point of view. To date such a theorem has been proved[4] only under

---

*This work was supported in part by the National Science Foundation and by Humboldt Foundation.

**Humboldt Foundation U.S. Senior Scientist Awardee. Permanent address: Dept. of Physics, University of Maryland, College Park, Maryland 20742

assumptions (such as spherical symmetry) which severely restrict its generality. However, at least for the case of a fixed boundary, the set of spacelike surfaces has a simpler topology in a Lorentz manifold than in a Riemann manifold. Thus Seifert[6] has shown that there exists a M.S. bounded by the boundary of any compact set on a Cauchy hypersurface, provided the spacetime has no spacelike singularities. This leads to the conjecture that complete M.S. exist in any Lorentz manifold with suitable conditions (such as asymptotic simplicity) at conformal infinity $\mathcal{I}^+$. In this connection, Choquet-Bruhat[7] has shown that there exist M.S. in Lorentz manifolds which are close to Minkowski space (in the sense of a weighted Sobolev measure). Among well-known examples of M.S. in general relativity are totally geodesic surfaces ("plane" surfaces, on which the entire extrinsic curvature tensor vanishes) such as the surface t = const. in static spacetimes, or the surface of time-symmetry in time-symmetric spacetimes. Examples of non-plane M.S. are the surfaces t = const. in the Kerr spacetime and the maximal foliation of Schwarzschild-Kruskal space found by Estabrook et al[9] and by Reinhart.[10] This work can be immediately extended to discuss all spherically symmetric M.S. in any spherically symmetric static spacetime. We summarize the argument to illustrate some mathematical and physical points.

The "area" (3-volume) of a spacelike surface S is given by $A_1 = \int_S ({}^3g)^{1/2}\, d^3x^i$ where ${}^3g$ is the determinant of the induced metric components in the coordinates x^i on the surface. We could find the M.S. by varying the surface S in this coordinate-invariant integral. However, like in the analogous variational principle for geodesics, it is more convenient to use instead the non-invariant integral $A_2 = \int_S {}^3g\, d^3x^i$. Every spacelike M.S. corresponds also to an extremum* of A_2; however, extremizing A_2 also fixes the coordinates such that g = const**. After normalizing this "equal-area" map of R^3 into the space-time we have $A_1 = A_2$ at the extrema. The variational principle A_2, unlike A_1, also defines timelike and null M.S. (The proof of these statements is entirely analogous to the corresponding proof about geodesics).

We take a line element of the form

$$ds^2 = -B(r)dt^2 + C(r)dr^2 + r^2(d\theta^2 + \sin^2\theta d\phi^2) \tag{1}$$

Let s, θ, φ be coordinates on a spherically symmetric 3-surface t = t(s), r = r(s),

---

*We use "extremum" here in the sense of stationary point.

**However, this does not mean that a M.S. which is spacelike somewhere will be spacelike everywhere.

and let ' denote the s-derivative. Then we have

$$A_2/16\pi^2 = \int L ds = \int (Cr'^2 - Bt'^2)\, r^4 ds \qquad (2)$$

Since neither t nor s occurs in the integrand, the variational equations have the two first integrals (where a_1, a_2 are constants)

$$\partial L/\partial t' = 2r^4 Bt' = a_1 \qquad (3a)$$

$$L = r^4(Cr'^2 - Bt'^2) = a_2 \qquad (3b)$$

Eq. (3a) determines the preferred "equal-area" parameter s. It can be eliminated from Eqs. (3) to find the first integral of the variational equations for the "proper slope" $\phi = B^{1/2}dt/C^{1/2}dr = (B/C)^{1/2}\, t'/r'$ of the M.S.,

$$\phi = 0 \quad \text{or} \quad \phi = \pm\, (1 + ar^4B)^{-1/2} \qquad (4)$$

where $a = 4a_2/a_1^2$ = const. (For the Schwarzschild case, $B = C^{-1} = 1 - 2m/r$, this is Reinhart's solution .[10])The explicit description of the M.S. $t = t(r)$ can be obtained from this equation by quadrature. Reinhart has discussed the behaviour of the M.S. near the horizon of Schwarzschild-Kruskal spacetime and has shown that they can be smoothly continued from one r,t coordinate patch to the next.

Some general features of the M.S. follow directly from Eq. (4). The set of spherically symmetric M.S. form a two-parameter family: one parameter can be taken to be a, which determines ϕ of Eq. (4). Each ϕ in turn defines a one-parameter family of M.S. which are related by time-translations, $t \to t$ + const. For example, the family corresponding to $\phi = 0$ consists of the familiar surfaces t = const. Alternately, in the case of a two-sheeted "wormhole" solution like Schwarzschild-Kruskal[11] or extended Reissner-Nordström[12] spacetime we may think of the two parameters as the asymptotic t-values on the two sheets, provided the M.S. extends to both asymptotic regions. For example, in the Schwarzschild-Kruskal spacetime the usual surfaces of course have asymptotic values t,t ("antisymmetric" in u, v space), and the set of surfaces found by Estabrook et al.[9] have asymptotic values t,-t ("symmetric") on the two sheets. A time translation of the latter produces surfaces with general asymptotic values (t_1,t_2). These are all the spherically symmetric M.S. which are everywhere regular. In addition, there are M.S. which hit the singularity at r=0. In the description of Eq. (4), spacelike M.S. correspond to positive "area", hence $a_2 > 0$, $a > 0$. If a is small enough that $(1 + ar^4B) > 0$ for all r (i.e. $a \le 16/27m^4$ in the Schwarzschild case), the M.S. extends to r=0. If $1 + ar^4B \ge 0$ only for $r \ge r_o$ (or $r \le r_1$), the M.S. is of the Estabrook et al. type and does not hit the singularity (or hits the singularity in two places). Finally, we remark that the case a = 0 corresponds to lightlike M.S., and any spherically symmetric set of radial null lines (geodesics) forms such a M.S. Fig. 1 illustrates these M.S. in Schwarzschild-Kruskal spacetime.

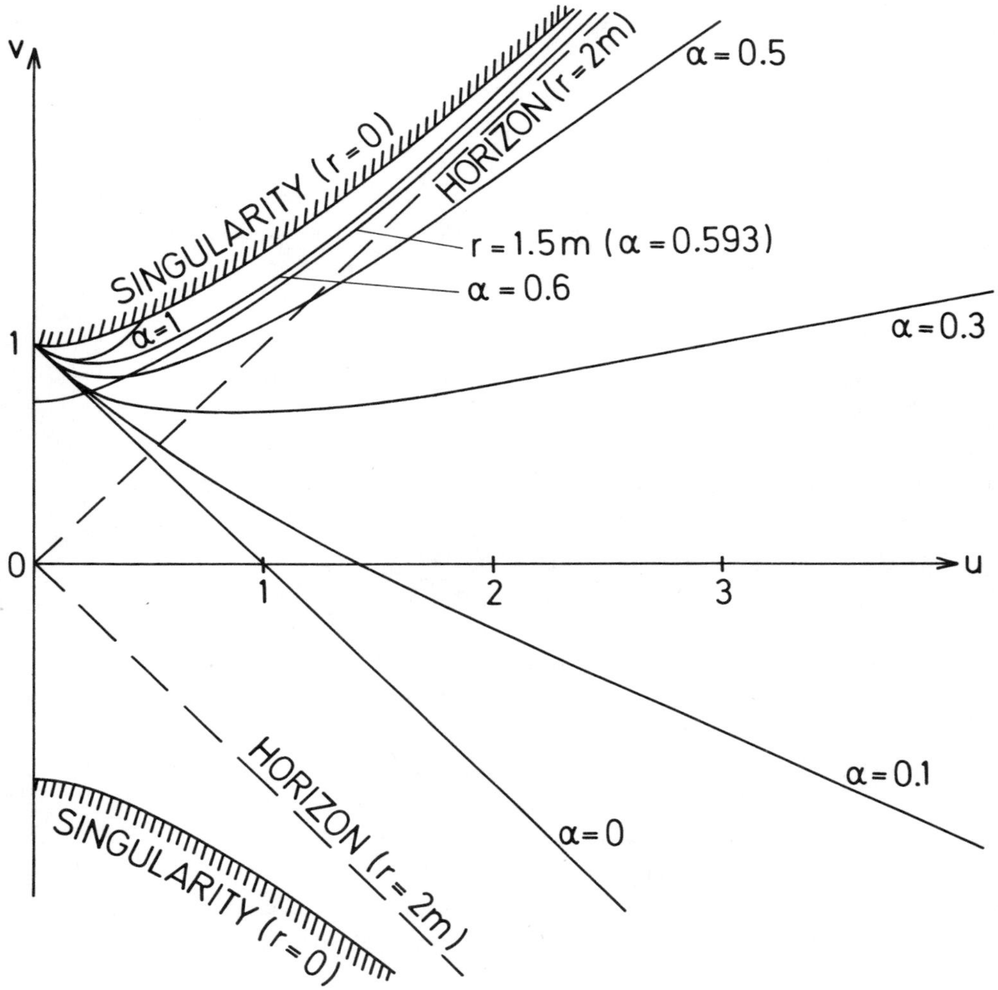

Figure 1. Extremal spacelike surfaces in Kruskal diagram. The plot is a numerical integration of Eq. (4) with $B = C_4^{-1} = 1 - 2m/r$, converted to Kruskal u, v coordinates. The parameter is $\alpha = a/m^4$. All surfaces start at u = 0, v = 1 and run to infinity ($\alpha < 0.593$) or back to the singularity ($\alpha > 0.593$). All other spherically symmetric M.S. with one or two intersections with the singularity at r = 0 are obtained from these by a time translation, t' = t + const. Except for the surface r = 1.5m we do not show the M.S. which do not intersect the singularity (see ref. 9 for a plot of those M.S.). All M.S. shown are spacelike except the limit $\alpha = 0$, which is null.

III. NON-SPHERICALLY SYMMETRIC M.S. AND BLACK HOLES IN SLOW MOTION

Schücking[4] has pointed out that, in analogy to Bernstein's theorem[5] for Euclidean space, one would expect the class of complete M.S. in general relativity to be quite restricted; for example, in Minkowski space Calabi[5] showed that the obvious hyperplanes are the only complete M.S. For a Schwarzschild-Kruskal or other wormhole type spacetime one similarly expects that complete M.S. are the asymptotically static ones (t = const) discussed above, and surfaces which are related asymptotically to t = const surfaces by Lorentz transformations, and no others. These asymptotically Lorentz-transformed surfaces would describe the black hole from the point of view of an asymptotically uniformly moving observer, i.e. they would represent a moving black hole.*

Exact solutions for moving black holes are not known, but if we assume that such M.S. indeed exist, then we can find them by approximation. One such class has been found by Smarr[13] in the linear approximation; we give here the general solution in that approximation. A convenient way to perturb the M.S. equation is to use the spacelike description ("3+1 decomposition") of Einstein's equations[14]. We start with a totally geodesic spacelike surface, such that the extrinsic curvature density $\pi^{ij} = g^{1/2} K^{ij}$ vanishes at t = 0. We want to propagate to another M.S. in a given spacetime; hence we perform the converse of the usual time development computation and ask what lapse function N (with shift N^i assumed zero) preserves the vanishing (trace π^{ij}) = π to first order, i.e. makes $\dot{\pi} = 0$. Since the spacetime geometry is uniquely determined by the initial data, it is not surprising that only the initial data enter into the first order answer. As these data we take a conformally flat+ initial metric, and a vanishing initial extrinsic curvature

$$ds^2 = \psi^4 ds_\flat^2 \qquad \pi^{ij} = 0 \qquad \text{at } t = 0 \tag{5}$$

It is well known[16] that the initial value equations then demand

$$\nabla_\flat^{\,2}\psi = 0, \tag{6}$$

that the Schwarzschild-Kruskal spacetime (in isotropic coordinates) is the time development of the solution $\psi = 1 + m/2r$, and that the solution $\psi = \sum_{i=1}^{n} 1+m_i/2r_i$ represents an initial configuration of n "wormholes", each with a separate, asymptotically flat interior.

---

*We do not distinguish here between "black hole" and "wormhole". Coordinate transformations of Schwarzschild spacetime which are asymptotically Lorentz transformations have been discussed by a number of authors[15], but the spacelike sections t = const of these solutions do not have any particular geometrical significance.

+The subscript $\flat$ denotes the flat metric, and operators formed in this metric.

From the equations of time development[14] it is easy to compute

$$\dot{\pi} = -2g^{1/2}\ \nabla^2 N = 0 \tag{7}$$

as the requirement for N. The Laplacian ∇^2 in the initial metric ds^2 can be expressed in terms of the flat space Laplacian $\nabla_{\flat}^2$, and in view of (6), Eq. (7) takes the form

$$\nabla_{\flat}^2\ (\psi N) = 0 \quad . \tag{8}$$

Recalling that N is related to a spacelike surface near the initial surface via N = proper time separation from initial surface, we recognize that for an asymptotic Lorentz transformation we must choose a solution of (8) corrsponding to $\ell = 1$. In particular, for motion along a polar axis ($\theta=0$) and for the case of a single Schwarzschild black holes we have

$$N = (v_1 r - 8v_2/m^3 r^2)\cos\theta/(1+m/2r) \tag{9}$$

and the spacelike surface is described by

$$\tau = (g^{oo})^{1/2} t = [(1-m/2r)/(1+m/2r)]\ t = N$$

or

$$t = (v_1 r - 8v_2/m^3 r^2)\cos\theta/(1-m/2r) \tag{10}$$

Since this is a first order approximation we must have v_1, $v_2 << 1$. We see that v_1 measures the velocity in the asymptotic region $r \to \infty$, and, by changing variables to $r' = m^2/4r$, we see that v_2 measures the velocity in the asymptotic region $r \to 0$. Again these M.S. form a 2-parameter (v_1, v_2) family, because there are two asymptotically flat regions in this spacetime. The solution with $v_1 = v_2$ has odd symmetry in Kruskal u, v coordinates, hence can be described as a "slowly moving version" of the t= constant spherically symmetric M.S. It was first found by Smarr[13]. The solution (10) with $v_1 = -v_2$ has even symmetry similar to the spherically symmetric surfaces of Estabrook et al.[9] In addition, the simple solution (which contains both $\ell = 0$ and $\ell = 1$ components)

$$t = N_o + v_1 r \cos\theta/(1-m/2r) \qquad (N_o = \text{const}) \tag{11}$$

is of interest; here the worm-hole is moving as seen from $r \to \infty$, but is static as seen from $r \to 0$. These statements about the velocity of the black hole could be made more precise by evaluating the whole extrinsic curvature of these slices, and computing from it the total momentum as given by a surface integral[17] in one or the other asymptotic region. We note that if the analogue of the Bernstein theorem is true in Schwarzschild-Kruskal spacetime, there should not exist any solutions of the exact M.S. equations corresponding to solutions of Eq. (8) with $\ell > 1$.

IV. EXAMPLE OF APPLICATION WITH MOVING BLACK HOLES

The method of section III applied to an initial value solution for several wormholes can only produce a M.S. with respect to which the whole collection of wormholes is slowly moving as a unit; it cannot yield an approximate solution to the more interesting problem of relative motion of wormholes with respect to each other. A solution of the initial value equations essentially different from (5) is needed, in which the extrinsic curvature tensor sufficiently near each moving wormhole has the approximate form determined by (9) for a single moving hole. Smarr[13] has suggested one procedure to obtain such a solution. We want to focus here on the simple situation provided naturally by the time-development of a two-wormhole solution of type(5). Initially the holes have vanishing relative velocity, but as we propagate into the future we should expect a gravitational acceleration of the holes toward each other, so that at later times a veloity should be aquired. Our slow motion approximation is only sufficiently accurate to compute the initial acceleration of the wormholes. The weak-field (large wormhole separation) limit of this problem should give the basic law of Newtonian gravity theory. In part because no suitable definition for the distance between wormholes is available,* this initial acceleration has never been analytically computed. The proposal here is to measure acceleration as a change of velocity, and velocity by comparison with the slowly moving wormhole solution.

Let r_1, r_2 denote the distances of a field point in Euclidean 3-space from two origins 1,2, respectively. The two origins themselves are separated by a Euclidean distance d. We consider an initial value solution of type (5) with

$$\psi = \psi_{(2)} = (1+m/2r_1 + m/2r_2)/(1+m/2d) \tag{12}$$

These data can be thought of as representing two wormholes located near $r_1 = 0$ and $r_2 = 0$ respectively with zero initial velocity (see Ref. 16). Here ψ is normalized such that it reaches the value 1 in the asymptotic regions $r_1 \to 0$ and $r_2 \to 0$, because we want to compare the time development of those regions with a single moving wormhole. If the single wormhole were not moving it could be described by initial values of type (5) with a ψ which asymptotically (say at $r_1 \to 0$) agrees with (12), namely

$$\psi = \psi_{(1)} = 1 + m_B/2r_1 \qquad m_B = m/(1+m/2d) \tag{13}$$

*Distance along some geometrically preferred path, such as the minimum distance D between the wormhole throats, is affected by the collapse of the individual wormholes. For example, D actually increases initially. More globally defined distances, such as that which measure the electrical capacity coefficients of the wormholes when a test electric field is threaded through them, may be more appropriate. The advantage of the present approach is that it is independent of any distance definition for evaluating the acceleration.

Here m_B is the "bare" mass[16] of one of the wormholes, measured, as above, by its asymptotically Schwarzschildian behaviour, or via a suitable surface integral. Incidentally, the total mass of the system is $2m/(1+m/2d)^2$, so that the parameter m measures neither the bare nor the total mass, except when $d \gg m$.

Now we let both the two wormholes (12) and the single wormhole (13) develop in time. To lowest order the spacelike geometry is unchanged and therefore is still described by two conformal factors ψ_1 and ψ_2. However, the extrinsic curvature tensor begins to differ from zero according to the equation of time development,

$$\dot{\pi}_{ij}\Big|_{t=0} = \sqrt{g}\,(N_{|ij} - NR_{ij})$$

$$= (\sqrt{g}/\psi^2)\,(M\psi_{ij} + M_{ij}\psi + 2M_\kappa\psi^\kappa g_{ij} - 3\psi_{(i}M_{j)}) \qquad (14)$$

Here $M = \psi N$, | denotes 3-dimensional covariant differentiation in the metric (5), R_{ij} is the Ricci tensor of that metric, sub and superscripts on scalars denote flat space derivatives, and g_{ij} is the flat-space metric. For the time development of the two wormholes we choose proper time as time parameter,

$$N_{(2)} = 1 \qquad (15)$$

From Eq. (7) we see that this does correspond to a M.S. to this order. For the time development of the single wormhole we choose a time coordinate which agrees with (15) at $r_1 \to 0$, and deviates from it at $r_1 > 0$ by acquiring whatever velocity on the other sheet, as in Eq. (11), is necessary that the resulting $\dot{\pi}_{ij}$ agree most closely with that of the double wormhole case. Thus we choose

$$N_{(1)} = 1 + vz/\psi_{(1)} \quad \text{or} \quad M_{(1)} = \psi_{(1)} + vz \qquad (16)$$

with v, the velocity in the z-direction at t=1, as yet undetermined.

Since the comparison is to be made near $r_1 \to 0$ we expand the term $m/2r_2$ in $\psi_{(2)}$ to lowest order, assuming the wormholes to lie along the z axis,

$$\psi_{(2)} = \psi_{(1)} - (m_B z/2d^2) \qquad (17)$$

We note that M and ψ enter Eq. (14) symmetrically, hence in order that $\dot{\pi}_{(1)ij} = \dot{\pi}_{(2)ij}$ we must have, as can be verified by explicit calculations*,

$$v = -\,m_B/d^2 \quad \text{at} \quad t = 1 \qquad (18)$$

Since the wormholes started with zero initial velocity, their acceleration toward each other is m_B/d^2. The agreement with the Newtonian limit is gratifying, but not too much should be read into the exact agreementfor all d, because d is a distance in the auxiliary Euclidean space conformally related to the initial spacelike surface. Its scale is well-defined by the requirement $\psi \to 1$ at $r_1 \to 0$, but another equally reasonable scale would be one for which $\psi \to 1$ at $r_1, r_2 \to \infty$. The present calculation does have the advantage that all the quantities have clear geometrical significance, even though one could argue

---

*For example,

$$\dot{\pi}_{(2)ij} = (2/\psi^2)(\psi\psi_{ij} + \psi_\kappa\psi^\kappa g_{ij} - 6\psi_i\psi_j) - (m_B/d^2\psi^2)(z\psi_{ij} + 2\psi_3 g_{ij} - 3\psi_{(i}\delta_{j)}{}^3).$$

whether they represent the "best" definition of velocity, distance etc. The use of M.S. in this calculation could probably be replaced by other equally meaningful geometrical conditions that allow making a correspondence between spacelike surfaces in the one- and two-wormhole spacetimes. At least M.S. seem very simple and natural for this purpose. They may similarly help with the physical interpretation of other spacetimes involving moving wormholes or moving matter.

IV. M.S. IN COSMOLOGY

Whereas in asymptotically flat spaces complete M.S. seem to be common and plentiful, closed M.S. in closed universes are exceptional and scarce. It is easy to give examples of closed universes - i.e. spacetimes of topology $M_3 \times R$, where M^3 is closed and spacelike - in which no closed spacelike M.S. exist. Consider the "uniformly expanding" universes with metric

$$ds^2 = -dt^2 + R^2(t)d\sigma^2 \tag{19}$$

where $d\sigma^2$ is some t-independent metric on M_3, and $R(t)$ is a monotonically increasing function. The "area" (3-volume) of any spacelike surface $(t = t(x^i))$ can be increased by performing a (constant) time translation Δt, because each area element will be changed by the factor $R^3(t(x^i) + \Delta t)/R^3(t(x^i))$. Thus no spacelike surface can be a M.S.

This simple example already shows that the existence of M.S. is closely connected with behavior of the radius of the universe $R(t)$, when defined. For models of type (19), which includes of course the Friedman models, M.S. can only exist if $dR/dt = 0$ at some t. The familiar moment of maximum expansion is an example of a M.S. in Friedman universes. In fact, in all homogenous universes, independent of field equations, each extremum of R corresponds to a M.S. This follows from Hsiang's theorem[18], that any extremal orbit of a closed group of isometries is a M.S. Static universes of course can have a whole family of spacelike M.S., and the deSitter universe shows that even universes with large-scale dynamics can have such a family of M.S. (namely any t=const surface in the "static frame"). However, if we use the Einstein field equations with a reasonable energy condition, the possibilities for M.S. are greatly reduced.

The simplest consequence that can be read out of the Einstein equation does not concern the first variation of the "area" (3-volume) functional (which vanishes for M.S.), but the second variation. The formula for the second variation of the area of a submanifold is well known[19] (for a reasonably simple derivation see Appendix): Suppose S_t is a family of spacelike surfaces parametrized by t. Let $\underline{n}$ be the field of unit normals to these surfaces, and choose a coordinate system such that the t-coordinate runs along integral curves of the normal, and the rest of the coordinates (x^i) run along the surface. In other words, $\partial/\partial x^i$ shall be tangential vectors and $\partial/\partial t$ shall be a normal vector of length N, so that

$$\partial/\partial t = N\underline{n} \tag{20}$$

Let $A(t)$ be the 3-volume of S_t, $<,>$ the scalar product and Ricci (,) the Ricci tensor in spacetime, then

$$d^2A/dt^2 = \int (N_i N^i <\underline{n},\underline{n}> - N^2 K_{ij} K^{ij} - N^2 \text{Ricci}(n,n))\, d^3A \tag{21}$$

Here $i, j = 1,\ldots 3$, subscripts and superscripts on scalars again denote differentiation, K_{ij} represents the extrinsic curvature of the surface, and d^3A the 3-volume element on the surface. In a Lorentz space we have $<n,n> = -1$, and Einstein's equations determine

$$\text{Ricci}(n,n) = \kappa(T(n,n) + \frac{1}{2}\,\text{Tr } T) \geq 0 \tag{22}$$

by the (weak) energy condition. Thus we find

$$d^2A/dt^2 \leq 0 \tag{23}$$

and equality is possible only if N is constant and K_{ij} vanishes. For this latter case one can then show [20]that $d^4A/dt^4 < 0$ unless spacetime is flat. Thus we find; *Any closed M.S. in a closed universe which satisfies Einstein's equations with the energy condition is a local maximal surface and is locally isolated.*

From Eq. (23) we can also conclude that the Hessian of the M.S. under these conditions is negative definite, that the index of the M.S. is 0, and that there are no Jacobi fields (unless $K_{ij} = 0$ and $T_{\mu\nu} = 0$). Hence, by the generalization of the Morse index theorem[21] , there are no conjugate boundaries on a M.S. Thus the situation for spacelike M.S. is just the opposite as that for timelike geodesics, where the energy conditions implies the existence of conjugate points and associated singularity theorems.

One would like to conclude from the above theorem that there can be in fact at most one M.S. in a closed universe. If the area were a function of a finite number of variables rather than a functional then the fact that all its stationary points (M.S.) are maxima (Eq. (23)) would imply by Morse theory that it has in fact only one stationary point, i.e. only one M.S. It is conceivable that Morse theory can be extended to the area functional on spacelike surfaces. Below we show a somewhat more restricted theorem valid for Cauchy M.S.[22]

Suppose S_o is a Cauchy M.S. and S_1 is another M.S. in a closed universe. Erect a family F of timelike geodesics from S_o to S_1, normal to S_o, parametrized by t (proportional to arclength) such that $t = 0$ on S_o and $t = 1$ on S_1, and denote the tangent to these geodesics by ∂_t. Consider the "length" (>0 for timelike curves)

$$L = - \int <\partial_t, \partial_t> dt \tag{24}$$

as a function of position on S_o. Since S_o is compact, L reaches a maximum. Let $c(t)$ be this maximizing geodesic; it must meet S_1 orthogonally. Let $c(t,s)$ be a variation of $c(t)$ within the family F. Then $c(o,s)$ and $c(1,s)$ are curves on S_o

and S_1 respectively, and ∂_s is parallel along c(t). The standard second variation formula[23] gives

$$(1/2)d^2L/ds^2 = -\langle D_s\partial_s,\partial_t\rangle \Big|_o^1 - \int (\langle R(\partial_s,\partial_t)\partial_s,\partial_t\rangle + \langle D_s\partial_t, D_s\partial_t\rangle)dt \tag{25}$$

(all at s=0). Now sum Eq. (25) over an orthonormal set of directions ∂_s in S_o. The boundary terms are proportional to the extrinsic curvature components $K(\partial_s,\partial_s)$, so after summation they become TrK and vanish because S_o, S_1 are M.S. The term in the Riemann tensor becomes $\mathrm{Ricci}(\partial_t, \partial_t) = R_{oo} > 0$ by the energy condition. The last term on the right vanishes because $D_s\partial_t = D_t\partial_s$ and ∂_s is parallely transported. The left side becomes $\frac{1}{2}\nabla^2_{S_o} L$, which is negative for a maximum L. But the right side is positive (except in the vacuum case, where L = const would be possible). This contradiction shows that S_o and S_1 must in fact be identical.

Thus the picture begins to emerge that closed universes in general relativity have either one absolute maximal spacelike surface or none at all. Physically this distinction corresponds to universes in which every family of observers finds their spacelike surfaces to have bounded 3-volume - i.e. when typically the universe reaches a maximum expansion phase and recollapses - vs. universes in which at least some family of observers finds that their 3-volume increases forever, in general without bound. The two types correspond to closed, and flat or "open", Friedmann universes, and it should be noted that models of the latter type with closed spacelike sections do exist (at the cost of losing global isotropy). Thus the two classes are non-empty, and this classification by behavior at large expansion is complementary to the more common one which focuses on the behavior near the singularity. For Friedmann universes the sign of the "Newtonian energy" determines the class (recollapse for energy < 0, expansion for energy ≥ 0). It can be evaluated at any epoch. It would be useful to have a similar quantity for more general universes. The next section reports a small step in this direction for a restricted set of universes.

V. PERTURBATIONS OF FLAT UNIVERSES

The most familiar example of a closed flat universe is the 3-torus universe, of topology $T_3 \times R$, obtained by identifying points in Minkowski space according to three independent spatial translations. Other closed flat universes differ only by the use of other discrete spacelike isometries as identification rules. These universes have a family of parallel M.S., and constitute an example where the L of the last section differs from zero (and is of course constant). It is therefore interesting to study the behavior of these M.S. when these spaces are perturbed. Choquet-Bruhat's theorem[7] guarantees the existence of H-surfaces in the perturbed spaces, but for this case cannot guarantee that H=0 is possible, which would correspond to M.S. We show that in fact there can be no M.S. in a nontrivial perturbation.

This result is proved by a straightforward application of the results of Brill and Deser[24]. In that paper the perturbations were decomposed into matter perturbations $\delta T^{o}{}_{o}$, TT parts of the metric and extrinsic curvature perturbations δg^{TT}_{ij}, $\delta\pi^{TT}_{ij}$, and longitudinal geometrical perturbations. The latter are determined by the initial value equations, whereas the former can be chosen freely and constitute the non-trivial perturbations. Among the quantities that are determined when the calculation is carried to second order is the space average of the mean extrinsic curvature,

$$c_\pi = V^{-1} \int \pi_i{}^i \, dV \tag{26}$$

namely one has

$$(1/6)Vc_\pi^2 = \int (g^{mn}\delta g^{TT}_{ij,m}\, \delta g^{TT}_{k\ell,n} + \delta\pi^{TT}_{ij}\, \delta\pi^{TT}_{k\ell})\, g^{ik}\, g^{i\ell}\, g^{1/2}\, d^3x + 2 \int \delta T_o{}^o \, d^3x \tag{27}$$

Since the right side is positive for nontrivial (non-coordinate) perturbations, we must have $c_\pi \neq 0$, hence $\pi_i{}^i \neq 0$ no matter what surface is chosen as the perturbed initial surface. Thus no surface can be a M.S.

Since this proof was carried out on the level of the initial value problem, one only needs to know that the perturbation is small on some initial surface. Since the absence of M.S. signals a continuing expansion, the perturbed solution will at late times in general differ considerably from the unperturbed flat universe. Occurrence of a M.S. at these late times is not excluded by the above local argument. However, if the matter and gravitational wave contents adiabatically decompresses as the expansion proceeds, the universe at later times will be a perturbation of a larger flat universe, and the argument can be repeated.

Examples of such perturbations of flat universes are some Kasner universes (with parameters close to the values 1,0,0), and the Gowdy T_3 universe[25] . Both show the characteristic continued expansion without formation of M.S. Thus, for nearly flat T_3 universes one can tell whether a M.S. exists by computing the expression (27) at an arbitrary time. A similar energylike expression added to the "Newtonian" energy balance in nearly Friedmann universes would enable one to tell, at any time, whether a M.S. will be formed in these universes. Whether one can write a workable criterion for M.S. in more general universe models in terms of initial data on an arbitrary spacelike surface is of course still an open question.

Acknowlegements. I thank J. Cavallo, F. Flaherty, J. Nester, B. Reinhart, T. Roman, E. Schücking, H. J. Seifert and L. Smarr for helpful discussions.

APPENDIX

THE SECOND VARIATION FORMULA FOR M.S. AREA

In the notation of section IV, we compute the surface integral $A(t) - A(0) = \int_{S_t - S_o} \underline{n} \cdot \underline{dA}$ as the integral over the volume V bounded by S_t, S_o, $\int_V \mathrm{div}\, \underline{n}\, dV$.

But

$$dV = N dt dA \tag{A1}$$

hence

$$dA/dt \Big|_{t=0} = \lim_{t \to 0} \int_o^t \int_{S_o} n^\alpha{}_{;\alpha} N dt dA = \int_{S_o} n^\alpha{}_{;\alpha} N dA. \tag{A2}$$

For a M.S. we demand $dA/dt = 0$ for arbitrary N, and obtain

$$n^\alpha{}_{;\alpha} = 0 \tag{A3}$$

as the usual M.S. condition. By a similar argument we find

$$dA/dt \Big|_o^t = \int_V n^\alpha{}_{;\alpha\beta} N n^\beta dV \tag{A4}$$

Now use the Riemann tensor to exchange the covariant differentiations, integrate by parts, use (A3) and note that the definition of the extrinsic curvature implies

$$n^\alpha{}_{;\beta} n^\beta{}_{;\alpha} = K_{ij} K^{ij}, \tag{A5}$$

and that

$$n^\alpha{}_{;\alpha} n^\beta N_\alpha = - (N^i N_i / N) \langle n,n \rangle .$$

Finally use (A1) and take the limit $t \to 0$ to obtain directly Eq. (21)

REFERENCES

[1.] See, for example, R. Courant, Dirichlet's principle, conformal mapping and minimal surfaces, Interscience (1950).

[2.] Arnowitt, Deser and Misner in L. Witten (editor), Gravitation, Wiley (1962) D. Brill and S. Deser, Ann. Phys. (N.Y.) 50 (1968) 548

[3.] J. York and N. O'Murchadha, J. Math. Phys. 14, (1973) 1551

[4.] E. Schücking, paper presented at this meeting

[5.] For Euclidean space see J.C.C. Nitsche, Bull. Am. Math. Soc. 71 (1965); for Minkowski space see E. Calabi, Global Analysis Part II, S.S. Chern and S. Small (editors), Am. Math. Soc. 1970, p. 223; S. -Y Chang.and S. -T Yan, Preprint Courant Institute 1975

[6.] H. -J. Seifert, Thesis, Hamburg 1968

[7.] Y. Choquet-Bruhat, C.R. Acad. Sc. Paris, 280 (1975) 169

[8.] D. Brill, Ann. Phys. (N.Y) 7 (1959) 466

[9.] Estabrook, Wahlquist, Christensen, DeWitt, Smarr and Tsiang, Phys. Rev. D7 (1973) 2814

[10.] B. L. Reinhart, J. Math. Phys. 14, (1973) 719

[11.] M. Kruskal, Phys. Rev. 119, (1960) 1743

[12.] J. C. Graves and D. Brill, Phys. Rev. 120, (1960) 1507

[13.] L. Smarr, Bull. Am. Phys. Soc. 18, (1973) 643

[14.] See, for example, ref. 2 or Misner, Thorne, Wheeler, Gravitation, Freeman (1973), Chapter IIV.21

[15.] See, for example, N. Rosen, Ann. Phys. (N.Y.), 17, (1962) 269

[16.] D. Brill and R. Lindquist, Phys. Rev. 131. (1973) 471

[17.] N. O'Murchadha and J. York, Jr., Phys. Rev. (1975)

[18.] W. Hsiang, Proc. Nat. Acad. Sci. (USA) 56, (1966) 5

[19.] A. Duschek, Math. Z. 40, (1936) 279

[20.] D. Brill, Nuovo Cim. Suppl. II, Ser. 1, (1964) 1

[21.] S. Smale, J. Math. Mech. 14, (1965) 1049

[22.] D. Brill and F. Flaherty, Comm. Math. Phys. (to appear)

[23.] See, for example, N. Hicks, Notes on Differential Geometry, D. van Nostrand 1965

[24.] D. Brill and S. Deser, Comm. Math. Phys. 32, (1973) 291

[25.] R. H. Gowdy, Ann. Phys. (N.Y.) 83, (1974) 203

A NECESSARY CONDITION FOR KILLING HORIZON TO BE EVENT HORIZON

P. Hajicek

Institute for Theoretical Physics, University of Berne, Berne, Switzerland

We consider a Killing horizon [1], that is a null hypersurface $\mathcal{H}^+$ embedded in a spacetime $\mathcal{M}$ which allows an isometry group $\mathcal{G}$ such that

1) $g\mathcal{H}^+ = \mathcal{H}^+$, for all $g \in \mathcal{G}$,

2) the null geodesic generators of $\mathcal{H}^+$ are trajectories of some one-dimensional subgroup $\mathcal{G}_1$, say, of $\mathcal{G}$. Let ξ^i be a Killing vector field generating $\mathcal{G}_1$.

We suppose:

1) $\mathcal{H}^+$ has the topology $S^2 \times R^1$, is geodesically complete (every non-diverging null hypersurface is totally geodesic) and non-degenerate [2], or, at $\mathcal{H}^+$:

$$\xi^i{}_{;j}\xi^j = \kappa\xi^i \ , \kappa \neq 0.$$

It follows [3] that there is another Killing horizon, $\mathcal{H}^-$, in $\mathcal{M}$ intersecting $\mathcal{H}^+$ at the two-submanifold where $\xi^i = 0$.

2) There is a neighbourhood $\mathcal{U}$ of $\mathcal{H}^+ \quad \mathcal{H}$ in $\mathcal{M}$ such that the Einstein-Maxwell equations are satisfied in $\mathcal{U}$.

Let $\mathcal{S}$ be a compact space-like section of $\mathcal{H}^+$. $\mathcal{S}$ is a Riemann 2-manifold with the topology S^2 and with induced positive definite metric γ_{AB} (in some coordinate chart x^A on). One can show that all such $\mathcal{S}$ are isometric (e.g. [4]).

Consider a vector field ℓ^i along $\mathcal{S}$ which is tangential to the generators of $\mathcal{H}^+$. ℓ^i is given up to the transformation

$$\ell^{i'} = f\ell^i \tag{1}$$

where $f(x^A)$ is any positive C^∞ function on $\mathcal{S}$. The covariant derivative of ℓ^i along the coordinate curve x^A on $\mathcal{S}$ is denoted by $\ell^i{}_{;A}$. One can show [5] that

$$\ell^i{}_{;A} = \Omega_A\ell^i \ ,$$

where Ω_A is a vector field on $\mathcal{S}$. Ω_A changes as follows

$$\Omega'_A = \Omega_A + (\ell g\, f)_{,A} \ , \tag{2}$$

if ℓ^i is transformed according to (1). It is well-known that the divergence-free element of the class of vectors Ω'_A given by Eq. (2) for all f's exists and is uniquely determined ($\mathcal{S}$ is compact). Moreover, it does not depend on the section $\mathcal{S}$ chosen. Let Ω_A denote, from now on, only this vector field:

$$\Omega^A{}_{;A} = 0 \tag{3}$$

Such Ω_A describes the amount of rotation in ξ^i: it has been shown [5] that, in case $F_{ij} = 0$,

$$\xi_{[i}\xi_{j;\ell]} = 0 \text{ in } \mathcal{U} \leftrightarrow \Omega^A = 0 \text{ at } \mathcal{S}$$

Finally, following Carter [6], we introduce the radial electric and magnetic fields on $\mathcal{H}^+$: an observer flying through $\mathcal{H}^+$ at p can use an orthonormal frame whose time vector is his velocity u^i, and whose space-like vectors e_1^i, e_2^i are tangential to $\mathcal{H}^+$. Then, e_3^i is uniquely determined by the condition to be oriented outwards of $\mathcal{H}^+$ and can be called radial direction for the observer. The components E and H of the electric and magnetic field as seen by the observer in the radial direction can be shown to be constant along the generators and independent of u^i (this follows from the fact that $\mathcal{L}_\xi F = 0$). They define, in such a way, two invariant functions $E(x^A)$, $H(x^A)$ on $\mathcal{S}$.

One can show [5], that these four quantities $\mathcal{J}_{AB}$, Ω_A, E, H given on $\mathcal{S}$ determine uniquely the gravitational and electromagnetic field in $\mathcal{U}$.

Not every Killing horizon is an event horizon. A necessary condition for this has been given by Hawking and Ellis [7]: there should be no outer trapped surfaces near $\mathcal{H}^+$. This implies

$$\int_{\mathcal{S}} (\Omega^2 + E^2 + H^2).\sqrt{\mathrm{Det}(\mathcal{J}_{AB})}\; d^2x < 4\pi \;, \tag{4}$$

where

$$\Omega^2 = \mathcal{J}_{AB}\Omega^A\Omega^B \;[5].$$

Example: Kerr outer ($r = r_+$) and i nner ($r = r_-$) horizons $\mathcal{H}_+$ and $\mathcal{H}_-$: Eq. (4) is satisfied at $\mathcal{H}_+$, but it is not satisfied at $\mathcal{H}_-$. Thus, Eq. (4) is not just a generalization of the Kerr-Newman inequality

$$a^2 + e^2 + h^2 < m^2 \;, \tag{5}$$

because this does not distinguish between $\mathcal{H}_+$ and $\mathcal{H}_-$. Moreover, the source of the fields Ω_A, E and H can also be outside charge and matter currents, not only the black hole itself. Eq. (4) says that the fields Ω_A, E and H cannot be arbitrarily strong at the horizon of a black hole. This need not mean that outside fields can destroy black holes, even in principle (the necessary fields are in any case too high for any reasonable astrophysical situation). Rather, if we try to produce such strong fields at the horizon in order to violate the Eq. (4), then we had to collect so much energy and it had to be brought so near to the horizon, that another horizon will be formed enclosing all this energy together with the original black hole. On the other hand, if we try to "destroy" the black hole from far away, then the fields, however strong at their sources, can be modified by the spacetime curvature in the neighbourhood of the horizon so that the Eq. (4) is always satisfied. A family of exact solutions of Einstein-Maxwell equations which could probably be used to construct a model of such a situation has been found by Ernst [8]. Another confirmation of this hypothesis in a special case is [9]. More work, however, is needed to clarify the matter.

We can derive an analogon of Kerr-Newman inequality (5) in case of axial symmetry, generated by the Killing vector η^i. Then, the total angular momentum J of the hole is well-defined [10]. On the other hand, the total electric and magnetic charges Q_e and Q_m are well-defined in general case by Carter [6]. According to these definitions, we can calculate J, Q_e and Q_m by means of $\mathcal{J}_{AB}$, Ω_A, E and H:

$$\mathcal{J} = -\frac{1}{8\pi}\int_{\mathcal{S}} \mathcal{J}_{AB}\Omega^A\eta^B \sqrt{\mathrm{Det}(\mathcal{J}_{AB})}\ d^2x \ ,$$

$$Q_e = \frac{1}{4\pi}\int_{\mathcal{S}} E\sqrt{\mathrm{Det}(\mathcal{J}_{AB})}\ d^2x \ ,$$

$$Q_m = -\frac{1}{4\pi}\int_{\mathcal{S}} H\sqrt{\mathrm{Det}(\mathcal{J}_{AB})}\ d^2x \ .$$

Using Schwartz inequality for the squares J^2, Q_e^2, Q_m^2 and denoting

$$\mathcal{A} = 16\pi\ M_{irr}^2 = \int_{\mathcal{S}} \sqrt{\mathrm{Det}(\mathcal{J}_{AB})}\ d^2x \ ,$$

$$\overline{R^2} = \frac{1}{\mathcal{A}}\int_{\mathcal{S}} r^2\sqrt{\mathrm{Det}(\mathcal{J}_{AB})}\ d^2x \ ,$$

where $2\pi r(p)$ is the circumference of the η^i-trajectory going through p on $\mathcal{S}$, we obtain from (4)

$$\frac{4J^2}{\overline{R^2}} + Q_e^2 + Q_m^2 < 4M_{irr}^2 \ .$$

This inequality holds for a stationary, axisymmetric, but otherwise arbitrarily deformed black hole.

The squared mean circumference radius $\overline{R^2}$ can, in general, differ from the effective Schwarzschild radius $R_0^2 = \mathcal{A}/4\pi$: for a pancake-shaped black hole, $\overline{R^2} > R_0^2$ and, for a cigar-shaped one, $\overline{R^2} < R_0^2$. We notice that, in the latter case, the inequality (6) would be weakened by replacing $\overline{R^2}$ by R_0^2.

The author is indebted to Dr. B. Carter and to Dr. G. Gibbons for useful conversations.

REFERENCES

[1] B. Carter, J. Math. Phys. 10,70 (1969).

[2] B. Carter, in Black Holes. Ecole d'ete de Physique theorique, Les Houches, 1972. Ed. by C. DeWitt, B. DeWitt. Gordon and Breach, 1973.

[3] R. H. Boyer: Proc. Roy. Soc. A311, 245 (1969).

[4] L. Smarr: Phys. Rev. D7, 289 (1973).

[5] P. Hajicek: J. Math. Phys. 16, 518, 523 (1975).

[6] B. Carter: "The Electrical Equilibrium of a Black Hole". Preprint, Cambridge, 1972.

[7] S. W. Hawking, G. F. R. Ellis: The Large Scale Structure of Spacetime. Cambridge University Press, Cambridge, 1973.

[8] F. J. Ernst: "Black Holes in a Magnetic Universe." Preprint, Chicago, January 1975.

[9] A. R. King, J. P. Lasota, W. Kundt: "Black Holes and Magnetic Fields." Preprint, Hamburg, 1975.

[10] J. M. Bardeen, B. Carter, S. W. Hawking, Commun. Math. Phys. 31, 161 (1973).

GRAVITATIONAL ENERGY FROM A QUADRATIC LAGRANGIAN WITH TORSION*

Carlos A. Lopez

Departamento de Física, Facultad de Ciencias Físicas y Mathematicas
Universidad de Chile, Casilla 5487, Santiago, Chile

SUMMARY

Gravitation is considered as a gauge field within the formalism of Utiyama [1] and Kibble [2]. The potentials in this approach correspond to the local affine connexion $A^{ij}{}_{\mu}$ and the tetrad $h_k{}^{\mu}$. The associated gauge fields are the local curvature $R^{ij}{}_{\mu\nu}$ and the local torsion $C^i{}_{k\ell}$. To write the invariant Lagrangian in empty space-time, we follow the analogy with electrodynamics. We thus propose to adopt as the free gravitational Lagrangian, the quadratic scalar density

$$Lg = L_R + L_C$$

$$= \frac{1}{4}(\det h_k{}^{\mu})^{-1}[R^i{}_{jk\ell}R_i{}^{jk\ell} + C^i{}_{jk}C_i{}^{jk}] \tag{1}$$

The corresponding equations of motion are

$$\partial_\nu(h^{-1}R_{ij}{}^{\mu\nu}) + h^{-1}A_{ja\nu}R_i{}^{a\mu\nu} - h^{-1}A_{ia\nu}R_j{}^{a\mu\nu}$$

$$= (\Sigma_C)^{\mu}{}_{ij} \tag{2}$$

$$\partial_\nu(h^{-1}C_\alpha{}^{\mu\nu}) - C_\nu{}^{\varepsilon\mu}(b^k{}_\alpha h_{k,\varepsilon}{}^{\nu} - A^{\nu}{}_{\alpha\varepsilon}) = \tag{3}$$

$$= (T_R)^{\mu}{}_{\alpha} + (T_C)^{\mu}{}_{\alpha} .$$

where, via Noether's theorem, Σ_C is identified as the spin of the torsion field, T_R represents the energy of the curvature field and T_C is the energy of the torsion field. When these equations are cast in covariant language (see [2]), we obtain

$$R^{\alpha\beta\mu\nu}{}_{;\nu} - 1/2\ C^{\mu}{}_{\varepsilon\nu}R^{\alpha\beta\varepsilon\nu} - 1/2\ C^{\nu}{}_{\varepsilon\nu}R^{\alpha\beta\mu\varepsilon} = (\Sigma_C)^{\mu\alpha\beta} , \tag{4}$$

$$C^{\beta\mu\nu}{}_{;\nu} - 1/2\ C^{\mu}{}_{\varepsilon\nu}C^{\beta\varepsilon\nu} - 1/2\ C^{\nu}{}_{\varepsilon\nu}C^{\beta\mu\varepsilon} = (T_R)^{\beta\mu} + (T_C)^{\beta\mu} . \tag{5}$$

The new feature we note here at once is the presence of torsion in empty space. Its source is the sum of the energy tensors of the curvature and torsion fields. In covariant notation they become

$$(T_R)^{\beta\mu} = R^{\varepsilon}{}_{\eta\gamma}{}^{\beta}R_{\varepsilon}{}^{\eta\gamma\mu} - 1/4\ g^{\beta\mu}R^{\varepsilon}{}_{\eta\gamma\delta}R_{\varepsilon}{}^{\eta\gamma\delta} , \tag{6}$$

$$(T_C)^{\beta\mu} = C^{\varepsilon}{}_{\eta}{}^{\beta}C_{\varepsilon}{}^{\eta\mu} - 1/4\ g^{\beta\mu}C^{\varepsilon}{}_{\eta\gamma}C_{\varepsilon}{}^{\eta\gamma} . \tag{7}$$

* Supported in part by the Multinational Project in Physics from the Organization of American States.

Note the similarity between these expressions with Maxwell's energy tensor of electrodynamics. They are quadratic in the fields, symmetric and traceless. On the other hand, the source of the curvature field is the torsion spin, which is given by

$$(\Sigma_C)^{\mu\alpha\beta} \equiv 1/2(C^{\alpha\beta\mu}-C^{\beta\alpha\mu}) \ . \tag{8}$$

In Riemannian spaces (no torsion) the energy tensors (6) and (7) vanish identically. This implies that a gravitational wave is characterized, besides the curvature, by the torsion of space-time.

REFERENCES

[1] R. Utiyama, Phys. Rev. 101 (1956), 1597.
[2] T. W. Kibble, J. Math. Phys. 2 (1961), 219.

POSITIVE ENERGY IN U_4 THEORY*

Eckehard W. Mielke

Institut für Reine und Angewandte Kernphysik
der Universität 23 Kiel, Ohlshausenstrasse 40/60,
Federal Republic of Germany

ABSTRACT

The positive energy conjecture for gravitational fields on maximal hypersurfaces has been extended to general relativity with spin and torsion. Some consequences for cosmological spin fluid models without singularities are discussed.

It is the purpose of this paper to extend the well-known theorem that in empty space-time the change of the total gravitational energy [1]**

$$16\pi E^{\Delta} = -8\int_{M^3} \Delta\phi \, d\mu_g \tag{1}$$

of conformally related asymptotically flat 3-metrics defined on maximal hypersurfaces is strictly positive--or zero in the time-symmetric case--to U_4 theory in the presence of matter.

For convenience, the field equations for general relativity with spin and torsion (U_4 theory) will be adopted in the pseudo-Einsteinian form [2]:

$$^{(4)}R^{\mu\nu} = 8\pi\,(\tilde{\sigma}^{\mu\nu} - \tfrac{1}{2}g^{\mu\nu}\,\tilde{\sigma}^{\kappa}{}_{\kappa}), \tag{2}$$

$$\begin{aligned}\tilde{\sigma}^{\mu\nu} = \sigma^{\mu\nu} &+ 8\pi\{\, \tau^{\alpha\beta\mu}\,\tau_{\alpha\beta}{}^{\nu} - 4\tau^{\mu\alpha}{}_{[\beta}\,\tau^{\nu\beta}{}_{\alpha]} \\ &-2\tau^{\mu\alpha\beta}\,\tau^{\nu}{}_{\alpha\beta} + \tfrac{1}{2}g^{\mu\nu}(4\tau_{\delta}{}^{\alpha}{}_{[\beta}\,\tau^{\delta\beta}{}_{\alpha]} \\ &+ \tau^{\delta\alpha\beta}\,\tau_{\delta\alpha\beta})\,\}.\end{aligned} \tag{3}$$

The metric on U_4 will be assumed in the form

$$ds^2 = -dt^2 + g_{ab}\,dx^a\,dx^b \tag{4}$$

with respect to a Gaussian normal coordinate system. One of the initial value

---

*Work supported in part by the Studienstiftung des deutschen Volkes.

**Absolute units where $G = c = \hbar = 1$ are used throughout. All geometrical objects are calculated with regard to the unique symmetric Riemannian connection.

equations, the so-called Hamiltonian constraint, reads [1]

$$S^{ab} S_{ab} - S^2 - R = - 16\pi \tilde{\sigma}_*^* \tag{5}$$

where

$$S_{ab} \equiv \frac{1}{2} \frac{\partial}{\partial t} g_{ab} \tag{6}$$

is the second fundamental form and

$$\tilde{\sigma}_*^* \equiv \tilde{\sigma}_{\mu\nu} u^\mu u^\nu \tag{7}$$

is the local energy density measured with respect to a timelike unit normal u of the initial hypersurface M^3. For the goals to be achieved here it is enough to specify the following independent subset [3] of the complete initial data on M^3:

a) the "conformal metric" $\tilde{g}_{ab} \equiv (g)^{-\frac{1}{3}} g_{ab}$

b) the "kinetic energy" scalar

$$\hat{S} \equiv g^{ab} g^{ce} S_{ac} S_{be} \geq 0$$

and

c) the "extrinsic time" $S \equiv g^{ab} S_{ab}$.

These initial values are <u>conformally invariant</u> under the conformal deformations

$$g_{ab} \rightarrow \bar{g}_{ab} = \phi^4 g_{ab}, \; S_{ab} \rightarrow \bar{S}_{ab} = \phi^4 S_{ab}, \; \phi > 0 \tag{8}$$

whereas

$$\bar{R}\,\phi^5 = R\phi - 8\,\Delta\phi\,. \tag{9}$$

Therefore, the requirement that a conformally related metric $\bar{g}_{ab}$ satisfies the constraint (5) is equivalent to

$$(\hat{S} - S^2)\,\phi^5 - R\phi + 8\Delta\phi = - 16\pi\, \bar{\tilde{\sigma}}_*^*\, \phi^5\,. \tag{10}$$

It follows that the change E^Δ of the total energy corresponding to conformally related metrics (8), which are free of singularities and defined on open, simply connected manifolds M^3, is obtained from

$$16\pi\, E^{\Delta} = \int_{M^3} (\hat{S} - S^2 + 16\pi\overline{\tilde{\sigma}}{}^*_*)\, \phi^{-1} d\mu_{\bar{g}}$$

$$- \int_{M^3} R\phi \; d\mu_g \; . \qquad (11)$$

With regard to this expression, an interesting connection between a recent generalization of Aubin's theorem [4] and the positive energy conjecture can be pointed out. Due to this theorem, it is always possible to choose an initial metric within the equivalence class $\tilde{g}_{ab}$ ("conformal superspace") such that $R = -\beta^2$, β = const., on an open manifold M^3 which is diffeomorphic to an open submanifold of a compact manifold M^3_c. Hypersurfaces admitting metrics with $R > 0$ as a result of given values in (5) do not cause trouble, because in that case $\tilde{g}_{ab}$ contains as well a metric with $R = 0$. Consequently, the last term in (11) can always be arranged to be non-negative and for $S = 0$, the case of a maximal foliation, the variation of the total energy of conformally related 3-geometries will always be non-negative, if

$$\int_{M^3} \tilde{\sigma}^*_* \, \phi^5 \, d\mu_g \geq 0 \quad . \qquad (12)$$

The minimum necessarily occurs in the time-symmetric case, i.e. $\overline{S}_{ab} = 0$, and it will be zero in empty space-time, i.e. $\tilde{\sigma}^{\mu\nu} = 0$.

If, as favoured by dimensional considerations, a scaling of $\overline{\tilde{\sigma}}{}^*_*$ according to $\overline{\tilde{\sigma}}{}^*_* = \tilde{\sigma}^*_* \, \phi^{-8}$ will be assumed, (12) can be replaced by the sufficient condition

$$\tilde{\sigma}^*_* \geq 0 \; . \qquad (12')$$

As a semiclassical example, the intrinsic spin angular momentum tensor [5]

$$\sqrt{g}\; \tau_{\alpha\beta}{}^{\mu} = \frac{\partial \mathcal{L}}{\partial \psi_{,\mu}} \; M_{\alpha\beta}\, \psi \qquad (13)$$

will be calculated with respect to a matter Lagrangian of an arbitrary spin field expressed in the first order formalism [6]. In the asympotically flat region [7], the Lagrange density including a coupling to an external field $B(x)$ reads as follows:

$$L = \psi^+ (\beta^\mu \partial_\mu + m)\psi + \psi^+ B(x) \psi . \tag{14}$$

In (13) the infinitesimal operators $M_{\alpha\beta}$ have been associated [8] with a finite dimensional irreducible representation of the Lorentz group SL(2, C) specified by (l_o, l_1).

If, as in spin fluid models, the matter is assumed to be transported along the velocity u and the condition

$$\tau_{\alpha\kappa}{}^{\kappa} = 0 \tag{15}$$

will be imposed, the spin squared terms in (7) turn out to be proportional to the first Casimir operator

$$C_1 \equiv \frac{1}{2} M_{\alpha\beta} M^{\alpha\beta} = l_o^2 + l_1^2 - 1 \tag{16}$$

of the Lorentz group. (Without (15) the algebra [9] of the β^μ can be employed to reduce the calculation.) Thus, for a representation with only one spin content $s = l_o$, $l_1 = l_o + 1$, the evaluation* of the local energy density results in:

$$\tilde{\sigma}_* = \rho - 16\pi n^2 s(s+1) + 2n\langle M^{\alpha\beta}\rangle \nabla_{[\alpha} u_{\beta]} \tag{17}$$

In order to prevent singularities in such cosmological models, it has been suggested [2] that the <u>strong energy condition</u>

$${}^{(4)}R_{\mu\nu} u^\mu u^\nu \geq 0 . \tag{18}$$

has to be violated.

In the case of a fluid with no vorticity this is even possible without violating (12') in the range

$$64\pi s(s+1) n^2 - 3p \geq \rho \geq 16\pi s(s+1) n^2 \tag{19}$$

*The details of these calculations will be presented elsewhere.

of the rest mass density. This depends also on the equation of state $p = p(\rho)$ of the pressure.

ACKNOWLEDGEMENT

It is a pleasure for the author to acknowledge several enlightening discussions with Prof. F. W. Hehl on U_4 theory and critical comments on the first version of this note. Furthermore, he would like to thank Prof. R. Ruffini for much encouragement and support.

REFERENCES

[1] N. O'Murchadha and J. W. York, Jr., Phys. Rev. D10, 428 (1974).

[2] F. W. Hehl, P. von der Heyde, G. D. Kerlick, Phys. Rev. D10, 1066 (1974).

[3] E. W. Mielke, Conformal Changes of Metric in General Relativity, subm. to Nuovo Cimento (1975).

[4] J. L. Kazdan and F. W. Warner, J. Diff. Geometry 10, 113 (1975).

[5] F. W. Hehl, GRG 5, 491 (1974).

[6] A. S. Wightman, Proc. Symposia Pure Math. 23, 441 (1973).

[7] T. Regge and C. Teitelboim, Phys. Lett. 53B, 101 (1974).

[8] A. Trautmann, Bull. Acad. Polon. Sci. 20, 503 (1972).

[9] A. J. Bracken, J. Phys. A8, 800 (1975).

ABSORPTION OF MASSIVE SCALAR FIELD BY A CHARGED BLACK HOLE

Takashi Nakamura
Department of Physics, Kyoto University, Kyoto, Japan
and
Humitaka Sato
Research Institute for Fundamental Physics, Kyoto University, Kyoto, Japan

ABSTRACT*

Absorption and reflection of charged, massive scalar field by the Reisner-Nordstrom black hole are investigated through a numerical computation. The absorption is suppressed when (Schwarzschild radius)<(Compton wave length) and the amplification of the wave occurs when the level crossing condition is satisfied.

*Work published in Phys. Lett. 61B, 371 (1976).

Inertia and Gravity in General Relativity

Andreas Quale

Institute of Physics, Oslo University, Oslo 3, Norway.

The work I am going to present here is based on a mathematical formalism developed in [1]. Briefly, what I shall do here is to try to indicate for you: (i) what the new formalism consists of; (ii) why it is introduced into General Relativity; (iii) how this is done; and (iv) some of the main results that have come out of this approach.

The metric tensor g_{ik} of a spacetime region $\mathcal{S}$ is split into two symmetric matrices (latin indices taking the values 0 to 3)

$$g_{ik} = \gamma_{ik} + h_{ik} \tag{1}$$

in such a way that γ_{ik} may be physically identified as the metric which would have been observed in $\mathcal{S}$ if there had been no gravitational interactions taking place in this region. The meaning of this proposition will be clarified later in this talk.

The idea is to regard γ_{ik} as representing the background geometry of the region $\mathcal{S}$, which is independent of the dynamics going on in this region, and h as a dynamical variable describing the gravitational field in $\mathcal{S}$. Thus, γ appears in a role somewhat similar to that of the Minkowski metric η in Special Relativity and linearized General Relativity. We shall refer to γ_{ik} and h_{ik} as the inertial background and the gravity potential, respectively. We stress two points right here: (a) γ_{ik} need not describe a flat geometry--its physical properties will be discussed later; and (b) the dynamical field h_{ik} need not be small. We now take any (local or nonlocal) field quantity f, depending on the Riemannian metric g_{ik} and (possibly) matter fields μ in General Relativity, and rewrite f as a function f' of γ, h and μ :

$$f(g, \mu) = f(\gamma + h, \mu) \equiv f'(\gamma, h, \mu) . \tag{2}$$

The right hand side of (2) is then interpreted physically as a function of the dynamical fields h and μ, defined on the background γ-space. Thus for instance, the split (1) induces a similar splitting of the Christoffel affinity Γ and Riemann tensor R of the g-metric:

$$\Gamma^i_{k\ell} = C^i_{k\ell} + \Delta^i_{k\ell}\,, \tag{3}$$

$$R^i_{k\ell m} = P^i_{k\ell m} + S^i_{k\ell m}\,, \tag{4}$$

where C and P are the Christoffel affinity and Riemann tensor of the γ-metric, and the remainders Δ, S are tensor functions of h in γ-space. In this way, the whole of General Relativity may be rewritten as a physical theory of interacting dynamical fields of gravity (h) and matter (μ), all defined on an externally fixed geometrical background--the inertial γ-space. Thus, for example, the Einstein field eqs. now take the form

$$G_{ik}(\gamma;h) = \kappa\, T_{ik}(\gamma;\mu)\,, \tag{5}$$

where the left hand side is a complicated nonlinear differential "wave operator", acting on the field h in γ-space. An obvious special case of this theory is linearized General Relativity (obtained by taking $\gamma = \eta$, h = small); however, we emphasize that the formalism is in general exact, involving no approximations or perturbation expansions. Now, why go to all this trouble--what can one hope to gain from such an alternative formulation of General Relativity? In Special relativity all matter fields and interactions are described on a fixed geometrical background (usually Minkowski space); and in this space, one can describe classical and quantum field theory. It is then a notable fact that many of the results of this treatment depend crucially on the existence and properties of the background. For instance, the symmetry group of Minkowski space yields weak conservation laws for the energy and momentum of any field system, while internal gauge groups yields conserved charge-currents. The whole formalism is easily made generally covariant; and there is no trouble with "coordinate effects", since all physical objects are defined on a fixed background space.

When this theory is extended to include gravitation--i.e. to general relativity--then very many of these field-theoretical methods and concepts break down, for the obvious reason that now there is no fixed background: the observed geometry of spacetime is identified with the dynamical field of gravitation. This leads to all the well-known difficulties: the conservation laws become non-covariant (pseudotensors), and the notions of gravitational field energy-momentum and gravitational waves become physically rather obscure (to say the least); while the quantitation of gravity is at present beset by very vexing problems, as has been reviewed at this meeting already by Dr. Nieuwenhuizen. Hence, the present attempt to formulate general relativity within a theoretical framework which makes it easier to compare gravity with other (matter) interactions, applying to them all the same basic field-theoretical methods and comcepts. The "philosophy", so to speak, may be summarized as follows: Gravity is a fundamental interaction, and should be treated in line with other interactions. The present background-space formalism is devised to accomplish this in a physically more satisfactory way than does the standard (no background) formalism of general relativity. The split (1) is the crucial part of our formalism--once it has been defined, the rest follows in a more or less straightforward manner. How, then, can it be justified physically? In other words: how can one perform the split (1) in such a way as to ensure that no local dynamics will appear into the "fixed" inertial background γ ? Briefly, the argument runs as follows: Take any given region $\mathcal{S}$ with a Riemannian geometry determined by the metric tensor g_{ik}. Does it make sense to ask what would be the metric tensor γ_{ik} of $\mathcal{S}$, if there had been no local gravitational interactions taking place in $\mathcal{S}$? This latter condition obviously implies, among other things, that the local density of matter energy-momentum must be negligible as gravitational field sources--i.e. the γ_{ik} of (1) must satisfy the vacuum Einstein field eqs. But, even so: how do we know that no free gravitational fields--i.e. gravitational waves, or locally bound semistable gravitational field configurations (geons) - will appear into our γ ? The trouble is that such free gravitational fields are not well understood: it is not known whether geons actually exist, as exact solutions of the vacuum field eqs., and there is no consensus as to exactly what one means by a "free gravitational wave". However, it is clear that if such objects are not present in $\mathcal{S}$, then the only agents that can influence physically the geometry γ in $\mathcal{S}$ are masses external to this region. Well-known cosmological arguments, based on Machs Principle, then indicate that the dominant part of this influence must come from the distant masses of the Universe. This means, in turn, that the geometry of $\mathcal{S}$ (in so far as it is determined by external sources) must be spacetime-homogeneous--i.e. admit rigid space and time translations as a group of motions. On the other hand, if geons or gravitational waves (whatever they are) should be present in $\mathcal{S}$, one would not expect this region to have such a translational symmetry. This gives us the possibility of eliminating geons and gravitational waves in $\mathcal{S}$. Moreover, the translational symmetry also makes it possible to observe (indirectly) the background geometry of any region $\mathcal{S}$ by comparing the observable geometry of $\mathcal{S}$ with that of any nearby vacuum region $\mathcal{S}_0$ which is of a size comparable to that of $\mathcal{S}$ and has a translation-invariant geometry.

In this way, one may deduce from astronomical observations that the background geometry of the spacetime region containing the Solar System is flat to a very high accuracy. On the other hand, if $\mathcal{S}$ is of cosmological dimensions--containing, say, very many clusters of galaxies--one would expect its background geometry to reflect any large-scale curvature that may be present in the Universe. Hence, one expects the background geometry of a region to be in general dependent on the scale of the region; this scale dependence must then be determined by cosmological considerations.

Let me just remark here that this notion of background geometry is only physically well-defined if certain restrictions are imposed on the cosmology of the Universe is too close to a "fireball" state. On the other hand, it is not mandatory to adopt Mach's Principle in its strict form; one may well allow absolute elements into the geometry of spacetime, if only their influence on the local geometry is also assumed to have translational invariance--an assumption which is always made, tacitly or explicitly. Even when the background geometry of a region is determined, this does not fix the functional form of γ and h in (1), as functions $\gamma(x)$, $h(x)$ of a given coordinate system (x) -- it follows from the Principle of Equivalence that γ and h are fixed only up to an arbitrary coordinate transformation. Thus, the fields $\gamma(x)$, $h(x)$ are not locally observable; while $g(x)$, of course, is. This constitutes a major difference between the present approach and the Bimetric formulation of general relativity, as developed by N. Rosen and others: in the latter, the background metric tensor γ is assumed to be observable, thus violating the Principle of Equivalence ("ether drift" experiments.) This non-observability of h, is interpreted as an internal gauge degree of freedom in the gravity potential--in the special case of linearized general relativity, the appearance of such a gauge is, of course, well known. When the split (1) is plugged into the action principle of general relativity with the curvature scalar of the g-metric as the gravitational Lagrangian $\mathcal{L}_g$ we obtain two kinds of conservation laws:

(a) The geometrical symmetry of the background γ_{ik}-space gives rise to a parameter-group of invariance in the action principle, and thus (by Noether's theorem) a family of tensors of gravitational field energy-momentum, covariantly and weakly conserved together with the matter energy-momentum.

(b) The internal gauge of h_{ik} gives rise to a function group of invariance (isomorphic to the general coordinate group) in the action principle, leading to a family of strongly concerved internal "charge-currents" of gravity. It can then be shown that these currents are physically identical to the energy-momentum tensors already mentioned--in other words: the "charge" of gravity is just the "energy" of this field, as determined by the symmetry of the background. It is remarkable that this formalism is wholly covariant--all physical fields that appears in it are tensors, all derivatives are covariant, and it is even possible to define (on the background space) an operation of covariant integration, as the inverse of covariant differentation. The well-known "coordinate effects" of standard general relativity are then avoided--on the other hand, the noncovariance of the pseudotensors is reflected here as a gauge-invariance appearing in the gravitational energy-momentum tensor family. Finally, we ask whether it is possible to find a member of this family which is gauge-invariant, and thus may be regarded as representing the energy-momentum of gravity, analogous to matter energy-momentum in SRT. There turns out to be precisely one such tensor, namely the variational derivative

$$T^{ik}_{(g)} \equiv \frac{\delta \mathcal{L}_g}{\delta \gamma_{ik}} , \tag{6}$$

which is, of course, what one would expect from matter physics in SRT. However, this tensor $T_{(g)}$ is then always proportional to the corresponding energy-momentum tensor

$$T^{ik}_{(m)} \equiv \frac{\delta \mathcal{L}_m}{\delta \gamma_{ik}} \tag{7}$$

of matter. In other words: the gravitational energy of any region is always totally determined by the matter in the region:

One consequence of this result is that the notion of a "free gravitational wave" seems to be rather inappropriate: signals transmitted through the gravitational field should be thought of as having an inductive character, in that they require both an emitter and an absorber for their existence. Thus, "gravitational waves" should be expressible wholly as functionals of their matter sources, somewhat analogous to the Wheeler-Feynman formulation of electrodynamics in special relativity. Another consequence is that a separate quantum theory of gravity is no longer called for--the quantum behaviour of the gravitational field h_{ik} is wholly determined by that of the quantized local matter field courses. Thus, for instance, on the particle level one would expect to have only tree diagrams, and no loop diagrams. This, of course, accords well with the present situation in covariant quantum gravity theory, as already reviewed here: graviton trees are good (can be dealt with satisfactorily by standard methods of quantum field theory), while graviton loops are bad (amazingly resistant to such treatment). In conclusion, these results suggest that perhaps our whole understanding of the nature of the gravitational interactions, and its relation to other interactions, needs some rethinking.

Reference

[1] Forts der Physik 21, 265-325 (1973).

METRIC SOLUTIONS OF SPINNING MASSES

Akira Tomimatsu

Research Institute for Theoretical Physics, Hiroshima University, Hiroshima, Japan

and

Humitaka Sato

Research Institute for Fundamental Physics, Kyoto University, Kyoto, Japan

The only known rotational solutions of vacuum Einstein equation satisfying an asymptotical flatness are the Kerr metric and the Tomimatsu-Sato metric. A list of the papers on the T-S metric is given in the references: finding of the solution in [1], [2], [3], [12] and a general review in [3]; discussion on an event horizon and a naked singularity in [5], [6], [7], [8]; discussion on a pole structure in [10], [11]; weak limit and the extreme Kerr limit in [2], [12], [14]; geodesics in [2], [15], [16], [17]; transformation to the Einstein-Maxwell solutions in [18], [19], [20], [21] and to the NUT like solution in [22].

REFERENCES

[1] A. Tomimatsu and H. Sato, "New Exact Solution for the Gravitational Fields of a Spinning Mass," Phys. Rev. Letters 29 (1972), 1344.

[2] A. Tomimatsu and H. Sato, "New Series of Exact Solutions for Gravitational Fields of Spinning Masses," Prog. Theor. Phys. 50 (1973), 95.

[3] H. Sato, "Gravitational Field of Spinning Mass - Kerr metric and Tomimatsu-Sato metric," Lecture at "Enrico Fermi" Summer School, July, 1975 (RIFP-224).

[4] C. Reina and A. Treves, "Axisymmetric Gravitational Fields" (Preprint 1975) (Univ. Milano).

[5] G. W. Gibbons and R. A. Russell-Clark, "Note on the Sato-Tomimatsu Solution of Einstein's Equations," Phys. Rev. Letters 30 (1973), 398.

[6] E. N. Glass, "Structure of the Tomimatsu-Sato Gravitational Field," Phys. Rev. D7 (1973), 3127.

[7] M. A. Abramovicz, J. P. Lasota and M. Demianski, "A Note on Tomimatsu-Sato Metric" Institute of Astronomy (Warsaw) preprint (1973).

[8] A. Tomimatsu and H. Sato, "Event Horizon of the Tomimatsu-Sato Metrics" Lettere Nuovo Cimento 8 (1973), 740.

[9] J. E. Economou and F. J. Ernst, "Weyl Conform Tensor of $\delta = 2$ Tomimatsu-Sato Spinning Mass Gravitational Field," J. Math. Phys. (in press).

[10] F. J. Ernst, "New Representation of the Tomimatsu-Sato Solution," IIT (Chicago) Preprint, 1975 August.

[11] J. E. Economou, "Approximate form of the Tomimatsu-Sato $\delta = 2$ Solution near the poles $x = 1$, $y = \pm 1$," IIT (Chicago) Preprint, 1975, August.

[12] F. J. Ernst, "Complex potential formulation of the axially symmetric gravitational field problem," J. Math. Phys. 15 (1974), 1409.

[13] W. Kinnersley and E. F. Kelly, "Limits of the Tomimatsu-Sato Gravitational Field," J. Math. Phys. 15 (1974), 2121.

[14] Y. Tanabe, "Multipole Moments in General Relativity," Prog. Theor. Phys. 55 (1976), 106.

[15] S. K. Bose and M. Y. Wang, "Geodesic Motions in the Tomimatsu-Sato Metric," Phys. Rev. D8 (1973), 361.

[16] J. D. Alonso, "Numerical Study of timelike geodesics in the first Tomimatsu-Sato metric," (Observ. de Meudon preprint, 1974).

[17] M. Calvani and R. Catenacci, "Geodesic Motion in the Tomimatsu-Sato Space-times" (Padova University preprint, 1975).

[18] F. J. Ernst, "Charged Version of Tomimatsu-Sato Spinning-Mass Field," Phys. Rev. D7 (1973), 2520.

[19] M. Y. Wang, "Class of Solutions of Axial-Symmetric Einstein-Maxwell Equations," Phys. Rev. D9 (1974), 1935.

[20] K. C. Das and S. Banerji, "Charged Tomimatsu-Sato Metrics," Phys. Letters 50A (1975), 409.
[21] G. Onengiit and M. Serdaroglu, "Two-Parameter Static and Five-Parameter Stationary Solutions of the Einstein-Maxwell Equations," Nuovo Cimento 27B (1975), 213.
[22] C. Reina and A. Treves, "NUT-like generalization of axisymmetric gravitational fields," J. Math. Phys. 16, 834 (1975).

ON THE GENERAL SOLUTION OF EINSTEIN EQUATIONS NEAR THE COSMOLOGICAL SINGULARITY

E. M. Lifshitz

Institute for Physical Problems, Academy of Sciences, Moscow, USSR

Professor E. M. Lifshitz has presented in two lectures a detailed discussion of the properties of solutions of Einstein field equations near the cosmological singularity. The material presented is contained in the following references:

[1] E. M. Lifshitz, I. M. Khalatnikov, Investigations in relativisitc cosmology. Adv. in Phys. 12, 185 (1963).

[2] V. A. Belinskii, I. M. Khalatnikov, E. M. Lifshitz, Oscillatory approach to a singular point in the relativistic cosmology. Adv. in Phys. 19, 525 (1970).

[3] E. M. Lifshitz, I. M. Lifshitz, I. M. Khalatnikov. An asymptotical analysis of the oscillatory approach to a singular point in the homogeneous cosmological models. JETP, 59, 322 (1970).

[4] V. A. Belinskii, I. M. Khalatnikov, E. M. Lifshitz. Oscillatory approach to a singular point in homogeneous models with rotating axes. JETP, 60, 1969 (1971).

[5] V. A. Belinskii, I. M. Khalatnikov, E. M. Lifshitz. On the construction of a general cosmological solution of the Einstein equations with a singularity in time. JETP, 62, 1606 (1972).

A short summary of this work (together with the relevant references) is given also in Sections 118, 119 of the latest edition of the "Classical Theory of Fields" (Pergamon Press, 1975, Editori Riuniti, 1975).

SPATIALLY HOMOGENEOUS UNIVERSES*

A. H. Taub

Mathematical Department, University of California
Berkeley, California

ABSTRACT: In this paper spatially homogeneous universes are defined as space-times admitting a three parameter group whose generating Killing vectors are space like and such that the group acts transitively on a three dimensional space-like hypersurface. The geometry of such hypersurfaces is discussed in terms of a triad of vectors which generate the reciprocal group. Expressions for the curvature two forms of these three spaces are given in terms of the constants of structure of the group, the triad of vectors and an arbitrary 3 x 3 matrix. The geometry of a space-time admitting such invariant hypersurfaces is next treated as are the Einstein equations for such a space-time. The derivation of the field from a variational principle is then given. The conditions under which the operation of deriving the Einstein field equations from the usual variational principles commutes with the imposition of symmetry conditions are discussed.

The classification of the singularities that can occur in fluid filled homogeneous space-times given by Ellis and King is discussed. The paper concludes with remarks about methods of dealing with space-times containing singularities that would be relevant in various physical applications.

I. Introduction

The field equations of the Einstein theory of gravitation, namely the equations(1)

$$G^{\mu\nu} = R^{\mu\nu} - \frac{1}{2}\,{}^4g^{\mu\nu}R = T^{\mu\nu} \tag{1.1}$$

express the fact that the energy content of space-time as described by the stress energy tensor $T^{\mu\nu}$ determines the geometry of space-time through the requirement that its metric tensor ${}^4g_{\mu\nu}$ be such that the Einstein tensor determined by it and its first and second derivatives be such that these equations are satisfied. In some problems the tensor $T^{\mu\nu}$ is determined as a function of the ${}^4g_{\mu\nu}$ and various fields which are described by differential equations involving these fields and the ${}^4g_{\mu\nu}$. In case the source of the gravitational field is a perfect fluid, the stress energy tensor is of the form

$$T^{\mu\nu} = (w+p)U^{\mu}U^{\nu} + p\,{}^4g^{\mu\nu} \tag{1.2}$$

where p is the pressure, w the energy density and U^{μ} with

$$U^{\mu}U_{\mu} = -1 \tag{1.3}$$

the fluid velocity.

In this case the equations describing the motion of the matter are

$$T^{\mu\nu}_{\ ;\nu} = 0 \tag{1.4}$$

*Work on this paper was done in part while the author was on sabbatical leave from the University of California, Berkeley and in residence at the College de France, Paris and at the Departement de Mecanique, University Pierre et Marie Curie (Paris VI).

These equations are a consequence of equations (1.1) in view of the Bianchi identities. Thus in dealing with the Einstein equations for this case we are confronted with a situation where we are to solve equations (1.1) without any knowledge of the left-hand sides (the ${}^4g_{\mu\nu}$) or the right hand sides (the thermodynamic variables p and w and the velocity U^μ). The nature of the independent variables, the coordinates χ^μ, is also not specified completely for the topology of space-time is another thing to be determined.

This situation is not as hopeless as it seems. One may obtain equations involving the metric tensor alone in a variety of ways. Thus we note that equations (1.2) are equivalent to

$$(T^\mu{}_\nu - p\delta^\mu_\nu)\,(T^\nu_\rho + w\delta^\nu_\rho) = 0 \tag{1.5}$$

where

$$\begin{aligned} p &= \frac{1}{4}T + \frac{1}{12}\left(12\,S - 3T^2\right)^{\frac{1}{2}} \\ w &= -\frac{1}{4}T + \frac{3}{12}\left(12\,S - 3T^2\right)^{\frac{1}{2}} \\ T &= T^\mu{}_\mu\,, \quad S = T^\mu{}_\nu\,T^\nu{}_\mu \end{aligned} \tag{1.6}$$

In view of equations (1.1), equations (1.5) are equations to be satisfied by the Einstein tensor. Equations (1.6) relate the pressure and energy density of the fluid to the trace of the Einstein tensor (the scalar curvature) and the trace of its square (which of course is related to the invariant $R_{\mu\nu}R^{\mu\nu}$. Thus singularities in p and w are related to singularities in the geometry. Further unless the geometry is such that equations (1.5) are satisfied with p and w, real positive quantities given by equations (1.6), the source of the gravitational field cannot be a perfect fluid.

Another method of reducing the problem of solving equations (1.1) to one involving only the metric is to integrate equations (1.2). We shall outline this method for the case in which there exists an equation of state of the form

$$w = w(p) \tag{1.7}$$

A similar technique may be used where the fluid is characterized by a caloric equation of state in which the rest internal energy per unit mass is given as a

(1) The following conventions are employed below: The signature of space-time is +2 and units are such that $8\pi G = c = 1$. The metric of space-time is denoted ${}^4g_{\mu\nu}$, the curvature and Einstein tensors by the usual $R^\mu_{\nu\sigma\tau}$, $G_{\mu\nu}$ etc. and covariant differentiation by a semicolon separating indices. The corresponding quantities for an embedded three-space will be denoted by g_{ij}, $R^{*i}{}_{jk\ell}$, G^*_{ij} etc; and a bar separating indices. Greek indices run from 1 to 4. Latin indices from 1 to 3, χ^4 being a time coordinate. A comma separating indices denotes partial derivative. The sign conventions for the Riemann and Ricci tensors are : for an arbitrary vector b^μ

$$b^\mu{}_{;\sigma\tau} - b^\mu{}_{;\tau\sigma} = -\,b^\rho\,R^\mu{}_{\rho\sigma\tau}\,, \quad R_{\sigma\tau} = R^\mu{}_{\sigma\mu\tau}\,, \quad R = R^\sigma{}_\sigma$$

function of two thermodynamic variables and the conservation of mass is required. We define the function S(w) (or S(p)) by the equation

$$\frac{ds}{S} = \frac{dw}{w+p} \tag{1.8}$$

Then we may choose co-moving coordinates so that [1]

$$U^{\mu} = e^{-\phi}\,\delta^{\mu}_{4} \tag{1.9}$$

where

$$Se^{-\phi} = w + p = f(x^{i})\,(-\,{}^{4}g)^{-\frac{1}{2}}$$

$$U_{\mu} = e^{-\phi}\,{}^{4}g_{4\mu} = e^{\phi}\,V_{\mu}(x^{i}) \tag{1.10}$$

$$V_{4} = -\,1\,,\quad V_{i} = V_{i}(x^{j}),\quad i,j = 1,2,3$$

Equations (1.1) then become partial differential equations involving the functions $\phi(x^{i},\,x^{4})$, $V_{i}(x^{i})$ and the

$${}^{4}g_{ij} = g_{ij}(x^{4},\,x^{k})\,,\quad k = 1,2,3 \quad .$$

Still another method for dealing with equations (1.1) is to assume a form for the metric tensor ${}^{4}g_{\mu\nu}$ and compute $T^{\mu\nu}$ from the equations. For the Robertson-Walker (Friedmann) cosmological models of General Relativity it is assumed that space-time is spatially homogeneous and isotropic. That is there exist space-like hypersurfaces which admit a six parameter group of motions and hence are of constant curvature. These symmetry requirements imply that the metric tensor may be written as

$$ds^{2} = R^{2}(t)\,d\sigma^{2} - dt^{2}$$

where $d\sigma^{2}$ is the line element in a three space of constant curvature and hence we may introduce coordinates such that

$$d\sigma^{2} = (dx^{2} + dy^{2} + dz^{2})/(1 + kr^{2}/4)$$

with $r^{2} = x^{2} + y^{2} + z^{2}$ and $k = +1,\ 0,\ -1$.

The stress-energy tensor determined from this metric is that of a perfect fluid whose pressure and energy density depend on the function R(t) and its first second derivatives. These space-times are conformally flat, that is, have a vanishing conformal (Weyl) tensor. They also have the property that a physical singularity necessarily develops if w and p obey the inequalities

$$\mu + 3p \geq 0$$

or

$$1 \geq \frac{dp}{dw} \geq 0,\quad w_{o} + p_{o} > 0,\quad w_{o} + 3p_{o} > 0$$

where w_{o} and p_{o} on the energy density and pressure at some arbitrary time [2].

The nature of and classification of the singularities that occur in the space-time determined by the Einstein field equations continues to be one of the most important problems in General Relativity even though the work of Penrose and Hawking (cf. [3], [4]) has thrown much light on the subject. One may attempt to gain further insight by examining exact solutions of equations (1.1). The problem of determining such solutions is eased considerably if one makes some assumptions on the symmetry of space-time as was done in the Robertson-Walker models. We have seen however that the assumptions made there led to extremely restricted models.

If one relaxes these assumptions by restricting the group of motions to be a three-parameter group instead of a six parameter one and further requires that this group have as invariant varieties, three dimensional hypersurfaces in space-time, one obtains the so-called homogeneous but anisotropic models. The symmetry properties enable one to write the Einstein equations as ordinary differential equations and thus considerably reduce the complexity of the problem of determining their solutions. Unfortunately the problem is still quite difficult.

In the following I shall review the characterization of the three parameter groups which act transitively on a three-space. The Bianchi groups and show how the Einstein equations simplify when the space-time admits such a group of motions. I shall also discuss some geometric properties of such space-times and techniques for studying them by variational and other methods.

2. SPATIALLY-HOMOGENEOUS UNIVERSES

These Universes are required to admit a three-parameter simply-transitive group of motions acting on space-like hypersurfaces. We denote a basis of Killing vectors of the simply-transitive group by ξ_A $(A = 1,2,3)$. Then

$$[\xi_A, \xi_B] = C^C_{AB}\, \xi_C \tag{2.1}$$

where [,] is the usual commutation operation and the structure constants C^C_{AB} satisfy

$$C^A{}_{B[C}\, C^B{}_{DE]} = 0 \tag{2.2}$$

The matrix

$$M = ||\xi^\mu_A||$$

is of rank 3. The minimum invariant varieties are then geodesically parallel hypersurfaces and may be taken to be the surfaces $x^4 = t =$ constant. Then $\xi^4_A = 0$ and ξ^i_A $(i = 1,2,3)$ are independent of t. We may take for the curves $x^1 =$ constant $x^2 =$ constant, $x^3 =$ constant, the geodesics orthogonal to one of the surfaces $t =$ constant. Then the line-element of space-time becomes

$$ds^2 = -\,dt^2 + g_{ij}\, dx^i\, dx^j \tag{2.3}$$

The ξ^j_A need only be specified in the initial surface and this has been done in [5] where the cannonical forms of the structure constants given by Bianchi [6] were used.

Reciprocal group generators B_a tangent to the group orbits which satisfy

$$[B_a, \xi_A] = 0 \qquad \text{all } a, \text{ all } A \tag{2.4}$$

in each hypersurfaces of transitivity can be chosen to satisfy the initial condition $\xi_A = B_a\,\delta^a_A$ at one point. In the initial hypersurface the vectors then obey

$$[B_a, B_b] = -\,C^c_{ab}\,B_c \tag{2.5}$$

where the C^c_{ab} are the same as the C^A_{BC}. If we propagate the B_a by the requirement that

$$[B_a, n] = 0 \tag{2.5a}$$

where $n^\mu = \delta^\mu_4$ is the normal to the hypersurface of transitivity then all the conditions given above including (2.5) are preserved, the $B^4_a = 0$ and B^i_a are independent of t. Explicit forms for the B^i_a are to be found in [5].

The geometry of the spaces t = constant may be studied in terms of the basis vectors B^i_a and the reciprocal basis of one forms B^a_i. We have

$$g_{ij} = \gamma_{ab}(t)\;B^a_i\;B^b_j \tag{2.6}$$

The Ricci rotation coefficients

$$\Lambda^c_{ab} = B^c_j\;B^j_{a|i}\;B^i_b \tag{2.7}$$

where the symbol "|" denotes the covariant derivative with respect to the g_{ij} satisfies

$$\Lambda^c{}_{ab} - \Lambda^c{}_{ba} = C^c_{ab} \tag{2.8}$$

If we use the quantities γ_{ab} and γ^{ab} defined by

$$\gamma_{ab}\,\gamma^{bc} = \delta^c_a$$

to raise and lower the indices a,b,c··· we see that

$$\Lambda_{gab} = \gamma_{gc}\;\Lambda^c{}_{ab}$$

satisfies

$$\Lambda_{gab} + \Lambda_{agb} = \gamma_{ga|i}\;B^i{}_b = 0 \tag{2.9}$$

$$\Lambda^c_{ab} = -\frac{1}{2}\gamma^{cg}(C_{gab} + C_{bga} - C_{abg}) \tag{2.10}$$

The quantities

$$\Lambda^c_{ai} = \Lambda^c_{ab} B^b_i = B^c_j B^j_{a|i} = - B^j_a B^c_{j|i} \tag{2.11}$$

may be considered as connection one-forms for the invariant hypersurfaces. The Cartan structure equations then become

$$B^c_{j|k} - B^c_{k|j} = B^a_k \Lambda^c_{aj} - \Lambda^c_{ak} B^a_j \tag{2.12}$$

The curvature two-forms are

$$\begin{aligned} R^{*c}_{aij} &= B^k_a B^c_\ell R^{*\ell}_{kij} \\ R^{*c}_{aij} &= \Lambda^c_{aj|i} - \Lambda^c_{ai|j} + \Lambda^c_{bi}\Lambda^b_{aj} - \Lambda^c_{bj}\Lambda^b_{ai} \end{aligned} \tag{2.13}$$

However, in view of equations (2.10) to (2.12) and the fact that the Λ^c_{ab} are independent of the X^i, we have

$$\Lambda^c_{aj|i} - \Lambda^c_{ai|j} = \Lambda^c_{ab}\Lambda^b_{de}(B^d_i B^e_j - B^d_j B^e_i)$$

Thus equation (2.13) may be written as

$$R^{*c}_{aij} = (\Lambda^c_{ab}\Lambda^b_{de} + \Lambda^c_{bd}\Lambda^b_{ae})(B^d_i B^e_j - B^d_j B^e_i) \tag{2.14}$$

It is then evident that the geometry of the invariant hypersurfaces is determined by the constants of structure of the group and the $\gamma_{ab}(t)$.

The Bianchi identities derived from equations (2.14) are

$$\varepsilon^{ijk} R^{*c}_{aij|k} = 0 \quad .$$

These are satisfied as a consequence of the constants of structure satisfying the Jacobi identity equation (2.2).

It is useful to write [7]

$$C^a_{bc} = \eta_{bct} n^{at} + \delta^a_c a_b - \delta^a_b a_c \tag{2.15}$$

or

$$C^a{}_{bc} = \varepsilon_{bct}\, m^{at} + \delta^a_c\, a_b - \delta^q_b\, a_c$$

where

$$\eta_{bct} = \sqrt{\gamma}\, \varepsilon_{bct}\ , \quad \gamma = \det ||\gamma_{ab}||, \quad n^{at} = n^{ta}\ , \quad m^{at} = m^{ta} \quad .$$

Then

$$C^a{}_{ba} = 2\, a_b = -\, \gamma_{bc}\, B^c{}_{i|j}\, g^{ij}.$$

The curvature tensor $R^{*i}{}_{jk\ell}$ may then be expressed as a quadratic expression in the quantities n^{ab} and a^a .

3. THE GEOMETRY OF SPACE-TIME

In the coordinate system in which equations (2.3) hold the only non-vanishing components of the Christoffel symbols are

$$\Gamma^4_{ij} = K_{ij} = \frac{1}{2}\, g_{ij,4}$$

$$\Gamma^i_{4j} = g^{ik}\, K_{kj} = K^i{}_j \qquad (3.1)$$

$$\Gamma^i_{jk} = \{^{\,i}_{jk}\}$$

where the $\{^{\,i}_{jk}\}$ are the Christoffel symbols of the invariant hypersurface with metric g_{ij}. The K_{ij} are the components of the second fundamental form of the hypersurface, the exterior curvature.

In this coordinate system, the non-vanishing components of the Riemann tensor are

$$R^\ell{}_{ijk} = R^{*\ell}{}_{ijk} + K^\ell{}_j\, K_{ik} - K^\ell{}_k\, K_{ij}$$

$$R^o{}_{ijk} = K_{ik|j} - K_{ij|k} \qquad (3.2)$$

$$R^\ell{}_{ook} = \dot{K}^\ell{}_k + K^\ell{}_m\, K^m{}_k$$

where the dot denotes the derivative with respect to t and the starred quantities are the components of the curvature tensor on the invariant hypersurface.

The non-vanishing components of the Ricci tensor formed from the $g_{\mu\nu}$ are

$$R^i_{\ j} = R^{*i}_{\ \ j} - \dot{K}^i_{\ j} - K\,K^i_{\ j}$$

$$R^o_{\ k} = (K^\ell_{\ k} - \delta^\ell_{\ k}\,K)_{|\ell} \tag{3.3}$$

$$R^o_{\ o} = \dot{K} + K^i_{\ j}\,K^j_{\ i}$$

Hence

$$R = R^* + K^i_{\ j}\,K^j_{\ i} - K^2 \ .$$

When one is concerned with homogeneous universes in which the source of the gravitational field is a perfect fluid, it is often convenient to use coordinates co-moving with the fluid [8]. In that case the metric of spacetime may be shown to take the form given by equations (1.9) and (1.10). The generators of the reciprocal group may be propagated by the requirement

$$[B_a, u] = 0$$

instead of by equation (2.5a). The relations between the components of the Ricci tensor of space-time and that of the invariant hypersurface may be found in [8].

It is evident from what has been said above that the Einstein equations reduce to ordinary differential equations in spatially homogeneous universes. The independent variable is t, the proper time along curves orthogonal to the invariant hypersurfaces (or the proper time along a fluid elements world line) and the dependent variables include the $\gamma_{ab}(t)$ (and the thermodynamic variables characterizing the fluid in case the sole source of the gravitational field is a perfect fluid). The explicit form of these equations may be found in [8] for the fluid-filled case. In the next section we examine the conditions under which they may be derived from a variational principle.

4. VARIATIONAL PRINCIPLES

It is known [1] that when an equation of state exists the Einstein field equations for a self-gravitating fluid may be derived from the variational principle based on the action integral

$$^4I = \int_V (R + 2p(\phi))\sqrt{-{}^4g}\ d^4x \tag{3.1}$$

where R is the scalar curvature of space-time and ϕ is a function of the metric tensor given in co-moving coordinates by equations (1.10). When the space-time is spatially homogeneous this action integral becomes

$$^4I = \int \mathscr{L}(x^o)\, ds^o \int L(x^i)\, d^3x \ .$$

In reference [8] condition were found under which the Einstein equations for $\phi(t)$ and $\gamma_{ab}(t)$ are the Euler equations derived from the action integral

$$I = \int \mathscr{L}(x^o)\, dx^o \ . \qquad (3.2)$$

That is, that paper examined when the operation of deriving the Euler equations commutes with the imposition of the spatial homogeneity symmetry condition.

It is not necessarily valid to impose the symmetry before taking the variations. To see this consider varying a Lagrangian of the form

$$I = \int L(\phi^A, \phi^A_{,\mu})\, d^4x$$

where $\phi^A (A = 1, \cdots, n)$ are some fields and $\phi^A_{,\mu}$. their derivatives. One obtains for the variation of I^*

$$I^{*'} = \int \left(\frac{\partial L}{\partial \phi^A} - \frac{\partial}{\partial x^\mu} \frac{\partial L}{\partial \phi^A_{,\mu}} \right) \phi'^A\, d^4x + \int \left(\frac{\partial L}{\partial \phi^A_{,\mu}} \phi'^A \right)_{,\mu} d^4x$$

where the variation has been taken by introducing a one-parameter family $\phi^A(x,e)$ and differentiating with respect to the parameter e. The last term in this expression is eliminated by imposing boundary conditions on the ϕ'^A. One then deduces the Euler-Lagrange equations

$$\frac{\partial L}{\partial \phi^A} - \frac{\partial}{\partial x^\mu} \frac{\partial L}{\partial \phi^A_{,\mu}} = 0 \ ,$$

from $I^{*'} = 0$ by considering the set of all ϕ'^A satisfying these boundary conditions. However when the variations are required to satisfy some symmetry conditions, the boundary conditions may, by the symmetry, restrict the variations ϕ'^A in the interior of the region of integration. Alternatively one can say that for an arbitrary variation obeying the symmetry restriction, the Euler-Lagrange equations may not follow from the expression for $I^{*'}$ for the last term in this expression might not vanish.

In this cited paper it was shown that if and only if the spatially homogeneous universes are of the type called class A by Ellis and MacCallum [9] the action integral given by (3.1) may be replaced by one of the form (3.2) in deriving the Einstein field equations for $\phi(t)$ and $\gamma_{ab}(t)$. The same result holds for the

A D M Lagrangian. Class A spatially homogeneous universes are those for which

$$C^{a}_{ba} = 2\ a_{b} = 0. \tag{3.3}$$

Those which violate this condition are said to be of class B. The paper also examines the possibility of obtaining the correct Einstein field equations for class B universes by first imposing the symmetry conditions on the integral in equation (3.1) and then adding additional conditions.

Thus spatially homogeneous universes of class A may be treated in the same manner as a classical mechanical system with a configuration space described by the variables ϕ and γ_{ab}.

The hamiltonian for this system may readily be determined from the Lagrangian given in [8] and the Einstein field equations may be written in Hamiltonian form.

If one is interested in the stability of such universes under perturbations which also have the same symmetry properties as the unperturbed space-time, one studies the linear equations satisfied by the perturbations. These equations may be obtained by either perturbing the non-linear Einstein field equations or by using the fact that such equations may also be derived from a variational principle - namely that given by the so-called second variation [cf. 10]. This latter variational principle may be derived from the action principle given by equation (3.2).

If however, the perturbations do not have the same symmetry properties as the unperturbed metric tensor, the second variation integral still exists but must be derived from the action integral given by equation (3.1).

5. SINGULARITIES

References was made earlier to the singularity theorems of Penrose and Hawking [3, 4]. These theorems which show that under very plausible assumptions on the nature of space-time the space-time must be geodesically incomplete. That is, there is a geodesic $\lambda(v)$ defined for an affine parameter $o \leq v < v_{+}$ which cannot be extended to the parameter value v_{+}. The space-time is then said to contain a singularity. However these theorems do not describe the nature of the singularity in term of the behaviour of the Riemann curvature tensor near the singularity or in terms of invariants formed from the components of the curvature tensor.

Ellis and King [2] classify the nature of the singularity a curve runs into in terms of the behaviour of the components of the Riemann tensor $R_{abcd}(v)$ with respect to an orthonormal basis $\ell_{a}(v)$ (a tetrad) defined along $\lambda(v)$:

(I) If all the components R_{abcd} tend to finite limits with respect to a tetrad parallely propagated along $\lambda(v)$ the singularity is said to be locally extendible;

(ii) If some component $R_{abcd}(v)$ with respect to a parallely propagated tetrad

does not go to a limit but there is some other tetrad in which they do all go to a finite limit as $v \to v_+$, the singularity is said to be an intermediate one.
(iii) If there is no orthonormal basis along $\lambda(v)$ such that the components $R_{abcd}(v)$ go to finite limits as $v \to v_+$ the singularity is called a curvature singularity.

These latter singularities may be further characterized as (a) matter singularities if some Ricci tensor components do not go to finite limits for any orthonormal frame along $\lambda(v)$ or (b) conformal singularities if the conformal (Weyl) tensor components $C_{abcd}(v)$) do not all go to finite limits.

Ellis and King show that fluid filled Bianchi type Universes cannot in general remain spatially homogeneous and discuss cases in which matter singularities occur. One result is that if the velocity vector is orthogonal to the invariant hypersurface then both the energy density and the curvature invariant $R_{ab}\,R^{ab}$ diverge on a particular invariant hypersurface. They also discuss cases in which there is an intermediate singularity when the spatial homogeneity breaks down but the density does not become infinite and the space-time can be extended across a Cauchy horizon.

Clarke [11] has shown that a singularity reached on a time-like curve in a globally hyperbolic space-time must be a point which is either a curvature or intermediate singularity or the curvature tensor is of Petrov type D and electrovac. This result does not apply only to Bianchi type Universes.

The results given in references [2] and [11] are of importance. However much more needs to be known about singularities in space-time. If for example the infinities in the curvature tensors which occur at singularities are such that certain integrals over submanifolds of space-time containing the singular points exist, then it may be possible to give generalizations of the Einstein field equations which obtain at these points. I illustrate this remark by recalling that in classical hydrodynamics and in general relativistic hydrodynamics when heat conductivity and viscosity are set equal to zero, shock waves occur. These hypersurfaces are singular ones in space-time. We have no real difficulty in determining the equations that describe the space-time containing them and the jump conditions that must hold across the singularity. Indeed they may all be obtained from a variational principle property interpreted.

Another example in which a singularity is made "inoccuous" is the following. Cahill and Taub have shown [12] that a static spherically symmetric distribution of a self-gravitating field in which the energy density and pressure have a singularity at the origin r = o in that they are each proportional to r^{-2} may be fitted to a time dependent similarity solution for t > o. This solution holds in the region $o \leq r/t < z_1$, and is a Robertson-Walker space-time with w = 3p. For

$r > z_1 t$ and $t > o$ the solution is static and separated from the Robertson-Walker one by a shock-wave. Thus the matter singularity at the origin is traded for a shock-wave by this extension procedure.

This example suggests that methods of extending space-times beyond the singularities which involve "mild types" of singularities be investigated.

A tool for accomplishing this task may be provided by using the techniques that play a role in relating the integrals of curvature invariants to topological properties of space-time. These techniques may be based on the identities which follow from the Bianchi identities and which relate quadratic expressions formed from combination of the connection one-forms and the curvature two forms. Such equations may be used to determine the Euler and Pontrjagin classes of space-times. The spinor form of these identities are treated in [13].

REFERENCES

[1] Taub, A. H.: Proceedings of the 1967 Colloque on "Fluides et champs gravitationnels en relativite generale" n° 170 Paris, Centere National de la Recherche Scientifique (1969).
[2] Ellis, G. F. R. and King, A. R.: Commun Math. Phys. 38 (1974) 119-156.
[3] Hawking, S. W. and Penrose, R.: Proc. Roy. Soc. (London) A314 (1970) 529.
[4] Hawking, S. W. and Ellis, G. F. R.: The Large Scale Structure of Space-Time. (1973) Cambridge.
[5] Taub, A. H.: Ann. of Math. 53 (1951) 472-490.
[6] Bianchi, L.: Lezioni sulla teoria dei gruppi continui finite di transformazioni. Sperni, Pisa (1918) 550.
[7] Estabrook, F., Wahliquist, H. D., Behr, C. G.P J. Math. Phys. 9 (1968) 491.
[8] Mac Callum, M. A. H., Taub, A. H.: Commun. Math. Phys. 24 (1972) 173-189.
[9] Ellis, G. F. R., Mac Callum, M. A. H.: Commun. Math. Phys. 12 (1969) 108.
[10] Taub, A. H.: Commun. Math. Phys. 15 (1969) 235-254; and Relativistic Fluid Dynamics, ed. C. Cattaneo (Lectures at the Centro Internazionale Matematico Estivo, Bressanone, 1970) Edizioni Cremonese, Rome (1971).
[11] Clarke, C. J. S., Commun. Math. Phys. 41 (1975) 65-78.
[12] Cahill, M.E., Taub, A.H., Commun. Math. Phys. 21 (1971) 1-40.
[13] Taub, A. H., Curvature Invariants, Characteristic Classes and the Petrov Classification of Space-times. To appear in A. Lichnerowicz "Festschriff".

THE VACUUM BLACK HOLE UNIQUENESS THEOREM AND ITS CONCEIVABLE GENERALISATIONS

Brandon Carter

Groupe d'Astrophysique Relativiste
Observatoire de Paris, 92 Meudon

ABSTRACT

A brief outline is given of the mathematical results (due to A. Papapetrou, F. J. Ernst, B. Carter and D. C. Robinson) that together provide a completely water-tight proof of the truth of the conjecture that the Kerr solutions are the only topologically spherical stationary axisymmetric black hole solutions of the vacuum Einstein equations. Further results (due to A. Lichnerowicz, W. Israel, S. W. Hawking and others) provide in combination a fairly convincing but not quite complete demonstration that all stationary black hole vacuum solutions satisfy these conditions (of axisymmetry and spherical topology).

Work on the extension of these results to allow for the presence of electromagnetic and other conceivable external fields is discussed.

OVERVIEW

The purpose of this paper is to give a brief review of the progress that has been made so far towards proving what has frequently been referred to loosely as the "Israel-Carter conjecture."

Pioneer work on static (non-rotating) vacuum solutions of Einstein's equations by Doroshkevich, Zeldovich and Novikov (1965) [1], Mysak and Szekeres [2] (1966) and others, which culminated in the theorem of Israel (1967) [3] lead to the following fairly precise conjecture, which I quote in a form published in 1968 [4] at a time when the term "black hole" was not yet in general use: ". . . the Schwarzschild solution is unique among asymptotically flat static-vacuum solutions in being bounded by a simple non-singular Killing horizon ("simple" meaning that the constant time cross sections are topologically spherical) which suggests that the family of stationary axisymmetric assymptotically flat vacuum solutions with the same property may also be very restricted. It may be conjectured that the Kerr fields may be the only examples."

An alternative version was also formulated (by Israel [3],[5]) in terms of what various authors have referred to as the "infinite red shift surface" or "surface of the ergosphere" instead of in terms of the "Killing horizon" which is what would be described today as the "surface of the black hole." (In the static case the two coincide so no ambiguity arises). This variant is definitely false, however, as may be seen by considering a construction originally suggested by Bardeen (unpublished) in which one imagines a small perturbing ring of matter rotating outside the horizon but within the ergosphere in a Kerr solution background; it is evident that this would give rise to a perturbation of the vacuum field which would inevitably extend outside the ergosphere.

On the other hand, the truth of the first version quoted above (whose formulation was greatly influenced by discussions with R. Penrose) may now be considered to be fully confirmed.

A fairly complete proof subject to a mathematically interesting but physically marginal technical reservation was obtained in 1971 [6] in the form of a series of lemmas (whose detailed proofs were described in a paper entitled "The stationary axisymmetric black hole problem" (1972), of which the contents was published in the proceedings of the (1973) les Houches summer school [7]). The only remaining technical reservation has recently been removed by an additional lemma given by D. Robinson (1975) [8].

This scope of the original and now completely established conjecture has been greatly extended by several authors of whom the boldest is J. A. Wheeler who is responsible for the conjecture that, in the words of his own colourful catch phrase, "a black hole has no hair," because "anything that can be radiated away will be radiated away."

This all embracing conjecture really contains two distinct parts. The first is the conjecture that a gravitationally collapsing body really will in fact form a well behaved black hole in the strict sense (as defined e.g. by Carter (1971) [6], Hawking (1972) [9]) (instead of forming what Penrose (1969) [10] has called a "naked singularity") and that the region exterior to the horizon will settle down ultimately to a stable stationary final state. The stability of the Kerr solutions as final equilibrium states can now be considered fairly well established due to a great deal of work by many authors, but the question of naked singularities remains wide open. These dynamic aspects of the problem will not be considered further in the present discussion.

The second part of the conjecture that "a black hole has no hair" concerns the possible forms of the final stationary equilibrium states (when and if they are attained). The idea is that the only classical degrees of freedom of these states should be those corresponding to the very lowest asymptotic multipole moments--namely those to which there does not correspond any direct radiation mode--of the various classical (i.e. integer spin) fields that may be present in any particular physical situation.

In the pure gravitational case one has only a zero rest mass tensor (spin-2) field which can radiate only in the quadrupole and higher modes, so the conjecture would therefore allow only a monopole (mass) and dipole (angular momentum) degree of freedom. This is precisely consistent with what has been proved to be the case, since these are just the degrees of freedom possessed by the Kerr solutions.

The additional degrees of freedom one would expect to find associated with an electromagnetic field are even more restricted since a zero rest mass vector (spin-1) field can also radiate in the dipole as well as higher modes. Hence, according to the conjecture, the only extra degree of freedom associated with an electromagnetic field should be that corresponding to an electromagnetic monopole, i.e. once the mass angular momentum and charge of an electromagnetic black hole have been specified there should be no further freedom to specify even the magnetic dipole moment (although a magnetic monopole moment could be allowed if the physical boundary conditions permitted). One can make the even stronger conjecture that the only black hole equilibrium solutions of the Einstein-Maxwell equations are the Kerr-Newman solutions, which in fact have just these degrees of freedom, but this, while probably true, would be going beyond the strict no-hair conjecture as I understand it. If the no hair conjecture for Einstein-Maxwell black holes is taken to mean that the equilibrium solutions come in continuous families characterised only by mass, angular momentum and charge (electric and, if one wishes, magnetic monopole) which are discrete (without bifurcations) then it has in fact been proved to be correct by the recent work of Carter [7] and Robinson [11] subject to the same restrictions--of axisymmetry and spherical topology--as were assumed in the original conjecture concerning the pure vacuum case as stated above. Although it makes the existence of a uniqueness theorem seem very likely, this "no-hair theorem" does not itself exclude the existence of other presently unknown families of solutions in addition to that of Kerr-Newman. (If they exist, however, such families would be physically pathological because the previous work of Israel [12] and Muller zum Hagen et al. [13],[14] establishes that their angular velocity could not be continuously reduced to zero, and nor, by the work of Carter [7] and Robinson [8] could their charge.) Thus the extent of our knowledge concerning the Einstein-Maxwell equilibrium states today is about the same as that of our knowledge concerning pure vacuum equilibrium states in 1973: the electromagnetic analogue of Robinson's final (1975) step [7] to a complete uniqueness theorem has not yet been obtained.

It is generally understood that the conjecture that "a black hole has no hair" includes the implication that the equilibrium states satisfy the axisymmetry condition that is assumed in all mathematical work referred to so far (except that of Israel) since any deviation from axisymmetry would correspond to the presence of a radiatable multipole moment. An almost complete proof that this condition must necessarily be satisfied, at least in the pure vacuum and electromagnetic cases is obtainable from a theorem of Hawking [15],[16] to the effect that a stationary non-axisymmetric black hole cannot (in a certain precise technical sense) rotate. (The work of Hawking [15],[16] and Israel [3] in the pure vacuum case and of Carter [7] in the electromagnetic case goes most of the way towards an appropriate generalisation of a well known theorem of Lichnerowicz to the effect that the non-rotating equilibrium states must be static, whence it follows from the work of Israel [3],[12] and Muller zum Hagen et al [13],[14] that they must in fact be not merely axisymmetric but, as one would have guessed, spherical.)

Even though their existence as discrete classes of black hole equilibrium states would not be incompatible with the basic idea of the "no-hair" conjecture, Hawking [15],[16] has in fact also been able to virtually rule out the possibility of topologically non-spherical (e.g. toroidal) black hole equilibrium states.

(The only possibility not yet rigourously excluded, except in the static case [17], is a compound topology consisting of several individually spherical black holes on a common axis of symmetry). This means that the previously cited results already provide a virtually complete proof of the truth of the no-hair conjecture as far as pure vacuum and electromagnetic equilibrium states are concerned, and that if Robinson's (1975) [7] lemma could be extended to the electromagnetic case the absolute uniqueness (as opposed merely to "baldness") of the Kerr-Newman solutions as electromagnetic black hole equilibrium states would be effectively established.

In what follows I shall present reasonably precise statements (but not proofs) of the key lemmas and theorems necessary for establishing the mathematical results quoted above insofar as they apply to black hole equilibrium states satisfying the basic simplifying conditions of axisymmetry and spherical topology, i.e. the conditions whose necessity (for rotating black holes surrounded by fields satisfying any of the standard simple classical equations--Klein Gordon, Maxwell, etc.) has been established by Hawking [15],[16].

One of the purposes of listing these theorems which cover the case of the only two physically known long range classical fields (i.e. gravitation and electromagnetism) is to show more clearly what is needed for the results to be extended to other hypothetical (e.g. scalar or massive vector) fields to which the "no-hair" or "generalised Israel-Carter" conjecture may be applied.

In the case of a simple scalar field--for which even monopole radiation exists--it is clear that the no-hair conjecture would allow no extra degrees of freedom at all. It is extremely easy to show (as was first remarked by Hawking [18] in the context of a discussion of the Brans Dicke theory) that in a black hole equilibrium cannot have an external scalar field.

The case of a massive vector field satisfying the Proca Field equations has been considered by Beckenstein [19],[20],[21]. Since a massive (unlike a massless) vector field has not only dipole but even (longitudinal) monopole radiation modes, the no-hair conjecture would imply that a black hole equilibrium state should not be able to have an external massive vector field either (c.f. Teitleboim [22]). However, although Beckenstein's work makes the truth of this conjecture seem very likely, much remains to be done before the question can be considered to be as completely settled as in the electromagnetic and simple scalar cases, because in the massive vector case more of the crucial steps (including the proof of "circularity") in the line of argument are still missing.

The case of (massless-vector) Yang-Mills type fields other than electromagnetism is interesting because for these the no-hair conjecture does not forbid the existence of additional monopole degrees of freedom which might be associated e.g. with baryon and lepton conservation (c.f. Carter [23]). In fact, the number of additional degrees of freedom with such a field could be quite large.

To start with (as has been emphasized by Perry [24]) for any Maxwell-Einstein vacuum solution there are exactly corresponding solutions of the Einstein-Yang-Mills fields equations. In addition to these (as has been pointed out by Wang [25]) there are also black hole analogues of the rather less trivial Yang-Mills fields which t'Hooft [26] has shown how to construct in flat space any Yang-Mills field generated by a group (such as SU [3]) containing a subgroup covering the orthogonal group SO [3]. It would be interesting to make a systematic study of this question.*

THE CIRCULARITY CONDITION

The results described below apply to a stationary, axisymmetric, assymptotically flat space-time manifold M. We shall primarily be concerned with the domain of outer communications (the region outside the black hole) which will be denoted by <I> and which is defined as consisting of all events in M that from which there can be constructed both future directed and past directed timelike lines extending out to arbitrarily large distances in the assymptotically flat region. (If we were considering dynamic gravitational collapse from an initially well behaved space, the requirement that there exist past directed connections to infinity would be superfluous, but the familiar example of the Kruskal extension of the spherical Schwarzschild solution reminds us that the ultimate stationary state need not be well behaved when extended back into the past.) It will be supposed that the domain of communications is bounded in the future by a well behaved horizon H^+ which is topologically the product of a 2-sphere (representing a section of the black hole boundary) with the real line (representing a null generator of the horizon). The domain of communications itself will be supposed to have the topology of the product of a 2-sphere with (which may be he supposed to have co-ordinates θ, ϕ) with a 2-plane (which may be supposed to have co-ordinates r, t). Nothing is assumed about the past boundary of <I> nor about the interior of the black hole which may be highly singular.

The condition that the space is stationary-axisymmetric means that it is possible to choose co-ordinates x^0, x^1, x^2, x^3 on <I> in such a way that two of them, $x^0 = t$ say and $x^3 = \phi$ say are ignorable, i.e. so that the metric and other physical field components depend only on $x^1 = r$ and $x^2 = \theta$ say but are independent of t and ϕ. The vector kields k^a, m^a respectively generating the stationary and axisymmetry group actions may be represented in such a system by

$$k^a = \delta^a_0 \quad , \quad m^a = \delta^a_3$$

*A recent preprint by P. B. Yasskin (Solutions for Gravity Coupled to Massive Gauge Fields, Univ. of Maryland, 1975) shows that an (W-1) parameter family of Einstein-Yang-Mills solutions (including the trivial solutions and Wang-t'Hooft type solutions as special cases) for an N-parameter gauge group can be constructed automatically from any Einstein-Maxwell solution (without altering the geometry). Yasskin conjectures that these are the only Einstein-Yang-Mills black hole solutions.

(where latin indices a,b run from 0 to 3) and the metric tensor will have the form

$$ds^2 = g_{\alpha\beta}\, dk^\alpha dk^\beta + 2\, g_{3\alpha}\, dk^\alpha d\phi + 2\, g_{0\alpha}\, dk^\alpha dt + X d\phi^2 + 2\, W d\phi dt - V dt^2 \tag{1}$$

where Greek indices α, β run from 1 to 2 and where $X = m^a m_a$, $W = m^a k_a$ and $V = - k^a k_a$. It is possible to achieve great simplifications in the study of spaces such as this if one can establish what I refer to generally as <u>circularity conditions</u>. The simplest example is the case of a stationary axisymmetric vector field u^a say, said to be <u>circular</u> if it has no r or θ components, i.e., if it is a linear combination of the vector fields generating the symmetries, of the form

$$u^a = u^0 k^a + u^3 m^a \tag{2}$$

(cf. figure; the field would be said to be <u>static</u> in the even more restricted case when u^3 is zero). The concept of circularity can be naturally extended [7] from vector to other tensor fields, and in particular to the metric tensor. A stationary axisymmetric space will be said to have a <u>circular metric</u> if the symmetry action is orthogonally transitive in the sense that the 2-surfaces of constant t, ϕ may be chosen orthogonal to those of constant r, θ so that one obtains

$$g_{\alpha 0} = 0 \quad , \quad g_{\alpha 3} = 0 \tag{3}$$

in the metric form (1), i.e., the metric tensor has the form

$$g_{ab} \longleftrightarrow \begin{pmatrix} -V & 0 & 0 & W \\ 0 & g_{11} & g_{12} & 0 \\ 0 & g_{21} & g_{22} & 0 \\ W & 0 & 0 & X \end{pmatrix}$$

(the metric would be said to be <u>static</u> in the even more restricted case when $W = 0$ also).

If this condition were not satisfied it would be very difficult to calculate the exact location of the horizon H^+ bounding the hole even for an explicitly known metric, but when the circularity condition is satisfied the situation of the horizon can easily be pinned down as forming part of the locus where the quantity

$$\rho^2 = VX + W^2 \quad (\equiv - g_{00}\, g_{33} + g_{03}^2) \tag{4}$$

vanishes. Formally we have the following:

<u>Lemma 1</u> (Carter [27],[7]

If the circularity condition (3) and also the causality condition (no closed timelike lines) are satisfied then the quantity ρ^2 defined by (4) is strictly positive everywhere on <I> (except on the symmetry axis where $X = W = 0$ so that ρ^2 is zero) and ρ^2 <u>is zero</u> on the horizon H^+ bounding <I> .

(In the restricted static case when W is zero the horizon occurs where $V = 0$ as was pointed out by Vishveshwara [28]. However in general the "ergosphere," i.e. domain of negative V, will extend outside the horizon.)

In order to be able to use this result we need to know that the circularity condition is in fact satisfied. Fortunately this can be established without too much difficulty in many simple situations. We have

Lemma 2a (Papapetrou [29]) if the pure vacuum Einstein field equations are satisfied everywhere then so also is the circularity condition.

Lemma 2b (Carter [27],[7]) if the vacuum Einstein-Maxwell equations are satisfied everywhere then so also is the circularity condition.

The necessity of the circularity condition can in fact be established under considerably more general conditions e.g. in the presence of external perfect fluid matter rings [30],[7] provided that the fluid flow vector (and the electric charge current vector if any) themselves satisfy the vector circularity condition (2) (but not otherwise).

It is also easy (in fact much easier than in the electromagnetic case) to extend the original Papapetrou theorem (Lemma 1a) to apply in the presence of a scalar (Klein-Gordon) field. What I have not been able to do is to prove that the circularity condition must hold in the presence of a massive vector (Proca) field. (Nor do I have a counterexample, so the question is open.) The reason why the approach that works for the electromagnetic field breaks down for a massive vector field is described in an appendix. It would be desirable to settle the question one way or the other because the circularity condition [2] was used as a basic postulate both in Beckenstein's original [20],[21] attack on the Proca no-hair problem and in a more recent reexamination of the problem by Perry [31].

THE PROOF OF UNIQUENESS

The next step, when circularity has been established, is to fix the choice of the coordinates completely i.e. to specify r and θ. In practice it is convenient to work rather with ellipsoidal polar type coordinates λ, μ which are related to radial and angular coordinates r, θ with the standard assymptotic properties by

$$\lambda = r - M, \qquad \mu = \cos\theta$$

(where M is the assymptotically defined mass of the black hole).

It is possible to take advantage of the well known fact that the quantity ρ is harmonic when the pure vacuum or vacuum Einstein-Maxwell equations are satisfied to prove the following:

Lemma 3 (Carter [6],[7]) if the source free Einstein or Einstein-Maxwell equations are satisfied everywhere in the domain of outer communications ≪ ≫ then (subject to the conditions of Lemmas 1 and 2) it can be covered completely (apart from the standard degeneracy on the symmetry axis) by a system of coordinates t, λ, μ, ϕ such that the metric has the form

$$ds^2 = \frac{d\lambda^2}{\lambda^2 - c^2} + \frac{d\mu^2}{1-\mu^2} + X d\phi^2 + 2W d\phi dt - V dt^2 \qquad (5)$$

where ϕ is periodic (with period 2π) μ runs from -1 to +1 (the limits being reached on the north and south polar axes) and where λ is bounded below by the positive constant valve C (which it reaches on the horizon $^{+}$) and runs to $+\infty$ in the assymptotically flat far distance. Moreover ρ, X, W, V are functions of λ, μ only, and such that $\rho^2 = (\lambda^2 - c^2)(1 - \mu^2)$.

The next step, which relies heavily on the use of the Ernst [32],[33] transformations of the field equations, is to obtain an elliptic boundary condition for the smallest possible number of unknowns. We have Lemma 4 (Carter [6], [7]): Under the conditions established by Lemma 3, a solution of the full set of field equations and asymptotically flat black hole boundary conditions is uniquely determined by a solution in the 2-dimensional domain $-1 < \mu < 1$, $C < \lambda < \infty$ with metric

$$ds^2 = \frac{d\lambda^2}{\lambda^2 - c^2} + \frac{d\mu^2}{1 - \mu^2} \tag{6}$$

of the system given by the variation principle derived from an integral of the form $\int L \, d\lambda d\mu$ subject to boundary conditions as follows.

(a) In the pure vacuum case the Lagrangian has the form

$$L = \frac{|\nabla X|^2 + |\nabla Y|^2}{2X^2} \tag{7}$$

where the unknowns X (which has already been introduced in the metric, and which by causality must be positive except on the axis) and Y satisfy the conditions that they and their derivatives are bounded, that on the axis $\mu \to \pm 1$, $\partial Y/\partial\lambda$, $\partial Y/\partial\mu$ and X (but not $\partial X/\partial\mu$) must tend to zero, and that in the limit $\lambda \to \infty$,

$$\lambda^{-2} X = (1 - \mu^2)(1 + 0(\tfrac{1}{\lambda}))$$

$$Y = 2J\mu(3 - \mu^2) + 0(\tfrac{1}{\lambda})$$

where J is the assymptotically defined angular momentum.

(b) In the vacuum Einstein-Maxwell case

$$L = \frac{|\nabla X|^2 + |\nabla Y + 2(E\nabla B - B\nabla E)|^2}{2X^2} + 2\,\frac{|\nabla E|^2 + |\nabla B|^2}{X} \tag{8}$$

where the four quantities X, Y, E, B (of which the latter determine the electric and magnetic parts of the field) satisfy the conditions that they and their derivatives are bounded, that $\partial E/\partial\lambda$, $\partial B/\partial\lambda$, $\partial Y/\partial\lambda$, $\partial Y/\partial\mu + 2(E\,\partial B/\partial\mu - B\,\partial E/\partial\mu)$, and X (but not $\partial X/\partial\mu$) must tend to zero on the axis $\partial \to \pm 1$, and that as $\lambda \to \infty$

$$E = -Q\mu + 0(\tfrac{1}{\lambda})$$

$$B = -P\mu + 0(\tfrac{1}{\lambda})$$

$$Y = 2J\mu(3 = \mu^2) + 0(\tfrac{1}{\lambda})$$

$$\lambda^{-2} X = (1 - \mu^2)(1 + 0(\tfrac{1}{\lambda}))$$

where J, Q, P are the assymptotic angular momentum, electric charge, and magnetic monopole moment (if any).

The final step to establishing the absolute uniqueness of the Kerr solutions is provided by the following

Lemma 5a (Robinson [7])

For given values of the parameters J and C, the solution X, Y to the boundary value problem specified by Lemma 4a is unique.

The unique solution is of course the representative of a Kerr solution--namely that Kerr solution with angular momentum J and with mass M given in terms of J and C by

$$M = \tfrac{1}{2}\left((C^2 + 2J)^{\frac{1}{2}} + \quad (C^2 - 2J)^{\frac{1}{2}} \right)$$

The proof of this last Lemma is very elevant and worth describing. Suppose one has two distinct pairs of fields X, Y an the domain, and let us for convenience use a dot for the difference, a bar for the arithmetic average and angle brackets for the geometric average of corresponding functions, i.e.

$$\dot{f} \equiv (f(X_1,Y_1) - (f(X_2,Y_2))$$

$$\bar{f} \equiv (f(X_1,Y_1) + f(X_2,Y_2))$$

$$\langle f\rangle = f(X_1,Y_1)\ f(X_2,Y_2)$$

where f is any function of X and Y and where the two pairs of fields under consideration are distinguished by the suffices 1,2. Then it can easily be checked that (for any pairs of functions X_1,Y_2 and X_2,Y_2) the relation

$$\rho|\bar{X}(\tfrac{\nabla Y}{X})^{\cdot} - \dot{Y}(\overline{\tfrac{\nabla X}{X}})|^2 + \rho|\dot{X}(\overline{\tfrac{\nabla Y}{X}}) - \dot{Y}(\overline{\tfrac{\nabla X}{X}})|^2 + \overline{\rho|\dot{Y}\,\tfrac{\nabla Y}{X} + X(\tfrac{\nabla X}{X})^{\cdot}|^2}$$

$$\equiv \langle X\rangle^2\ \nabla\{\rho^{\nabla}(\frac{\dot{X}^2 + \dot{Y}^2}{\langle X\rangle^2}) - \dot{Y}(X^2G_Y)^{\cdot} + \dot{Y}^2(\overline{XG_X}) - \bar{X}\,\dot{X}\ (XG_X)^{\cdot} \qquad (9)$$

is identically satisfied, where we use the further abbreviations

$$G_X = \frac{\delta L}{\delta X}\quad , \quad G_Y = \frac{\delta L}{\delta Y}$$

for the functional derivatives of the pure vacuum Lagrangian function (7). Since these must vanish when the field equations are satisfied, it can be seen that the identity gives rise to an equation between a non-negative quantity on the left hand side and a divergence on the right hand side. Hence integrating over the domain, and using Green's theorem and the boundary conditions, one can establish that the quantity on the left hand side is zero everywhere, which leads on the conclusion that $\dot{X}$ and $\dot{Y}$ must be zero--i.e. the two hypothetical solutions must coincide after all.

The identity (9) discovered by Robinson (1975) [8] is a generalisation of a simpler identity (Carter (1971 [6]) to which it reduces when the two pairs of functions are supposed to be at most infinitesimally distinct, so that the dot is interpreted no longer as a finite difference but as a differential--or as a derivative with respect to a continuous variation parameter--and the averaging operations become redundant. This original identity--which can be read off from (9) simply by ignoring the presence of the bars and angle brackets--was sufficient to establish a no-hair theorem to the effect that vacuum black hole solutions occur in families characterised by only two parameters (C and J, or equivalently M and J) without bifurcations. Robinson [11] has succeeded in generalising the original (1971) [6] identity in a different direction so as to include the electromagnetic field scalars E and B in addition to X and Y, but only within the framework of infinitesimal variations not finite differences. This has made it possible to establish a rigorous no-hair theorem for the Einstein-Maxwell black holes, i.e. to prove that the solutions come in discrete non-bifurcating families characterised only by the three parameters mass, angular momentum and electric charge (or by four parameters if one is willing to consider magnetic monopoles). This generalises the earlier work of Ipser [34], Wald [35], and Carter [7] which had already ruled out electromagnetic bifurcations other than those uniquely characterised by the charge (and magnetic monopole) parameter starting from a pure vacuum (Kerr) solution.

I shall not write out Robinson's electromagnetic identity because it would occupy half a page and the probability of including at least one typographical error would be of order unity. Suffice it to say that it equates a divergence (when the field equations are satisfied) to the sum of no less than eight distinct positive definite terms, some of them quite complicated compared with only three comparatively simple positive definite terms in the original pure vacuum case. If one could generalise this identity appropriately to allow for finite differences by the insertion of averaging bars and brackets as in the pure vacuum identity (9), then it would be possible to establish the absolute uniqueness of the Kerr Newman electromagnetic vacuum solutions. In view of the remarkable success of this whole mathematical program so far, I would be strongly inclined to bet that such a finite difference electromagnetic identity does in fact exist. However, in view of the enormous number of ways in which averaging bars and brackets may be inserted, and the effort required to check each distinct possibility, it is not surprising that no one has yet found it. In the meanwhile, the conceivable existence of pathological (undischargeable and unstoppably rotating) families of electromagnetic black hole equilibrium states other than those of Kerr Newman is not completely excluded.

If and when this question is settled, there will remain the mathematically more difficult (but physically much less important) problems of clearing up the analogous questions concerning Proca and Yang-Mills black holes.

APPENDIX

The extension of the theorem to the effect that the metric must be circular from the pure vacuum to the vacuum Einstein-Maxwell case depends on first showing that the Maxwell tensor Fab itself satisfies an appropriate circularity condition. Circularity of the metric tensor has been defined as the absence of cross components between the $t\phi$, and r,θ directions, and Papapetrou's proof was based on the demonstration that a globally sufficient (as well as locally necessary) condition for this is that the Ricci tensor Rab should satisfy an exactly analogous condition of absence of cross components (which is automatically satisfied in the pure vacuum case when Rab is zero). The circularity condition for the antisymmetric tensor Fab--which automatically guarantees that of Rab (which is a quadratic function of Fab) is somewhat different in that it consists of a requirement not that the cross components be absent but rather than Fab should <u>only</u> have cross components. This requirement can be expressed briefly in terms of contractions (denoted by a dot) with the symmetry generators k^a, m^a as

$$k.F.m = 0 \quad , \quad k.{}^*F.m = 0$$

where ${}^*F$ denotes the dual electromagnetic field (i.e. ${}^*F_{ab} = \varepsilon_{ab}{}^{ed}F_{ed}$). Now it can easily be shown to follow merely from the condition that F be invariant under the symmetry group actions generated by k and m that

$$\partial(k.F.m) = k.\partial F.m$$
$$\partial(k.{}^*F.m) = k.\partial{}^*F.m$$

where ∂ is the Cartan exterior differentiation operator, which when acting on a scalar (such as k F.m or $k.{}^*F.m$) is the same as the gradient.

When the ordinary source free Maxwell equations, which can be expressed as

$$\partial F = 0 \quad , \quad \partial{}^*F = 0$$

are satisfied, it thus follows immediately that k.F.m and $k.{}^*F.m$ are <u>constants</u> and hence, since they necessarily vanish on the symmetry axis (where m^a is zero), that they vanish everywhere as required. (It can also be shown that this is a sufficient as well as obviously necessary condition for it to be possible to choose a vector potential A satisfying $F = \partial A$ which satisfies the original <u>vector</u> circularity condition (2) which is equivalent to $k.{}^*A.m = 0$.)

In the case of a vector field with mass μ, the field equations are

$$\partial F = 0 \quad , \quad \partial{}^*F = \mu^{2*}A$$

where

$$F = \partial A$$

Under these conditions we still obtain

$$\partial(k.F.m) = 0$$

and hence that k.F.m. itself must be zero, but for the dual scalar we now have

$$\partial(k.{}^*\partial A.m) = \mu^2 k.{}^*A.m$$

What one would like to be able to prove is that both sides of this equation are zero, but I cannot at present see how to deduce this.

Acknowledgements: I wish to thank Gary Gibbons and Malcolm Perry for informative discussions.

REFERENCES

[1] A. G. Doroshkevich, Ya B Zeldovich, and I. D. Novikov (1965) Z.E.T.P. 49 170 (Eng. Trans. Sov. Phys. JETP 22, 122).
[2] L. A. Mysah and G. Szekeres (1966) Can. J. Phys., 44, 617.
[3] W. Israel (1967) Phys. Rev. 164, 1776.
[4] B. Carter (1968) Phys. Rev. 174, 1559.
[5] W. Israel (1971) G.R.G. Journal, 2, 53.
[6] B. Carter (1971) Phys. Rev. Lett. 26, 331.
[7] B. Carter (1974) in Black Holes, Ed. B. and C. Dewitt, Gordon and Breach.
[8] D. C. Robinson (1975) Phys. Rev. Lett. 34, 905.
[9] S. W. Hawking (1974) in Black Holes, Ed. B. and C. Dewitt, Gordon and Breach.
[10] R. Penrose (1969) Riv. Nuovo Cimento I, 1, 252.
[11] D. C. Robinson (1974) Phys. Rev. D10, 458.
[12] W. Israel (1968) Common. Math. Phys. 8, 245.
[13] H. Muller Zum Hagen, D. C. Robinson and H. J. Seifert (1973), G.R.G. Journal, 4, 53.
[14] H. Muller Zum Hagen, D. C. Robinson and H. J. Seifert (1974), G.R.G. Journal, 5, 59.
[15] S. W. Hawking (1972) Common. Math. Phys. 25, 152.
[16] S. W. Hawking and G. F. R. Ellis (1973), The Large Scale Structure of Space Time, Cambridge University Press.
[17] G. W. Gibbons (1974) Common. Math. Phys. 35, 13; H. Muller zum Hagen and H. J. Siefert (1973), Int. J. Th. Phys. 8, 443.
[18] S. W. Hawking, Common. Math. Phys. 26, 167.
[19] J. D. Beckenstein (1971) Phys. Rev. Lett. 28, 452.
[20] J. D. Beckenstein (1972) Phys. Rev. D5, 1239.
[21] J. D. Beckenstein (1972) Phys. Rev. D5, 2403.
[22] C. Teitelboim (1972) Phys. Rev. D5, 2941.
[23] B. Carter (1974) Phys. Rev. Lett. 33, 558.
[24] M. Perry (1975) Yang Mills Fields in General Relativity (unpublished).
[25] M. Y. Wang (1975) preprint (Miami).
[26] G. 't Hooft (1974) Nucl. Phys. B79, 276.
[27] B. Carter (1969) J. Math. Phys. 10, 70.
[28] Vishveshwara (1968) J. Math. Phys. 9, 1319.
[29] A. Papapetrou (1966) Ann. Inst. H. Poincarre A4, 83.
[30] Kundt, W. and Trumper, M. (1966) Z Physik, 192, 419.
[31] M. Perry (1975) preprint (Cambridge).
[32] F. J. Ernst (1968) Phys. Rev. 167, 175.
[33] F. J. Ernst (1968) Phys. Rev. 167, 1175.
[34] J. R. Ipser (1971) Phys. Rev. Letters 27, 529.
[35] R. Wald (1971) Ph.D. Thesis, Princeton.

ON GRAVITATION IN MICROPHYSICS

IS EINSTEIN'S FORM OF GENERAL RELATIVITY STILL VALID FOR CLASSICAL FIELDS WITH SPIN?

Paul von der Heyde and Friedrich W. Hehl

Institut für Theoretische Physik der Universität zu Köln
D-5000 Köln 41, Fed. Rep. Germany

SUMMARY

Einstein's theory of gravitation contains elements which contradict an appropriate description of the gravitational interaction of elementary, massive fields with spin. Neither the equivalence principle, nor the success of the macroscopic theory enforce a symmetric connection on spacetime. The equivalence principle rather suggests a Riemann-Cartan (U_4) geometry of the world. Accordingly, the field equations of general relativity cannot always be extrapolated to the level of elementary fields.

Conventionally matter with elementary spin is forced into Einstein's formalism and hence a spin concept is given up which is independent of orbital angular momentum. We uphold an independent spin concept and point out the intimate relation existing between the field equations of gravity sought-after and the localization of the energy-momentum and the spin angular momentum of the matter under consideration. Energy-momentum and spin angular momentum are acted on by gravitation and consequently should be sources of gravitation, too. We localize them by their respective canonical tensor. For spinning matter the resulting U_4 theory coincides only in linear approximation with the extrapolated Einstein theory.

In an Appendix we present several theses in order to comment on some controversial issues in U_4 theory.

1. PRELIMINARY REMARKS

It is generally assumed that the gravitational interaction of elementary fields basically follows the principles laid down in general relativity (GR) for the case of tangible matter. The spirit of our attempt is to take for granted Einstein's fundamental ideas on gravitation, like the equivalence principle and the structure of his field equation. We will show that the outcome of his construct, however, contains elements which are not appropriate to a microscopic theory of gravity. Such considerations seem to be a legitimate undertaking at a Conference on the Fundamentals of GR.

We will end up with a general relativistic field theory enriched by a torsion of spacetime (U_4 theory). This theory was proposed by Sciama [27], Kibble [15], and others. For new results in U_4 theory and detailed references to the literature see ref. [11], e.g. We will consider only massive, local matter fields possessing a special relativistic Lagrangian density. For simplicity we restrict ourselves to the discussion of tensor matter ("bosons"). The corresponding treatment of spinor matter ("fermions") is strictly analogous and leads to the same general results. The mathematical notions and the terminology we generally use are taken from Schouten's Ricci Calculus [25].

2. EQUIVALENCE PRINCIPLE

The equivalence principle (EP) links up the special relativistic gravity free description of matter with the general relativistic description which includes gravitation. It was proposed by Einstein as a heuristic principle and was based on the following observation: If we have a reference frame in a gravitational field of a certain strength, then, within the limits of measurability, any massive test particle will be equally accelerated relative to that reference frame. This fact is explained by the equivalence principle which states more generally: Locally the properties of special relativistic matter in a non-inertial frame of reference are equal to the properties of the same matter in a corresponding gravitational field. Here locally means within an infinitesimal surrounding of a fixed point of spacetime.

Let us explain this in more detail. First, the postulated impossibility to distinguish the behavior of matter can only apply to such material properties which can be described by locally defined quantities. This will be treated in Sect. 4. Secondly, at least in macrophysics (and probably in microphysics, too) the notion "infinitesimal" has to be considered as an idealization of "sufficiently small, but not too small", which allows for a simple mathematical description of material objects.

It is senseless, for example, to ask for the change in the properties of a gas over a spatial distance dl if dl is smaller than the average distance of the particles constituting the gas. After all, the local values of the properties of the gas are computed by averaging over a region much bigger than dl. Analogously, it is senseless to ask for the change of these quantites within the time interval dt = dl/c (c = velocity of light) since, within this interval, the averaging regions are practically causally disconnected. With respect to the electron field, however, the same dl can be understood as a finite length and hence as a non-local quantity. Accordingly, the notion "local" strongly depends on the level of matter description.

The EP implies two things:

a.) In the (pseudo-) Euclidean Minkowski spacetime the influence of the gravitational field on matter can be simulated by going over to a non-inertial reference frame within the infinitesimal surrounding of the point under consideration.

b.) In an arbitrary gravitational field within the infinitesimal surrounding of each point of spacetime, there exists a reference frame relative to which the description of matter takes its special relativistic form (as far as it concerns locally defined properties).

As Einstein has shown, the EP can be understood to express a geometrical property of the spacetime continuum. For this purpose (b.) has to be interpreted not as a statement about matter, but about the geometry of the world:

c.) The world is locally Euclidean, but, if gravity is present, the infinitesimal Euclidean spaces cannot be fitted together to a global Minkowski space, but rather constitute a non-Euclidean space.

One comes back to (b.) if one assumes (c.) and considers matter with locally defined properties and known special relativistic behavior in each infinitesimal Euclidean space.

Accordingly, the EP traces back the phenomenon of gravity to properties of the geometry of the world and requires this geometry to be locally Euclidean. The successful application of this principle to tangible matter, its purely geometrical nature and remarkable simplicity, lead us to the belief that it is sensible to speak about an infinitesimal Euclidean space. In that domain of validity in which a special relativistic theory of matter without gravity is successful (down to at least about 10^{-15} cm), the EP should describe correctly the action of gravitation on matter.

3. MINIMAL COUPLING

In our formulation the EP can be applied to elementary matter since it does not contain macroscopical properties of matter. Let be given a matter field $\psi_R(x^\alpha)$ in a Minkowski spacetime R_4. We take Cartesian coordinates x^α, $\eta_{\alpha\beta}$ denotes the Minkowski metric. The matter field is supposed to have a special relativistic, locally defined Lagrangian density $L_R = L_R\ (\psi_R, \partial_\alpha\psi_R, \eta_{\alpha\beta})$.

Now we carry out at each point an arbitrary, unique, and differentiable transformation ε which transforms the constant orthonormal tetrads $\underset{\sim}{\xi}_\alpha$ into spacetime dependent tetrads $\underset{\sim}{\xi}_i(x^\beta)$:

$$\begin{aligned}&\underset{\sim}{\xi}_\alpha \rightarrow \underset{\sim}{\xi}_i(x^\beta) = \varepsilon_i^{\alpha}(x^\beta)\underset{\sim}{\xi}_\alpha\ ;\quad \psi_R(x^\beta) \rightarrow \psi(x^\beta)\,;\\ &\eta_{\alpha\beta} \rightarrow \eta_{ij}(x^\gamma) = \varepsilon_i^{\alpha}\varepsilon_j^{\beta}\eta_{\alpha\beta};\ \partial_\alpha \rightarrow \overset{\varepsilon}{\nabla}_i = \partial_i + \varepsilon_\beta^{k}(\partial_i\varepsilon_j^{\beta})\ f_{k}^{\cdot j}\ .\end{aligned} \tag{1}$$

Here we have assumed the transformation ε to be invertible everywhere. Hence we have $\det \varepsilon_i^{\alpha} = (-\det \eta_{ij})^{\frac{1}{2}} \neq 0$ and the coefficients of the inverse transformation are defined by $\varepsilon_\alpha^{i}\varepsilon_i^{\beta} = \delta_\alpha^{\beta}$. The partial derivatives are related by $\partial_i = \varepsilon_i^{\alpha}\partial_\alpha$ and $f_k^{\cdot j}$ is the generating operator of the infinitesimal transformation of the field ψ, the transformation property of which is assumed to be known. The constancy of the Minkowski metric yields

$$\overset{\varepsilon}{\nabla}_i\ \eta_{jk} = 0\qquad , \tag{2}$$

as can be seen easily from $(1)_3$ and $(1)_4$. With (2) we can cast the covariant derivative into the form $\overset{\varepsilon}{\nabla}_i = \partial_i + \overset{\varepsilon}{\Gamma}{}_{ij}^{\cdot k}\, f_k^{\cdot j}$ with [1])

1) In future Latin (Greek) indices will be raised and lowered by the metric with the corresponding indices.

$$\overset{\varepsilon}{\Gamma}{}_{ij}^{\ \ k} := \left\{ {k \atop ij} \right\} + \overset{\varepsilon}{\Omega}{}_{ij}^{\ \ ..k} - \overset{\varepsilon}{\Omega}{}_{j.i}^{\ .k} + \overset{\varepsilon}{\Omega}{}^{k}_{\ .ij} \quad . \tag{3}$$

Here we have $\{{k \atop ij}\} := \eta^{kl}(\partial_i\eta_{jl} + \partial_j\eta_{il} - \partial_l\eta_{ij})/2$ as Christoffel's symbol with respect to the metric η, and $\overset{\varepsilon}{\Omega}{}_{ij}^{\ \ ..k} := \varepsilon^k_\beta\, \partial_{[i}\varepsilon^\beta_{j]}$ is a measure of the anholonomity of the tetrads $\underset{\sim}{\xi}_i$. This means that the transformation ε of the tetrads is equivalent to a unique (holonomic) coordinate transformation $x^i = x^i(x^\alpha)$ if and only if the local relation $dx^i = \varepsilon_\alpha{}^i dx^\alpha$ is integrable, i.e. if $\partial_{[\beta}\varepsilon^i_{\alpha]} = \varepsilon^j_\alpha\, \varepsilon^k_\beta\, \overset{\varepsilon}{\Omega}{}_{jk}^{\ \ ..i} = 0$. So in general it will not be possible to introduce a coordinate system x^i.

With respect to $\underset{\sim}{\xi}_i$, The Lagrangian density takes the minimally coupled form [2])

$$L_R \rightarrow \mathcal{L} = (-\det \eta_{ij})^{\frac{1}{2}}\, L_R(\psi,\ \overset{\varepsilon}{\nabla}_i\psi,\ \eta_{ij}) . \tag{4}$$

The EP, as discussed in Sect. 2, now tells us in a.) that (4) is locally identical with the Lagrangian density of the field ψ in a gravitational field. In particular, we have to interprete $\overset{\varepsilon}{\Gamma}{}_{ij}^{\ \ k}$ as (pseudo-) field strength of the gravitational field.

As yet, according to a.), we have only simulated locally the gravitational field in flat spacetime R_4. This enables us now to introduce a genuine gravitational field, in which we can carry out the inverse procedure b.). For this purpose we start with a Lagrangian density of the field ψ, which is supposed to be given in a non-Euclidean spacetime with respect to coordinates x^i (compare c.)). This spacetime has to be linearly connected and metric in order to guarantee local Euclidicity everywhere. Infinitesimal changes of functions will be controlled by the covariant derivative $\nabla_i = \partial_i + \Gamma_{ij}^{\ \ k}(x^l)\, f_k^{\ .j}$, the metric $g_{ij}(x^l)$ obeys $\det g_{ij} \neq 0$, and, in analogy to (2), we have

$$\nabla_i g_{jk} = 0 \quad . \tag{5}$$

Resolved with respect to the connection $\underset{\sim}{\Gamma}$ of spacetime, we get

$$\Gamma_{ij}^{\ \ k} = \left\{ {k \atop ij} \right\} + S_{ij}^{\ \ ..k} - S_{j.i}^{\ .k} + S^k_{\ .ij} \quad . \tag{6}$$

---

[2]) It is known that L_R is determined by the matter equation only up to a divergence $\partial_\alpha D^\alpha$. Generally the resulting Lagrangian densities are not equivalent for different vector densities $\underset{\sim}{D}$. Therefore $\partial\underset{\sim}{D}$ has to be treated as an "observable" quantity in gravitational theory. This is consistent with the fact that the distributions of energy-momentum and spin angular momentum, as sources of gravity (see below), depend on the choice of $\underset{\sim}{D}$.

{} is Christofel's symbol with respect to the metric g. The tensor $S_{ij}^{\cdot\cdot k}(x^l) := \Gamma_{[ij]}^{k}$ is called torsion and introduces 24 independent components into the gravitational field strength. The 40 components of $\underset{\sim}{\Gamma}$ left over are determined by the potential g with its 10 independent components. A space with the connection (6) is called U_4 or Riemann-Cartan space. For S = 0 it degenerates into a Riemannian space V_4. The general material Lagrangian density has the minimally coupled form (4) since $\underset{\sim}{\eta}$ and $\underset{\sim}{g}$, and $\underset{\sim}{\overset{\varepsilon}{\Gamma}}$ and $\underset{\sim}{\Gamma}$ are equivalent locally:

$$\mathcal{L} = (-\det g_{ij})^{1/2}\, L_R\, (\psi\, (x^l),\ \nabla_i\psi,\ g_{ij}). \tag{7}$$

In order to verify b.), we carry out an arbitrary transformation of the (holonomic) $\underset{\sim}{\xi}_i$ of the U_4. Let be ε again unique, invertible, and differentiable:

$$\underset{\sim}{\xi}_i(x^l) \rightarrow \underset{\sim}{\xi}_\alpha(x^l) = \varepsilon_\alpha^i(x^l)\ \underset{\sim}{\xi}_i(x^l)\ ;\quad \psi(x^l) \rightarrow \psi_\varepsilon(x^l)\ ;$$
$$g_{ij}(x^l) \rightarrow g_{\alpha\beta}(x^l) = \varepsilon_\alpha^i\, \varepsilon_\beta^j\, g_{ij};\quad \nabla_i \rightarrow \nabla_\alpha = \partial_\alpha + \Gamma_{\alpha\beta}^{\gamma}\ f_\gamma^{\cdot\beta} \tag{8}$$

We defined $\Gamma_{\alpha\beta}^{\gamma} := \varepsilon_\alpha^i \varepsilon_\beta^j \varepsilon_k^\gamma (\Gamma_{ij}^{k} - \varepsilon_\delta^k \partial_i \varepsilon_j^\delta)$ and have again $\partial_\alpha = \varepsilon_\alpha^i \partial_i$.

Let us now consider a fixed point X^l. At this point we are able to choose the 16 components of ε_α^i and its 4x16 derivatives independently in the sense of initial values. We require $g_{\alpha\beta}$ to be the Minkowski metric at the point $\underset{\sim}{X}$ and as such to be locally constant:

$$g_{\alpha\beta}\ (X^l) = \eta_{\alpha\beta}\ ;\quad \partial_\gamma g_{\alpha\beta}\ (X^l) = 0. \tag{9}$$

By (9), 10 components of ε and 40 components of ∂ε are fixed at the point $\underset{\sim}{X}$. In particular, (9) allows to express Christoffel's symbol in terms of the derivatives of the transformation coefficients alone. For the transformed connection we get

$$\Gamma_{\alpha\beta}^{\gamma}(X^l) = \varepsilon_\alpha^i\ \varepsilon_\beta^j\ \varepsilon_k^\gamma (S_{ij}^{\ \ k} - S_{j\cdot i}^{ok} + S_{\cdot ij}^{k} - \overset{\varepsilon}{\Omega}_{ij}^{\cdot\cdot k} + \overset{\varepsilon}{\Omega}_{j\cdot l}^{\cdot k} - \overset{\varepsilon}{\Omega}_{\cdot ij}^{k})(X^l) \tag{10}$$

with $\overset{\varepsilon}{\Omega}_{ij}^{\ \ k} := \varepsilon_\delta^k\, \partial_{[i}\ \varepsilon_{j]}^\delta$. If we fix the remaing 24 components of ∂ε by

$$S_{ij}^{\cdot\cdot k}\ (X^l) = \overset{\varepsilon}{\Omega}_{ij}^{\cdot\cdot k}\ (X^l), \tag{11}$$

we have

$$\Gamma_{\alpha\beta}{}^{\gamma}(X^1) = 0. \tag{12}$$

Then the covariant derivative ∇_α at the point $\underset{\sim}{X}$ reduces to the partial derivative ∂_α. As required by the EP, spacetime turns out to be locally Euclidean with respect to metric and connection. With $\det g_{\alpha\beta}(X^1) = -1$, we get back the original special relativistic form of the Lagrangian density $\mathcal{L}(X^1) = L_R(\psi_\varepsilon, \partial_\alpha\psi_\varepsilon, \eta_{\alpha\beta})$, where ψ_ε is identified locally with ψ_R.

The sytems $\underset{\sim}{\xi}_\alpha$ is not totally fixed by $\Gamma_{\alpha\beta}{}^{\gamma} = 0$. Any constant (homogeneous) Lorentz transformation with its 6 parameters $\xi_\alpha \to \Lambda_\alpha{}^{\cdot\beta}\xi_\beta$ leaves (9) and (11) invariant. Clearly constant Lorentz transformations cannot imitate a gravitational field in the local R_4 since they link inertial frames of references with each other.

It is not possible in general to transform the connection in the whole U_4 to zero by the same coefficients ε, since the derivatives $\partial\varepsilon$ cannot be prescribed independently at $\underset{\sim}{X}+d\underset{\sim}{X}$ if they are fixed in $\underset{\sim}{X}$. Equation (9), i.e. the introduction of orthonormal tetrads, can be extended to the whole spacetime, however. This is essential for embedding spinor matter in a U_4 since spinors are defined with respect to orthonormal tetrads.

4. Locality

The spacetime of Einstein's gravitational theory is the V_4 without torsion. As we have shown, there is no reason for this restriction from the EP alone. The EP rather suggests a U_4 for the description of the gravitational field. Technically the difference in our discussion, as compared to the usual reasoning in Einstein's theory, is that we take tetrads seriously from a physical point of view. They are fundamental as reference frames for making realistic measurements. In these in general anholonomic tetrads the connection of a U_4 locally can be made to vanish and the Minkowski metric to appear, as we have seen.

Often in textbooks it is stated that torsion $\underset{\sim}{S}$ should vanish in order to allow for the local transformation $\underset{\sim}{\Gamma} \to 0$. In the proof, however, only transformations $x^i \to x^\alpha = x^\alpha(x^i)$ are considered which yield only (holonomic) coordinates again with $\varepsilon_i^\alpha = \partial_i x^\alpha$ and hence $\overset{\varepsilon}{\Omega} = 0$. The assumption that the local relation $dx^\alpha = \varepsilon_i^\alpha dx^i$ is integrable, is a non-local concept, however, and contradicts the local character of the EP. Naturally it is possible to define coordinates x in the tangent R_4 spanned by the Euclidean parallel displacement of the $\xi_\alpha(X^1)$. In the case $S \neq 0$, these x^α cannot be mapped uniquely on the coordinates x^i in the U_4.

We have applied the EP to the Lagrangian density of matter and not to the special relativistic laws following from it. Sometimes it is assmmed that the EP can be applied to all laws. This is only possible for such laws, however, which are formulated in local quantities. In Einstein's V_4 theroy for macroscopic matter the notion of locality is different from the one in microphysics. Hence applications of the EP which make sense macroscopically, may be illegitamate in microphysics.

Consider the energy momentum theorem for a special relativistic ideal gas, e.g.,

$$\partial_\beta T_\alpha^{\cdot\beta} = \partial_\beta[(\rho\ c^2 + \Pi)\ u_\alpha u^\beta - \Pi\delta_\alpha^\beta] = 0\ . \tag{13}$$

Here ρ is the matter density, $c\cdot u^\alpha$ the 4-velocity of the gas and Π its pressure. This is a local law if we consider u^α and the energy-momentum density $p_\alpha = cu^\alpha$ as fundamental dynamical variables of the material continuum. Accordingly, (13) should have the form $\nabla_j T_i^{\ j} = 0$ in the gravitational field.

From a microphysical point of view, $T_\alpha^{\cdot\beta}$ is the average of the energy-momentum distributions of elementary fields. The microphysical version of $\underset{\sim}{T}$ and of the conservation theorem (13) resemble the macrophysical expressions. The energy-momentum density $\underset{\sim}{p}$ is proportional to the gradient of the field variables, however, which is covariant for non-scalar matter only after minimal coupling to the gravitational field. Accordingly, on that level, the energy-momentum theorem is a non-local law. It may lose its simple structure in the gravitational field without violation of the EP. The same holds true for Euler-Lagrange equations and other laws, which generally depend on second derivatives of the matter fields. Thus only in the case of local laws does the application of the EP lead to the same results as the minimal coupling in the special relativistic Lagrangian.

So far we have only studied the action of a given gravitational field on matter. Now we turn to the question, how a gravitational field is produced by a certain distribution of matter.

5. Einstein's Two Field Equations

Expressed in the independent variables $\underset{\sim}{\psi}$, $\underset{\sim}{g}$, $\underset{\sim}{S}$, we have $\mathcal{L}=\mathcal{L}(\psi,\partial\psi, \underset{\sim}{g}, \partial\underset{\sim}{g}, \underset{\sim}{S})$. In order to close the system, metric and torsion have to depend on the distribution of matter via field equations. In Einstein's theory, the energy-momentum tensor of (13) serves as a source of the metric field:

$$\widetilde{G}^{ij}(g,\ \partial g,\ \partial\partial g,) = kT^{ij} \tag{14}$$

$\widetilde{G}$ is the Einstein tensor [3] with respect to the V_4 connection and k the relativistic gravitational constant. The second field equation, which determines torsion, simply reads

$$S_{ij}^{\cdot\cdot k} = 0 \quad . \tag{15}$$

In Einstein's theory this equation is not understood as a field equation, it rather appears as an a priori requirement in the context of the discussion of the EP. This justification we have already rejected above. As we will see later-on, the constraint (15) cannot be justified a posteriori by the success of Einstein's theory either. We get such an idea already from (7). Torsion S only enters the Lagrangian density of non-scalar fields, i.e. torsion acts on nothing but on matter with spin. Consequently only spinning matter should produce torsion. Macroscopical matter usually carries only energy-momentum but no elementary spin angular momentum, mainly because the former is an additive quantity of a monopole type, whereas the latter is a dipole quantity which averages out. Hence the gravitational interaction of normal macroscopic matter is determined to a good approximation by the energy-momentum distribution and the metric, even if (15) is not fulfilled.

6. The Source of Gravity and its Special Relativistic Limit

The success of Einstein's theory does not give an argument that the field equations (14) and (15) should keep their simple form on the level of elementary fields. We can learn from (14), however, thaty a certain true energy-momentum tensor of the matter field should exist which represents a source (not necessarily the only one) of

---

3) $\widetilde{G}_{ij} := \widetilde{R}_{kij}{}^{k} - 1/2 \times g_{ij} g^{lm} \widetilde{R}_{klm}{}^{k}$. For the definition of the curvature tensor $\widetilde{R}$ of the V_4 compare footnote 7 for S = 0.

the gravitational field. This tensor, not yet specified, will be denoted by $\hat{\Sigma}$. So we have

$$G^{ij}(g,\ \partial g,\ \partial\partial g,\ S,\ \partial S) = k\ \hat{\Sigma}^{ij}. \tag{16}$$

In the tensor G, which has to be determined, there cannot enter 2nd derivatives $\partial\partial S$ of the torsion if we demand plausibly that the theory should not contain a further universal constant and that G is polynomial in its variables and their derivatives. As energy-momentum tensor of matter, $\hat{\Sigma}$ should not depend substantially on the gravitational fields. To be more positive, $\hat{\Sigma}$ ought to be a local quantity which can be transformed into its special relativistic form by means of a suitable transformation of the tetrads. This can be achieved the other way round by application of the EP if the local energy-momentum tensor of the R_4 theory is given [4]):

$$\hat{\Sigma}^{\alpha\beta}(\psi_R, \partial\psi_R, \eta) \rightarrow \hat{\Sigma}^{ij}\ (\psi, \nabla\psi, g)\ . \tag{17}$$

Which tensor now does determine the true energy-momentum distribution of matter in an R_4? This question has been discussed already for quite a time. Let us consider the facts emerging from special relativistic field theory.

Dynamical observables in special relativity are energy-momentum P and angular momentum J. For a closed system they obey the conservation theorems

$$P_\alpha = \text{const.};\quad J_{\alpha\beta} = \text{const.} \tag{18}$$

If the Lagrangian density of the system is known, P and J can be expressed by integrals over their corresponding canonical densities[5]):

$$P_\alpha = \int_\infty \Sigma_\alpha^{\cdot\beta} dS_\beta;\quad J_{\alpha\beta} = \int_\infty (\tau_{\alpha\beta}^{\cdot\cdot\gamma} + m_{\alpha\beta}^{\cdot\cdot\gamma})\ dS_\gamma\quad ; \tag{19}$$

$\Sigma_\alpha^{\cdot\beta} := L_R\delta_\alpha^\beta - (\partial L_R/\partial\partial_\beta\psi_R)\partial_\alpha\psi_R$ canonical energy-momentum tensor:

$\tau_{\alpha\beta}^{\cdot\cdot\gamma} := (\partial L_R/\partial\partial_\gamma\psi_R) f_{[\alpha\beta]}\psi_R$ canonical spin angular momentum tensor;

[5]) Here we don't consider the requirements of second quantization.

[4]) For local quantities we take the same symbol irrespective of the space which they are embedded in.

$$m_{\alpha\beta}^{\cdot\cdot\gamma} := x_{[\alpha}\Sigma_{\beta]}^{\cdot\ \gamma}$$ canonical orbital angular momentum tensor. (20)

We have used Cartesian coordinates and have integrated over an arbitrary spacelike hypersurface of the R_4 with an area element dS_β. The differential conservation theorems

$$\partial_\beta \Sigma_\alpha^{\cdot\beta} = 0 \quad ; \qquad \partial_\gamma \tau_{\alpha\beta}^{\cdot\cdot\gamma} - \Sigma_{[\alpha\beta]} = 0 \tag{21}$$

follow from translational and rotational invariance via Noether's theorem. They lead back to (18). In $(21)_2$ we have carried out the differentiation $\partial_\gamma m_{\alpha\beta}^{\cdot\cdot\gamma} = -\Sigma_{[\alpha\beta]}$ since m, in contrast to τ, is not a tensor with respect to a Poincare transformation. Accordingly, the tensor equation $(21)_2$ is the appropriate field theoretical formulation of the conservation theorem of angular momentum.

Evidently the system (21) is not sufficient in order to determine uniquely the distribution of energy-momentum and spin angular momentum of the system. We can substitute the canonical tensors Σ, τ by a whole class of tensors $\hat{\Sigma}, \hat{\tau}$ (compare ref. [9] for details).

$$\hat{\Sigma}_\alpha^{\cdot\beta}[U] = \Sigma_\alpha^{\cdot\beta} - \partial_\gamma(U_\alpha^{\cdot\beta\gamma} - U^{\beta\gamma}_{\cdot\cdot\alpha} + U^{\gamma\cdot\beta}_{\cdot\alpha}) \quad ; \quad [U_{(\alpha\beta)}^{\cdot\cdot\ \gamma} = 0] \quad ;$$

$$\hat{\tau}_{\alpha\beta}^{\cdot\cdot\gamma}[U,Z] = \tau_{\alpha\beta}^{\cdot\cdot\gamma} - U_{\alpha\beta}^{\cdot\cdot\gamma} + \partial_\delta Z_{\alpha\beta}^{\cdot\cdot\gamma\delta}; \quad [Z_{(\alpha\beta)}^{\cdot\cdot\ \gamma\delta} = Z_{\alpha\beta}^{\cdot\cdot(\gamma\delta)} = 0] \ , \tag{22}$$

which fulfill the same conservation theorems (21) and lead to the same observables P and J.

This could make one believe that the distributions of energy-momentum and spin are only secondary concepts not to be taken too seriously. One should not forget, however, that the regauging in (22) is essentially connected with the relation DivCurl = 0, which is valid in an R_4, but broken in a curved space. Accordingly, the considerable freedom in the choices of U and Z is partially a consequence of the special relativistic description of matter. As soon as certain distributions $\hat{\Sigma}$, $\hat{\tau}$ are embedded in curved space, the fields U, Z have to fulfill additional differential equations which constrain the choices considerably. And finally, at least for energy-momentum, there is no localization problem if we assume the correct gravitational theory to be known, since the source of gravitation ought to be described by a well localized quantity. Hence it seems sensible to speak of genuine distributions of energy-momentum and spin even in the special relativistic limit. The question is now: Which choices of U and Z do yield this genuine distribution?

7. Symmetric Energy-Momentum Tensor and the Extrapolated Einstein Theory

According to a widely used assumption, the energy-momentum tensor of matter should be symmetric. Let us inquire into this assumption. By putting $U_{\alpha\beta}^{\cdot\cdot\gamma} = \tau_{\alpha\beta}^{\cdot\cdot\gamma} - \partial_\delta X_{\alpha\beta}^{\cdot\cdot\gamma\delta}$ in (22), we arrive at the most general set of tensors $\hat{\Sigma}$, $\hat{\tau}$ with a symmetric $\hat{\Sigma}$:

$$\hat{\Sigma}^{\alpha\beta}[\tau+\partial X] = \Sigma^{\alpha\beta} - \partial_\gamma(\tau^{\alpha\beta\gamma} - \tau^{\beta\gamma\alpha} + \tau^{\gamma\alpha\beta}) + 2\partial_{\gamma\delta} X^{\gamma(\alpha\beta)\delta} \; ;$$
$$\hat{\tau}_{\alpha\beta}^{\cdot\cdot\gamma}[\tau, Z-X] = \partial_\delta(Z_{\alpha\beta}^{\cdot\cdot\gamma\delta} - X_{\alpha\beta}^{\cdot\cdot\gamma\delta}) \; ; \; [X_{(\alpha\beta)}{}^{\gamma\delta} = X_{\alpha\beta}{}^{(\gamma\delta)} = 0] \; . \tag{23}$$

If we want to avoid, according to (17), non-local terms in $\hat{\Sigma}$, we ha have to choose $X = 0$ [6]). In this special case we recover the so-called Belinfante-Rosenfeld symmetrization [3, 24] of the canonical energy-momentum tensor.

As Rosenfeld [24] has shown, $\hat{\Sigma}^{\alpha\beta}[\tau]$ is the special relativistic limit of the tensor

$$\hat{\Sigma}^{ij}[\tau] = \Sigma^{ij} - \tilde{\nabla}_k(\tau^{ijk} - \tau^{jki} + \tau^{kij}) \; , \tag{24}$$

where the canonical tensors Σ, τ, as defined in (20), are local quantities minimally coupled to the V_4 of Einstein's theory. $\hat{\Sigma}[\tau]$ turned out to be identical with the variational derivative of the Lagrangian density $\tilde{\mathcal{L}} = \tilde{\mathcal{L}}(\psi, \nabla\psi, g)$ with respect to the metric, provided the matter equation $\delta\tilde{\mathcal{L}}/\delta\psi = 0$ is fulfilled:

$$\hat{\Sigma}^{ij}[\tau] \quad \tilde{\sigma}^{ij} := (2/e) \times \delta\tilde{\mathcal{L}}/\delta g_{ij} \; ; \; e := (-\det g_{ij})^{1/2} . \tag{25}$$

For the so-called metric energy-momentum tensor $\tilde{\sigma}$ (and for $\hat{\Sigma}[\tau]$), we simply have the conservation theorem

$$\tilde{\nabla}_j \; \hat{\Sigma}_i^{\cdot j}[\tau] = \tilde{\nabla}_j \; \tilde{\sigma}_i^{\cdot j} = 0 \quad . \tag{26}$$

Consequently, the symmetry assumed for $\hat{\Sigma}$, leads to (16) as the microphysical extrapolation of Einstein's equation and to a Riemannian

---

6) A non-vanishing tensor X is useful for the description of the energy-momentum distribution of massles scalar matter; compare Callan, Coleman, and Jackiw [5] and Schwinger [26]. Since we concentrate on spinning matter, this procedure is of no relevance in our context.

connection of spacetime:

$$\tilde{G}^{ij} = k \;\tilde{\sigma}^{ij} \quad ; \quad S_{ij}^{\cdot\cdot k} = 0 \; . \tag{27}$$

Equation $(27)_1$ can be derived, as in the macrophysical case, by variation of the total Lagrangian density $\tilde{\mathcal{L}}_{total} = \tilde{\mathcal{L}} + \tilde{\mathcal{R}}/2k$ with respect to the metric, where $\mathcal{R} := e g^{lm} R_{klm}{}^{k}$ denotes the density of the curvature scalar of the V_4. Equation $(27)_2$ again has the character of a postulate.

This extension of Einstein's theory to microscopical domains looks so convincingly simple that it is often considered as the final answer and believed to yield the correct gravitational interaction for any matter field. One of the fundamental assumptions, however, is the general symmetry of the material energy-momentum tensor.

This assumption is justified for the case of a matter field without spin. In (22) we then put $\tau = \hat{\tau} = 0$. For local spinless matter fields, apart from the massless case (see footnote 6), we furthermore exclude non-local terms in $\hat{\Sigma}$ by requiring $Z = 0$, $U = 0$. Consequently the genuine energy-momentum tensor for a massive field is identified uniquely with the canonical tensor, which, in this case, is symmetrical according to (21) and coincides with the metric energy-momentum tensor $\tilde{\sigma}$. Hence we accept that the gravitational interaction of local massive matter fields without spin is governed by the extrapolated Einstein equation (27) and by minimal coupling of the Lagrangian density to the V_4.

8. The Emancipation of Spin

If the massive matter field carries spin angular momentum, however, then the localization of energy-momentum by the metrical tensor does not seem to be appropriate any longer and is open to doubt in several directions:

a.) There seems to be no reason why the energy-momentum tensor of spinning matter should be symmetric. The belief in its general symmetry has its roots in Planck's [23] field theoretical formulation of $E = Mc^2$, namely in:

$$\text{c x momentum density} = (1/c) \text{ x energy flux density} \qquad (\hat{\Sigma}^{\alpha o} = \hat{\Sigma}^{o\alpha}) \tag{28}$$

Hence the velocity which the field energy is transported with, should be always proportional to the local momentum of the field. Planck's considerations were based essentially on Newtonian particle dynamics and on classical continuum mechanics, that is on the behavior of tangible, spinless matter. With that background, Planck postulated the relation $(28)_1$, implying the symmetry of the energy-momentum tensor of any matter distribution, and realized that postulate by means of a suitable definition of the momentum density.

Without doubt, Planck also took as a guiding line, as well as Einstein later-on, the only elementary field theory known at that time, namely Maxwell's vacuum electrodynamics also containing the relation $(28)_1$. The Maxwell field is a very special field since it has mass 0, and this fact disqualifies it, in spite of carrying spin, to serve as a model theory for massive spinning matter. The fact that Einstein's gravitational theory only works with a symmetrical energy-momentum tensor is not convincing either, since we know that the geometry of the world is not necessarily Riemannian. We will see that a U_4 allows for an embedding of asymmetrical energy-momentum tensors into spacetime in a most natural way.

b.) Dirac's electron theory points to an asymmetrical energy-momentum tensor. As is well known, there the velocity operator u^α, represented by the γ-matrices $u^\alpha = c\gamma^\alpha$, uncouples from the energy-momentum operator $p_\alpha = i\hbar\partial_\alpha$. According to the Dirac equation, we rather have the operator relation

$$Mu^\alpha = P^\alpha - 2i\ s^{\alpha\beta}P_\beta \quad , \tag{29}$$

where $s^{\alpha\beta}=(i/2)\times\gamma^{[\alpha}\gamma^{\beta]}$ is the spin operator and M the mass of the electron. Integration shows that p leads to the macroscopically observable group velocity, whereas the second term in (29) represents an internal, non-convective zitterbewegung caused by the spin of the field. This contradicts Planck's claim ofa local proportionality of velocity and momentum. The energy-moment distribution should be represented by the asymmetrical operator $p_\alpha u^\beta$, which turns out to be related directly to the canonical energy-momentum tensor of the Dirac field.

c.) Consider for a moment the metrical tensor $\tilde{\sigma}$ to represent the genuine distribution of energy-momentum. Then, in a local tangent space, the conservation theorem (26) yields no force density on matter in accordance with the assumption in the special relativistic theory. Hence the simple structure of (26) could be taken as an argument in favor of the localization (24). This argument, which is claimed to originate from the equivalence principle, may be applicable to matter the energy-momentum distribution of which is of a pure monopole type. There the gravitational force density $f_\sigma \sim \sigma \cdot \Gamma$ the monopole density σ is mediated by the field strength Γ, which can be transformed to zero.

Spin, however, as a synonym for an internal dipole moment of the energy-momentum distribution, should couple to the gradient of the field strength, i.e. to the curvature R of spacetime, which cannot be transformed to zero. Accordingly, we expect that in the tangent space there acts a force density $f_\tau \sim \tau \cdot R$ on spinning matter. A force of this type has been already discovered by Mathisson [18] in the framework of Einstein's theory. There it couples to the (orbital) angular momentum of extended rotation bodies and leads to a deviation from the "free" geodesic motion. It is easy to show that a force density of the Mathisson type is hidden in (26). For that purpose we substitute in (26) $\hat{\Sigma}[\tau]$ by means of the right hand side of (24) and apply the Ricci identity:

$$\tilde{\nabla}_j \Sigma_i^{\cdot j} = \tau^{jkl} \tilde{R}_{iljk} \quad . \tag{30}$$

If we take the force density on the right hand side for granted,(30) points to the canonical tensors Σ,τ as giving the genuine distributions of energy-momentum and spin of matter and excludes the "force-free" distribution right away. As already stated, this interpretation does not contradict the equivalence principle since the energy-momentum theorem is a non-local law in this context.

d.) According to its definition (20), the canonical tensor Σ is a local quantity since we supposed the Lagrangian density to be local. The equations (24) and (25) show, however, that the metrical tensor $\tilde{\sigma}$ may be non-local if the canonical spin tensor τ depends on first derivatives of the material field.

e.) The most serious objection against the localization by (24) is, that is neglects the fundamental difference between spin and orbital angular momentum. Similar as energy-momentum, spin should be a localizable and local quantity. Not accidentally, both conservation theorems can be expressed exclusively in terms of these dynamical quantities. A localization of orbital angular momentum $\hat{m}_{\alpha\beta}^{\cdot\cdot\gamma}[U] = x_{[\alpha}\hat{\Sigma}_{\beta]}^{\cdot\gamma}[U]$ is excluded right from the beginning since it is not invariant under shifting of the coordinate origin. Hence from a field theoretical point of view, energy-momentum and spin are the fundamental dynamical observables. If we split the total angular momentum J = S + M in a spin S and an orbital part M with

$$S_{\alpha\beta}[U] := \int_\infty \hat{\tau}_{\alpha\beta}{}^{\cdot\cdot\gamma}[U, Z]dS_\gamma \; ; \; M_{\alpha\beta}[U] := \int_\infty x_{[\alpha}\hat{\Sigma}_{\beta]}{}^{\cdot\gamma}[U]\, ds_\gamma \; , \qquad (31)$$

the symmetrization (23) with X = 0 leads to

$$S_{\alpha\beta}[\tau] = 0 \; ; \; M_{\alpha\beta}[\tau] = M_{\alpha\beta}[0] + \int_\infty \tau_{\alpha\beta}{}^\gamma dS_\gamma \; . \qquad (32)$$

Consequently, (23), and hence also the Belinfante-Rosenfeld symmetrization, gauge spin to zero and interprete it as an orbital angular momentum. This procedure is not justified by field theory. It contradicts the important role which spin plays, besides mass, in the classification of elementary particles.

In accordance with our considerations, it suggests itself to take the canonical tensors Σ,τ for the description of the genuine distributions of energy-momentum and spin of any massive matter field. There is additional justification. According to their dimension and their appearance in (22), U should be connected with the dipole moment and Z with the quadrupole moment of the energy momentum distribution of the matter field. The quadrupole moment does not appear in a local field, which suggests Z=0. The identification of U with spin leads to the mentioned difficulties. So $\hat{\Sigma}[0] = \Sigma$, $\hat{\tau}[0,0] = \tau$ remain as the most natural choices.

This localization takes the spin concept seriously and treats the spin S[0] as fundamental observable, which is not amenable to a regauging procedure. As demanded Σ, is local and allows the application of (17) as an expression of the equivalence principle. The same applies to τ. And, after all, the canonical tensors are very simple quantities. One of the most attractive feature, however, is the fact that this localization leads to a very natural and simple generalization of Einstein's field equations.

9. U_4 Theory with Canonical Sources

Let the dynamical properties of matter be completely determined by the 16+24 components of the canonical tensors. Both fundamental properties, mass and spin, are subject to the gravitational interaction, as can be read off from the minimal coupling or the Mathisson force. Consequently, mass and spin should also pay an active role in the gravitational interaction and should leave imprints on spacetime. For this purpose we have to drop the unjustified condition (15) since in a V_4 with the 10 independent components of its metric tensors there is no place for an active role of spin. Let then the geometry of the world be one of a Riemann-Cartan space U_4, as already suggested by the equivalence principle. The U_4 geometry should be determined by the two field equations (compare (27))

$$G^{ij} = k\, \Sigma^{ij} \; ; \quad T_{ij}^{\cdot\cdot k} = k\, \tau_{ij}^{\cdot\cdot k} \quad . \tag{33}$$

Tha canonical tensors $(20)_1$ and $(20)_2$, which are minimally coupled to the U_4, serve as sources in the field equations. Hence the tensors G, T comprise 16+24 components, too. They depend only on the 10+24 independent components of the metric g and torsion S, however. Consequently, G and T have to fulfill a purely geometrical identity with 6 independent components. Obviously, this identity has to be compatible with the angular momentum theorem. It follows, without reference to field equations, via the Rosenfeld identities as applied to the Lagrangian density $\mathcal{L}$ which is minimally coupled to the U_4 [7]):

$$\overset{*}{\nabla}_k \tau_{ij}^{\cdot\cdot k} - \Sigma_{[ij]} = 0 \quad . \tag{34}$$

Similarly, we require the existence of a further identity of the type of (30) which should be the geometrical image of the energy-momentum theorem. Also independently from (33), the covariance of the Lagrangian under coordinate transformations yields

$$\overset{+}{\nabla}_j \Sigma_i^{\cdot j} = \tau^{klj} R_{ijkl} \quad . \tag{35}$$

This restricts the choices of G and T heavily. Additionally, for spinless matter, (33) should pass over into the V_4 field equations (27). For matter with spin, both sets of equations should coincide within the linear approximation [8]).

Let us start with the last requirement. In $(27)_1$ we neglect the non-linearities of the left hand side and the minimal coupling on the right hand side. With (25) and (24) we get

$$2\, \partial_{[k} \left\{ {}^{\;k}_{i]j} \right\} - g_{ij}\, g^{lm} \partial_{[k} \left\{ {}^{\;k}_{l]m} \right\} = k\, \Sigma_{ij} - k \partial_k \left(\tau_{ij}^{\cdot\cdot k} - \tau_{j\cdot i}^{\cdot k} + \tau^k_{\cdot ij} \right) . \tag{36}$$

Here indices can be raised and lowered by means of the Minkowski

---

7) We define $\overset{*}{\nabla}_k := \nabla_k + 2S_{kl}^{\cdot\cdot l}$ and $\overset{+}{\nabla}_j \Sigma_i^{\cdot j} := \overset{*}{\nabla}_j \Sigma_i^{\cdot j} + 2S_{ji}^{\cdot\cdot l} \Sigma_l^{\cdot j}$. $R_{ijk}^{\cdot\cdot\cdot l} := 2\partial_{[i} \Gamma_{j]k}^{\;\;l} + 2\Gamma_{[i|m}^{\;\;\;l} \Gamma_{|j]k}^{\;\;m}$ is the curvature tensor of the U_4.

8) For the linearization procedure in U_4 theory compare Arkuszewski, Kopczynski, and Ponomariev [2] and Adamowicz [1].

metric. In order to put (36) into the form $(33)_1$, we shift the spin terms to the left hand side. The physical meaning is that we absorb again that part of the energy-momentum tensor into the source which arises from the material spin distribution and which has been subtracted out from Σ by means of the Belinfante-Rosenfeld procedure. A short calculation reveals that then the left hand side of (36) can be expressed in terms of the U_4 connection Γ according to

$$2\,\partial_{[k}\Gamma_{i]j}{}^{k} - g_{ij}\, g^{lm}\, \partial_{[k}\Gamma_{l]m}{}^{k} = k\,\Sigma_{ij}\,, \tag{37}$$

provided we put $S_{ij}{}^{k} = k(\tau_{ij}{}^{k} + \delta^{k}_{[i}\tau_{j]l}{}^{l})$. Equation $(33)_2$ shows this to be equivalent to the choice

$$T_{ij}{}^{k} := S_{ij}{}^{k} + \delta^{k}_{i}\, S_{jl}{}^{l} - \delta^{k}_{j}\, S_{il}{}^{l}\,. \tag{38}$$

An algebraic relation between spin and torsion appears desirable. For spinless matter it guarantees that spacetime degenerates to a V_4. Furthermore torsion averages out for normal macroscopical matter. Thus we do not run into contradictions with Einstein's theory.

The left hand side of (37) is just the linear approximation of the Einstein tensor of the U_4. Hence we identify G according to

$$G_{ij} := R_{kij}{}^{k} - g_{ij}\, g^{lm}\, R_{klm}{}^{k}/2. \tag{39}$$

Together with (38), this guarantees the theory to pass over into the extrapolated Einstein theory in the case of spinless matter. A closer look on U_4 geometry reveals that these choices of G and T are compatible with the conservation theorems indeed.

The field equations are coupled by the conservation theorems. So it is clear that they cannot be deduced from a total Lagrangian density by independent variations. They can be uncoupled, however. We subtract suitable terms from both sides of the first field equation and apply the second field equation (compare (36)):

$$G^{ij} - \overset{*}{\nabla}_k (T^{iji} - T^{jki} + T^{kij}) = k\Sigma^{ij} - \overset{*}{\nabla}_k (\tau^{ijk} - \tau^{jki} + \tau^{kij}) =: k\sigma^{ij}. \tag{40}$$

Now, according to the Rosenfeld identities, the sources σ and τ turn out to be independent variational derivatives of the matter Lagrangian minimally coupled to the U_4:

$$(2/e)\times\delta\mathscr{L}/\delta g_{ij}\equiv\sigma^{ij} \quad ; \quad (1/e)\times\delta\mathscr{L}/\delta S_{ij}^{\cdot\cdot k}\equiv-\tau_{k}^{\cdot ji}+\tau^{ji}_{\cdot\cdot k}-\tau^{i\cdot j}_{\cdot k} \quad . \tag{41}$$

The the uncoupled field equations (40) and $(33)_2$ can be derived from the total Lagrangian density $\mathscr{L}_{total}=\mathscr{L}+\mathscr{R}/2k$ by variation with respect to metric and torsion. Here $\mathscr{R}:=eg^{lm}R_{klm}^{\cdot\cdot\cdot k}$ is the density of the curvature scalar of the U_4. Hence we get $\mathscr{L}_{total}$ from the corresponding Lagrangian density in the V_4 simply by substituting the U_4 connection Γ in place of the Christoffels {}.

10. Concluding Remarks

U_4 theory seems to us a serious competitor of Einstein's V_4 theory. Our aim was to show that, for elementary spin, U_4 theory is a necessary and natural extension of Einstein's theory. For more details of the apparatus of U_4 theory and for simple applications compare, for instances, the references [8, 10, 14, 17, 28, 29].

A short remark about massless matter seems appropriate. U_4 theory is only applicable to matter which possesses a well-defined spin distribution. Certain fields have the property of losing some of their independent degrees of freedom in the limit mass$\to$0. Take the four-component Proca field (spin - $1\hbar$). In the limit of vanishing mass, i.e. in the case of the Maxwell field, a one parameter gauge group arises. The canonical distributions of energy-momentum and spin are not gauge invariant, and spin even may be transformed to zero. This disqualifies the photon spin as a source of torsion. The mentioned redistribution of angular momentum allows for an interpretation of spin as orbital angular momentum and, accordingly, the symmetry (28) has to be accepted in the Maxwell case. Thus V_4 theory seems appropriate for "degenerated" massless fields. This is also suggested by the fact that the special relativistic Lagrangian density of such fields is already generally covariant and needs only the substitution $\eta\to g$. In the case of the electron (spin = $\hbar/2$), however, we do not lose any degree of freedom for mass $\to$ 0, correspondingly, neutrinos do produce torsion.

Acknowledgements

One author (F. W. Hehl) would like to thank Professor Remo Ruffini and Professor G. Denardo for the invitation to the Conference. Furthermore he gratefully acknowledges valuable discussions with Prof. Jürgens Ehlers, Prof. P. van Nieuwenhuizen, and Prof. H. K. Urbantke.

Appendix

Thesis A:

U_4 theory is the local gauge theory of the Poincare group. So it is expected to possess a better renormalization behavior than Einstein's theory.

The Poincare group consisting of rigid rotations (6 parameters) and rigid translations (4 parameters) in the Minkowski spacetime may be generated by infinitesimal, constant-parameter motions of the local orthonormal tetrad ε^i_{α}, where Latin indices refer to the coordinates x^i. This means parallel transport of the tetrads along a constant vector field λ^i and additional constant Lorentz rotations $\lambda_{\alpha}^{\cdot\beta}$.

Non-constant motions of the tetrads, corresponding to a non-rigid motion of matter, lead to the "interactions" $(\partial_j\lambda^{\alpha})\,\Sigma_{\alpha}^{\cdot j} + (\partial_j\lambda_{\alpha}^{\cdot\beta})\tau_{\beta}^{\ \alpha j}$. This identifies canonical energy-momentum spin angular momentum as independent currents of the corresponding translational and rotational gauge fields. Invariance of the system under local Poincare transformations is achieved by introducing a new tetrad $e^{\alpha}_i(x)$ (10 independent components) and an affine tetrad connection $\Gamma_{i\alpha}^{\ \ \beta}(x)$ (24 independent components) as the translational and rotational gauge potentials, respectively. Observe that $\Gamma_{i\alpha}^{\ \ \beta}$ is already a metric connection (U_4 connection) because of its antisymmetry in the last indices. Geometrically by $g_{ij} = e^{\alpha}_i e^{\beta}_j \eta_{\alpha\beta}$ and $D_i = \partial_i + \Gamma_{i\alpha}^{\ \ \beta} f_{\beta}^{\cdot\alpha}$ lengths, angles, and parallel orientations are connected between different points in a curved spacetime. Compare for this whole development Sciama [27] and Kibble [15] and the detailed discussions of Kerlick [14] and Nester [20].

Taking the curvature scalar $\mathcal{R}(e,\Gamma)$ as the free Lagrangian of the independent fields e,Γ, one immediately arrives at the field equations (33) with (38, 39). Thus U_4 theory turns out to be the gauge theory of the Poincare group. It is now expressed with respect to the fields e and Γ and not, equivalently, with respect to metric g and torsion S. The relations between the different field pairs are similar to the equs. $(8)_3$, $((9)_1$, and (10) (replace ε by e). Hence a variation of the tetrad e corresponds to a change in the metric g and, as Γ is fixed, to a certain related change in the torsion S; a variation of the affine tetrad connection Γ, keeping the tetrads fixed, corresponds to a variation of torsion S alone.

The gauge group of U_4 theory is the Poincare group and not the group of general coordinate transformations plus the Lorentz group, as it is sometimes stated. General coordinate transformations in flat spacetime don't correspond to rigid motions, nor do they include Lorentz rotations. In our view, general coordinate transformations arise in a different way: The U_4 theory is locally Poincare invariant; that is parallel propagation $e'^{\alpha}_i(x) = [\delta^k_{\ i}\ (1+\lambda^j(x)\ D_j) + \partial_i\lambda^k(x)]\ e^{\alpha}_k(x)$ and local rotation $e'^{\alpha}_i(x) = (\delta^{\alpha}_{\beta} + \lambda_{\beta}^{\cdot\alpha}(x))e^{\beta}_i(x)$ of the tetrads do not change physics. The translational part, however, may be equivalently generated by a general coordinate transformation $x'^j = x^j - \lambda^j(x)$, taking the tetrads of the point $x'=x$, plus an additional local rotation $\lambda^j(x)\ \Gamma_{j\beta}{}^{\alpha}\ e_i^{\beta}$. Hence local Poincare invariance implies general coordinate invariance.

Gauge theories of interaction have a preferred status with respect to covariant quantization and renormalization. See the review article of Veltman [30]. Applying the background field method in the framework of Einstein's theory, Deser and van Nieuwenhuizen [6] recently have shown that the coupled Dirac-Einstein system is non-renormalizable on the 1-loop level. Compare also 't Hooft [13] and van Nieuwenhuizen [21]. The trouble arises from an residual axial-vector interaction which cannot be compensated in the framework of the Einstein field equations, but which is exactly of the same type as the additional spin-torsion interaction occuring in U_4 theory. Since furthermore U_4 theory is the true gauge theory of gravitation, we formulate the following hypothesis: The coupled U_4-Dirac system is 1-loop renormalizable.

Thesis B:

A metric and affine space (Riemann-Cartan space) has to be postulated in U_4 theory. It cannot be derived by varying independently the metric and an unconstrained affine connection.

In U_4 theory, as compared to GR, one looks for independent rotational or rather contortional degrees of freedom of spacetime, which are then related to spin angular momentum of matter. U_4 theory is a dualistic theory comprising matter in interaction with spacetime (or field proper). This dichotomy in matter and space implies, according to the idea of GR, that spacetime is, in some sense, the image of matter and vice versa. Consequently the way we describe matter has its implications for the geometry to be chosen for spacetime and the other way round.

Now, matter fields are introduced via their transformation properties under a Poincare transformation. Thus in the geometry to be deduced by means of the EP, we have to find elements taken over from the Poincare group. As we have seen in Sect. 3, the Riemann-Cartan geometry U_4, characterized by the metric postulate $\overset{\Gamma}{\nabla} g=0$, is the appropriate geometry for spinning massive "Minkowski" or "Poincare" matter.

Consequently, in a dualistic theory of matter and space, the geometry is prescribed by the properties of matter. Hence, from a physical point of view, it doesn't make sense to leave open the relation between affine connection Γ and metric g and to pretend that this relation should be derived by an independent variation with respect to the connection. Accordingly, the so-called "Palatini Principle" (compare the interesting discussion of El-Kholy, Sexl, and Urbantke [7], see also Misner, Thorne, and Wheeler [19]) is to be rejected in a dualistic theory of matter [9]).

matter [9]).

Incidentally, this point of view is consistent with the analysis of Trautman [28] and Kopczynski [16].

In his pioneering work Sciama [27] used what he calls a Palatini technique. What he really does, however, is to implicitly assume a U_4 beforehand and to vary the orthonormal tetrads and the affine tetrad connection of a U_4 independently. As we have seen in Thesis A, this is equivalent to varying independently metric and torsion. So also Sciama's tetrad connection is constrained to be of a U_4 type a priori.

Thesis C:

Torsion is produced by the spin of elementary particles and not by the so-called spin of stars or galaxies.

In Sect. 8 and Sect. 9 we have tried to show that it is the concept of an irreducible elementary spin which leads to U_4 theory. Any so-called spin, which can be resolved into orbital angular momentum, is appropriately described within the framework of Einstein's V_4 theory. The antisymmetric part of the energy-momentum tensor and the spin tensor "will have a direct physical meaning if the particles must be considered as point particles, this being equivalent to the statement that it has no physical sense to speak of an internal structure of these particles... The only possible contributions would come from the spins of the elementary particles..." (Papapetrou [22]). The same is true for the physical meaning of torsion. For a contrasting point of view see, for example, Adamowicz [1].

The unitary representations of the Poincare group are numbered by mass and spin showing the mutual independence of mass and spin or energy-momentum and spin angular momentum, respectively. Furthermore, there is only one fundamental constant in nature with the dimension of spin, namely Planck's constant. This constant, multiplied only by small numbers, should appear in Lagrangian density of matter and, accordingly, in the spin angular momentum tensor of the 2nd field equation.

If by some later developments in experimental physics the spin of all spinning elementary particles could be resolved somehow into orbital angular momentum, then U_4 theory would lose its meaning as a fundamental theory of gravity. Such an outcome seems improbable, however, because of the far-reaching importance of the Poincare group and due to the independence of translations and rotations in Minkowski's spacetime. Observe that even quarks and partons carry intrinsic spin!

[9]) In unified field theories of the general relativistic type, for Lagrangians other than R, the Palatini Principle is even inconsistent. See Buchdahl [4].

Thesis D:

Tordions could exist. But the Lagrangian of U_4 Theory is appreciably constrained by a theorem of the Lovelock type.

One can arrive at the field Lagrangian of U_4 theory, the density of the curvature scalar

$$\mathcal{R} = eg_{ij}\, R^{ij} = eg_{ij}\, R^{(ij)} \quad ,$$

by means of the usual simplicity arguments. The curvature tensor R_{ijkl} is, as in GR, antisymmetric in respectively the first and the last two indices, but the ri Ricci tensor R_{jk} now is asymmetric in general. $\mathcal{R}$ leads to the algebraic relation $(33)_2$ between spin and torsion. Thus in vacuum there is no torsion.

The uniques position of R in GR is guaranteed by Lovelock's theorem. One of us has shown (von der Heyde [12]) that there is a generalized Lovelock theorem in U_4 theory: The U_4 field equations are uniquely determined up to three parameters if they are assumed to be polynomial in metric and torsion and their derivatives, respectively, and if they contain no fundamental constants of dimension $\neq 1$ except the gravitational constant K. The three parameters can be fixed by means of the EP, i.e. with the help of equ. (12). We then arrive at the two field equations (33).

(So the price we have to pay for allowing torsion to propagate into the vacuum, is to introduce a new fundamental constant, say an elementary length. The Planck length. $l = \sqrt{hkc}$ would do with the supplementary Lagrangian

$$e\, l^2 R_{[ij]}\, R^{[ij]} = e\, l^2 (\overset{*}{\nabla}_k\, T_{ij}^{\cdot\cdot k})\, (\overset{*}{\nabla}_\ell\, T^{ij\ell}).$$

In a sense l isn't a real new constant if we suppose h to be known.

Now torsion could propagate in vacuum. Besides the usual gravitational waves of the metric type ("gravitons"), we would have contortional waves ("tordions"). The tordions would be rather heavy, certainly not massless. But this leads to a more speculative theory...

References

[1]. W. Adamowicz. Equivalence between Einstein-Cartan and General Relativity Theories in the Linear Approximation for a Classical Model of Spin. Preprint Warsaw University (1974).

[2]. W. Arkuszewski, W. Kopczynski, and V. N. Ponomariev. On the Linearized Einstein-Cartan Theory. Ann.Inst. H. Poincare A21, 89-95 (1974).

[3]. F. J. Belinfante. On the Spin Angular Momentum of Mesons. Physica 6, 887-898 (1939).

[4]. H.A. Buchdahl: Non-Linear Lagrangians and Palatini's Device. Proc. Cambridge Phil. Soc. 56, 396-400 (1960).

[5]. C. G. Callan, S. Coleman, and R. Jackiw. A New Improved Energy-Momentum Tensor. Ann. Phys. (N.Y.) 59, 42-73 (1970).

[6]. S. Deser and P. van Nieuwenhiuzen. Nourenormalizability of the quantized Dirac-Einstein system. Phys. Rev. D10, 411-420 (1974).

[7]. A. A. El.Kholy, R. U. Sexl, and H. K. Urbantke. On the so-called "Palatini method" of variation in covariant gravitational theories. Ann. Inst. H. Poincare A18, 121-136 (1973).

[8]. F. W. Hehl. Spin und Torsion in der allgemeinen Relativitäts-theorie oder die Riemann-Cartansche Geometrie der Welt. Habilitation thesis (mimeographed) Techn. Univ. Clausthal (1970). See also Gen. Rel. Grav. J. 4, 333-349 (1973) 5, 491-516 (1974).

[9]. F. W. Hehl. On the Energy Tensor of Spinning Massive Matter in Classical Field Theory and General Relativity. Reports on Mathematical Physics (Torun) to appear 1976.

[10]. F. W. Hehl, P. von der Heyde, and G. D. Kerlick. General Relativity with Spin and Torsion and its Deviations from Einstein's Theory. Phys. Rev. D10, 1066-1069 (1974).

[11]. F. W. Hehl, P. Von der Heyde, G. D. Kerlick, and J. M. Nester. General Relativity with Spin and Torsion: Foundations and Prospects. Rev. Mod. Physics, to appear 1976.

[12]. P. von der Heyde. A Generalized Lovelock Theorem for the Gravitational Field with Torsion. Phys. Letters 51A, 381-382 (1975).

[13]. G. 't Hooft. Quantum Gravity. Kyoto Conference on Mathematical Physics (1975).

[14]. G. D. Kerlick. Spin and Torsion in General Relativity: Foundations, and Implications for Astrophysics and Cosmology: Ph.D. thesis. Princeton Univ. (1975).

[15]. T. W. B. Kibble, Lorentz Invariance and the Gravitational Field. J. Math. Phys. 2, 212-221 (1961).

[16]. W. Kopczynski. The Palatini Principle with Constraints. Bull. Acad. Polon. Sci., Ser. sci. math. astr. phys., 23, 467-473 (1975).

[17]. B. Kuchowicz. A Spin-Dependent Geometry of the Universe. Lecture given at Int. School Cosmol. Grav. in Erice. Preprint Warsaw University (1973).

[18]. M. Mathisson. Neue Mechanik materieller Systeme. Acta Phys. Polon. 6, 163-200 (1937).

[19]. C. W. Misner, K. S. Thorne, and J. A. Wheeler. Gravitation. Freeman, San Francisco (1973).

[20]. J. M. Nester. Gravity as a Gauge Theory. Preprint Univ. of Maryland (1975).

[21]. P. van Nieuwenhuizen. Review of Covariant Quantization of Gravitation. Marcel Grossmann Meeting, Trieste (1975).

[22]. A. Papapetrou. Non-symmetric Stress-Energy-Momentum Tensor and Spin-Density. Phil. Mag. 40, 937-946 (1949).

[23]. M. Planck. Bemerkungen zum Prinzip der Aktion und Reaktion in der allgemeinen Dynamik. Physik. Zeitschr. 9, 828-830 (1908).

[24]. L. Rosenfeld. Sur le tenseur d'impulsion-energie. Mem. Acad. Roy. Belgique, cl. sc., tome 18, fasc. 6 (1940).

[25]. J. A. Schouten. Ricci Calculus, 2nd ed. Springer, Berlin (1954).

[26]. J. Schwinger. Particles, Sources, and Fields. Addison-Wesley, Reading, Mass. (1970).

[27]. D. W. Sciama. On the Analogy between Charge and Spin in General Relativity. In "Recent Developments in General Relativity", p. 415-439. Pergamon + PWN, Oxford (1962).

[28]. A. Trautman. On the Einstein-Cartan Equations. I-IV. Bull. Acad. Polon. Sci., Ser. sci. math. astr. phys., 20, 185-190, 503-506, 895-896; 21, 345-346 (1972).

[29]. A. Trautman. On the Structure of the Einstein-Cartan Equations. Symposia Mathem. 12, 139-162 (1973).

[30]. M. Veltman. Gauge Field Theory. Intern. Symp. on Electron and Photon Interactions at High Energies, Bonn (1973).

THE HULSE-TAYLOR PULSAR AND GRAVITATIONAL SPIN PRECESSION

N. D. Hari Dass
The Niels Bohr Institute, Copenhagen, Denmark

and

V. Radhakrishnan
Raman Research Institute, Bangalore, India

The recent discovery of the pulsar 1913+16 by Hulse and Taylor [1] has led to considerable excitement among physicists, as, for the first time a study of the gravitational interactions of close binary systems is possible. Considerable literature has accumulated since on various topics of astrophysical and physical interest; the use of the pulsar as a laboratory for relativistic and gravitational phenomena have been studied here [2].

In this note [3] we explore the gravitational spin-precession effects in the binary system and suggest ways of observing such effects. The information one gets from a pulsar is data on pulse shape, pulse duration, spectral composition, polarisation etc. The relevant physical effects have to be disentangled from these. Apart from testing gravitational theories, for which there are many other tests, one is given the exciting opportunity to possibly test pulsar models too.

We leave all details concerning the calculations but remark that these have been made within the magnetic pole models for pulsars. The derivations of the spin-precession effects are also not given here. We refer the interested reader to our papers on these subjects [4] as well as the already mentioned literature [2].

There are three types of spin-precession effects that are possible in the binary system: the first of these is due to spin-orbit coupling induced by gravitational interactions. For the binary system characterized by masses m_1, m_2 and orbit parameters, a, e, the average rate of spin-precession is [4] (m_1 is spinning)

$$\Omega_{s.o} = \frac{3}{2}\left(\frac{Gm_2}{ac^2}\right)^{3/2}\left(1+4\,\frac{m_1}{m_2}\right)\left(1+\frac{m_1}{m_2}\right)^{-1/2}(1-e^2)\,\frac{c}{a} \tag{1}$$

for the binary pulsar this amounts to about 2° per year. It should be noted that if $m_2 >> m_1$, $\Omega_{s.o}$ is exactly half the Perihelion advance, and in fact does not constitute an independent test. See ref. [4] for a discussion of this interesting connection.

The second spin-precession effect is due to spin-spin coupling between the pulsar and its companion if the latter is also spinning. In the case of the binary pulsar this is expected to be small as orbital angular momentum of the pulsar increases as $R^{1/2}$. If the parent is spinning very rapidly, one has to consider this effect, too.

The last contribution comes from the fact that there could be a quadrupole interaction. Once again, for the binary pulsar this effect is expected to be smaller than (1) by a factor of 10^{-3}. Interestingly, in the earth-sun system this quadrupole effect is 10^3 times bigger than $\Omega_{s.0}$ and is in fact responsible for the precession of the equinoxes [4]. The question that now faces us is, how can we observe $\Omega_{s.o}$. In the currently popular pulsar models the pulsar radiation emanates from a cone (say of half-angle α) whose axis is tilted with respect to the axis of rotation by an angle ξ. The pulse duration then is the duration for which our line of sight to the pulsar is contained within this cone. Likewise, the sweep in the polarisation angle observed within the pulse will depend upon the inclination of the line of sight with the magnetic axis. The simple point is that this inclination varies systematically in the binary system due to spin-precession effects. Thus a study of the varying pulse width and polarisation sweep should provide us with a direct means of observing these precession effects. Our results are:

Pulse width ΔP:

$$\cos \frac{\Delta P(t)}{2} = \frac{\cos \alpha - \cos \xi \cos \bar{\theta}_{LS}(t)}{\sin \xi \sin \theta_{LS}(t)} \tag{2}$$

Polarisation Sweep $\Delta\phi$:

$$\sin \frac{\Delta\phi(t)}{2} = \frac{\cos \alpha - \cos \xi \cos \bar{\theta}_{LS}(t)}{\sin \xi \sin \alpha} \tag{3}$$

$\bar{\theta}_{LS}(t)$ is given by

$$\cos \bar{\theta}_{LS}(t) = \cos \eta \cos \theta_{LS} + \sin \eta \sin \theta_{LS} \cos \tilde{\omega} t \tag{4}$$

where η is the angle between the pulsar axis of rotation of the orbit normal, θ_{LS} is the inclination of the line of sight with the orbit normal, and $\tilde{\omega}$ is the spin-precession frequency.

We make a few concluding remarks. The pulse shapes are in general very complicated and an accurate determination of pulse width may at present be difficult. It is hoped that such measurements can be made feasible soon. The interesting feature of this present analysis is that if cone models are correct, the pulses should disappear entirely at some time. This also implies that binary pulsars not seen so far may in future precess into view. For a number of other interesting possibilities see ref. [3].

REFERENCES

[1] R. A. Hulse and J. H. Taylor, 1975, Ap.J (Letters) 195, L51.

[2] T. Damour and R. Ruffini, C. R. Acad. Sc. Paris, 279, A, 971 (1974); A. R. Masters and D. H. Roberts, Ap.J. Letters, 195, 107 (1975); K. Brecher, Ap.J. Letters, 195, 113 (1975); L. W. Esposito and E. R. Harrison, Ap.J. Letters, 196, 1 (1975); C. M. Will, Ap.J. Letters, 196, 3 (1975); D. M. Eardley, Ap.J. Letters, 196, 59 (1975); R. V. Wagoner, Ap.J. Letters, 196, 63 (1975); J. C. Wheeler, Ap.J. Letters, 196, 67 (1975); B. M. Barker and R. F. O'Connell (to be published in Phys. Rev.) and many more.

[3] The details of this note are presented elsewhere; see N. D. Hari Dass and V. Radhakrishnan (to be published in Astrophysical Letters, 1975).

[4] C. F. Cho and N. D. Hari Dass "The gravitational two body problem. A source view point" to appear in Annals of Physics. See also G. Börner, J. Ehlers and E. Rudolph "Relativistic effects in the binary pulsar PSR 1913+16 Max Planck Inst., Munich.

TIME-DEPENDENT MAXWELL FIELDS IN STATIONARY GEOMETRY

B. Lukács and Z. Perjés

Central Research Institute for Physics, Budapest, Hungary
High Energy Physics Department

ABSTRACT

The following question is posed: to what extent does the stationarity of the space-time restrict the time dependence of the electromagnetic field in the absence of currents? We derive the field equations of the system. These equations contain, as a special case, the stationary electrovac problem. We show that a phase factor in the Maxwell field can depend on time. Either this factor is of the form $\vartheta = \varepsilon t$ or the electromagnetic field is type N. We find an exact particular solution of the problem belonging to Kundt's metrics.

1. PROBLEM STATEMENT

There are matter fields in general relativity which must be time-independent whenever the space-time itself is stationary. A well-known example is the perfect fluid [1]. In the present paper we are investigating the behavior of the current-free Maxwell field in a stationary geometry.

A quick reference to the energy-momentum spinor $\Phi_{ABC'D'} = \Phi_{AB}\bar{\Phi}_{C'D'}$ suffices to see that it is only a phase factor in Φ which may depend on time under these conditions. More stringent conclusions have been reached by Witten [2]. He has shown that in a quite arbitrary electrovac domain, the metric determines the form of the Maxwell field up to a real constant of integration ε (giving rise to the phase factor $e^{i\varepsilon t}$ of Φ_{AB}), unless the field is type N.

In the 3-dimensional formalism which we present in sections 2 and 3, the electromagnetic field is conveniently represented by a complex 3 vector $\underline{H} = \underline{E} - i\underline{B}$ with $\underline{E}$ and $\underline{B}$ being the electric and magnetic vectors, respectively. The field equations are rewritten in terms of $\underline{H}$. We give the full details of this method in the hope that we may even clarify some results in preceding papers [3,4]. In section 4 we show that the only time-varying quantity is θ, the pase of $\underline{H}$, as we would expect from the above spinor reasoning. Witten's condition finds its expression in the result that either $\theta = \varepsilon t$ or $\underline{H}^2 = 0$ must hold for these fields.

In section 5 we apply the complex triad formalism [5] to the $\underline{H}^2 = 0$ class and obtain particular solutions depending on three arbitrary functions. These solutions are characterized by parallel rays hence they belong to Kundt's metrics [6]. Other time-dependent fields available by the complex triad formalism are currently studied.

2. GEOMETRIC CONSIDERATIONS

Coordinate systems adapted to the time-like Killing vector (such that $\xi^\mu = \delta^\mu_0$) allow for transformations of the space coordinates

$$x'^i = x'^i(x^k) \qquad i,k,\ldots = 1,2,3. \tag{2.1}$$

Allowable also are the choices of the time coordinate $t = x^0$

$$t' = t + u(x^i) \tag{2.2}$$

with $u(x^i)$ an arbitrary function, and the changes of time-scale by a constant factor.

The decomposition of world tensors into space-like and time-like parts can be done covariantly under transformations (2.1) and (2.2) if space-like indices are all chosen contravariant and the zero indices covariant. As an example, the energy-momentum tensor $T_{\mu\nu}$ is properly represented by its parts T^{ik}, T^i_0 and T_{00}.

Let us arrange the metric variables $\tilde{g}_{\mu\nu}$ in the form

$$d\tau^2 = f(dt + \omega_i dx^i)^2 - f^{-1} g_{ik} dx^i dx^k \tag{2.3}$$

where f, ω_i and g_{ik} are functions of the space coordinates. In this notation, the product of vectors a^μ and b^μ may be decomposed as follows:

$$a^\mu b^\nu g_{\mu\nu} = f^{-1}(a_0 b_0 - a^i b^k g_{ik}) = f^{-1}(a_0 b_0 - \underline{a}.\underline{b}) \tag{2.4}$$

From this the rule for tensors is obvious. Take the energy-momentum tensor for the Maxwell field,

$$T_{\mu\nu} = F_{\mu\alpha} F^\alpha{}_\nu + \frac{1}{4} g_{\mu\nu} F_{\rho\sigma} F^{\rho\sigma} \tag{2.5}$$

We introduce the electric vector

$$E^i \stackrel{d}{=} f^{-3/2} F^i{}_0 \tag{2.6}$$

and magnetic vector

$$B_i \stackrel{d}{=} \frac{1}{2} f^{-3/2} \varepsilon_{ijk} F^{jk} \sqrt{g} \tag{2.7}$$

where $g = \det[g_{ik}]$, and obtain the decomposition

$$\begin{aligned} T_{00} &= \frac{1}{2} f^2 (\underline{E}.\underline{E} + \underline{B}.\underline{B}) \\ T^i_0 &= +f^2 \varepsilon^{irs} E_r B_s . g^{-1/2} \\ T^{ik} &= -f^2 (E^i E^k + B^i B^k - \frac{1}{2} g^{ik} (\underline{E}.\underline{E} + \underline{B}.\underline{B})). \end{aligned} \tag{2.8}$$

Note that all covariant operations with three-vectors are performed by the "metric" g_{ik} and g^{ik} satisfying $g_{ir} g^{kr} = \delta^k_i$.

Einstein's gravitational equations $R_{\mu\nu} - \frac{1}{2} g_{\mu\nu} R = -kT_{\mu\nu}$ may then be written [3] as

$$(\nabla - \underline{G})\, \underline{G} = \overline{\underline{H}}\, \underline{H} - \overline{\underline{G}}\, G \tag{2.9a}$$

$$\nabla \times \underline{G} = \overline{\underline{H}} \times \underline{H} - \overline{\underline{G}} \times \underline{G} \tag{2.9b}$$

$$R_{ik} + \overline{G}_i G_k + G_i \overline{G}_k - \overline{H}_i H_k - H_i \overline{H}_k = 0, \tag{2.9c}$$

where the units are chosen to have $k = 2$ and the complex gravitational and electromagnetic vectors are defined, respectively, by

$$\underline{G} \stackrel{d}{=} \nabla \ln f - i f \nabla \times \underline{\omega}; \quad \underline{H} \stackrel{d}{=} \underline{E} - i \underline{B} \tag{2.10}$$

3. MAXWELL'S EQUATIONS

In the preceding section the complex vectors $\underline{G}$ and $\underline{H}$ have been used for decomposing Einstein's equations. We now extend this procedure to the current-free field equations

$$\left(\sqrt{-\tilde{g}}\ F^{\mu\nu}\right)_{,\nu} = 0 \tag{3.1}$$

where $\tilde{g} \overset{d}{=} \det[\tilde{g}_{\mu\nu}]$. These equations may be written in a detailed form using line element (2.3)

$$(f^{-1}\sqrt{g}\ F^{ij})_{,j} + (f^{-1}\sqrt{g}\ F^{i0})_{,0} = 0, \tag{3.2}$$

$$(f^{-1}\sqrt{g}\ F^{i0})_{,i} = 0$$

The comma denotes partial derivative. In terms of the electric and magnetic vectors (2.6) and (2.7) one has

$$(\varepsilon^{ijk}B_k f^{1/2})_{,j} - (\sqrt{g}\ E^i f^{-1/2} + \varepsilon^{ijk}\omega_j B_k f^{1/2})_{,0} = 0 \tag{3.3a}$$

$$(\sqrt{g}\ E^i f^{-1/2} + \varepsilon^{ijk}\omega_j B_k f^{1/2})_{,i} = 0 \tag{3.3b}$$

The remaining field equations are

$$F_{\mu\nu} = A_{\nu,\mu} - A_{\mu,\nu} \tag{3.4}$$

with A_μ, the vector potential. In three dimensions, the Bianchi identities amount to

$$\varepsilon^{ijk}g^{-1/2}A_{k;j;i} = -\ (\varepsilon^{ijk}g^{-1/2}F_{jk})_{;i} = 0 \tag{3.5}$$

the covariant derivatives referring to the 3-metric g_{ik}. Here the quantities F_{jk} may be expressed in the form

$$F_{jk} = f^{-1/2}\varepsilon_{jkr}B^r g^{1/2} + (E_j\omega_k - E_k\omega_j)f^{1/2} \tag{3.6}$$

Substituting back in (3.5) we get

$$(B^i f^{-1/2} + \varepsilon^{ijk}E_j\omega_k f^{1/2}g^{-1/2})_{;i} = 0. \tag{3.7a}$$

Note the structural correspondence with Eq. (3.3b). In order to obtain the analog of equation (3.3a), we consider

$$g^{-1/2}\varepsilon^{ijk}(E_k\sqrt{f})_{;j} = g^{-1/2}\varepsilon^{ijk}(A_{0,k,j} - A_{k,0,j}) = -g^{-1/2}\varepsilon^{ijk}A_{k,j,0}$$

$$= -\frac{1}{2}\,g^{-1/2}\varepsilon^{ijk}F_{jk,0}.$$

Again, on the right end we make use of Eq. (3.3a) and get

$$(g^{-1/2}\varepsilon^{ijk}E_k f^{1/2})_{;j} + (B^i f^{-1/2} + \varepsilon^{ijk}E_j\omega_k f^{1/2}g^{-1/2})_{,0} = 0. \tag{3.7b}$$

Field equations (3.3) and (3.7) are rewritten in the complex notation as

$$\nabla \times \underline{H} + \frac{1}{2}(\underline{G} + \overline{\underline{G}}) \times \underline{H} - \underline{\omega} \times \dot{\underline{H}} + i\dot{\underline{H}}f^{-1} = 0 \tag{3.8a}$$

$$\nabla.\underline{H} - \frac{1}{2}(\underline{G} + \overline{\underline{G}}).\underline{H} - (\underline{G} - \overline{\underline{G}}).\underline{H} - \underline{\omega}.\dot{\underline{H}} = 0. \tag{3.8b}$$

Time transformations (2.2) result in a gauge freedom of the second kind for the field quantities. In this context we note that the operators $\partial/\partial t$ and $\nabla - \underline{\omega}\partial/\partial t$, furthermore the complex vectors $\underline{G}$ and $\underline{H}$ are invariant under transformations (2.2). Hence the form of the Einstein-Maxwell equations (2.9) and (3.8) is invariant with respect to the time-gauge transformations.

4. PROPERTIES OF THE FIELDS

Einstein's equations (2.9c) imply that the only field quantity which can have time variance is the phase θ of the complex electromagnetic vector. Therefore we take the liberty to switch to a new notation

$$\underline{H} \to \underline{H}(x^i)e^{i\theta(x^i,t)}, \tag{4.1}$$

where ϑ is determined up to a time-independent additive term. Field equations (2.9) and (3.8) are transcribed as

$$(\nabla + i\nabla\theta - i\underline{\omega}\dot{\theta} - \tfrac{3}{2}\underline{G} + \tfrac{1}{2}\overline{\underline{G}}).\underline{H} = 0 \tag{4.2a}$$

$$(\nabla + i\nabla\theta - i\underline{\omega}\dot{\theta} + \tfrac{1}{2}\underline{G} + \tfrac{1}{2}\overline{\underline{G}})\times \underline{H} = f^{-1}\dot{\theta}\underline{H} \tag{4.2b}$$

$$(\nabla - \overline{\underline{G}}).\underline{G} = \overline{\underline{H}}.\underline{H} - \overline{\underline{G}}.\underline{G} \tag{4.2c}$$

$$\nabla \times \underline{G} = \overline{\underline{H}} \times \underline{H} - \overline{\underline{G}} \times \underline{G} \tag{4.2d}$$

$$R_{ik} + G_i\overline{G}_k + \overline{G}_iG_k - H_i\overline{H}_k - \overline{H}_iH_k = 0. \tag{4.2e}$$

The time derivative of Eq. (4.2b),

$$i(\nabla\dot{\theta} - \underline{\omega}\ddot{\theta}) \times \underline{H} = f^{-1}\ddot{\theta}\,\underline{H}, \tag{4.3}$$

upon contracting with $\underline{H}$ yields,

$$f^{-1}\ddot{\theta}\,\underline{H}^2 = 0. \tag{4.4}$$

The quantity f is the magnitude of the Killing vector, hence is positive. Either $\ddot{\theta}$ or $\underline{H}^2$ must vanish. When $\ddot{\theta} = 0$, from Eq. (4.3) it follows $\nabla\dot{\theta} = 0$. This gives, after removing the arbitrary time-independent phase term, $\theta = \varepsilon t$ where ε is a real constant.

In the flat-space Maxwell theory, solutions with $\theta = \varepsilon t$ are readily generated. In contrast with this, we have not, as yet, been able to cope with the nonlinear system (4.2). The complexity of the problem is striking if we notice that for $\varepsilon = 0$ we obtain the general stationary electrovac equations

$$\begin{aligned}(\nabla - \underline{G}).\underline{H} &= \tfrac{1}{2}(\underline{G} - \overline{\underline{G}}).\underline{H}\\ \nabla \times \underline{H} &= -\tfrac{1}{2}(\underline{G} + \overline{\underline{G}}) \times \underline{H}\\ (\nabla - \underline{G}).\underline{G} &= \overline{\underline{G}}.\underline{G} - \overline{\underline{H}}.\underline{H}\\ \nabla \times \underline{G} &= \overline{\underline{G}} \times \underline{G} - \overline{\underline{H}} \times \underline{H}\end{aligned} \tag{4.5}$$

$$R_{ik} + \overline{G}_i G_k + G_i \overline{G}_k - \overline{H}_i H_k - H_i \overline{H}_k = 0.$$

Fields with $\underline{H}^2 = 0$ present a less hard problem. From $E^2 - B^2 = 0$ and $\underline{E}.\underline{B} = 0$ we conclude that the electromagnetic field is type N in the Petrov sense. Hence the theorem of Mariot and Robinson [7] ensures that there exists a shear-free geodesic null congruence in the space-time. We may then easily extend the proof of the Goldberg-Sachs theorem given by Newman and Penrose [8] to the present system and show that the gravitational field is algebraically special, and that the Weyl and Maxwell tensors share a multiple principal null congruence. Also the Killing field defines principal null directions belonging to the bivector $X_{\mu\nu} = \xi_{[\mu,\nu]} - \frac{i}{2}\varepsilon_{\mu\nu\alpha\beta}\xi^{\alpha\beta}\sqrt{-\tilde{g}}$.

In the 3+1 formalism we introduce a complex triad of basis vectors $(\underline{\ell},\underline{m},\underline{\overline{m}})$ with nonvanishing products $\underline{\ell}.\underline{\ell} = \underline{m}.\underline{\overline{m}} = 1$. The real 3-vector $\underline{\ell}$ is conveniently oriented along a principal congruence. For the bivector $X_{\mu\nu}$ we ensure this by setting

$$\underline{G}.\underline{m} = 0. \tag{4.6}$$

With the choice (4.6) of the triad, curves with tangent $\underline{\ell}$ are eigenrays of the gravitational field [5]. Similarly, the electromagnetic eigenrays are given by the condition

$$\underline{H}.\underline{m} = 0. \tag{4.7}$$

For type N fields, the electromagnetic eigenrays are geodesics, and (4.7) entails $\underline{H}.\underline{\ell} = 0$.

When the electromagnetic and gravitational vectors possess a common eigenray congruence, conditions (4.6) and (4.7) may be fulfilled simultaneously. Field equations (4.2) take the triad form [5].

$$D\rho = \rho^2 + G_0\overline{G}_0 \tag{4.8a}$$

$$DG_0 = (2\rho + G_0 - \overline{G}_0)G_0 \tag{4.8b}$$

$$D\tau = \rho\tau + \overline{G}_0 G_- \tag{4.8c}$$

$$\delta\rho = \overline{G}_+ G_0 \tag{4.8d}$$

$$-\overline{\delta}G_0 + DG_- = \rho G_- - \overline{G}_0 G_- \tag{4.8e}$$

$$\delta\tau + \overline{\delta\tau} = 2\tau\overline{\tau} + \rho\overline{\rho} - G_0\overline{G}_0 - H\overline{H} + G_-\overline{G}_+ \tag{4.8f}$$

$$-\delta G_- = (\rho - \overline{\rho})G_0 - \overline{\tau}G_- - H\overline{H} + G_-\overline{G}_+ \tag{4.8g}$$

$$-\delta G_0 = \overline{G}_+ G_0 \tag{4.8h}$$

$$D\ \ln(Hf^{1/2}) = \rho \tag{4.8i}$$

$$\delta\ \ln H = \overline{\tau} - i\delta\theta - \frac{1}{2}\overline{G}_+ + i\omega_+\dot{\theta} \tag{4.8j}$$

In the above equations the subscripts 0, + and - denote $\underline{\ell}$,$\underline{m}$ and $\underline{\overline{m}}$ components, respectively, and $H \stackrel{d}{=} H_-$. Gauge transformations (2.2) allow us to set $\omega_0 = f^{-1}$. Choosing the space coordinates such that $D = \partial/\partial r$ and $\delta = \Omega\partial/\partial r + \xi^2\partial/\partial x^2 + \xi^3\partial/\partial x^3$ hold, we have the further equations

$$D\xi^a = \bar{\rho}\xi^a \qquad (a = 2,3) \tag{4.9a}$$

$$D\Omega = \bar{\rho}\Omega \tag{4.9b}$$

$$\delta\bar{\xi}^a - \bar{\delta}\xi^a = \overline{\tau\xi^a} - \tau\xi^a \tag{4.9c}$$

$$\delta\bar{\Omega} - \bar{\delta}\Omega = \overline{\tau\Omega} - \tau\Omega + \bar{\rho}-\rho . \tag{4.9d}$$

$$D\omega_+ = \bar{\rho}\omega_+ \tag{4.9e}$$

$$\bar{\delta}\omega_+ - \delta\omega_- = \tau\omega_+ - \bar{\tau}\omega_- - f^{-1}(G_0 + \rho - \bar{G}_0 - \bar{\rho}) . \tag{4.9f}$$

The system consisting of Eqs. (4.8) and (4.9) governs the type N fields with common eigenrays. It can be explicitly solved as will be shown in the next section.

5. THE KUNDT-TYPE SOLUTIONS

We now elaborate on the integration method of field equations (4.8) and (4.9) and obtain the Kundt-type metrics with parallel multiple principal null directions [6]. These fields have the property [5]

$$\rho + G_0 = 0 \tag{5.1}$$

equivalent to vanishing of the spin coefficient $\rho_{\text{Newman-Penrose}}$

From (4.8a) we obtain [9]

$$\rho = -G_0 = -\frac{1+i\rho^0}{2r} , \quad \rho^0 = \bar{\rho}^0 \tag{5.2}$$

where we use a degree sign for functions of x^2 and x^3. The origin of the coordinate r has been fixed by absorbing an integration "constant" into r. Eqs. (4.9a) yield

$$\Omega/\Omega^0 = \xi^a/\xi^{a0} = \frac{1}{\sqrt{r}}\, e^{\frac{i\rho^0}{2}\ln r} . \tag{5.3}$$

We transform to a coordinate system in which $\xi^{20} = -i\xi^{30} \stackrel{d}{=} P$ and rotate the vector $\underline{m}$ to make P real. The remaining coordinate freedom is written in terms of the complex coordinate $z = x^2 + ix^3$ as $z' = F(z)$ where $F(z)$ is an arbitrary analytic function. It will prove convenient to introduce the complex quantity Q by $\Omega^0 = 2PQ$.

The structure (2.10) of the gravitational vector $\underline{G}$ is expressed in the triad relations

$$D \ln f = G_0 + \bar{G}_0 \tag{5.4}$$

$$\bar{\delta} \ln f = G_- .$$

Hence

$$f = f^0 r; \qquad f^0 = \overline{f^0}; \tag{5.5}$$

$$G_- = \frac{2P}{\sqrt{r}} e^{-\frac{i\rho^0}{2}\ln r}\left(\frac{\overline{Q}}{r} + \frac{\partial}{\partial z}\ln f^0\right) \tag{5.5}$$

Equation (4.8d) yields

$$\partial/\partial\overline{z}\,[(1+i\rho^0)f^0] = 0. \tag{5.6}$$

For later convenience, we write the solution of (5.6) in the form

$$(1+i\rho^0)f^0 = \frac{1}{2}\frac{dw(z)}{dz} \tag{5.7}$$

where $w(z)$ is an arbitrary analytic function. From (5.7) we have

$$f^0 = w_{,z} + \overline{w}_{,\overline{z}}\,. \tag{5.8}$$

Next we integrate Eq. (4.8c) and get

$$\tau = \frac{2P}{\sqrt{r}} e^{-\frac{i\rho^0}{2}\ln r}\left[\tau^0 + \frac{1-i\rho^0}{2}\left(\ln r\ \ln_{,z} f^0 - \frac{\overline{Q}}{r}\right)\right]. \tag{5.9}$$

The integration "constant" τ^0 is determined by Eq. (4.9c),

$$\tau^0 = \ln_{,z} P. \tag{5.10}$$

Eliminating the term $H\overline{H}$ from equations (4.8f) and (4.8g) and using the freedom in the choice of the z coordinate, we obtain

$$2P^2 = 1/f^0. \tag{5.11}$$

From (4.9d) a linear inhomogeneous equation for Q follows,

$$2(\overline{Q}_{,\overline{z}} - Q_{,z}) = w_{,z} - \overline{w}_{,\overline{z}}. \tag{5.12}$$

This has the general solution

$$Q = R(z,\overline{z})_{,\overline{z}} - \frac{1}{2} w(z) \tag{5.13}$$

where $R(z,\overline{z})$ is a real function.

Having chosen $\omega_0 = 1/f$ in the preceding section, we can still add a function of z and $\overline{z}$ to the time coordinate. The solution of Eqs. (4.9e) and (4.9f) is transformed to $\omega_+ = 0$. Maxwell's equations (4.8i) and (4.8j) can now be integrated to find

$$H = \frac{\chi_{,z}(z)}{f^0 r} e^{-\frac{i\rho^0}{2}\ln r + i\theta(t)} \tag{5.14}$$

where $\chi(z)$ is an analytic function. From (4.8g) we extract the equation

$$\chi_{,z}\overline{\chi}_{,\overline{z}} = \left[2(w_{,z} + \overline{w}_{,\overline{z}})R - ww_{,z}\overline{z} - \overline{ww_{,\overline{z}}z}\right]_{,z\overline{z}}\,. \tag{5.15}$$

Solution of (5.15) yields

$$2(w_{,z} + \overline{w}_{,\overline{z}})R = \chi\overline{\chi} + ww_{,z}\overline{z} + \overline{ww_{,\overline{z}}z} + q(z) + \overline{q}(\overline{z}) \tag{5.16}$$

with $q(z)$ an arbitrary analytic function. We may use (5.16) for eliminating R.

Finally we put together the above results and calculate the 3-metric from the relation $g^{ik} = \ell^i \ell^k + m^i \bar{m}^k + \bar{m}^i m^k$. Also, in the present coordinate system, we have $\omega^i = \ell^i$ and $f = (w,_z + \bar{w},_{\bar{z}})r$. Our results can be summarized in the line element

$$d\tau^2 = dt\left[(w,_z + \bar{w},_{\bar{z}})rdt + 2dr + (w-2R,_{\bar{z}})d\bar{z} + (\bar{w} - 2R,_z)dz\right] - dzd\bar{z}. \tag{5.17}$$

We may choose a new coordinate $v = r - R$ and obtain a form in terms of the analytic functions $w(z)$, $g(z)$ and $\chi(z)$:

$$d\tau^2 = dt\left\{\left[(w,_z + \bar{w},_{\bar{z}})v + \frac{1}{2}(\chi\bar{\chi} + q + \bar{q} + ww,_z\bar{z}) + \overline{ww,_z z}\right]dt + 2dv + wd\bar{z} + \bar{w}dz\right\} - dzd\bar{z}. \tag{5.18}$$

This, together with the electromagnetic vector (5.14) is an explicit example of a time-dependent Maxwell field coupled to a stationary space-time.

ACKNOWLEDGMENTS

We have benefited from discussions with Dr. A. Sebestyén and Dr. H. Urbantke.

REFERENCES

[1] J. L. Synge: Relativity - The General Theory, North-Holland, 1960.
[2] L. Witten, in Gravitation, ed. by L. Witten, Wiley, 1962.
[3] Z. Perjés, Phys. Rev. Letters 27, 1668 (1971).
[4] B. Lukács and Z. Perjés, GRG Journal 4, 161 (1973).
[5] Z. Perjés, J. Math. Phys. 11, 3383 (1970).
[6] W. Kundt, Z. für Physik 163, 77 (1961).
[7] F. A. E. Pirani, in Lectures on General Relativity, Prentice Hall, 1964.
[8] E. Newman and R. Penrose, J. Math. Phys. 3, 566 (1962).
[9] The metrics arising from the alternative solution $\rho = -G_0 = i\rho^0$ can also be transformed to the form (5.18).

SOLUTIONS OF EINSTEIN-CARTAN EQUATIONS

A. R. Prasanna*

Institut für theoretische Kernphysik, Universität Bonn,
Nussallee 16, Bonn, Federal Republic of Germany

Following Trautman's [1] reformulation of the Einstein-Cartan theory, Kopczynski [2] was the first to obtain a solution, wherein he considered the problem of studying the geometry of the space-time supporting the gravitational field produced by a spherically symmetric distribution of incoherent matter composed of spinning particles. Assuming a classical description of spin, i.e. the spin density tensor s^{i}_{jk}

$$s^{i}{}_{jk} = U^{i} S_{jk} , \qquad U^{k} S_{jk} = 0$$

Wherein U^{i} is the four-velocity vector and S_{jk} is the intrinsic angular momentum tensor, he showed the existence of a two parameter family of world models of the Friedman type without singularities. Following this Trautman [3] showed that by starting from a Robertson-Walker type of line element, again for a classical description of spin, the Friedman equation takes the form

$$\frac{\dot{R}^2}{2} - \frac{GM}{R} + \frac{3G^2S^2}{2C^4R^4} = 0 \quad ,$$

solving which he obtained a non-zero minimum radius at t = 0, as given by $R \approx (3GS^2/2MC^4)^{1/3}$, Stewart and Hajicek [4] commenting on this work showed that the singularity in Trautman's model was avoided mainly because of the perfect isotropy introduced. However as Hehl, von der Heyde and Kerlick [5] have pointed out, one can always rewrite Einstein-Cartan equations such that the torsion effects are included in the energy momentum tensor of matter and in principle singularity might be avoided by violating the positive energy condition of Penrose-Hawking theorems.

Instead of making only qualitative statements it is preferable to consider

---

*Alexander von Humboldt Research Fellow. Address after September 1975: Physical Research Laboratory, Navrangpura, Ahmedabad-380009, India.

regular fluid distributions with non-zero pressure and try to solve the system of Einstein-Cartan equations and then study what effects the spin density or torsion will have on physically plausible situations. Further, it has recently been pointed out by Florides [6], in an approximate way, that a rigidly rotating perfect fluid would not match with the exterior Kerr solution. It is our desire to see whether by introducing torsion this situation could be improved.

We start with simple situations of spherical and cylindrical symmetry [7] [8]. The field equation of the Einstein-Cartan theory are given by:

$$R^i_{\ j} - \frac{1}{2} R\, \delta^i_j = - \kappa\, t^i_{\ j} \ ,$$

$$Q^i_{\ jk} - \delta^i_j\, Q^\ell_{\ \ell k} - \delta^i_k\, Q^\ell_{\ j\ell} = - \kappa\, S^i_{\ jk}$$

where t^i_j is the canonical energy-momentum tensor as given by

$$t^i_j = \hat{T}^i_{\ j} + \frac{1}{2}\, g^{im}\, \nabla_k\, (S^k_{\ jm})$$

$\hat{T}^i_j$ being the symmetric energy-momentum tensor. As mentioned earlier, S^i_{jk}, the spin density tensor, for a classical description of spin, takes the form $S^i_{jk} = U^i S_{jk}$, with $U^k\, S_{jk} = 0$.

1. Spherical symmetry

$$ds^2 = - e^{2\mu}\, dr^2 - r^2 d\theta^2 - r^2 \sin^2\theta\, d\phi^2 + e^{2\nu}\, dt^2 \ .$$

Assuming the spins of the individual particles to be aligned along the radial direction, for a static sphere, the equations are given by:

$$8\pi p = 16\pi^2 K^2 = \frac{1}{r^2} + e^{-2\mu}\left(\frac{2\nu'}{r} + \frac{1}{r^2}\right)$$

$$8\pi\rho = 16\pi^2 K^2 + \frac{1}{r^2} + e^{-2\mu}\left(\frac{2\mu'}{r} - \frac{1}{r^2}\right)$$

$$e^{-2\mu}\left[\left(\frac{\nu''}{r} - \frac{\nu'}{r^2} - \frac{1}{r^3}\right) - \mu'\left(\frac{2\nu'}{r} + \frac{1}{r^2}\right) + \nu'\left(\frac{\mu' + \nu'}{r}\right)\right] + \frac{1}{r^3} = 0$$

and

$$\frac{dp}{dr} + (p + \rho)\,\nu' + \frac{\kappa K}{2}\,(K' + K\nu') = 0$$

Redefining pressure and density as

$$\tilde{p} = p - 2\pi K^2 , \qquad \tilde{\rho} = \rho - 2\pi K^2 ,$$

we then obtain the equations just as in Einstein's theory and one can then use any of the well-known solutions and work back the equation of state connecting p, ρ and K.

(1) $$ds^2 = -(1-\frac{r^2}{R^2})^{-1} dr^2 - r^2 d\theta^2 - r^2 \sin^2\theta \, d\phi^2 + \{B_2 - B_1 (1-\frac{r^2}{R^2})^{1/2}\}^2 dt^2 .$$

$$B_2 = \frac{3}{2} (1 - \frac{a^2}{R^2})^{1/2} \qquad B_1 = \frac{1}{2} , \qquad \frac{2m}{a} = \frac{a^2}{R^2} ,$$

$$8\pi p = 8\pi\rho - \frac{6}{R^2} + \frac{B2}{2\pi A R^2} (8\pi\rho - \frac{3}{R^2})^{1/2}$$

$$K = A \{B_2 - B_1 (1 - \frac{r^2}{R^2})^{1/2}\}^{-1} , \quad A = \frac{1}{8\pi} (8\pi\rho_o - \frac{3}{R^2})^{1/2} [3(1 - \frac{a^2}{R^2})^{1/2} - 1]$$

(2) $$ds^2 = - \frac{(D+2Cr^2)}{(1+B_1 r^2)(D+Cr^2)} dr^2 - r^2 d\theta^2 - r^2 \sin^2\theta \, d\phi^2 + (D+Cr^2) dt^2$$

$$B_1 = -\frac{m}{a^3} , \qquad C = \frac{m}{a^3} , \qquad D = (1 - \frac{3m}{a})$$

$$K = A (D + Cr^2)^{-1/2} , \qquad A = \frac{1}{4\pi} [8\pi\rho_o (1 - \frac{3m}{a}) - \frac{6m}{a^3} + \frac{9m^2}{a^4}]^{1/2}$$

$$8\pi\rho_o = 8\pi p_o + \frac{6m}{a^3} (1- \frac{2m}{a}) (1 - \frac{3m}{a})^{-1}$$

(3) $$ds^2 = - \frac{7}{4 - \frac{1}{2}(r/a)^{7/3}} dr^2 - r^2 d\theta^2 - r^2 \sin^2\theta \, d\phi^2 + \frac{1}{2} (r/a) \, dt^2$$

$$K = A(a/r)^{1/2} , \qquad A < (1/8\pi) (2/3)^{1/2} (1/a)$$

$$8\pi p (1 \pm Z)^2 = \frac{2}{7} (3\pi\rho + 5\pi p) (6 + 2Z^2 \pm 7Z) - \frac{(2 \pm 2Z)^{7/3} (\pi A)^{2/3}}{14 a^2 (3\pi\rho + 5\pi p)^{1/3}}$$

$$Z = [1 + (3\pi\rho + 5\pi p)/1024 \, a^2 \pi^4 A^4]^{1/2}$$

In fact in the above treatment we have determined K purely by an artificial spin continuity equation introduced ad hoc in the general equation of continuity, viz.

$$K' + K\nu' = 0 \quad \Rightarrow \quad K = Ae^{-\nu} ,$$

as the field equations do not determine K. This is highly unsatisfactory. Further a perfect radial alignment of spins implies the presence of a magnetic monopole which is also not desirable. However we find that relaxing the symmetry helps solve these two problems.

2. Cylindrical symmetry

$$ds^2 = - e^{2\mu-2\nu}(dr^2 + dz^2) - r^2 e^{-2\nu} d\phi^2 + e^{2\nu} dt^2 .$$

Assuming the spin to be aligned along the axis of symmetry, i.e. $S_{12} = K \neq 0$, the field equations turn out to be:

$$e^{2(\nu-\mu)} (2\nu'' - \mu'' + \frac{2\nu'}{r} - \nu'^2) + \frac{\kappa^2 K^2}{4} = - \kappa\rho ,$$

$$e^{2(\nu-\mu)} (\nu'^2 - \frac{\mu'}{r}) - \frac{\kappa^2 K^2}{4} = \kappa p_r ,$$

$$e^{2(\nu-\mu)} (- \mu'' - \nu'^2) - \frac{\kappa^2 K^2}{4} = \kappa p_\phi ,$$

$$e^{2(\nu-\mu)} (- \nu'^2 + \frac{\mu'}{r}) - \frac{\kappa^2 K^2}{4} = \kappa p_z ,$$

and $K' + K\mu' = 0 \Rightarrow K = Be^{-\mu}$

Using a similar technique as in the case of spherical symmetry, we obtain a solution

$$\nu = [Xr^{(n+1)} / (n+1)] + C_1$$

$$\mu = [2xr^{(n+1)} / n(1+r)] - [x^2 r^{2(n+1)} / (2n+1) (2n+2)] + D_1 r + D_2$$

with the exterior solution

$$ds^2 = - A^2 r^{-2C(1-C)} (dr^2 + dz^2) - r^{2(1-C)} d\phi^2 + r^{2C} dt^2$$

and the equation of state

$$(\rho - \Gamma p_\phi) = 2\pi K^2 (1 - \Gamma)$$

The constants B and C are connected with the mass of the cylinder per unit length

$$M_g = \frac{C}{2} + 16\pi^2 B^2 a^{2(1-C)} (\frac{1}{2} + \frac{C}{n+3})$$

However for a more realistic situation one should consider a co-existing magnetic field which would induce spin polarization. Hence one needs to first generalize Maxwell's equation for a space with torsion [9]. Starting from the invariant form of Maxwell equations

$$F = d\phi \ , \qquad dF = 0 \ , \qquad dF^* = J^* \ , \qquad dJ^* = 0 \ ,$$

we obtain in a Riemann-Cartan space, referred to arbitrary frame θ^i in the cotangent bundle, the Maxwell equations as given by:

$$\nabla_k F_{ij} + \nabla_j F_{ki} + \nabla_i F_{jk} = Q^m{}_{ij} F_{km} + Q^m{}_{ki} F_{jm} + Q^m{}_{jk} F_{im} \ ,$$

$$2 \nabla_j F^{kj} + (Q^k{}_{ij} - \delta^k_i Q^\ell{}_{\ell j} - \delta^k_j Q^\ell{}_{i\ell}) F^{ij} = J^k \ ,$$

$$\nabla_i J^i + Q^k{}_{ik} J^i = 0$$

wherein ∇_i denotes the covariant derivative taken with respect to the asymmetric connection Γ^i_{jk} . Using these equations along with the Einstein-Cartan equations we have obtained [10] a solution for a cylinder of perfect fluid with a magnetic field as given by:

$$\nu = \text{Ln}\ (A_1 + A_2\, r^2/a^2)$$

$$\mu = 2\nu - \left(\frac{2\Gamma}{\Gamma-1}\right) \left\{ \left(\frac{A_2}{A_1} \frac{r}{a} \right) \tan^{-1} \left(\sqrt{\frac{A_2}{A_1}}\, \frac{r}{a} \right) \right\} - \frac{C_1 r}{(\Gamma-1)} - \frac{(\Gamma \text{Ln}\ A_1 + C_2)}{(\Gamma-1)} \ ,$$

with

$$8\pi p_r = e^{(2\nu-2\mu)} \left\{ \left(\frac{-2\Gamma}{\Gamma-1}\right) \left[\sqrt{\frac{A_2}{A_1}}\, \frac{1}{ar} \tan^{-1} \left(\sqrt{\frac{A_2}{A_1}}\, \frac{r}{a} \right) - \frac{A_2}{a^2} \frac{(A_1 - A_2 r^2/a^2}{(A_1 + A_2 r^2/a^2)} \right] - \frac{C_1}{(\Gamma-1)r} \right\}$$

$$8\pi p_z = -\, 8\pi p_r + 32\, \pi^2 A^2 e^{-2\mu}$$

$$8\pi\rho = (8\Gamma\, A_1 A_2\, e^{-2\mu}) \ /\ a^2\ (\Gamma - 1)$$

$$H^2 = e^{-2\mu} \left\{ 16\, \pi^2 A^2 - \frac{4\, A_1 A_2}{a^2 (\Gamma-1)} \right\} \ , \qquad J^i = 0$$

$$A = (1/4\pi) \left\{ (1/\Gamma-1) \left[\Gamma c^2 + \frac{\lambda(2-\lambda)}{a^2} \left(a^\lambda + \frac{c^2\, a^{2-\lambda}}{4(1-\lambda)^2} \right)^2 \right] \right\}^{1/2}$$

satisfying an equation of state

$$\rho = \Gamma(\, p_r + p_z - H^2/_{4\pi} \,) \ .$$

and going over smoothly to the solution of Ghosh and Sen Gupta [11] in the absence of matter.

There is one interesting feature to be noticed in the case of a cylindrically symmetric distribution. From the equations for pressure one notices that when $K \neq 0$, both p_r and p_z cannot be identically zero simultaneously. In fact we have

$$(p_r + p_z) = \left(\frac{4\pi G}{c^2}\right) K^2 \quad ,$$

indicating that even a slight increase in pressure would increase the spin density enormously and this fact needs consideration.

Finally, I should like to just say that we are now considering the most general case of axisymmetrical distributions in the form

$$ds^2 = - e^{2\alpha} dr^2 - e^{2\beta} d\theta^2 - e^{2\Gamma} [adt - (r^2+a^2) d\phi]^2 + e^{2\sigma} [dt - a\sin^2\theta \, d\phi]^2 \; ,$$

$$\alpha, \beta, \Gamma, \sigma \quad \text{functions of } (r, \theta, t) \; .$$

However, as I mentioned earlier, since we have in mind to see whether one could obtain a solution for the Kerr metric in the interior, with torsion, we could restrict this study first to the stationary case in the form

$$ds^2 = - \Delta e^{2\alpha} dr^2 - \Delta e^{2\beta} d\theta^2 - \frac{e^{2\Gamma}\sin^2\theta}{\Delta} [adt - (r^2+a^2) d\phi]^2 + \frac{e^{2\sigma}}{\Delta} [dt - a\sin^2\theta \, d\phi]^2 \; ,$$

$$\Delta = (r^2 + a^2\cos^2\theta), \quad \alpha, \beta, \Gamma, \sigma, \quad \text{functions of } r \; ,$$

and we have first found that assuming the spins to be aligned along the axis of symmetry, the spin density is determined in terms of the symmetric energy-momentum tensor T^i_j (which we have not yet specified) as follows:

$$\frac{\partial}{\partial\theta} (K \sin\theta) = 2 \, \Delta^{1/2} e^{\beta} \sin\theta \quad (U^3\hat{T}_{13} + U^4 \hat{T}_{14})$$

$$\frac{\partial}{\partial r} (K e^{\Gamma+\sigma}) = - 2 \, \Delta^{1/2} e^{(\alpha+\Gamma+\sigma)} \quad (U^3\hat{T}_{23} + U^4\hat{T}_{24}) \; .$$

Further work is in progress.

It is a pleasure to thank the Alexander von Humboldt-Stiftung for the award of a Research Fellowship and for supporting my participation in this conference. The hospitality extended to me at the ICTP by Professor Abdus Salam, UNESCO and the IAEA is very much appreciated.

REFERENCES

[1] A. Trautman, Bull. Acad. Pol. Sci. 20, 185, 503 (1972).

[2] W. Kopczynski, Phys. Letters A39, 219 (1972).

[3] A. Trautman, Nature 242, 7 (1973).

[4] J. Stewart and P. Hajicek, Nature 244, 96 (1973).

[5] F. W. Hehl et al., Phys. Rev. D10, 1066 (1974).

[6] P. S. Florides, Nuovo Cimento 25B, 251 (1975).

[7] A. R. Prasanna, Phys. Rev. D11, 2076 (1975).

[8] A. R. Prasanna, Phys. Rev. D11, 2083 (1975).

[9] A. R. Prasanna, to appear in Phys. Letters A (1975).

[10] A. R. Prasanna, to appear in Phys. Rev. D.

[11] R. Ghosh and R. Sen Gupta, Nuovo Cimento 38, 1579 (1965).

ROTATING MODEL FOR RADIATION AND RADIATION-REACTION FORCE OF A FREELY-FALLING CHARGE

by

Daniel C. Wilkins*†
Department of Physics, University of California
Santa Barbara, California 93106

Two related questions that dogged the heels of gravitation theory for a long time are: does a freely-falling charge radiate, and does it follow the same path as a neutral test body? If gravitation were like any other force, the answer to the first question would be "yes," and to the second "no." According to Larmor's formula any accelerated charge in flat space-time loses energy at a rate proportional to the square of its acceleration [1]. And, following Lorentz, the electromagnetic field of an accelerated charge is so deformed from its Coulomb value, that the net self-force does not vanish [2]. This additional force gives rise to what is called "radiation reaction" or "damping".

The problem when we come to the General Theory of Relativity is that gravitation is not like any other force. It occupies a unique position. The (strong) principle of equivalence, built into the foundations of Einstein's Theory, states that all local physical effects of the gravitational field vanish at a point. The effects of space-time curvature only reveal themselves over regions of finite extent. If we assume that the charge follows a geodesic the field in the immediate neighborhood is apparently Coulomb-like. Consistent with the original assumption, this field exerts no self-force, so geodesic motion should result. But if there is no radiation reaction, the power radiated ought to vanish too, at least for a quasi-periodic orbit. (The case of a uniformly accelerated charge in flat space-time with vanishing radiation reaction but finite power loss shows however, that radiation reaction and power loss do not go hand-in-hand for a general, non-periodic motion [3]. Otherwise, energy loss without orbital decay would be a patent violation of energy conservation.

This heuristic argument leads to conclusions just the opposite of flat-space physics. The resulting "Equivalence Principle paradox,"

*Supported in part by the National Science Foundation.
†Present address: Tata Institute of Fundamental Research, Homi Bhabha Road, Bombay 400-005, India.

which haunted the corridors of relativity theory for decades was put to rest apparently only as recently as the early 1960's [4]. Those investigations show that in the weak field, slow-motion approximation the flat-space formulae are more or less correct. But, as we shall explain presently, radiation reaction due to gravitational acceleration was found to follow from a completely novel, non-local mechanism.

To study these issues in the simplest possible way, we use a vacuum geometry consisting of two vacuum regions separated by a thin spehrical shell. The shell is assumed constructed in some specified (and admittedly artifical) manner so that both regions are exactly flat, and with the inner inertial frame (fixed relative to the sphere's center) rotating uniformly at angular velocity Ω relative to the outer. For shell radius r_o, the vacuum metric is:

$$ds^2 = - dt^2 + dr^2 + r^2 d\theta^2 + r^2 \sin^2\theta \; [d\phi - \Omega(r)dt]^2, \tag{1}$$

where $\Omega(r) = 0$ in $r>r_o$ but $\Omega(r) = \Omega$ in $r<r_o$.

As a very simple source of radiation, we let a scalar point dipole at the center of the shell rotate in the equatorial plane. The dipole couples to a weak, massless scalar field Φ satisfying the usual wave equation with source:

$$\Phi^{;\mu}{}_{;\mu} = - 4\pi T \tag{2}$$

A semicolon denotes a covariant derivative.

T is the scalar charge density. There are two matching conditions on Φ across the shell: Φ and $n^\mu \, \partial\Phi/\partial x^\mu = \partial\Phi/\partial r$ are continuous. (n^μ is a unit normal outward from the sphere). The latter condition results from integrating eq. (2) across the neutral ($T = 0$) shell. In addition these waves must be purely outgoing outside the shell. Using these conditions, one can solve for the field everywhere.

The stress-energy tensor of the scalar field is

$$T_{\mu\nu} = (1/4\pi)(\Phi_{;\mu} \, \Phi_{;\nu} - \tfrac{1}{2} g_{\mu\nu} \, \Phi^{;\alpha} \, \Phi_{;\alpha}) \; . \tag{3}$$

Fluxes of energy and angular momentum radiated outward through a sphere of radius r are

$$\dot{E} = - r^2 \int T^r{}_\mu K^\mu{}_{(t)} \, d\Omega \quad ,$$
$$\dot{J} = r^2 \int T^r{}_\mu \, K^\mu{}_{(\phi)} d\Omega \quad . \tag{4}$$

$K^\mu{}_{(j)}$ $(j = t, \phi)$ denotes the timelike or axial killing vector and $\cdot = d/dt$. Inside the shell, the locally non-rotating polar coordinates $X^{\mu'}$ are the same as eq. (1) except that $\phi' = \phi - \Omega t$. (We use a prime on all quantities pertaining to $r<r_o$.) It follows that $\partial/\partial\phi' = \partial/\partial\phi$ and $\partial/\partial t' = \partial/\partial t + \Omega \, \partial/\partial\phi$. The Killing vectors inside the shell are thus related to those outside by

$$\underset{\sim}{K}'_{(t)} = \underset{\sim}{K}_{(t)} + \Omega \underset{\sim}{K}_{(\phi)} \quad ,$$
$$\underset{\sim}{K}'_{(\phi)} = \underset{\sim}{K}_{(\phi)} \tag{5}$$

A source rotating uniformly at a rate $d\phi/dt = \Omega_d$ as seen from outside, rotates locally at $\Omega'_d = \Omega_d - \Omega$. Then $\Phi_{,t} = - \Omega_d \, \Phi_{,\phi}$, which implies $\dot{E} = \Omega_d \, \dot{J}$. Likewise $\dot{E}' = \Omega'_d \, \dot{J}'$. Since $\dot{J}' = \dot{J}$ follows from (3), (4), (5) and the junction conditions for arbitrary motion, we have for a uniformly rotating source,

$$\dot{E}' = (\Omega'_d/\Omega_d) \; \dot{E} \quad . \tag{6}$$

A charge subject possibly to some non-gravitational force has different accelerations relative to a local freely-falling observer and relative to infinity. The Larmor values of the power as calculated locally and far away must also differ. Eq. (6) represents the precise rotational analog of this result. (Here $r<r_o$ corresponds to the local observer, $r>r_o$ to the observer at infinity and centripetal acceleration takes the place of linear acceleration). Moreover, eq. (6) is just right to insure "photon" conservation. The scalar quanta emitted by a steadily rotating source are monochromatic with frequencies Ω'_d, Ω_d in the two regions. Writing $N_{i,r,t}$ for the numbers of incident, reflected and transmitted quanta, and $\dot{E} = \hbar\Omega_d \, dN_t/dt$, $\dot{E}' = \hbar\Omega'_d(\frac{dN_i}{dt} - \frac{dN_r}{dt})$ then eq. (6) leads to the desired conservation condition.

One finds that the outgoing energy flux outside the shell is

$$\dot{E} = P_M \left\{ (\frac{v}{u} \sin u)^2 + \left[\frac{\sin u}{u} \, (1-\frac{v^2}{u^2}) + \frac{v^2}{u^2} \cos u\right]^2 \right\}^{-1} , \tag{7}$$

where $u \equiv \Omega'_d r_o$, $v \equiv \Omega_d r_o$. $P_M = \frac{1}{3} p^2 \Omega_d^4$ is the power from a dipole of magnitude p in shell-free Minkowski space.

It is notable that the power $\dot{E}$ can differ substantially from the flat-space result P_M, being either much larger or smaller. Investigations of radiation from charges falling into black holes show how dramatic are the effects of large spacetime curvature which intervene between source and observer [5].

In discussing the system's time development, we can suppose that Ω remains constant because there is neither direct coupling to the neutral shell nor indirect, gravitational coupling to the shell in the test-field approximation.

Rahter than coming to rest locally, the dipole tends toward $\Omega_d = 0$ (when $\dot{E} = 0$). This has a curious consequence. If initially the dipole rotates in the same sense as the dragging but more rapidly (say $0 < \Omega < \Omega_d$), then as the rotation slows below $\Omega_d = \Omega$, the local rotation, $|\Omega'_d|$, starts increasing. The dipole is absorbing rotational kinetic energy even though the outside observer says energy is being carried away. Inside the shell, ingoing waves predominate over outgoing waves.

A locally non-rotating dipole is like a charge following a geodesic. Such a motion will not be maintained because the flux of angular momentum $\dot{J}' = \dot{J} = \dot{E}/\Omega_d$ does not vanish. There is thus radiation reaction. It is due to the backscattered field, the field arising from the shell's presence. The scattering of the field occurs inside the material of the shell. This mechanism of radiation reaction is decidedly non-local, involving regions of much larger extent than the charge. DeWitt and DeWitt [4] described this remarkable effect in their paper of 1964.

To counteract the force of radiation reaction, we can constrain the dipole to remain locally at rest, $\Omega'_d = 0$. Then, as we saw above, t the energy and angular momentum loss outside do not vanish. Chitre, Price and Sandberg [6] discussed this phenomenon of "radiation from an unmoving charge" for a charge orbited by two equal masses on opposite sides at the same distance. Here again, coupling between the metric and the electromagnetic field (distortions of the Coulomb field) is responsible for the field "breaking off" to propagate away as radiation.

Further details on the work described here will be presented elsewhere [7].

REFERENCES

[1]. See, e.g., J. D. Jackson, Classical Electrodynamics (John Wiley, New York, 1962), pp. 469-70.

[2]. Ref. 1, Section 17.3.

[3]. F. Rohrlich, Classical Charged Particles, (Addision-Wesley, Massachusetts, 1965).

[4]. T. Fulton and F. Rohrlich, Ann. Phys. 9, 499 (1960); F. Rohrlich, Nuovo. Cim. 21, 811 (1961); A. Peres, Ann. Phys. 12, 86 (1961), R. Brehme and B. S. DeWitt, Ann. Phys. 9, 220 (1960); C. M. DeWitt and B. S. DeWitt, Phys. 1, 3 (1964); for a readable review of the Equivalence Principle paradox in the prerelativistic approximation, see V. L. Ginzburg, Sov. Phys. Uspekhi 12, 565 (1970), and References cited therein.

[5]. R. Ruffini, Phys. Letters 41B- 334 (1972); M. Johnston, R. Ruffini, and F. Zerilli, Phys. Letters 49B, 185 (1974).

[6]. D. M. Chitre, R. H. Price and V. D. Sandberg, Phys. Rev. Letters 31, 1018 (1973).

[7]. D. C. Wilkins, to be published.

ELECTROMAGNETIC PERTURBATIONS OF A ROTATING BLACK HOLE

Steven L. Detweiler

Center for Theoretical Physics
University of Maryland
College Park, Maryland 20742

ABSTRACT: The equations governing the electromagnetic perturbations around a rotating black hole are examined and found to yield a simple, one dimensional wave equation with a short ranged and purely real potential.

I. INTRODUCTION

In an examination of the perturbations of a rotating black hole, Teukolsky (1973) made the important discovery of a set of separable equations governing the Newman-Penrose quantities Φ_o and Φ_2 for electromagnetic perturbations and Ψ_o and Ψ_4 for gravitational perturbations. Each of the equations governing the radial dependence of a perturbation may be cast into the form of a one dimensional wave equation with a potential. But, unfortunately, both conceptual and numerical difficulties are encountered in an examination of any one of these radial equations. In general, each potential is both complex and long range. To emphasize the situation these difficulties persist even in the Schwarzschild limit, where it is known that there do exist radial wave equations with purely real and short range potentials describing both electromagnetic and gravitational perturbations.

In an examination of the gravitational perturbations of the Schwarzschild metric, Chandrasekhar (1975) found transformations relating the perturbation equations based upon the Newman-Penrose formalism (Bardeen and Press 1973) with those well behaved equations originally based upon gauge dependent metric perturbations (Regge and Wheeler 1957, Zerilli 1970, Friedman 1973). And Chandrasekhar and Detweiler (1975), limiting themselves to axisymmetric gravitational perturbations, used similar transformations on Teukolsky's equation for spin two fields to find a set of four radial wave equations each of which reduces either to Regge and Wheeler's equation or to Zerilli's equation in the Schwarzschild limit.

In this seminar I will describe a technique for treating the electromagnetic perturbations of the Kerr metric in a manner similar to that in which the axisymmetric gravitational perturbations may be treated. Specifically I will describe a transformation of Teukolsky's radial wave equation for electromagnetic fields to a form which has a purely real and short ranged potential and which reduces to the simple form available in the Schwarzschild limit.

II. ELECTROMAGNETIC FIELDS IN THE SCHWARZSCHILD GEOMETRY

In the static and spherically symmetric geometry of a Schwarzschild black hole the partial differential equations governing an arbitrary electromagnetic perturbation separate in the usual Schwarzschild coordinates. The coordinate dependence of a quantity Φ may be written as

$$\Phi(t,r,\theta,\phi) = e^{-i\sigma t} R(r) Y^{m}{}_{\ell}(\theta,\phi) , \tag{1}$$

where R(r) obeys the simple radial wave equation

$$d^2R/dr_*^2 + [\sigma^2 - \frac{\ell(\ell+1)}{r^2}(1 - \frac{2M}{r})] R = 0. \tag{2}$$

The new radial coordinate r_* is the Regge and Wheeler "tortoise" coordinate defined by

$$dr_*/dr = (1 - 2M/r)^{-1} \tag{3}$$

In principle such a Φ contains all of the information in a test electromagnetic field with frequency σ and spherical harmonic indices ℓ and m. This equation was first found by Wheeler (1955) using a straightforward analysis of Maxwell's equations in a spherically symmetric background symmetry. Later Price (1972), using the formalism developed by Newman and Penrose (1962), found that the Newman-Penrose quantity Φ_1 also had radial dependence determined by equation (2).

III. TEUKOLSKY'S EQUATION FOR ELECTROMAGNETIC FIELDS IN THE KERR GEOMETRY

Using the Newman-Penrose formalism in conjunction with Kinnersley's choice of a null tetrad (Kinnersley 1969) Teukolsky found separable equations for two different linear combinations of the components of the electromagnetic field tensor - the Newman-Penrose Φ_0 and Φ_2. He assumed that the electromagnetic field was only a test field and thus was able to ignore its influence on the geometry. In the Boyer and Lindquist (1967) coordinate system the time and the azimuthal angular dependence of the test field may be assumed to be of the form exp(-iσt + imϕ), where σ is the frequency and m an integer. An electromagnetic perturbation, Ψ_s, may then be separated as

$$\Psi_s = \exp(pi\sigma t + im\phi)\ S_s(\theta)\ R_s(r)\ , \tag{4}$$

where Ψ_s is Φ_0 for the choice s = + 1 and $-(r-ia\cos\theta)^2\ \Phi_2$ for the choice s = - 1.

The separated equation for R_s takes the form

$$\Delta^{-s}\ \frac{d}{dr}\ (\Delta^{s+1}\ \frac{dR_s}{dr}) - V_s\ R_s = 0\ , \tag{5}$$

where

$$V_s = -\ K^2/\Delta + isK\Delta'/\Delta - 2isK' + \lambda_s \tag{6}$$

with

$$K = (r^2 + a^2)\sigma - am \tag{7}$$

and

$$\lambda_s = E - 2a\sigma m + a^2\sigma^2 - s(s+1)\ . \tag{8}$$

The quantity Δ is a function familiar from the Boyer and Lindquist representation of the Kerr metric.

$$\Delta = r^2 - 2\ Mr + a^2 \tag{9}$$

The quantities M and a are the mass and the angular momentum parameter of the black hole; and the event horizon is located at the larger value of r, r_+, where Δ = 0. It is clear that equation (5) does not reduce to equation (2), the simple form available, when a = 0.

The quantity E appearing in equation (8) is an eigenvalue in the equation

$$\frac{1}{\sin\theta}\ \frac{d}{d\theta}\ (\sin\theta\ \frac{d\ S_s}{d\theta}) + (a^2\sigma^2\cos^2\theta - \frac{m^2}{\sin^2\theta} - 2a\sigma s\ \cos\theta - 2ms\ \frac{\cos\theta}{\sin^2\theta} - s^2\cot^2\theta + E - s^2)S_s = 0 \tag{10}$$

for the θ dependence of Ψ_s. In the Schwarzschild limit this eigenvalue problem may be solved explicitly and the angular dependence, $e^{im\phi}\ S_s(\theta)$, reduces to the spin-weighted spherical harmonics and E takes on the value $\ell(\ell+1)$. For the more general case, when $a \neq 0$, Teukolsky and Press (1974) have tabulated the results of claculations of E for various values of $a\sigma$ and m.

IV. A TRANSFORMATION OF THE RADIAL WAVE EQUATION

Teukolsky and Press show that there is a curious relationship between any solution of equation (5) with $s = +1$ and a solution with $S = -1$. In particular,

$$\left(\frac{d}{dr} - \frac{iK}{\Delta}\right)^2 R_{-1} = -\frac{1}{2} R_{+1} \tag{11}$$

We have found a similar operation which, when performed on R_{-1} yields not a solution to equation (5) but rather a solution to a completely different wave equation which has a purely real potential and which reduces to equation (2) when a goes to zero.

Our search for such an operator was guided by two major facts. First, the operation described by equation (11) effectively goes from one complex potential to its complex conjugate potential. Thus we desire an operation which goes "half" this distance and leaves us with a real potential. Second, Teukolsky and Press describe the conserved energy flux of an electromagnetic field in terms of R_{-1} and R_{+1}. We assume that this conserved energy flux must be proportional to the Wronskian of two solutions of our sought for wave equation with a real potential.

With the advantage of hindsight we define a new dependent field variable χ by

$$\chi = \Delta p[(q + k/\Delta)R_{-1} + b\ dR_{-1}/dr] \tag{12}$$

where

$$q = 2K^2/\Delta^2 - \lambda_{-1}/\Delta - iK'/\Delta \tag{13}$$

and

$$b = 2iK/\Delta \tag{14}$$

The quantity K is a constant

$$K = (\lambda_{-1}^2 + 4a\sigma m - 4\,a^2\sigma^2)^{\frac{1}{2}} \tag{15}$$

and the function p is given by

$$p = (2\,K^2/\Delta - \lambda_{-1} + K)^{-\frac{1}{2}}. \tag{16}$$

From equations (5) and (11) it can be shown that χ obeys a wave equation of the form

$$\Delta\, d^2\chi/dr^2 + U\chi = 0 \tag{17}$$

where the new potential U is

$$U = -K^2/\Delta + \lambda_{-1} + \Delta(Kp')'/Kp. \tag{18}$$

In this formulation U is clearly real and, as one would expect, in the axisymmetric case, when m = 0, U depends on the frequency only via σ^2 not via σ.

The only difficulty that might arise with this transformation occurs if

$$2K^2/\Delta - \lambda_{-1} + K = 0 \tag{19}$$

for some r. If this were to happen p, from equation (16) would be undefined and the above transformation would break down. Fortunately, for the values of E tabulated by Teukolsky and Press equation (14) fails to hold for any value of r.

Due to the poor behaviour of U on the event horizon it is advantageous to change the dependent field variable to

$$Y = (r^2 + a^2)^{\frac{1}{2}}\,\chi\Delta^{-\frac{1}{2}} \tag{20}$$

and the independent variable to r_* defined by an extension of (3) to be

$$dr_*/dr = (r^2 + a^2)/\Delta$$

The wave equation now reduces to

$$d^2Y/dr_*^2 - WY = 0 \tag{21}$$

with

$$W = \frac{(-K^2 + \lambda\Delta)}{(r^2 + a^2)^2} + \frac{\Delta^2 (Kp')'}{(r^2 + a^2)^2\, Kp} - \left(\frac{\Delta}{r^2 + a^2}\right)^{3/2} \left[\frac{\Delta^{\frac{1}{2}}}{(r^2 + a^2)^{\frac{1}{2}}}\right]'' \tag{22}$$

after being calculated explicitly, W is given by

$$W = \frac{(-K^2 + \lambda_{-1}\Delta)}{(r^2 + a^2)^2} - \frac{\Delta r(\Delta r + 4\, Ma^2)}{(r^2 + a^2)^4} - \frac{\Delta[\Delta(10r^2 + 2\nu) - (r^2 + \nu)\ (11r^2 - 10rM + \nu)]}{(r^2 + a^2)^2\ [(r^2 + \nu)^2 + \eta\Delta]} + \frac{12\Delta r(r^2 + \nu)^2\ [\Delta r - (r^2 + \nu)\ (r - M)]}{(r^2 + a^2)^2\ [(r^2 + \nu)^2 + \eta\Delta]^2} + \frac{\Delta(r - M)^2 \eta\ [\Delta\eta - (r^2 + \nu)(4r^2 + 4\nu + 2)]}{(r^2 + a^2)^2\ [(r^2 + \nu)^2 + \eta\Delta]^2}\,, \tag{23}$$

where ν and η are two constants defined by

$$\nu = a^2 - am/\sigma \tag{24}$$

and

$$\eta = (K - 2\lambda_{-1})/4\sigma\ . \tag{25}$$

In the Schwarzschild limit the last two terms in equation (22) cancel and W is given by

$$\lim_{a\to 0} W = -\sigma^2 + (1 - 2M/r)\ \lambda_{-1}/r^2 \tag{26}$$

In this limit λ_{-1} is just $\ell(\ell+1)$; and, thus equation (21) reduces to the standard wave equation governing electromagnetic perturbations around a Schwarzschild black hole, equation (2).

Now the potential W in equation (21) is purely real and short ranged, thus we are immediately led to a conservation law for the flux of energy of an electromagnetic perturbation about a rotating black hole. The Wronskian of two independent solutions of equation (21) is independent of r_*, and thus we have

$$Y^* dY/dr_* - Y dY^*/dr_* = \text{constant}. \tag{27}$$

For large r_* a general solution of equation (21) is of the form

$$Y \sim A_{out}\, e^{i\sigma r_*} + A_{in}\, e^{-i\sigma r_*} \tag{28}$$

and near the horizon

$$Y \sim B_{out}\, e^{ikr_*} + B_{in}\, e^{-ikr_*} \tag{29}$$

where

$$k = \sigma - am/2Mr_+ \tag{30}$$

Equation (27) now implies that

$$(k/\sigma)\ (|B_{out}|^2 - |B_{in}|^2) = |A_{out}|^2 - |A_{in}|^2 \tag{31}$$

The terms in this equation are proportional to the energy of the ingoing and outgoing radiation at both the horizon and at infinity. This fact may be shown rigorously by a direct examination of the electromagentic stress energy tensor.

Equation (31) clearly exhibits the process of superradiant scattering. A wave packet sent in from infinity and scattered off of the black hole has $B_{out} = 0$. Equation (31) shows that such a scattered wave is amplified, $|A_{out}|^2 > |A_{in}|^2$, if $(k/\sigma) < 0$ (for a more general discussion of superradiant scattering see Press and Teukolsky 1972).

V. CONCLUSIONS

In this seminar I have given a transformation of Teukolsky's radial wave equation for electromagnetic perturbations about a rotating black hole to a form which reduces to the standard equation, equation (2), in the Schwarzschild limit. And, since the new potential is real, it is a simple matter to formulate the concepts of superradiant scattering and the conservation of energy of the electromagnetic perturbations. In addition, this new wave equation is short ranged which leads to easy numerical analysis.

Techniques similar to these described here may be applicable when $s = \pm 2$. Thus, it seems most likely that there exists a "real Potential" formulation of the problem of gravitational perturbations around a Kerr black hole. Work in this direction is in progress.

I am grateful to Professor S. Chandrasekhar for many fruitful discussions.

The research reported in this paper has in part been supported by the Center for Theoretical Physics of the University of Maryland and by the National Aeronautics and Space Administration under grant NGR-21-002-010 with the University of Maryland.

REFERENCES

Bardeen, J. M. and Press, W. H. 1973 J. Math. Phys. 14, 7.

Boyer, R. H. and Lindquist, R. W. 1967 J. Math. Phys. 8, 265.

Chandrasekhar, S. 1975 Proc. R. Soc. Lond. A 343, 289.

Chandrasekhar, S. and Detweiler, S. 1975 Proc. R. Soc. Lond. A, in press.

Fackerell, E. D. and Ipser, J. R. 1972 Phys. Rev. D5, 2455.

Friedman, J. L. 1973 Proc. R. Soc. Lond. A335, 163.

Kinnersley, W. 1969, J. Math. Phys. 10, 1195.

Newman, E. and Penrose, R. 1962, J. Math. Phys. 3, 566.

Press, W. H. and Teukolsky, S. A. 1972, Nature, 238, 211.

Price, R. H. 1972, Phys. Rev., D5, 2439.

Regge, T. and Wheeler J. A. 1957, Phys. Rev. 108, 1063.

Teukolsky, S. A. 1973, Astrophys. J. 185, 635.

Teukolsky, S. A. and Press, W. H. 1974 Astrophys. J. 193, 443.

Wheeler, J. A. 1955 Phys. Rev. 97, 511.

Zerilli, F. J. 1970 Phys. Rev. D2, 2141.

SPIN COEFFICIENTS AND ELECTROMAGNETISM IN REISSNER-NORDSTRØM SPACETIME

Mark D. Johnston
Joseph Henry Laboratories, Department of Physics
Princeton University, Princeton New Jersey*

ABSTRACT: The problem of the electromagnetic fields created by azimuthally symmetric charge and current sources in a background Reissner-Nordstrøm geometry is solved in the framework of the spin coefficient formalism of Newman and Penrose.

Considerable interest is attached to the problem of the accretion cloud expected to surround a "black hole" due to matter trapped from an orbiting companion star. We here address the problem of the electromagnetic field created by such a differentially rotating and possibly charged cloud, considered a perturbation on the background electric field of the charged black hole. The mass of the cloud is assumed to have negligible effect on the gravitational field of the black hole, described by the Reissner-Nordstrøm metric:

$$ds^2 = g_{ab}\, dx^a\, dx^b = \sum dt^2 - \sum^{-1} dr^2 - r^2(d\theta^2 + \sin^2\theta\, d\phi^2)$$
$$\sum \equiv 1 - \frac{2M}{r} + \frac{Q^2}{r^2} \tag{1}$$

M is the mass and Q the charge of the black hole. The background electric field is:

$$F_o = \frac{Q}{r^2}\, dt \wedge dr \tag{2}$$

Our problem is that of solving Maxwell's equations in curved spacetime,

$$\nabla_b F^{ab} = - 4\pi j^a$$
$$\nabla_b {*F}^{ab} = 0 \, , \quad {*F}^{ab} = \frac{1}{2} E^{abcd} F_{cd} \tag{3}$$

where the source j^a is specified. For simplicity we assume j^a is azimuthally symmetric and stationary (we thus neglect both the magnetohydrodynamic evolution of the cloud and radiation processes.)

We formulate and solve the problem in the framework of the spin coefficient formalism of Newman and Penrose as developed in their seminal paper [1]; see also Pirani [2]. This provides a useful and interesting example of the power of the spin coefficient approach.

The spin coefficients (scalars) are defined through the null tetrad of vectors

*Present address: Department of Physics, Massachusetts Institute of Technology, Cambridge, Massachusetts

ℓ^a, n^a, m^a, $\bar{m}^a$ which satisfy the following relations:

$$\ell_a n^a = 1 \ , \ m_a \bar{m}^a = -1 \qquad (4)$$

All other inner products between pairs vanish. All vectors and tensors are then projected onto this tetrad. Spin coefficients arise through the appearance of projected covariant derivatives of the tetrad vectors, and are defined as follows:

$$\begin{aligned}
\pi &= -\bar{m}^a \ell^b n_{a;b} & \alpha &= \tfrac{1}{2}(n^a \bar{m}^b \ell_{a;b} - \bar{m}^a \bar{m}^b m_{a;b}) \\
\rho &= m^a \bar{m}^b \ell_{a;b} & \kappa &= m^a \ell^b \ell_{a;b} \\
\lambda &= -\bar{m}^a \bar{m}^b n_{a;b} & \varepsilon &= \tfrac{1}{2}(n^a \ell^b \ell_{a;b} - \bar{m}^a \ell^b m_{a;b}) \\
\mu &= -\bar{m}^a m^b n_{a;b} & \gamma &= \tfrac{1}{2}(n^a n^b \ell_{a;b} - \bar{m}^a n^b m_{a;b}) \\
\tau &= m^a n^b \ell_{a;b} & \sigma &= m^a m^b \ell_{a;b} \\
\nu &= -\bar{m}^a n^b n_{a;b} & \beta &= \tfrac{1}{2}(n^a m^b \ell_{a;b} - \bar{m}^a m^b m_{a;b})
\end{aligned} \qquad (5)$$

The projected covariant derivative operators which will appear are denoted by:

$$\begin{aligned}
D &= \ell^a \nabla_a & \Delta &= n^a \nabla_a \\
\delta &= m^a \nabla_a & \bar{\delta} &= \bar{m}^a \nabla_a
\end{aligned} \qquad (6)$$

We also make use of the correspondence between a spinor dyad basis $\zeta_o^A = 0^A$, $\zeta_1^A = i^A$, and the null tetrad above through:

$$\begin{aligned}
\ell^a &= \sigma^a_{A\dot{x}} 0^A \bar{0}^{\dot{x}} & n^a &= \sigma^a_{A\dot{x}} i^A \bar{i}^{\dot{x}} \\
m^a &= \sigma^a_{A\dot{x}} 0^A \bar{i}^{\dot{x}} & \bar{m}^a &= \sigma^a_{A\dot{x}} i^A \bar{0}^{\dot{x}}
\end{aligned} \qquad (7)$$

Here 0^A and i^A are basis 2-spinors satisfying

$$0^A i^B - 0^B i^A = \varepsilon^{AB} \quad ,$$

and $\sigma^a_{A\dot{x}}$ are the spinor connections. The dyad components (scalars) of a spinor are then defined, e.g.,

$$\phi_{a\dot{x}} = \zeta_a^A \bar{\zeta}_{\dot{x}}^{\dot{x}} \phi_{A\dot{x}} \quad .$$

The spin coefficients are easily related to the projected covariant derivatives of the basis spinors.

Maxwell's equations (3) assume the spinor form

$$\nabla_B{}^{\dot{W}} \phi^{AB} = J^{A\dot{W}} \qquad (8)$$

where $\phi^{AB} = F^{B}{}_{\dot{Z}}{}^{A\dot{Z}}$; $F^{A\dot{W}B\dot{Z}} = \sigma_a^{A\dot{W}} \sigma_b^{B\dot{Z}} F^{ab}$, (9)

and $J^{A\dot{W}} = -4\pi \sigma_a^{A\dot{W}} j^a$. (10)

In terms of the spin coefficients we find:

$$D\phi_{01} - \bar{\delta}\phi_{00} = (\pi-2\alpha)\,\phi_{00} + 2\rho\,\phi_{01} - \kappa\phi_{11} + J_{0\dot{0}} \quad (11a)$$

$$\delta\phi_{01} - \Delta\phi_{00} = (\mu-2\nu)\,\phi_{00} + 2\tau\,\phi_{01} - \sigma\phi_{11} + J_{0\dot{1}} \quad (11b)$$

$$D\phi_{11} - \bar{\delta}\phi_{01} = (\rho-2\varepsilon)\,\phi_{11} + 2\pi\,\phi_{01} - \lambda\phi_{00} + J_{1\dot{0}} \quad (11c)$$

$$\delta\phi_{11} - \Delta\phi_{01} = (\tau-2\beta)\,\phi_{11} + 2\mu\,\phi_{01} - \nu\phi_{00} + J_{1\dot{1}} \quad (11d)$$

where ϕ_{ab} are the dyad components of ϕ_{AB}:

$$\phi_{ab} = \zeta_a^{A} \zeta_b^{B} \phi_{AB} \quad (12)$$

Similarly $$J_{a\dot{\omega}} = \zeta_a^{A} \bar{\zeta}_{\dot{\omega}}^{\dot{W}} J_{A\dot{W}} \quad (13)$$

We now make the specific and most convenient choice of tetrad for Reissner-Nordstrøm spacetime:

$$\begin{aligned} \ell^a &= 2^{-\frac{1}{2}}(\Sigma^{-\frac{1}{2}}, \Sigma^{\frac{1}{2}}, 0, 0) \\ n^a &= 2^{-\frac{1}{2}}(\Sigma^{-\frac{1}{2}}, -\Sigma^{\frac{1}{2}}, 0, 0) \\ m^a &= 2^{-\frac{1}{2}}(0, 0, r^{-1}, i(r \sin\theta)^{-1}) \\ \bar{m}^a &= 2^{-\frac{1}{2}}(0, 0, r^{-1}, -i(r \sin\theta)^{-1}) \end{aligned} \quad (14)$$

ℓ_a and n_a are tangent to geodesics corresponding respectively to outgoing and incoming spherical waves centered on r = 0. The spin coefficients are thus:

$$\begin{aligned} &\pi = \lambda = \tau = \nu = \kappa = \sigma \\ &\rho = -\frac{1}{\sqrt{2}\, r}\Sigma^{\frac{1}{2}} = \mu \\ &\alpha = -\frac{1}{2\sqrt{2}\, r}\cot\theta = -\beta \\ &\varepsilon = \frac{1}{4\sqrt{2}}\Sigma^{-\frac{1}{2}}\Sigma' = \gamma \quad (\Sigma' = \frac{d}{dr}\Sigma) \end{aligned} \quad (15)$$

To proceed with the solution of the system (11) we use the explicit form of the tetrad vectors and the spin coefficients. Note that the only nonzero components of j_a are j_0 and j_3, so we define

$$J_0 = J_{0\dot{0}} = J_{1\dot{1}} = -\frac{4\pi}{\sqrt{2}}\Sigma^{\frac{1}{2}} j_o$$
$$J_1 = J_{0\dot{1}} = -J_{1\dot{0}} = \frac{4\pi\ ir\sin\theta}{\sqrt{2}} j_3 \tag{16}$$

In like manner we define the scalars

$$\Phi_0 = \frac{1}{2}\phi_{00} = \ell^a m^b F_{ab}$$
$$\Phi_1 = \frac{1}{2}\phi_{01} = \frac{1}{2}(\ell^a n^b F_{ab} - m^a \bar{m}^b F_{ab}) \tag{17}$$
$$\Phi_2 = \frac{1}{2}\phi_{11} = \bar{m}^a n^b F_{ab}$$

The projected derivatives become

$$D = \frac{1}{\sqrt{2}}\Sigma^{\frac{1}{2}}\frac{\partial}{\partial r} = -\Delta$$
$$\delta = \frac{1}{\sqrt{2}r}\frac{\partial}{\partial\theta} = \bar{\delta} \tag{18}$$

where we have taken advantage of the symmetry $\partial/\partial\phi = 0 = \partial/\partial t$.

Maxwell's equations become:

$$D\Phi_1 - \delta\Phi_0 = -2\alpha\Phi_0 + 2\rho\Phi_1 + \frac{1}{2}J_0$$
$$D\Phi_2 - \delta\Phi_1 = (\rho - 2\varepsilon)\Phi_2 - \frac{1}{2}J_1$$
$$\delta\Phi_1 + D\Phi_0 = (\rho - 2\varepsilon)\Phi_0 + \frac{1}{2}J_1 \tag{19}$$
$$\delta\Phi_2 + D\Phi_1 = 2\rho\Phi_1 + 2\alpha\Phi_2 + \frac{1}{2}J_0$$

The conditions on the field F_{ab} are:

$F_{ab} \to 0$ as $r \to \infty$, is regular as $r \to r_+$ (r_+ and r_- are the larger and smaller roots of $r^2\Sigma = 0$;; r_+ is the location of the event horizon) and at the poles $\theta = 0, \pi/2$, and is continuous.

The system (19) is equivalent to

$$\begin{aligned}
&\delta(\Phi_0 + \Phi_2) = 2\alpha\ (\Phi_0 + \Phi_2) \\
&D(\Phi_0 + \Phi_2) = (\rho - 2\varepsilon)\ (\Phi_0 + \Phi_2) \\
&\delta(\phi_2 - \phi_0) + 2D\Phi_1 = 4\rho\ \Phi_1 + 2\alpha\ (\Phi_2 - \Phi_0) + J_o \\
&D(\Phi_2 - \Phi_0) - 2\delta\ \Phi_1 = (\rho - 2\varepsilon)\ (\Phi_2 - \Phi_0) - J_1
\end{aligned} \tag{20}$$

which may be solved in pairs. The first two equations have the solution

$$\Phi_0 + \Phi_2 = \frac{c}{r \sin\theta}\ \Sigma^{\frac{1}{2}}$$

but the boundary and regularity conditions force c = 0, thus $\Phi_0 = -\Phi_2$. Substituting into the second pair yields.

$$\begin{aligned}
&D\Phi_1 - \delta\Phi_0 = 2\rho\ \Phi_1 - 2\alpha\ \Phi_0 + \frac{1}{2} J_0 \\
&D\Phi_0 + \delta\Phi_1 = (\rho - 2\varepsilon)\ \Phi_0 + \frac{1}{2} J_1
\end{aligned} \tag{21}$$

The homogeneous system from (21) is

$$\begin{aligned}
&\frac{\partial}{\partial r}\ (r\Sigma^{\frac{1}{2}}\ \Phi_0) = -\frac{\partial}{\partial\theta}\ \Phi_1 \\
&r^{-1}\ \Sigma^{\frac{1}{2}} \frac{\partial}{\partial r}\ (r^2\ \Phi_1) = \frac{1}{\sin\theta}\ \frac{\partial}{\partial\theta}\ (\sin\theta\ \Phi_0)\ .
\end{aligned} \tag{22}$$

This system is separable: writing

$$\begin{aligned}
&\Phi_0 = R^o(r)\ \Theta^0(x) \\
&\Phi_1 = R^1(r)\ \Theta^1(x)\ ,
\end{aligned} \tag{23}$$

where $x \equiv \cos\theta$, we find:

$$\begin{aligned}
&\frac{d}{dx}\ [(1 - x^2)\ \frac{d}{dx}\ \Theta^1] = -\ n(n + 1)\ \Theta^1 \ \rightarrow\ \Theta^1 = P_n(x) \\
&\frac{\sin\theta}{\Theta^o}\ \frac{d}{dx}\ \Theta^1 = \text{const.} \ \rightarrow\ \Theta^o = \sin\theta\ \frac{dPn}{dx}\ .
\end{aligned} \tag{24}$$

Thus Φ_0 and Φ_1 may be written as series:

$$\Phi_o = \sum_n R_n^o(r) \sin\theta\, P_n'(\cos\theta) ,$$

$$\Phi_1 = \sum_n R_n^1(r)\, P_n(\cos\theta) . \qquad (25)$$

The equations for the radial functions are

$$\frac{d}{dr}\left(r\Sigma^{\frac{1}{2}} R_n^o\right) = R_n^1$$

$$\frac{1}{r}\Sigma^{\frac{1}{2}} \frac{d}{dr}\left(r^2 R_n^1\right) = n(n+1)\, R_n^o . \qquad (26)$$

The substitutions

$$U_n = r^2 \Sigma^{\frac{1}{2}} R_n^o$$

$$G_n = r^2 R_n^1$$

convert the system to

$$G_n = r^2 \frac{d}{dr}\left(r^{-1} U_n\right)$$

$$\frac{dG_n}{dr} = n(n+1) r^{-1} \Sigma^{-1} U_n \qquad (27)$$

which may be solved as a power series in $Z = r - r_+$. There is one solution of (27) in which U_n is a polynomial of degree $(n + 1)$ in Z, but there is another solution (called V_n) expressible as a nonterminating power series in Z^{-1}. So we have as a result:

$$G_n = r^2 \frac{d}{dr}\left(r^{-1} U_n\right) \qquad \text{(polynomial of degree } (n+1))$$

$$H_n = r^2 \frac{d}{dr}\left(r^{-1} V_n\right) \qquad \text{(power series in } Z^{-1})$$

(see Appendix). The boundary and regularity conditions require R_n^o and R_n^1 be expressed in terms of U_n and G_n near the horizon $r = r_+$, and in terms of V_n and H_n as $r \to \infty$:

$$\Phi_o = \frac{\sin\theta}{r^2} \Sigma^{-\frac{1}{2}} \sum_n \begin{Bmatrix} \alpha_n U_n \\ \beta_n V_n \end{Bmatrix} P_n'(\cos\theta) \qquad \begin{matrix} r \to r_+ \\ r \to \infty \end{matrix}$$

$$\Phi_1 = \frac{n}{r^2} + \frac{1}{r^2} \sum_n \begin{Bmatrix} \gamma_n G_n \\ \delta_n H_n \end{Bmatrix} P_n(\cos\theta) \qquad \begin{matrix} r \to r_+ \\ r \to \infty \end{matrix} \qquad (28)$$

where η, α_r, β_n, γ_n, and δ_n are complex constants.

Returning to (21) and including the source terms we have

$$\frac{\partial}{\partial r}(r\Sigma^{\frac{1}{2}}\Phi_o) = \sin\theta\,\frac{\partial\Phi_1}{\partial x} + \frac{\sqrt{2}}{2}\,r\,J_o$$
$$\frac{1}{r}\Sigma^{\frac{1}{2}}\frac{\partial}{\partial r}(r^2\Phi_1) = -\frac{\partial}{\partial x}(\sin\theta\,\Phi_o) + \frac{\sqrt{2}}{2}\,r\,J_1 \qquad (29)$$

Substituting the series (25) and integrating against $\sin\theta\,P'_m$ or P_m yields:

$$\frac{d}{dr}(r\Sigma^{\frac{1}{2}}R^o_n) = R^1_n + \frac{\sqrt{2}\,(2n+1)r}{4r(n+1)}\,J_{on}$$
$$\frac{1}{r}\Sigma^{\frac{1}{2}}\frac{d}{dr}(r^2R^1_n) = n(n+1)\,R^o_n + \frac{\sqrt{2}\,(2n+1)r}{4}\,J_{1n} \qquad (30)$$

where $J_{on} = \int J_o P_n dx;\quad J_{1n} = \int J_1 \sin\theta\, P'_n\,dx$.

Defining $\tilde{R}^o_n = r^2\Sigma^{\frac{1}{2}}R^o_n$, $\tilde{R}^1_n = r^2 R^1_n$, (31)

and converting to second order equations yields:

$$r^2\Sigma\,\frac{d^2\tilde{R}^o_n}{dr^2} = n(n+1)\tilde{R}^o_n + S_{on}$$
$$r^2\Sigma\,.\,\frac{d^2\tilde{R}^1_n}{dr^2} + r^2\Sigma'\,\frac{d\tilde{R}^1_n}{dr} = n(n+1)\tilde{R}^1_n + S_{1n} \qquad (32)$$

where

$$S_{on} = \frac{\sqrt{2}(2n+1)}{4}\,r\,\Sigma\left[r^2\Sigma^{-\frac{1}{2}}J_{on} + \frac{1}{n(n+1)}\frac{d}{dr}(r^3 J_{1n})\right]$$
$$S_{1n} = \frac{\sqrt{2}(2n+1)}{4}\,r^2\left[r\,J_{1n} + \frac{d}{dr}(r^2\Sigma^{\frac{1}{2}}J_{on})\right] \qquad (33)$$

We can immediately write the Green's function for this system as follows:

$$\tilde{G}^o_n(r,s) = \begin{cases} \dfrac{U_n(r)\,V_n(s)}{s^2\,\Sigma(s)\,W^o_n(s)} & r_+ < r < S \\[2ex] \dfrac{U_n(s)\,V_n(r)}{s^2\,\Sigma(S)\,W^o_n(S)} & S < r < \infty \end{cases}$$

$$\tilde{G}^1_n(r,s) = \begin{cases} \dfrac{G_n(r)\ H_n(S)}{S^2\ \Sigma(S)\ W^1_n(S)} & r_+ < r < S \\[2ex] \dfrac{G_n(S)\ H_n(r)}{S^2\ \Sigma(S)\ W^1_n(S)} & S < r < \infty \end{cases}$$

where

$$W^o_n \equiv U_n V'_n - U'_n V_n$$

$$W^1_n \equiv G_n H'_n - G'_n H_n$$

And so the solutions are:

$$\tilde{R}^o_n(r) = \int_{r_+}^{\infty} G^o_n(r,S)\ S_{on}(S)\ dS$$

$$\tilde{R}^1_n(r) = \int_{r_+}^{\infty} G^1_n(r,S)\ S_{1n}(S)\ dS\ .$$

From the definitions (14) and (17) we can extract the field components:

$$F_{01} = -\frac{\sqrt{2}}{r^2}\sum_n \mathrm{Re}(\tilde{R}^1_n)\ P_n(\cos\ \theta)$$

$$F_{02} = \frac{\sqrt{2}\ \sin\theta}{r}\ \sum_n \mathrm{Re}(\tilde{R}^o_n)\ P'_n(\cos\ \theta)$$

$$F_{23} = \sqrt{2}\ \sin\ \theta \sum_n\ \mathrm{Im}(\tilde{R}^1_n)\ P_n(\cos\ \theta)$$

$$F_{13} = \frac{\sqrt{2}\ \sin^2\theta}{r\Sigma}\ \sum_n \mathrm{Im}(\tilde{R}^o_n)\ P'_n(\cos\ \theta)$$

The solution in this form is amenable to computer calculation, given a suitable choice of charge and current.

APPENDIX

The radial functions: the coefficients of the polynomial solution of (27),

$$U_n = \sum_k a^n_k\ z^k, \quad z = r - r_+ \quad ,$$

satisfy the recursion relation

$$dK(K+1)\ a^n_{k+1} = [n(n+1) - K(K-1)]a^n_k \ , \quad d \equiv r_+ - r_-$$

whence

$$U_o = 1 + Z$$

$$U_1 = Z(1 + Z/d)$$

$$U_3 = \frac{Z}{d^2}\ (2Z + d)(Z + d)$$

The corresponding G_n are given by

$$G_n = r^2 \frac{d}{dr}\ (r^{-1}\ U_n) = r\ \frac{dU_n}{dr} - Un \quad .$$

The power series solution

$$V_n = \sum_k b^n_k\ Z^{-(S+K)}$$

gives:

$$[S(S+1) - n(n+1)]b^n_o = 0 \rightarrow S=n$$

$$[n(n+1) - (n+K)\ (n+K+1)]b^n_k = d(n+K-1)(n+K)b^n{}_{k-1}$$

thus

$$V_o = 1$$

$$V_1 = \frac{1}{Z}\ [1 - \frac{d}{2}\cdot\frac{1}{Z} + \frac{3d^2}{10}\cdot\ \frac{1}{Z^2} - \ \ldots]$$

$$V_2 = \frac{1}{Z^2}\ [1 - d\cdot\frac{1}{Z} + \frac{6d^2}{7}\cdot\frac{1}{Z^2} - \ \ldots]$$

From this H_n follows by

$$H_n = r\ \frac{dVn}{dr} - Vn \quad .$$

ACKNOWLEDGMENTS

I am grateful to Professor Remo Ruffini for suggesting this problem and to Debbie Brown for aid with the final manuscript.

REFERENCES

[1]. E. Newman and R. Penrose, J. Math. Phys. 3 (1962) p.566.

[2]. F. A. E. Pirani, in Lectures on General Relativity: 1964 Brandeis Summer Institute in Theoretical Physics, Vol. I (Prentice-Hall, Englewood Cliffs, New Jersey, 1965), pp. 249-374.

Perturbations of an Analytic Background Metric

Mark A. Peterson
Amherst College
Amherst, Massachusetts, U. S. A.

and

Remo Ruffini
Princeton University
Princeton, New Jersey, U. S. A.

Introduction

Perturbation calculations for the propagation of waves through a curved background space have been reported by many authors. In this paper we describe a method for doing such calculations which does not require the background metric to have high symmetry, but only to be analytic. It is basically the method of Hadamard [1] as described by DeWitt and Brehme [2]. This method is usually considered an existence proof rather than an effectively computable method, but we offer some suggestions to make it practical for computation. That Hadamard's method can be used effectively in computations was shown in an example by K. P. Chung [3].

§I. Perturbation equations in a fixed background

The linearized equation for a perturbation $h_{\mu\nu}$ away from a solution $g_{\mu\nu}$ of the Einstein equations

$$R_{\mu\nu} - \frac{1}{2} g_{\mu\nu} R = - KT_{\mu\nu} \tag{1}$$

is well known to be [4]

$$h^{\alpha}{}_{\alpha;\beta\mu} - (f_{\beta;\mu} + f_{\mu;\beta}) + h_{\beta\mu;\alpha}{}^{;\alpha} + 2R^{\nu}{}_{\beta\mu\alpha} h_{\nu}{}^{\alpha} + (f_{;\gamma}{}^{;\gamma} - h^{\gamma}{}_{\gamma;\alpha}{}^{;\alpha}) g_{\mu\beta} = -2K\delta T_{\mu\beta} \tag{2}$$

where $f_{\beta} = h_{\beta}{}^{\alpha}{}_{;\alpha}$, $\delta T_{\mu\beta}$ is the perturbation of the source, and the derivatives are covariant with respect to the background metric $g_{\mu\nu}$. Under an infinitesimal change of coordinates (gauge transformation) in the vacuum region ($R_{\alpha\beta}=0$, $T_{\alpha\beta}=0$)

$$x^{\mu} \rightarrow x^{\mu} + \xi^{\mu} \tag{3}$$

the components of the perturbation $h_{\mu\nu}$ change according to Killing's equation

$$h_{\mu\nu} \rightarrow h_{\mu\nu} - \xi_{\mu;\nu} - \xi_{\nu;\mu} \tag{4}$$

$$f_\mu \to f_\mu - \xi_\mu{}^{;\nu}{}_{;\nu} - \xi^\nu{}_{;\nu;\mu} \tag{5}$$

$$h^\alpha{}_\alpha \to h^\alpha{}_\alpha - 2\xi^\alpha{}_{;\alpha} \tag{6}$$

Equation (2) simplifies greatly if we can find a ξ^μ satisfying the following relations

$$\xi_\mu{}^{;\nu}{}_{;\nu} = f_\mu - \frac{1}{2} h^\alpha{}_{\alpha;\mu} \tag{7}$$

$$\xi^\mu{}_{;\mu} = \frac{1}{2} h^\alpha{}_\alpha$$

since in that case $f_\mu = 0$, $h^\alpha{}_\alpha = 0$ after the transformation. (7) is prima facie an overdetermined system of equations for ξ_μ. The integrability condition that a solution exist is

$$f_\mu{}^{;\mu} = h^\alpha{}_{\alpha;\mu}{}^{;\mu} \tag{8}$$

In fact, the equation (2) itself guarantees that this integrability condition holds, as one sees by contracting the two free indices. Therefore there exists a gauge in which $f_\mu = 0$, $h^\alpha{}_\alpha = 0$, and in this gauge (2) becomes

$$h_{\beta\mu;\alpha}{}^{;\alpha} + 2R^\nu{}_{\beta\mu\alpha} h^\alpha{}_\nu = -2K\delta T_{\mu\beta} \tag{9}$$

The perturbation is seen to obey a covariant wave equation.

§II. The wave equation in flat space

Our method for attacking equation (9) is best understood if we first consider the analogous equation in flat space

$$\left(\nabla^2 - \frac{\partial^2}{\partial t^2}\right) h_{\alpha\beta} = -2K\delta T_{\alpha\beta} \tag{10}$$

We seek a Green's function solution

$$h_{\alpha\beta} = -2K \int G_{\alpha\beta\mu\nu}(\mathbf{x}, \mathbf{x}')\delta T^{\mu\nu}(\mathbf{x}')d^4\mathbf{x}' \tag{11}$$

Hence the Green's function $G_{\alpha\beta\omega\nu}(\mathbf{x}, \mathbf{x}')$ satisfies

$$\left(\nabla^2 - \frac{\partial^2}{\partial t^2}\right) G(\mathbf{x}, \mathbf{x}') = \delta^4(\mathbf{x}-\mathbf{x}') \tag{12}$$

where

$$G_{\alpha\beta\mu\nu}(\mathbf{x}, \mathbf{x}') = G(\mathbf{x}, \mathbf{x}')\frac{1}{2}\left[\delta_{\alpha\mu}\delta_{\beta\nu} + \delta_{\alpha\nu}\delta_{\beta\mu}\right] \tag{13}$$

If we analytically continue (12) in the complex variable $t + i\tau$ to the imaginary axis we find

$$(\nabla^2 + \frac{\partial^2}{\partial \tau^2}) G(x, x') = \delta^4(x - x') \tag{14}$$

the Laplace equation. But a Green's function solution for the 4-dimensional Laplacian is clearly

$$G(x, x') = \frac{-1}{4\pi^2 \left[|\underset{\sim}{r} - \underset{\sim}{r}'|^2 + (\tau - \tau')^2 \right]} \tag{15}$$

because it solves the homogeneous elliptic problem if $x \neq x'$ and has the right singularity at $x = x'$. To find the Green's function for the original problem we need only analytically continue back to the real axis. The analytic continuation can be achieved in Equation (11) simply by rotating the contour in the complex t-plane as shown in Figure 1. The rotation introduces an explicit $-i$, but since we are seeking a real solution given a real source it is sufficient to take the imaginary part of G on the real axis. Also the two singularities on the real axis are approached in the manner shown in Figure 1, associated with half-retarded, half-advanced boundary conditions.

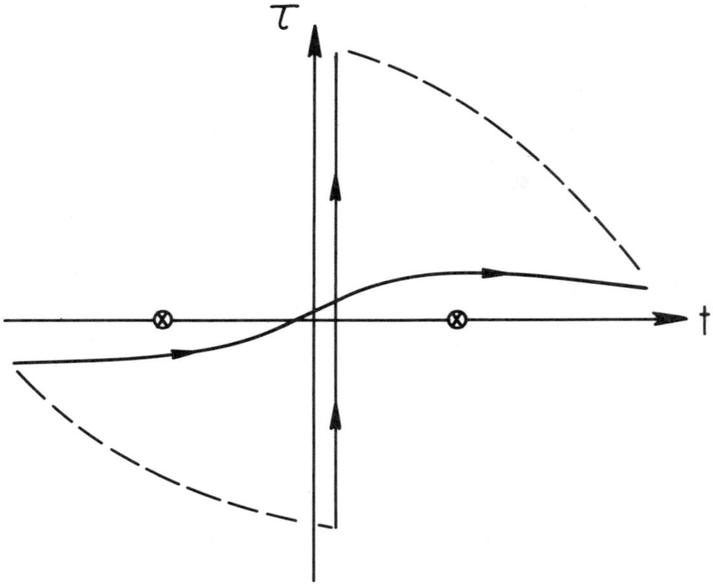

Figure 1

Thus we have in all

$$G(x,x') = \lim_{\varepsilon \to 0} \operatorname{Im} \frac{-1}{4\pi^2 \left[|\underset{\sim}{r}-\underset{\sim}{r}'|^2 - (t-t')^2 - i\varepsilon \right]} = -\frac{1}{4\pi} \delta \left(|\underset{\sim}{r}-\underset{\sim}{r}'|^2 - (t-t')^2 \right) \tag{16}$$

This is of course the Lienard-Wiechert potential, the well known solution to the problem.

§III. Hadamard's method

The method of Hadamard for constructing a Green's function solution to equation (9) has been described by DeWitt and Brehme [2] and by Chung [3]. The basic idea is just that of Section II. We analytically continue everything in the time variable to the imaginary axis, so that the problem becomes elliptic. There we find Hadamard's elementary solution, which is just the Green's function for the elliptic problem analogous to (15). Finally, the Green's function for the hyperbolic problem is found by continuing the elementary solution back to the real t-axis as in Section II, and taking the imaginary part.

We therefore consider the elliptic problem, which is simpler than the hyperbolic since the Green's function singularity occurs at an isolated point and the metric is definite. The role of the distance function in flat space is played by the geodesic distance function

$$s(x,x') = \int_\gamma \sqrt{g_{\alpha\beta} \frac{dx^\alpha}{d\tau} \frac{dx^\beta}{d\tau}}\, d\tau \tag{17}$$

where γ is the geodesic connecting x and x' (assumed to be unique). A slightly more convenient function is

$$\sigma(x,x') = \frac{s^2}{2} \tag{18}$$

It obeys the Hamilton-Jacobi equation

$$g^{\alpha\beta} \frac{\partial\sigma}{\partial x^\alpha} \frac{\partial\sigma}{\partial x^\beta} = 2\sigma \tag{19}$$

It is also useful to introduce the quantity

$$\Delta(x,x') = \det\left(\frac{\partial^2\sigma}{\partial x^\alpha \partial x'^\mu}\right) \left[g(x)g(x')\right]^{-1/2} \tag{20}$$

which satisfies

$$\frac{\partial \Delta}{\partial x^\alpha}\frac{\partial \sigma}{\partial x^\beta} g^{\alpha\beta} = 4 - g^{\alpha\beta}\frac{\partial^2 \sigma}{\partial x^\alpha \partial x^\beta} \tag{21}$$

as can be seen by differentiating (19) once at x and once at x'. To prove (21) we must assume $\sigma_{;\alpha\mu}$ has an inverse (as a matrix). The locus of points where this fails to hold, the caustic set, marks the boundary of the region within which our construction is valid.

With this notation established the elementary solution of Hadamard can be written

$$G_{\alpha\beta\mu\nu}(x,x') = \frac{1}{8\pi^2}\left[\frac{u_{\alpha\beta\mu\nu}(x,x')}{\sigma(x,x')} + v_{\alpha\beta\mu\nu}(x,x')\ln\sigma(x,x') + w_{\alpha\beta\mu\nu}(x,x')\right] \tag{22}$$

The first term is the analogue of the elementary solution in flat space, and it has the right singularity as $x \to x'$ if

$$\lim_{x\to x'} u_{\alpha\beta\mu\nu}(x,x') = \frac{1}{2}(g_{\alpha\mu}(x)g_{\beta\nu}(x)+g_{\alpha\nu}(x)g_{\beta\mu}(x)) \tag{23}$$

The succeeding terms have no analogue in flat space, but are necessary if $G_{\alpha\beta\mu\nu}$ is to solve the elliptic homogeneous equation away from $x = x'$. Let us demand that $u_{\alpha\beta\mu\nu}$, $v_{\alpha\beta\mu\nu}$, and $w_{\alpha\beta\mu\nu}$ be regular analytic functions and

$$v_{\alpha\beta\mu\nu}(x,x') = \sum_{n=0}^{\infty} v^{(n)}_{\alpha\beta\mu\nu}(x,x')\sigma^n(x,x')$$

Then plugging the form (22) into the elliptic homogeneous equation obtained from (9) we get terms in σ^{-2}, σ^{-1}, and $\sigma^n \ln\sigma$ (where $n = 0,1,\ldots$) with coefficients involving $u_{\alpha\beta\mu\nu}$, $v^{(n)}_{\alpha\beta\mu\nu}$. Since the whole expression must vanish and the coefficient functions are regular, the coefficient of each singular term must vanish separately. Thus we obtain an infinite system of equations for $u_{\alpha\beta\mu\nu}$, $v^{(n)}_{\alpha\beta\mu\nu}$ to express the vanishing of these coefficients,

$$\left[2u_{\alpha\beta\mu\nu;\gamma} - \Delta^{-1}\Delta_{,\gamma}u_{\alpha\beta\mu\nu}\right]\sigma_{,\delta}\, g^{\gamma\delta} = 0 \tag{24}$$

$$\left[2v^{(0)}_{\alpha\beta\mu\nu;\gamma} - \Delta^{-1}\Delta_{,\gamma}v^{(0)}_{\alpha\beta\mu\nu}\right]\sigma_{,\delta}\, g^{\gamma\delta} + 2^{(0)}v_{\alpha\beta\mu\nu} = -u_{\alpha\beta\mu\nu;\gamma}{}^{;\gamma} - 2R^{\gamma\delta}_{\alpha\beta}\, u_{\gamma\delta\mu\nu} \tag{24a}$$

$$\left[2nv^{(n)}_{\alpha\beta\mu\nu;\gamma} - n\Delta^{-1}\Delta_{\gamma}v^{(n)}_{\alpha\beta\mu\nu}\right]\sigma_{,\gamma}\, g^{\gamma\delta} + 2n(n+1)v^{(n)}_{\alpha\beta\mu\nu} = -v^{(n-1)}_{\alpha\beta\mu\nu\delta;\gamma}{}^{\gamma} - 2R^{\gamma\delta}_{\alpha\ \beta}\, v^{(n-1)}_{\gamma\delta\mu\nu} \quad (n\geq 1) \tag{24b}$$

By Hamilton-Jacobi theory

$$\frac{\partial \sigma}{\partial x^{\alpha}} = \tau \frac{dx^{\beta}}{d\tau} g_{\alpha\beta}(x) \tag{25}$$

where $\frac{dx^{\beta}}{d\tau}$ is tangent to the geodesic from x' to x at x and τ is any affine parameter along this geodesic. Thus the equations (24) are seen to be ordinary differential equations along the geodesics emanating from x'. To emphasize this we can rewrite the system as

$$\frac{D}{D\tau}(\Delta^{-1/2} u_{\alpha\beta\mu\nu}) = 0 \tag{26}$$

$$\frac{D}{D\tau}(\tau \Delta^{-1/2} v^{(0)}_{\alpha\beta\mu\nu}) = -\frac{1}{2}\Delta^{-1/2}(u_{\alpha\beta\mu\nu;\gamma}{}^{;\gamma} + 2R^{\gamma}{}_{\alpha}{}^{\delta}{}_{\beta} u_{\gamma\delta\mu\nu})$$

$$\vdots$$

$$\frac{1}{\tau^{n}}\frac{D}{D\tau}(\tau^{n+1}\Delta^{-1/2} v^{(n)}_{\alpha\beta\mu\nu}) = -\frac{1}{2n}\Delta^{-1/2}(v^{(n-1)}_{\alpha\beta\mu\nu;\gamma}{}^{;\gamma} + 2R^{\gamma}{}_{\alpha}{}^{\delta}{}_{\beta} v^{(n-1)}_{\gamma\delta\mu\nu})$$

etc. $(n \geq 1)$

The initial value of $u_{\alpha\beta\mu\nu}(x, x')$ at $x = x'$ is given by (23). The initial values for the $v^{(n)}_{\alpha\beta\mu\nu}$ are determined recursively by taking the limit as $x \to x'$ in each equation above. Thus the above system of ordinary differential equations completely determines $u_{\alpha\beta\mu\nu}$ and $v_{\alpha\beta\mu\nu}$.

The analytic continuation of the elementary solution to the real t-axis leads to the Green's function

$$G_{\alpha\beta\mu\nu}(x, x') = \operatorname{Im} \lim_{\varepsilon \to 0} \frac{1}{8\pi^2}\left[\frac{u_{\alpha\beta\mu\nu}}{\sigma - i\varepsilon} + v_{\alpha\beta\mu\nu}\ln(\sigma - i\varepsilon) + w_{\alpha\beta\mu\nu}\right] \tag{27}$$

The functions $u_{\alpha\beta\mu\nu}$ and $v_{\alpha\beta\mu\nu}$ still obey the differential equations (26) by analytic continuation to real time, so they are real, and $w_{\alpha\beta\mu\nu}$ may be taken real. Hence

$$G_{\alpha\beta\mu\nu}(x, x') = \frac{1}{8\pi}\left[u_{\alpha\beta\mu\nu}(x, x')\delta(\sigma) - v_{\alpha\beta\mu\nu}(x, x')\theta(-\sigma)\right] \tag{28}$$

It is apparent from the first of equations (26) that

$$u_{\alpha\beta\mu\nu} = \Delta^{-1/2}\frac{1}{2}(\bar{g}_{\alpha\mu}(x, x')\bar{g}_{\beta\nu}(x, x') + \bar{g}_{\alpha\nu}(x, x')\bar{g}_{\beta\mu}(x, x')) \tag{29}$$

where

$$\frac{D}{D\tau}\bar{g}_{\alpha\mu}(x, x') = 0 \tag{30}$$

and

$$\lim_{x \to x'} \bar{g}_{\alpha\mu}(x, x') = g_{\alpha\mu}(x) \tag{31}$$

The significance of $\bar{g}_{\alpha\mu\nu}(x, x')$ is that it accomplishes parallel transport from x' to x.

In all, the Green's function admits a pretty interpretation. Signals are parallel transported from x' to x along the light cone (the $\delta(\sigma)$ term), and in addition there are terms (from $\theta(-\sigma)$) which come along geodesics inside the light cone, arriving later than might be expected. These terms, it may be imagined, express in the language of ray optics waves which have scattered from the curvature tensor to reach the observer, so do not take the most direct path. The factor Δ describes the focussing of the wave. It blows up at caustic points.

§IV. Numerical considerations

The Green's function can be found explicitly by integrating the system (26) along geodesics. To do this, however, one appears to need not only the functions $u_{\alpha\beta\mu\nu}$, $v_{\alpha\beta\mu\nu}$ as one proceeds along a given geodesic, but also their derivatives. Previous applications of this method [3] have found these derivatives numerically by constructing additional geodesics close to the original one. But the problem of finding the geodesic which connects two given points is difficult and time consuming in practice, since it is a global problem. We have used a method instead which eliminates the necessity of finding nearby geodesics. The idea is to enlarge (26) to be a system of equations for $u_{\alpha\beta\mu\nu}$, $v_{\alpha\beta\mu\nu}$ and their derivatives along the geodesic of interest. To do this we simply adjoin to (26) the derivatives of these same equations. In the process derivatives of the Riemann tensor appear, but these are known functions. Thus we replace unknown functions by known ones at the expense of enlarging the system. The problem is a purely local one once the geodesic of integration has been selected.

We also need all derivatives of σ and Δ on the geodesic, and similar remarks apply to these functions. To illustrate the procedure, the derivatives of σ are determined by

$$\begin{aligned} g^{\alpha\beta}\sigma_{,\alpha}\sigma_{,\beta} &= 2\sigma \\ g^{\alpha\beta}\sigma_{;\alpha\gamma}\sigma_{,\beta} &= 2\sigma_{,\gamma} \end{aligned} \tag{32}$$

$$g^{\alpha\beta}\sigma_{;\alpha\gamma\delta}\sigma_{,\beta} + g^{\alpha\beta}\sigma_{;\alpha\gamma}\sigma_{;\beta\delta} = \sigma_{;\gamma\delta}$$

etc. (32 continued)

which can be rewritten using (25) as ordinary equations along the geodesic of integration

$$\tau \frac{D\sigma}{D\tau} = 2\sigma \tag{33}$$

$$\tau \frac{D\sigma}{D\tau}{}_{,\gamma} = \sigma_{,\gamma}$$

$$\tau \frac{D}{D\tau}\sigma_{;\gamma\delta} = R^{\varepsilon\alpha}{}_{\gamma\delta}\sigma_{,\varepsilon}\sigma_{,\alpha} + \sigma_{;\gamma\delta} - g^{\alpha\beta}\sigma_{;\alpha\gamma}\sigma_{;\beta\delta}$$

etc.

where τ is any affine parameter.

Similarly we can interpret (21) as an equation for

$$\tau \frac{D\Delta}{D\tau} = (4-\sigma_{;\alpha\beta}g^{\alpha\beta})\Delta \tag{34}$$

and by differentiating it we obtain equations for $\Delta_{,\alpha}$, $\Delta_{;\alpha\beta}$, etc.

The result of all this is a large system of ordinary differential equations for determining $G_{\alpha\beta\mu\nu}(x, x')$ starting at the source position x' and moving out along a geodesic to the observer's position x. The initial values for all quantities are explicitly determined by taking the limit $x \to x'$ in the system of equations itself, as noted in DeWitt and Brehme [2]. One may now integrate the system numerically.

The first term in the Green's function beyond geometrical optics, $v^{(0)}_{\alpha\beta\mu\nu}(x, x')$, is easily accessible by this method. However the system of equations grows so rapidly that for higher terms one must take advantage of the many symmetries and algebraic identities in the problem to reduce it to a practical size. The following observations are useful for reducing the system.

1) Since nearly all indices in the system arise from repeated differentiation, they may be put into ascending order with the help of the Riemann tensor.

2) $g_{\mu\nu}(x')\bar{g}^{\mu}{}_{\alpha}(x, x')\bar{g}^{\nu}{}_{\beta}(x, x') = g_{\alpha\beta}(x)$

$$\bar{g}^{\mu}{}_{\alpha}(x, x')\frac{\partial\sigma}{\partial x'^{\mu}} = \frac{\partial\sigma}{\partial x^{\alpha}}$$

$$g^{\mu\nu}(x')v^{(n)}_{\alpha\beta\mu\nu} = 0$$

are algebraic identities expressing parallel transport. Of course the derivatives of these identities are also identities.

3) The indices $\mu, \nu, \ldots$ at the source point $\mathbf{x}'$ label uncoupled systems. For example, if the source trajectory is radial in spherical coordinates, only the indices $\mu, \nu, \ldots = 0, 1$ are needed.

The method is something of a brute force method, but with the suggestions given in this section, it is useable. Moreover, since it is still exact, it may form the basis for approximation schemes taking advantage of the peculiarities of individual problems.

References

1 J. Hadamard, Lectures on Cauchy's Problem in Linear Partial Differential Equations, Dover Publications, Inc., New York, 1952.

2 B. DeWitt and R. Brehme, Annals of Physics 9 (1960), 220.

3 K. P. Chung, Il Nuovo Cimento 14B (1973), 293.

4 F. Zerilli, Phys. Rev. D2 (1970), 214.

FORCED PERTURBATIONS OF THE REISSNER-NORDSTRØM GEOMETRY

Joseph Weinstein*

Department of Physics, Princeton University
Princeton, New Jersey 08540 U.S.A.

ABSTRACT: Equations derived by the author governing the coupled electromagnetic-gravitational perturbations induced by a charged test particle upon a Reissner-Nordstrøm background geometry are presented in outline form and a sketch given of the method of derivation. A detailed presentation of these results will be given elsewhere.

The presenceof coupling between the gravitational and electromagnetic fields represented by the appearance of the matter-energy tensor in the Einstein equations and of metric-dependent terms in the Maxwell equations can lead to qualitative modifications in the properties of gravitational or electromagnetic radiation propagating through, or generated in, regions with strong background fields. Gravitational radiation may be induced by an electromagnetic wave propagating through a region of strong space-time curvature and, conversely, may itself induce an electromagnetic wave when propagating through a region with strong electromagnetic fields. This effect is a first-order phenomenon which may be investigated by ordinary perturbation methods [1, 2].

We have investigated this coupling for the case of perturbations induced by a (possibly charged) test particle upon the Reissner-Nordstrøm geometry. This background metric represents a charged black hole and hence clearly meets the criterion of strong background fields. For each of the multipole modes in a spherical-harmonic expansion of the perturbation field (designated by harmonic indices L, M (= -L, -L+1,..., L-1, L) and parity index P (= 1, -1)) the perturbation field may be characterized completely by a pair of radial functions $F^P_{LM}(r)$, $G^P_{LM}(r)$ representing the electromagnetic and gravitational fields, respectively [1]. These functions satisfy a pair of coupled, Schroedinger-like differential equations to be solved under boundary conditions of purely-outgoing radiation at infinity and purely-ingoing radiation at the black-hole boundary [3] of the form

$$\frac{d^2}{dr_*^2}\begin{pmatrix} F^P_{LM} \\ G^P_{LM}\end{pmatrix} + (\omega^2 - V^P_L(r))\begin{pmatrix} F^P_{LM} \\ G^P_{LM}\end{pmatrix} + S^P_L(r)\begin{pmatrix} 3m & (-1)^{P+1}2e\sqrt{(L-1)(L+2)} \\ (-1)^{P+1}2e\sqrt{(L-1)(L+2)} & -3m \end{pmatrix}\begin{pmatrix} F^P_{LM} \\ G^P_{LM}\end{pmatrix} = \begin{pmatrix} A^{(F)P}_{LM} \\ A^{(G)P}_{LM}\end{pmatrix}$$

in which V^P_L serves as the effective potential, $A^{(F)P}_{LM}$ and $A^{(G)P}_{LM}$ are source terms, and ω is the frequency of the radiation. The coupling between the two fields appears in the matrix coefficients in the third term and in the source terms. The equation may be decoupled by the linear transformation which diagonalizes the matrix coefficient [4]. These equations without the source term have already been

*Current address: 10807 Lombardy Road, Silver Spring, Maryland

given for the purpose of analysis of the stability of the Reissner-Nordstrøm metric by Vincent Moncrief [4] and are believed to be equivalent under a transformation of the radial functions to the equations previously derived by another method by Zerilli [1] but which could not be decoupled [5]. The exact form of the coefficients appearing in the equations is complicated and will be given elsewhere[6].

The above equations were derived through an extension of the method developed by Moncrief for analyzing the stability of the Reissner-Nordstrøm geometry [7]. The total action of the system was decomposed into the sum of two terms, I_{field} representing the action of the electromagnetic and gravitational fields including their mutual interaction, and I_{inter} representing the action arising from the interaction of these two fields with the perturbing test particle. The ADM Hamiltonian formulation [8] was used to express the action integrals. In this formulation, the gravitational field is represented by the spatial components of the metric tensor g_{ij} and their conjugate momenta π^{ij}, the electromagnetic field is represented by the spatial components of the vector potential A_i and their conjugate momenta $\mathcal{E}^i$, and the test particle is represented by its coordinates x^i and momenta p_i. Besides these variables, the lapse function N and shift function N^i appear as Lagrange multipliers determining the coordinate system, and the scalar potential ϕ appears as a Lagrange multiplier determining the electromagnetic gauge. The second term I_{inter} was then treated as a small perturbation to the first I_{field} and a series expansion introduced for the field variables. It is then found that the first-order perturbations δg_{ij}, $\delta\pi^{ij}$, δA_i, $\delta\mathcal{E}^i$, δN, δN^i, and $\delta\phi$ can all be determined from a variational principle based upon the action integral $I = \delta^2 I_{field} + \delta I_{inter}$.

At this point an expansion in Regge-Wheeler tensor spherical harmonics [3,9] is introduced to eliminate the radial variables from the Hamiltonian and the field variables re-expressed in terms of their Regge-Wheeler-Zerilli parameterization coefficients. As a consequence of the spherical symmetry and parity invariance of the problem the Hamiltonian splits into a sum of independent terms, each characterized by distinct harmonic indices L, M, and parity index P. Of the new field variables five arise from an expansion of the Lagrange multipliers δN, δN_i, and $\delta\phi$, and continue to play a similar role in the reduced system. Variation of these quantities leads to five constraint equations which must be satisfied by the remaining eighteen variables. Specification of a coordinate system and gauge provides another five equations, reducing the number of variables per harmonic mode to eight, corresponding to the four degrees of freedom of the electromagnetic and gravitational fields (one degree of freedom per field per parity mode).

The reduction proceeds by introducing the canonical transformation given by Moncrief [7] from the original parameterization to a new set of variables chosen so that precisely five of the variables appear in the constraint equations. These equations may then be solved explicitly for these variables in terms of the source functions and the result substituted back into the Hamiltonian. The five variables conjugate to the constrained quantities are the gauge and coordinate dependent quantities and do not affect this reduced Hamiltonian. Consistency of the reduction process requires that the expressions for the constrained variables obtained from the constraint equations and from the dynamical equations obtained by variation of the quantities conjugate to them be equivalent; this is guaranteed by the Bianchi identities, provided that the motion of the test particle obeys the laws of conservation of momentum and energy. In the final Hamiltonian only eight gauge-independent quantities appear, variation of which leads to the above system of equations for each of the two parity cases.

ACKNOWLEDGEMENT

I would like to thank Dr. Remo Ruffini of the Institute for Advanced Study, Princeton, N. J. for assisting me and encouraging me in this work, and Dr. Vincent Moncrief of the University of Utah for providing me with information concerning his method for analyzing the perturbations of a spherically-symmetric geometry. A travel grant from Princeton University is also gratefully acknowledged.

NOTES

[1] F. Zerilli, Phys. Rev. D7 (1974) 860.
[2] M. Johnston, R. Ruffini, and F. Zerilli, Phys. Rev. Lett. 31 (1972) 1377.
[3] F. Zerilli, Phys. Rev. D2 (1970) 2141; R. Ruffini, in Black Holes (ed.DeWitt) Gordon and Breach Science Publishers, New York (1972) 451-546.
[4] V. Moncrief, Phys. Rev. D9 (1974) 2207 and Phys. Rev. D10 (1974) 1056.
[5] M. Johnston, R. Ruffini, and M. Peterson, Lett. Nuovo Cimento 9 (1974) 217.
[6] J. Weinstein, to be submitted for publication.
[7] V. Moncrief, Phys. Rev. in press.
[8] R. Arnowitt, S. Deser, and C. Misner, in Gravitation: an Introduction to Current Research, ed. Louis Witten; John Wiley and Sons, Inc., New York (1961) 227-265.
[9] T. Regge and J. Wheeler, Phys. Rev. 108 (1957) 1063.

SHOCK WAVES IN RELATIVISTIC MAGNETOHYDRODYNAMICS

André Lichnerowicz
Collège de France, Paris (France)

INTRODUCTION

The theory of relativistic fluids and plasmas plays an important role in theoretical astrophysics. Magnetohydrodynamical shock waves can appear in the physics of the sun, of the stars and also of the galaxies. However, relativistic M. H. D. is simpler than the classical theory from some aspects, since relativistic dynamics present a natural harmony with Maxwell equations.

The main purpose of this talk is to develop a rigorous study of magnetohydrodynamical shock waves. It is reasonable to use, as background, a curved given space-time, which is convenient for various astronomical applications. But nothing is changed if we assume that the space-time is flat.

We consider here a perfect gas with an infinite conductivity; we give, for this gas, the main differential system of the relativistic M. H. D. and the corresponding shock equations. For our study, the relativistic form of the compressibility conditions for a fluid is a main tool. It is known now that it is possible to deduce these relativistic conditions from considerations of Statistical Mechanics.

From the shock equations and from the compressibility conditions, it is possible to deduce mathematically
1) The timelike character of the magnetohydrodynamical shock waves.
2) The main inequalities between the different speeds and also the Thermodynamics of the shocks.
3) The existence and uniqueness of a non trivial solution of the shock equations.

For these purposes, a generalized relativistic Hugoniot function is introduced. It is possible to find a detailed treatment in [7].

1 - Thermodynamical Perfect Fluid

a) Let $(V_4, \underline{g})$ be a given space-time. The so-called metric tensor $\underline{g}$ is Lorentzian, of signature +--- and can be written on the domain U of a local chart $\underline{g}|_u = g_{\alpha\beta}\, dx^\alpha \otimes dx^\beta$ (α, β, each greek index = 0, 1, 2, 3).

Consider in V_4 a perfect fluid described by the energy tensor:

$$T_{\alpha\beta}^{(f)} = (\rho + p)\, u_\alpha\, u_\beta - p g_{\alpha\beta} \qquad (1\text{-}1)$$

where ρ is the proper energy density, p the pressure, u the unit 4-velocity, oriented towards the future; ρ includes a matter density and an internal energy density. We put according to old considerations of Taub

$$\rho = c^2\, r\, (1 + \varepsilon/c^2) \quad (r > 0)$$

where r is the matter density and ε the specific internal energy. Let us consider the scalar:

$$\rho + p = c^2 r \; (1 + \varepsilon/c^2 + p/(c^2 r))$$

We put:

$$i = \varepsilon + p/r = \varepsilon + pV \quad (\text{where } V = 1/r) \qquad (1\text{-}2)$$

i is the specific enthalpy. I have introduced systematically, long-time ago, the following thermodynamical variable (index of the fluid) $f = 1 + i/c^2$ (where $f \geqslant 1$). The energy tensor can be written

$$T^{(f)}_{\alpha\beta} = c^2 rf \; u_\alpha \; u_\beta - pg_{\alpha\beta} \qquad (1\text{-}3)$$

b) The proper temperature Θ and the specific entropy S of the fluid satisfy, as in classical thermodynamics, the differential relation:

$$\Theta \; dS = d\varepsilon + pdV = di - Vdp = c^2 df - Vdp \quad (\Theta > 0)$$

Therefore

$$c^2 df = Vdp + \Theta dS \quad (V > 0, \; \Theta > 0) \qquad (1\text{-}4)$$

The thermodynamical variable $\tau = fV$ (the dynamical volume) plays an important role in the relativistic frame work and corresponds to the classical specific volume. We consider τ as a given function of p and S, $\tau = \tau\,(p, S)$ which is an equation of state for the fluid.

c) Let Σ be a regular hypersurface, $\phi = 0$ its local equation. We put $\ell = d\phi$. The speed v^Σ of Σ with respect to the fluid is given by:

$$(v^\Sigma/c)^2 = (u^\alpha \; \ell_\alpha)^2 \; [(u^\alpha \; \ell_\alpha)^2 - \ell^\alpha \; \ell_\alpha]^{-1} \qquad (1\text{-}5)$$

v^Σ is $< c$ if and only if $\ell^\alpha \; \ell_\alpha < 0$ (Σ is timelike). If v is the speed of sound for the fluid, we obtain easily that $v^2/c^2 = 1/\gamma$), where:

$$c^2 \; \tau'_p = - V^2(\gamma - 1) \qquad (1\text{-}6)$$

We assume, in the following part, that $\tau(p, S)$ satisfies the following compressibility conditions:

$$\tau'_p < 0 \qquad\qquad \tau'_S > 0 \qquad (H_1)$$

and the convexity condition

$$\tau''_{p^2} > 0 \qquad (H_2)$$

These conditions are the relativistic extensions of the classical conditions of Hermann Weyl; $\tau'_p < 0$ expresses that $v < c$ (or $\gamma > 1$). It is possible to prove these conditions for a relativistic Maxwell-Boltzmann gas or for a quantum gas((H_1) for a relativistic Boltzmann gas by W. Israel, general case by Lucquiaud).

2 - Energy Tensor For Relativistic Magnetohydrodynamics

a) An electromagnetic field is represented by two skew 2-tensors H and G; the one H is the electric field-magnetic induction tensor. If $*H$ is the dual tensor, the spacelike vectors orthogonal to u:

$$e_\beta = u^\alpha H_{\alpha\beta} \qquad b_\beta = u^\alpha (*H)_{\alpha\beta}$$

are respectively the electric field and the magnetic induction with respect to the timelike direction defined by u. If μ (a given constant) is the magnetic permeability of the fluid, $b = \mu h$, where h is the magnetic field. In first approximation, the electric current is the sum of two terms:

$$\underline{J} = \nu\, \underline{u} + \sigma\, \underline{e}$$

where ν is the proper density of electric charge and σ the conductivity of the fluid.

b) Magnetodydrodynamics is here the study of the properties of a perfect fluid with an infinite conductivity $\sigma = \infty$; $\underline{J}$ and thus the product $\sigma\, \underline{e}$ being essentially finite, we have necessarily $\underline{e} = 0$. The electromagnetic field is reduced to the magnetic field h with respect to the fluid. (That is with respect to u). It is well-known that the energy tensor of this field is:

$$\tau_{\alpha\beta} = \mu\{|h|^2 (u_\alpha u_\beta - \tfrac{1}{2} g_{\alpha\beta}) - h_\alpha h_\beta\}$$

where $|h|^2 = -h^\rho h_\rho$ is strictly positive. The complete energy tensor of the system (fluid + field) is:

$$T_{\alpha\beta} = (c^2 rf + \mu|h|^2)\, u_\alpha u_\beta - q g_{\alpha\beta} - \mu\, h_\alpha h_\beta \qquad (2\text{-}1)$$

with

$$q = p + \tfrac{1}{2}\mu|h|^2$$

3 - The Main System for Relativistic Magnetohydrodynamics

a) The differential system of relativistic Magnetohydrodynamics is given by the following considerations. First, we assume that the matter density r (which corresponds to the specific number of particules) is conservative: there is no creation or annihilation of particles. If ∇ is the operator of covariant differentiation:

$$\nabla_\alpha (r\, u^\alpha) = 0 \tag{3-1}$$

Maxwell equations give only here:

$$\nabla_\alpha (h^\alpha\, u^\beta - u^\alpha\, h^\beta) = 0 \tag{3-2}$$

and the equations of relativistic dynamics are:

$$\nabla_\alpha\, T^{\alpha\beta} = 0 \tag{3-3}$$

(3-1), (3-2), (3-3) is the main system. A straightforward calculus shows that this system implies the so-called equation of adiabatic flow

$$u^\alpha\, \partial_\alpha S = 0 \tag{3-4}$$

b) The characteristic manifolds (or wave fronts) of our system are the tangential waves ($u^\alpha \ell_\alpha = 0$) generated by the streamlines, the magnetosonic waves, solutions of:

$$P(\ell) \equiv c^2 rf(\gamma-1)(u^\alpha \ell_\alpha)^4 + (c^2 rf + \mu|h|^2\gamma)(u^\alpha \ell_\alpha)^2 \ell^\beta \ell_\beta - \mu(h^\alpha \ell_\alpha)^2 \ell^\beta \ell_\beta = 0 \tag{3-5}$$

and the Alfven waves solutions of:

$$D(\ell) \equiv (c^2 rf + \mu|h|^2)\ (u^\alpha \ell_\alpha)^2 - \mu(h^\alpha \ell_\alpha)^2 = 0 \tag{3-6}$$

Under the assumption $\tau'_p < 0$ (or $v < c$), (3-5) and (3-6) define speeds v^{MS}, v^{MF}, v^A with the inequalities:

$$v^{MS} \leqslant \begin{matrix} v \\ v^A \end{matrix} \leqslant v^{MF} < c$$

If we put $\beta = \sqrt{(c^2 rf + \mu|h|^2)/\mu}$, (3-6) shows that the Alfven waves are generated by the integral curves of the two vector fields

$$\underline{A} = \beta\, \underline{u} + \underline{h} \qquad\qquad \underline{B} = \beta\, \underline{u} - \underline{h}$$

where $\underline{A}$ and $\underline{B}$ are timelike and oriented towards the future.

c) The mathematical analysis of the main system shows that this system is not strictly hyperbolic in the sense of Gårding-Leray, but that the differential operators which occur are products of strictly hyperbolic operators. It is possible to prove by different ways that the regular local Cauchy problem for our system admits an existence theorem and that the solution is geometrically unique. We see that there exists in this case a notion of domain of influence.

4 - General Shock Equations

We assume always that the $g_{\alpha\beta}$ and their first derivatives are continuous:

a) A state Y of the system (fluid + field) at a point x of V_4 is defined by the values of p, S, u, h (8 parameters). A magnetohydrodynamical shock wave is a solution of the main system in a weak sense, such that there exists a hypersurface Σ - the wave front - with the following conditions:

1) On both sides of Σ, the states are continuous functions of x and the main system is satisfied in the usual sense.

2) The variables defining the states are regularly discontinuous across Σ and, in the neighborhood of Σ, the main system is satisfied in the sense of the distributions.

We will show that under the compressibility conditions $\tau'_p < 0$, $\tau'_S > 0$, Σ is necessarily timelike. If we decompose $\underline{u}$ and $\underline{h}$ into a tangential component and a normal component with respect to Σ ($\phi = 0$; $\ell = d\phi$), we obtain

$$u^\beta = v^\beta + (u^\alpha \ell_\alpha / \ell^\alpha \ell_\alpha)\ell^\beta, \quad h^\beta = t^\beta + (h^\alpha \ell_\alpha / \ell^\alpha \ell_\alpha)\ell^\beta \quad (v^\beta \ell_\beta = t^\beta \ell_\beta = 0)$$

We denote as y_0 the state at $x \in \Sigma$ before the shock, as y_1 the state after the shock; [Q] is the discontinuity $Q_1 - Q_0$ of a quantity across Σ. It follows from the main system, by a classical argument, the general shock equations:

$$[r\ u^\alpha]\ell_\alpha = 0 \quad [h^\alpha\ u^\beta - u^\alpha\ h^\beta]\ell_\alpha = 0 \quad [T^{\alpha\beta}]\ell_\alpha = 0 \qquad (4-1)$$

we add to (4-1) the assumption

$$[S] > 0 \qquad (4-2)$$

which is the corresponding formulation of the so-called Clausius-Duhem inequality. (4-1) expresses the invariance of the scalar:

$$a(Y) = ru^{\alpha} \ell_{\alpha}$$

the invariance of the vector tangent to Σ:

$$v^{\beta}(Y) = (h^{\alpha}\ell_{\alpha})u^{\beta} - (a/r)h^{\beta}$$

and the invariance of the vector:

$$w^{\beta}(Y) = (c^2\tau + \mu|h|^2/r^2)aru^{\beta} - q\ell^{\beta} - \mu(h^{\alpha}\ell_{\alpha})\ h^{\beta}$$

If $a = 0$ (tangential shock), we have $u_0^{\alpha}\ \ell_{\alpha} = u_1^{\alpha}\ \ell_{\alpha} = 0$ and Σ has a null speed with respect to the fluid in the two states. The study of this case is trivial and we assume $a \neq 0$ in the following part.

b) If we decompose w^{β} into a tangential and a normal component, we obtain a scalar e and a tangent vector which are invariants of the shock. There different invariants give the following results: first the two thermodynamical variables and the three scalars $|h|^2$, $u^{\alpha}\ell_{\alpha}$, $h^{\alpha}\ell_{\alpha}$ satisfy the five scalar relations:

$$r_1\ u_1^{\alpha}\ \ell_{\alpha} = r_0\ u_0^{\alpha}\ \ell_{\alpha} = a \tag{4-3}$$

$$f_1\ h_1^{\alpha}\ \ell_{\alpha} = f_0\ h_0^{\alpha}\ \ell_{\alpha} = b \tag{4-4}$$

$$(h_1^{\alpha}\ \ell_{\alpha})^2/a^2 - |h_1|^2/r_1^2 = (h_0^{\alpha}\ \ell_{\alpha})^2/a^2 - |h_0|^2/r_0^2 = H \tag{4-5}$$

$$q_1 - [c^2\ a^2/(\ell^{\alpha}\ell_{\alpha})]\tau_1 = q_0 - [c^2\ a^2/(\ell^{\alpha}\ell_{\alpha})]\tau_0 = e \tag{4-6}$$

$$\chi_1\ \alpha_1^2 = \chi_0\ \alpha_0^2 = L \tag{4-7}$$

where:

$$\alpha = c^2\tau - \mu\#\!\!\# = D(\ell)/a^2\ , \quad \chi = |h|^2 + a^2\#\!\!\#/(\ell^{\alpha}\ell_{\alpha}) \tag{4-8}$$

$\alpha = 0$ expresses that Σ is an Alfven wave front for the state Y. The quantities (4-8) have the following properties: $(\ell^{\alpha}\ell_{\alpha})\ \chi \leqslant 0$ and $(\ell^{\alpha}\ell_{\alpha})\chi = 0$ if and only if ℓ is in the 2-plane ($\underline{u}$, $\underline{h}$); $\ell^{\alpha}\ell_{\alpha} \geqslant 0$ implies $H \leqslant 0$ and thus $\alpha > 0$.

Secondly, the tangential components of the velocity and of the magnetic field satisfy

$$(h_1^{\alpha}\ \ell_{\alpha})v_1^{\beta} - (u_1^{\alpha}\ \ell_{\alpha})t_1^{\beta} = (h_0^{\alpha}\ \ell_{\alpha})v_0^{\beta} - (u_0^{\alpha}\ \ell_{\alpha})t_0^{\beta} \tag{4-9}$$

$$(c^2 r_1 f_1 + \mu |h_1|^2)(u_1^\alpha \ell_\alpha) v_1^\beta - \mu (h_1^\alpha \ell_\alpha) t_1^\beta = (c^2 r_0 f_0 + \mu |h_0|^2)(u_0^\alpha \ell_\alpha) v_0^\beta - \mu (h_0^\alpha \ell_\alpha) t_0^\beta \tag{4-10}$$

The determinant of the left members of (4-9), (4-10) with respect to v_1^β, t_1^β can be written

$$D_1(\ell) = (c^2 r_1 f_1 + \mu |h_1|^2)(u_1^\alpha \ell_\alpha)^2 - \mu (h_1^\alpha \ell_\alpha)^2 = a^2 \alpha_1$$

If $\alpha_1 \neq 0$, (4-9), (4-10) give v_1^β, t_1^β in terms of quantities which satisfy the five scalar equations.

c) Suppose $\alpha_1 = 0$; Σ is then a timelike Alfven wave front after the shock. It follows from (4-7) either $\alpha_0 = 0$ or $\chi_0 = 0$.

An <u>Alfven shock</u> is a shock such that $\alpha_0 = \alpha_1 = 0$. An Alfven shock can be of the type A or B, exactly as an Alfven wave. It is easy to prove that if $\tau_p' < 0$, the two thermodynamical variables and the three scalars $|h|^2$, $u^\alpha \ell_\alpha$, $h^\alpha \ell_\alpha$ are invariant across the shock. The direction of the tangential magnetic field after the shock is undetermined, but determines the direction of the tangential velocity according to the following condition: the vector A (resp. B) is invariant under an Alfven shock of type A (resp. B).

Moreover, it is possible to prove the following result: A shock wave such that $\alpha_0 \alpha_1 = 0$ is compatible with usual Alfven waves if and only if it is an Alfven shock ($\alpha_0 = \alpha_1 = 0$). The cases $\alpha_1 = 0$, $\chi_0 = 0$, $\alpha_0 \neq 0$ and symmetrically $\alpha_0 = 0$, $\chi_1 = 0$, $\alpha_1 \neq 0$ are forbidden as unstable.

5 - <u>Hugionot function for Magnetohydrodynamics</u>

a) An initial state Y_0 of the system (fluid + field) being given at $x \in \Sigma$, we consider in the following part the set of the possible states Y satisfying the two conditions:

$$H(Y) = H(Y_0) = H \qquad L(Y) = L(Y_0) = L \tag{5-1}$$

so that

$$\chi = L\,(c^2\tau - \mu H)^{-2} \qquad (\tau \neq \mu H/c^2)$$

It is also convenient to introduce the variable:

$$\bar{q} = p + \frac{1}{2}\mu\chi$$

and to substitute for the equation (4-6), the equation:

$$\bar{q}_1 - \bar{q}_0 = c^2 a^2 (\tau_1 - \tau_0)/(\ell^\alpha \ell_\alpha) \tag{5-2}$$

A thermodynamical state (τ, p) of the fluid defines, under the conditions (5-1) a point Z of the plane $(\tau, \bar{q})$. This plane plays in the following part an important role. We have between the variables τ, S and $\bar{q}$ the functional relationship:

$$\bar{q} = \frac{1}{2}\mu L(c^2\tau - \mu H)^{-2} + p(\tau, S) \qquad (5-3)$$

where $p = p(\tau, S)$ is obtained by inversion of $\tau = \tau(p, S)$.

b) Let us introduce the function $\mathcal{H}(Z_0, Z)$ which is a function of $Z = (\tau, \bar{q})$ for a given initial point Z_0:

$$\mathcal{H}(Z_0, Z) = c^2(f^2 - f_0^2) - (\tau+\tau_0)(p-p_0) + (\tau-\tau_0)\cdot\frac{1}{2}\mu(\chi+\chi_0-2\chi_0\alpha_0/\alpha) \qquad (5-4)$$

I call $\mathcal{H}(Z_0, Z)$ the Hugoniot function for magnetohydrodynamics. Clearly $\mathcal{H}(Z_0, Z_0) = 0$ and I have proved that we can substitute for (4-4) the <u>Hugoniot equation</u> $\mathcal{H}(Z_0, Z_1) = 0$ in the system of the five scalar relations. Differentiating $\mathcal{H}$ in Z, we obtain according to (1-4)

$$d\mathcal{H} = 2f\,\Theta\, dS + (\tau - \tau_0)d\bar{q} - (\bar{q} - \bar{q}_0)d\tau \qquad (5-5)$$

In particular, if we differentiate along a straight line Δ of the plane $(\tau, \bar{q})$ issued from Z_0, we have

$$d\mathcal{H} = 2f\,\Theta\, dS \qquad (5-6)$$

Moreover, a straightforward calculus gives that along Δ defined by (5-2):

$$\tau'_S\,\alpha dS = P(\ell)/(a^2\ell^\alpha\ell_\alpha)d\tau = P(\ell)c^{-2}a^{-4}\,d\bar{q} \qquad (5-7)$$

The relation (5-7) is important and is absent of the classical papers of Magnetohydrodynamics.

6 - <u>Orientation of the Shock Wave Fronts</u>

We assume the compressibility conditions (H_1). It is easy to show that, under these conditions, $\ell^\alpha\ell_\alpha \geqslant 0$ implies $P(\ell) > 0$. Moreover α is > 0 and, according to (5-7), $dS/d\bar{q}$ is > 0 along the ray (Z_0, Z_1) of the plane $(\tau, \bar{q})$. But since $\mathcal{H}(Z_0, Z_0) = \mathcal{H}(Z_0, Z_1) = 0$, the function $\mathcal{H}(Z_0, Z)$ is stationary at one point Z_s at least of (Z_0, Z_1) and the same thing is true for S, according to (5-6). We obtain a contradiction and $\ell^\alpha\ell_\alpha$ is < 0. We have

Theorem - If the fluid satisfies the conditions (H_1), the shock wave fronts Σ are necessarily timelike. If v_0^Σ and v_1^Σ are the speeds of Σ with respect to the fluid before and after the shock, we have $v_0^\Sigma < c$, $v_1^\Sigma < c$.

Now we know that $\chi = k^2 \geqslant 0$ and, for a convenient choice of the signs, (4-7) can be written:

$$k_1 \alpha_1 = k_0 \alpha_0$$

7 - Thermodynamics of the Shocks

The Hugoniot equation can be written:

$$\mathcal{H}(z_0, z_1) \equiv c^2(f_1^2 - f_0^2) - (\tau_0 + \tau_1)(p_1 - p_0) + (\tau_1 - \tau_0) \cdot \frac{1}{2} \mu (k_1 - k_0)^2 = 0 \qquad (7\text{-}1)$$

According to (1-4), we have:

$$c^2 f_p' = V > 0 \qquad c^2 f_S' = \Theta > 0$$

and differentiating:

$$(\partial/\partial p)(c^2 f^2) = 2\tau \qquad (7\text{-}2)$$

According to the Clausius-Duhem inequality (4-2), we have $S_0 < S_1$ for each point $x \in \Sigma$. It is easy to deduce from the compressibility conditions that $S_1 = S_0$ implies that we have an Alfven shock at x. It follows from (7-1), (7-2) and the compressibility conditions.

Theorem - Consider an effective shock which is not an Alfven shock we have under the compressibility conditions (H_1) and (H_2):

$$p_1 > p_0 \qquad f_1 > f_0 \qquad \tau_1 < \tau_0 \qquad (7\text{-}3)$$

Assuming $S_1 > S_0$, I will prove, for example, that $p_1 > p_0$. If we suppose $p_1 \leqslant p_0$, we have integrating (7-2) for $S = S_0$:

$$c^2\{f^2(p_0, S_0) - f^2(p_1, S_0)\} = 2\int_{p_1}^{p_0} \tau(p, S_0)\, dp \leqslant (p_0 - p_1)\{\tau(p_0, S_0) + \tau(p_1, S_0)\}$$

according to the convexity condition (H_2). It follows a fortiori:

$$c^2(f_1^2 - f_0^2) - (\tau_0 + \tau_1)(p_1 - p_0) > 0$$

we deduce from (7-1) that $\tau_1 < \tau_0$, which contradicts $p_1 \lesssim p_0$, $S_1 > S_0$ and (H_1). The other proofs are similar.

Therefore we have $\alpha_1 < \alpha_0$. It is possible to prove that a shock wave which is not an Alfven shock wave is compatible with the usual Alfven waves if and only if $\alpha_0 \alpha_1 > 0$. We obtain two types of shocks: the <u>slow shocks</u> for which $\alpha_1 < \alpha_0 < 0$ and the <u>fast shocks</u> for which $0 < \alpha_1 < \alpha_0$. We deduce easily from the relation $k_1^2 \alpha_1^2 = k_0^2 \alpha_0^2$, that for a slow shock, the magnitude field decreases and for a fast shock, it increases across the shock.

8 - Isentropic Curves and Speeds of the Shock Waves

a) An initial state Y_0 being given, we consider the set of the states Y satisfying:

$$H(Y) = H(Y_0) = H \qquad k\alpha = k_0 \alpha_0 \text{ (with } \alpha \alpha_0 > 0)$$

The straightline $\tau = \mu H/c^2$ is forbidden in the plane $(\tau, \bar{q})$. In this plane, the isentropic curve $\mathcal{S}$ corresponding to the value S of the entropy is defined by:

$$\bar{q} = \frac{1}{2} \mu k_0^2 \alpha_0^2 (c^2\tau - \mu H)^{-2} + p(\tau, S) \tag{8-1}$$

Differentiating twice along $\mathcal{S}$, we obtain

$$(d^2\bar{q}/d\tau^2)_{\mathcal{S}} = - M/\tau'^3_p \tag{8-2}$$

where:

$$M = \tau''_{p^2} - 3 c^4 \mu k^2 \tau'^3_p \alpha^{-2} > 0 \tag{8-3}$$

according to the conditions (H_1), (H_2). The curve $\mathcal{S}$ is therefore <u>convex</u>.

b) Let Δ be a ray of the plane $(\tau, \bar{q})$ issued from Z_0:

$$\bar{q} - \bar{q}_0 = m(\tau - \tau_0) \qquad d\bar{q} = m d\tau$$

we have the following lemma:

<u>Lemma - Under the conditions (H_1), (H_2), we have at each point $Z_{\mathcal{S}}$ of Δ where S is stationary</u>.

$$(d^2 S/d\tau^2)_{Z_{\mathcal{S}}} < 0 \tag{8-4}$$

In fact, differentiating (8-1) twice along Δ, we obtain:

$$(d^2S/d\tau^2)_{Z_{\mathcal{S}}} = \{(\tau_p'/\tau_S')(d^2\bar{q}/d\tau^2)_{\mathcal{S}}\}_{Z_{\mathcal{S}}} = -\{M/(\tau_p'^2\tau_S')\}_{Z_{\mathcal{S}}} < 0$$

c) Consider a shock such that $Z_0 \to Z_1$. It follows from the lemma that the point Z_S of the ray (Z_0, Z_1), where $S(\tau)$ is stationary, is unique and corresponds to a maximum for S. For a fast shock ($0 < \alpha_1 < \alpha_0$) we have, according to (5-7):

$$(dS/d\tau)_0 < 0 \rightarrow P(\ell)_0 > 0, \quad (dS/d\tau)_1 > 0 \rightarrow P(\ell)_1 < 0$$

If we interpret the signs of α and of $P(\ell)$, we obtain

Theorem - Under the conditions (H_1), (H_2), the speeds v_0^Σ and v_1^Σ of a shock front Σ satisfy the inequalities

1) for a fast shock

$$v_0^{MS} < v_0^A < v_0^{MF} < v_0^\Sigma \qquad v_1^{MS} < v_1^A < v_1^\Sigma < v_1^{MF}$$

2) for a slow shock

$$v_0^{MS} < v_0^\Sigma < v_0^A < v_0^{MF} \qquad v_1^\Sigma < v_1^{MS} < v_1^A < v_1^{MF}$$

d) Let us consider in the plane $(\tau, \bar{q})$, the Hugoniot curve defined by $\mathcal{H}(Z_0, Z) = 0$. It follows from (5-5):

$$2f\,\Theta\,(dS/d\tau)_{\mathcal{H}} + (\tau - \tau_0)\,(d\bar{q}/d\tau)_{\mathcal{H}} - (\bar{q} - \bar{q}_0) = 0 \tag{8-5}$$

and $(dS/d\tau)_{\mathcal{H}} = 0$ at Z_0. Differentiating (8-5), we obtain

$$2f\,\Theta\,(d^2S/d\tau^2)_{\mathcal{H}} + 2(d(f\Theta)/d\tau)_{\mathcal{H}}\,(dS/d\tau)_{\mathcal{H}} + (\tau-\tau_0)(d^2\bar{q}/d\tau^2)_{\mathcal{H}} = 0 \tag{8-6}$$

and $(dS/d\tau)_{\mathcal{H}} = 0$ at Z_0. Thus the Hugoniot curve and the isentropic curve corresponding to $S = S_0$ have a contact of second order at Z_0 and we have at this point:

$$(d^2\bar{q}/d\tau^2)_{\mathcal{H}} = (d^2\bar{q}/d\tau^2)_{\mathcal{S}} = -(M/\tau_p'^3)_{Z_0}$$

Differentiating once again (8-6), we obtain at Z_0:

$$(d^3S/d\tau^3)_{\mathcal{H}} = [M/(2f\,\Theta\,\tau_p'^3)]_{Z_0}$$

It follows that, in the neighborhood of Z_0, we have along $\mathcal{H}$:

$$S - S_0 = [M/(12f\Theta)]_{Z_0} (p - p_0)^3$$

we see that, across a weak shock, the increase of the entropy is of third order with respect to the shock strength $(p_1 - p_0)$.

9 - Hugoniot Curve and Solution of the Shock Equations

Solving $\tau = \tau(p, S)$ for S, we obtain a function $S = S(p, \tau)$. We suppose that the function $S(p, \tau)$ is such that the assumptions (H_1), (H_2) are satisfied.

I will prove that if $v_0^A < v_0^{MF} < v_0^\Sigma$ (resp. $v_0^{MS} < v_0^\Sigma < v_0^A$), the general shock equations have one and only one solution such that $v_1^A < v_1^\Sigma$ (resp. $v_1^\Sigma < v_1^A$).

a) Consider, in the plane $(\tau, \bar{q})$, the branch of the isentropic curve $\mathcal{S}$ corresponding to $S = S_0$, such that $\tau < \tau_0$, with $\alpha\, \alpha_0 > 0$. We will show that if $\alpha_0\, P(\ell)_0 > 0$, the equation of shock admit an unique non trivial solution such that $\alpha_1\, \alpha_0 > 0$.

We have along $\mathcal{S}$, according to (5-5)

$$(d\mathcal{H}/d\tau)_{\mathcal{S}} = (\tau - \tau_0)\ (d\bar{q}/d\tau)_{\mathcal{S}} - (\bar{q} - \bar{q}_0)$$

Differentiating a second time, we have:

$$(d^2\mathcal{H}/d\tau^2)_{\mathcal{S}} = (\tau - \tau_0)\ (d^2\bar{q}/d\tau^2)_{\mathcal{S}}$$

that is, according to (8-2):

$$(d^2\mathcal{H}/d\tau^2)_{\mathcal{S}} = -(\tau - \tau_0)\ M\ \tau_p'^{-3} < 0$$

Therefore, on the considered arc of $\mathcal{S}$:

$$(d\mathcal{H}/d\tau)_{\mathcal{S}} > 0 \qquad \mathcal{H}(Z_0, Z) < 0$$

b) Let Δ be a straight line issued from Z_0, with slope m. It follows from (5-6) that along Δ

$$(d\mathcal{H}/d\tau)_\Delta = 2f\ \Theta\ (dS/d\tau)_\Delta \tag{9-1}$$

Let m_0 be the slope of $\mathcal{S}$ at Z_0. It follows from our global assumptions that $\bar{q}$ can be arbitrarily large along our part of $\mathcal{S}$; $\mathcal{S}$ being convex, we see that if $m < m_0$, Δ intersects the considered branch of $\mathcal{S}$ in one and only one point $Z_A \neq Z_0$ and for this point

$$\mathcal{H}(Z_0, Z_A) < 0 \qquad (9\text{-}2)$$

But $S(Z_0) = S(Z_A) = S_o$ and $S(\tau)$, along Δ, is necessarily stationary between Z_0 and Z_A, at a point Z_S necessarily unique, according to the lemma, point which is a strict maximum for S along Δ.

The assumption $\alpha_0 \; P(\ell)_0 > 0$ gives $(\frac{dS}{d\tau})_\Delta \; (Z_0) < 0$ and so $(\frac{d\mathcal{H}}{d\tau})_\Delta \; (Z_0) < 0$. When Z goes from Z_0 to Z_A along Δ, $\mathcal{H}(Z_0, Z)$ begins being positive and it follows from (9-2) that there exists one at least point Z_1 on Δ, between Z_0 and Z_A, such that $\mathcal{H}(Z_0, Z_1) = 0$.

We know that, along Δ, S is increasing from Z_0 to Z_S, has a maximum at Z_S and is decreasing after this point; $\mathcal{H}$ has the same behaviour and thus the point Z_1 is necessarily unique.

Therefore <u>to each number $m < m_0$ corresponds an unique point Z_1 of the Hugoniot curve of Z_0, with $\tau_1 < \tau_0$ (and $\alpha_1 \; \alpha_0 > 0$) such that $m = (\bar{q}_1 - \bar{q}_0)/(\tau_1 - \tau_0)$</u> .

We note that we have, along the Hugoniot curve, according to (5-5):

$$2f \; \Theta \; dS + (\tau - \tau_0)^2 \; dm = 0$$

and that therefore <u>S is increasing when m is decreasing</u>.

For $m = c^2 a^2/(\ell^\alpha \ell_\alpha) < 0$, z_1 satisfies (5-2) and $\mathcal{H}(Z_0, Z_1) = 0$; Z_1 being known, τ_1 and p_1 are known and f_1 is given for example by the Hugoniot equation. We know χ_1, and so $|h_1|^2$; we deduce from (4-4) the value of $h_1^\alpha \; \ell_\alpha$, from (4-3) the value of $u_1^\alpha \; \ell_\alpha$. The shock equations haves thus an unique non trivial solution such that $\alpha_0 \; \alpha_1 > 0$.

It is easy to verify that the condition

$$c^2 a^2/(\ell^\alpha \ell_\alpha) < m_0$$

is equivalent to $\alpha_0 \; P(\ell)_0 > 0$. We have proved:

<u>Theorem - If the function S (p, τ) satisfies the compressibility conditions (H_1), (H_2) for $\tau \leqslant \tau_0$ and for arbitrarily large values of p, to each state Y_0 verifying $\alpha_0 \; P(\ell)_0 > 0$ (that is $v_0^{MF} < v_0^\Sigma$ for a fast shock and $v_0^{MS} < v_0^\Sigma < v_0^A$ for a slow shock) corresponds a non trivial unique solution of the shock equation such that $\alpha_0 \; \alpha_1 > 0$.</u>

We see that, though general shock equations (4-3) to (4-7) appear relatively complicated from the mathematical point of view, there equations are however good equations, according to the previous theorem.

REFERENCES

[1] F. Hoffman and E. Teller, Phys. Rev. t80 (1950) p. 692.
[2] A. H. Taub, Arch. Rat. Mech. Anal. 3 (1959) p. 312.
[3] Y. Choquet-Bruhat, Astron. Acta. 6 (1960) p. 354.
[4] W. Israel, Proc. Roy. Soc. A259 (1960) 129.
[5] A. Lichnerowicz, Relativistic Hydrodynamics and Magnetohydrodynamics (1967) Benjamin, New York.
[6] A. Lichnerowicz, Ann. Int. Poincare 7 (1967) 271; and Comm. Math. Phys. 12 (1969) 145.
[7] A. Lichnerowicz, Relativistic fluid Dynamics C.I.M.E. (1970) Cremonese, Roma (1971) p. 89-203.
[8] J. Lucquiaud, C. R. Acad. Sci. Paris 270 (1970) p. 85 et Sur l'équation de Boltzmann relativiste Thèse Paris (1975).
[9] A. Lichnerowicz, Phys. Scriptat 2 (1970) 221-225.

ON RELATIVISTIC MAGNETOHYDRODYNAMICS PROCESSES IN THE FIELDS OF BLACK HOLES

Remo Ruffini *

Institute for Advanced Study, Princeton, New Jersey, U.S.A.

and

Joseph Henry Physical Laboratories, Princeton, New Jersey, U.S.A.

ABSTRACT: Some experimental results giving evidence for the observations both of neutron stars and black holes in our galaxy and some classical theoretical results in black hole physics are summarized. Recent advances in relativistic magnetodydrodynamics are then reviewed with special emphasis on the analysis of simple models leading to a better understanding of the properties of the magnetosphere of a black hole. The stability of the magnetosphere against processes of vacuum polarization is discussed as well as its relevance for possible sources of detectable signals of gravitational radiation.

1. INTRODUCTION

It has been known since Friedman [1] and Lemaitre [2] presented analytic solutions of the Einstein field equations and Hubble [3] presented observations of the recession of galaxies and of the expansion of the Universe, that large effects of general relativity could be found in nature [4]. Since there is only one Universe, however, the example of relativistic cosmology, important as it is, scarsely seemed a solid ground from which to start further experimental verifications of the many possible and complex predictions of Einstein's theory of gravitation. Accordingly another entire branch of activity grew up devoted to testing general relativity by means of precision measurements in the weak gravitational field of the solar system [5].

General relativistic effects in the solar system are exceedingly small. It was therefore a natural temptation to suppose that they would be equally small anywhere in our galaxy, and if the only sizable effects of general relativity were those occurring on a cosmological scale, the opinion became widespread that one might as well neglect general relativity in all real physical situations. Einstein's theory, though beautiful and elegant, was then in danger of being neglected to purely philosophical discussions.

The theoretical works of Landau [6], Chandrasekhar [7], Oppenheimer and Volkoff [8] and Oppenheimer and Snyder [9] clearly pointed to the possible existence of gravitationally collapsed stellar objects requiring general relativity for

* Alfred P. Sloan Fellow

their description. Bade and Zwicky [10] were the first to understand the full relevance of these researches for astrophysics; they postulated that supernova were the transition from a normal star into a neutron star. In the meantime it became more and more clear [11] that the existence of a critical mass, theoretically predicted by Chandrasekhar [7] for the white dwarf stars, and by Oppenheimer and Volkoff [8] for the neutron stars, was the single most important factor in determining whether gravitational collapse to a "black hole" could occur [9].

In 1968 the discovery of Pulsars, and especially the discovery of the Pulsar NP0532 still pulsating in the center of the remnant of the supernova explosion of 1054, gave to all these theoretical works the first spectacular confirmation: the identification of pulsars with rotating neutron stars was compelled by much experimental evidence [12].

It became very clear, therefore, that unless the concept of critical mass against gravitational collapse could be shown to be unphysical then the existence of black holes in nature had to be considered unavoidable. Much work was spent in making the concept of critical mass of a neutron star free of any theoretical assumptions on the behaviour of matter at supranuclear densities (still largely unknown) [13]. The general conclusion was reached that within the framework of general relativity a neutron star could never have a mass larger than 3.2 $M_\odot$.

The fundamental experimental discovery which allowed at once to greatly increase our knowledge on neutron stars and also to identify the first black hole came with the discovery of binary x-ray sources by the Uhuru satellite in 1971 [14]. In close binary systems formed by a normal star and by either a neutron star or a black hole matter transferred from the normal star into the gravitationally collapsed companion will be so compressed and heated-up that radiation in the x-ray region will be emitted. This most important prediction had been advanced by Shklovsky [15], Zel'dovich and Novikov [16] and Hayakawa and Matsuoka [17]. The observations by Giacconi and his co-workers clearly demonstrated that this kind of binary system had indeed been discovered [12], [14].

A possible classification of these binary x-ray sources was proposed [18]: In one class we considered sources with: (1) x-ray emitted in sharp and regular pulses with time scales varying from seconds to minutes. (2) x-rays fluxes 10^{36} ergs/sec $\lesssim dE/dt \lesssim 10^{38}$ ergs/sec $\simeq 10^5\ L_\odot$. These sources were identified with accreting neutron stars: (a) The sources of energy would then simply be given by gravitational binding energy which is expected to be as large as 0.1 mc^2 at the neutron star surface. A very reasonable accretion rate 10^{14} gr/sec $\lesssim dM/dt \lesssim 10^{17}$ gr/sec could then explain the observed intensity in the x-ray flux. (b) The pulsational period could be explained by the rotational period of the neutron star transmitted to the accreting magnetosphere by a non-axially-symmetric magnetic field. We know quite a large number of sources belonging to this class:

Cen X3 with an intrinsic pulsational period of $P_o \simeq 4.84$ sec a binary period of 2.09 days and a neutron star mass $0.6 \lesssim Mx/M_\odot \lesssim 1.8$ [19].

Hercules X1 with an intrinsic pulsational period of $P_o \simeq 1.24$ sec a binary period of 1.7 days and a neutron star mass $M_x \sim 1.3\ M_\odot$ [20].

SMCX-1 with an intrinsic pulsational period of $P_o \simeq 0.716$ sec a binary period 3.893 days [21] and a neutron star mass $M_x \approx 2\ M_\odot$ [18] but still largely undetermined.

3U 0900-40 or Vela X1 with an intrinsic pulsational period of $P_o \simeq 283$ sec, a binary period of 8.95 days and a neutron star mass $1.3 \lesssim M_x/M_\odot \lesssim 2.2$ [19].

Very likely three additional sources 3U 1223-62 with an intrinsic pulsational period of $P_o \simeq 11.64$ min, 3U 1728-24 with a $P_o \simeq 4.31$ min, 3U 1813-14 with a period $P_o \simeq 31.9$ min [22] are members of the same class.

The fact that all the masses of the pulsating sources which have been measured are indeed smaller than $3.2\ M_\odot$ is extremely satisfactory in light of the theorem on the maximum mass of a neutron star [13].

In a different class we considered sources with:
(1) x-ray luminosities of $10^{36} \lesssim dE/dt \lesssim 10^{38}$ ergs/sec.
(2) x-ray fluxes without any regular pulsation, with burst-like structures down to a few msec or, any way, with no sharply repetitive features.
(3) Mass larger than $3.2\ M_\odot$.

These sources were identified with accreting black holes:
(a) The large binding energy that a particle can have near the horizon (up to 100% or to 42% as a function of the electromagnetic structure or rotation of the black hole, see §2) allows one to explain the energy source of the x-ray flux as originating from the gravitational energy of matter accreting into the black hole at a rate $dM/dt \sim 10^{14} - 10^{17}$ gr/sec.

(b) The absence of regular pulsations agrees with the results that only stationary and, therefore, solutions endowed with axial symmetry around the rotation axis, can be found for a black hole (see §2). The existence of burst-like structure on time scales of msec then agrees, in order of magnitudes, with the characteristic time scale $\tau \sim GM/c^3$ expected for matter spiralling around a black hole (see §2).

(c) The mass estimate larger then the absolute upper limit of the neutron star mass ($3.2\ M_\odot$) forces the identification of the collapsed object with a black hole. The best example of such binary x-ray sourcesis still represented by Cygnus X1.

This summarizes the situation for neutron stars, and for black holes of $\sim 10\ M_\odot$, however, much interest is being currently devoted to the possibility of detecting black holes of much larger size (in the range $10^2 - 10^{10}\ M_\odot$). There has been for a long time a proposal to look for black holes in the core of globular clusters [23] or in the galactic cores [24]. Since the early appearance of the Uhuru catalogue, Gursky [25] has repeatedly stressed that an abnormally large number of x-ray sources were coinciding with globular clusters. This general idea had been reproposed by Silk and Arons [26] and by Bahcall and Ostriker [27]. The real novel result has come from the very important observations by Heise, Brinkman, den Boggende, Parsignault, Grindlay and Gursky [28-30]. There is clear evidence of "activity" from the core of one of these globular clusters: sharp bursts have been observed by the Netherland's Astronomical Satellite from the source 3U 1820-30 which is located in the globular cluster NGC6624. The intensity of the x-ray source (usually $\sim 2 \times 10^{38}$ ergs/sec) increases by a factor of 30 in less then a second and then decays exponentially over the next 10 seconds. Observations from the SAS-3 satellite of the same source have indicated that the bursts are repetitive with an approximate time constant of 0.182 days. There is evidence for the existence of many more of these "Bursters" from observations from Uhuru, SAS-3, the Vela Satellites and the Dutch Satellite [30].

Some possible steps to build a model for these sources are given in §3, §4 and §5. What we would like to stress here, however, are some of the major difficulties in establishing the presence of a black hole in the core of a globular cluster and to indicate some possible way by which this identification could gain credibility:

(1) Now, in sharp contrast with Cygnus X1, a direct measurement of the mass of the collapsed object cannot be made exploiting the Keplerian laws of a binary system, since only a single massive object of $10^2 \sim 10^3\ M_{\odot}$ is expected in the core of a globular cluster.

(2) Since a mass of $10^3\ M_{\odot}$ is assumed for the collapsed object there is no serious constraint in explaining the energy requirement of the x-ray source of $\sim 10^{38}$ ergs/sec.

(3) The compactness of the emission region, was made mandatory in the case of Cygnus X1 both by the energy requirement (large gravitational binding energy) and by the very fast time variability $\tau \sim GM/c^3$, in the x-ray spectrum. In this case, however, the requirement of compactness is not yet essential and can only follow from a model based on a black hole to explain both the intensity and the spectrum of the emitted x-rays. Unless the uniqueness of this model is proved, however, the identification of a black hole in the core of a globular cluster is not compelling.

The search for very large black holes of $\sim 10^{10}\ M_{\odot}$ has mainly been centered on the core of the emitting region of radio lobes in extragalactic radio sources. Three basic requirements must be fulfilled by any model of these radio sources [31]:

(1) The energy contained in the radio emitting lobes is estimated to be of the order of 10^{61} ergs and originating in the central nucleus.

(2) The directionality of the emission processes of the pairs of relativistic plasmoids ejected from the nucleus.

(3) The repetitive character of ejection phenomenas which can occur as often as a few million years apart, keeping the same directionality.

These three requirements can be fulfilled by a model based on a rotating-magnetized black hole; for the energy requirement see e.g. the mass formula given by Eq.(8) in §2 and for a possible mechanism of extraction of rotational energy and storage in the magnetosphere see §3. More work is needed, however, in building a detailed model.

The many experimental results being received daily from x-ray satellites (like Uhuru, SAS3, Copernicus, Ariel, ANS and the Vela satellites) and the many ground observations on these x-ray sources made from optical and radio telescopes, have forced a large improvement in the sophistication of theoretical models.

We would like to reach an understanding of all the many variabilities in the spectra and in the intensity of these x-ray sources and to indeed allow to use them as a tool for probing some of the most novel and profound predictions and effects of general relativity in the strong field limit. I will report in my lecture some of the progress which has been made in the fundamentals of the "model making" process in binary x-ray sources with special emphasis on the physical processes occurring near the horizon of the gravitationally collapsed object. I will especially emphasize progress being made in the analysis of the relativistic magnetohydrodynamic treatment in the field of black holes made mandatory by the very high accretion rate observed in some of these sources. It is worthwhile to emphasize that an accretion rate of $\sim 10^{17}$ gr/sec or ten times the mass of Mont Blanc per second in a region of few kilometers in radius, is such that complex magnetohydrodynamical processes should be expected and predictions departing drastically from test particles approximation in the field of black holes will be needed. For completeness, however, some of these results on test particle considerations and a few classic results on black hole physics will be summarized.

2. BASIC RESULTS OF BLACK HOLE PHYSICS

Since the introduction by Kerr [32] of an exact solution of the Einstein equations and its generalization by Newman <u>et al.</u> [33] to the set of Einstein-Maxwell equations, it has usually been assumed that the metric

$$ds^2 = \Sigma\Delta^{-1}dr^2 + \Sigma\, d\theta^2 + \Sigma^{-1}\sin^2\theta[a dt - (r^2 + a^2)d\phi]^2$$
$$- \Sigma^{-1}\Delta[dt - a\sin^2\theta d\phi]^2 \qquad (1)$$

with $\Delta = r^2 + 2Mr + a^2 + Q^2$, $\Sigma = r^2 + a^2\cos^2\theta$ (here, and in the following, we choose $G = c = 1$ and Greek indices vary from 0 to 3) with the associated electromagnetic field tensor

$$F = 2Q\,\Sigma^{-2}(r^2 - a^2\cos^2\theta)\, dr\wedge(dt - a\sin^2\theta d\phi)$$
$$- 4Q\,\Sigma^{-2}\, ar\cos\theta\sin\theta d\theta\wedge[a dt - (r^2 + a^2)d\phi] \qquad (2)$$

should be expected to describe the most general stationary electrovacuum solution endowed with a regular horizon and asymptotically flat at infinity ("black hole"). The advances made toward the proof of the "uniqueness" of the Kerr-Newman solution have been summarized by Carter [34].

Here, the metric has been written in Schwarzschild-like coordinates; M represents the total mass as measured from infinity, $a = L/M$ is the angular momentum per unit mass, and Q equals the charge of the object. At any point outside the horizon, $r > M + (M^2 - Q^2 - a^2)^{\frac{1}{2}}$, it is possible to introduce an orthonormal tetrad and then, with respect to this tetrad, define the components E and B of the electromagnetic field [35]. It is of particular interest to consider the tetrad [36]

$$\omega^{(0)} = (\Delta/\Sigma)^{\frac{1}{2}}(dt - a\sin^2\theta d\phi), \quad (3.1)$$

$$\omega^{(1)} = (\Sigma/\Delta)^{\frac{1}{2}}dr, \quad (3.2)$$

$$\omega^{(2)} = \Sigma^{\frac{1}{2}}d\theta, \quad (3.3)$$

$$\omega^{(3)} = \sin\theta\, \Sigma^{-\frac{1}{2}}[(r^2 + a^2)d\phi - a dt], \quad (3.4)$$

which gives for the electric field E and the magnetic field B the following expressions:

$$E_{(1)} = Q\Sigma^{-2}(r^2 - a^2\cos^2\theta), \quad (4.1)$$

$$B_{(1)} = Q\Sigma^{-2}2ar\cos\theta; \quad (4.2)$$

the other components are identically zero.

From the expression of the electromagnetic field tensor F and its dual, we can also define the electric and magnetic lines of force [35] associated with the electromagnetic field, which is associated, in turn, with the given background geometry see Fig. 1.

The first major step in the study of the properties of the solution described by the metric (1) deals with the analysis of its stability against small perturbations. The classic work in this field was made, in the case of a Schwarzschild metric (a = Q = 0) by Regge and Wheeler as far back as 1957 [37]. By exploiting the symmetry of the background space Regge and Wheeler were able to separate the angular part of the perturbations using tensorial spherical harmonics. This treatment was further extended by Vishveshwara [38] and then Zerilli [39] was able to reduce the entire set of perturbations to two independent functions, fulfilling Schrődinger-like equations, governing their radial and time dependence. These works not only gave the proof of the stability of the Schwarzschild solution against small perturbations but they also gave a complete set of eigenfunctions in which any small perturbation in the background geometry could be expanded, this has lead to the possibility of computing a large set of processes of emission of electromagnetic and gravitational radiation in strong gravitational fields (see e.g. §6). The corresponding analysis of the stability of the Reissner-Nordstrøm metric

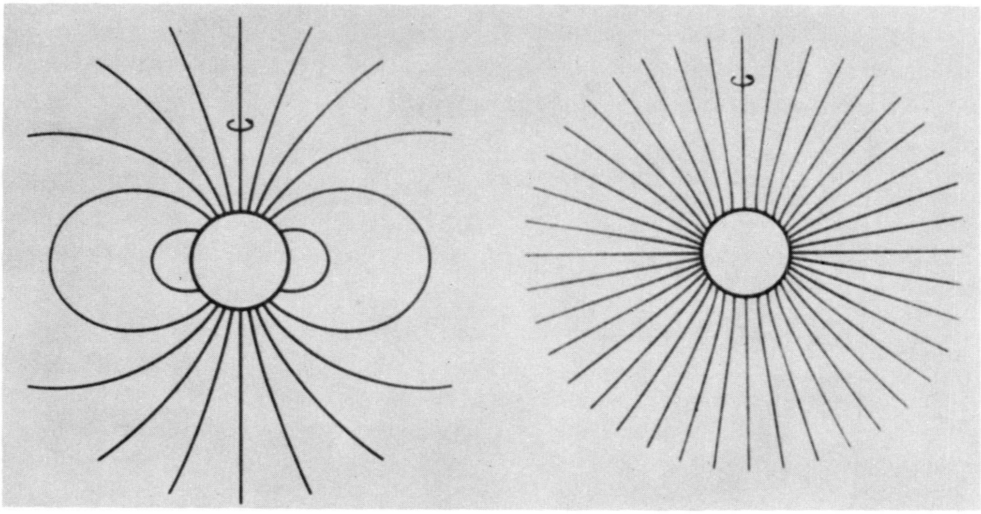

Figure 1: Magnetic (left and electric (right) lines of force of a Kerr-Newman geometry. For details see ref. [35].

(a = 0, Q ≠ 0) has been done by Zerilli [40] along the lines of the Regge-Wheeler-Zerilli formalism by the use of tensorial spherical harmonics. An alternative derivation using a gauge invariant formulation of the gravitational and electromagnetic perturbations has been done by Moncrief [41] and Weinstein [42]. The same problem has also been treated in the framework of the Newman-Penrose formalism by Lee [43]. Once again quite apart from establishing the stability of the Reissner-Nordstrøm solutions these works have lead to the detailed description of two new physical processes: the gravitationally induced electromagnetic radiation [44] and the electromagnetically induced gravitational radiation [45].

For the Kerr metric (a = 0 and Q = 0) Teukolsky [46] was able, using the spinorial formalism of Newman and Penrose, to separate the variables of the most general [47] perturbation of a Kerr metric and to reduce the entire problem to the solution of a single "master" ordinary differential equation. Still some question remains, however, if the stability of the Kerr solution has been satisfactorily proved [48].

No attempt has been made, up to date, in studying the complete perturbation of a Kerr-Newman geometry ($a \neq 0$, $Q \neq 0$) [49]. The second important step in analyzing the properties of collapsed objects described by Eq. (1) has been the study of the motion of test particles in this metric. The fundamental contribution in this field has been given by the work of Carter [50] in which he shows how the Hamilton-Jacobi equation governing the motion of a test particle in the metric (1) admits the separation of variables and leads to a set of normal differential equations. The equations of motion are then equivalent to the following set of normal differential equations

$$d\theta/d\lambda = \{C - \cos^2\theta\,[a^2(\mu^2 - E^2) + p_\phi^2 \sin^{-2}\theta]\}^{\frac{1}{2}}\Sigma^{-1} \tag{5.1}$$

$$dr/d\lambda = \{P^2 - \Delta[\mu^2 r^2 + (p_\phi - aE)^2 + C]\}^{\frac{1}{2}}\,\Sigma^{-1} \tag{5.2}$$

$$d\phi/d\lambda = [-(aE - p_\phi \sin^{-2}\theta) + a\,\Delta^{-1} P]\,\Sigma^{-1} \tag{5.3}$$

$$dt/d\lambda = [-a(aE \sin^2\theta - p_\phi) + (r^2 + a^2)\Delta^{-1} P]\,\Sigma^{-1} \tag{5.4}$$

with $P = E(r^2 + a^2) - p_\phi a - qQr$ and E the energy measured from infinity of a test particle with azimuthal angular momentum p_ϕ, rest mass μ, charge q, proper time $\tau = \mu\lambda$, and where C is the fourth constant of the motion introduced by Carter [51].

The analysis of the motion of the test particles has been done either by direct numerical integration of Eqs. (5) , for selected values of the constants of the motion [52], or by the use of the effective potential techniques [53]. A few results of particles orbits in the field of a Kerr and a Schwarzschild black hole are given in Figs. 2 and 3, compared and contrasted with the Newtonian results.

If electromagnetic fields are present the orbits of charged test particles are even more different qualitatively and quantitatively: both in the case of a Reissner-Nordstrøm solution [55] and in the case of a Kerr-Newman solution [56] we can find stable circular orbits with binding energies of up to 100%. Therefore, at least in principle, by the different motions of test particles we can identify the amounts of electromagnetic structure and of rotation in the background geometry.

The next step which has been important in giving a new understanding of black holes physics has been the gedanken process proposed by Penrose [57] and by Floyd and Penrose [58]. The new basic contri-

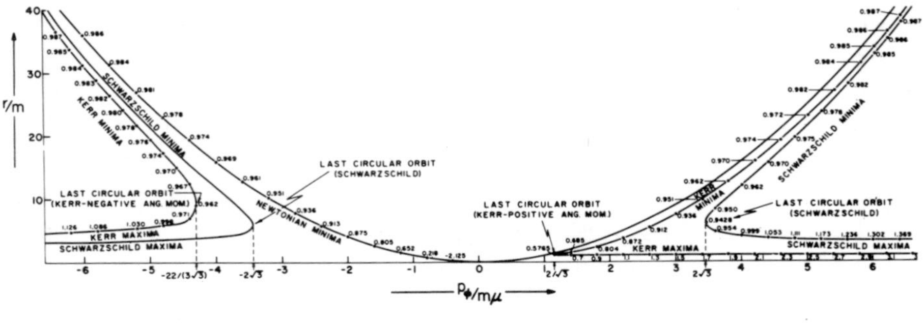

Figure 2: The minima (stable circular orbit) and maxima (unstable circular orbit) of the "effective potential" experienced by a test particle moving in the field of a Schwarzschild ($a = Q = 0$) and an extreme Kerr black hole ($a = M$, $Q = 0$) are here compared and contrasted with the corresponding Newtonian results. The distance r is given in units of the mass of the black hole and the angular momentum of the incoming particle p_ϕ in units of the product of the mass M of the black hole and of the mass μ of the incoming particle, both expressed in geometrical units $G = C = 1$. Numerical values of the effective potential E in units μ are displayed on the representative curves of the maxima and minima for selected values of the distance and of the angular momentum. Particularly impressive is the difference of binding energy between corrotating ($p_\phi/\mu M>0$) and counterrotating orbits in the case of a Kerr metric. Details in ref. [53].

bution of this work has been to point out the possibility of extracting energy from a black hole by processes occurring outside its horizon. The mechanism by which this process can occur is most transparent if the concept of effective potential, well familiar from other branches of physics, is introduced [53]. Particles counter-rotating with the black hole can acquire, in a limited region outside the horizon, a negative total energy as measured from infinity, their capture can therefore reduce the total mass-energy of the black hole (see Figs.4 and 5).

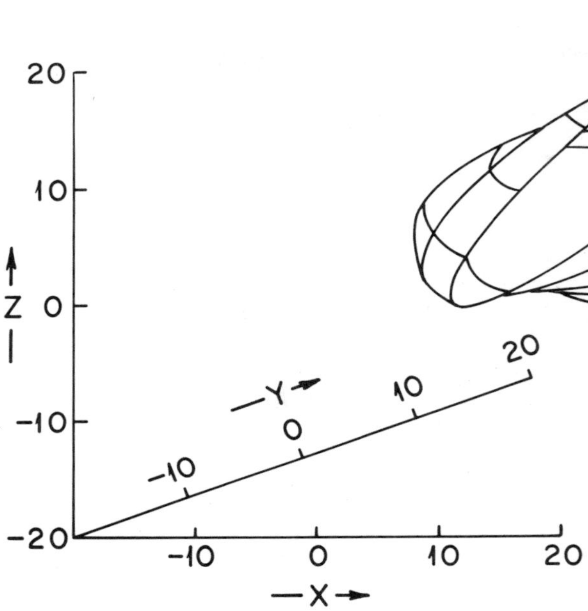

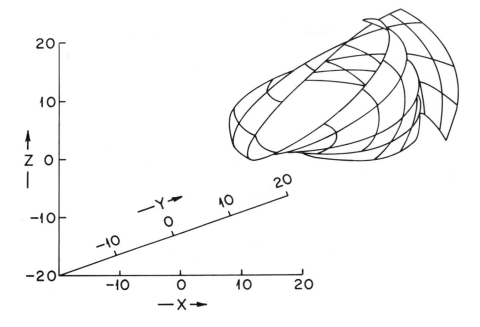

Figure 3: Motion of an uncharged cloud of test particles of mass μ specific angular momentum $p_\phi/\mu M = 1$, energy $E/\mu = 0.968$ corrotating around an extreme Kerr black hole with $a = M$. Points on a vertical lines are isochronous as seen from an observer at infinity. The typical "Klein bottle" behaviour of this cloud of particles illustrates one of the major differences between Newtonian orbits of test particles and orbits in the strong gravitational fields of a rotating object ("Wilkins effect" [54]): the orbit of the test particles is now no longer complanar but acquire an additional periodicity in the θ direction. Details in ref. [52].

These considerations were later extended to the extraction of electromagnetic energy from a black hole [56]. The region in which these negative energy states can be found, either due to the electromagnetic or rotational interaction, and therefore processes of energy extraction can occur, is defined as the "effective ergosphere" of a black hole and the radius of this "effective ergosphere" given by [59]

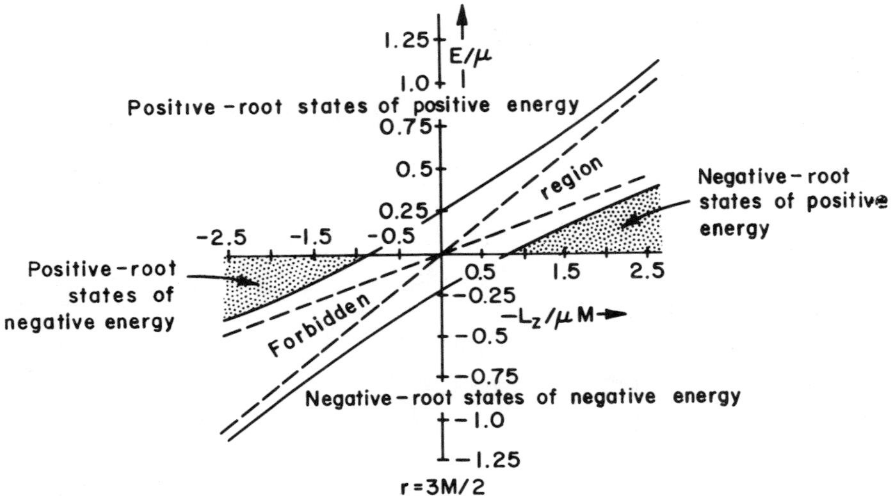

Figure 4: The energy E of states allowed to a particle of angular momentum p_ϕ and rest mass μ in the gravitational field of an extreme Kerr black hole (a = M) at a radius r = 3M/2 are plotted as a function of the angular momentum p_ϕ. The seas of positive and negative root states are shown [56]. The boundary of the seas (particles with $p_r = p_\theta = 0$) is given by the equation $E^2[r^3 + a^2\ (r + 2M)] - 4\ M\ Eap_\phi + (2M - r)\ p_\phi^2 - \mu^2 r^2(r - 2M) - a^2\mu^2 r = 0$. The forbidden region refers to particles with an immaginary momentum. Classically a particle can only attain the states of the positive root solutions, the negative root states acquire a meaning only in the quantum regime (see §6). The existence of negative energy states in the positive root solutions allows the possibility of extracting energy from a black hole. The separation between the positive and negative root solutions goes to zero for $r \to r_+$. Details in refs. [53], [55] and [56].

$$r_+ = M + (M^2 - Q^2 - a^2)^{\frac{1}{2}} \leq r \leq M + [M^2 + Q^2(q^2/M^2 - 1)]^{\frac{1}{2}}\ , \quad (6)$$

where, as usual a, Q and M, are the three parameters of the black hole, r_+ is the radius of the horizon and q/μ is the charge to mass ratio of the test particle by which the process of energy extraction occurs.

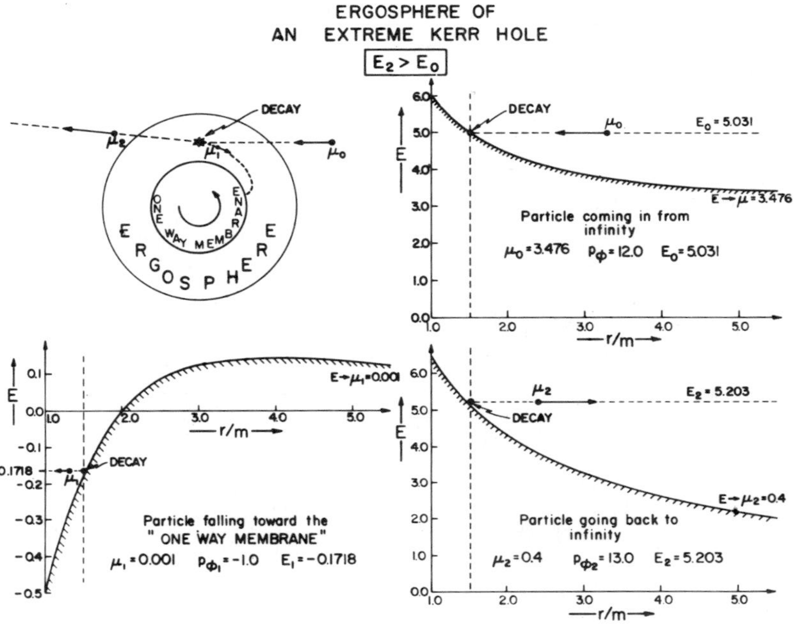

Figure 5: A particle of mass μ_0 coming from infinity with total energy E_0 and a positive value of the angular momentum p_{ϕ_0} can penetrate in the ergosphere of an extreme Kerr hole and here decay in two particles. One particle of mass μ_1, negative value of the angular momentum p_{ϕ_1} and a negative value of the total energy E_1, falls towards and penetrates the horizon r_+. The second particle of mass μ_2, positive value of the angular momentum $p_{\phi 2}$ and or positive value of the total energy E_2 goes back to infinity. The remarkable new feature in this process is that the energy E_2 of the particle coming in (Floyd and Penrose) [57], [58]. The numerical example here considered has been computed assuming the conservation of four momentum in the decay process. On the upper left side a qualitative diagram shows the main feature of the decay process in the equatorial plane of the ergosphere of a Kerr hole. In the upper right side is the effective potential for the incoming particle. The effective potential is plotted at the lower left and lower right side for the particle falling toward the horizon and for the one going back to infinity. This energy gain process critically depends on the existence and on the size of the ergosphere which in turn depends upon the value a/m of the hole. The particle falling towards the horizon will in general alter and reduce the ratio a/m of the black hole. Details in ref. [53].

Two new important and related concepts were introduced by analyzing the back effect on the black hole of the capture of test particles. Exploiting the concept of an effective potential it was possible to discover the existence of reversible and irreversible transformations in black hole physics [56] see Fig. 6.

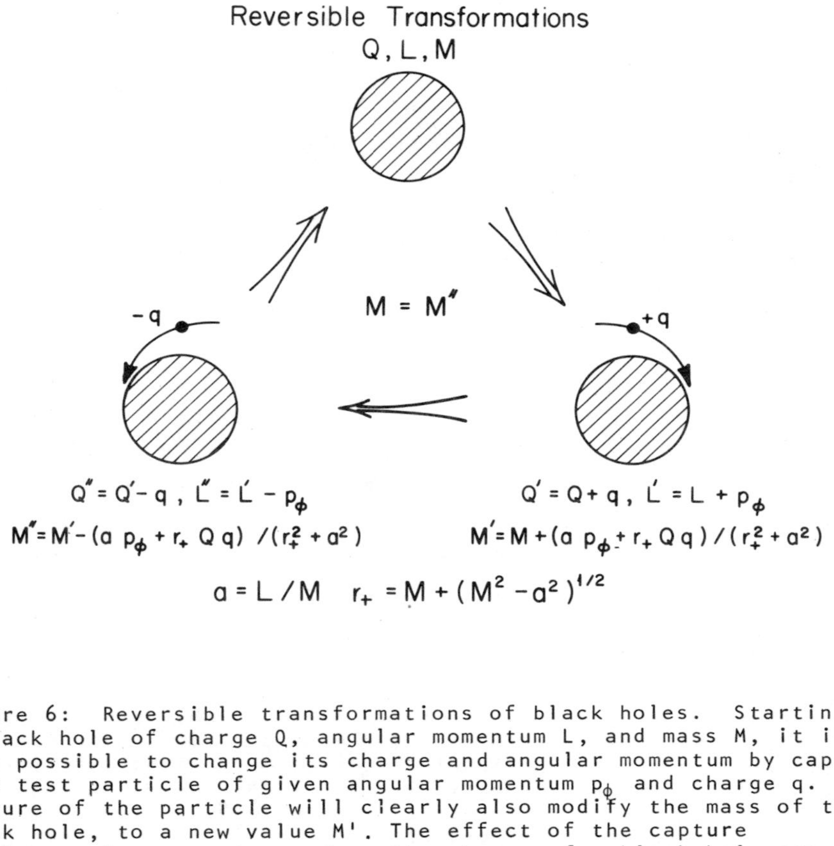

Figure 6: Reversible transformations of black holes. Starting with a black hole of charge Q, angular momentum L, and mass M, it is always possible to change its charge and angular momentum by capture of a test particle of given angular momentum p_ϕ and charge q. The capture of the particle will clearly also modify the mass of the black hole, to a new value M'. The effect of the capture on the angular momentum and on the charge of a black hole can be counterbalanced by the capture of an oppositely charged particle with opposite value of the angular momentum. In general, however, after the capture of the two particles, the black hole, though endowed with the same initial charge and angular momentum will have augmented its mass M'' > M. It has however been shown in ref. [56] that there exists a subclass of transformations (the reversible ones) for which the black hole can recover the <u>same</u> value of the initial mass M'' = M. To reach reversibility, the capture process must be performed infinitely slowly, particles with zero kinetic energy at the horizon (r_+). In this case the mass increment is the one given by the formula in the figure. Details in ref. [56].

In turn from the concept of reversible transformations it was possible to introduce the following differential equation [56]:

$$dM = [(L/M)dL + r_+ QdQ]/[r_+^2 + L^2/M^2] \tag{7}$$

this equation integrated gave the well-known formula [56]

$$M^2 = (m_{ir} + Q^2/4\, m_{ir})^2 + L^2/4m_{ir}^2 \tag{8}$$

which gives the total mass-energy of a black hole in terms of its rest mass (m_{ir} ≡ irreducible mass), its Coulomb energy and its rotational energy. The area of the black hole horizon is then given, most simply, by

$$\dot{A} = 4\pi(r_+^2 + a^2) = 16\,\pi\, m_{ir}^2 \quad . \tag{9}$$

From Eq. (8) it is then possible to see that the change of irreducible mass of a black hole by capture of a test particle is given by:

$$4\, m_{ir}\, \delta\, m_{ir} = [E(r_+^2 + a^2) - ap_\phi - r_+ qQ]/(M^2 - a^2 - Q^2)^{\frac{1}{2}} , \tag{10}$$

where E is the energy of the particle of charge q and angular momentum p_ϕ. From the formulae of the effective potential [53] and from Eq. (10) it is then possible to see that we always have

$$\delta\, m_{ir} \geq 0 \tag{11}$$

the equality to zero holding only for reversible transformations (capture of the particle occurring infinitely slowly with zero radial momentum at the horizon). This inequality implies that the area of the horizon for <u>any</u> arbitrary classical transformation must increase monotonically. This same result was obtained quite independently, with a different approach by Hawking [60].

The analogy between these concepts of reversible and irreversible transformations and the usual thermodynamical transformations as well as the analogy between the area of the horizon (monotonically increasing), and the usual thermodynamical concept of entropy were stressed by Bekenstein [61]. By a direct differentiation of Eq. (8) and from the surface area as given by Eq. (9) we have [61]

$$dM = \frac{K}{8\pi}\, dA + \Omega\, dL + \Phi\, dQ \tag{12}$$

with $K = (m^2-a^2-Q^2)^{\frac{1}{2}}/(r_+^2+a^2)$ "surface gravity" at the horizon [62], $\Omega = \frac{\partial M}{\partial L} = a/[r_+^2 + a^2]$ angular velocity of the horizon [35], $\Phi = Qr_+/[r_+^2 + a^2]$ the electrostatic potential at the horizon [63]. Eq. (12) is <u>formally</u> identical to the well—known thermodynamic relation

$$dM = T\, ds + \Omega\, dL + \Phi\, dQ \tag{13}$$

for the change in mass energy in a thermodynamic transformation, where now T is the temperature and S the entropy of the thermodynamical system rotating with an angular velocity Ω. As usual we should take care in going from conventional units to geometrical units. We then have for

$$\frac{G}{c^3}\, h_{conv} = h_{geom}\ (cm^2);\quad G\, k\ T_{conv}/c^4 = T(cm)$$

where k as usual is the Boltzmann constant and for the entropy

$$S_{conv}/k = S_{dimensionless} \quad .$$

Bekenstein suggested that the similarity between Eq. (12) and Eq.(13) is more than pure analogy and that indeed the quantity $K \times h_{geom}$ should be identified, within a numerical factor of order of unity, with the temperature of the black hole and the surface area S divided by h_{geom} should be identified again within a factor of order of unity with the black hole entropy. Hawking [64], has gone much further in proving the ansatz of Bekenstein. He has shown that if quantum processes of pair creation are taken into account then indeed a black hole spontaneously radiates a black body spectrum of temperature

$$T_{geom} = h_{geom}\ K/2\pi \tag{14}$$

and that the entropy of the black hole should be given by

$$S_{dimensionless} = (A/4h)_{geom} \quad . \tag{15}$$

Much work has still to be done to have a deeper and complete understanding of these important analogies and identities. More results on these processes of black holes evaporation are presented in the reports of Gibbons [65], Davies [66] and Unruh [67].

It is however by now clear that all these phenomena, important as they are from a first principle point of view, do not lead directly to any astrophysical observations unless small black holes are generated in primordial stages of cosmology. The reason is simply stated it is impossible to form small black holes and overcome the barrier against gravitational collapse in matter, due to electromagnetic interactions and the quantum principle. The only way to form black holes, we believe, is to overcome the critical mass of a neutron star: $M_{crit} \sim 2M_{\odot}$. The effective temperature of a black hole from Eq.(14) is

$$T_{conv} \simeq 0.62 \times 10^{-7} \; (M_{\odot}/M) \; {}^{o}K \tag{16}$$

implying a lifetime for the black hole against evaporation

$$\tau \simeq E/(dE/dt) \sim 2 \times 10^{63} \; (M/M_{\odot})^{3} \text{ years} \tag{17.1}$$

with and energy flux

$$dE/dt \simeq 10^{-22} \text{ erg/sec } (M_{\odot}/M)^{2} \quad . \tag{17.2}$$

Only in the case of black holes of mass $M \simeq 10^{15}$gr these effects of vacuum polarization could be very relevant. We turn therefore to a different aspect of Eq.(8) namely to the possibility of using black holes as <u>realistic</u> sources of energy for astrophysical processes. The point to stress here is that the only limitations on the amount of rotational and electromagnetic energy which can be stored in a black hole in the process of gravitational collapse is given by the following relation:

$$a^{2} + Q^{2} \leq M^{2}$$

and at least in principle a very large amount of energy could still be extracted from such a collapsed object. A few numerical examples are given in Table I.

Mass ($M_{\odot}$)	Initial Radius of Collapsed Object $r_{+} = M + \sqrt{M^{2} - a^{2} - Q^{2}}$ (cm)	Minimum Rotational Period $\omega = a/(r_{+}^{2} + a^{2})$ (sec)	Maximum Strength of the Magnetic Field (G)	Maximum Energy Extractable from the Collapsed Object (erg)
10	1.47×10^{6}	0.653×10^{-3}	1.05×10^{18}	0.689×10^{55}
10^{8}	1.47×10^{13}	0.653×10^{4}	1.05×10^{11}	0.689×10^{62}
10^{12}	1.47×10^{17}	0.653×10^{8}	1.05×10^{7}	0.689×10^{66}

Table I: Typical values of the parameters associted to an extreme collapsed object ($M^{2} = Q^{2} + a^{2}$ and $Q = a$) for selected values of the mass. Details in ref. [35].

3. STRUCTURE OF A BLACK HOLE UNDER REALISTIC ASTROPHYSICAL CONDITIONS

Under any realistic astrophysical condition a black hole, in order to be detected, must be accreting plasma (see Introduction). It then becomes natural to ask which one of the four analytic solutions with the metric given by Eq. (1), [Schwarzschild (a=Q=0), Reisner-Nordstrøm (a=0), Kerr(Q=0) or the Kerr-Newman] is the best approximation to describe realistically an accreting black hole.

We know that electromagnetic fields are extremely important in the description of either an accreting neutron star in a binary system or an isolated neutron star, like a pulsar. The reason being that the characteristic periodic emission in the radio, or in the optical or in the x-rays are explained by plasma phenomena occurring in the magnetosphere of a rotating neutron star. The main issue to clarify is the role a magnetosphere will play in the corresponding situation for a black hole. What became very clear since the first steps in analyzing this problem, was that a test particle approximation, discussed in previous paragraphs, could not lead even to first approximation analysis as soon as rates of accretion of $dM/dt \sim 10^{17}$ gr/sec were considered. A more complex magnetohydrodynamical analysis was needed.

For a rotating star endowed with a magnetic field with an arbitrary inclination angle with respect to the rotation axis, the exact analytic solution of its electromagnetic structure has been known in great detail due to the classical work of Deutsch [68]. Only much later starting from general considerations of energy minimization, Treves and I [69] pointed out the possibility of having charge separation playing an important role in the magnetosphere of a rotating collapsed object. The motivation at the time, was to have a physical argument in order to understand the very special electric to magnetic field ratio of the Kerr-Newman solution. It became however very clear that this phenomenon would be relevant in many realistic astrophysical conditions. I also presented the problem in the following way [70]:

"It has been known for a long time that objects of astrophysical interest ($M \sim 1 M_{\odot}$) can never be endowed with a net charge of large value. The reason is simply stated: The presence of a net charge on a star will produce selective accretion of oppositely charged material. The system will discharge extremely fast, under any reasonable assumption for the density of interstellar material. These very straightforward considerations can be largely modified if the object under examination is not only endowed with a net charge but also with rotation and strong magnetic fields. The accreting material could then be trapped in the magnetosphere of the rotating object (for simplicity we can assume

a dipole magnetic field aligned with the rotation axis), completely screen the net charge and consequently stop any process of selective accretion. Strictly speaking the system as a whole is now neutral, though the "bare" object and the magnetosphere are indeed endowed with large and opposite charges. The possible existence of such an object strongly depends upon two major assumptions:
a) that the lifetime of charged particles in the magnetosphere is long enough, (b) that the accretion through the "polar cup" must be negligible. In this region, in fact, particles will not be trapped any longer by the lines of force of the magnetic field and they are free to impinge on the object and discharge it.

Quite apart from these considerations, is there any physical reason why the "bare" object itself should have a net charge? This is the motivation for the analysis presented in the work by Ruffini and Treves".

For point (b) above see the concept of "Plasma Horizon", later developed [71]. To clarify this problem with a very simple model, consider a black hole endowed with a net charge and surrounded by a ring of equal and opposite charge [72]. See Fig.7. The problem can be solved nearly analytically by writing the covariant Maxwell equations in the background geometry [73]

$$F^{\mu\alpha}{}_{;\alpha} = 4\pi J^{\mu} \quad (18.1) \qquad {}^*F^{\mu\alpha}{}_{;\alpha} = 0 \quad (18.2)$$

with a four current vector

$$J^{r}=0;\ J^{\theta}=0;\ J^{\phi}=\Omega\ \delta(r-b)\ \delta(\cos\theta)/2\pi b^{2};\ J^{t}=q\delta(r-b)\delta(\cos\theta)/2\pi b^{2}, \quad (19)$$

where the charge q is the net charge on the ring of radius b in the equatorial plane of a black hole endowed with charge $Q=-q$ (global charge neutrality). The quantity Ω is a free parameter describing additional currents in the ring. Due to the high degree of symmetry of the problem the equations (18.2) can be fulfilled by introducing a vector potential with components

$A_r = 0;\ A_\theta=0;\ A_\phi=\Sigma_\ell\ f_\ell(r)\ T_\ell(\cos\theta);\ A_t=\Sigma_\ell\ g_\ell(r)\ P_\ell(\cos\theta)$ where $T_\ell = \sin\theta\ (dP_\ell/d\theta)$ and $P_\ell(\cos\theta)$ are the usual Legendre polynomials The radial functions $f_\ell(r)$ and $g_\ell(r)$ fulfill the two equations

$$\frac{d}{dr}\left(r^2\ \frac{dg_\ell}{dr}\right) = \ell(\ell+1)g_\ell r^2/\Delta \quad (20.1);\quad \frac{d}{dr}\left(\frac{\Delta}{r^2}\ \frac{df_\ell}{dr}\right)=\ell(\ell+1)\ f_\ell/r^2 \quad (20.2)$$

with the boundary conditions at the ring given by

$$\left[\frac{dg_\ell}{dr}\right]_{b-\varepsilon}^{b+\varepsilon}=(2\ell+1)q\ P_\ell(0)/b^2 \quad (21.1);\quad \left[\frac{df_\ell}{dr}\right]_{b-\varepsilon}^{b+\varepsilon} = \frac{(2\ell+1)b^2\Omega}{\ell(\ell+1)\Delta(b)}\ \frac{dP_\ell(0)}{d\theta} \quad (21.2)$$

results are given in Figs.7 and 8.

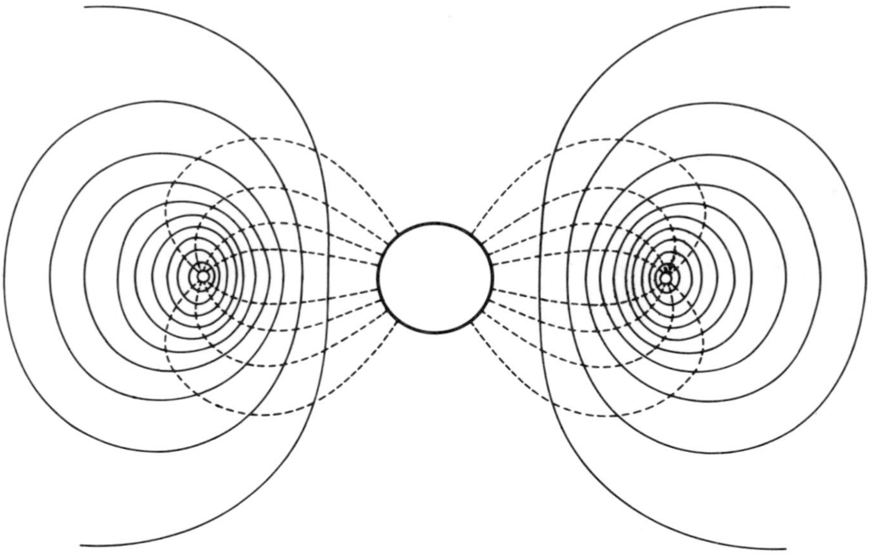

Figure 7: Electric (dashed lines) and magnetic lines of force of a ring endowed with a net charge and current in the background field of an oppositely charged non rotating black hole. Details of compretations in Ref. [72].

This elementary example allows us to understand some important aspects of the problem. Far away the electric field has only a quadrupole component while the magnetic field a dipole one. If we go far enough therefore, the strength of the magnetic field is always larger than the one of the electric field, and plasma, if we neglect special regions of alignment between the two fields, can always be suported by the magnetic field structure (see Fig. 8). Near the black hole or near the ring the electric field will always be larger in magnitude then the magnetic field and no confinement of plasma will be possible. To make more quanitative these considerations we have introduce the concept of plasma horizon (see Fig. 8). More analytic details are given in the report by Hanni [71] and in ref. (74). The simple model considered here completely neglects the effects of rotation. We expected, however, that rotation plays an essential role in giving rise to the charge separation process [70]. This further step has recently been made in the works of Wald [75], Peterson [76], Chitre and Vishveshwara [77], King, Lasota, Kundt [78] . Using a well-known result

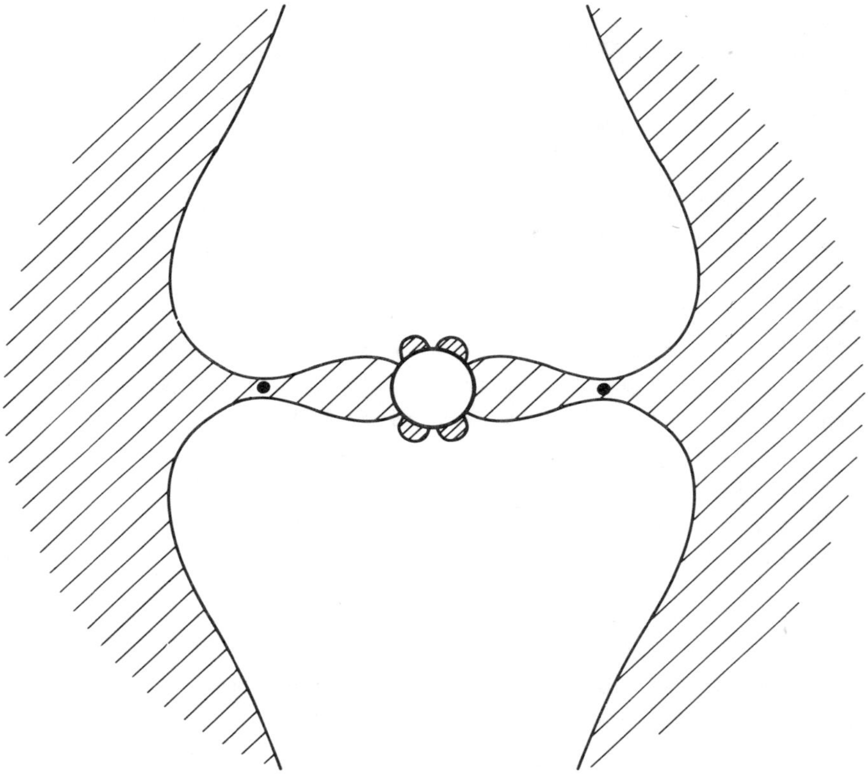

Figure 8: "Plasma Horizon" for the current ring considered in Fig. 7. The dashed region indicates the points were $|E \times B| > E^2$. Particles in this region, if moving with the appropriate velocites, can be trapped by the magnetic fields. Details in ref. [71], [72], [74].

by Papetrou [79] that a Killing vector in a vacuum spacetime can be used as a vector potential for a Maxwell test field in that background geometry, Wald has derived an analytic solution for the electromagnetic field occurring when a stationary, axisymmetric black hole is placed in an originally uniform magnetic field aligned along the symmetry axis of the black hole [75]. By considering the "injection energy" of test particles [61] he shows that a black hole with angular momentum aM imbedded in a uniform magnetic field B_o will tend to accrete a net charge $Q = 2\, B_o\, aM$. Similar conculsions have been reached in the work of King, Lasota and Kundt [78]. Quite different and more similar to the example of Fig.7 is the work by Peterson: using the Teukolsky's perturbation equation for electromagnetic perturbations he considers a ring with net charge q in the field of a rotating black hole and he studies the minimum energy configuration. Chitra and Vishveshavara [77] have also addressed their interest to a similar problem. This problem has been solved in the framework of the approximation given in ref. [69] by Damour [80]. Damour finds that indeed a minimal energy configuration is obtained for a charge on the black hole equal and opposite to the one in the magnetosphere and of value

$$Q = 2M\, a\Omega/(b-2M) \quad ,$$

where the notations are the same as in Eq. (19). All these works are helpful in clarifying some of the issues connected with the structure of a realistic magnetosphere for a rotating collapsed object. The following question however still remains to be answered: what will happen to all these idealized models (rings, uniform fields, etc.) in presence of an accreting plasma? Will they still be able to exist or the plasma accretion will not allow these minimal energy configurations ever to be reached? To answer these questions we must go to a higher degree of sophistication and treat the more realistic problem of plasma accreting into a rotating gravitationally collapsed object.

4. ON THE BASIC EQUATIONS OF RELATIVISTIC MAGNETOHYDRONAMICS

In the following we are going to consider the relativistic equations of magnetohydrodynamic of a plasma moving in a given background geometry. We shall neglect all the feed back of the plasma distribution on the metric itself under the well justified assumption that the mass of the accreting plasma can be considered negligible with respect to the mass energy of the collapsed object.

The classical results in this field have been given by Lichnerowicz [81][82] and in the following we are going to summarize only the main results:

To describe a plasma accreting in a curved background, we have most generally two tensors $F_{\mu\nu}$ and $G_{\mu\nu}$ which fulfill the following equations:

$$F_{(\mu\nu,\lambda)} = 0, \tag{22}$$

$$G^{\alpha\beta}{}_{;\beta} = 4\pi J^{\alpha} , \tag{23}$$

where the semicolon indicates, as usual, covariant differentiation, and J^a is the four-current

$$J^{\alpha} = \varepsilon u^{\alpha} + \sigma F^{\alpha\beta} u_{\beta} \quad , \tag{24}$$

where ε is the charge density, σ denotes the conductivity, and u_{β}

represents the four-velocity of the plasma. According to Lichnerowicz [81], [82] we can define the following four vectors:

$$E_\alpha = F_{\alpha\beta} u^\beta , \quad (25.1)$$

$$B_\alpha = - *F_{\alpha\beta} u^\beta , \quad (25.2)$$

$$D_\alpha = G_{\alpha\beta} u^\beta , \quad (25.3)$$

$$H_\alpha = - *G_{\alpha\beta} u^\beta , \quad (25.4)$$

with

$$D_\alpha = \lambda E_\alpha \quad (26.1)$$

and

$$B_\alpha = \mu H_\alpha , \quad (26.2)$$

where λ and μ are, respectively, the dielectric permittivity and the magnetic permeability of the medium, here considered isotropic. I shall restrict myself in the following text to the case of infinite conductivity ($\sigma = \infty$) and $\mu = \lambda = 1$. Under these limitations, we shall have

$$G_{\alpha\beta} = F_{\alpha\beta} \quad (27)$$

and

$$F_\alpha{}^\beta u_\beta = 0 \quad . \quad (28)$$

The total energy momentum tensor for the plasma is then given by

$$T_{\mu\nu} = (\rho + p) u_\mu u_\nu + p g_{\mu\nu} + (F_{\mu\alpha} F_\nu{}^\alpha - \tfrac{1}{4} g_{\mu\nu} F_{\alpha\beta} F^{\alpha\beta})/4\pi , \quad (29)$$

where

$$\rho = r(1 + \text{specific internal energy}), \quad (30)$$

and r is the proper rest mass density of particles. The following conservation laws must be fulfilled:

$$T^{\mu\nu}{}_{;\nu} = 0 , \quad (31.1)$$

the conservation of particles number,

$$(r u^\alpha)_{;\alpha} = 0 , \quad (31.2)$$

the energy conservation equation,

$$u_\mu (T^{\mu\nu}{}_{;\nu}) = 0, \text{ which implies } [(\rho + p) u^\nu]_{;\nu} = p_{,\nu} u^\nu, \text{or, for } p=0, \quad (31.3)$$

$$[ru^{\nu}]_{;\nu} = 0 \qquad (31.4)$$

and finally, the equation for the stream lines of the fluid,

$$(\rho + p)u^{\lambda}\nabla_{\lambda}u^{\alpha} = -(g^{\alpha\lambda} + u^{\alpha}u^{\lambda})p_{,\lambda} + F^{\alpha\lambda}J_{\lambda} . \qquad (31.5)$$

These equations have been solved numerically for any axially symmetric configuration and some results as well as some details of the very complex techniques of integration are given by Wilson [83]. What we shall try in the following, instead, is to solve some special case with $p=0$ and $\sigma=\infty$. Again the main reason for this is to have a simple model, solvable analytically, in order to gain an understanding of more complex and general configurations. We shall also require, again for the sake of simplicity, that the solution will satisfy both the condition of stationarity and axial symmetry.

5. ON AN ANALYTIC SOLUTION FOR THE MAGNETOHYDRODYNAMICS EQUATIONS

We assume [84] that the motion of the plasma is given by the geodetic in the background Kerr geometry. We shall also assume that at infinity $u_t=-1$ and $u_\phi=0$ which also implies that u_θ is a constant of the motion. The velocity field of the particles is then known and given by the following formulae [50]:

$$v^r = u^r/u^t = \frac{-\Delta\,[-\Delta u_\theta^2 + 2Mr(r^2 + a^2)]^{\frac{1}{2}}}{\Sigma(r^2 + a^2) + 2Mra^2\sin^2\theta} \qquad (32.1)$$

$$v^\theta = u^\theta/u^t = \Delta u_\theta/[\Sigma(r^2 + a^2) + 2Mr\,a^2\sin^2\theta] \qquad (32.2)$$

$$v^\phi = u^\phi/u^t = g^{\phi t}/g^{tt} \qquad (32.3)$$

From the condition of stationarity and axisymmetry, one finds for the component of the electromagnetic field:

$F_{\phi t} = 0$ (33.1) $F_{r\phi} = A_{\phi,r}$ (33.2) $F_{\theta\phi} = A_{\phi,\theta}$ (33.3)

$F_{rt} = A_{t,r}$ (33.4) $F_{\theta t} = A_{t,\theta}$ (33.5) and we define

$$F_{r\theta} \equiv H_\phi \qquad (33.6)$$

the condition of infinite conductivity (28) leads to the following relations:

$$F_{tr} = v^\theta\,H_\phi + v^\phi\,\frac{\partial A_\phi}{\partial r} \qquad (34.1)$$

$$F_{t\theta} = - v^r H_\phi + v^\phi \frac{\partial A_\phi}{\partial \theta} \qquad (34.2)$$

implying that the knowledge of A_ϕ and H_ϕ completely determine the electromagnetic field structure.

It has also been shown by Damour [85] that the condition of infinite conductivity [28] straightforwardly imply, even in this most general case, the classical "frozen in" condition for the magnetic lines of force in the matter flow.

The entire problem reduces now to the determination of the two functions A_ϕ and H_ϕ . From Eq. (22) $*F^{\phi\mu}{}_{;\mu} = 0$ and the condition of stationarity we have

$$\frac{\partial}{\partial r}(H_\phi v^r) + \frac{\partial}{\partial \theta}(H_\phi v^\theta) + \frac{\partial v^\phi}{\partial \theta}\frac{\partial A\phi}{\partial r} - \frac{\partial v^\phi}{\partial r}\frac{\partial A\phi}{\partial \theta} = 0 \qquad (35.1)$$

from the ϕ component of the equation of infinite conductivity and the condition of stationarity, we obtain

$$v^r \frac{\partial A\phi}{\partial r} + v^\theta \frac{\partial A\phi}{\partial \theta} = 0 \quad . \qquad (35.2)$$

We also have from the t component of Eq. (28)

$$v^r \frac{\partial At}{\partial r} + v^\theta \frac{\partial At}{\partial \theta} = 0 \quad . \qquad (36)$$

Equations (35.2) and (36) imply that both A_t and A_ϕ are constant along the trajectories. We can choose $A_t = 0$ identically which will imply an overall charge neutrality for the system. Eq. (35.2) implies simply that A_ϕ is an arbitrary function of θ or

$$A_\phi = A(\theta_\infty) \qquad (37.1)$$

where

$$\theta_\infty = \theta - u_\theta \int_r^\infty [-\Delta u_\theta^2 + 2Mr(r^2 + a^2)]^{-1/2} dr \quad . \qquad (37.2)$$

From Eqs. (35) and (36) and from the additional Maxwell equation $*F^{t\mu}{}_{;\mu}=0$ then follows, in complete generality

$$H_\phi = (v^\phi/v^r)\, F_{\theta\phi} \quad . \qquad (38)$$

To solve the entire problem, therefore, no differential equation must be solved but we know analytically an entire family of solutions which are determined up to an arbitrary function A_ϕ.

Once the electromagnetic field has been determined we can compute from Eq. (23), by direct differentiation, the four current j^{α}, and from that compute the torque of the magnetosphere on the black hole and the charge distribution in the magnetosphere.
We have chosen the following form of the function A_{ϕ}:

$$A_{\phi} = A_0 \quad (1 - |\cos\theta|) \quad . \tag{39}$$

We then have for the torque:

$$T = \int (J^r F_{r\phi} + J^{\theta} F_{\theta\phi}) \sqrt{-g}\, d^3x \tag{40.1}$$

or for our specific example

$$T = A_0^2 \left[\frac{r_+^2 + a^2}{a^2 r_+} \arctan\left(\frac{a}{r_+}\right) - \frac{1}{a} \right] . \tag{40.2}$$

The current distribution as well as the magnetic lines of force corresponding to Eq. (39) are given in Figs.9, 10 and 11.
The total charge in the magnetosphere is simply given by

$$Q_{mag} = \int \sqrt{-g}\, J^t\, d^3x \tag{41.1}$$

or using the Gauss theorem we can extend the integration over the horizon

$$Q_{Hole} = \int_{Hor} \sqrt{-g}\, [g^{tt} g^{rr} v^{\theta} H_{\phi} - g^{rr}(g^{tt} v^{\phi} - g^{t\phi}) \frac{\partial A_{\phi}}{\partial r}]\, d\theta d\phi \quad \text{or}$$

$$Q_{Hole} = - Q_{mag} = \frac{1}{4\pi} \int_{Hor} \sqrt{-g}\, F^{tr}\, d\theta d\phi \quad . \tag{41.2}$$

This result clearly implies that if the infalling gas:
(1) follows geodesics; (2) is at rest on infinity with zero angular momentum (3) has a motion in the θ direction such as $v^{\theta}(\theta) = v^{\theta}(\pi-\theta)$ then the net charge in the magnetosphere and on the horizon is null. Any deviation from these requirements of very high symmetry will generally induce a net charge on the surface of the black hole. In the example given in Eq. (39) by assuming

$$U_{\theta}(\theta) = - U_{\theta}(\pi-\theta) = \text{const.} \tag{42}$$

we obtain a net charge on the hole equal and opposite to the one in the magnetosphere given by:

$$Q_{Hole} = - Q_{mag} = \pi\, a\, A_o\, u_{\theta}/16\, M^2 \quad . \tag{43}$$

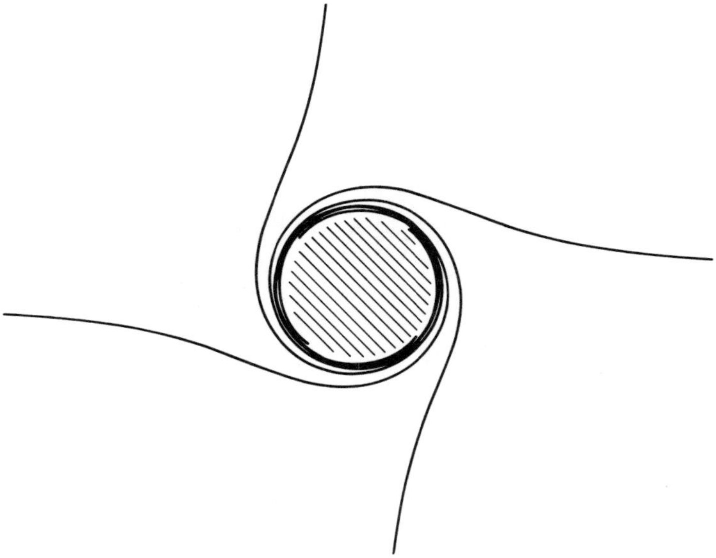

Figure 9: Magnetic lines of force [86] in the equatorial plane of the black hole corresponding to the vector potential given by Eq. (39). The winding of the lines of force due to the dragging of inertial frames is here very clearly illustrated. This is the fundamental mechanism by which rotational energy can be extracted from a black hole (see also Fig. 12). Here and in the following figures the black hole is assumed maximally rotating (a=M).

It is important to illustrate by a very direct analogy the physical processes by which the torque and the energy extraction process do occur around a rotating black hole (see Fig. 12). Of course, in the analytic solution here considered the angular momentum extracted from the black hole, is transmitted to the plasma by the torque of the magnetic field and given back to the black hole by the accreting plasma. This clearly follows from the requirement of stationarity imposed to our solution. In a more realistic situation, however, in which the magnetic field and gas pressure are strong enough to make the particle trajectories depart significantly from a geodesic motion, it has been shown by numerical calculations that a net outward flow of angular momentum is possible [87], [88].

The main point to stress, here, is that a very large amount of rotational energy can be stored into a black hole (up to 29% of its mass-energy, see Eq.(8)) and the mechanism here presented could be a realistic one by which this energy could be extracted and stored in

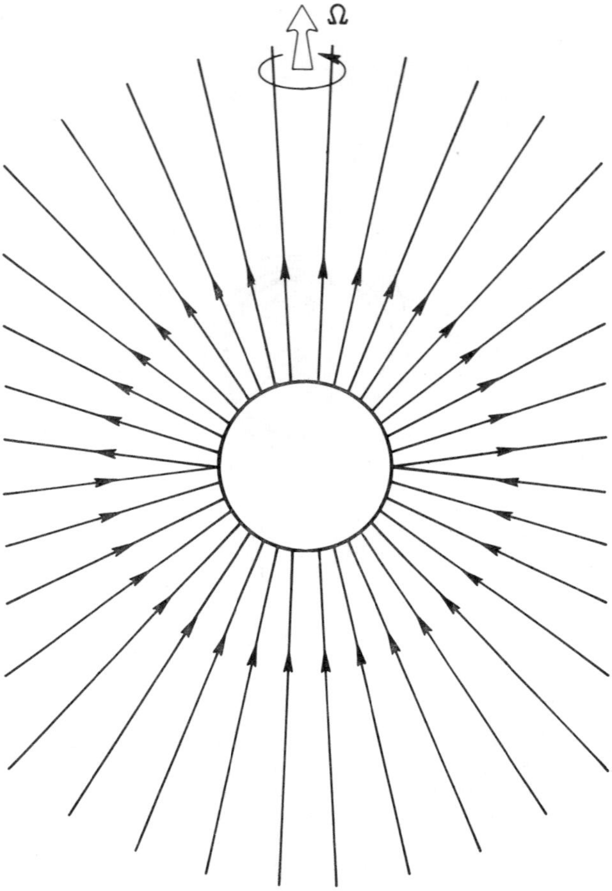

Figure 10: Magnetic lines of force in a plane ϕ=const. corresponding again to the vector potential given by Eq.(39) and with a $u_0(\theta) = -u_\theta(\pi-\theta)$=const. A three-dimensional plot will present the feature of magnetic lines of force ingoing in the southern emisphere and outgoing in the northern one, as well as the characteristic winding shown in Fig.9.

the magnetosphere [89]. One of the major questions still to be answered deals with the stability of this process of energy extraction. There is a macroscopic instability, magnetic pressure overcoming pull of gravitational attraction [90], rotational instabilities, etc. There is also a microscopic instability of quantum origin which we are going to discuss in the next paragraph.

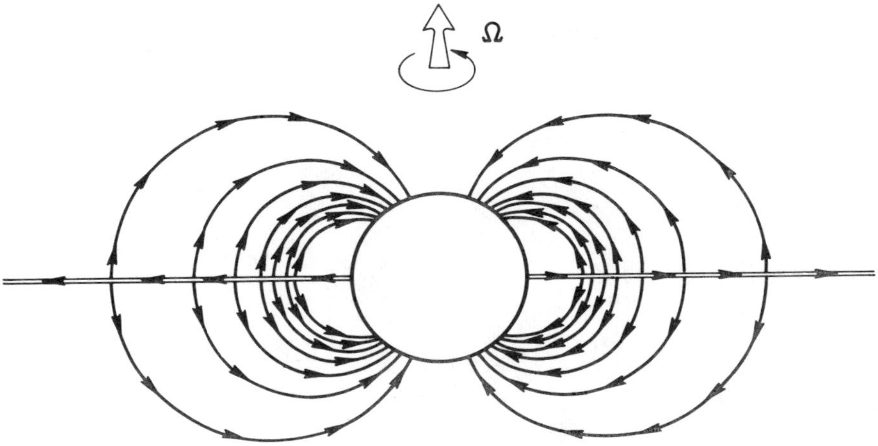

Figure 11: Lines of current calculated for the vector potential given by Eq. (39). The presence of an infinite sheet of current in the equatorial plane is due to the particularly simple angular dependence chosen for the vector potential in Eq.(39). This infinite current sheet could be eliminated by suitably choosing a smoother (θ) dependence of the vector potential A_ϕ .

6. KLEIN PARADOX AND BLACK HOLES

In 1923 Oscar Klein [91] pointed out the possibility of creating pairs of particles out of the vacuum if a static potential (electromagnetic or other) with a strong enough gradient exists over dimensions comparable to the Compton wavelength of a particle. Klein stressed that this process was clearly a consequence of relativistic wave equations. This physical process was studied in much detail by Sauter [92] and especially by Heisenberg and Euler [93] who gave a very beautiful treatment to take into account the effects of vacuum polarization in strong electric fields by introducing an "effective" dielectric and magnetic constant of the vacuum (see Fig. 13). Interesting are the work of Pauli on this subject [94] and later, in 1952, the work by Schwinger [95]. Schwinger was able to generalize these results to estimate the rate of pair creation by a uniform electric field.

The application of the Klein-Sauter-Heisenberg-Euler approach to black hole physics came through a direct analogy between the E^+ and E^- solutions of a quantum field and the positive and negative root solutions introduced by Christodoulou and myself [56] (see also §2 and Fig.4, where the level crossing between the positive and the negative root

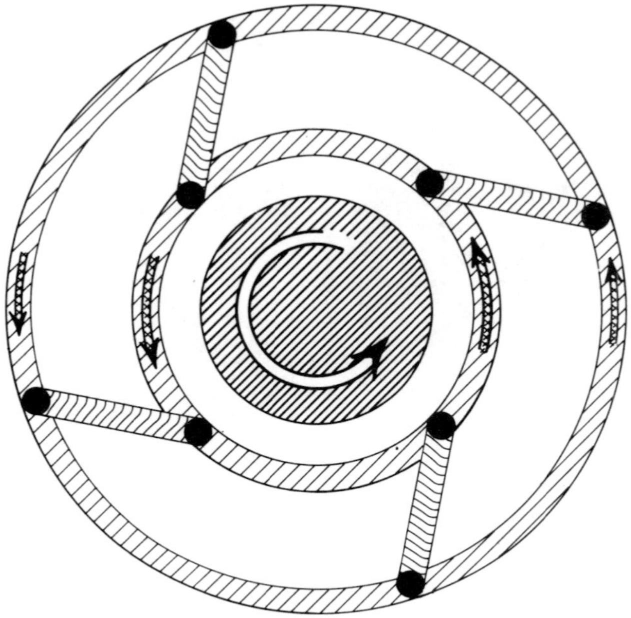

Figure 12: Idealized machine to extract rotational energy from a black hole. Two shells of different radii are dragged from the rotation of the black hole with the characteristic angular velocity $\Omega_{dragging} = 2aMr/[(r^2+a^2)^2 - \Delta a^2 \sin^2\theta]$. If connections between the two shells are made (ropes, springs, or in the case of the magnetosphere magnetic lines of force), rotational energy can be pumped out from the innermost shell to the external one, and rotational energy can be extracted from a black hole.

solutions is clear. The real meaning of these classical solutions were found by Deruelle and myself [97] by analyzing the classical limit of a relativistic field theory ($h \to 0$) and more specifically the classical limits of the positive and negative energy states of a quantized field of spin zero. It became manifest that the classical positive and negative root solutions were, very simply, the classical limits of the positive and negative energy states of the quantized field [97],[98],[100]. The negative energy solutions could then be

Die exakte Durchrechnung dieses Problems verdankt man Sauter¹). In Fig. 1 ist zunächst die potentielle Energie $V(x)$ als Funktion der Koordinate (das elektrische Feld wird parallel zur x-Achse angenommen) aufgetragen, ferner die Linien $V(x) + mc^2$ und $V(x) - mc^2$. Die Rechnungen von Sauter zeigen, daß die Eigenfunktionen, die etwa zum Eigenwert E_0 gehören, nur in den Gebieten I und III groß sind, daß sie jedoch im Innern des Gebietes II exponentiell abnehmen. Dies hat zur Folge, daß eine Wellenfunktion, die etwa zunächst nur in einem Gebiet, z. B. I, groß ist,

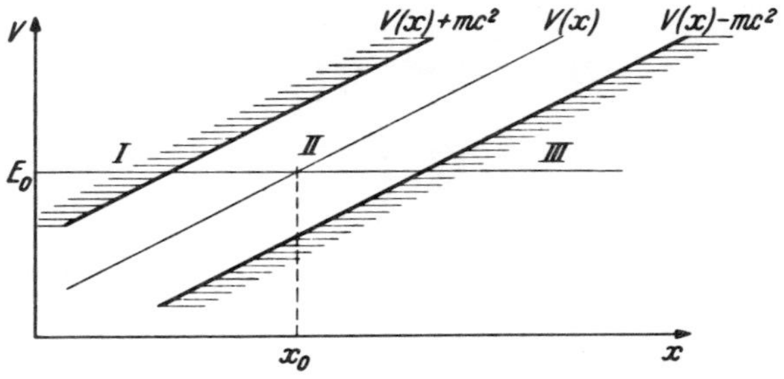

Fig. 1.

allmählich in das Gebiet III ausläuft, wobei der Durchlaßkoeffizient durch das Gebiet II, das hier die Rolle eines Gamowberges spielt, nach Sauter von der Größenordnung $e^{-\frac{m^2 c^3}{\hbar e |\mathfrak{E}|}\pi}$ ist. Bezeichnet man $|\mathfrak{E}_k| = \frac{m^2 c^3}{\hbar e}$

¹) F. Sauter, ZS. f. Phys. **69**, 742, 1931.

Figure 13: This is the fundamental diagram which governs the process of particle creation by a static potential (reproduced from ref.[93]). Regions I and III correspond to the solutions E^+ and E^- of the relativisitc quantum field. If the gradient of the potential is large enough, then a "level crossing" can occur between the regions I and III and, for a given value of the energy, a tunnelling of the wave function can occur from region I to III, through region II. The solution of the wave equation in region II, as shown by Sauter, is exponentially decreasing and the amplitude of the wave transmitted to region III critically depends from this exponential solution. The amplitude of the tunneling is directly related to the rate of pair creation, see ref.[95] and [96].

interpreted as describing particles once the usual transformation had been made: $t^c=-t$, $q^c=-q$, $E^c=-E$.

Since level crossing between the positive and the negative root solutions (E^+ and E^-) can exist near the surface of the black hole (see Fig. 14) we can apply the results of the Klein Paradox: processes of pair creation due to vacuum polarization should then occur. The pairs are created either at the expenses of the rotational energy or of the Coulomb energy of the black hole.

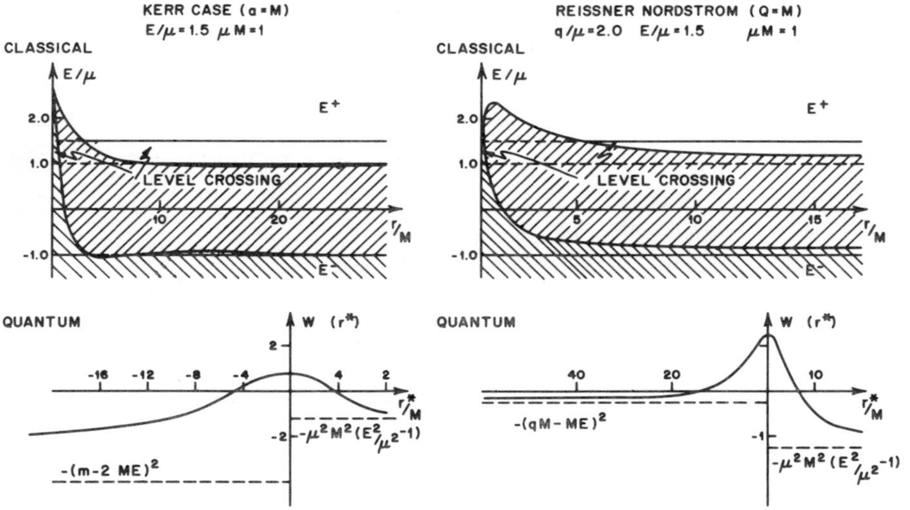

Figure 14: Effective potentials for particles of mass μ and charge q in an extreme Kerr black hole (a=M) and in an extreme Reissner Nordstrøm black hole (Q=M). Level crossing between the positive root solutions (E^+) and the negative root solutions (E^-) occurs due to the electromagnetic or rotational field of the black hole and of the test particle. The dashed region between the E^+ and E^- states corresponds to region II in fig. 13: no particles with real momentum can be found in this region which is forbidden to classical particles. Tunnelling from region E^+ to E^- can, however, occur on a quantum level, through this region, exactly in analogy to Fig. 13. To describe this tunnelling process a covariant Klein-Gordon equation $\nabla_\alpha \nabla^\alpha \Phi = \mu^2 \Phi$ has been solved in the given background geometry. The lower part of the figure shows the effective potential for this equation. Details in [97][100][101] and [102].

Once again we have studied the simplest possible case: the one of black holes in vacuum, not surrounded by any plasma or any additional electromagnetic or rotational field.

The case of a Kerr solution and creation of pair due to the shear of the dragging of the inertial frames was presented in pioneering works by Zel'dovich [101],[102], important analytic properties have been found by Starobinsky [103] and Starobinsky and Churilov [104] Unruh as formulated the problem in the framework of a second quantized field theory [105] and detailed results for the tunnelling of a massive scalar field have been given by Deruelle and myself [106]. The main results are summarized in the report of Deruelle [100]. The case of a Reissner Nordstrøm solution has been studied by three different approaches: by Zaumen [107] using some of the properties of the mass formula given in ref. [56], by Gibbons [108] using a second quantized formulation and by Nakamura and Sato [109] using the classical Sauter-Heisenberg-Euler approach. The main results are summarized in the reports by Gibbons [65], Sato [110] and Zaumen [111]. Finally the most general case of a Kerr–Newman solution has been treated by Damour and myself [112]. In this treatment we have followed the traditional approach in à la Sauter by studying the barrier penetration of a relativistic quantized scalar field and we have also used a powerful alternative technique. Since we are mainly interested in studying the process of vacuum polarization from macroscopic black holes ($M > M_\Theta$, $r_+ > 1.47$ km) and the pair creation occurs in a region of the approximate size of the Compton wavelength of the created particle ($\lambda < 10^{-9}$ cm), we can always choose at every point an inertial system of reference and proceed in our analysis exactly like in an usual flat space computations. We can in particular introduce at every point the tetrad given by Eq. (3) which indeed determines a Lorentz frame, and in that frame both electric and magnetic lines of force are parallel and given by Eq. (4). We can then locally apply the classical works of Heisenberg and Euler and the one by Schwinger and obtain not only the values of the critical fields but also the rate of particles created. This approach allows us to also treat the vacuum polarization processes leading to the creation of spin $\frac{1}{2}$ Dirac particles.

We then obtain for the transmission coefficient between the positive and the negative energy states for a Dirac field of spin $\frac{1}{2}$ and particles of mass μ and charge q the expression [113]:

$$T = \exp{-\rho} \quad \text{with} \quad \rho = \pi\mu^2/(qE_{(1)}) + 2\pi(n + \tfrac{1}{2} + \sigma_{(1)})\, B_{(1)}/E_{(1)}\,, \tag{44}$$

where n is the quantum number associated with the harmonic oscillation of the electron in the plane orthogonal to E and B. The minimum opacity is then obtained for $n=0$ and $\sigma_{(1)} = -\frac{1}{2}$ which means for a pair created along the common direction of the electric and magnetic fields. This implies that, as seen from infinity each particle will carry away an angular momentum in the ϕ direction given by $m=a\,\omega\sin^2\theta$ where ω represents the energy of the particle. Following Schwinger [113], [114] we can estimate the number of particles created by the formula

$$N = 2 \int \mathrm{Im}\,\mathcal{L} \sqrt{|g|}\, d^4x \qquad (45)$$

where

$$\sqrt{|g|} = \Sigma \sin\theta \qquad (46)$$

and

$$2\,\mathrm{Im}\mathcal{L} = (4\pi)^{-1} (E_{(1)}\, q/\pi)^2 \sum_{n=0}^{\infty} n^{-2} (n\pi B_{(1)}/E_{(1)}) \cot h(n\pi B_{(1)}/E_{(1)}) \exp -(n\pi\mu^2/qE_{(\qquad (47)$$

from these expressions we can compute the rate of particle creation per unit volume outside the horizon. Because for every pair produced one of the particles is expelled at infinity and one with opposite charge and angular momentum is captured by the black hole, it follows that in these processes of vacuum polarization the charge, the total mass-energy M, and the angular momentum of a black hole are reduced. The limitations imposed by the vacuum polarization process on the electromagnetic field of a Kerr-Newman geometry are given in Table 2; it is clear that the electromagnetic field cannot become larger than 2×10^{12} g. More details are given by Damour [113] and Davis [114].

$M_{ir}/M_{\odot}$	Maximum strength of electromagnetic field in Gauss	Maximum net Charge in Electron Charge	Maximum energy Extractable in erg
1	B 1.18×10^{19} / 2.00×10^{12} P	B 2.14×10^{39} / 3.63×10^{32} P	B 1.79×10^{54} / 5.15×10^{40} P
10^2	1.18×10^{17} / 1.88×10^{12}	2.14×10^{41} / 3.41×10^{36}	1.79×10^{56} / 4.55×10^{46}
10^4	1.18×10^{15} / 1.77×10^{12}	2.14×10^{43} / 3.21×10^{40}	1.79×10^{58} / 4.03×10^{52}
10^6	1.18×10^{13} / 1.67×10^{12}	2.14×10^{45} / 3.03×10^{44}	1.79×10^{60} / 3.59×10^{58}
10^8	1.18×10^{11} / 1.18×10^{11}	2.14×10^{47} / 2.14×10^{47}	1.79×10^{62} / 1.79×10^{62}

It is very important to understand how the presence of a plasma will affect all these considerations. In particular these processes of vacuum polarization together with the mechanism of energy extraction presented in §6 can offer a very attractive model for the explanation of the recently observed x and γ-ray bursts (see §1) [115]. The electromagnetic field could be largely amplified by the accretion of material with the mechanism described in §5, if an instability is then reached a flare could be generated and the process of vacuum polarization would then occur. A natural prediction of this model [115] is the possibility of having repetitive outbursts at a temporal distance which will be determined uniquely by the strength of the magnetic field in the accreting material, by the rate of matter accretion and by the angular momentum of the black hole. It is also interesting to notice that the critical field defined by Eq. (44) automatically gives $\sim 10^{20}$ eV for the energy of created pairs Possible coincidences, therefore, should be analyzed between the very energetic cosmic rays showers and γ or x-ray bursts.

7. X-RAY SOURCES AND GRAVITATIONAL RADIATION

The observation of more than 200 pulsars and 40 binary x-ray sources [12] and the clear evidence for a finite lifetime their processes of emission of radiation, suggests the existence inside our own galaxy of $\sim 10^8 - 10^9$ gravitationally collapsed stars. There exists therefore a good possibility of observing the process of gravitational collapse itself.

There are different processes in which, at least in principle, we should be able to observe this phenomenon:

(a) from observations of supernova inside our galaxy and in nearby galaxies we conclude that processes of gravitational collapse occur approximately once every thirty years in every galaxy. From the pulsar observations (Crab Nebula, Vela) we conclude that some of these processes give rise to neutron stars [12].

(b) We are still far from understanding the process of gravitational collapse leading to a black hole; we do not know if black holes are directly formed by the implosion of a massive degenerate core of white dwarf material, or if instead, they are formed by a two step process. First a neutron star forms and then by accretion of additional matter the neutron star reaches the critical mass and undergo total gravitational collapse to a black hole [116] [12].

What has become clear in recent years [117] is that no matter what the details of the gravitational collapse are of the order of a few percent of the mass energy of the original star is likely to be radiated in a burst of gravitational waves during the occurrence of the collapse process and this in turn would give a detectable signal of gravitational radiation [117], [118], [119].

There is however quite a different process which appears to be extremely interesting, particularly in light of the possibility of storing a large amount of energy in the magnetosphere of a collapsed object: The coupling between electromagnetic and gravitational radiation. Some aspects of this coupling have been discussed by Choquet-Bruhat [120], its relevance as a tool to detect gravitational radiation signals has been discussed by Partridge and Ruffini [121].

The basic formalism to analyze the coupling between electromagnetic and gravitational radiation in the field of a collapsed object has been given by Zerilli [40] (see also refs. [41], [42] and [43]). Once again for the sake of simplicity, this coupling has been studied in the field of a Reissner-Nordstrøm geometry.

Zerilli considers the equations governing first-order perturbations both in the gravitational and electromagnetic field of a given Reissner-Nordstrøm geometry ($g^{(0)}_{\mu\nu}$, $F^{(0)}_{\mu\nu}$)

$$g_{\mu\nu} = g^{(0)}_{\mu\nu} + h_{\mu\nu} \quad (48.1) \qquad F_{\mu\nu} = F^{(0)}_{\mu\nu} + f_{\mu\nu} \quad (48.2)$$

the corresponding set of Einstein-Maxwell equations can then be written

$$G_{\mu\nu} = 8\pi(T_{\mu\nu} + E_{\mu\nu}) \quad (49.1)$$

$$[(-g)^{1/2} F^{\mu\nu}]_{,\nu} = 4\pi(-g)^{1/2} J^{\mu} , \quad (49.2)$$

where $T^{\mu\nu}$ is the matter energy-momentum tensor and J^{μ} is the four current driving the perturbation. By expanding the perturbations $h_{\mu\nu}$ and $f_{\mu\nu}$ in terms of tensor and vector spherical harmonics, following the classic Regge-Wheeler-Zerilli approach [37], [39], Zerilli was able to reduce the entire problem of the perturbation analysis to four master equations

$$\frac{d^2R^{(e.m)}_{LM}}{dr^{*2}} + (\omega^2 - V^{grav}_L)\, R^{(e.m)}_{LM} = a^{(e.m)}_{LM}\, f_{LM} + \text{grav source} , \quad (50.1)$$

$$\frac{d^2f^{(e.m)}_{LM}}{dr^{*2}} + (\omega^2 - V^{elm}_L)\, f^{(e.m)}_{LM} = b^{(e.m)}_{LM}\, R^{(e.m)}_{LM} + C^{(e.m)}_{LM}\, \frac{dR^{(e.m)}_{LM}}{dr^*} + \text{source}, \quad (50.2)$$

two for the electric (e) parity solutions and two for the magnetic (m) parity solutions. Here r^* is the usual tortoise radial coordinate [118]. This highly coupled system of equations has been numerically integrated [122] and two limiting cases of relevant physical processes have been considered:

(a) Gravitationally induced electromagnetic radiation [44].

In this process a gravitational perturbation, generated for sake of an example, by the radial infall of an uncharged object, is considered in the background geometry and in the electric field of the Reissner-Nordstróm solution. It is found that due to the high coupling of Eq. (50) an amount of electromagnetic radiation

$$(\Delta E)_{em} \sim 0.03\ m\left(\frac{m}{M}\right)\ \left(\frac{Q}{M}\right)^2 \qquad (51)$$

and an amount of gravitational radiation

$$(\Delta E)_{grav} \sim 0.01\ m\left(\frac{m}{M}\right) \qquad (52)$$

are radiated by the system of the black hole of the infalling test particle. Here m is the mass of the test particle and Q and M, as usual, are the charge and the mass of the black hole.

The qualitative explanation of this phenomenon is described in Fig. 15. Due to the very large difference of cross section between electromagnetic and gravitational radiation this result is of the greatest interest for possible observational consequences.

Similar analysis, with a range of validity limited to cases in which $Q \ll M$, have been considered in the framework of the Newman-Penrose formalism by Chitre, Price and Sandberg [123],[125]. Particularly interesting is the work by Chitre [124] who used a Kerr brackground still in the approximation $Q \ll M$.

(b) Electromagnetically induced gravitational radiation [45],[126].

This is the extreme opposite case of the one considered in the previous example (a), in which a test particle of charge q and mass m with $q/m \gg 1$ is moving in a background of a Reissner–Nordstrøm solution of charge Q and mass M. Due to the high coupling of the system of Eq. (50) the electromagnetic field of the test particle generates an interference term with the background electromagnetic field which can generate a considerable amount of gravitational radiation. This effect was first analyzed, in the framework of a relativistic treatment in flat space, by Khriplovich and Sushkov [127]. Our treatment generalizes their results to the case of strong gravitational fields

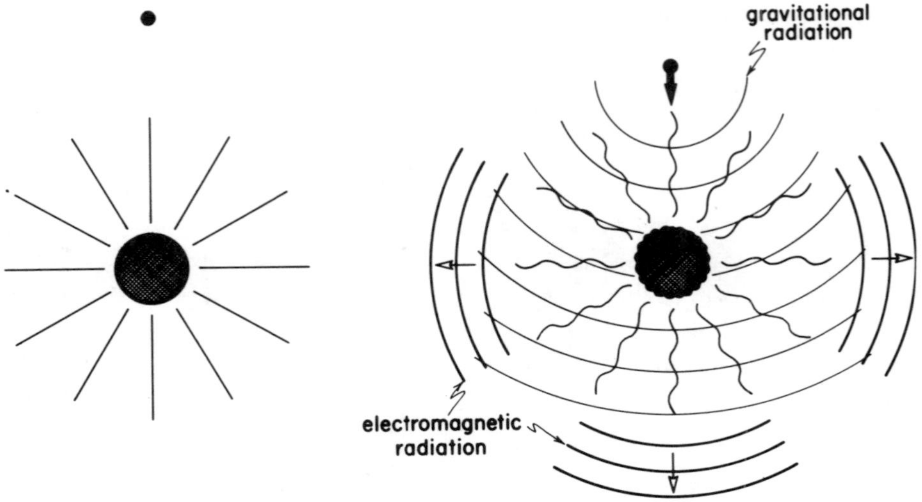

Figure 15: Schematic representation of the gravitationally induced electromagnetic radiation. We have considered [44] the electromagnetic and gravitational radiation emitted in the infall of an uncharged test particle of mass m in the field of a Reissner-Nordstrøm black hole of mass M and charge Q. Similar effects have clearly to be also expected to occur in the case of black holes surrounded by a magnetosphere (see §3). The infalling particle perturbs the background electromagnetic field by the gravitational field of the test particle and scattering of gravitational waves. The major result is simply stated: in the limit of $Q/M \sim 1$ the amount of electromagnetic radiatiom emitted by the infall of an uncharged object can be larger than the corresponding amount of gravitational radiation.

and takes into proper account the entire and complex coupling between electromagnetic and gravitational radiation. Details are given in refs. [45] and [126]. In the specific case we have considered, radial infall of a test particle with $q/m=-2$ falling in a Reissner–Nordstrøm black hole with $Q/M = 0.5$ the total amount of electromagnetic radiation emitted

$$(\Delta E)_{elec} \simeq 0.021\, q(q/M) \quad ,$$

while the gravitational radiation is

$$(\Delta E)_{grav} \simeq 0.015\, q(q/M)$$

it is interesting to remark that the back effect of the gravitational radiation on the electromagnetic one, contrary to what had been purported by some, is totally negligible (less than 2% of the total electromagnetic radiation).

Still along the lines of analyzing some of the physical consequences of the system of Eqs. (50) we must mention the work by Gerlach [128]. Studying the solutions of Eq. (50) in the W.K.B. approximation limit he has been able to find evidence for the existence of "beating" modes between the electromagnetic and the gravitational radiation.

Looking toward the future it appears very likely that if the expected accuracy of the new generation gravitational waves detectors will be reached [129], then signal of gravitational waves will be detected [130]. One of the most powerful tools for identifying their source will certainly be the detection of an associated burst of high energy particles and of electromagnetic radiation.

REFERENCES

[1] A. Friedman, Zeit. fur Physik 10, 377 (1922); Zeit. fur Physik 21, 326 (1924).

[2] G. Lemaitre, Ann. Soc. Sci. Bruxelles 47A, 49 (1927).

[3] E. Hubble, Proc. Nat. Acad. 15, 168 (1929).

[4] For a review on recent observations in cosmology see the report by R.B. Partridge in these proceedings.

[5] For a review on recent precision measurements for testing gravitational theories see the report by F. Everitt, for theoretical interpretation of experimental results see the report by K. Nordtvedt, both in these proceedings.

[6] L.D. Landau. Phys. Z. Sow 1, 285 (1932).

[7] S. Chandrasekhar, Atrophys. J. 74, 81 (1930); S. Chandrasekhar, Mon. Not. R.A. 91, 456 (1931); S. Chandrasekhar, Z. Astrophys. 5, 321 (1932); S. Chandrasekhar. Mon. Not. R.A.S. 95, 207 (1935).

[8] R.J. Oppenheimer and G.M. Volkoff, Phys. Rev. 55, 374 (1938).

[9] R.J. Oppenheimer and R. Snyder, Phys. Rev. 56. 455 (1939).

[10] W. Bade and F. Zwicky, Phys. Rev. 45, 138 (1934).

[11] J.A. Wheeler in Gravitation and Relativity, Eds. H.Y. Chiu and W.F. Hoffman, (W.A. Benjamin Inc., New York 1963), 195 ff.

[12] For a discussion on this entire topic see, e.g. Neutron Stars, Black Holes and Binary x-ray Sources, Eds. H. Gursky and R. Ruffini (D. Reidel Publ., Co., Dordrecht-Holland 1975).

[13] C.E. Rhoades, Ph.D. Thesis (1971), Ph.D. Dissertation, Princeton Univ., C.E. Rhoades and R. Ruffini, Phys. Rev. Letters 32, 324 (1974), have proved

under the assumptions (a) knowledge of an equation of state at nuclear and subnuclear densities, (b) equilibrium configurations computed in the framework of general relativity, (c) $(dp/d\rho) < c^2$, that the neutron star mass must always be smaller than 3.2 $M_\odot$. M. Nauenberg and G. Chapline, Ap. J. 179, 277 (1973), have given a variational principle to reach under similar assumptions very similar results. A.G. Sabbadini and J.B. Hartle, Ap. and Space Science 25, 117 (1973) have proven that without condition (c) an upper limit can still be found to be 5.05 $M_\odot$. R.V. Wagoner and R.C. Malone, Ap. J. Letters 189, L75 (1974) have purported very considerable effects in the computations of the maximum mass of a neutron star due to the theory of gravitation used: their treatment, however, leads directly to the violation of the conservation law of the four momentum of an isolated system and contains basic analytic mistakes.

[14] For a discussion of this entire topic, see. e.g. X-ray Astronomy, Eds. R. Giacconi and H. Gursky (D. Reidel Publ., Co., Dordrecht-Holland 1974). see also ref. [12].

[15] I.S. Shklovskii, Societ Atron. 11, 749 (1968).

[16] Ya.B. Zel'dovich and I.D. Novikov, Suppl. Nuovo Cimento IV, 810 (1966).

[17] S. Hayakawa and M. Matsuda, Progr. Theoret. Phys. Suppl. Japan 30, 204 (1964).

[18] R. Leach and R. Ruffini, Ap. J. Letters 180, L15 (1973).

[19] See, e.g. Y. Avni and J.N. Bahcall, Ap. J. Letters 202, L131 (1975).

[20] See, e.g. J. Nelson in the Proceedings of the "E. Fermi Summer School 1975", Eds. R. Giacconi and R. Ruffini (North Holland, Amsterdam 1975).

[21] R. Lucke, D. Yeutis, H. Friedman, G. Fritz and S. Shulamn, Ap. J. Letters 206, L25 (1976).

[22] N.W. White et al. I.A.U. Circular No.2870 (1975).

[23] For an interesting discussion of this topic see, e.g. P.J E. Peebles, Gen. Rel and Grav. 3, No.1, 63 (1972).

[24] See, e.g. J.A. Wheeler, "Mechanics for jets", Pontifica Academia Scientiarum, Scripta Varia No.35, 539 (1971).

[25] See, e.g. H. Gursky, p.182 in ref. [12].

[26] J. Silk and J. Arons, Ap. J. Letters 200, 131 (1975).

[27] J.N. Bahcall and J.P. Ostriker, Nature 256, 23 (1975).

[28] J. Grindlay et al. Ap. J. Letters 205, L127 (1976).

[29] J. Grindlay and H. Gursky, Ap. J. Letters 205, L131 (1976).

[30] J. Heise, A.C. Brinkman, A.J.F. den Boggende, D.R. Parsignault, J. Grindlay and H. Gursky, Nature 261, 562 (1976).

[31] See, e.g. Galactic and Extragalactic Radio Astronomy, Eds. G.L. Verschurr and J.I. Kellerman (Springer-Verlag, Berlin 1974).

[32] R.P. Kerr, Phys. Rev. Letters 11, 237 (1963).

[33] E.T. Newman, E. Couch, R. Chinnapared, A. Exton, A. Prakash and R. Torrence, J. Math. Phys. 6, 918 (1965).

[34] B. Carter, in these proceedings.

[35] D. Christodoulou and R. Ruffini in Black Holes, Eds. B. DeWitt and C. DeWitt (Gordon and Breach, New York 1973), R151-R160.

[36] B. Carter, Phys. Rev. 174, 1559 (1968).

[37] T. Regge and J.A. Wheeler, Phys. Rev. 108, 1063 (1957).

[38] C.V. Vishveshwara, Phys. Rev. D1, 2870 (1970).

[39] F. Zerilli, Phys. Rev. Letters 24, 727 (1970);
F. Zerilli, J. Math. Phys. 11, 2203 (1970);
F. Zerilli, Phys. Rev. D2, 2141 (1970).

[40] F. Zerilli, Phys. Rev. D9, 860 (1974).

[41] V. Moncrief, Phys. Rev. D9, 2707 (1974);
V. Moncrief, Phys. Rev. D10, 1057 (1974).

[42] J. Weinstein, these proceedings and senior thesis, Princeton Univ., June 1975.

[43] C.H. Lee, J. Math. Phys, in press.

[44] M. Johnston, R. Ruffini and F. Zerilli, Phys. Rev. Letters 37, 1317 (1973).

[45] M. Johnston, R. Ruffini and F. Zerilli, Phys. Letters B49, 185 (1974)

[46] S.A. Teukolsky, Phys. Rev. Letters 29, 1114 (1972);
S.A. Teukolsky, Ap. J. 185, 635 (1973).

[47] That indeed the Teukolsky equation describes all the gravitational perturbations of a Kerr black hole has been shown by R.M. Wald, J. Math. Phys. 14, 1453 (1973).

[48] See, e.g. J.M. Stewart, Proc. Roy. Soc. (London) A344, 65 (1975);
See also W. Press and S. Teukolsky, Ap. J. 185, 649 (1973);
J.B. Hartle and D.C. Wilkins, Comm. Math. Phys. 38, 47 (1974);
J.L. Friedman and B.F. Schutz, Phys. Rev. Letters 32, 243 (1974).

[49] An alternative approach to the perturbations analysis of a general stationary metric based on the Hadamard technique is presented in these proceedings by M. Peterson and R. Ruffini.

[50] B. Carter, Phys. Rev. 174, 1559 (1968);
B. Carter, Comm. Math. Phys. 10, 280 (1968).

[51] See also L.P. Hughston, R. Penrose, P. Sommers and M. Walker, Comm. Math. Phys. 27, 303 (1972).

[52] See, e.g. M. Johnston and R. Ruffini, Phys. Rev. D10, 2324 (1974).

[53] R. Ruffini and J.A. Wheeler, Cosmology from Space Platforms, E.S.R.O. Book SP52, Paris 1971. Reproduced in Black Holes, Gravitational Waves and Cosmology, M. Rees, R. Ruffini and J.A. Wheeler (Gordon a d Breach, New York, London 1974).

[54] D.C. Wilkins, Phys. Rev. D5, 814 (1972).

[55] See, e.g. R. Ruffini in Black Holes, Eds. B.DeWitt and C. DeWitt (Gordon and Breach, New York, London 1972), p.499 and following.

[56] D. Christodoulou and R. Ruffini, Phys. Rev. D4, 3552 (1971).

[57] R. Penrose, Rivista Nuovo Cimento 1, 252 (1969).

[58] R. Penrose and R. Floyd, Nature 229, 193 (1971).

[59] G. Denardo, L. Hively and R. Ruffini, Phys. Letters 49B, 185 (1974).

[60] S.W. Hawking, Phys. Rev. Letters 26, 1344 (1971).

[61] J.D. Bekenstein, Ph.D. Thesis, Princeton Univ., 1972;
J.D. Bekenstein, Phys. Rev. D7, 2333 (1973);
J.D. Bekenstein, Phys. Rev. D9, 3292 (1974).

[62] J.M. Bardeen, B. Carter and S.W. Hawking, Comm. Math. Phys. 31, 170 (1973).

[63] B. Carter in Black Holes, Eds. B. DeWitt and C. DeWitt (Gordon and Breach, New York, London 1973).

[64] S.W. Hawking, Nature 238, 30 (1974);
S.W. Hawking, Comm. Math. Phys. 43, 199 (1975);
S.W. Hawking, Phys. Rev. D13, 191 (1976).

[65] G. Gibbons, these proceedings.

[66] P.C.W. Davies, these proceedings.

[67] W. Unruh, these proceedings.

[68] A.J. Deutsch, Ann. Astrophys. 18, 1(1955).

[69] R. Ruffini and A. Treves, Astrophys. Letters 13, 109 (1973).

[70] R. Ruffini in Black Holes, Eds. B. DeWitt and C. DeWitt (Gordon and Breach, New York, London 1973), see p.525.

[71] See, e.g. R. Hanni, these proceedings.

[72] R. Ruffini, Invited talk delivered at the Seventh Texas Symposium on Relativistic Astryphysics, Dallas, December 1974, Published by New York Acad. of Sciences, 1975, p.95.

[73] Much work has been done in the analysis of electromagnetic test fields in a curved background. For a Schwarzschild brackground see, J.A. Wheeler, Phys. Rev. 97, 511 (1955). R. Ruffini, J. Tiomno and C.V. Visheshwara,Nuovo Cimento Letters 3, 211 (1972). For a Reissner-Nordstrøm background see refs.[40]-[43]. For a Kerr background see ref.[46] and E.D. Fackerell and J.R. Ipser, Phys. Rev. D5, 2455 (1972); J.M. Cohen and L.S. Kegeles, Phys. Rev. D10, 1070 (1974); S. Chandrasekhar, Proc. Roy. Soc. (London) A349, 1 (1976); S. Detweiler, these proceedings.

[74] T. Damour, R. Hanni, R. Ruffini and J. Wilson, to be submitted for publication.

[75] R.M. Wald, Phys. Rev. D10, 1860 (1974).

[76] J.A. Peterson, Phys. Rev. D12, 8 (1975).

[77] D.M. Chitre and C.V. Visheshwara, Phys. Rev. D12, 1538 (1975).

[78] A.R. King, J.P. Lasota and W. Kundt, Phys. Rev. D12, 3037 (1975); See also report by W. Kundt in these proceedings.

[79] A Papapetrou, Ann. Inst. H. Poincare 4, 83 (1966).

[80] T. Damour, to be submitted for publication.

[81] A. Lichnerowicz, Relativistic Hydrodynamics and Magnetohydrodynamics (W.A. Benjamin Inc., New York, Asmterdam 1967).

[82] A. Lichnerowicz, "Ondes de choc, ondes infinitesimales, etc." Proceedings of the Centro Italiano Matematico Estivo 1970. See also A. Lichnerowicz's report in these proceedings.

[83] J. Wilson, in these proceedings.

[84] R. Ruffini and J. Wilson, Phys. Rev. D12, 2959 (1975).

[85] T. Damour, Invited talk presented at the Seventh Texas Symposium on Relativistic Astrophysics, 1974, proceedings published by New York Academy of Sciences, 1975, p.120.

[86] For the definition of lines of force in a curved background, see, e.g. D. Christodoulou and R. Ruffini in Black Holes, Eds. B. DeWitt and C. DeWitt (Gordon and Breach, New York 1973), p.R151 and following, see also R. Hanni and R. Ruffini, Nuovo Cimento Letters 15, 189 (1976).

[87] J. Wilson, Invited talk presented at the Seventh Texas Symposium on Relativistic Astrophysics, 1974. Proceedings published by the New York Academy of Sciences, 1975, p.123.

[88] J. Wilson, lectures delivered at the "E. Fermi Summer School" 1975, Eds. R. Giacconi and R. Ruffini (North Holland, Amsterdam 1976).

[89] This leads to the definitions of "black holes: the largest storehouse of energy in the Universe", D. Christodoulou and R. Ruffini, Gravity Research Award, Third prize.

[90] For a classic example of this instability see, e.g. J.M. Leblanc and J.R. Wilson, Ap. J. 161, 541 (1970).

[91] O. Klein, Zeit. fur Physik 53, 157 (1929).

[92] F. Sauter, Zeit. fur Physik 69, 742 (1931).

[93] W. Heisenberg and H. Euler, Zeit. fur Physik 98, 714 (1936).

[94] W. Pauli, 1935-36, "The theory of the positron and related topics", Princeton Seminar, Princeton University.

[95] J. Schwinger, Phys. Rev. 82, 664 (1951).

[96] T. Damour, these proceedings.

[97] N. Deruelle and R. Ruffini, Phys. Letters 52B, 437 (1974).

[98] This identification was made transparent by choosing a special system of coordinates by T. Damour, Nuovo Cimento Letters 12, 315 (1975).

[99] N. Deruelle, These de Doctorat Troisieme Cycle-Ecole Normale Superieure, Paris 1975.

[100] N. Deruelle, these proceedings.

[101] Ya. B. Zel'dovich, Sov. Phys.-JETP Letters 14, 180 (1971).

[102] Ya.B. Zel'dovich, Sov. Phys.-JETP Letters 35, 1085 (1972).

[103] A.A. Starobinski, Sov. Phys.-JETP 37, 28 (1973).

[104] A.A. Starobinski and S.M. Chrilov, Sov. Phys.-JETP 38, 1 (1974).

[105] W. Unruh, Phys. Rev. D10, 3194 (1974).

[106] N. Deruelle and R. Ruffini, Phys. Letters 57B, 248 (1975).

[107] W.T. Zaumen, Nature 247, 530 (1974).

[108] G. Gibbons, Comm. Math. Phys. 44, 245 (1975).

[109] T. Nakamura and H. Sato, Phys. Letters 61B, 371 (1976).

[110] M. Sato, these proceedings.

[111] W.T. Zaumen, these proceedings.

[112] T. Damour and R. Ruffini, Phys. Rev. Letters 35, 463 (1975).

[113] T. Damour, these proceedings.

[114] L. Davis, these proceedings.

[115] See ref. [72], p. 110 and on.

[116] R. Ruffini and J.A. Wheeler, Physics Today, 30 January 1971.

[117] For a review of this field see R. Ruffini, "Les ondes gravitationnelles", La Recherche 6, 907 (1975).

[118] For the study of sources of gravitational radiation in the field of black holes and the analysis of their spectral and polarization distribution, see e.g. R. Ruffini in Black Holes, Eds. B. DeWitt and C. DeWitt (Gordon and Breach, New York, London 1973).

[119] For the computation of the cross section of a gravitational wave detecter see Chap. 7 of M. Rees, R. Ruffini and J.A. Wheeler, Black Holes, Gravitational Waves and Cosmology (Gordon and Breach, New York, London 1974).

[120] Y. Choquet Bruhat, these proceedings.

[121] R.B. Partridge and R. Ruffini, "Gravitational waves and a search for the associated microwave radiation", Gravity Research Award, third award, 1970.

[122] M. Johnston, R. Ruffini and M. Peterson, Nuovo Cimento Letters 9, 217 (1974).

[123] D.M. Chitre, R.H. Price and V. Sandberg, Phys. Rev. Letters 31, 1018 (1973).

[124] D.M. Chitre, Phys. Rev. D11, 760 (1975).

[125] D.M. Chitre, R.H. Price and V. Sandberg, Phys. Rev. D11, 747 (1975).

[126] R. Ruffini, Phys. Letters 41B, 334 (1972).

[127] O.P. Sushkov and I.B. Khriplovich, Soviet Phys.-JETP 39, 1 (1974).

[128] U. Gerlach, Phys. Rev. Letters 32, 1023 (1974).

[129] See, e.g. W. Fairbank, Invited lecture delivered at the 1975 "E. Fermi Summer School",Eds. R. Giacconi and R. Ruffini (North-Holland, Amsterdam 1976).

[130] See, e.g. R. Ruffini, Invited talk delivered at the 1975 "E. Fermi Summer School", Eds. R. Giacconi and R. Ruffini (North-Holland, Amsterdam, 1976).

MAGNETOHYDRODYNAMICS NEAR A BLACK HOLE*

James R. Wilson

Lawrence Livermore Laboratory, University of California
Livermore, California U. S. A.

INTRODUCTION

We present in this paper a numerical computer study of hydromagnetic flow near a black hole. First, the equations of motion are developed to a form suitable for numerical computations. Second, the results of calculations describing the magnetic torques exerted by a rotating black hole on a surrounding magnetic plasma and the electric charge that is induced on the surface of the black hole are presented.

THE EQUATIONS

We start from the divergence of the energy momentum tensor for a perfect fluid

$$T^{\nu}_{\mu;\nu} = \frac{1}{\sqrt{-g}} \frac{\partial}{\partial x^{\nu}}(T^{\nu}_{\mu} \sqrt{-g}) + \frac{1}{2} \frac{\partial g^{\alpha\beta}}{\partial x^{\mu}} T_{\alpha\beta} \tag{1}$$

where

$$T_{\mu\nu} = (\rho + \varepsilon + P)U_{\mu} U_{\nu} + g_{\mu\nu} P. \tag{2}$$

To put this in a form suitable for computations, a momentum density $S_{\mu} = (\rho + \varepsilon + P)U_{\mu} U^{t}$, a time four velocity $V^{\nu} = U^{\nu}/U^{t}$, and a number density $D = \rho U^{t}$ are introduced, where ρ is the proper number density of the fluid, ε is the proper thermal energy density, P is the pressure, and U^{μ} is the usual four velocity.

With manipulation, the energy and particle conservation equations become

$$\frac{1}{\sqrt{-g}} \frac{\partial}{\partial x^{\nu}}(S_{\mu} V^{\nu} \sqrt{-g}) + \frac{\partial P}{\partial x^{\mu}} + \frac{1}{2} \frac{\partial g^{\alpha\beta}}{\partial x^{\mu}} \frac{S_{\alpha} S_{\beta}}{S^{t}} = 0 \tag{3}$$

and

$$\frac{1}{\sqrt{-g}} \frac{\partial}{\partial x^{\nu}}(D V^{\nu} \sqrt{-g}) = 0. \tag{4}$$

While these equations are sufficient to determine hydrodynamic flow, it is preferable to use only the three space-like momentum elements of Eq. (3) and the time-projected part of $T^{\mu}_{\nu;\mu}$ to find the time behavior of ε.

Consider $U^{\nu}T^{\mu}_{\nu;\mu} = 0$. Then, using the condition $U^{\nu}U_{\nu} + 1 = 0$ to evaluate and simplify the divergence terms, and introducing an energy density $E = \varepsilon U^{t}$, the energy equation becomes:

$$\frac{\partial}{\partial x^{\mu}}(E V^{\mu} \sqrt{-g}) + P \frac{\partial}{\partial x^{\mu}}(U^{t} V^{\mu} \sqrt{-g}) = 0. \tag{5}$$

These equations have now been made as similar as possible to the Newtonian Eulerian hydrodynamic equations, for which a large body of numerical methodologies exist.

---

*This work was performed under the auspices of the U. S. Energy Research & Development Administration.

Next, we add the equations for the magnetic fields. At the start we specialize to axial symmetry. The magnetic field is described by two independent variables: H_ϕ, the component of the magnetic field about the axis of symmetry; and A_ϕ, the vector potential component about the symmetry axis. Consider the plasma as a perfect conductor expressed by setting the comoving electric field $U^\mu F_{\mu\nu}$ equal to zero, where $F_{\mu\nu}$ is the electromagnetic field tensor. With the condition of a zero electric field the components of the electromagnetic tensor are

$$F_{RZ} = H_\phi,\ F_{R\phi} = \frac{\partial A_\phi}{\partial R},\ F_{Z\phi} = \frac{\partial A_\phi}{\partial Z}, \tag{6}$$

$$F_{tZ} = V^\phi F_{Z\phi} - V^R F_{RZ},\ F_{tR} = V^\phi F_{R\phi} + V^Z F_{RZ},$$

and

$$F_{t\phi} = -V^R F_{R\phi} - V^Z F_{Z\phi}.$$

This last equation can be reinterpreted as the time-evolution equation for the vector field component A_ϕ

$$\frac{\partial A_\phi}{\partial t} = - V^R \frac{\partial A_\phi}{\partial R} - V^Z \frac{\partial A_\phi}{\partial Z}. \tag{7}$$

The time evolution of H_ϕ is found from the Maxwell equation $F_{\mu\nu;\eta} + F_{\eta\mu;\nu} + F_{\nu\eta;\mu} = 0$, with $\mu = R$, $\nu = Z$, $\eta = t$. (Other components of the equation are solved identically by choosing A_ϕ and H_ϕ to describe the magnetic field.)

The equation for H_ϕ is

$$\frac{\partial H_\phi}{\partial t} = \frac{\partial}{\partial Z}\left(V^\phi \frac{\partial A_\phi}{\partial R}\right) - V^Z H_\phi - \frac{\partial}{\partial R}\left(V^\phi \frac{\partial A_\phi}{\partial Z}\right) + V^R H_\phi . \tag{8}$$

Finally, the electromagnetic force $F_{\mu\nu} J^\nu$ must be added to the momentum equation where $J^\nu = \frac{1}{\sqrt{-g}} \frac{\partial}{\partial \chi^\mu} (F^{\mu\nu} \sqrt{-g})$ is the electric current. Because the plasma is a perfect conductor, no additional terms occur in the energy equation.

In the rest of the paper we are concerned only with a static metric. To recapitulate, the equations in this situation are

$$\frac{\partial D}{\partial t} + \frac{1}{\sqrt{-g}} \frac{\partial}{\partial \chi^i} (\sqrt{-g} V^i D) = 0 \qquad \begin{matrix} i = R,\ Z \\ j = R,\ Z,\ \phi \end{matrix} \tag{9}$$

$$\frac{\partial E}{\partial t} + \frac{1}{\sqrt{-g}} \frac{\partial}{\partial \chi^i} (\sqrt{-g} V^i E) + P \left[\frac{\partial U^t}{\partial t} + \frac{1}{\sqrt{-g}} \frac{\partial}{\partial \chi^i} (\sqrt{-g} V^i U^t)\right] = 0$$

$$\frac{\partial S_j}{\partial t} + \frac{1}{\sqrt{-g}} \frac{\partial}{\partial \chi^i} (S_j V^i \sqrt{-g}) + \frac{\partial P}{\partial \chi^j} + \frac{1}{2} \frac{\partial g^{\alpha\beta}}{\partial \chi^j} \frac{S_\alpha S_\beta}{S^t} + J^\nu F_{\nu j} = 0$$

$$\frac{\partial H_\phi}{\partial t} = \frac{\partial H_\phi}{\partial Z} \quad V^\phi \frac{\partial A_\phi}{\partial R} - V^Z H_\phi \quad - \frac{\partial}{\partial R} \quad V^\phi \frac{\partial A_\phi}{\partial Z} + V^R H_\phi$$

$$\frac{\partial A_\phi}{\partial t} = - V^R \frac{\partial A_\phi}{\partial R} - V^Z \frac{\partial A_\phi}{\partial Z} .$$

The V^i are found from S_j, D, and E by using the velocity normalization condition $U^\mu U_\mu + 1 = 0$. A perfect gas equation-of-state with an adiabatic coefficient of 1.5 completes the hydrodynamic description of the system. A coefficient of 1.5 was chosen because the electrons are completely relativistic while the ions are weakly relativistic at the level where shock waves are formed.

We use the Boyer-Lindquist form of the Kerr metric expressed in cylindrical R, Z coordinates.

$$g^{RR} = \frac{\Delta}{\rho^2}\left(\frac{Z^2}{\Delta} + \frac{R^2}{r^2}\right) ; \; g^{RZ} = \frac{RZ}{\rho^2}\left(\frac{\Delta}{r^2} - 1\right) ; \; g^{ZZ} = \frac{\Delta}{\rho^2}\left(\frac{R^2}{\rho^2} + \frac{Z^2}{r^2}\right) \tag{10}$$

$$g^{\phi\phi} = \frac{r^2}{\rho^2 R^2} - \frac{a^2}{\rho^2 \Delta} ; \; g^{\phi t} = \frac{2amr}{\rho^2 \Delta} ; \; g^{tt} = -\left(1 + \frac{2mr(a^2 + r^2)}{\rho^2 \Delta}\right) ; \; \sqrt{-g} = \frac{R\rho^2}{r^2} ,$$

where $r^2 = R^2 + Z^2$, $\rho^2 = r^2 + a^2 Z^2/r^2$, $\Delta = r^2 - 2mr + a^2$, m is the hole mass, and a is the hole angular momentum.

To solve these differential equations, a finite, rectangular mesh is set up in R, Z space. The variables S_j, A_ϕ, and V_i are considered to be centered at the mesh corners. The variables D, E, P, H_ϕ, and $g^{\alpha\beta}$ are thought to be centered between the mesh corners (Fig. 1). The grid is drawn for a = 0 so that 20 equal zones span the hole in each direction. Ten zones span the hole for a = 1. Outside the hole, zone size increases by a constant factor zone-to-zone so that the outside boundary is 10 times the hole radius. Forty zones span each direction. See Fig. 2 for overall calculational grid.

Many methods exist in the literature of numerical techniques for solving these equations. Several second-order schemes were tried, but were found to give very poor results near the horizon of the hole because of sharp changes in the variables from one zone to the next. The scheme adopted is about the simplest.

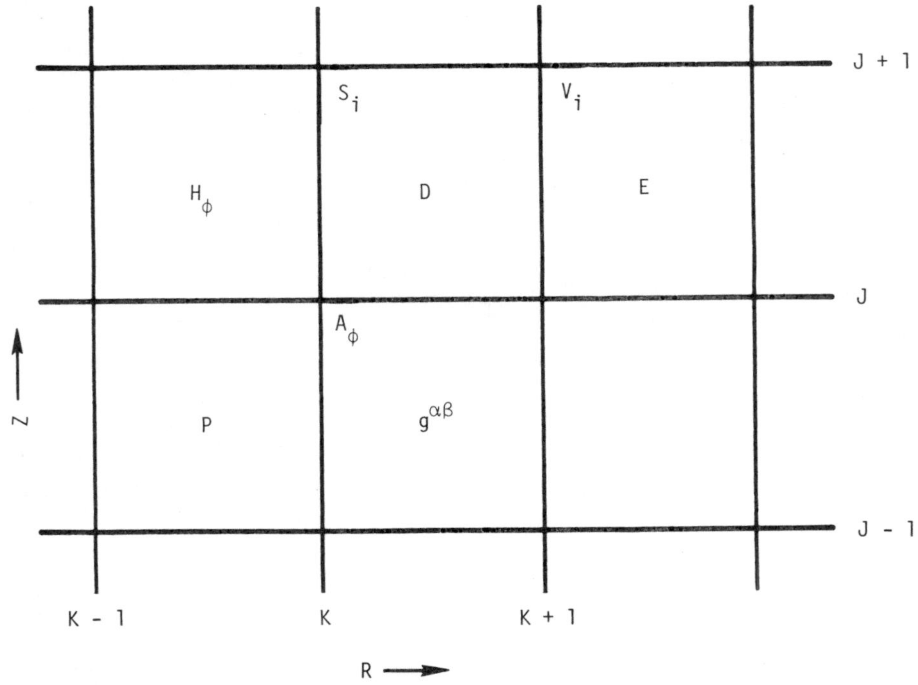

Fig. 1. Mesh for interpreting difference equations. Quantities S_i, V_i, and A_ϕ are centered at intersections of solid lines. Quantities D, E, P, H_ϕ, and $g^{\alpha\beta}$ are centered as indicated between lines.

In convection terms such as $\frac{1}{\sqrt{-g}} \frac{\partial}{\partial \chi^i} (DV^i \sqrt{-g})$, the flux of mass in the Z direction across a zone boundary is represented by $D_b\ (V^i_{K,J} + V_{K,J+1}) \cdot (\sqrt{-g}_{K,J} + \sqrt{-g}_{K-1,J}) \cdot (R_{J+1} - R_J)/4$.

Where D_b is the density in the zone behind the zone boundary with respect to the velocity, i.e., if $V_{K,J} + V_{K,J+1}$ is greater than zero, $D_b = D_{K-1,J}$, otherwise $D_b = D_{K,J}$. In all but convection terms, straight arithmetic averages are used to make all terms in any equation center at the same point in space. The non-transport part of the momentum equation is time-centered one-half time-step from the transport-type equations. Pressure in the energy equation is time-centered with respect to the new and old energies, otherwise all equations use old values to compute the new values. An artificial viscosity, or Richtmyer-Von Neumann dissipation pressure is introduced to provide the correct entropy change in shock waves. Dissipation pressure is

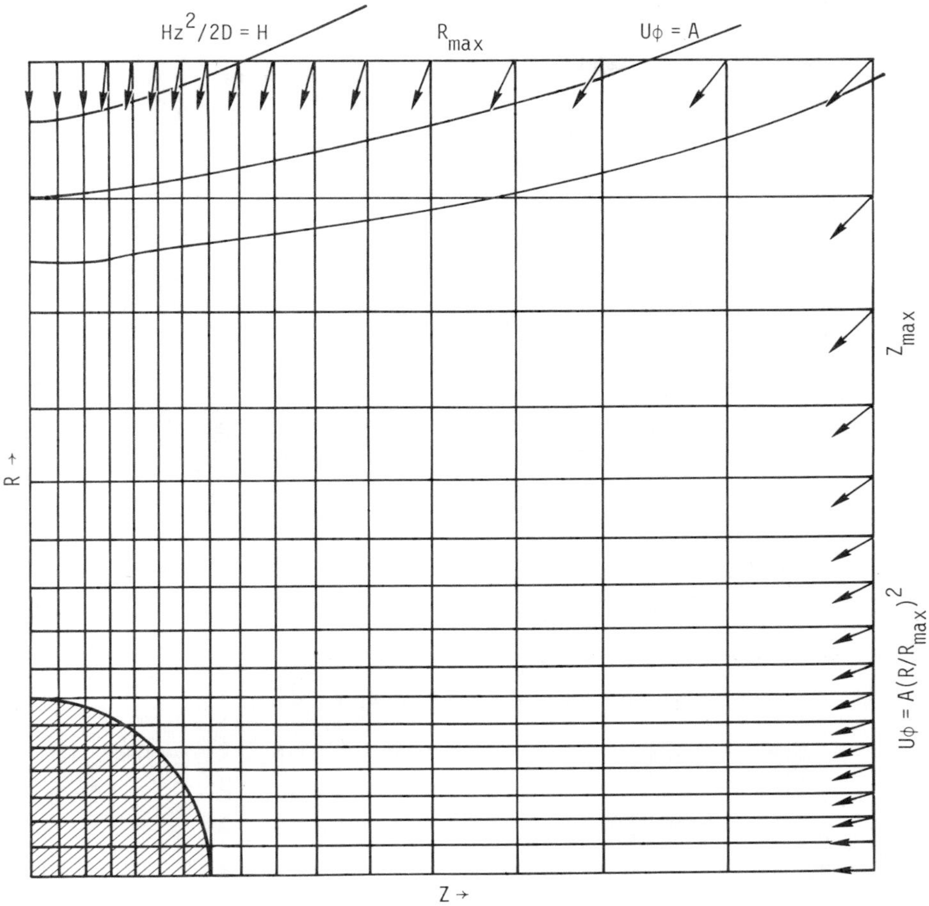

Fig. 2. Overall view of mesh used in calculations. Hatched zones are considered as interior to the black hole. Curved lines indicate how magnetic field lines look near the boundary.

$$Q = D\left[\left(\Delta_R V^R\right)^2 + \left(\Delta_Z V^Z\right)^2\right],$$

$\Delta_R V^R$ is the change in R velocity across a zone in the R direction. Here, Q is equal to zero unless $\frac{\partial \rho}{\partial t} > 0$ and $\Delta_R V^R < 0$ or $\Delta_Z V^Z < 0$. The Q pressure is added to the P pressure everywhere that P occurs in the equations of motion.

We made test calculations to study the accuracy of the shock calculations in flat space. The error in shock compression was about 5% for nonrelativistic shocks and rose to about 20% for a shock velocity of 0.995, which is much higher than encountered in these calculations. Shock waves form a few black hole radii out, and hence the velocities are about one-half to one-third of light velocity.

The metric is singular on the horizon because the Δ term disappears on the horizon. Numerical problems very close to the horizon are avoided when a minimum of 0.025 m^2 is placed on Δ. This allows material to actually fall into the hole. The self-gravitational effects of the plasmas are assumed to be negligible.

A test calculation was also made for material free-falling into a Schwartzchild hole with no internal, rotational, or magnetic energy. The magnitude of the velocities was good, but the directions were inclined slightly from the 45° line toward the equator and the axis. Just outside the horizon, the density at the waist and axis was 2 to 3 times the density at 45°. The density among the 45° line is very close to the true solution. If the present simple differencing scheme is used in a Newtonian calculation, the same divergence toward the waist and axis is observed. This is because the zones are not square. In the Newtonian calculation this divergence can be eliminated by higher order corrections. However, these more complicated difference methods, when used with the relativistic equations, are unstable near the horizon. Although a density factor of three may seem large, the density of free-fall, pressureless gas varies as

$$D \simeq 1/r^{3/2}\ (1 - 2m/r),$$

so that it diverges at the horizon. Also, the value of S_t near the horizon rises about three to six times that of the true solution in these calculations. In this case S_t varies in the same manner as D, however S_t is found by the velocity renormalization condition, and the momentum density varies as

$$S_r \simeq 1/(r - 2m)^2.$$

Indirectly, S_t is more divergent than D near the horizon. (Most of the interesting action in the following calculations occurs a few hole radii from the hole so that large errors close to the horizon will hopefully not overly affect the results.) A too large S_t increases the inward gravitational forces relative to the centrifugal and magnetic forces.

EXAMPLE CALCULATIONS

For these calculations, a grid mesh is set up in an R, Z space extending from zero to about 10 to 20 times the gravitational radius, $m(G = c = 1)$. Material is introduced at the outer boundary with a density and an R, Z velocity that uniform, cold, nonrotating material falling from infinity would have.

$$D = 1/\ r\sqrt{2mr}\ (1 - 2m/r) \quad ,$$

$$S_R = -R/\ r^3(1 - 2m/r)^2 \ , \ S_Z = -Z/\ r^3(1 - 2m/r)^2$$

The magnetic field and angular momentum about the Z axis are added to this cold material. The boundary is far enough from the origin, and the added fields and angular momentum are small enough so that if the material had fallen from infinity with those values of magnetic flux and angular momentum, the flow would have been only slightly affected. These boundary conditions are rather specialized, but they could represent the inside flow in black hole accretion from a solar wind. The range of angular momentum values considered here may be too low for most real astronomical situations, i.e., disk accretion. A main reason for considering this type of accretion is that it develops quickly on the characteristic hydrodynamic time scale.

It may be useful to consider a few elementary, relevant facts of orbits about a black hole [1]. In a Schwartzchild metric, the minimum radius and minimum specific angular momentum for stable circular orbits in the equator are 6 m and $2\sqrt{3}m$, respectively. Angular momentum values in the examples are always less than the above values so that material cannot be stopped by rotation alone. However, the shear in angular velocity produces large toroidal (H_Φ) fields from an initial poloidal field (H_R, H_Z). If enough field pressure is developed this way, material is slowed sufficiently to produce a shock wave, even with relatively low angular momentum. In an extreme Kerr metric, the last stable circular orbit is at the horizon and has an angular momentum of only $2\sqrt{3}$. Even without angular momentum, particles rotate about a Kerr hole and, if any poloidal field is present, the total magnetic field is strongly amplified and retards the inward flow. Also, relatively small angular momentum (of the order of unity) will affect the flow if the plasma is corotating with an extreme Kerr hole. The dragging of inertial frames in the Kerr metric can be thought of as adding Coriolis force $\omega U_\phi/R$ where ω is the angular velocity of the inertial frame.

The calculations are limited by three variables: a, the angular momentum of the black hole (henceforth the mass m of the hole will be taken as one); A, the specific angular momentum of the infalling gas at the equator boundary; and, H, the ratio of magnetic field energy density to Plasma mass density at the boundary. The magnetic field at the boundary is taken as totally poloidal. It is dominantly an H_Z field along the cylindrical boundary. Because the density decreases along the cylindrical boundary going from the equator, the field intensity also decreases; the field is not strictly an H_Z field. All points not on the equator are given an angular momentum of $A(R/R_e)$ where R_e is the cylindrical radius at the equator. The average angular momentum per unit mass is approximately 5/6 of the nominal value A, because the outside of the calculational mesh is usually square. Calculations are symmetric with respect to the equator, Z = 0, plane.

In a previous paper (Wilson 1971) [2] similar calculations were made for the case of no magnetic field. For in-fall into a Schwartzchild hole, very large angular momentum was needed to alter the flow from smooth adiabatic infall. Even a large angular momentum of A - 4 formed only a weak shock wave, and it arose primarily from the material colliding at the equator. The centrifugal force deflects the material from straight in the equator. To really stop material it is probably necessary to use an A appreciably greater than 2 3, the value of angular momentum for the last stable orbit around a Schwartzchild hole. When the material slows down in a shock wave, for example, it acquires more energy per particle, but does not change its angular momentum. It decreases its angular momentum and falls inside the radius of the last stable orbit, even though it began with an angular momentum greater than the angular momentum of the last stable orbit. It can also be pushed into the hole by fluid thermal pressures. Kerr hole calculations with a = 0.7, A = 3 and a = 1, A = 2 were marginal for the development of a shocked region. With a = 1, A = 3 a strong clear shock developed in the waist region, forming a hot toroidal doughnut of gas in the equatorial region. The internal energy was about 0.1 when the shock region radius had grown to about 8. The radius of the shocked region grew with a velocity of about 0.1 until the shock wave was out to about 10. Whether it slows down at some moderately small radius or keeps growing outward indefinitely is not known. With $a = 1$, $A = -4$ the centrifugal forces were large far out, and strongly deflected the material toward the waist. In close, the lessening of the centrifugal force by antirotation pulled so strongly on the material that no shock waves were formed. In general, pure hydrodynamic flows tend to be simple.

The first calculation is for free fall into a Schwartzchild hole with a very low angular momentum and magnetic field ($a = 0$, $A = 0.0001$, and $H = 0.0001$). The flow is only slightly perturbed from spherical infall. However, enough field and nonradial flow is produced to induce a small electric charge on the hole. The electric charge is given by $e = 2A^0 A$ where A^0 is the vector potential on the equator of the hole and equal to the magnetic moment of the hole. Figure 3 shows the magnetic field structure at 90. The poloidal field lines tend toward a radial field configuration, but the field lines are not radial because the material with field in it has only been falling in for a finite time. Also, the field started at a radius of 20 (with the configuration described in the last paragraph) so the field will never become strictly radial. In a radial field, a current sheet forms on the equator. In this calculation the current is concentrated near the equator. The electric charge density is likewise concentrated in the equator region. The electric field lines could be thought of as arising mostly from the horizon of the hole and near the equator. Consider an ideal case of a purely radial magnetic field given by $A = A\ (1 - |Z|/r)$. The toroidal field can be found easily from Eq. (8) and the free-fall velocity $V^r = (1 - 2m/r)\ 2m/r$:

$$H_\phi = \frac{1}{r^2}\ \frac{r}{2m}\ A_0\ \frac{V\phi(\theta)}{\sin\theta}\ .$$

In the computer calculations $V_\phi = \sin^2\theta$, for θ small because of the boundary condition on A. Hence H_ϕ is small near the axis in Figure 3. The smallness of H_ϕ near the equator arises because the poloidal field is not radial near the equator.

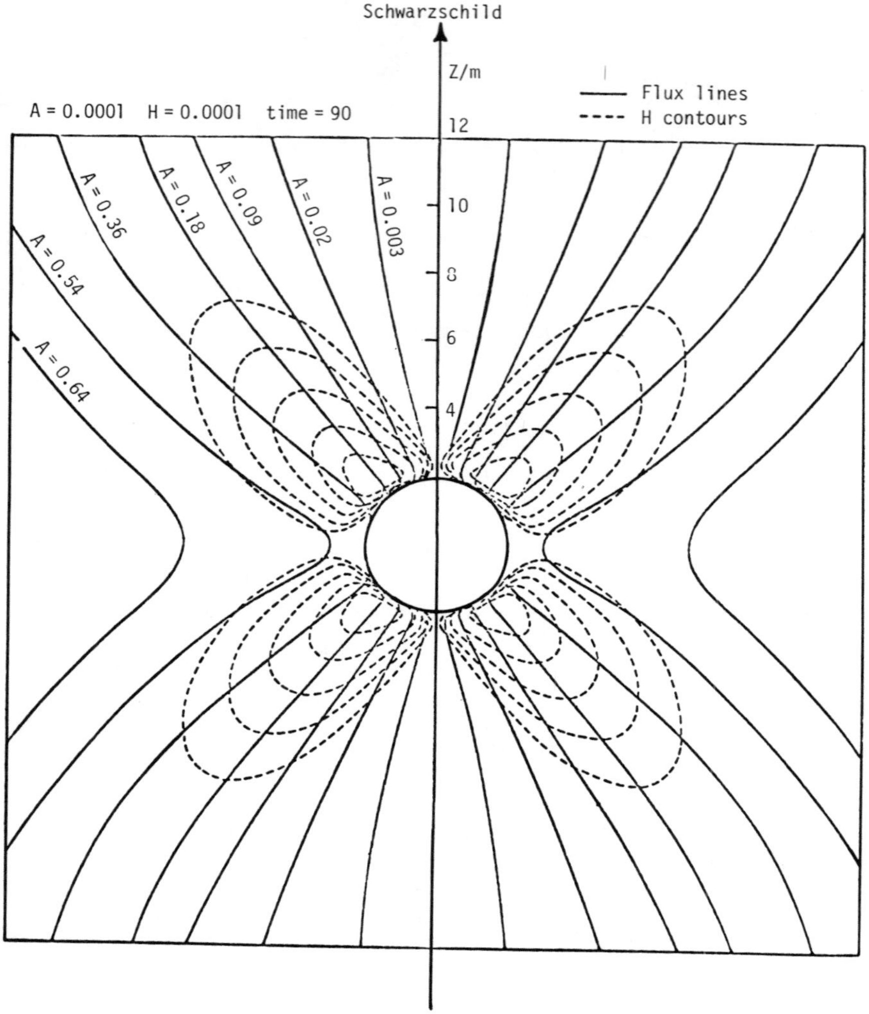

Fig. 3. Field configuration for in-fall onto a Schwarzchild hole at a time of 90 with A = 0.0001 and H = 0.0001. Solid lines are poloidal field lines, labels represent flux contained within lines, but scale is arbitrary. Dashed lines are lines of constant H_ϕ values. H_ϕ differs by a factor of two from one dashed line to the next.

The charge and A_ϕ^0 are increasing linearly with time. The radial electric field has two terms, $U_\phi H_\theta$ and $U_\theta H_\phi$ (θ is the spherical polar angle), arising from the condition of perfect conductivity. Breaking the spherical symmetry by either having H_θ or U_θ not equal to zero produces a charge. In this low-field, low-rotation example, the charge arises primarily from the $U_\phi H_\theta$ term. If true steady-state accretion exists, with radial infall, the magnetic field would be purely radial and no charge could arise.

The charge on the hole is calculated by the integral of the radial electric field at the inner-most zones over the surface of the hole. Whether the charge belongs to the hole or the infalling gas is obscure. For a star accreting cold material at a rate of m, the density near the horizon is given by

$$D = \frac{\dot{m}}{8\pi m} \quad \frac{2m}{r} \quad \frac{1}{(r - 2m + 4mm)} .$$

Integrating this density out from the horizon to a small fraction of the radius, the mass near the hole is

$$\delta m = 2m\dot{m} \log \frac{(r - 2m + 4m\dot{m})}{4m\dot{m}} .$$

Just outside the hole is a moderately large mass with essentially the same mean coordinate density D as the hole density. With the large charge-to-mass ratio of electrons (2×10^{21}), this exterior layer can sustain as large a charge as is interesting. For $m = 10^{17}$ g/s, $m = M$, and $\delta m/m = 2.2 \times 10^{-22} \log (r - 2m + 4m\dot{m})/4mm$. Analytical solutions presented elsewhere in the proceedings show that charge can be induced onto black holes by external fields, but these are static, or steady-state, solutions that have had infinite time for the field to penetrate the hole.

For the simple solution of steady radial infall, whose A_ϕ is a function of θ only, $H_\theta = 0$ everywhere. For nonsteady radial flow, $(\partial A_\phi/\partial t) = - V^r(\partial A_\phi/\partial r)$. Since V^r approaches zero at the horizon, H_θ diverges near the horizon. In steady flow the field lines can be thought of as all lying on top of each other at the horizon. That is, while $H_\theta = 0$ up to the horizon, it is singular at the horizon. In the real world the H_θ just becomes large, of the order of $1/(1 - 2m/r)$, which tends to $1/4m$. The magnetic field in a local tetrad has the physical component

$$H_{(\theta)} = \frac{H_\theta}{r^2 \sin^2\theta} \quad 1 - \frac{2m}{r}^{\frac{1}{2}} .$$

The ratio of densities $D/H^2_{(\theta)}$ thus tends to a constant value, i.e., it is neither zero nor infinity.

The next calculation uses a = 0, A = 0.0001, and H = 0.01. This magnetic field is large enough to strongly deflect the material from its free-fall motion. Figure 4 shows how the charge and equatorial magnetic flux vary with time.

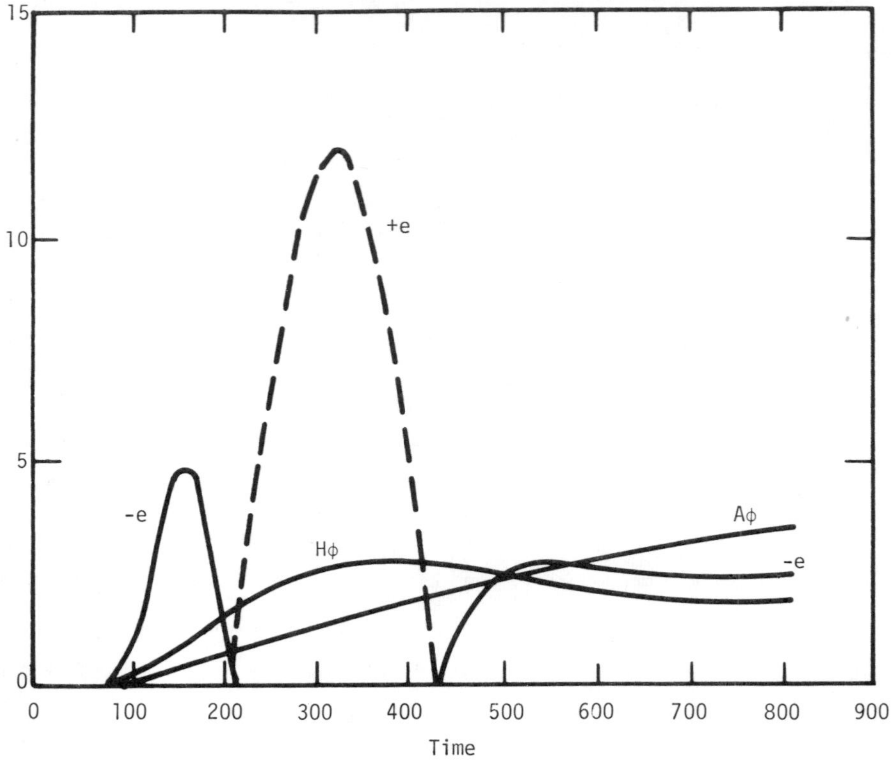

Fig. 4. Charge e on the hole and vector potential A_{ϕ} on the equator of the hole as functions of time for the example of in-fall onto a Schwarzchild hole with A = 0.0001 and H = 0.01.

The charge changes sign because both H_{ϕ} and U_{θ} are reversing signs near the horizon. The magnetic field configuration is shown in Fig. 5. The velocity vectors are closely parallel to the field lines, and the material is only modestly slowed from its free-fall velocity.

For free-fall into a black hole, the angular velocity

$$V^{\phi} = - 1 - \frac{2m}{r} \frac{V_{\phi}}{r^2 \sin^2_{\theta}} .$$

Therefore, if V_{ϕ} is a function of angle θ, only the shear

$$\frac{V}{r} = -\frac{2}{r^3}\quad 1 - \frac{3m}{r}\quad \frac{V}{\sin^2\theta}$$

reverses sign at a radius of 3 m. In the first calculation the free-fall velocity V^r goes to zero so that the field H does not reverse sign. In the more general case it is not clear how the field grows near the horizon.

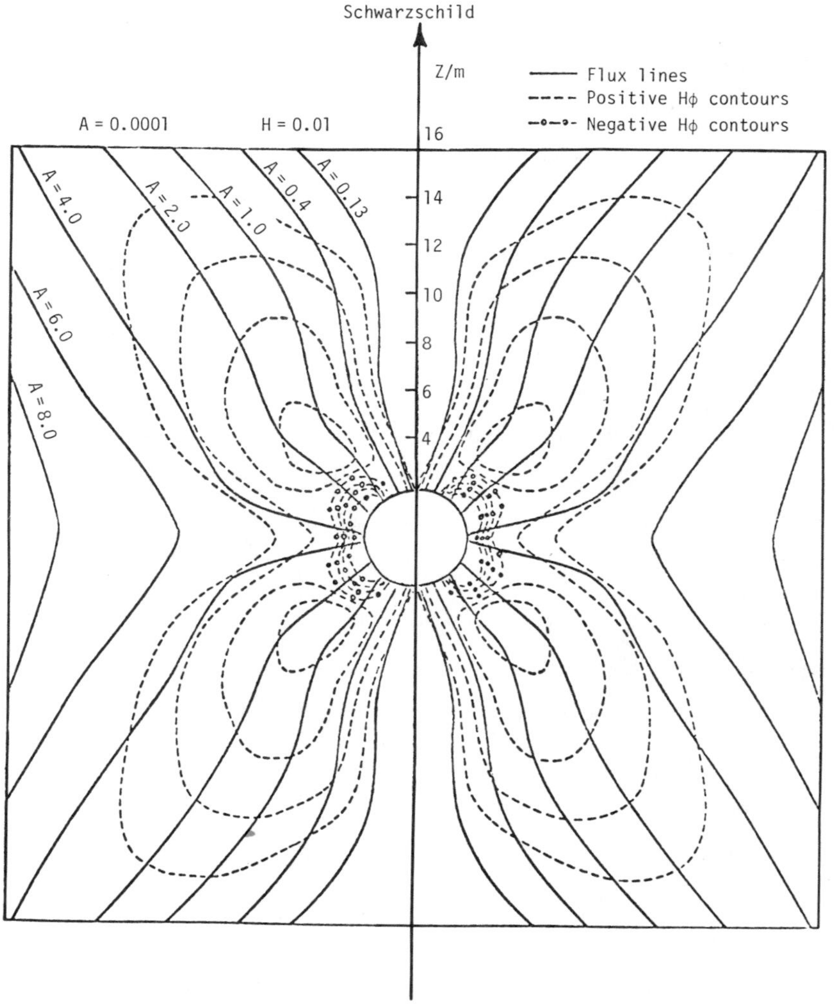

Fig. 5. Magnetic field configuration for in-fall to a Schwarzchild hole with A = 0.0001 and H = 0.01. Solid lines are poloidal field contours. Dashed lines are toroidal field contours. H_ϕ differs by a factor of two from one dashed line to the next.

In the last Schwarzchild calculation, the rotational momentum is given a value large enough to modestly affect the flow if no field were present [2]. Consider A = 2 and a small value of magnetic field, H = 0.0001. Here, the centrifugal field deflects material to the waist, enhancing the magnetic field. The rotational shear also builds up the H_ϕ magnetic field to substantial values. Flow without a magnetic field and with an angular momentum of two forms no shock. However enough additional field generation occurs with this relatively low field to slow the material enough to produce a shock and greatly slow the infall. The field acquires the complicated form shown in Fig. 6. The shock front is at the surface where the poloidal field has kinks. The H_ϕ field has a steep gradient at a radius of about 6 to 10. The thermal energy behind the shock is about 0.03 units.

Material falling in a Kerr metric experiences a rotational dragging. This dragging is resisted by the rigidity of magnetic fields. This combination of frame dragging and field rigidity leads to the possibility of extracting energy from the rotating hole. To study this effect in its simplest form, two calculations were run with A = 0, H = 0.01, with a a = 0.7 or 1. With this strong a field, the torque on the infalling gas is quite significant. In Fig. 7, the specific angular momentum averaged over the total mass of material in the calculational grid is plotted as a function of time (the calculational grid extends out to R, Z = 10). Initially, the material in the grid had no magnetic field. The material has to fall from the outside nearly to the horizon before the torques are large enough to have effect. Thus, the angular momentum of the plasma does not start to rise until a time of 30. Only in the case of the extreme Kerr is the dragging effect strong. In Fig. 8, the angular velocity of the material and of the inertial frames are compared for the two values of a. The material angular velocity curves result from averaging the angular velocities between 45° and the equator. The speed-up only affects a volume of space rather close to the horizon. When material is corotating with the hole, angular momentum is still transferred out. The dashed line in Fig. 6 shows the net increase of angular momentum for a calculation with a = 1, A = 1, H = 0.0001. Because of the weaker

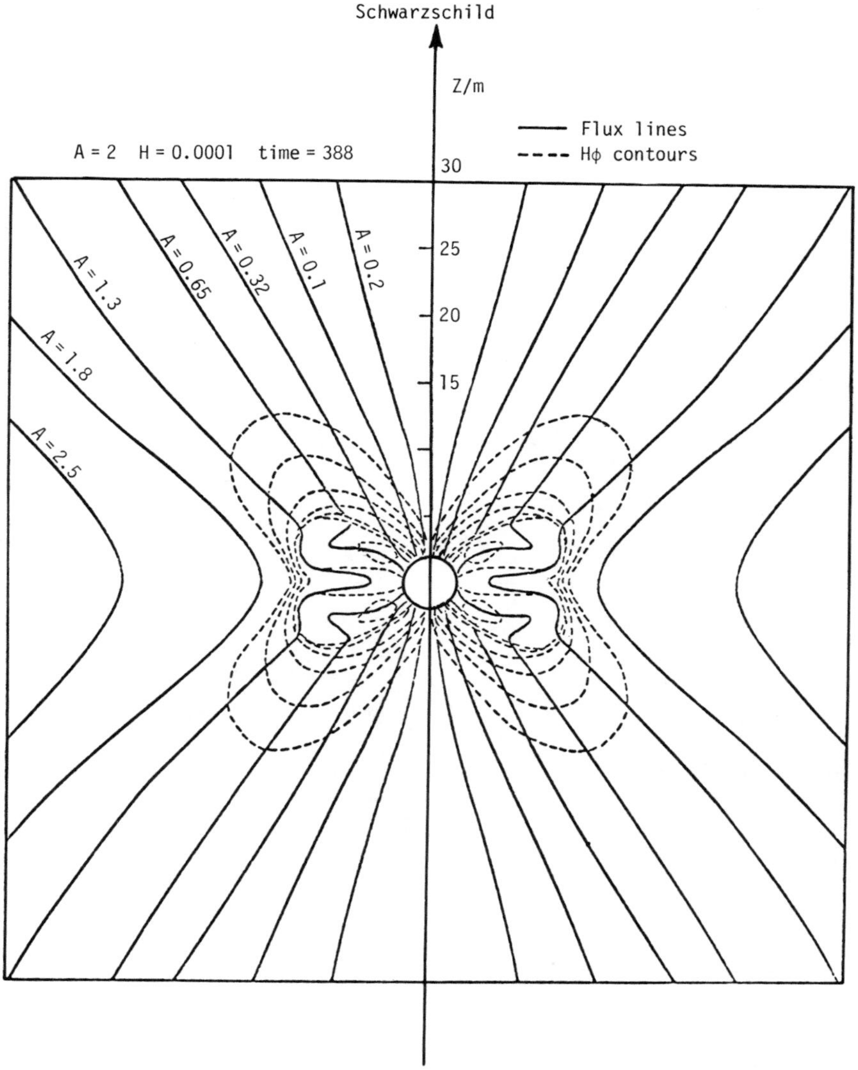

Fig. 6. Magnetic field configuration for in-fall to a Schwarzchild hole with $A = 2$ and $H = 0.0001$. Labels are the same as Fig. 4.

magnetic field, the field takes longer to build to where it can affect the material motions. For Fig. 6, the dashed line was shifted about 100 units in time to the left. This case is more interesting because the fields and momentum forces are large enough to partially stagnate the in-falling material and produce a ring of material around the hole with an augmented angular momentum. Figure 9 shows

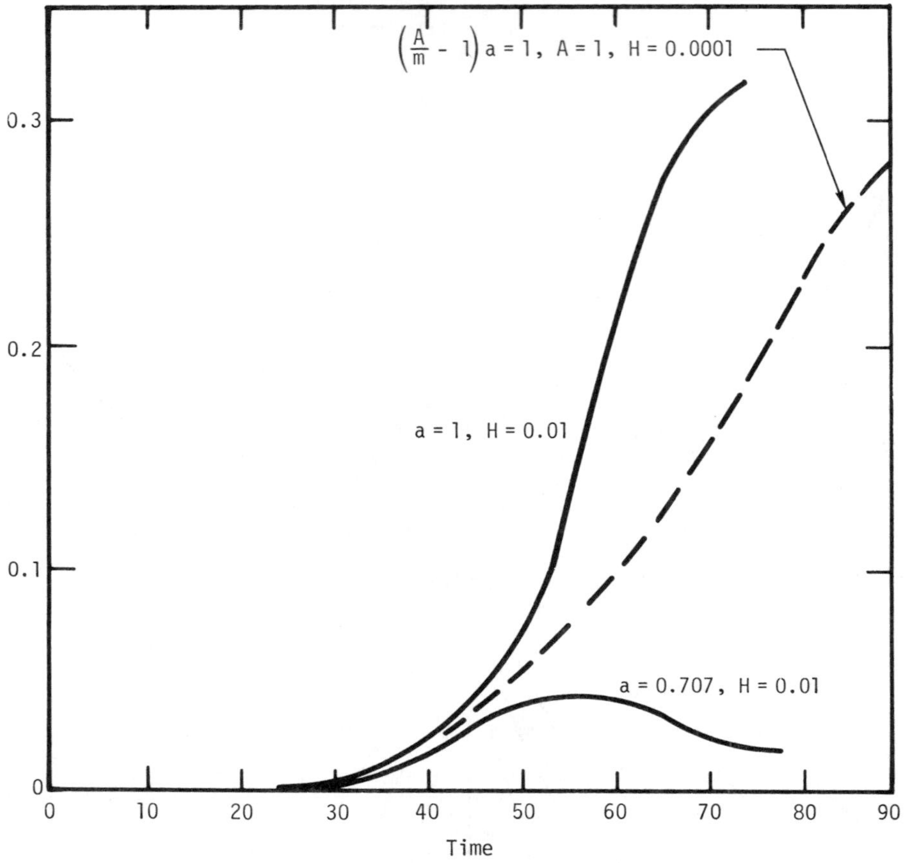

Fig. 7. Angular momentum per unit mass averaged over the calculational grid versus time. Dashed curve has been shifted forward in time about 100 units.

the isoangular momentum contours as well as the field lines and velocity vectors. Outside the A = 1 contour, A = 1 except near the axis. The velocity pattern is very typical of all flows resulting in shock formation. A swirling region is formed several hole-radii out by the inside of the swirl is always falling into the hole because of lack of support. Material at a radius of 2 to 4 has an angular momentum of 2. However, as it slides down field lines to the hole it will lose that angular momentum. The stagnation region continues to grow slowly with time.

Figures 10 and 11 show the time evolution of the electric charge, the toroidal magnetic field, and the vector potential on the hole for the cases a = 1,

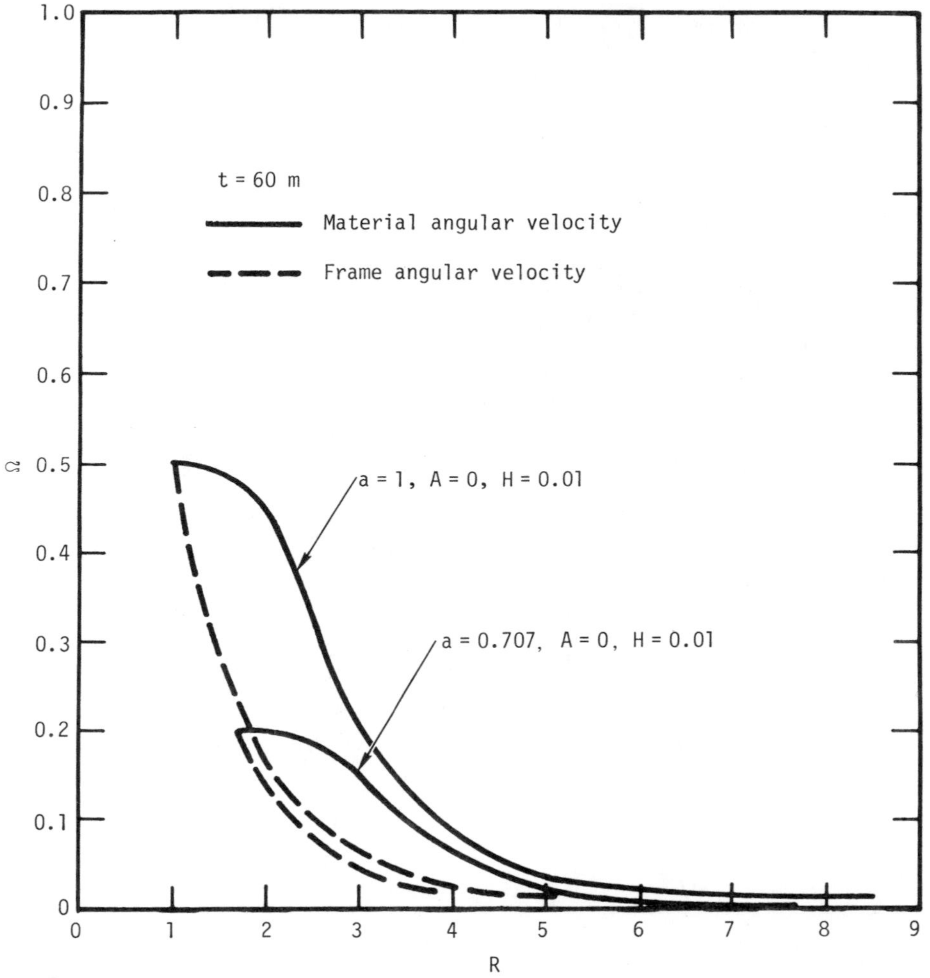

Fig. 8. Angular velocity Ω versus radius. Solid curves are material angular velocity. Dashed curves are frame angular velocity. For the upper curves a = 1 and for the lower curves a = 0.7. In both cases A = 0 and H = 0.01.

H = 0.0001, and A = 0 or 1. Very large charges are generated on the scales of the magnetic fields. The plasma has a charge of sign opposite to the hole charge. It is not quite equal because nonradial field lines leave the calculational mesh with some rotational velocity. The charge from the field lines exiting the calculational region is $e = \int d\vec{A}\,(\vec{v} \times \vec{H})$, which is small but not negligible. For infalling material with A = 1, the plasma charge is almost twice the hole charge. The system would be electrically neutral only at very large distances.

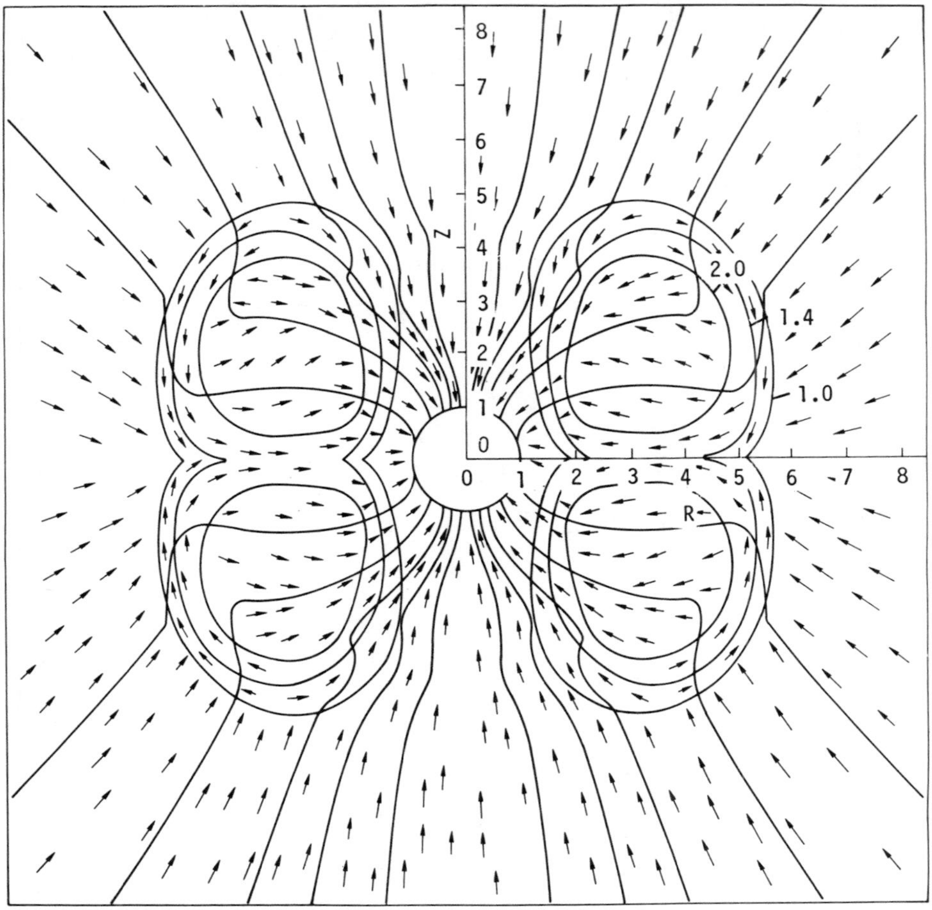

Fig. 9. Poloidal field lines for the in-fall of material onto an extreme Kerr metric, a = 1, with A = 1 and H = 0.0001. Closed looping lines are contours of A = 1, 1.4, and 2. Arrows indicate direction of material flow. In the region of high A the magnitudes of the velocity are two to five times less than the length of the vectors would indicate. Also the magnitude of the velocity goes to zero near the hole.

Calculations were made for angular momentum A = 2, 3, and 4 in the Kerr metric with a magnetic field parameter of H = 0.0001. The behavior qualitatively is similar to the A = 1 example discussed above. Figure 12 shows the field lines and isotoroidal field contours plotted for a = 1, A = 2, and H = 0.0001. The higher rotation rates simply speeded up the formation of the stagnation region. Calculations with a = 1, A = 2, H = 10^{-5} and 10^{-6} were also done. The H = 10^{-5} behaved similarly to H = 10^{-4}, just taking more time to develop. A very limited

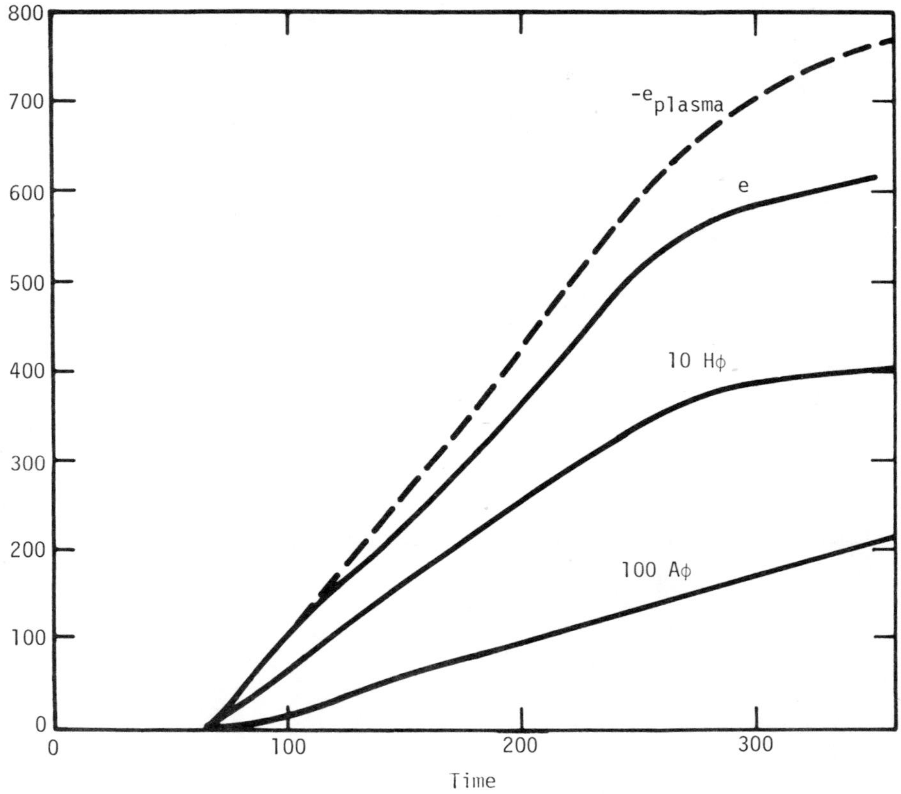

Fig. 10. Hole charge, vector potential at the equator of the hole, and toroidal field near the hole as functions of time for a Kerr metric with a = 1, A = 0, and H = 0.0001. Dashed curve is the negative of the charge in the plasma outside the hole.

shocked region was formed for $H = 10^{-6}$ even after a time of 400. The field did not multiply enough to have any discernible effect on the plasma flow. A calculation was also performed with counter-rotating plasma a = 1, A = -4, and H = 0.01. Even with these large values of A and H, the plasma flowed smoothly, though not straight into the hole. No shock wave was formed.

SUMMARY

We looked at very specialized cases of accretion onto a black hole and noted two main effects: the induction of charge on a black hole and the extraction of energy from rotating black holes by the magnetic rigidity. At present it is hard to see what observable effects these processes have. At the moment, the interest is mostly in principle. What is needed is an analysis of the fields to

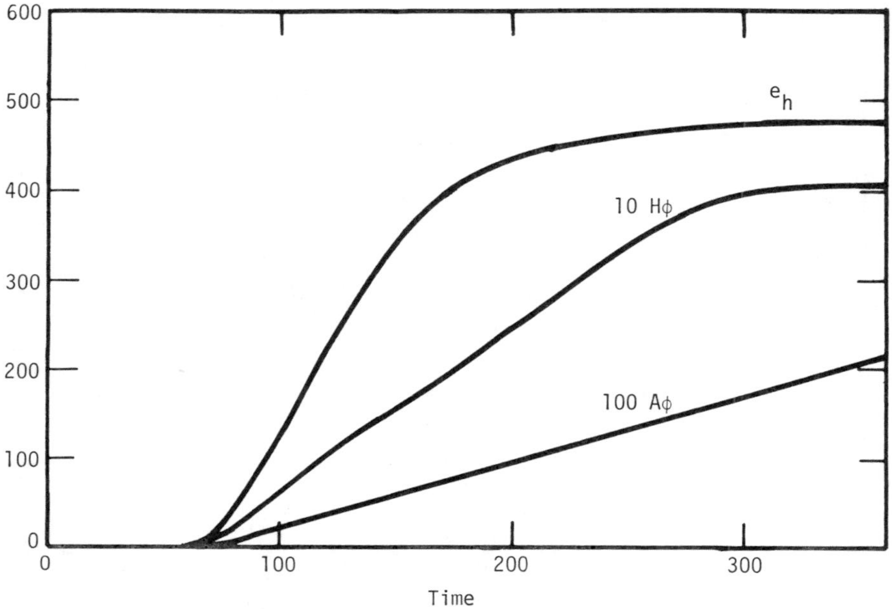

Fig. 11. Hole charge, vector potential at the equator of the hole, and toroidal yield near the hole as functions of time for a Kerr metric with a = 1, A = 1, and H = 0.0001.

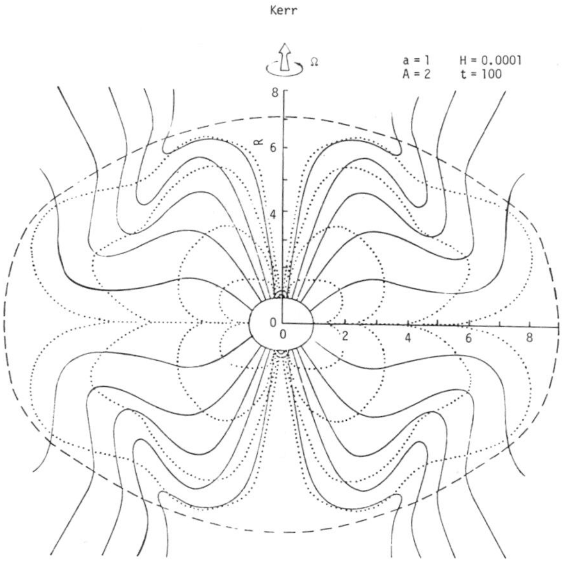

Fig. 12. Magnetic field configuration at a time of 100 for in-fall onto a Kerr metric with a = 1, A = 2, and H = 0.0001. Solid lines are poloidal field lines, dotted lines are contours of H_ϕ differing by factor of two, and the dashed curve is the position of the shock wave.

see if they grow into catastrophically magnetohydrodynamically unstable configurations, and thus produce short bursts of high-energy particles [3][4]. There is a tendency to form current sheets and concomitant charge sheets on the equator. In the present calculations there is also a build up of density along the equator, which argues against any sudden release of field energy near the waist, by, say, field reconnection, because inertial effects would dampen it. When a shock develops, the density near the equator builds up almost an order of magnitude above free-fall density. We might envisage material falling in smoothly with a simple field configuration, building up a complicated field near the hole, and extracting some energy from the hole. It would then become very unstable, releasing a lot of field energy and allowing the material stacked near the hole to fall in. The whole process would then repeat itself [3][4][5]. This is very speculative.

Acknowledgments

The author would like to thank R. Hanni, R. Leach, and J. LeBlanc for checking the computer program and making several helpful suggestions.

REFERENCES

[1] R. Ruffini, J.A. Wheeler, European Space Research Organization Colloquium, Vol. SP52-1969; J.M. Bardeen, W.H. Press, S.A. Teukolosky, Ap.J. 178: 347, 1972.
[2] J.R. Wilson, Ap.J. 173: 431, 1972.
[3] R. Ruffini, Proceedings of the Seventh Texas Tymposium on Relativisitc Astrophysics, New York Academy of Science, p. 95, 1975.
[4] T. Damour and R. Ruffini, Phys. Rev. Letters 35, 463 (1975).
[5] J.R. Wilson, Proceedings of the Seventh Texas Symposium on Relativistic Astrophysics, New York Academy of Science, p. 123 (1975); T. Damour, Proceedings of the Seventh Texas Symposium on Relativistic Astrophysics, New York Academy of Science, p. 113 (1975); R. Ruffini and J.R. Wilson, Phys. Rev. D 12 2959 (1975).

COUPLING OF HIGH FREQUENCY GRAVITATIONAL AND ELECTROMAGNETIC WAVES

Y. Choquet-Bruhat

Département de Mécanique Analytique
Université Pierre et Marie Curie - Paris 75005

1. Introduction

There is a very general and powerful method to study high frequency waves. It originates from the W.K.B. method, which was used in the construction of approximate solutions of the Schrödinger equation under the form

$$u(x) = f(x)\, e^{i\omega\varphi(x)}$$

where x is the point in space-time X (four coordinates), φ is a real valued smooth phase function, ω a dimensionless parameter which will be large in applications, and f a slowly varying amplitude. Such a type of approximation is also a particular case of the approximate solutions called in mechanics "progressive waves," with wave-fronts $\varphi(x)$ = cte, that is the physical quantities u(x) which vary much faster in a direction transversal to the wave front than on the wave front.*

A progressive wave on the n-dimensional manifold X is defined by a function of $X \times \mathcal{R}$, and a phase function on X. It may be written (ω large parameter)

$$\{ u(x, \xi) \}_{\xi = \omega\varphi}$$

The method has been used and adapted to a large variety of physical situations, by general considerations or ad hoc procedures. It has also received interest from mathematicians, and has been given its full generality in the linear and generic case, by Leray, in collaboration with Garding and Kotaké [1], more than ten years ago. They consider a formal serie

$$u = \sum_{p=0}^{\infty} \omega^{\mu-p} u_p(x, \xi)\big|_{\xi = \omega\varphi(x)} \qquad (1)$$

They compute formally its derivatives using the obvious rule

$$\frac{\partial u_p}{\partial x^i} = \{\partial_i u_p + \omega u'_p \varphi_i\}_{\xi = \omega\varphi}$$

*Because of the appearance of the two rates of variation the study of progressive waves is also called the two timing, or the multiple scale, method.

where the notations are:

$$\frac{\partial u_p}{\partial x^i} = \frac{\partial}{\partial x^i} u_p \ (x, \omega\varphi(x))$$

$$\partial_i u_p = \frac{\partial}{\partial x^i} u_p \ (x, \xi)$$

$$u'_p = \frac{\partial}{\partial \xi} u_p \ (x, \xi)$$

and they say that u is an asymptotic solution of the partial differential equation

$$a \ (x, \frac{\partial}{\partial x}) \ u = 0 \qquad (2)$$

if the formal serie obtained by replacing u by (1):

$$\sum_{p=0}^{\infty} \omega^{\mu' - p} F_p \ (x, \xi) \ | \ \xi = \omega \varphi(x)$$

is such that

$$F_p = 0 \quad \text{for all} \quad x \in X \quad , \quad \xi \in \mathcal{R} \ .$$

The same definitions apply to a system of partial differential equations, with u a set of unknowns--the indexes μ and μ' may also have to be replaced by sets of indexes depending on the order of the equations of the system.

A remarkable fact is that the first condition $F_o = 0$ imposes that the phase φ satisfies the eikonal (characteristic) equation. Once the phase has been chosen verifying this equation, the equations [system] $F_p = 0$ can be solved inductively for p = 1, 2, ... by the resolution of a [system of] linear differential equation of propagation along the rays for the unknown u_o, u_1, ... , considered as functions on X, depending on the parameter ξ, at least if φ = cte is a "simple characteristic,"* of the partial differential equation [system].

I have given in [5] an extension of these general results to non-linear partial differential equations when the first term u_o ("background") is independent of ξ: the propagation equations for u_1, ... are now not differential equations on X for functions u_p on X depending on the parameter ξ, but differential equations

*If φ is a multiple characteristic there may, or may not be propagation. The results depend not only on the principal terms, but also on the lower order terms of the partial differential operator (cf [5]).

on $X \times \mathcal{R}$ for functions $(x, \xi) \to u_p(x, \xi)$: the functional dependance of u_p on ξ (ξ has to be replaced afterwards by $\omega\varphi$) varies during the propagation; it is the physically well known phenomena of distortion of signals--stiffening of signals in hydrodynamics--which is linked with the apparition of shocks.

A fundamental problem for these asymptotic solutions to be physically relevant is for them to be true asymptotic solutions in the usual sense, when ω is large. We will require that the finite series $v_N = \sum_{p=0}^{N} \omega^{\mu - p} u_p(x, \omega\varphi)$ is an asymptotic solution of order p in the sense that

$$|| a(x, \frac{\partial}{\partial x}) v_N || < M \omega^{\mu' - N - 1} \qquad (3)$$

where M is some constant and $|| \; ||$ some convenient norm, for instance the maximum norm.

The above requirement is easy to satisfy in the linear case: $a(x, \frac{\partial}{\partial x}) v_N$ is some linear combinations (with coefficients functions on X) of the u_p and their derivations in x and ξ. These derivatives have all to be uniformly (i.e. independently of ω), bounded for the expansion in inverse powers of ω to make sense. If $F_p = 0$ for $p = 0, \ldots, N$ then automatically the condition (3) is satisfied. Remark that the condition u_p uniformly bounded for large ω implies

$$\int_0^T u_p'(x, \xi)\, d\xi \qquad \text{bounded}$$

condition which will give directly results obtained sometimes by indirect and not always rigorous averaging arguments.

In the non-linear case polynomials of the u_p and their derivatives appear after the first two steps and the conditions of boundedness will not be satisfied (except for special types of equations) unless one makes additional restrictions, sometimes severe, on the expansions. In most cases it will be impossible to go beyond the third step.

We shall in the following apply the general method sketched above to the coupled Einstein Maxwell system. Indeed the W.K.B. method has long ago been applied to the linear system of Maxwell equations in Minkowski space time: the first two steps give the geometrical optics approximation.

The qualitative properties of W.K.B. type solutions for the gravitational field have already been introduced by Wheeler in his study of geons [2]. Later Y. Choquet-Bruhat [3-6] and Gomes [7] have applied the general method of progressive waves to Einstein equations when the magnitude of the perturbation with respect to the background metric is of order ω^{-2}. In a well known paper Isaacson [8] has treated the case where this perturbation is of order ω^{-1}: there appears a new important phenomenon, called "back reaction" (terminology of Taub [9]), and the paper contains many deep and interesting physical comments. However there is some arbitrariness in the way Isaacson extracts from Einstein equations a wave equation to which he applies the W.K.B. method, treating the other terms by physical considerations on averages.

We shall follow here the method given in our papers [3-6], applying in a straightforward manner the generalization of the W.K.B. method sketched above to the full set of Einstein-Maxwell equations, and using averages on the family of wave fronts, only as it is required by the validity of the expansions.*

We shall treat both the cases where the magnitude of the perturbation is of order ω^{-2} or ω^{-1}: we shall see that in both cases electromagnetic and gravitational high frequency waves are linked. Except in special situations, a gravitational h.f. wave automatically generates an electromagnetic h.f. wave, and vice versa.

Note that an analogous result is obtained by the study of coupled electromagnetic and gravitational shock waves** (Lichnerowicz, [11]).

2. Definitions and general properties

The Einstein-Maxwell equations in vacuo read, in local coordinates:

$$R_{\alpha\beta} - \tau_{\alpha\beta} = 0 \qquad (4)$$

$$\nabla_\alpha F^{\alpha\beta} = 0 , \qquad \oint \nabla_\alpha F_{\beta\gamma} = 0 \qquad (5)$$

---

*See also, for the average lagrangian techniques, Taub [9].

**Note also that the electromagnetic h.f. has the properties of a pure radiation field Lichnerowicz [10].

$g_{\alpha\beta}$ hyperbolic metric on the space time V_4, $R_{\alpha\beta}$ Ricci tensor of g, $F_{\alpha\beta}$ exterior 2-form (electromagnetic field), $\tau_{\alpha\beta}$ Maxwell tensor of $F_{\alpha\beta}$.

We look for a perturbation of a given metric and electromagnetic field of the type (progressive wave):

$$g_{\alpha\beta}(x,\ \omega\varphi(x)) = \bar{g}_{\alpha\beta}(x) + \varepsilon h_{\alpha\beta}(x,\ \omega\varphi(x)) \tag{6}$$

$$F_{\alpha\beta}(x,\ \omega\varphi(x)) = G_{\alpha\beta}(x) + \omega\varepsilon\, H_{\alpha\beta}(x,\ \omega\varphi(x)) \tag{7}$$

where ε and ω are dimensionless parameters (in applications ε and ω^{-1} will be small).

In fact the magnitude of a tensor--or the comparison of magnitudes of two of them--can only be defined in a coordinate system, or through the use of a background metric. Since all our study in this paper will be made in local coordinates we shall simply set, for instance if T is a 2-tensor

$$|T| = \sup_{\alpha,\beta} |T_{\alpha\beta}| \tag{8}$$

and if it is a tensor field on a domain U of a chart in V_4:

$$||T|| = \sup_{x\in U}\ \sup_{\alpha,\beta} |T_{\alpha\beta}| \quad . \tag{9}$$

We shall deduce from (6),

$$g^{\alpha\beta}(x,\ \omega\varphi(x)) = \bar{g}^{\alpha\beta}(x) - \varepsilon\bar{h}^{\alpha\beta}(x,\ \omega\varphi(x)) + 0(\varepsilon^2) \tag{10}$$

$$\bar{h}^{\alpha\beta} = \bar{g}^{\alpha\lambda}\bar{g}^{\beta\mu}h_{\lambda\mu} \tag{11}$$

We shall compute the derivatives in a straightforward manner, for instance:

$$\frac{\partial h_{\alpha\beta}(x,\ \omega\varphi(x))}{\partial x^\lambda} = \partial_\lambda h_{\alpha\beta}(x,\ \omega\varphi(x)) + \omega h'_{\alpha\beta}(x,\ \omega\varphi(x))\,\ell_\lambda \tag{12}$$

$$h'_{\alpha\beta} = \frac{\partial h_{\alpha\beta}}{\partial(\omega\varphi)}\ ,\quad \partial_\lambda h_{\alpha\beta} = \frac{\partial\{h_{\alpha\beta}\}\omega\varphi = \text{cte}}{\partial x^\lambda}$$
$$\ell_\lambda = \frac{\partial\varphi}{\partial x^\lambda} \tag{13}$$

and analogous formulas for $F_{\alpha\beta}$ and the higher derivatives. We shall always suppose that all the coefficients of the expansion in powers of ω^{-1} we shall write on a domain U are bounded in the norm defined above (i.e. uniform).

We shall say that the perturbed metric and electromagnetic field (6), (7) are an asymptotic solution of order p of Einstein-Maxwell system on U V_4 if there exists a constant M, independent of ω, such that, in the considered coordinates the norms of the first members of (4), (5) are smaller than $M\omega^{-p}$.

3. Case I, $\varepsilon = \omega^{-2}$

We shall see that, in this case, there is a complete splitting (no back reaction) between each step of approximation.

Let us set:

$$R_{\alpha\beta} = \overset{0}{R}_{\alpha\beta} + \frac{1}{\omega}\overset{1}{R}_{\alpha\beta} + \dots \tag{14}$$

$$\tau_{\alpha\beta} = \overset{0}{\tau}_{\alpha\beta} + \frac{1}{\omega}\overset{1}{\tau}_{\alpha\beta} + \dots \tag{15}$$

with ($\bar{R}_{\alpha\beta}$ Ricci tensor of $\bar{g}$)

$$\overset{0}{R}_{\alpha\beta} \equiv \bar{R}_{\alpha\beta} + \frac{1}{2}\overline{\ell^{\lambda}}\ell\lambda h''_{\alpha\beta} + \frac{1}{2}(\bar{g}_{\alpha\lambda}\ell_{\beta} + \bar{g}_{\beta\lambda}\ell_{\alpha})\overset{1}{F}{}'^{\lambda} \tag{16}$$

F^{λ} is here the left hand side of the harmonicity condition:

$$F^{\lambda} \equiv g^{\alpha\beta}\Gamma^{\lambda}_{\alpha\beta} = \bar{g}^{\alpha\beta}\bar{\Gamma}^{\lambda}_{\alpha\beta} + \frac{1}{\omega}\overset{1}{F}\lambda + 0\left(\frac{1}{\omega^{2}}\right) \tag{17}$$

and

$$\overset{0}{\tau}_{\alpha\beta} = \tau_{\alpha\beta}(G_{\lambda\mu}) = \bar{\tau}_{\alpha\beta} \tag{18}$$

1) Conditions of order zero (algebraic)

A necessary and sufficient condition for

$$g_{\alpha\beta} = \bar{g}_{\alpha\beta} + \frac{1}{\omega^{2}} h_{\alpha\beta} \tag{19}$$

$$F_{\alpha\beta} = G_{\alpha\beta} + \frac{1}{\omega} H_{\alpha\beta} \tag{20}$$

to be an asymptotic solution of order zero of Einstein equations is, under the hypothesis of boundedness that we have made:

$$\overset{0}{R}_{\alpha\beta} - \bar{\tau}_{\alpha\beta} = 0 \tag{21}$$

but, under these same hypotheses (in particular $h'_{\alpha\beta}$ uniformly bounded), we have

$$\lim_{T=\infty} \frac{1}{T}\int_{0}^{T} \overset{0}{R}_{\alpha\beta}(x,\xi)\, d\xi = \bar{R}_{\alpha\beta}(x) \tag{22}$$

and

$$\frac{1}{T}\int_{0}^{T} \bar{\tau}_{\alpha\beta}(x)\, d\xi = \bar{\tau}_{\alpha\beta}(x) \tag{23}$$

We therefore conclude that the background metric $\bar{g}$ and electromagnetic field G must be a solution of the empty space equations

$$\bar{R}_{\alpha\beta} - \bar{\tau}_{\alpha\beta} = 0 \quad .$$

This conclusion is the first example of the very strong requirements on our expansions that are a straightforward consequence of their meaning and that we deduce by "averages on ξ," that is, on the family of wave fronts.

The equations (21) are now, when φ is chosen, a linear homogeneous system for $h''_{\alpha\beta}$. Two different situations occur:

1) $\bar{\ell}^{\lambda} \ell_{\lambda} \neq 0$: φ = cte is not a wave front of the background.

The general solution for (21) is then (we use once more also the "averaging in ξ" argument)

$$h_{\alpha\beta} = a_{\alpha}\ell_{\beta} + a_{\beta}\ell_{\alpha} \tag{24}$$

Such a perturbation may be canceled by a change of coordinates conserving $\bar{g}$ (cf [4]) thus is considered as physically meaningless.

2) $\bar{\ell}^{\lambda} \ell_{\lambda} = 0$, φ = cte is a wave front (i.e. isotropic) for $\bar{g}$.

The equation (21) are then equivalent to

$$\overset{1}{F}{}^{\lambda} = 0 \tag{25}$$

that is:

$$\ell_{\alpha} h - 2\bar{\ell}^{\lambda} h_{\alpha\lambda} = 0 \quad , \qquad h = \bar{g}^{\lambda\mu} h_{\lambda\mu} \quad . \tag{26}$$

Note that these relations, obtained as gauge conditions in other contexts, are here necessary conditions to be satisfied.

A similar study for the Maxwell equations shows that the background field should satisfy empty space equations

$$\nabla_{\alpha} G^{\alpha\beta} = 0 \quad , \qquad \oint \nabla_{\alpha} G_{\beta\gamma} = 0 \tag{27}$$

and the perturbation be a pure radiation field in the sense of Lichnerowicz-Mariot:

$$H_{\alpha\beta} = \ell_{\alpha} b_{\beta} - \ell_{\beta} b_{\alpha} \quad , \quad \ell^{\alpha} b_{\alpha} \, (x, \omega\varphi(x)) = 0 \tag{28}$$

In radiatives coordinates, such that $x^{o} = \varphi(x)$, $\ell_{o} = 1$, $\ell_{i} = 0$, $\bar{\ell}^{o} = 0$, the algebraic relations we have obtained read

$$H_{oi} = b_1 \quad , \quad H_{ij} = 0 \quad , \quad \bar{\ell}^i b_i = 0 \tag{29}$$

$$\bar{\ell}^i h_{ij} = 0 \quad , \quad \bar{g}^{ij} \, h_{ij} = 0 \tag{30}$$

h_{ij}, significant part of the perturbation, is, at each point, in a 2-plane.

2) Propagation equations: they will be obtained by writing that $R_{\alpha\beta} - \tau_{\alpha\beta}$, $\nabla_\alpha G^{\alpha\beta}$ and $\oint \nabla_\alpha F_{\beta\gamma}$ are of order ω^{-2} in the sense defined above. One finds, without any further hypothesis, the remarkable fact that the significant part of the gravitational perturbation, and the perturbed electromagnetic field, propagate without deformation along the rays of the wave fronts φ = cte [13]. Namely, by a straightforward computation in radiative coordinates (cf. Y. Choquet-Bruhat [4]), using only the equations $R^1_{ij} - \tau^1_{ij} = 0$, and the algebraic conditions obtained above one finds that:

$$- \ell^\lambda \bar{\nabla}_\lambda h'_{ij} - \frac{1}{2} h'_{ij} \, \bar{\nabla}_\lambda \bar{\ell}^\lambda = \bar{g}_{ij} \bar{G}^{oh} b_h - \bar{G}^o_i b_j - G^o_j b_i \tag{31}$$

Using Maxwell equations and the algebraic condition $\bar{\ell}^i \, b_i = 0$, we find

$$2\bar{\ell}^\lambda \nabla_\lambda b_i + b_i \nabla_\lambda \bar{\ell}^\lambda = \bar{G}^{oj} h'_{ij} \tag{32}$$

in both cases it may be proved that the algebraic conditions are conserved by the differential equations, hence it is coherent to take them as initial conditions.

An important property which appears in these propagation equations is the coupling of electromagnetic and progressive (or high frequency) waves: it results from (31) that an electromagnetic wave b_i generates a gravitational wave h_{ij}. The only exceptional case is when b_i satisfies the following relation:

$$\tau^1_{ij} = \bar{g}_{ij} \bar{G}^{oh} b_h - \bar{G}^o_i b_j - \bar{G}^o_j b_i = 0 \, , \tag{33}$$

The converse statement, generation of an electromagnetic wave by a gravitational wave is a consequence of (32), the only exceptional case being this time

$$\bar{G}^{oj} \, h'_{ij} = 0 \, . \tag{34}$$

The same exceptional cases of uncoupling have been found by Lichnerowicz [10] in his study of coupled gravitational and electromagnetic shock waves. He showed that they are satisfied if and only if the ray vector ℓ_λ is an eigenvector of the background field $G_{\lambda\mu}$

$$\bar{G}^{\lambda\mu} \ell_\lambda = k\bar{\ell}^\mu \tag{35}$$

From (31) and (32), by contracted product with $\bar{h}'^{ij}$ and b_i, one deduces a propagation equation for

$$E = - \bar{b}^i b_i + \frac{1}{4} \bar{h}'^{ij} h'_{ij} \tag{36}$$

which may be written as a conservation law. It reads, in arbitrary coordinates

$$\bar{\nabla}_\lambda (E\ell^\lambda) = 0 \tag{37}$$

with

$$E = - \bar{b}^\lambda b_\lambda + \frac{1}{4} (\bar{h}'^{\lambda\mu} h'_{\lambda\mu} - \frac{1}{2} (\bar{g}^{\lambda\mu} h'_{\lambda\mu})^2) \tag{38}$$

E which reads in radiative coordinates $E = - \bar{b}^i b_i + \frac{1}{4} \bar{h}'^{ij} h'_{ij}$ is > 0, it is to be interpreted as the energy of the coupled wave: $- \bar{b}^i b_i > 0$ is well known to be the electromagnetic energy of a pure radiation field.

We see from equation (38) that the total energy, gravitational plus electromagnetic, of the wave is conserved. Obviously each one is not conserved separately, and in the course of the propagation there will be a continual exchange between the two types of energy. We see moreover from equations (31) and (32) giving the propagation of the perturbed fields, that the polarizations of these fields will not be parallel translated as it is in the uncoupled case. To have a more concrete physical insight into the results let us set

$$- b^i b_i = \beta^2 \qquad \text{(electromagnetic energy)}$$

$$\frac{1}{4} h'^{ij} h'_{ij} = \alpha^2 \qquad \text{(gravitational energy)}$$

The equations (31) and (32) give immediately:

$$\nabla_\lambda (\alpha^2 \ell^\lambda) = X \tag{39}$$

$$\nabla_\lambda (\beta^2 \ell^\lambda) = -X \tag{40}$$

where X, function characteristic of the coupling is, in radiative coordinates

$$X = G^{oi} h'_{ij} b^j$$

that is, in arbitrary coordinates

$$X = G^{\alpha\beta}\ \ell_\beta\ h'_{\alpha\lambda}\quad b^\lambda$$

(remark that X is invariant by the transformation $h'_{\alpha\lambda} \to h'_{\alpha\lambda} + \tau_\lambda \ell_\alpha + \tau_\alpha \ell_\lambda$, $b^\lambda \to b^\lambda + k\ell^\lambda$, as it should).

To study the relative variation of the gravitational and electromagnetic energy, and also of the polarizations of the fields, let us set

$$\beta^2 = E\sin^2\theta \quad , \quad \alpha^2 = E\cos^2\theta$$

or, more precisely:

$$\frac{1}{2} h_{ij} = e_{ij}\ E^{1/2}\cos\theta \quad , \ e^{ij} e_{ij} = 1$$

$$b_i = e_i\ E^{1/2}\sin\theta \quad , \quad e^i e_i = 1$$

Then we deduce from (39) and (38):

$$\ell^\lambda \partial_\lambda \theta = \psi$$

with

$$\psi = -\frac{1}{2} G^{oi} e_{ij} e^j \qquad \left(= -\frac{X}{2E\sin\theta\cos\theta}\right)$$

we remark that

$$|\psi| < \frac{1}{2}\ |G_\ell|$$

where $|G_\ell|$ is the length of the spatial, or isotropic, vector $G^{\alpha\beta}\ell_\beta$.

The equation (41) is not, properly speaking, a differential equation for θ, since it has to be coupled with differential equations for e_{ij}, e_i, deduced also from (39), (40). These equations are:

$$\ell^\lambda \nabla_\lambda e_{ij} = \tau g\theta(\psi\ e_{ij} - \frac{1}{2} g_{ij} G^{oh} e_h + \frac{1}{2} G^o_j\ e_i + \frac{1}{2} G^o_i\ e_j)$$

$$\ell^\lambda \nabla_\lambda e_i = \text{cotg}\theta\ (-\psi\ e_i + G^{oj} e_{ij})$$

Some qualitative features can be easily deduced. For instance we see from (41) that if the background electromagnetic field tends to zero away from the source fast enough to be integrable on a ray then, on this ray, θ together with the radio of gravitational and electromagnetic energy, tends to a constant value. (Cf. also Gerlach [12], for physical comments.)

From the last differential equations we see, as could have been anticipated already in (39) and (40) that the polarizations are no more parallel transported--there is a Faraday transport--when and only when there is a background electromagnetic field which is not exceptional ($G^{\alpha\lambda}\ell_\lambda \neq k\ell^\alpha$). A more detailed study of these equations, using spinor formalism, has been done recently by Gerlach [12].

All the conditions have not been satisfied in the above construction, namely

$$\overset{1}{R}_{oi} - \overset{1}{\tau}_{oi} = 0 \tag{42}$$

$$\overset{1}{R}_{oo} - \overset{1}{\tau}_{oo} = 0 \tag{43}$$

a straightforward computation shows that these equations are linear in h_{oi} and h_{oo}, and can be made to vanish by choice of these components of $h_{\alpha\beta}$: note that $h_{o\alpha}$ may be made zero by a change of coordinates conserving $\bar{g}$ but such a change introduces an additional term $\frac{1}{\omega^3} k_{ij}$ in $\bar{g}$, which would give a contribution of order ω^{-1} to $R_{o\alpha}$. In fact it is somewhat simpler for actual computations (and we have seen, legitimate) to take $h_{o\alpha} = 0$ and

$$g_{ij} = \bar{g}_{ij} + \frac{1}{\omega^2} h_{ij} + \frac{1}{\omega^3} k_{ij} \quad . \tag{44}$$

Equations (42) and (43)--called henceforth "supplementary conditions"--appear then as linear (algebraic equations) for k''_{ij} which read

$$\bar{g}^{ij} k''_{ij} = f_o \tag{45}$$

$$\bar{\ell}^i k''_{ij} = f_j \tag{46}$$

4. Case II, $\varepsilon = \omega^{-1}$

It is the case considered by Isaacson [8] in his well known paper on high frequency gravitational waves. We now have

$$g_{\alpha\beta}\ (x,\omega\varphi) = \bar{g}_{\alpha\beta} + \frac{1}{\omega} h_{\alpha\beta}\ (x,\omega\varphi) \tag{47}$$

$$F_{\alpha\beta}\ (x,\omega\varphi) = G_{\alpha\beta}\ (x) + H_{\alpha\beta}\ (x,\omega\varphi) \tag{48}$$

The expansions of the Maxwell and Ricci tensors are:

$$R_{\alpha\beta} = \omega\overset{-1}{R}_{\alpha\beta} + \overset{0}{R}_{\alpha\beta} + \frac{1}{\omega}\overset{1}{R}_{\alpha\beta} + \ldots \tag{49}$$

$$\tau_{\alpha\beta} = \overset{0}{\tau}_{\alpha\beta} + \frac{1}{\omega}\overset{1}{\tau}_{\alpha\beta} + \ldots \tag{50}$$

1) The metric g and field F are asymptotic solutions of order -1 (with $h_{\alpha\beta} \neq a_\alpha \ell_\beta + a_\beta \ell_\alpha$) if and only if φ is a wave front of the background $h_{\alpha\beta}$, $H_{\alpha\beta}$ satisfy the algebraic conditions (26), (28) as in the case I. However, in the case $\varepsilon = \omega^{-1}$, we do not find, at this stage, any restriction on the backgrounds g, G.

We note, as a consequence of (26), that the gravitational waves are exceptional in the sense of Lax-Boillat:

$$\overset{1}{g}{}^{\lambda\mu} \ell_\lambda \ell_\mu = 0 \ , \quad \text{with} \quad g^{\alpha\beta} = \bar{g}^{\alpha\beta} + \frac{1}{\omega} \overset{1}{g}{}^{\alpha\beta} + \ldots \tag{51}$$

This result, trivial in the case I, was not obvious here, it implicates that the waves will propagate without distortion [13].

2) Propagation equations:

To have an asymptotic solution of order zero we must satisfy the non-linear equations:

$$\overset{0}{R}_{\alpha\beta} - \overset{0}{\tau}_{\alpha\beta} = 0 \tag{52}$$

It turns out again that these equations decompose in two sets.

a) In radiative coordinates $\overset{0}{R}_{ij} - \overset{0}{\tau}_{ij}$ is identical with the corresponding expression found in case I, except for the addition of $\bar{R}_{ij} - \bar{\tau}_{ij}$: by averaging in ξ we see that this term must be zero, thus we have the same propagation equation (31). On the other hand the verification to order zero of Maxwell equations imposes (after averaging in ξ) that the background G satisfies Maxwell empty space equations, and the perturbation H (which is again a pure radiation) the propagation equation (32).

The energy E has the same definition as in case I and satisfies the same conservation law.

3) Supplementary condition.

The difference--essential--between cases I and II comes from the supplementary conditions.

We still have to satisfy

$$\overset{0}{R}_{oi} - \overset{0}{\tau}_{oi} = 0 \tag{53}$$

which imposes $\bar{R}_{oi} - \bar{\tau}_{oi}$ and determines h_{oi} through a linear equation and, at last

$$\overset{0}{R}_{oo} - \overset{0}{\tau}_{oo} = 0 \tag{54}$$

only equation where the nonlinearity of Einstein equation will play a role. Indeed

$$\overset{0}{R}_{oo} - \overset{0}{\tau}_{oo} \equiv \frac{1}{4}(\bar{h}^{ij}h_{ij})'' + \bar{g}^{im}\bar{\nabla}_i h'_{om} + \frac{1}{2}h'_{oo}\bar{\nabla}_\lambda \bar{\ell}^\lambda$$
$$- \bar{g}_{oo}G^{oh}b_h + 2G^i_o b_i + \bar{b}^\lambda b_\lambda - \frac{1}{4}\bar{h}'^{ij}h'_{ij} + \bar{R}_{oo} - \bar{\tau}_{oo} = 0 \tag{55}$$

Therefore, by averaging in ξ

$$\bar{R}_{oo} - \bar{\tau}_{oo} = a \equiv \lim_{T=\infty} \frac{1}{T}\int_0^T E(x, \xi)\, d\xi \tag{56}$$

that is, in arbitrary coordinates

$$\bar{R}_{\lambda\mu} - \bar{\tau}_{\lambda\mu} = a\ell_\lambda \ell_\mu \tag{57}$$

where a, average energy on a family of wave fronts is

$$a(x) = \lim_{T=\infty} \frac{1}{T}\int_0^T E(x, \xi)\, d\xi \tag{58}$$

and satisfies the conservation law

$$\bar{\nabla}_\lambda (a\bar{\ell}^\lambda) = 0 \tag{59}$$

We conclude that, in the type of solutions we are looking for, the background $\bar{g}$ cannot satisfy empty space equation: if $a = 0$, then $E = 0$, therefore $b^i = 0$ and $h_{ij} = 0$.

Note also that a background verifying (57) can be perturbed only by very special progressive waves to give an approximate solution of Einstein equations: these waves where the wave front is normal to ℓ.

An application to perturbation of the Reissner Nordstrom solution has been given in [6] and [14].

REFERENCES

[1] L. Garding, T. Kotake, J. Leray, Bull. Soc. Math 92, 263, 1964.

[2] J. A. Wheeler, Geometrodynamics, New York, Academic Press, 1962.

[3] Y. Choquet-Bruhat, C. R. Acad. Sci. Paris, 258, 3809-12, 1964.

[4] Y. Choquet-Bruhat, Comm. Maths. Phys. 12, 16-35, 1969.

[5] Y. Choquet-Bruhat, J. Maths. pures et appl. 48, 117-158, 1969.

[6] Y. Choquet-Bruhat in "Gravitation," papers in honor of N. Rosen, edited by

Kuiper and Peres, Gordon and Breach, 1971, et Colloque "on des et radiations gravitationnelles" publication CNRS, 1973; see also Fatton, Thèse 3e cycle, Paris, 1970.

[7] A. Gomes, C.R. Acad. Sci. Paris, 262, 412, 1966; 262, 603, 1966 et thèse université Coimbra 1966.

[8] R. A. Issacson, Phys. Rev. 166, 1263, 1969.

[9] A. H. Taub, Comm. Maths. Phys. 31, 310, 1973.

[10] A. Lichnerowicz, Ann. di Mat. pura ed. appl. 50, 1, 1960.

[11] A. Lichnerowicz, Colloque "Ondes et radiations gravitationneles" publication CNRS, 1973.

[12] W. Gerlach, Phys. Rev. Lett 32, 1033, 1974 and preface to 1975.

[13] See also J. Madore, Comm. Maths. Phys. 27, 297, 302, 1972.

[14] For the coupling of electromagnetic and gravitational waves by the eigenfunction expansion method of Regge-Wheeler-Zerilli see:

F. Zerilli, Phys. Rev. D7, 860, 1974.

And for physical applications:

M. Johnston, R. Ruffini, F. Zerilli, Phys. Rev. Lett. 31, 1317, 1973.

M. Johnston, R. Ruffini, F. Zerilli, Phys. Rev. Lett. 49B, 185, 1974.

PLASMA HORIZONS OF A CHARGED BLACK HOLE*

Richard Squier Hanni†

Department of Physics
Stanford University
Stanford, California 94305

INTRODUCTION

The most promising way of detecting black holes seems to be through electromagnetic radiation emitted by nearby charged particles. The nature of this radiation depends strongly on the local electromagnetic field, which varies with the charge of the black hole. It has often been purported that a black hole with significant charge will not be observed, because the dominance of the Coulomb interaction forces its neutralization through selective accretion.

This paper will show that it is possible to balance the electric attraction of particles whose charge is opposite that of the black hole with magnetic forces and (assuming an axisymmetric, stationary solution) covariantly define the regions in which this is possible. A Kerr-Newman hole in an asymptotically uniform magnetic field and a current ring centered about a Reissner-Nordström hole are used as examples, because of their relevance to processes through which black holes may be observed.

DEFINITION OF THE PLASMA HORIZON

In a stationary axisymmetric system, there are two Killing vectors, ξ_t and ξ_φ. The definition of the electric and magnetic fields is determined uniquely by their relation to the electric and magnetic fluxes. [1]

$$E^\alpha = F^{\alpha\beta}\eta_\beta \qquad\qquad B^\alpha = -{}^*F^{\alpha\beta}\eta_\beta \tag{1}$$

where Greek indices run from 0 to 3,

$$\eta = C\left[\xi_t - \frac{(\xi_\varphi, \xi_t)}{(\xi_\varphi, \xi_\varphi)}\xi_\varphi\right] \tag{2}$$

and C is a constant of normalization. In the Kerr-Newman solution, this definition reduces to that of Christodoulou and Ruffini [2] and η reduces to the 4-velocity of the zero angular momentum observer. [3] η approaches ξ_t far from the hole, and is timelike everywhere outside the event horizon of a Kerr-Newman hole.

A unit charge will experience a Lorentz-force equal to the contraction of its timelike 4-velocity, u_β, with the electromagnetic field:

$$K^\alpha = F^{\alpha\beta}u_\beta \qquad\qquad u_\beta u^\beta < 0 \tag{3}$$

Are there trajectories for a particle (timelike 4-velocities) such that the Lorentz-force exerted on it is orthogonal to the electric field?

$$E^{\alpha} K_{\alpha} = E^{\alpha} F_{\alpha\beta} u^{\beta} = 0 \tag{4}$$

Only if the 4-velocity of the unit charge is orthogonal to the 4-vector V_{γ} defined by:

$$V_{\gamma} = E^{\alpha} F_{\alpha\gamma} = \eta_{\beta} F^{\alpha\beta} F_{\alpha\gamma} \tag{5}$$

Since u_{β} is timelike, V_{γ} must be spacelike or zero. Conversely, when V_{γ} is spacelike or zero, there always exists a timelike 4-velocity u_{β} orthogonal to V_{γ}. Thus, the magnetic force can balance the electric force on a charged particle if and only if V_{γ} is spacelike or V_{γ} is zero:

$$V_{\gamma} V^{\gamma} > 0 \qquad \text{or} \qquad V_{\gamma} = 0 \tag{6}$$

We define the plasma horizon [4,5] as the boundary of the region in which (6) is satisfied. It is the surface on which V_{γ} is lightlike and nonzero:

$$V_{\gamma} V^{\gamma} = 0 \qquad \text{and} \qquad V_{\gamma} \neq 0 \tag{7}$$

This concept is new and should not be confused with the null hypersurfaces also referred to as horizons. Physically, the plasma horizon is the boundary of the region in which the magnetic field can support an infinitely thin plasma against Coulomb attraction.

Projected onto the orthonormal tetrad with 4-velocity η, V_{γ} has the form:

$$V_{\hat{\gamma}} = F^{\hat{\alpha}\hat{\beta}} F_{\hat{\alpha}\hat{\gamma}} \eta_{\hat{\beta}} = F^{\hat{t}\hat{\alpha}} F_{\hat{\alpha}\hat{\gamma}} \tag{8}$$

In this local Lorentz frame the temporal components of the electric and magnetic fields are zero, so the requirement that the Lorentz-force on the test charge be orthogonal to the electric field (4) reduces to a condition on the spatial 3-vectors. The magnetic force is maximized as the 3-velocity of the charge approaches the speed of light and is orthogonal to the magnetic field. Only that part of the magnetic field which is orthogonal to the electric field can balance the electric field. Thus the cross product of the electric and magnetic fields must have at least the same magnitude as the square of the electric field, for the Lorentz-force to be orthogonal to the electric field. The scalar defining the plasma horizon (7) takes the simple form:

$$V_{\hat{\gamma}} V^{\hat{\gamma}} = |E_{\hat{\theta}} B_{\hat{r}} - E_{\hat{r}} B_{\hat{\theta}}| - |E_{\hat{r}}^2 + E_{\hat{\theta}}^2| \tag{9}$$

BLACK HOLE IN AN ASYMPTOTICALLY UNIFORM MAGNETIC FIELD

The galactic magnetic field and that of a magnetic star are essentially uniform on the scale of a black hole formed by stellar collapse. The effect of both fields on the spacetime background is small. Therefore, to calculate the electromagnetic radiation produced by a black hole in a binary system, it is good to understand the effect of a black hole on a weak asymptotically uniform magnetic field.

The corresponding weak field solution in the Reissner-Nordström background follows from the usual separation of variables: [6]

$$A_\alpha dx^\alpha = B \frac{r^2 - Q^2}{2} \sin^2\theta \, d\varphi - \frac{Q}{r} dt \tag{10}$$

If the term second order in the charge of the black hole is parametrized and the potential (10) is substituted into Zerilli's [7] linearization of the Einstein-Maxwell equations, the resulting geometry is:

$$ds^2 = -\left(1 - \frac{2M}{r} + \frac{Q^2}{r^2}\right) dt^2 + \left(1 - \frac{2M}{r} + \frac{Q^2}{r^2}\right)^{-1} dr^2 + r^2 d\theta^2 + r^2 \sin^2\theta \left(d\varphi + \frac{2BQ}{r} dt\right)^2 \tag{11}$$

and the nonzero tetrad components of the electromagnetic field are:

$$E_{\hat{r}} = Q/r^2 \qquad B_{\hat{r}} = B\left(1 - \frac{3Q^2}{r^2}\right)\cos\theta \tag{12.1}$$

$$B_{\hat{\theta}} = -B \sin\theta \left(1 - \frac{2M}{r} + \frac{Q^2}{r^2}\right)^{\frac{1}{2}} \tag{12.2}$$

Again the asymptotic strength of the magnetic field, B, must be small, but the charge, Q, is not restricted as it was in the weak field approximation. Equations (9) and (12) determine the plasma horizon, whose separation from the axis:

$$r \sin\theta = \frac{Q}{Br}\left(1 - \frac{2M}{r} + \frac{Q^2}{r^2}\right)^{-\frac{1}{2}} \tag{13}$$

decreases asymptotically as 1/r. Therefore, the volume of the region of forced accretion is finite.

The asymptotically uniform magnetic field in the Kerr solution was first published by Wald. [8] That electromagnetic field also follows immediately from Teukolsky's [9] separation of the wave equation. The latter method is easily extended to give the weak field approximation in the Kerr-Newman background; [10]

$$A_\alpha dx^\alpha = B\,\frac{r^2+a^2-Q^2}{2}\,\sin^2\theta\, d\varphi + \frac{aB}{\Sigma}\,[(Mr - Q^2)(1+\cos^2\theta) + Q^2]\,(dt - a\sin^2\theta\, d\varphi) \qquad (14)$$

It may be possible to generalize this Newman-Penrose analysis to solve the linearized Einstein-Maxwell equations for a black hole with charge and angular momentum. Alternatively, that solution might be derivable from the weak field solution, as in the Reissner-Nordström case, but it appears that electronic symbol manipulation would be required.

In the limit that the magnetic field does not affect the geometry, neither does the charge of the black hole, so the metric is Kerr:

$$ds^2 = \Sigma\Delta^{-1}\, dr^2 + \Sigma d\theta^2 + \Sigma^{-1}\sin^2\theta\,[a dt - (r^2+a^2) d\varphi]^2 - \Sigma^{-1}\,\Delta\,[dt - a\sin^2\theta\, d\varphi]^2 \qquad (15)$$

where $\Sigma = r^2 + a^2\cos^2\theta$ and $\Delta = r^2 - 2Mr + a^2$. The tetrad components of the electromagnetic field are the superposition of those of a Kerr-Newman hole and an asymptotically uniform magnetic field:

$$E_{\hat r} = F_{\hat r\hat t} = \Sigma^{-2} A^{-\frac{1}{2}} \{BaM\,[2r^2\sin^2\theta\Sigma - (r^2+a^2)(r^2 - a^2\cos^2\theta)(1+\cos^2\theta)] + Q(r^2+a^2)(r^2 - a^2\cos^2\theta)\} \qquad (16.1)$$

$$E_{\hat\theta} = F_{\hat\theta\hat t} = \Sigma^{-2} A^{-\frac{1}{2}} \Delta^{\frac{1}{2}}\, 2ra^2\sin\theta\cos\theta\,\{BaM(1+\cos^2\theta) - Q\} \qquad (16.2)$$

$$B_{\hat r} = F_{\hat\theta\hat\varphi} = \Sigma^{-2} A^{-\frac{1}{2}} \cos\theta\{B[(r^2+a^2)\,\Sigma^2 - 2Mra^2(2r^2\cos^2\theta + a^2(1+\cos^4\theta))] + 2Qar(r^2+a^2)\} \qquad (16.3)$$

$$B_{\hat\theta} = F_{\hat\varphi\hat r} = \Sigma^{-2} A^{-\frac{1}{2}} \Delta^{\frac{1}{2}} \sin\theta\,\{Qa(r^2 - a^2\cos^2\theta) - B\,[Ma^2(r^2 - a^2\cos^2\theta)(1+\cos^2\theta) + r\,\Sigma^2]\} \qquad (16.4)$$

where $A = (r^2+a^2)^2 - \Delta a^2\sin^2\theta$.

Figure 1(a) shows the lines of force where the magnetic dipole moment of the Kerr-Newman hole ($a = 3M/4$, $Q = M/10$) is directed opposite to the applied magnetic field. Figure 2(a) shows the lines of force when the magnetic dipole moment of the black hole is aligned with the external field. The concentration and rarefaction of the magnetic field at the poles and the waist are of particular interest, as they determine where the plasma can be supported.

The plasma horizon is specified by equations (9) and (16). Figure 1(b) shows plasma horizons corresponding to asymptotic magnetic field strengths proportional to the adjacent integers. Figure 2(b) is identical, except that the orientation of the Kerr-Newman hole with respect to the applied magnetic field has been reversed. The two inner plasma horizons of the system with the magnetic dipole moment of the black hole aligned with the applied field (Figure 2(b)) are manifestly unstable, because the Lorentz-force on a charge supported against the Coulomb attraction of the hole is directed toward the equatorial plane. If the

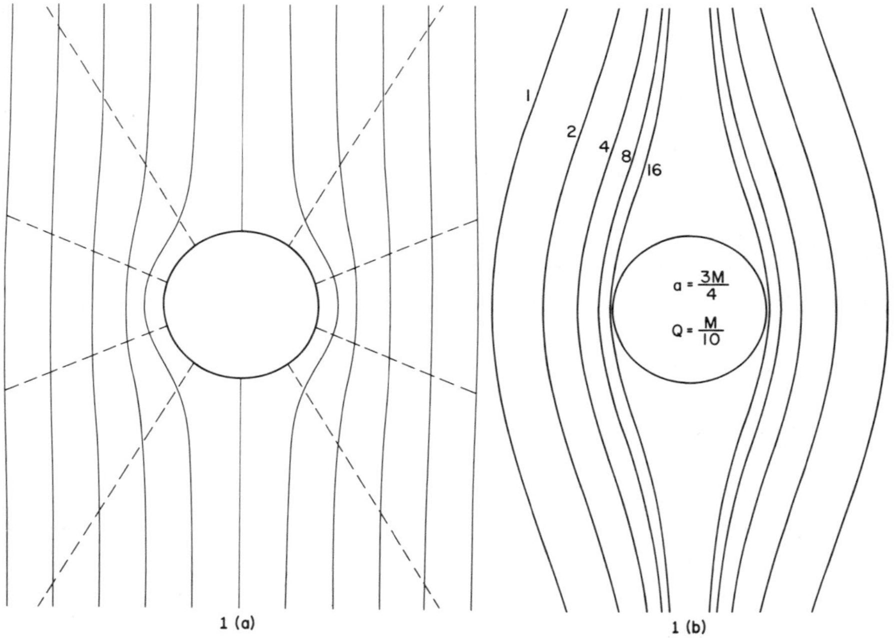

FIGURE 1: The circles represent the event horizon of a Kerr-Newman hole whose magnetic dipole moment is anti-parallel to an asymptotically uniform magnetic field in Boyer-Lindquist coordinates. The dashed lines in 1(a) are the electric lines of force; the continuous lines the magnetic lines of force. The open curves in 1(b) are the plasma horizons corresponding to asymptotic magnetic field strengths $B = nQ/M^2$.

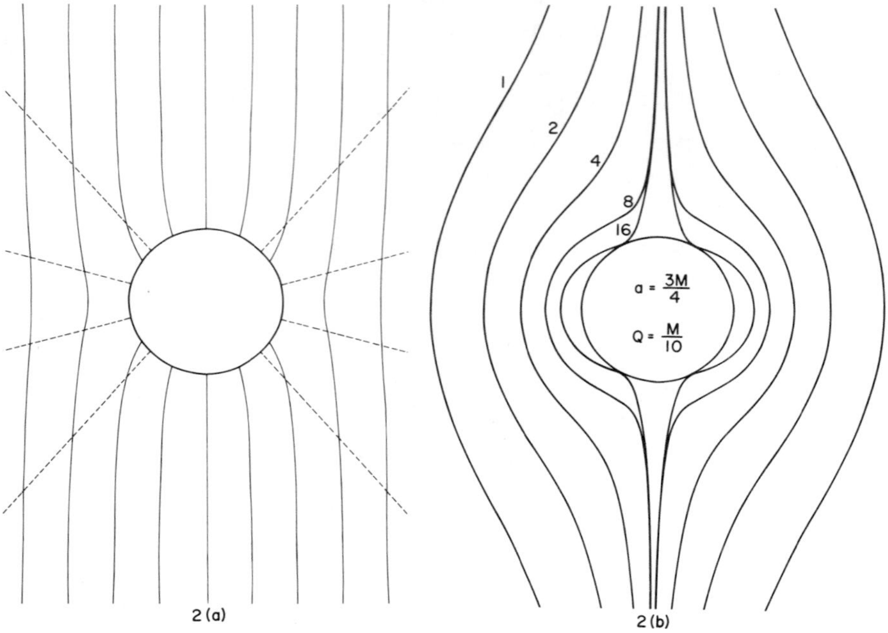

FIGURE 2: Same as Figure 1, except that the magnetic dipole moment of the black hole is parallel to the applied magnetic field.

self interaction of the plasma may be neglected, plasma near such a horizon will move through it obliquely and then accrete directly.

While the preceeding results suggest that the plasma horizon receeds from the hole as the asymptotically uniform magnetic field is decreased, there remains a question of principle. Can an isolated black hole have a plasma horizon? Since there is no applied field, the scalar defining the plasma horizon reduces to:

$$V_\gamma V^\gamma = Q^4 \Sigma^{-8} \mathbf{A}^{-1} \{16 r^4 a^6 \cos^4\theta \sin^2\theta\, \Delta - (r^2+a^2)^2 (r^2 - a^2\cos^2\theta)^4\} \tag{17}$$

$\mathbf{A}$ is positive outside the event horizon. The expression in brackets reduces to $-(r^2+a^2)^2\,(r^2-a^2)^4$ along the axis and $-(r^2+a^2)^2\,r^8$ in the equatorial plane. Its derivative with respect to $\cos^2\theta$ is a polynomial cubic in $\cos^2\theta$ with no roots between zero and one. Therefore $V_\gamma V^\gamma < 0$, and there is no plasma horizon for an isolated black hole.

STATIC BLACK HOLE IN A CIRCULAR CURRENT

Wilson's [11] calculations of the collapse of a magnetic star have shown that a ring of current with net charge opposite that of the collapsing star may form in the equatorial plane. As a model for the resulting system, consider a Reissner-Nordström hole centered in a circular current with opposite net charge.

Maxwell's equations were solved [4,5] in the limit that the effect of the current ring on the background geometry, and thus that of the charge of the black hole, can be neglected. A similar problem was investigated by Petterson. [12] The scalar defining the plasma horizon is:

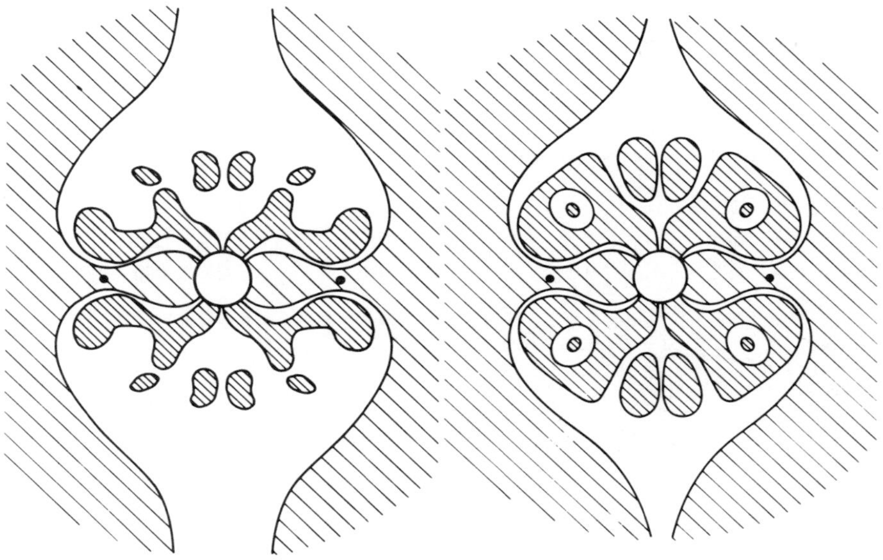

3 (a) 3 (b)

FIGURE 3: The regions in which an infinitely thin plasma can be supported by the electromagnetic field of a current ring centered about a Reissner-Nordström hole against the Coulomb attraction of the hole are shaded. Their boundaries constitute the plasma horizon. J^{φ}/J^{t} equals 2 in 3(a), 3 in 3(b).

$$V_\gamma V^\gamma = [(1 - \frac{2M}{r})A_{t,r} A_{\varphi,r} + A_{t,\theta} A_{\varphi,\theta}/r^2]^2 / r^2\sin^2\theta$$
$$- [(1 - \frac{2M}{r})(A_{t,r})^2 + (A_{t,\theta})^2 / r^2]^2 / (1 - \frac{2M}{r}) \qquad (18)$$

Plasma horizons corresponding to different currents are illustrated in Figure 3. Far away from the black hole the electric quadrupole and magnetic dipole moments dominate, so the separation of the plasma horizon from the axis has a finite asymptotic limit. The astrophysical implications of these results and the related stability problem have been discussed by Ruffini [4] and others. [10]

ACKNOWLEDGEMENTS

The concept of the plasma horizon was conceived and refined in collaboration with T. Damour, R. Ruffini, and J. R. Wilson. We thank D. M. Haskin and F. Thorne for their valuable assistance in providing computational facilities.

---

*Supported in part by the National Science Foundation Grant GP-38610.

†Dupont Graduate Fellow.

[1] Hanni, R. S., Senior Thesis, Princeton University,unpublished (1973).

[2] Christodoulou, D. and R. Ruffini, "On the Electrodynamics of Collapsed Objects" in Black Holes,Gordon and Breach, N. Y. (1973).

[3] Bardeen, J. M., Ap.J. 162 (1970) 71.

[4] Ruffini, R., Seventh Texas Symposium on Relativistic Astrophysics, New York Acad. of Sciences, N. Y. (1975).

[5] Hanni, R. S., Seventh Texas Symposium on Relativistic Astrophysics, New York Acad. of Sciences, N. Y. (1975).

[6] Hanni, R. S. and R. Ruffini, To be published.

[7] Zerilli, R., Phys. Rev. D. 9 (1974) 860.

[8] Wald, R. M., Phys. Rev. D. 10 (1974) 1070.

[9] Teukolsky, S. A., Ap. J. 185 (1973) 635.

[10] Damour, T., R. S. Hanni, R. Ruffini and J. Wilson. To be published.

[11] Wilson, J. R., Seventh Texas Symposium on Relativistic Astrophysics, New York Acad. of Sciences, N. Y. (1975).

[12] Petterson, J. A., Phys. Rev. D. 10 (1974) 3166.

MAGNETIC BLACK HOLES

W. Kundt

I. Institut fur Theoretische Physik der Universitat
Hamburg, West Germany

Black holes are commonly thought of as stationary even in astrophysical context, and hence unrelated to regularly pulsating sources. This view is suggested by Price's work indicating that black holes embedded in electrovac adjust their multipole moments to those of the 3-parametric Kerr-Newman family (of charged black holes) via radiation, and also due to the no-hair-theorems by Israel, Carter, Hawking and Robinson proving that there can be at most 3-parametric families of stationary black hole spacetimes which moreover reduce to the 2-parametric Kerr family in the limit of vanishing electric charge. The purpose of this communication is to point out that realistic black holes may be endowed with a rather large magnetic flux corresponding initially to $\sim 10^{13}$G, which is anchored within a surrounding plasma of high conductivity. This flux may well be inclined with respect to the hole's angular momentum, and therefore give rise to phenomena similar to those attributed to rotating magnetic neutron stars, the greatest difference being a shorter lifetime and a different current geometry. The plasma dynamics near the black hole are governed by electromagnetic forces until almost up to the horizon, and the lifetime of the magnetic moment is determined by diffusion of the plasma that confines the flux (rather than by gravitational collapse), which may well be of astrophysical importance.

A detailed analysis of these problems has been started in [1]. The following results, partially still unpublished, are worth mentioning:

1) Magnetic fields can penetrate black holes; they can be treated as test fields for

$$B_{18}(M/M_{\odot}) < 1, \qquad (B_{18} := B/10^{18}\mathrm{G}).$$

2) The magnetic flux ϕ of an asymptotically uniform field going down one hemisphere of the horizon (can be defined uniquely and) is given by

$$\phi = \pi r_+^2 \, B(1 - a^4/r_+^4)$$

where r_+ = horizon radius, B = magnetic field strength at infinity, a = specific angular momentum in geometrical units. The hole thus appears to expel the flux increasingly with increasing a/r_+ (< 1).

3) Black holes appear to have zero polarizability because no induced dipole field is seen near infinity.

4) A rotating magnetic black hole creates an electric (unipolar induction) field $\vec{E}$ with nonvanishing invariant $\vec{E}\cdot\vec{B}$. It therefore tends to selectively accrete charged particles. As a result, charged particles will redistribute such as to achieve $\vec{E}\cdot\vec{B} = 0$, c.f. [2], and there will be a surplus volume charge in the

the surrounding plasma.

5) In a steady state situation, it is unlikely that the total electric current down the hole should vanish without a net charge in the hole. This charge Q cannot be obtained from the principles of minimum field energy, or zero injection energy (which result in just one sign of $\vec{E}\cdot\vec{n}$ at the horizon ($\vec{n}$ being its normal), and will be of the sign of, but smaller than the charge calculated in [3]: $Q = \alpha B J (G/c^3)$ where α is of order 2/5, J = angular momentum of the hole, G = Newton's constant.

6) Near the horizon, electromagnetic forces are stronger than gravitational forces for

$$B_2 \beta_{-2} (M/M_{\odot})(m/m_p) > 1$$

where β = toroidal velocity in units of c, and m_p = proton mass. This shows that gravity is only a minor perturbation for magnetic fields about 10^2 Gauss.

7) Magnetic torques will force the plasma near the horizon to essentially corotate with the hole. Corotation of the plasma must stop near the velocity-of-light cylinder.

8) An alignment of the magnetic moment with the axis of rotation (or conceivably even a counter-alignment) may be expected on the slowdown timescale of the hole, and might depend on details, c.f. [4].

9) The magnetic flux will leak out from the hole (via infall of an oppositely oriented flux) due to instabilities of the surrounding plasma. Such are
a) a neutral rain down the hole, or wind leaving the corotating magnetosphere,
b) a finite plasma conductivity (due to particle collisions at relatively high density, and due to scattering on photons at relatively low density, and
c) limited toroidal conductivity (in a strong magnetic field).

REFERENCES

[1] King, A., Lasota, J. P., Kundt, W. Black Holes and Magnetic Fields, Phys. Rev. D to appear (1975).

[2] Goldreich, P., Julian, W.H. Ap. J. 157, 869 (1969).

[3] Wald, R. M. Phys. Rev. D10, 1680 (1974).

[4] Goldreich, P. Ap. J. 160, L11 (1970).

RELATIVISTIC ELASTICITY APPLIED TO NEUTRON STARS

H. Quintana*

Department of Astronomy-University of Chile
Casilla 36-D, Santiago, Chile

ABSTRACT: A short account is presented of the development of general relativistic elasticity. The theory is applied to slowly rotating relativistic solid stars. Explicit expressions are derived for the shear, pressure and energy momentum tensors, valid up to third order in the angular velocity for a solid star that has a spherical relaxable structure.

I. INTRODUCTION

The general relativistic theory of elasticity has been the subject of several unrelated and only partially successful papers. The first (largely unknown) formulation of the theory was due to Souriau (1958, 1964), who discussed the basic mathematical concepts. More attention was paid to the work of Synge (1959) in which he (mistakingly) assumed that only time rates of strain, but not absolute strain states, could be described in general relativity. A suitable formalism for the description of a continuous medium was developed by Oldroyd (1970), who introduced the concept of convected derivative. A more complete and independent formulation of relativistic elasticity theory has been given by Carter and Quintana (1972, hereafter called Paper I), who also discuss some of the earlier work. They developed elasticity theory for simple solids, including high-pressure materials, with a view to applications to neutron star models. Ehlers (1973) has reviewed and extended this work to include entropy as a variable.

One interesting application of relativistic elasticity, namely, the description of the interaction of gravitational waves with elastic solid bodies, has been discussed in the literature. Some of this work has been based on the ideas of Synge (1959) or on the article of Rayner (1963) (Papapetrou 1972, Lamoureux 1974; see also the reviews by Maughin 1973, 1974). A preliminary discussion of this problem based on Paper I has been reported elsewhere (Carter and Quintana 1974). The dynamical nature of the interaction means that only time rates of strain are considered, allowing Synge's formulation to be applicable. Furthermore, due to the approximations made when discussing a test body, the differences between the various formulations turn out to be negligible.

A second application of astrophysical interest concerns the description of the elastic structure of neutron star models. Carter (1973) has proposed a variational principle for a relativistic solid star and Carter and Quintana (1975 b, Paper III) have derived the general elasticity formulae as applied to a rotating relativistic solid star (for any angular velocity). The present work, which is based

*Present Address: European Southern Observatory, c/o CERN, 1211 GENEVE 23

on this last paper, gives explicit expressions for the various elastic tensors for a slowly rotating star (endowed with a spherical relaxable elastic structure) in a specific coordinate system (a more complete account of this work is being published elsewhere [Quintana 1976]).

The observed glitches and noise in the radio emission from pulsars have been interpreted as evidence for the existence of a solid component in neutron stars. Sudden rotation induced quakes change the moment of inertia, which implies associated sudden changes in the angular velocity. In not too massive neutron stars ($0.1 - 0.5M_{\odot}$) there exists one solid component, a crust on top of a fluid core.

The heavier core dynamically dominates the shape of the star, forcing it to be very nearly that of a perfect fluid star. For more massive stars also a solid core is likely to exist in which the lattice forces derive from the repulsive nuclear forces. The rigidity of this solid material is five to six orders of magnitude higher than that corresponding to the crust material. Since stars with solid cores are more massive it is expected that general relativistic corrections will be more important for them. Consequently, a general relativistic treatment of the elastic structure is desirable. For a discussion of the (newtonian) starquake theory and a review of the general properties of neutron stars see Ruderman (1972). Canuto (1975) has reviewed the equations of state for solid nuclear matter and Krotscheck et al (1975) discuss in detail the starquake theory and the physics of the crust material.

A general relativistic description of the starquake theory has been given by Carter and Quintana (1975 a). It was shown that the relevant parameters in the theory are the third order coefficients in the expansion of the angular momentum in terms of the angular velocity. These coefficients are determined by the third order equations of the structure of a slowly rotating star. The second order and the third order structure equations for a slowly rotating perfect fluid star have been derived and integrated by Hartle (1967, 1973) and Hartle and Thorne (1968), among others. For perfect fluid stars it is found that the changes in the moment of inertia for different angular velocities depend on the $\ell=1$ third order correction to the $g_{t\phi}$ metric element, but not on the $\ell=3$ function. For solid stars a semi-Newtonian treatment of the starquake theory has been given by Baym and Pines (1970) (for an incompressible material) and by Heintzmann et al. (1973). A more exact calculation should properly consider the elastic material of the solid component of the star and general relativistic elasticity theory should be used to describe its structure. The material of the star is taken to be a perfect solid, defined by two equations of state that give the pressure and rigidity as functions of density (Paper I).

For rotating solid stars a three parameter family of relaxable states can be defined by the total baryon number, the actual angular velocity Ω and the initial angular velocity $\bar{\Omega}$ of the star in the state for which the star would be fully unsheared, which one might presume to be the state in which it froze. This initial angular velocity $\bar{\Omega}$ determines a particular relaxable structure. To compute the coefficients of the moment of inertia for any relaxable structure it is only necessary to solve the perfect fluid and the solid star equations in what is called the spherical relaxable structure (that is to say, when $\bar{\Omega}=0$), as was shown by Carter and Quintana [1975 a]. Because, in practice, the perfect fluid and the solid structure equations should be solved simultaneously, it is convenient in the solid case to perform a similar treatment to the perfect fluid analysis. In particular, we shall follow the work of Hartle [1967, 1973]. The slowly rotating state is considered as a perturbation of the spherical state with the same baryon number, both for the perfect fluid and the solid star, since the state of the spherical relaxable structure with vanishing angular velocity is identical to the state of the spherical perfect fluid star. Therefore, one has a common unperturbed state.

II. THE SLOWLY ROTATING SOLID STAR MODEL

It is convenient to use a coordinate system in which the coordinates $x^0=t$ and $x^3=\phi$ are defined by the symmetries of the problem (stationarity and axisymmetry). The two other coordinates are chosen as spherical coordinates, $x^1=r$ and $x^2=\theta$ (asymptotically flat at infinity), with the scale of the radial coordinate to be fixed later on. Provided the elastic structure is ϕ- reversible, the metric for a stationary equilibrium state of an axisymmetric rigidly rotating solid star can [Carter 1975] be expressed in the Papapetrou form,

$$ds^2=g_{00}\, dt^2+2g_{03}\, dt\, d\phi+g_{33}\, d\phi^2+g_{11}(dx^1)^2+2g_{12}\, dx^1 dx^2+g_{22}(dx^2)^2 \tag{1}$$

where the metric elements are functions of x^1 and x^2. With the above choice of coordinates this metric can be written as a perturbation of the spherical metric, taking $g_{12} = 0$ (spherical coordinates), in a similar fashion to Hartle's [1967],

$$ds^2=-e^{\nu}(1+2h)dt^2+e^{\lambda}(1+\frac{2m}{r-2M})dr^2+r^2(1+2k)[d\theta^2+\sin^2\theta\{d\phi-(\omega+W)dt\}^2] \tag{2}$$

where M (representing the mass inside r radius), ν and λ are the radial functions describing the spherical solution, and ω is the purely radial first order function [Hartle 1967] that represents the rotation. The functions h, m, k, represent the second order perturbations in the angular velocity, and finally, W is the third order perturbation in the angular velocity, all of these being functions of r and Θ .

A material point 4-velocity has components u^{ϕ}, u^t

$$u^{\phi} = \Omega u^t \tag{3}$$

$$u^t = [-(g_{oo} + 2\Omega g_{o3} + \Omega^2 g_{33})]^{-1/2} \quad , \tag{4}$$

which in terms of the perturbation functions of the metric defined above, give

$$u^t = e^{-\nu/2}[1 - h + \tfrac{1}{2}e^{-\nu}\bar{\omega}^2 r^2 \sin^2\theta] + O(\Omega^4) \tag{5}$$

where $\bar{\omega} = \Omega - \omega$. Besides, u^t represents the redshift factor introduced in Paper III.

If one denotes the unperturbed spherical metric by $\bar{g}_{ab}$, the Eulerian perturbations $\delta g_{ab} = h_{ab}$ of the metric

$$g_{ab} = \bar{g}_{ab} + h_{ab} \tag{6}$$

are given explicitly by equation (2). The fundamental projection tensor

$$\gamma_{ab} = g_{ab} + u_a u_b \tag{7}$$

that projects on the plane orthogonal to the 4-velocity u^a, has components which can be immediately derived from (2) and (5).

A material point q at $x^a - \zeta^a$ in the unperturbed state, suffers a displacement ζ^a (a=r, θ), plus a rotation with angular velocity Ω. It is convenient to define a semilagrangian variation D as the variation induced by ζ^a (a true Lagrangian variation will include the translation in the φ-direction induced by the rotation). The relation between the semilagrangian variation D (variation of a quantity around the same material ring) and the Eulerian variation δ (with respect to a spacetime point) is the Lie derivative of the relevant quantity:

$$D - \delta = \mathscr{L}_{\zeta} \tag{8}$$

In particular, for the metric one has

$$Dg_{ab} = h_{ab} + 2\zeta_{(a;b)} \tag{9}$$

It is natural to choose the unperturbed state as the reference state for the strain and shear tensors of the elastic deformed state. At every point of the unperturbed star there exists a projection $\bar{P}$ from the spacetime to the 3-dimensional manifold X representing the body. After the perturbation generated by the displacement $\Delta x^a = \zeta^a$, the same material point q will be projected by a different projection P on the same point in X .

There exist material functions $\bar{n}(x)$, $\rho(\bar{n})$, $\bar{K}_{ab}$,...etc., that describe the unperturbed state. We can use these same functions to describe the reference state for the perturbed state, i.e. we take the same functional forms for the corresponding reference functions $n_o(x)$, $\rho(n_o)$, $K_{ab}(n_o)$,... etc. The actual perturbed state is described by corresponding functions n, $\rho(n)$, $K_{ab}(n)$, ... etc., that satisfy

$$n(x) - \bar{n}(x) = \delta n(x) \tag{10}$$

$$n(x) - n_o(x-\zeta) = Dn(x) \tag{11}$$

with similar expression holding for other quantities.

The strain tensor is defined by

$$e_{ab} = \frac{1}{2}(\gamma_{ab} - K_{ab}) \tag{12}$$

where K_{ab} is the reference quasi-Hookean tensor (value of γ_{ab} for the relaxed state: when $n=n_o$). The shear tensor at constant volume is defined by

$$s_{ab} = \frac{1}{2}(\gamma_{ab} - \eta_{ab}) \tag{13}$$

where the shear reference tensor is given in terms of the reference tensor K_{ab} by (isotropic solid)

$$\eta_{ab} = \left(\frac{n}{n_o}\right)^{-2/3} K_{ab} \tag{14}$$

In the small deformation approximation we keep terms of first order in the strain or the shear. The shear, from (12), (13) and (14), can be written as,

$$s_{ab} = e_{ab} - \frac{1}{3} e_c{}^c \gamma_{ab} + 0(e^2) \tag{15}$$

where $0(e)$ is the order of the strain or shear: $0(e) = 0(s)$.

For the unperturbed state $\bar{s}_{ab} = \bar{e}_{ab} = 0$, so that $e_{ab} = \delta e_{ab}$ and $s_{ab} = \delta s_{ab}$. The trace of the strain tenson $e_c{}^c$ gives the change in the density around a fixed point: the volume dilatation,

$$Dn = - n_o e_c{}^c \tag{16}$$

so that the Eulerian variationof the particle density from (8) and (16) is

$$\delta n = - n_o e_c{}^c - n_{o,a} \zeta^a \tag{17}$$

In the order of approximation used, for a perfect solid as defined in Paper I, the density is given by

$$\rho=\rho(n) + 0(e^2) = \rho(n_o) - n_o e_c{}^c \frac{d\rho}{dn} + 0(e^2) \tag{18}$$

and the pressure tensor is given by

$$p_{ab} = p(n)\gamma_{ab} - 2\mu(n)s_{ab} + 0(e^2)$$
$$= p(n_o)\gamma_{ab} - e_c{}^c[n_o \frac{dp}{dn} - \frac{2}{3}\mu(n_o)]\gamma_{ab} - 2\mu(n_o)s_{ab} + 0(e^2) \qquad (19)$$

where $\mu(n)$ is the rigidity as function of density and where $p(n_o)$ is the isotropic unsheared value of the pressure. Therefore the Eulerian variations of the density and pressure tensor are given by:

$$\delta\rho = -\rho_{,a}\zeta^a - n_o\frac{d\rho}{dn}e_c{}^c + 0(e^2) \quad , \qquad (20)$$

$$\delta p_{ab} = -[p_{,a}\zeta^a + \beta(n)\ e_c{}^c]\gamma_{ab} + p(n)\delta\gamma_{ab} - 2\mu(n)\ s_{ab} + 0(e^2) \qquad (21)$$

where the bulk modulus β is defined by

$$\beta(n) = n\frac{dp}{dn} \quad . \qquad (22)$$

The energy momentum tensor variation can be deduced from (18) and (19) and is given by the expression (Paper III)

$$\delta T_{ab} = (u_a u_b \frac{d\rho}{dn} + \gamma_{ab}\frac{dp}{dn})\delta n + (\rho + p)\delta\gamma_{ab}$$
$$- \rho(n)h_{ab} - 2\mu(n)\delta s_{ab} + 0(e^2) \quad . \qquad (23)$$

In the coordinate system being used, the covariant derivatives of the displacement $\zeta^a = \zeta^r, \zeta^\theta$ can be computed up to third order in the angular velocity. One obtains the following formulae:

$$\zeta_{(o;o)} = -\frac{1}{2}e^\nu\ \nu_{,r}\ \zeta^r + 0(\Omega^4)$$
$$\zeta_{(o;3)} = -\frac{1}{2}[(\omega_{,r} + \frac{2\omega}{r})\zeta^r + 2\omega\cot\theta\ \zeta^\theta]r^2\sin^2\theta + 0(\Omega^5) \qquad (24)$$
$$\zeta_{(3;3)} = (\frac{1}{r}\zeta^r + \cot\theta\ \zeta^\theta)r^2\sin^2\theta + 0(\Omega^4)$$
$$\zeta_{(r;r)} = (\zeta^r_{,r} + \frac{\lambda_{,r}}{2}\zeta^r)e^\lambda + 0(\Omega^4)$$
$$\zeta_{(r;\theta)} = \frac{1}{2}(e^\lambda\zeta^r_{,\theta} + r^2\zeta^\theta_{,r}) + 0(\Omega^4)$$
$$\zeta_{(\theta;\theta)} = (\frac{1}{r}\zeta^r + \zeta^\theta_{,\theta})r^2 + 0(\Omega^4)$$

It was shown quite generally in Paper III that for a rotating solid star the shear tensor is given in terms of the variations $\delta(w^2)$ of the determinant w^2 (variations induced by a change in the angular velocity $\delta\Omega$)

$$w^2 = (g_{o3}^{\;2} - g_{oo}\, g_{33})(u^t)^2 \tag{25}$$

by the expressions (α, β = r, θ)

$$s_{\alpha\beta} = \frac{1}{2}\,[Dg_{\alpha\beta} - \frac{1}{3}\,\{g^{\gamma\delta}\,Dg_{\gamma\delta} + w^{-2}D(w^2)\}\,g_{\alpha\beta}] \tag{26}$$

$$s_{33} = -\frac{1}{3}\,[\frac{1}{2}\,w^2\,g^{\gamma\delta}\,Dg_{\gamma\delta} - D(w^2)] \quad , \tag{27}$$

and the volume dilatation is given by

$$e_c^{\;c} = \frac{1}{2}\,[g^{\alpha\beta}\,Dg_{\alpha\beta} + w^{-2}\,D(w^2)] \quad . \tag{28}$$

For vanishing initial angular velocity $\bar{\Omega} = 0$, $\Omega = \delta\Omega$ (small), the variation of w^2 can easily be shown to be

$$D(w^2) = h_{33} + 2\zeta_{(3;3)} + \frac{(h_{o3} + \Omega g_{33})^2}{-\,g_{oo}} \quad , \tag{29}$$

so that the volume dilatation takes the form

$$\begin{aligned} e_c^{\;c} &= \frac{1}{2}\,\bar{\gamma}^{ab}(h_{ab} + 2\zeta_{(a;b)}) + \frac{1}{2}\,\frac{(h_{o3} + \Omega\, g_{33})^2}{-\,\bar{g}_{oo}\,\bar{g}_{33}} + 0(\Omega^4) \\ &= 2k + \frac{m}{r-2M} + \zeta^r_{\;,r} + (\frac{2}{r} + \frac{\lambda_{,r}}{2})\zeta^r + \zeta^\theta_{\;,\theta} + \cot\theta\;\zeta^\theta + \frac{1}{2}e^{-\nu}\bar{\omega}^2 r^2 \sin^2\theta + 0(\Omega^4) \end{aligned} \tag{30}$$

from which, using (17), the explicit expression for δn follows. In a similar fashion, it follows easily from (26), (27) and (28) that the shear tensor is given by,

$$\begin{aligned} s_{\alpha\beta} &= \frac{1}{2}\,[Dg_{\alpha\beta} - \frac{2}{3}\,e_c^{\;c}\,\bar{\gamma}_{\alpha\beta}] + 0(\Omega^4) \\ s_{33} &= \frac{1}{2}\,[Dg_{33} - \frac{2}{3}\,e_c^{\;c}\,\bar{\gamma}_{33} + \frac{(h_{o3} + \Omega g_{33})^2}{-\,\bar{g}_{oo}}] + 0(\Omega^4) \end{aligned} \tag{31}$$

or explicitly, using (9), (6), (24) and (30):

$$s_{rr} = e^{\lambda}[\frac{m}{r-2M} + \zeta^r_{\;,r} + \frac{\lambda_{,r}}{2}\,\zeta^r - \frac{1}{3}\,e_c^{\;c}] + 0(\Omega^4)$$

$$s_{r\theta} = \frac{1}{2}\,r^2[\frac{e^{\lambda}}{r}\,\zeta^r_{\;,\theta} + \zeta^\theta_{\;,r}] + 0(\Omega^4)$$

$$s_{\theta\theta} = r^2[k + \frac{1}{r}\zeta^r + \zeta^\theta_{,\theta} - \frac{1}{3}e_c{}^c] + O(\Omega^4)$$

$$s_{33} = r^2\sin^2\theta\ [k + \frac{1}{r}\zeta^r + \cot\theta\ \zeta^\theta - \frac{1}{3}e_c{}^c + \frac{1}{2}e^{-\nu}\bar{\omega}^2 r^2 \sin^2\theta] + O(\Omega^4)$$

Replacing the variation formulae in the expression of the energy momentum tensor, it is simple to derive explicit expressions for it. The variations of the density, the projection tensor $\delta\gamma_{ab} = \gamma_{ab} - \bar{\gamma}_{ab}$ and the shear tensor are known. The derivatives with respect to n can be combined with the expression for the variation of n to obtain,

$$\delta p = \frac{dp}{dn}\delta n = - p_{,r}\zeta^r - \beta e_c{}^c \tag{33}$$

$$\delta\rho = \frac{d\rho}{dn}\delta n = - \rho_{,r}\zeta^r - (\rho + p)e_c{}^c \tag{34}$$

Finally, for the mixed components of the energy momentum tensor, up to second order in the angular velocity, one obtains the following explicit expressions, in which

$$T_a{}^b = \bar{T}_a{}^b + \delta T_a{}^b$$

$$\bar{T}_i{}^k = p\delta_i^k \quad \bar{T}_o{}^o = -\rho \tag{35}$$

so that, with $\Lambda = \beta - \frac{2}{3}\mu$

$$\delta T_r{}^r = - [p_{,r}\zeta^r + \Lambda e_c{}^c + 2\mu(\frac{m}{r-2M} + \zeta^r_{,r} + \frac{\lambda_{,r}}{2}\zeta^r)] + O(\Omega^4) \tag{36}$$

$$\delta T_\theta{}^r = - \mu e^{-\lambda}[e^\lambda\zeta^r_{,\theta} + r^2\zeta^\theta_{,r}] + O(\Omega^4) \tag{37}$$

$$\delta T_\theta{}^\theta = - [p_{,r}\zeta^r + \Lambda e_c{}^c + 2\mu(k + \frac{1}{r}\zeta^r + \zeta^\theta_{,\theta})] + O(\Omega^4) \tag{38}$$

$$\delta T_\phi{}^\phi = - [p_{,r}\zeta^r + \Lambda e_c{}^c + 2\mu(k + \frac{1}{r}\zeta^r + \cot\theta\zeta^\theta + \frac{1}{2}e^{-\nu}\bar{\omega}^2 r^2\sin^2\theta) - (\rho+p)\bar{\omega}\Omega e^{-\nu} r^2\sin^2\theta] + O(\Omega^4) \tag{39}$$

$$\delta T_t{}^t = \rho_{,r}\zeta^r + (\rho+p)e_c{}^c - (\rho+p)\bar{\omega}\ \Omega\ e^{-\nu} r^2\sin^2\theta + O(\Omega^4) \tag{40}$$

The $T^t{}_\phi$ component, that contributes in first and third order, is found to be, in first order

$$T_\phi{}^t(\Omega) = (\rho + p)\bar{\omega}\ e^{-\nu} r^2\sin^2\theta \tag{41}$$

and in third order

$$T_\phi{}^t(\Omega^3) = e^{-\nu} r^2 \sin^2\theta\{-(\rho + p)W + \bar{\omega}[\delta\rho + \delta p + + (\rho + p)\ (2k - 2h + e^{-\nu}\bar{\omega}^2 r^2 \sin^2\theta) - 2\mu\ s_{\phi\phi}/r^2\sin^2\theta]\} \tag{42}$$

The above expressions can be used to write down the Einstein equations perturbed around the spherical solution,

$$\delta G_a{}^b = 8\pi\ \delta T_a{}^b \tag{43}$$

These equations can be integrated numerically to obtain the values of the strain and stresses throughout the star for a specific model. It is convenient to expand these equations into Legendre polynomials. The equations separates for different ℓ values. The second order equations separate into ℓ=0 and ℓ=2 parts. Similarly to the perfect fluid analysis [Hartle 1967] one takes advantage of the freedom of choice to scale the radial coordinate to set k_o=0, so that the ℓ=0 variables are reduced to three: m_o, h_o and the spherical displacement $\zeta^r{}_o$. The three needed ℓ=0 equations can be chosen to be the r-r, the t-t components and the hydrostatic equation, but contrarywise to the fluid case this last equation has no first integral, so all three are differential. There are five ℓ=2 variables: $h_2, m_2, k_2, \zeta^r{}_2, \zeta^\theta{}_2$. Besides the hydrostatic equation one needs four other Einstein equations, one of which turns out to be algebraic. So that there are four differential equations for ℓ=2 instead of two as the fluid system can be reduced to. The third order t-ϕ equation separates into ℓ=1 and ℓ=3 parts, in complete analogy to the perfect fluid star equations [Hartle 1973].

We have shown how to derive the structure equations for a slowly rotating solid star using the general relativistic elasticity theory. The largest uncertainties remaining for an actual computation derive from the uncertainties in the equation of state for densities higher than 3×10^{14} gcm^{-3}, in particular on the values of the rigidity. The solid neutron star material can be suitably described by the perfect solid idealisation. The structure of the star depends solely on the stationary state underconsideration and not on the dynamical mode or history of reaching that state (for a given elastic structure).

It is a pleasure to thank Brandon Carter for many stimulating discussions.

REFERENCES

G. Baym and D. Pines, Ann. Phys. 66 (1971) 816.
V. Canuto, Ann. Rev. Astron. Astrophys. 13 (1975)335.
B. Carter, Commun. Math. Phys. 30 (1973) 261.
B. Carter, Ann. Phys. 95 (1975)53.
B. Carter and H. Quintana, Proc. R. Soc. Lond. A331 (1972) 57.
B. Carter and H. Quintana, in "Ondes et Radiations Gravitationnelles", CNRS Colloque 220 (1974) 265 (Paris, CNRS ed.).

B. Carter and H. Quintana, Ann. Phys. 95 (1975 a) 74.
B. Carter and H. Quintana, Ap. J. 202 (1975 b) 511.
J. Ehlers, in "Relativity, Astrophysics and Cosmology" (1973) ed. W. Israel, (D. Reidel Publishing Company, Dordrecht).
J. B. Hartle, Ap. J. 150 (1967) 1005.
J. B. Hartle, Astrophys. Space Sci. 24 (1973) 385.
J. B. Hartle and K. S. Thorne, Ap. J. 153 (1968) 807.
H. Heintzmann, W. Hillebrandt, E. Krotscheck and W. Kundt, Ann. Phys. 81 (1973) 625.
E. Krotscheck, W. Kundt and H. Heintzmann (1975) preprint, submitted to Ann. Phys.
L. Lamoureux, in "Ondes et Radiations Gravitationnelles" CNRS Colloque 220 (1974) (Paris, ed. CNRS).
G. A. Maughin, Gen. Rel. Grav. 4 (1973) 251.
G. A. Maughin, Gen. Rel. Grav. 5 (1974) 13.
J. G. Oldroyd, Proc. R. Soc. Lond. A316 (1970) 1.
A. Papapetrou, Ann. Inst. H. Poincare A16 (1972) 63.
H. Quintana, Ap.J. (1976) in press.
C. B. Rayner, Proc. Soc. Lond. A272 (1963) 44.
M. Ruderman, Ann. Rev. Astron. Astrophys. 10 (1972) 427.
J. M. Souriau, Alger Math. 2 (1958) 103.
J. M. Souriau, "Géometrie et Relativité" (1964) (Paris, Hermann ed.).
J. L. Synge, Math. Zeit. 72 (1959) 82.

QUANTUM PROCESSES NEAR BLACK HOLES

G. W. Gibbons

University of Cambridge, D.A.M.T.P., Silver Street, Cambridge, ENGLAND

ABSTRACT

A general review is given of quantum processes near black holes with especial emphasis on the Hawking Thermodynamic Emission Process. Astrophysical applications are not discussed.

I wish in this talk to summarize recent work on quantum effects near black holes. In doing so I wish to confine myself to giving an outline of what principles go into the calculations and the results. I shall not discuss any astrophysical applications (for which see e.g. [40]). I have tried to put the various results in some sort of perspective and I hope that in doing so I have not given insufficient weight to anyone's contribution or incorrectly judged it. If I have done so I apologize in advance. In my talk I hope to indicate what parts of the theory look satisfactory and which require more work and also I shall try to indicate parallelswhich other parts of physics - especially the theory of quantum processes in strong external electromagnetic fields.

The first indication of potentially interesting effects arose when Penrose pointed out the existence of what has come to be known as the "Penrose Effect" [1]. This arises because of the existence of negative energy orbits inside the "ergosphere" of a rotating black hole (the region where the Killing vector which is timelike at infinity becomes spacelike). Given a region of negative energy orbits it is possible to extract energy - in this case the rotational energy of the black hole. One simply drops in a particle with positive energy E_1 and lets it split (inside the region) into 2 particles one with positive energy E_2 which emerges and the other with negative energy E_3 which remains inside. Since $E_1 = E_2 + E_3$ we have $E_2 > E_1$. This situation also occurs in electromagnetism near a point charge in special relativity or indeed in any deep enough potential well [2]. It also occurs near charged black holes [3].

In fact in any electro-magnetic background which is stationary, axisymmetric and invariant under simultaneous inversion of time and angle coordinates one finds that the energy E and angular momentum L of a particle of mass m and charge e must satisfy $(E\,dt + L d\phi + eA)^2 > m^2$ (1)
where A is the vector potential which falls to zero at infinity. This expression (or a simple generalization of it if there is a third constant of the motion) determines two surfaces $E^{\pm}(r,\theta)$ in the (E, r, θ) space between which a classical particle cannot exist. If the surface E^+ can fall below - m we have just the required situation referred to sometimes as "level crossing". The region r, space is referred to as a "generalized ergosphere" for the mode in question. It is easy to check the existence of such a region in "superheavy" atoms. If

$$A = \Phi dt + Bd\phi; \; \Omega = g_{\phi t}(g_{tt})^{-1} \tag{2}$$

the rate of rotation of inertial frames and $\sigma^2 = (g_{\phi t})^2 - g_{\phi\phi}\, g_{tt}$ we have

$$E^{\pm} = e\Phi + (L + eB)\Omega \pm \sigma[m^2 + (L + eB)^2]^{\frac{1}{2}} \tag{3}$$

On a horizon $\sigma \to 0$ and $\Omega \to \Omega_H$, $\Phi + \Omega B \to \Phi_H$ $E \pm \to e\Phi_H + L\Omega_H = \mu_H$ which may be thought of as a chemical potential for the mode in question.

From the duality between wave and particle one expects a similar phenomenon to occur for waves and indeed this turns out to be the case (Misner [4] and Zeldovich [5], and one has here the phenomenon of "super radiance." For a classical scalar field this arises because the conserved flux vector

$$J_\mu = (\bar{\phi}\nabla_{\mu\phi} - \phi\nabla_\mu\bar{\phi})/2i \tag{4}$$

need not necessarily be future directed timelike. An incident wave carrying positive flux can send negative flux down the hole and the reflected positive flux can be greater than the incident flux. All of this is very reminiscent of the well known "Klein Paradox" situation [6] and indeed in the most general case of a charged, rotating black hole we have a rather close analogy to the Klein Paradox.

In our previous notation we find the ϕ can be written as $\phi = e^{iEt} e^{iL\phi} \chi$ and χ obeys

$$1/\sigma(\nabla_A \sigma \nabla^A \chi) + \{(Edt + Ld\phi + eA)^2 - m^2\}\chi = 0 \tag{5}$$

where ∇_A denotes covariant differentiation in the r,θ plane. The conserved flux is

$$J = (Edt + Ld\phi + A)|\chi|^2 + (\bar{\chi}d\chi - \chi d\bar{\chi})/2i \tag{6}$$

The null generator of the horizon is $\ell = \partial/\partial t - \Omega_H \partial/\partial\phi$. The flux through the horizon is $\propto \langle J,\ell\rangle \propto E - \mu_H$. Thus it $E < \mu_H$ but $E^2 < m^2$ we have superadiance. This is of course just the previous criterion.

For classical spin ½ fields the situation is different the conserved flux vector

$$J_\mu = \bar{\psi}\gamma_\mu\, \psi \tag{7}$$

is always future directed timelike and so simple super radiance is not possible [7]. However it is still possible for negative energy to fall down the hole since the stress tensor of a spin ½ field does not obey the positive energy condition. Note that in both these cases a "hole" is necessary. Super radiance cannot occur unless a particle or energy can be trapped inside a certain region. Having seen how super radiance is possible, the analogy with "stimulated emission" is very close. On rather general grounds--Dirac [8] Feynman [9] Einstein [10] one expects --at least for bosons a related "spontaneous emission." Further each mode should be emitted with a coefficient just given by the super radiant coefficient (Starobinsky [11]). Note that while these physical arguments seem quite compelling one possible objection is that they seem to imply that a black hole can be some sort of thermal equilibrium with a surrounding heat bath. This as we shall see will turn out to be the case but at the early stages of this subject this seemed rather puzzling. Before I go on, it seems worthwhile here to point out that interesting as these speculations seem, the motivation for following them up would have been rather low had it not been in one's mind that rather small black holes (masses > planck mass ~ 10^{-5}g) had been postulated earlier by Hawking [12] as possibly arising in the early stages of a chaotic big bang universe, although the idea of black holes smaller than the Chandrasekhar limit had been suggested earlier by Zeldovich [13]. In this connection these early speculations brought to light an amusing coincidence Starobinsky [11] pointed out that the order of magnitude for the time for spontaneous loss of all of its angular momentum by a black hole of mass must be (in units such that $G = c = \hbar = k = 1$)

$$t \sim M^3 \tag{8}$$

Thus a hole would lose all of its angular momentum in less than 10^{10} year if its mass were less than 10^{-13}cm - a number not without significance in other contexts. The extension of these ideas to the charge of a black hole [14] showed similarly that unless the hole had a mass of this order, e^2/m_e it would be energetically favorable for it to discharge itself even if it possessed a single electron charge. The rate was expected to depend on the field strength and in a Schwinger [15] type way. Thus unless the electric field is less than the critical field m_e^2/e the rate is very fast. This implied that to have a charge comparable with its mass the black hole mass must exceed e/m_e^2 (which is coincidentally the least mass of a "classical geon" [16]. Essentially the same ideas seem to have occurred to Zaumen [17] independently. The story has also been taken up by Ruffini and Deruelle [19] and Ruffini and Damour [20]. These estimates made it very unlikely that mini black holes possessed charge.

Having seen the physical ideas which enter it remained to give them a more rigorous expression. The first person to tackle this problem was Unruh [21]. Since there is at present no well worked out candidate for a quantum theory of gravity Unruh adopted an approach in which the gravitational field was treated as a classical background--the so called external field approach. Thus one takes the equations describing a free quantum field in flat space and minimally couples them to the external field by the replacement $\partial_\mu \to \nabla_\mu - ieA_\mu$. This does not always yield a sensible theory [41] but in the case of spin 0, ½ and 1 a workable theory results.

The next problem one encounters is the definition of particle states or a vacuum state. This may be summarized as follows: the basic strategy of the quantum theory of fields is to resolve a field into normal modes. The coefficients of these normal modes obey the familiar bose einstein/fermi-dirac commutation/anticommutation relations. This gives field commutation/anticommutation relations which are independent of the choice of normal modes--provided they are properly normalized with the natural sesqui-linear form available:

$$\frac{1}{2i}\int(\overline{\phi}\nabla_\mu\phi - \phi\nabla_\mu\overline{\phi})d\Sigma^\mu \quad \text{for spin } 0 \tag{9.1}$$

$$\int\overline{\psi}\gamma^\mu\psi d\Sigma^\mu \quad \text{for spin } \tfrac{1}{2} \tag{9.2}$$

What is not independent of the choice is the vacuum state. Any transformation of the normal modes (Bogoliubov transformation) which mixes up particle and antiparticle modes (or positive and negative frequencies to use a conventional expression) will give an inequivalent definition of the vacuum state. Indeed it--as seems to occur in most practical examples the number of "created particles" diverges the two different state vectors may not even be connected by a unitary transformation [22]. Unruh made a particular choice--essentially that the particle modes be positive frequency with respect to the Killing vector $\partial/\partial t$ in the Kerr solution. He then computed the stress tensor expectation value $\langle T_{\mu\nu}\rangle$ in this state and found that although $\langle T_{00}\rangle$ was infinite, $\langle T_{0r}\rangle$ was finite and corresponded to an outward flux of super radiant modes at the expected rate. Similar results were subsequently found by Ford [23]. It should be mentioned that the gravitational background used was the maximally extended Kerr solution. We see that in general we meet three generic types of problem:

(1) choice of vacuum state

(2) infinities in $T_{\mu\nu}$

(3) breakdown for higher spins.

All of these problems occur and are familiar in the corresponding electromagnetic case. The next advance came with the work of Hawking [24]. He realized that

(1) One can only satisfactorily define particle states at infinity

(2) One must for a satisfactory treatment take collapse into account.

To take the first point--Hawking decided--in the spirit of the -matrix approach to define two vacua, the in vacuum and the out vacuum $|0_->$, $|0_+>$. Provided past infinity constitutes a cauchy surface (thus excluding the mixed white hole/black hole situation considered by Unruh) one may define an initial no particle state by the usual prescription of associating positive frequencies with particle states and conversely negative frequencies with anti-particle states. Since the idea of positive frequency is invariant under the asymptotic symmetry group--the B.M.S. group this remains valid even in the presence of gravitational radiation. Indeed one can show that the gravitational field of a plane wave (like a plane electromagnetic wave) is incapable of producing particles [25]. Since at infinity any ingoing radiation will be effectively plane it is clear that as far as past infinity is concerned we have a reasonable definition of what it means to say that there are initially no ingoing particles. Similarly we can identify outgoing particles at future infinity. The task of identifying particle states in the interaction region is much less clear. This is perhaps not unreasonable--physically a particle is really a certain sort of normal mode with high symmetry--

mathematically it is connected with irreducible representations of the Poincare Group [26]. Neither of these concepts is applicable near the black hole. These remarks do <u>not</u> apply to the strong electromagnetic field around a large black hole since here the typical wavelength of the created particles is much smaller than the horizon size.

Now to turn to the next point. The problem is to count the number of outgoing particles in a state which initially contained no ingoing particles. (We are working in the Heisenberg picture.)

$$N_i^{out} = \langle 0- | (a_i^{out})^+ a_i^{out} | 0 - \rangle \tag{10}$$

Taking the collapse into account and the very high redshifts associated with the formation of the event horizon Hawking was able to show that this number diverges which he also showed corresponded to a steady emission at a rate

$$R_i = \Gamma_i \left(\exp \frac{E_i - N_i \Omega_H - e_i \Phi_H}{T} \mp 1 \right)^{-1} \quad \begin{matrix} - \text{ for bosons} \\ + \text{ fermions} \end{matrix} \tag{11}$$

where E_i is the energy of the outgoing particles; N_i the angular momentum; the angular velocity of the hole; e_i the charge of the outgoing particle; Φ_H the electrical potential of the hole. Γ_i is the absorbtion coefficient of the hole for classical waves of energy E_i etc. and T is related to a constant K ("the surface gravity") which plays an important role in black hole physics. For a black hole of mass M, charge Q these constants are: $\Omega_H = J/M^2$; $K = r_+ - r_-/(2r^2)$; $r_0^2 = r_+^2 + J^2M^{-2}$; $\Phi_H = Qr_+/{}_0^2$; $r_\pm = M \pm [M^2 - J^2M^{-2} - Q^2]^{\frac{1}{2}}$; $K = 2\pi T$. In terms of an "irreducible mass" $M_0 = r_0$ and the event horizon area A_H we have:

$$A_H = 16\,\pi M_0^2 = 4\pi r_0^2 \tag{12.1}$$

$$M^2 = (M_0 + Q^2/4M_0)^2 + J^2/4M_0^2 \tag{12.2}$$

There are several remarkable features of this result. Firstly, observe that while it was necessary to include the collapse in order to give meaning to the calculation the details of the collapse do not enter at all into the answer. If one does not consider the collapse then the most natural way of doing the calculations (i.e. in the fully extended Kruskal manifold) give no production [27]. It should be pointed out that the result one gets in the Kruskal manifold depends crucially on what boundary conditions one sets on the past horizon. A suitable choice could in principle yield any desired result. Unruh has pointed out that one may obtain Hawking's result by chosing as part of one's set of normal modes a set entering the external region of the black hole from the past horizon and behaving like a complex exponential of the affine parameter on the post horizon (see his article). Secondly in order to obtain the result it is necessary to take into account normal modes of arbitrarily high energy (way above the Planck frequency) essentially because of the redshifting effect of the horizon. Indeed, if one imposes a cut off of frequence ω_c the emission will stop after a time $t \sim K^{-1} \log(\omega_c/K)$. This in fact is likely to be true in any calculation since it comes from considering how close to the horizon (in terms of an affine parameter) you have to be in order that you can send signals which reach infinity at a retarded time. Thirdly we have here just the thermal emission which was lacking in our previous physical arguments--although the analogy between atomic levels and a black hole is not absolutely precise. It should be mentioned here that the idea that a black hole has associated with it a temperature and an entropy (which is of course what these results imply) had been suggested previously by Beckenstein [28] on the basis of certain analogies between black hole physics and thermodynamics (cf. [29]) --the non-decreasing event horizon area playing the part of entropy. The precise relation is now seen to be $S = A_H/4$.

One could of course spend a great deal of time describing the implications of these spectacular results. In what follows I shall bring out just a few points. Before proceeding, however, I should note that Hawking's work encompasses all the previous results of Unruh and Starobinsky as limiting cases. The way in which this comes about for charged black holes and the relation of all this to the well known formulae due to Schwinger [15] for particle creation in uniform electric fields has been described by myself [30] cf also [20]. There are of course obvious analogies with thermionic emission from metal surfaces, $e\Phi_H$ acting as a "work function."

It is sometimes said that the super-radiant processes like particle creation in a uniform static electric field do not require a time dependent field in contradistinction to the Hawking process. This is not really correct. The stationary (and in general non physical) backgrounds can only be used in conjunction with boundary conditions. These boundary conditions relate to what particles enter the space from the past horizon or from the infinite past respectively. This decision is made (via the choice of boundary conditions for Feynman propagators or a choice of a complete set of normal modes) in a way appropriate for a situation in which the interaction is "turned off" in the inifnite past. From this point of view the "method of level crossing" amounts to:

1. a particular choice of normal modes (proportional to exp $i\omega t$).
2. a method of deciding whether the ingoing antiparticles overlap with outgoing particles (i.e. whether the Bogoliubov coefficients vanish or not)
3. an elegant representation of how large the effects will be by computing transmission coefficients in a certain effective potential. This is explained in more detail by Dr. Damour in his article and in [30] but the essential point is that what is a particle or antiparticle determined by the flux it carries through a cauchy surface. As I explained earlier this depends, for horizon modes, on $E - \mu_H$. Thus roughly speaking, a wave can appear as an antiparticle near the horizon ($(E - \mu_H) < 0$) but as a particle at infinity ($E > 0$).

From what has been said above it is clear that for all its spectacular successes, Hawking's derivation of his result is not entirely satisfactory. (For an alternative attempt see [31], for skeptical remarks [32], [33].) This it seems to me is not due to any fault on Hawking's part but due to the intrinsic unsatisfactories of the whole external field theory method. For instance we still have the divergences in $T_{\mu\nu}$ (and in the total number of particles) which need to be taken into account before we begin to feed back this expectation value into a Hartree-Fock version of Einstein's equations--if indeed that is appropriate. Some of these infinities can be absorbed as infinite renormalizations of the cosmological constant and Newton's constant G. Some, however, are of an essentially new king, as has been pointed out by DeWitt [34]. In terms of an effective Lagrangian they correspond to terms of the form $R_{ab} R^{ab}$ and $(R_{ab} g^{ab})^2$. No detailed calculations of these have been given in the black hole case (where in fact $R_{ab} g^{ab} = 0$) but there seems to be little ground for doubting their existence. Thus even the external field problem is unrenormalizable which is closely related to the fact that the only quantum theory of gravity which is amenable to calculations suffers from the same defect [35]. One might also worry that particle interactions have not been taken into account. This is presumably valid for large holes but for holes of size 10^{-13}cm this considerable flux of particles in such a small volume should surely require the use of interacting field theory.

All of these problems are deep and difficult, what I want to do in the final part of the talk is to cut through the Gordian Knot as it were, disregard the original field theory derivation and hang everything on the thermodynamic idea.

Before doing so it is perhaps worthwhile trying to see why a black hole should give a thermal spectrum at all. The most natural interpretation is that the Hawking process consists of very many individual events consisting of the creation of a virtual pair near the hole and the tunnelling of a member (presumably carrying negative energy) through the horizon, the other reaching infinity. What is seen is a statistical ensemble of such independent events each happening with a probability proportional to the phase space available. The factor Γ_i arises because the particles are created (in some sense) near the horizon and have to tunnel out through a combined curvature and centrifugal barrier. In all such calculations we need a constraint and in fact if we constrain the rate of dissipation of entropy or irreducible mass we obtain the Hawking formulae--with of course the temperature undetermined. Thus a black hole emits thermally because it is the most likely thing for it to do. Support for the idea comes from the work of Wald [22] and Parker [36] who have shown that the statistics of outgoing particles are those of a black body.

Given that we have a small body whose temperature decreases the more energy it loses. This is, of course, a manifestation of the well known fact that gravitating systems can have negative specific heats. The evolution of such a system will be towards a hotter and hotter state--the black hole presumably disappearing altogether, and presumably with it the baryons that make it up. To my knowledge now one has a mechanism whereby black holes can emit more baryons than antibaryon without making use of long range fields carrying baryonic charge [37].

If we adopt the thermodynamic viewpoint even when particle interactions are taken into account, then presumably a black hole will emit precisely what it would accrete from a heat bath at the same temperature. In terms of our simple arguments before the probability of emission is now no longer simply proportional to the available phase space. Carter, Lin and Perry and myself [38] have recently made some rough calculations on the basis. We represented a heat bath of strongly interacting particles as a perfect fluid with an equation of state. The accretion problem is straight forward and the corresponding emission is described by a sort of stellar wind. The interesting thing about our calculation is that provided the high density gas has a reasonably hard equation of state the wind is essentially transparent. This indicates that interactions might not be very important. It is perhaps worth reemphasizing here that for higher spin ($S > \frac{3}{2}$) the external field theory approach breaks down.

In this thermodynamic vein it is amusing to note that in this light the old theory of black body radiation requires revision. Consider a cavity of fixed volume V. They fill it with more and more energy.

At first the cavity will contain an ordinary black body gas and its temperature will rise. As it does, however, a qualitatively different behavior sets in. We have to maximize the entropy, S, of the configuration subject to the energy, E, being held constant.

$$S = 4\pi M^2 + \frac{4}{3} aVT^3 \tag{13.1}$$

$$E = M + aVT^4 \tag{13.2}$$

a is Stefans constant.

if $x = M/E$ and $y = (aV/E^5)\ /3\pi$

This amounts to maximizing $F = x^2 + y(1-x)^{3/4}$ on $[0,1]$. For $y > 2^5 3^{-1} 5^{-5/4} = 1.4266$ has no turning points and its greatest value is attained at $x = 0$ (pure radiation). For $1.4266 > y > 1.01440$ there is a local minimum at $x < \frac{4}{5}$ and a local maximum at $x < \frac{4}{5}$ but that $x = 0$ is still the global maximum. For $y < 1.01440$ the local maximum at $x > 0.97702$ is also a global maximum. Thus for a box with sufficient energy a black hole of mass $M = [8\pi T]^{-1}$ will condense out. As the energy of the enclosure is further increased the temperature will drop. Thus for any volume V there is a maximum temperature $T_m = (8\pi)^{-1}\, x_c^{3/4} (aV)^{-1/5} (3\pi y_c)^{4/5} = x_c^{-1/4} T_c$; $x_c = 0.97702$ $y_c = 1.01440$.

The point of the example is to indicate how basic Hawking's discovery is in relation to our views of fundamental physics. What part the sort of speculations contained in [38] on the deeper role that black holes have to play in the scheme of things remains to be seen. With that comment I shall close this review --in doing so it is appropriate for me to thank my many colleagues in this field but most especially S. W. Hawking for many discussions.

ADDENDUM

After completing the review I received the following preprints. The first of which is most pertinent to the final section:

S. W. Hawking, "Black Holes and Thermodynamics"
T. Damour "On the Correspondence between Classical and Quantum Energy States in Stationary Geometries"
N. Deruelle, R. Ruffini "Klein Paradox in a Kerr Geometry"
W. Unruh "Notes on Black Hole Evaporation"

REFERENCES

[1] R. Penrose. Riv. Nuovo Cimento 1 252 (1969).
[2] Cf. e.g. G. Denardo & A. Treves. Lett. al. Nuovo Cimento 8 295 (1973).
[3] G. Denardo & R. Ruffini. Phys. Lett. 45B 259 (1973); G. Denardo, L. Hively & R. Ruffini. Phys. Lett. 50B 270 (1974); cf. also Ya. B. Zeldovich & V. S. Popov. Uspekhi 14 673 (1972); L. I. Schiff, H. Snyder & J. Weinberg, Phys. Rev. 57 315 (1940).
[4] C. Misner. Bull. Amer. Phys. Soc. 17 472 (1972).
[5] Ya Zeldovich J.E.T.P. 35 (1972).
[6] O. Klein, Zeit. fur Phys. 53 157 (1929).
[7] W. Unruh. Phys. Rev. Lett. 31 1265 (1973).
[8] P. A. M. Dirac. Quantum Mechanics O.U.P.
[9] R. D. Feynman. Lectures on Physics, Vol. III.
[10] A. Einstein. Phys. Zeit 18 121 (1917).
[11] A. Starobinsky. J.E.T.P. 37 28 (1973).
[12] S. W. Hawking. M.N.R.A.S. 152 75 (1971); B. J. Carr & S. W. Hawking, M.N.R.A.S. 168 399 (1974).
[13] Ya. Zeldovich. J.E.T.P. 446 (1962).
[14] G. W. Gibbons & S. W. Hawking. Work reported at Warsaw Conference 1973 see "Gravitational Radiations and Gravitational Collapse" C. deWitt (ed.) Reidel (1974).
[15] J. Schwinger. Phys. Rev. 82 664 (1951).
[16] J. Wheeler. "Geometrodynamics" Academic Press (1972).
[17] W. T. Zaumen. Nature 247 530 (1974).
[19] N. Deruelle & R. Ruffini. Phys. Lett. 52B 437 (1975).
[20] T. Damour & R. Ruffini. "Quantum electrodynamic effects in Kerr-Newman Geometries" Princeton Preprint. Dec. 1974.
[21] W. Unruh. Phys. Rev. D 10 3194 (1974).
[22] Cf. M. Castagnino, A. Verbeuri & R. A. Weder. Nuovo Cimento 26B 396 (1975), Phys. Lett. 48A 99 (1974); R. Wald. Commun. Math. Phys. 46 (1975.
[23] L. Ford "Quantization of a Scalar Field in the Kerr Spacetime" Milwaukee Preprint UWM-4867-74-17 (1974).
[24] S. W. Hawking. Nature 248 30 (1974); S. W. Hawking. Commun. Math. Phys. 43 199 (1975).
[25] G. W. Gibbons. Commun. M. Phys. 45 191 (1975).
[26] E. Wigner. Ann. of Math. 40 (1939).
[27] D. Boulware. Phys. Rev. D 11 1406 (1975); D. Boulware. "Spin 1/2 Quantum Field Theory in Schwarzschild Space" Seattle Preprint RL-1388-689 (1975).
[28] J. D. Beckenstein. Phys. Rev. D7 2333 (1972).
[29] J. Bardeen, B. Carter & S. W. Hawking. Commun. Math. Phys. 31 162 (1973).
[30] G. W. Gibbons. Commun. Math. Phys. 44 245 (1975).
[31] U. Gerlach. "Mechanism of Black Body Radiation from an incipient Black Hole" Ohio Preprint.

[32] J. G. Taylor & P. Davies. Nature 250 37 (1974).
[33] P. W. Davies. J. Phys. A 8 609 (1975).
[34] B. deWitt. Physics Reports 29C. 295 (1975).
[35] E.g. S. Deser and P. van Nieuwenhuizen. Phys. Rev. d 10 411 (1974).
[36] L. Parker Phys. Rev. D 12 (1975).
[37] B. Carter. "Phys. Rev. Lett. 33 558 (1974).
[38] B. Carter, G. W. Gibbons, D. Lin & M. Perry "The Black Hole emission process in the high energy limit" in preparation.
[39] J. Sarfatt. Nature 240 101 (1972).
[40] B. J. Carr. "The Primordial Mass Spectrum" Caltech Preprint (1975).
[41] G. W. Gibbons. J. Phys. A 9 145 (1976).

KLEIN PARADOX AND VACUUM POLARIZATION

Thibaut Damour*

Department of Physics, Joseph Henry Physical Laboratories
Princeton University, Princeton, N.J. 08540

INTRODUCTION: This contribution reviews some of the methods which can be used when studying quantum fields in a given stationary classical external field. The attention is mainly directed towards cases where real pair creation can occur in such a stationary background. The paradigm of this situation is the Klein paradox. This paradox is best approached by the introduction of some energy diagrams (see the Figures) whose direct extension to black holes physics has proven to be very useful. Finally processes of real pair creation around a Kerr-Newman (charged and rotating) black hole and their feedback on the geometry are briefly discussed. It is also shown how the Hawking process can be recovered in this approach.

I KLEIN PARADOX IN FLAT SPACE

It is well known that every relativistic wave equation (e.g. Klein-Gordon's or Dirac's) admits symmetrically "positive frequency" as well as "negative frequency" solutions. Namely a relativistic wave of mass μ moving in free space and having a frequency ω and a wave vector $\underset{\sim}{k}$ must fullfill (here and in the following we choose $G = c = \hbar = 1$):

$$\omega^2 = \mu^2 + \underset{\sim}{k}^2 \tag{1a}$$

or

$$\omega = \pm (\mu^2 + \underset{\sim}{k}^2)^{\frac{1}{2}} \tag{1b}$$

This gives rise to the familiar spectrum of allowed states represented in Fig. I.

These states are the possible states of the quantum particle described by the wave. In the present situation (flat space, no external field) one can ignore the existence of these "negative" states because the "positive" states are stable; that is, there exists no posibility of decay of a "positive" state into a "negative" state. But if, for instance, the particle bears an electric charge ε and is embedded in a constant electric potential V, ω will be shifted by the amount ε V. More precisely in the quasi-classical (W.K.B.) approximation the particle must fulfill:

$$(\omega - \varepsilon V)^2 = \mu^2 + \underset{\sim}{k}^2 \tag{2a}$$

or

$$\omega = \varepsilon V \pm (\mu^2 + \underset{\sim}{k}^2)^{\frac{1}{2}} \tag{2b}$$

In the case of an electric field $\underset{\sim}{E} = (o,o, E(z))$ uniform between z_1 and z_2 and null outside, the Fig. II represents the corresponding sketch of allowed states.

*E.S.A. International Fellow

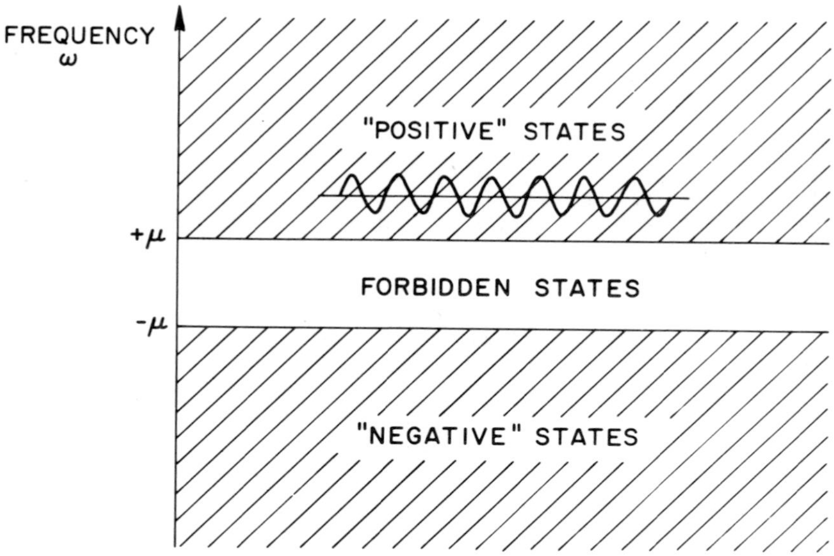

FIGURE I: The spectrum of allowed states for a free wave in flat space describing a quantum particle of mass μ is here represented in function of the frequency ω. In such a free field situation all the states are stable; that is, there is no possibility of a "positive" ("negative") wave decaying into a "negative" ("positive") one.

The key point now, which is the essence of the Klein paradox [1], is that the above mentioned stability of the "positive" states is lost for sufficiently strong electric fields. The same is true for "negative" states: for instance (see fig. II) a "negative" wave incident from the left will be both reflected back by the electric field and partly transmitted to the right as a "positive" wave. This transmission is nothing else but a Gamow tunnelling of the wave function through the classically forbidden states [2, 3]. Mathematically it is due to the fact that in Eq. (2a) k_z^2 will become negative in the barrier which means that the wave function will have an exponential behaviour ($\exp - \int |k_z| dz$ in W.K.B. approximation) instead of its usual oscillatory behaviour. The quantity of interest associated with this process will evidently be the transmission coefficient $|T|^2$ of the wave through the one-dimensional barrier between the "negative" and the "positive" states:

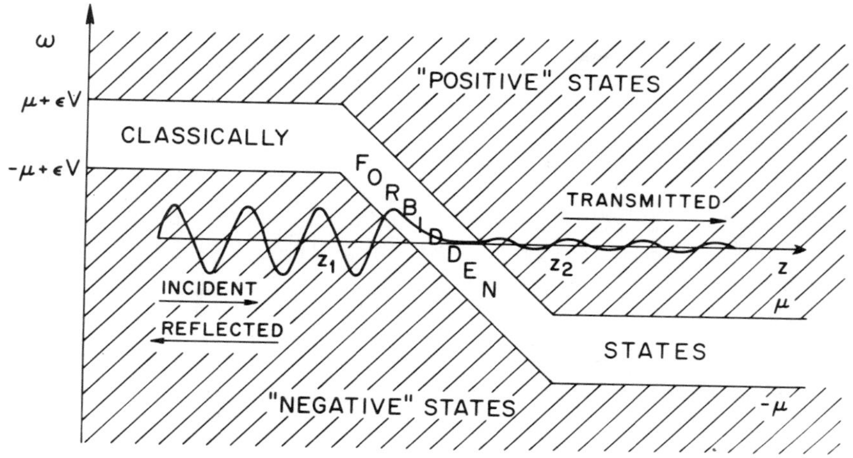

FIGURE II: In presence of a strong enough electric field the boundaries of the classically allowed states ("positive" or "negative") can be so tilted that a "negative frequency" is at the same level as a "positive frequency" (level crossing). Therefore a "negative" wave-packet incident from the left will be partially transmitted, after an exponential damping due to the tunnelling through the classically forbidden states, as a "positive" wave-packet outgoing to the right.

$$|T|^2 = |\text{transmitted flux}|/|\text{incident flux}| \tag{3}$$

In W.K.B. approximation one has simply $|T|^2 = \exp -2 \int_{\text{barrier}} |k_z|\, dz$. There will also be a reflection coefficient $|R|^2$ for the wave whose link to $|T|^2$ depends on the spin of the particle.

For fermions, the probability density is always positive and the probability flux directed as the group velocity. Therefore, the reflected flux will be smaller than the incident one and more precisely.

$$|R|^2 = 1 - |T|^2 \qquad \text{(fermions)} \tag{4a}$$

For bosons, "negative" waves have a negative density and therefore a flux oppositely directed to their group velocity. Consequently one sees immediately that the reflected flux will be larger than the incident one and that more precisely

$$|R|^2 = 1 + |T|^2 \qquad \text{(bosons)} \tag{4b}$$

But this last phenomenon: amplification of the wave by reflection that is "superradiance" though an interesting consequence of the spin-statistics connection, (see below) does not constitute the essence of the Klein paradox; what is important is the possibility of decay:

ingoing "positive" state → outgoing "negative" state

or

ingoing "negative" state → outgoing "positive" state

as exemplified by the scattering process described in Fig. II. And if such a decay process can take place, one can easily predict that there will be spontaneous real pair creation by the vacuum. For instance a heuristic reasoning using Dirac's Ocean of "negative" states (for electrons, but see II below for charged bosons) shows directly that the leakage of the "negative" states towards "positive" states will give rise to a continuous creation of electrons with a rate proportional to $|T|^2$ (see II below for details). Let us check this, using a second quantized approach [4].

II RATES OF PARTICLE CREATION

Along the lines of the scattering process described in Fig. II let us introduce [5-11] the ingoing and the outgoing states of the quantized field that we are considering. That is to say, if we want to investigate particle creation, by a given external field (electromagnetic and/or gravitational), in a given volume Ω of space-time, we have to introduce the solutions of the wave equation, written in that given background, from which one can build localized wave packets purely ingoing into Ω from the past. These states are the generalization of the ingoing stationary waves of a potential scattering process. In the following we will suppose that it has been possible to define meaning fully these ingoing states. We have to do the same thing for outgoing states, that is, states from which one can build localized wave packets purely outgoing from Ω into the future. Moreover it is essential that it has been possible to separate those ingoing (outgoing) states in "positive" and "negative" states. This choice is clear in the case of Fig. II where we have an effective potential with two asymptotic regions. See V below for the analogous situation in black holes physics.

Therefore we suppose that we have a complete basis of "positive" and "negative" ingoing modes $p_i^{in}(x)$ and $n_i^{in}(x)$ fullfilling the orthonormality relations

$$(p_i^{in}, p_k^{in}) = \delta_{ik} = \pm (n_i^{in}, n_k^{in}) \tag{5a}$$

$$(p_i^{in}, n_k^{in}) = 0 \tag{5b}$$

where (p, q) denotes the natural scalar product for our field, δ_{ik} is a general Kronecker symbol which may contain δ-functions in case of continuous normalization, and where, here and in the following, the upper sign is valid for fermions and the lower one for bosons.

Then we expand the quantized field as:

$$\phi(x) = \sum_i a_i^{in} p_i^{in}(x) + (b_i^{in})^+ n_i^{in}(x) \tag{6a}$$

with the (anti) commutation relations:

$$\left[a_i^{in}, (a_k^{in})^+\right]_\pm = \left[b_i^{in}, (b_k^{in})^+\right]_\pm = \delta_{ik} \tag{6b}$$

Finally the in-vacuum is defined by

$$a_i^{in} \,|{}^{in}_{vac}\rangle = b_i^{in} \,|{}^{in}_{vac}\rangle = 0 \tag{7}$$

that is, it is cancelled by the annihilation operators a^{in} and b^{in} for the in-particles and in-antiparticles of our field.

Evidently one can do the same thing with out-states and the phenomenon of particle creation will consist in that the in-vacuum contains out-states. More precisely, the mean number $\langle N_i\rangle = \eta_i$ of out-particles described by p_i^{out} that one will find in the in-vacuum is given by

$$\eta_i = \langle{}^{in}_{vac}\,|\,(a_i^{out})^+\, a_i^{out}\,|\,{}^{in}_{vac}\rangle \tag{8}$$

With a corresponding expression for the mean number of out-antiparticles. But from

$$\phi = \sum_k a_k^{in} p_k^{in} + (b_k^{in})^+ n_k^{in} = \sum_i a_i^{out} p_i^{out} + (b_i^{out})^+ n_i^{out} \tag{9}$$

and the orthonormality relations (5) one gets

$$a_i^{out} = \sum_k \quad (p_i^{out} \ , \ p_k^{in}) \ a_k^{in} + (p_i^{out} \ , \ n_k^{in}) \ (b_k^{in})^+ \tag{10}$$

hence

$$n_i = \sum_k \quad |(p_i^{out} \ , \ n_k^{in})|^2 \tag{11}$$

So that, as predicted above, to each possible channel for a decay $n_k^{in} \rightarrow p_i^{out}$ with what we may call, a non vanishing transmission amplitude:

$$T_{ik} = (p_i^{out} \ , \ n_k^{in}) \tag{12}$$

corresponds a mean number $|T_{ik}|^2$ of particles created in the mode p_i^{out}. The mean total number of particles created will be:

$$<N> = \sum_i \ n_i \ = \sum_{\substack{\text{channels} \\ i \ , \ k}} |T_{ik}|^2 \tag{13}$$

Of special interest in the applications, is the case where it is possible to choose the in-basis and the out-basis in such a correspondence that a n_i^{in} decays only in a p_i^{out} with the same i, that is when there is only one channel possible. This case occurs for instance when enough symmetries are present for allowing the use of factorized partial waves.

In such a case one can write:

$$T_{ik} = T_i \ \delta_{ik} \tag{14}$$

and therefore

$$n_i = |T_i|^2 \tag{15}$$

Eq. (14) means that the ingoing state n_i^{in} contains the outgoing part $T_i \ p_i^{out}$. As we will consider in the following particle creation in a stationary background, we will be able to separate the time t in the modes by a factor $e^{-i\omega t}$. Therefore the modes will be defined by a continuous index ω as well as by some other indices α. Hence the transmission amplitude will read:

$$T_{ik} = T_{\omega_i \alpha_i} \quad \delta(\omega_i - \omega_k) \quad \delta_{\alpha_i \alpha_k} \tag{16}$$

and we must use the familiar "golden rule" that [6]:

$$[\delta(\omega_i - \omega_k)]^2 = \delta(\omega_i - \omega_k)\frac{1}{2\pi}\int e^{i(\omega_i - \omega_k)t}\,dt = \delta(\omega_i - \omega_k)\frac{1}{2\pi}\int dt \tag{17}$$

which is easily justified by using nearly monochromatic wavepackets.

Therefore one gets a continuous rate of particle creation:

$$\frac{d}{dt} \langle N\rangle = \int d\omega \sum_{\alpha} \frac{1}{2\pi} |T_{\omega\alpha}|^2 \tag{18}$$

and in this formula $|T_{\omega\alpha}|^2$ is precisely the transmission coefficient previously introduced because the normalization in ω implies the normalization of the flux to $(2\pi)^{-1}$ which is just another phrasing of the preceding "golden rule". Finally it is of interest to compute the quantum fluctuations in the number of particles created. The number of particles in the state p_i^{out} is represented by the operator $N_i = (a_i^{out})^+ a_i^{out}$ which yields a mean number n_i of particles with a dispersion σ_i given by:

$$\sigma_i^2 = \left\langle {}^{in}_{vac} \middle| N_i^2 \middle| {}^{in}_{vac} \right\rangle - \left\langle {}^{in}_{vac} \middle| N_i \middle| {}^{in}_{vac} \right\rangle^2 \tag{19a}$$

which yields, using eq. (10), the usual formula for fermions or bosons:

$$\sigma_i^2 = n_i(1 \mp n_i) \tag{19b}$$

The total fluctuation is:

$$\sigma_N^2 = \sum_i \sigma_i^2 \tag{19c}$$

Therefore if each individual transmission coefficient is small that is if $n_i \ll 1$ (which does not mean necessarily that $\langle N\rangle$ is small due to the large number of possible channels) one gets the familiar $\langle N\rangle^{1/2}$ fluctuation.

It is interesting to interpret the preceding results on particle creation obtained by second quantization with the simple picture of the Dirac's ocean. In order to extend this picture to charged bosons let us begin with second quantization. We start from the expansion (6a) where the symbol Σ is understood as containing an integration over the frequency ω. The charge of the field ϕ is:

$$Q = \varepsilon(\phi, \phi) \tag{20a}$$

and its energy is

$$E = (\phi, i\partial_t\phi) \tag{20b}$$

where we use the scalar product in its usual order: $\phi^+ \times \phi$. Denoting by ω_i^+ the frequency of the "positive" state p_i^{in} and by ω_i^- the one of the "negative" state n_i^{in}, the sesquilinearity of the scalar product yields:

$$Q = \sum_i \varepsilon (a_i^{in})^+ a_i^{in} \pm \varepsilon b_i^{in} (b_i^{in})^+ \tag{21a}$$

$$E = \sum_i \omega_i^+ (a_i^{in})^+ a_i^{in} \pm \omega_i^- b_i^{in} (b_i^{in})^+ \tag{21b}$$

where, as before, the upper sign holds for fermions and the lower one for bosons. The use of the (anti)commutation relations (6a) yields:

$$Q = \sum_i \varepsilon (a_i^{in})^+ a_i^{in} - \varepsilon (b_i^{in})^+ b_i^{in} + \sum_{n_i^{in}} \pm \varepsilon \tag{22a}$$

$$E = \sum_i \omega_i^+ (a_i^{in})^+ a_i^{in} - \omega_i^- (b_i^{in})^+ b_i^{in} + \sum_{n_i^{in}} \pm \omega_i^- \tag{22b}$$

The last sums in (22a) and (22b) can be readily interpreted by saying that all the "negative" states n_i^{in} are filled by a wave normalized to unity which bears a charge $\pm \varepsilon$ and an energy $\pm \omega_i^-$. But in the case of Fig. II, or any similar situation, these waves will leak out of the "negative" sea and appear as an outgoing positive wave. More precisely in the case where only one channel is possible one can write:

$$n_i^{in} = R_i \, n_i^{out} + T_i \, p_i^{out} \tag{23a}$$

with

$$|R_i|^2 = 1 \mp |T_i|^2 \tag{23b}$$

The continuum normalization in ω implies that all the waves have a flux normalized to $(2\pi)^{-1}$ (an immediate corollary is that $|T_i|^2$ is identical with the transmission coefficient defined by Eq. (3)). Therefore in the case of <u>fermions</u>, the scattering process (23a) will be seen as an outgoing flux $|T_i|^2/2\pi$ of particles of charge $+\varepsilon$ and energy $+\omega_i^-$ associated to a defect of flux (hole) of $|T_i|^2/2\pi$ over the background sea which will appear as a $|T_i|^2/2\pi$ flux of antiparticles of charge $-(+\varepsilon)$ and energy $-(+\omega_i^-)$. For <u>bosons</u> the reflected part $R_i \, n_i^{out}$ will yield an excess of flux of $|T_i|^2/2\pi$ over the background sea and will appear directly as a $|T_i|^2/2\pi$ flux of antiparticles of charge $-\varepsilon$ and energy $-\omega_i^-$. (For bosons, antiparticles are directly "negative" waves and not holes). Therefore in both cases this simple picture yields the same quantitative predictions as the previous calculation (see Eq. (18)).

Let us now consider an application of what precedes to the study of pair creation by uniform electromagnetic fields in flat space.

III UNIFORM ELECTROMAGNETIC FIELDS

If we first consider following Sauter [2] a uniform electric field E in the z direction the electric potential will be (see Fig. II)

$$V = - E.z \tag{24}$$

We can use separated wave functions:

$$\phi = e^{i(k_x x + k_y y - \omega t)} \psi(z) \tag{25}$$

with continuous normalization in k_x , k_y and ω and discrete normalization for spin variables. In W.K.B. approximation, the equation for $\psi(z) \sim \exp i \int k_z \, dz$ is:

$$(\omega + Ez)^2 = \mu^2 + k_x^2 + k_y^2 + k_z^2 \tag{26}$$

independently of the spin.

Inside the barrier k_z^2 will be negative and the transmission coefficient will be

$$|T|^2 = e^{-\zeta} \tag{27a}$$

with the "opacity" ζ given by

$$\zeta = 2 \int_{\text{barrier}} |k_z| \, dz \tag{27b}$$

Therefore using Eq. 26 one gets

$$\zeta = \oint [\mu^2 + k_x{}^2 + k_y{}^2 - (\omega + \varepsilon Ez)^2]^{\frac{1}{2}} dz \tag{27c}$$

that is, trivially (with $\varepsilon E > o$):

$$\zeta = \pi \, (\mu^2 + k_x^2 + k_y^2)/\varepsilon E \tag{28}$$

Looking back to Eq. 25 we can write the transmission amplitude as:

$$T_{ik} = e^{i\alpha} \, e^{-\zeta/2} \, \delta(\omega_i - \omega_k) \, \delta(k_i^x - k_k^x) \, \delta(k_i^y - k_k^y) \, \delta_{\sigma_i \sigma_k} \tag{29}$$

where $e^{i\alpha}$ is a phase factor and $\delta_{\sigma_i \sigma_k}$ arises from the conservation of the projection σ of the spin on the z-direction.

The mean number of particles created is then obtained by using Eq. (13) and the "golden rule" (17) extended as usual to the other conjugated variables (x, k_x) and (y, k_y):

$$\langle N\rangle = \int \frac{dt\ d\omega}{2\pi} \frac{dx\ dk^x}{2\pi} \frac{dy\ dk^y}{2\pi} \sum_\sigma e^{-\zeta} \tag{30}$$

the integration over k_x and k_y is easily made by the substitution: $d\,k_x\, d\,k_y = \pi\, d\, k^2$ with $k^2 = k_x^2 + k_y^2$, the integration over ω is performed by noting that the frequency is linked to the position of the barrier by $d\omega = \varepsilon E\, dz$, with a result:

$$\langle N\rangle = \int d^4x \sum_\sigma \frac{1}{8\pi}\left(\frac{\varepsilon E}{\pi}\right)^2 e^{-\pi\mu^2/\varepsilon E} \tag{31a}$$

where d^4x = dxdydz dt and where the numerical value of $\mu^2/\varepsilon = \mu^2 c^3/\varepsilon\hbar$ ("critical electric field") is 4.414×10^{13} c.g.s. (for the production of electrons). In the case of spin 0 there are no supplementary channels due to the spin so that one gets a density of pair creation per unit space-time volume of:

$$\text{spin 0:}\quad (8\pi)^{-1}\ (\varepsilon E/\pi)^2\ e^{-\pi\mu^2/\varepsilon E} \tag{31b}$$

In the spin ½ case one has twice as many states due to $\sigma = \pm\frac{1}{2}$ which yields a spacetime density of pair creation:

$$\text{spin } \tfrac{1}{2}:\quad (4\pi)^{-1}\ (\varepsilon E/\pi)^2\ e^{-\pi\mu^2/\varepsilon E} \tag{31c}$$

These two results were obtained very simply by using a W.K.B. approximation but it happens surprisingly that they are exact because the opacity (28) is exact. This can be seen as follows: in the case of spin 0 the equation for $\psi(z)$ is:

$$d^2\psi(z)/dz^2 = \left[\mu^2 + k_x^2 + k_y^2 - (\omega + \varepsilon E z)^2\right] \psi(z) \tag{32a}$$

introducing

$$\xi = (\varepsilon E)^{-\frac{1}{2}}\ (\omega + \varepsilon E z) \tag{32b}$$

$$\lambda = (\mu^2 + k_x^2 + k_y^2)/\varepsilon E \tag{32c}$$

one gets
$$d^2\psi/d\xi^2 = (\lambda - \xi^2)\psi \tag{32d}$$

whose solution is the parabolic cylinder function [12, 13]

$$\psi = D_n(u) \tag{33a}$$

with
$$u = 2^{\frac{1}{2}}\ e^{-i\,\pi/4}\xi \tag{33b}$$

$$n = -\tfrac{1}{2} - i\,\lambda/2 \tag{33c}$$

when $\xi \to +\infty$ ψ is equivalent to $e^{-u^2/4}\, u^n$ which contains the phase factor $e^{i\xi^2/2}$, which implies a positive group velocity both for $\xi \to +\infty$ and $\xi \to -\infty$. Therefore Eq. (33a) describes a wave incident from the left and its transmission amplitude is obtained by rotating ξ of $e^{+i\pi}$ from $+\infty$ to $-\infty$ which transforms the transmitted wave into the incident one, hence: $T^{-1} = e^{in\pi} = e^{-i\frac{\pi}{2}} e^{\pi\lambda/2}$ which yields $|T|^2 = e^{-\pi\lambda}$ in perfect agreement with (28). The same conclusion is reached in the case of spin ½ simply because the iterated equation for $\psi(z)$ is obtained by the replacement $\lambda \to \lambda \pm i$ which does not alter the transmission coefficient.

Therefore the two expressions (31b) and (31c) yield exactly the rate of pair creation by a uniform electric field and yet they are different from Schwinger's expression [14]. This will be explained below, but let us first generalize these expressions to the situation where both an electric and a magnetic field (supposed uniforms) are present [3].

In such a case one can always use a frame where $\underset{\sim}{E}$ and $\underset{\sim}{B}$ are parallel (except when $|\underset{\sim}{E}| = |\underset{\sim}{B}|$ and $\underset{\sim}{E} \cdot \underset{\sim}{B} = 0$). In such a frame we can use separated wave functions:

$$\text{spin}\tfrac{1}{2}: \quad \phi = e^{i(k_x x - \omega t)}\, U_n\left[(\varepsilon B)^{\frac{1}{2}}(y + k_x/\varepsilon B)\right]\psi(z) \tag{34}$$

where U_n is a harmonic oscillator wave function of order n ($n = 0,1,2,\ldots$) which describes the motion of the electron in the plane orthogonal to the common direction of E and B. We then get the same first order equations for $\psi(z)$ as before with, instead of Eq. (32c):

$$\lambda = \mu^2/\varepsilon E + (n + \tfrac{1}{2} + \sigma)\, 2B/E \tag{35}$$

where σ is the projection of the spin along the common direction of E and B. And as before the opacity is exactly given by

$$\zeta = \pi\lambda \tag{36}$$

The spin 0 case is obtained when $\sigma = 0$ instead of $\sigma = \pm\tfrac{1}{2}$. Then the pair creation rate is given by a sum over the channels:

$$\langle N\rangle = \int \frac{dx\, dk_x}{2\pi}\, \frac{d\omega\, dt}{2\pi} \sum_n \sum_\sigma e^{-\zeta} \tag{37}$$

which yields:

$$\text{spin 0:} \quad \langle N\rangle = \int d^4x\, \frac{1}{8\pi}\left(\frac{\varepsilon E}{\pi}\right)^2 \frac{\pi B/E}{\operatorname{sh}(\pi B/E)}\, e^{-\pi\mu^2/\varepsilon E} \tag{38a}$$

$$\text{spin } \tfrac{1}{2}: \quad \langle N \rangle = \int d^4x \, \frac{1}{4\pi} \left(\frac{\varepsilon E}{\pi}\right)^2 \frac{\pi B/E}{\operatorname{th}(\pi B/E)} \, e^{-\pi\mu^2/\varepsilon E} \tag{38b}$$

IV COMPARISON WITH THE EFFECTIVE LAGRANGIAN APPROACH

This approach was pioneered by Heisenberg and Euler [3] and further developed by Schwinger [14]. Its main idea is that the polarization of the vacuum induced by (real or virtual) pair creation is equivalent to a non-linear modification of the field equations (Maxwell equations for electromagnetism) this modification being similar to the one used when working with dielectrics. We will not expound this method here but just quote the result: the effective Lagrangian for electromagnetism as modified by the creation of pairs of electrons and positrons in a uniform electromagnetic field is:

$$W = -\frac{F}{4\pi} - \frac{1}{8\pi^2} \int_0^\infty \frac{ds}{s^3} \, e^{-\mu^2 s} \left[(\varepsilon s)^2 \, G \, \frac{\operatorname{Re} \operatorname{ch} \varepsilon s X}{\operatorname{Im} \operatorname{ch} \varepsilon s X} - 1 - \frac{2}{3} (\varepsilon s)^2 \, F \right] \tag{39}$$

where $F = F_{\mu\nu} F^{\mu\nu}/4$, $G = F_{\mu\nu} {}^*F^{\mu\nu}/4$ and $X = [2(F + iG)]^{\frac{1}{2}}$

In presence of an electric field the integral in W has poles on the s-axis. If the integration path is considered to lie above this axis we obtain a positive imaginary contribution which is for parallel and uniform fields:

$$2 \operatorname{Im} W = \frac{1}{4\pi} \left(\frac{\varepsilon E}{\pi}\right)^2 \sum_{n=1}^{\infty} n^{-2} \frac{n\pi B/E}{\operatorname{th}(n\pi B/E)} \, e^{-n\pi\mu^2/\varepsilon E} \tag{40}$$

This is the probability per unit spacetime volume that at least one pair is created by the constant electromagnetic field or more precisely $\exp - \int d^4x \, 2 \operatorname{Im} W$ is the probability that no actual pair creation occurs during the history of the field [14, 11]. How is this probability related to the mean number of pairs created:

$$\langle N \rangle = \sum_i n_i = \int \frac{d^4x}{4\pi} \left(\frac{\varepsilon E}{\pi}\right)^2 \frac{\pi B/E}{\operatorname{th}(\pi B/E)} \, e^{-\pi\mu^2/\varepsilon E} \tag{41}$$

The answer can be found in the works of Nikishov [15], and, Narozhnyi and Nikishov [16] and uses the approach of Stueckelberg and Feynmann [17] [18]. We can expound it using the notation of II above and assuming for simplicity that T_{ik} is diagonal, therefore we have simply:

$$n_i^{in} = R_i \, n_i^{out} + T_i \, p_i^{out} \qquad \text{(no summation)} \tag{42}$$

where T_i is the transmission amplitude and R_i the reflexion amplitude. Let us denote:

$$|T_i|^2 \equiv n_i \tag{43a}$$

where n_i is now understood as a simple notation not yet related to the mean number of particles created.

We have therefore (see Eq. (4)):

$$|R_i|^2 = 1 \mp n_i \tag{43b}$$

where as before the upper sign is valid for fermions and the lower one for bosons. Now let us reinterpret the scattering process (42) "à la Stueckelberg" [17]:

$$n_i^{out} = R_i^{-1}\, n_i^{in} - (T_i/R_i)\, p_i^{out} \tag{44}$$

considering n_i^{out} as incident from the future and both refracted in the past and reflected in the future by the constant background field. The amplitude of "refraction" is R_i^{-1} and the one of "reflection" $- T_i/R_i$.

Therefore we can consider the new "reflection" coefficient.

$$|T_i/R_i|^2 = n_i/(1 \mp n_i) \tag{45}$$

as the relative probability for the creation of one pair: (n_i^{out}, p_i^{out}). The corresponding absolute probabilities are obtained by multiplying the relative ones by the probability $p_{i,o}$ that no pairs have appeared in the state i. Then the absolute probability for the creation of n pairs in the state i is:

$$p_{i,n} = p_{i,o}\, n_i^{\,n}\, (1 \mp n_i)^{-n} \tag{46}$$

For fermions n = 0,1 and for bosons n = 0,1,2, ... Writing that the total probability is unity, one obtains

$$p_{i,o} = (1 \mp n_i)^{\pm 1} \tag{47}$$

The associated generating functions are therefore:

$$p_i(\lambda) = \sum_n p_{i,n}\, \lambda^n = (1 \mp n_i \pm \lambda\, n_i)^{\pm 1} \tag{48}$$

with, correspondingly, for all the states:

$$P(\lambda) = \prod_i p_i(\lambda) = \sum_N P_N \lambda^N \tag{49}$$

Therefore one reaches simply the probabilities for the creation of N pairs throughout space-time.

For our purpose it will be sufficient to compute the mean number of pairs created in the state i by simply differentiating Eq. (48) with respect to λ and then setting $\lambda = 1$; this yields:

$$\langle n \rangle_i = \eta_i \tag{50}$$

in perfect agreement with Eq. (15) and (43a) above. And differentiating (49) yields the mean total number of pairs created:

$$\langle N \rangle = \sum_i \eta_i \tag{51}$$

which we can now compare to the absolute probability of the vacuum remaining a vacuum:

$$P_o = \prod_i p_{i,o} = \prod_i (1 \mp \eta_i)^{\pm 1} \tag{52a}$$

or $$P_o = \exp - 2 \text{ Im } W \tag{52b}$$

with $$2 \text{ Im } W = \sum_i \mp \ln(1 \mp \eta_i) = \sum_i \sum_{n=1}^{\infty} (\pm)^{n+1} n^{-1} \eta_i^{\,n} \tag{52c}$$

Equation (52c) which gives the imaginary part of the effective action provides us with the general connection we were looking for between the "transmission coefficient approach" and the "effective lagrangian" one.

When the individual transmission coefficients η_i are small, one will have 2 Im W $\simeq \langle N \rangle$. In the case of uniform parallel electric and magnetic fields the transmission coefficients are exactly known (see Eqs. (35) and (36)) and a simple calculation shows that Eq. (52c) reduces to Eq. (40).

Finally the fluctuations (19) are easily obtained from (48).

V APPLICATIONS TO BLACK HOLES PHYSICS

A black hole provides us with a stationary external field (gravitational and electromagnetic) in which we can study the behaviour of various quantum particles. According to the approach used in I above pair creation will occur if a "negative" state has in fact a "positive" frequency as seen from infinity (where we detect the particles see Fig. 11). In W.K.B. approximation we have a well defined wave 4-vector k = d (phase) which fullfills:

$$(k - \varepsilon A)^2 = - \mu^2 \qquad (53)$$

where A is the electric 4-potential and where the square is understood with respect to the background metric (signature +2).

Denoting by ξ_t the Killing vector expressing the stationarity of the spacetime (with $\xi_t^2 = -1$ at infinity) the frequency as seen from infinity is given by the scalar product:

$$\omega = - \xi_t \cdot k \qquad (54)$$

Equation (53) is equivalent to:

$$k = \varepsilon A \pm \mu . u \qquad (55)$$

where u is a unit <u>future</u> directed vector, (except when $\mu \to 0$ in which case $\mu . u$ is a null <u>future</u> directed vector). By comparison with Eq. (2) one sees that in the W.K.B. approximation the sign ± in Eq. (55) determines whether we are dealing with a "positive" or a "negative" state, therefore the frequency of a "negative" state as seen from infinity is:

$$\omega^- = \xi_t \cdot (\mu . u - \varepsilon A) \qquad (56)$$

So that if there exists a region in space ("effective ergosphere" [19]) where μu can be chosen such that:

$$\omega^- > \mu \qquad (57)$$

spontaneous pair creation will occur.

In the case of an uncharged black hole (A = 0) and taking $\mu \to o$ this condition is:

$$\xi_t \cdot \ell > o \qquad (58)$$

where ℓ is a future directed null vector. And this can be satisfied in a region where ξ_t ceases to be timelike for becoming space-like. Such a region, called ergosphere, exists outside the horizon for a rotating black hole. The crucial importance of the ergosphere for the energetics of blackholes was first noticed by Penrose and Floyd [20]. The corresponding pair creation process was predicted by Zel'dovich [21] and studied by Starobinsky [22] and Unruh [7] for massless fields. A detailed analysis of this process for massive particles has been given by Deruelle and Ruffini [23-25] using the generalization [19] to black holes physics of the one dimensional "effective potential" which was used in I and III above to compute the transmission coefficient from "negative" to "positive" states Inequality (57) can be satisfied around a charged blackhole implying pair creation as predicted by Gibbons and Hawking [26] and by Zaumen [27] and studied by Zaumen [27] and Gibbons [10]. A detailed treatment of this phenomenon using the above quoted "effective potential" approach has been given by Nakamura and Sato [28]. The more general case of a charged rotating black hole has been studied by Damour and Ruffini [29] with special emphasis on the possible astrophysical consequences [29-31] of the pair creation process [32].

Finally even in the case of an uncharged non-rotating vacuum black hole the condition (58) will be satisfied inside the horizon where ξ_t becomes space-like. Therefore one foresees the possibility of spontaneous particle creation by a Schwarzschild black hole. Actually this effect has been discovered by Hawking [8, 9] (see also Boulware [33] and Gerlach [34]). But the proof makes an explicit use of the time-dependent phase (collapse) leading to the formation of the hole. It is therefore interesting to see how it is possible to generalize the purely static barrier penetration treatment to this case. See below and [40].

Let us now define more precisely the "effective potential" for a particle of mass μ and charge ε moving around a Kerr-Newman hole of mass M, specific angular momentum a and charge e. The background geometry is:

$$ds^2 = \Sigma[\Delta^{-1} dr^2 + d\theta^2] + \Sigma^{-1} \sin^2\theta \, [(r^2 + a^2)d\phi - a\, dt]^2 - \Sigma^{-1} \Delta \, [dt - a \sin^2\theta \, d\phi]^2 \tag{59a}$$

with $\Delta = r^2 - 2Mr + a^2 + e^2$ and $\Sigma = r^2 + a^2 \cos^2\theta$ and the background electromagnetic potential is

$$A = - er\Sigma^{-1} (dt - a \sin^2\theta \, d\phi) \tag{59b}$$

As was shown by Carter [35] the corresponding Hamilton-Jacobi equation, that is the W.K.B. approximation (53) with $k_\alpha = S_{,\alpha}$, is separable. Therefore choosing the action (that is, the phase) S as:

$$S = - \omega t + m\phi + \Theta(\theta) + R(z) \tag{60a}$$

where $dz = - dr/\Delta$ (60b)

one gets $(d\Theta/d\theta)^2 + (m/\sin\theta - a\,\omega \sin\theta)^2 + \mu^2 a^2 \cos^2\theta = K$ (60c)

$$(dR/dz)^2 = - Z \tag{60d}$$

where $Z = \Delta[\mu^2 r^2 + K] - [\omega(r^2 + a^2) - am - e\varepsilon r]^2$ (60e)

K being Carter's fourth constant (35), generalization of the total angular momentum.

According to Eq. (60d) the classical region of accessibility is defined by:

$$Z(r;\omega,m;\ K) \leq o \tag{61}$$

but Z is quadratic in ω therefore the condition (61) can be written as:

$$\omega \geq E_o^+\ (r;\ m;\ K) \tag{62a}$$

or

$$\omega \leq E_o^-\ (r;\ m;\ K) \tag{62b}$$

with

$$E_o^\pm = (r^2 + a^2)^{-1}\ [am + e\varepsilon r \pm [\Delta(\mu^2 r^2 + K)]^{\frac{1}{2}}] \tag{62c}$$

(62a) defines the positive-root states E^+ and (62b) the negative-root states E^- introduced by Christodoulou and Ruffini [19] for their important role with respect to reversible and irreversible transformations of a black hole. It is easy to see that they correspond with the ambiguity of sign of Eq. (55). Some instances of them, displaying the phenomenon of level crossing (57), are represented in Fig. III.

Now let us consider the corresponding quantum problem. For simplicity's sake we will consider a scalar particle of mass μ and charge ε. It must fulfill the Klein-Gordon equation in the external field (59):

$$(\nabla^\alpha - i\varepsilon A^\alpha)(\nabla_\alpha - i\varepsilon A_\alpha)\ \Phi = \mu^2\ \Phi \tag{63}$$

where the wave function Φ can be separated [36] as:

$$\Phi = e^{i(m\phi - \omega t)}\ \chi(\theta)\ \psi(z) \tag{64a}$$

with [37] $dz = -\ dr/\Delta$ and

$$[-(\sin\theta)^{-1}\ (d/d\theta)\ \sin\theta\ d/d\theta + (m/\sin\theta - a\omega\sin\theta)^2 + \mu^2 a^2 \cos^2\theta]\chi = K\chi \tag{64b}$$

$$d^2\psi/dz^2 = Z\ \psi \tag{64c}$$

where Z is the same expression (60e) as before, K being now understood as an eigenvalue of the Eq. (64b) which is equivalent to the equation for spheroidal harmonics. This fact implies a close correspondence [37] between the quantum motion of the particle, described by Eq. (64c) and its classical limit: Eq. (60d) which is useful, for instance, in the search for resonances. Anyway in the case of level crossing as represented in Fig. III both the neighborhood of the horizon and spatial infinity are asymptotic regions where the local wave length is infinitely smaller than the local length scale and therefore where the W.K.B. ap-

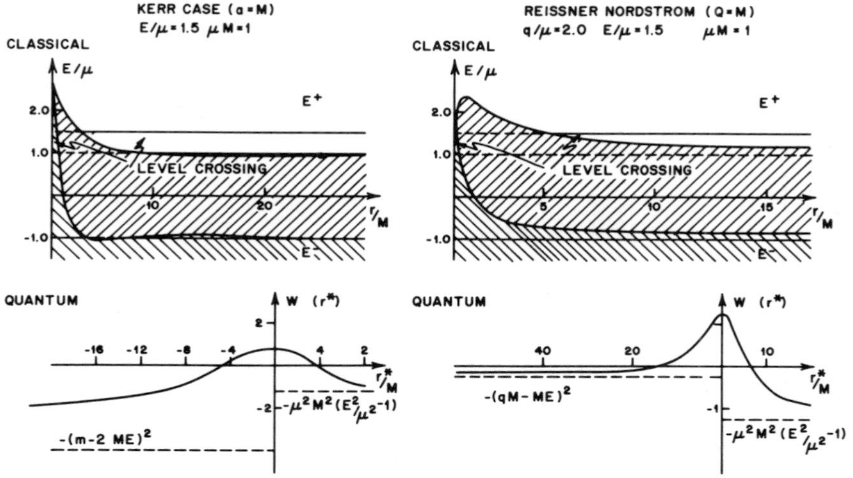

FIGURE III: In direct analogy with the flat-space situation depicted in Fig. II, the neighborhood of a black-hole can exhibit the phenomenon of level crossing between the "negative" and "positive" frequencies and therefore lead to spontaneous real pair creation. The upper part of the figure represents the quasi-classical diagrams of positive and negative-root states [19] in the two cases of a rotating (Kerr) or charged (Reissner-Nordstrom) black-hole. The lower part represents the explicit shape of the quantum potential which determines the behaviour of the wave function (using the coordinate $r_* = \int dr\, r^2/\Delta$). This shows the two asymptotic regions, the horizon ($r_* \to -\infty$) and spatial infinity ($r_* \to +\infty$), connected by a barrier where $W > o$.

proximation becomes exact. Hence, restricting ourselves to the study of pair creation processes occurring outside the horizon, we can apply the analysis of II above, where the choice of positive and negative states is provided by the asymptotic behaviour of the effective potential (see figure III and compare it to figure II). The rate of pair creation will be given by (see eq. (18)):

$$R \equiv \frac{d}{dt} \langle N \rangle = \int \frac{d\omega}{2\pi} \sum_{m,K} |T|^2_{\omega,m,K} \tag{65}$$

where $|T|^2$ is the transmission coefficient between "negative" and "positive" states computed by numerical integration of Eq. (64c) from the horizon to spatial infinity for all the modes (74a) fulfilling the condition for level crossing:

$$\mu < \omega < m\,\Omega + \varepsilon V \tag{66}$$

where Ω is the angular velocity and V the electric potential of the black hole [38]. Inequality (66) is easily obtained by taking the limit on the horizon either of (56) or of (62b).

It is possible too to use Eq. (64c) for studying the quantum resonances around a black hole.

It was found [39] that when the phenomenon of resonance is combined with the one of level crossing one obtains unstable modes growing exponentially with time as a consequence of a continuous pair creation of bosons.

Finally let us consider vacuum polarization around a Kerr-Newmann hole [29] and its possible astrophysical consequences. The key point is, here, to realize that as we will be interested in black holes of current astrophysical interest, that is $M \geq M_{\odot}$ and in the creation of pairs electrons-positrons, the compton wave length $1/\mu$ of the electron is much smaller than the characteristic dimension M of the hole. Therefore the only region where the behaviour of the particle is non-classical: that is during the tunnelling through the effective potential barrier, is very localized in the background geometry. There is no need for investigating the transmission coefficient by numerical means, a W.K.B. approximation is excellent:

$$\zeta = 2 \int_{\text{barrier}} Z^{\frac{1}{2}}\, dz \tag{67}$$

This can be done and agrees perfectly with what follows; but it is much simpler to exploit thoroughly the locality of the tunnelling by constructing in the neighborhood of any point (t, r, θ, ϕ) a local Lorentz frame. More precisely we can use the orthogonal tetrad defined by Carter [36]:

$$\omega^{(o)} = (\Delta/\Sigma)^{\frac{1}{2}}\,(dt - a\sin^2\theta\, d\phi) \tag{68a}$$

$$\omega^{(1)} = (\Sigma/\Delta)^{\frac{1}{2}}\, dr \tag{68b}$$

$$\omega^{(2)} = \Sigma^{\frac{1}{2}}\, d\theta \tag{68c}$$

$$\omega^{(3)} = \sin\theta\, \Sigma^{-\frac{1}{2}}[(r^2 + a^2)\, d\phi - a\, dt] \tag{68d}$$

In the inertial frame defined by (68) the background electric and the magnetic fields are parallel and directed along $\omega^{(1)}$. Their invariantly defined values are:

$$E_{(1)} = e\Sigma^{-2}\,(r^2 - a^2\cos^2\theta) \tag{69a}$$

$$B_{(1)} = e\Sigma^{-2}\, 2\, ar\, \cos\theta \tag{69b}$$

Locally $\mu^{-2} R_{\alpha\beta\gamma\delta}$ and $\mu^{-1} F_{\alpha\beta;\gamma}$ are negligible therefore we can apply the flat space, uniform fields results of III above. This yields a transmission coefficient $|T|^2 = \exp -\zeta$ valid for spin ½ , with

$$\zeta = \pi\, \mu^2/\varepsilon E_{(1)} + 2\pi(n + \frac{1}{2} + \sigma_{(1)})\, B_{(1)}/E_{(1)} \tag{70}$$

The transmission is then maximum when n = o and $\sigma_{(1)} = -\frac{1}{2}$. The first condition means that the particles move along the common direction of E and B (in the frame (68)). Going back to the coordinate frame, this implies the following link between the energy and the angular momentum of the particle as seen from infinity:

$$m = a\omega \sin^2\theta \tag{71}$$

Moreover the second condition $\sigma_{(1)} = -\frac{1}{2}$ shows that the pairs created are polarized, the branching ratio between a spin $\sigma_{(1)} = -\frac{1}{2}$ and a spin $\sigma_{(1)} = +\frac{1}{2}$ being $\exp\,(2\pi\, B_{(1)}/E_{(1)})$.

The mean number of pairs created is given by Eq. (38) which yields:

$$\langle N\rangle = \int g^{\frac{1}{2}}\, d^4x\, \frac{1}{4\pi} \left(\frac{\varepsilon E_{(1)}}{\pi}\right)^2 \frac{\pi B_{(1)}/E_{(1)}}{\mathrm{th}(\pi B_{(1)}/E_{(1)})}\; e^{-\pi\, \mu^2/\varepsilon E_{(1)}} \tag{72}$$

with $g^{\frac{1}{2}} = \Sigma \sin\theta$. Taking into account the charge ε, energy $\omega \simeq e\varepsilon r/\Sigma$ and angular momentum $m = a\omega \sin^2\theta$ carried away by the particle expelled to infinity ($e\varepsilon > o$) we get explicit estimates of the decrease of charge, mass and angular momentum of the black hole:

$$-\dot{e} = R\varepsilon \tag{73a}$$

$$-\dot{M} = R \langle\omega\rangle \tag{73b}$$

$$-\dot{J} = R \langle m\rangle \tag{73c}$$

where $\langle\omega\rangle$ and $\langle m\rangle$ are suitable spatial mean values of ω and m.$R = d\langle N\rangle/dt$ is the total rate of pair creation.

Let us state some of the results obtained [29]: If the mass of the black hole is larger than $7.2 \times 10^6\, M_\odot$ particle creation processes are never important, but if $M < 7.2 \times 10^6\, M_\odot$ and if the electric field reaches a critical strength $\sim 2 \times 10^{12}$ c.g.s. then an abrupt discharge of the hole via pair creation could occur releasing an energy $\Delta M \sim 5 \times 10^{40}\, (M/M_\odot)^3$ erg with the emission of electrons (or positrons) of energy $\omega \sim 10^{20}(M/M_\odot)$ eV.

The possible astrophysical relevance of this event as well as the conditions for its occurence are discussed in references [29-31]. Let us point out that the process of particle creation studied here always increases the irreducible mass of the black hole [19], in contrast with the phenomenon studied by Hawking. However it has been shown how it is possible to extend the barrier penetration treatment to study the evaporation of the horizon itself, which occurs even for an uncharged, non-rotating black-hole. For technical details see reference [40].

It is first necessary to realize why the preceding treatment excluded from the start the study of the evaporation of the horizon of the black hole: this came from the fact that the modes were defined and studied only in the open region outside the horizon (see e.g. Fig. III which uses a "tortoise" coordinate r_* going to $-\infty$ at the horizon). This situation is easily cured by using a coordinate system regular on the future horizon (the only existing one for a blackhole formed by collapse). This allows to study the behaviour of the modes from the singularity to infinity. As we saw before, the states, locally of negative frequency, inside the horizon can have a positive frequency as seen from infinity, but these states are classically confined in the black hole due to the "one-way" character of the horizon. However, relativistic wave mechanics allows the "leakage" of these states through the light barrier (i.e. the horizon). Physically this comes from the fact, well known in flat space, that a wave packet built only with one half of the frequency spectrum can never have a spatially compact support; therefore the wave functions describing the negative states inside the blackhole will necessarily have a small tail outside the horizon. And this tail, after peeling off the horizon and crossing the potential and centrifugal barrier, will appear at infinity as a positive wave, giving rise as usual to a flux of particles at infinity. (See Fig. IV).

Mathematically the result is very simply obtained by noticing that in Eddington-Finkelstein coordinates $v=t+r_*$, r, θ, ϕ (for simplicity we consider the Schwarzschild case, see [40] for details in the general case) a wave describing an antiparticle inside the horizon with a positive frequency ω at infinity has a logarithmic singularity $(2M-r)^{i4M\omega}$ at the horizon. But, using a small spread $\Delta\omega$ around ω, one can build with these states nearly monochromatic wave packets sharply localized on the horizon (because locally these states have infinitesimal wavelengths). Introducing new coordinates tangent to the old ones but locally geodesic at one point of the horizon we get a localized wave packet with a behaviour $\sim (2M-\hat{r})^{i4M\omega}$ where $\hat{r}$ is the locally geodesic coordinate associated to r. Then one can work as in flat space, Fourier-analyze this packet and apply the well known flat space result that a negative frequency wave is analytically continuable to the complex points of the forward tube (points $x + iy$ where y lies in the forward light cone). As the vector $\partial/\partial r$ is null and past directed in Finkelstein coordinates so is $\partial/\partial\hat{r}$, this implies that our negative wave must be

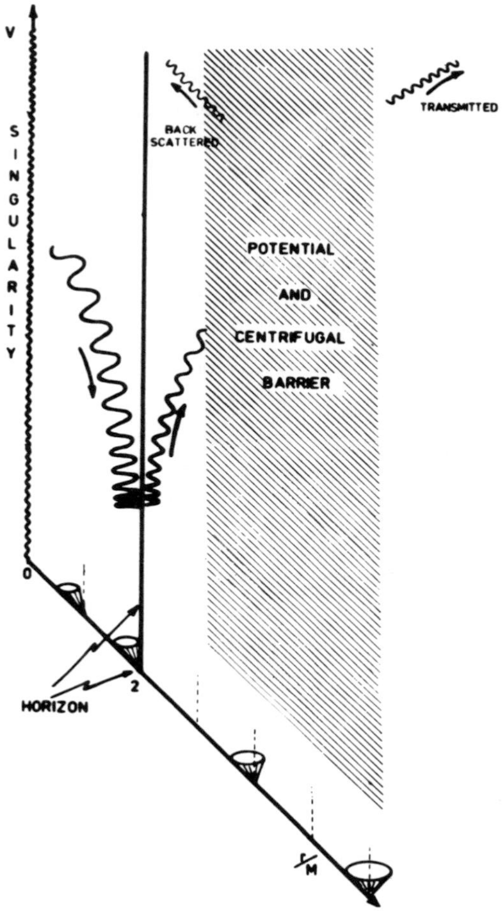

Figure IV: In usual Eddington-Finkelstein coordinates a qualitative representation is given of the splitting of a negative state into its inner component with negative energy flux falling on the singularity and its outer component (tail) peeling off the horizon and partially transmitted to infinity (if $\omega > \mu$) and partially back scattered into the hole (taken from ref. [40]).

analytic in the lower complex $\hat{r}$ plane. This fact tells us how to continue the logarithmic singularity from inside the horizon by an $e^{+i\pi}$ rotation in 2M-r yielding an outside tail term of

$$e^{-4\pi M\omega}(r-2M)^{i4M\omega}$$

necessarily associated to a state inside the horizon. After normalization of this negative state together with its associated tail and taking into account the transmission coefficient Γ of the wave through the combined potential and centrifugal barrier, one gets from formula (18) a continuous flux of particles at infinity. This flux is $(\Gamma/2\pi)\ (e^{8\pi M\omega}-1)^{-1}$ per unit of time and per unit range of frequency (which is Hawking's result [8]).

The same method can be straight forwardly applied to the general case of a Kerr-Newman black hole [40].

REFERENCES

[1] O. Klein, Zeit. fur Phys. 53 (1929) 157.
[2] F. Sauter, Zeit. fur Phys. 69 (1931) 742.
[3] W. Heisenberg and H. Euler, Zeit fur Phys. 98 (1935) 714.
[4] F. Hund, Zeit. fur Phys. 117 (1940) 1.
[5] W. Thirring, Principles of Quantum Electrodynamics (1958) Academic Press, N.Y.
[6] E. M. Lifshitz and L. P. Pitaevskii, Relativistic Quantum Theory, II, (1974) Pergamon press, New York.
[7] W. G. Unruh, Phys. Rev. D10 (1974) 3194.
[8] S. W. Hawking, Nature 248 (1974) 30.
[9] S. W. Hawking, Commun. Math. Phys. 43 (1975) 199.
[10] G. W. Gibbons, Commun. Math. Phys. (in press) see also G. W. Gibbons, in these proceedings.
[11] B. S. DeWitt, Physics Reports 19C (1975) 297.
[12] E. T. Whittaker and G. N. Watson, Modern Analysis (1965) Cambridge University Press.
[13] J. A. Wheeler in Fast Neutron Physics, J. B. Marion and J. L. Fowler ed., N.Y. Interscience publishers, (1960); Note that the transmission coefficient is not the same for the Klein-Gordon and for the Schroedinger equation.
[14] J. Schwinger, Phys. Rev. 82 (1951) 664.
[15] A. I. Nikishov, Sov. Phys. JETP 30 (1969) 660.
[16] N. B. Narozhnyi and A. I. Nikishov, Sov. Journal of Nuclear Physics 11 (1970) 596.
[17] E. C. G. Stueckelberg, Helv. Phys. Acta 14 (1941) 588; see particularly Fig. 2 in this reference. See also E. C. G. Stueckelberg: Helv. Phys. Acta 15 (1942) 23.
[18] R. P. Feynman, Phys. Rev. 76 (1949) 749.
[19] D. Christodoulou and R. Ruffini, Phys. Rev. D4 (1971) 3552; G. Denardo and R. Ruffini, Phys. Lett. 45B (1973) 259; G. Denardo, L. Hively and R. Ruffini Phys. Lett. 50B (1974) 270. The definition used by these authors is in fact only $\omega^- > o$.
[20] R. Penrose and R. Floyd, Nature 229 (1971) 193.
[21] Ya. B. Zel'dovich, Sov. Phys. JETP 35 (1972) 1085.
[22] A. A. Starobinsky, Sov. Phys. JETP 37 (1973) 28.
[23] N. Deruelle and R. Ruffini, Phys. Lett. 52B (1974) 437.
[24] N. Deruelle and R. Ruffini, Phys. Lett. 57B (1975) 248.
[25] N. Deruelle in these proceedings.
[26] G. W. Gibbons and S. W. Hawking in "Gravitational Radiation and Gravitational Collapse" C. DeWitt ed. Reidel (1974).
[27] W. T. Zaumen, Nature 247 (1974) 530; see also W. T. Zaumen in these proceedings.
[28] T. Nakamura and H. Sato, Phys. Lett. 61B, (1976), 371.
[29] T. Damour and R. Ruffini, Phys. Rev. Lett. 35 (1975) 463.
[30] R. Ruffini in Seventh Texas Symposium on Relativistic Astrophysics, P. G. Bergmann, E. J. Fenyves and L. Motz ed. N.Y. Academy of Sciences, NY (1975).
[31] S. Ames in these proceedings.
[32] For a general review of this approach see: T. Damour, N. Deruelle and R. Ruffini: to be submitted to Phys. Rev.
[33] D. G. Boulware in Seventh Texas Symposium on Relativistic Astrophysics, P. Bergmann, E. J. Feyves and L. Motz ed. NY Academy of Sciences, NY (1975).
[34] U. H. Gerlach, Phys. Rev. D15 (1975) (in press); see also: U. H. Gerlach in these proceedings.

[35] B. Carter, Phys. Rev. 174 (1968) 1559.
[36] B. Carter, Commun. Math. Phys. 10 (1968) 280.
[37] T. Damour, Nuovo Cimento Lett. 12 (1975) 315.
[38] See e.g. Black Holes, C. and B. DeWitt ed., Gordon and Breach, NY (1973).
[39] T. Damour, N. Deruelle and R. Ruffini, Nuovo Cimento Lett. 15, (1976),257.
[40] T. Damour and R. Ruffini, "Black-hole Evaporation in the Klein-Sauter-Heisenberg-Euler Formalism", Phys. Rev. D14 (in press).

CLASSICAL AND QUANTUM STATES IN BLACK HOLE PHYSICS

Nathalie Deruelle*
Institute of Astronomy, Cambridge (England)

ABSTRACT: In this report a quantum significance to the classical positive and negative energy states of a relativistic particle in the field of a black hole is presented. We show briefly how the possible level crossing between these positive and negative energy states can be interpreted as a Klein paradox. We then solve the Klein-Gordon equation in the Schwarzschild and Kerr cases according to this approach.

I. POSITIVE AND NEGATIVE ENERGY STATES

A black hole of mass M, charge Q, specific angular momentum a can be described, in Boyer-Lindquist coordinates, by the Kerr-Newmann metric:

$$ds^2=[1-(2Mr-Q^2)/\rho^2]dt^2-(\rho^2/\Delta)dr^2-\rho^2 d\theta^2-[r^2+a^2+a^2\sin^2\theta(2Mr-Q^2)/\rho^2]\sin^2\theta d\phi^2+$$
$$+[2(2Mr-Q^2)\ a\ \sin^2\theta/\rho^2]d\,\phi\,dt \tag{1}$$

where $\Delta = r^2-2Mr+Q^2+a^2$; $\rho^2=r^2+a^2\cos^2\theta$; $A_\mu dx^\mu=(Qr/\rho^2)(dt-a\sin^2\theta\ d\phi)$

Such a metric has two pseudo-singularities:

$$r_\pm = M \pm (M^2-a^2-Q^2)^{1/2} \quad (M^2>a^2+Q^2) \tag{2}$$

This coordinate system does not therefore describe correctly the complete manifold but since, according to the classical extension "à la Kruskal" of the manifold for example, nothing can emerge from the event horizon r_+, we shall disregard the region $o<r<r_+$ and consider only the region outside the event horizon where the metric (1) is well behaved.

The equations of motion of a classical test particle of mass μ and charge e can be deduced from the Hamilton-Jacobi equation written in the metric (1):

$$g^{\alpha\beta}(\partial_\alpha S+e\ A_\alpha)(\partial_\beta S+e\ A_\beta)=\mu^2 \tag{3}$$

The separation of the variables was carried out by Carter [1]:

$$S = - E\mu t + \Phi\phi + \Theta(\theta) + S(r) \tag{4}$$

where E is the specific energy and Φ the azimuthal angular momentum of the particle as measured at spatial infinity. The Hamilton-Jacobi equation then reduces to the set:

$$(d\Theta/d\theta)^2= K - (\Phi-\mu Ea\ \sin^2\theta)^2/\sin^2\theta - a^2\mu^2\ \cos^2\theta$$
$$(dS/dr)^2= \mu^2(r^2+a^2)^2\ [E^2-2AE+B]/\Delta^2 \tag{5}$$

*E. S. A. International Fellow

$$\text{with:}\quad A = \frac{a\Phi + eQr}{\mu(r^2+a^2)} \quad ; \quad B = \frac{(a\Phi + eQr)^2 - \Delta(\mu^2 r^2 + K)}{\mu^2(r^2+a^2)^2} \tag{5}$$

K is Carter's fourth constant of the motion [1] which reduces, in the Schwarzschild case, to $K=L^2\mu^2$ where L is the total specific angular momentum.

The radial equation of the motion then reads:

$$(dr/d\tau)^2 = [(r^2+a^2)/(r^2+a^2\cos^2\theta)]^2[E^2-2AE+B] \tag{6}$$

This equation is valid for $o<r<\infty$. But since we only consider here what happens outside the event horizon, we can rewrite eq. (5-6) in the following form, valid in the range $r_+<r<\infty$

$$\Delta^2(dS/dr)^2 = \mu^2(r^2+a^2)^2\ (E-E_o^+)(E-E_o^-) \tag{7}$$

$$(dr/d\tau)^2 = [(r^2+a^2)/(r^2+a^2\cos^2\theta)]^2(E-E_o^+)(E-E_o^-) \tag{8}$$

$$\mu\, E_o^{\pm}(r) = \frac{(a\Phi + eQr) \pm [\Delta(\mu^2 r^2+K)]^{1/2}}{r^2 + a^2} \tag{9}$$

This factorisation of the quadratic form (E^2- 2AE + B) -a direct consequence of the relativistic relation $p_\alpha\, p^\alpha = \mu^2$-was first performed by Christodoulou and Ruffini [2] in order to study the motion of test particles in the given geometry. This, in turn, lead to the introduction of the effective ergosphere [2, 3,4] i.e. the region surrounding a black hole where a particle in a positive energy state ($E>E_o^+$) can have a negative energy ($E<o$) and where it is possible by a Penrose process [5, 6], to extract energy from the hole.

$E_o^{\pm}(r)$ are the effective potentials for the positive and negative energy solutions. Their meaning is clear: they are the turning points of the particle. Classically only particles in a positive energy state ($E>E_o^+$) can exist since:

1. The "forbidden states" ($E_o^-<E<E_o^+$) have no physical meaning: they would correspond to particles having an imaginary momentum.

2. The negative energy states ($E<E_o^-$)("classically forbidden states") have no classical meaning since they would correspond, in the asymptotic flat space-time to particles of negative mass. See fig. I.

These classical positive and negative energy states acquire a full significance in the framework of a quantum theory [7].

Let us then consider a spin zero quantum test particle imbedded in the classical background field of a black hole. Its behaviour is described by the Klein-Gordon

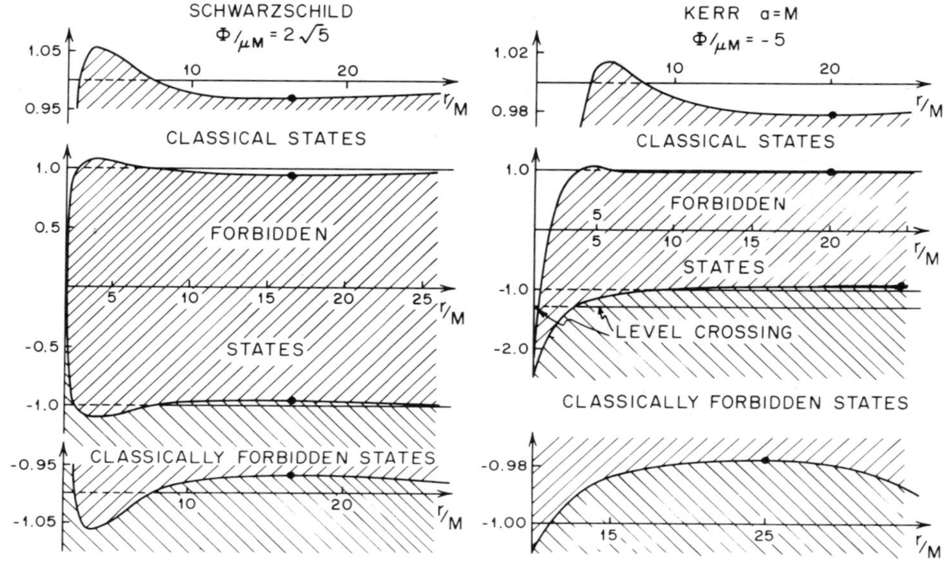

FIGURE I. The effective potentials for the positive and negative energy solutions $E_0^{\pm}(r)$ are here represented in the particular cases of the Schwarzschild and the extreme Kerr geometries.

In the Schwarzschild case:

$$E_o^{\pm}(\rho) = \pm\left[\left(1-\frac{2}{\rho}\right)\left(1+\frac{L^2}{M^2\rho^2}\right)\right]^{1/2} \qquad (\rho = r/M)$$

In the Kerr case ($a = M$, $\theta = \pi/2$):

$$E_o^{\pm}(\rho) = \frac{2(\Phi/\mu M) \pm (\rho-1)[\rho(\rho^3+\rho+2) + \Phi^2\rho^2/\mu^2 M^2]^{1/2}}{\rho^3 + \rho + 2}$$

The positive energy states ($E > E_o^+$) are the relativistic generalization of the Keplerian orbits. (The dots represent the last stable orbits). The forbidden states have no physical meaning since they would correspond to particles having an imaginary momentum. The negative energy states have no classical meaning since they would correspond to particles of negative mass.

These negative energy states acquire a significance in the framework of a quantum theory and correspond to the negative energy states of a relativistic wave equation [7]. In the case of a Kerr geometry (as well as in the Reissner-Nordstrom [8, 33] and the Kerr-Newman [9] geometries), a level crossing between the positive and negative energy states inside the effective ergosphere is possible (cf, fig). Pair production can then occur (Klein paradox). This cannot happen in the case of a "dead" Schwarzschild black hole which has no ergosphere.

equation written in the metric (1). (Here also we shall only consider what happens outside the event horizon):

$$g^{\alpha\beta} (D_\alpha + ie\, A_\alpha)(D_\beta + ie\, A_\beta)\, \Phi + \mu^2 \Phi = 0 \tag{10}$$

where $\Phi(x^\alpha)$ represents in the framework of a monoparticular interpretation the amplitude of the probability of finding the particle in x^α (and, in a second quantized theory, the modes by means of which the quantized field is expanded).

The conserved four-vector current associated with eq. (10) is:

$$J^\mu = \frac{i\sqrt{-g}}{2\mu} [\Phi^*(\partial^\mu + ieA^\mu)\Phi - \Phi(\partial^\mu - ieA^\mu)\Phi^*] \tag{11}$$

Separating the variables leads to [10]

$$\Phi = \exp i(m\phi - E\mu t) \times S_{m\ell}(\theta)\, R\,(r) \tag{12}$$

where E is the specific energy and m the magnetic quantum number of the particle. $S_{m\ell}(\theta)$ are the spheroidal harmonics of eigenvalues $\lambda_{m\ell}$, solutions of:

$$\frac{1}{\sin\theta} \frac{d}{d\theta} (\sin\theta \frac{d}{d\theta} S_{m\ell}) + [a^2\mu^2(E^2-1)\cos^2\theta - \frac{m^2}{\sin^2\theta} + \lambda_{m\ell}]\, S_{m\ell} = 0 \tag{13}$$

In the case a = o the spheroidal harmonics reduce to the Legendre polynomials and $\lambda_{m\ell}$ to $\ell(\ell+1)$.[11]

Outside the event horizon r_+, the radial Klein-Gordon equation now reads:

$$\begin{aligned} &\Delta \frac{d}{dr} \Delta \frac{dR}{dr} = -\mu^2(r^2+a^2)^2(E-E_o^+)(E-E_o^-)R \\ &\mu\, E_o^\pm(r) = \frac{(am + eQr) \pm [\Delta(\mu^2 r^2+K)]^{1/2}}{r^2 + a^2} \\ &K = \lambda_{m\ell} + \mu aE(\mu aE - 2m) \end{aligned} \tag{14}$$

(K is the quantum analogue of Carter's constant of the motion).

The separation of the variables (12) also gives:

$$J^o = \sqrt{-g}\quad \Phi^*\, \Phi[g^{oo}(E - \frac{e}{\mu}\, \frac{Qr}{\rho 2}) + \frac{g^{o3}}{\mu}\, (\frac{eQra\, \sin^2\theta}{\rho^2} - m)] \tag{15}$$

First one note the analogy between the radial Klein-Gordon equation (14) and the Hamilton-Jacobi equation (7). (This analogy can be made even clearer by introducing a new coordinate z such that $z = \int_r^\infty dr/\Delta$[12]). The shape of the effective potential felt by a quantum particle and therefore its behaviour can be immediately deduced from the classical potentials $E_o^\pm(r)$. Cf. fig. II.

The point to stress is that a quantum particle can exist (by exist we mean that the radial function has a oscillatory behaviour) not only in the positive state

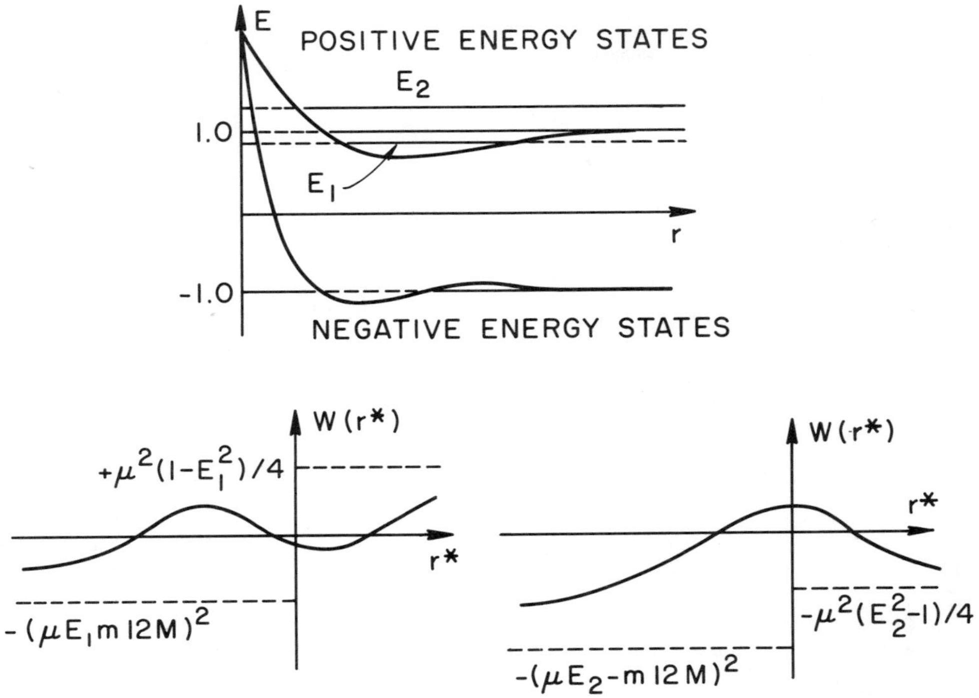

FIGURE II. These diagrams show in the particular case of the extreme Kerr geometry that the behaviour of a quantum particle in the field of a black hole can be immediately deduced from the classical diagrams $E_0^{\pm}(r)$. (b) The quantum potential $W(r^*)$ felt by a bound quantum particle of energy E_1 in the field of an extreme Kerr black hole is here represented as a function of the tortoise coordinate r^*. Since a leakage towards the horizon $r^*=-\infty$ of the wave function through the potential barrier is possible, the solutions of the wave equation will represent resonances. (Cf. & II). When the corresponding classical particle is not in a stable bound orbit (e.g. $E=E_3$ on (a)), $W(r^*)$ has only one zero and the Klein-Gordon equation has no stationary solutions. (c) Quantum potential felt by a quantum particle of energy E_2. A quantum particle can exist both in a positive and a negative energy state. A wave can then tunnel from an E^+ to an E^- region leading to the phenomenon of pair creation.

region $(E>E_0^+)$ but also in the classically forbidden region $(E<E_0^-)$. Moreover a quantum wave of energy E corresponding to a level crossing can tunnel through the forbidden states from the positive to the negative state region. This tunnelling which has no equivalent in the classical theory leads to the Klein paradox considered in the now classical papers of Klein [13], Sauter [14], Euler-Kockel [15], Heisenberg-Euler [16], Pauli [17], and examined in details in the context of curved space-time by Damour in these proceedings [see rf. [18-21]]. We shall only briefly outline it.

The equation $J^0=0$ (cf. eq. (15)) defines the curve:

$$\mu E_o(r) = \left(m - \frac{eQr\ a\ \sin^2\theta}{\rho^2}\right) \frac{g^{03}}{g^{00}} + \frac{eQr}{\rho^2} \tag{16}$$

with: $E_o^- \leq E_o \leq E_o^+$; $\mu E_o(r=r^+) = (am+eQr_+)/(r_+^2+a^2)$

This curve (cf. fig. II) delineates the frontier between the positive energy states where $J^o>0$ and the negative energy states where $J^o<0$. According to the usual interpretation [22-23] of the probability density J^o, the wave function Φ with eigenvalues (E,ℓ,m,e) describes, in the positive state region, a flux of particles of charge e energy E, quantum numbers ℓ and m; in the negative state region Φ describes a flux of antiparticles of charge -e, energy -E, quantum numbers ℓ,-m, thanks to the particle-antiparticle conjugation relations [22, 23]:

$$E_o^-(e, m) = - E_o^+(-e, -m)$$
$$J^\mu(\Phi, e) = - J^\mu(\Phi^*, -e) \tag{17}$$

These relations allows us to restrict the problem to the study of incident particles (E>o).

In the Schwarzschild case there can be no level crossing and the reinterpretation of the negative states in terms of antiparticles is straightforward. On the other hand if either of a and Q is non-zero, and if there is a level crossing between the E^+ and the E^- states, the same wave function Φ describes a flux of particles in the asymptotic flat space-time and a flux of antiparticles near the horizon. In the monoparticle interpretation here considered this paradox is resolved by saying that ingoing particles give rise to particles-antiparticles pairs [25]. In a second quantized theory one would resolve the paradox by saying that the gravitational and (or) the electromagnetic field of the black hole is strong enough to create pairs from the vacuum. (The same result is obtained in the case of spin ½. particles even though J^o is then positive definite [26].)

II. BOUND STATES IN THE SCHWARZSCHILD AND KERR GEOMETRIES

A. BOUNDARY CONDITIONS

As we already said we consider here the observable manifold to be limited to the region outside the event horizon; we can then introduce the "tortoise coordinate" r* which pushes the horizon r_+ back to $-\infty$:

$$dr^* = [(r_+^2+a^2)/r_+^2]\ r^2 dr/\Delta \tag{18}$$

The radial Klein-Gordon equation (14) then reads:

$$d^2u/dr^{*2} = [r_+^2/(r_+^2+a^2)]^2\ W(r^*)u;\ u = rR$$
$$r^4 W(r^*) = \Delta[\mu^2 r^2 + K + \frac{2M}{r} - \frac{2(a^2+Q^2)}{r^2}] - [\mu(r^2+a^2)E - (am + eQr)]^2 \tag{19}$$

where K is given by eq. (14).

It is important at this stage to stress the fact that since we restrict our attention to what happens outside the event horizon we shall not treat the phenomenon of black holes evaporation discovered by Hawking [27, 28].

The asymptotic forms of eq. (19) are:

$$\begin{aligned} r^* \to +\infty \quad &: d^2u/dr^{*2} \sim -[\mu r_+{}^2/(r_+{}^2+a^2)]^2(E^2-1)u \\ r^* \to -\infty \quad &: d^2u/dr^{*2} \sim -(\mu E - m\Omega - eV)^2 u \end{aligned} \tag{20}$$

where $\Omega = a/(r_+{}^2+a^2)$ is the angular velocity of the hole [29] and $V = Qr_+/(r_+2 + a^2)$ its electric potential. [30]

Since we study here the bound states (0<E<1) we must choose the decreasing exponential at $+\infty$:

$$u \overset{r^*\to\infty}{\sim} c \exp -[\mu r_+{}^2/(r_+{}^2+a^2)]\sqrt{1-E^2}\; r^* \tag{21}$$

and indeed we want the probability of finding the particle to tend to zero at infinity.

The general solution near the horizon has the form:

$$u(r^*) \overset{r^*\to-\infty}{\sim} A \exp i|(\mu E - m\Omega - eV)|r^* + B \exp -i|(\mu E - m\Omega - eV)|r^* \tag{22}$$

We shall choose the solution which represents a flux of particle or of antiparticles going <u>towards</u> the horizon. When there is no level crossing, the solution represents a particle everywhere and the quantization condition reads:

$$u(r^*) \overset{r^*\to-\infty}{\sim} B \exp -i|(\mu E - m\Omega - eV)|r^* \tag{23}$$

On the other hand when there is a level crossing, i.e. when:

$$\mu E < m\Omega + eV \; : \; \mu E < (ma + eQr_+)/(r_+{}^2+a^2) \tag{24}$$

an antiparticle falling towards the horizon is described by an outgoing wave (particle-antiparticle conjugation, see eq. (17)):

$$u(r^*) \overset{r^*\to-\infty}{\sim} A \exp + i\; |(\mu E - m\Omega - eV)|r^* \tag{25}$$

Therefore in all the cases we shall choose:

$$u(r^*) \propto \exp -i\; (\mu E - m\Omega - eV)r^*$$

B. SCHWARZSCHILD CASE

Eq. (19) reads in the Schwarzschild case (a = Q = 0):

$$d^2u/d\rho^{*2} = \left[(1 - \frac{2}{\rho})(\mu^2M^2 + \frac{\ell(\ell+1)}{\rho^2} + \frac{2}{\rho^3}) - \mu^2M^2E^2\right] u \tag{26}$$

The parameter μM is, in conventional units, the ratio of the radius of the black hole (GM/c^2) and the Compton wavelength of the particle ($\hbar/\mu c$).

The analogy between the radial Hamilton-Jacobi and Klein-Gordon equations tells us (cf. fig. II) that eq. (26) will have stationary solutions only if the corresponding classical particle can have a stable orbit. One must then have:

$$\ell(\ell+1) > 12\ \mu^2M^2 \quad ; \quad E^+_{o\ min} < E < E^+_{o\ max} \tag{27}$$

where $E^{+\ max}_{o\ min}$ are the extrema of $E^+_o(r)$.

When conditions (27) are satisfied one can solve eq. (26) using the W. K. B. approximation [25, 31]. Let a, b, c be the zeros of the effective quantum potential $W(\rho^*)$. The B.K.W. solution far from the hole ($\rho^* >> c$) is:

$$u(\rho^*) \sim (A/\sqrt{W(\rho^*)})\ e^{-|w|} \tag{28}$$

$$\text{where } w = \int_c^{\rho^*} \sqrt{W(\rho^*)}\ d\rho^*$$

The matching conditions imply that the solution near the horizon ($\rho^* << a$) is:

$$u(\rho^*) \sim [(Ae^{-i\pi/4})/\sqrt{-W}]\ \exp(-i\int_a^{\rho^*} \sqrt{-W}\ d\rho^*) \left[\frac{i \sinh e^{-\varepsilon}}{2} + 2 \cosh e^{\varepsilon}\right]$$
$$+ [(Ae^{i\pi/4})/\sqrt{-W}]\ \exp(i\int_a^{\rho^*} \sqrt{-W}\ d\rho^*) \left[-\frac{i \sinh e^{-\varepsilon}}{2} + 2 \cosh e^{\varepsilon}\right] \tag{29}$$

$$\text{where } h = \int_b^c \sqrt{-W(\rho^*)}\ d\rho^* \quad ; \quad \varepsilon = \int_a^b \sqrt{W}\ d\rho^*$$

This quantization condition (23) (no level crossing) then implies:

$$- i \sinh e^{-\varepsilon} + 4 \cosh e^{\varepsilon} = 0 \tag{30}$$

This equation only has solutions for complex values of the energy:

$$En = Eon - i\ \Gamma n/2$$

When $\Gamma n << Eon$, Eon is solution of:

$$h(E_{on}) \equiv \int_{b(Eon)}^{c(Eon)} \sqrt{-W(\rho^*, Eon)}\ d\rho^* = (n + \frac{1}{2})\pi \quad , n \in N \tag{31}$$

and Γn is given by:

$$\Gamma n = \left[e^{-2\varepsilon}/2(\partial h/\partial E) \right]_{E = Eon} \qquad (\Gamma n > 0) \tag{32}$$

Therefore the continuous spectrum of classical orbits is replaced by a discrete set of resonances of energy Eon and width Γn [7] of the kind first considered by Gamow in the problem of α - decay from a nucleus [32]: A quantum particle can tunnel through the potential barrier and is finally swallowed by the black hole. Table I gives some results of the numerical computation of Eon and Γn in the case $\mu M = 1$ (corresponding to a particle whose Compton wavelength is equal to the characteristic radius of the black hole).

TABLE I ($\mu M = 1$)

	Eon	Γn	
$\ell = 4$	0.976 - 0.977	$\sim 1.3 \times 10^{-6}$	1st resonance
$\ell = 4$	0.983 - 0.984	2.0×10^{-6}	2nd resonance
$\ell = 6$	0.989 - 0.990	$\sim 0.8 \times 10^{-15}$	1st resonance
$\ell = 6$	0.991 - 0.992	$\sim 1.3 \times 10^{-15}$	2nd resonance

The width Γ corresponds to a life-time:

$$\tau_{sec} = \frac{2}{\mu\Gamma} \frac{\hbar}{c^2} \tag{33}$$

If we take $\mu = m_e \sim 10^{-27}$g , a width $\Gamma \sim 10^{-6}$ corresponds to a lifetime $\tau \sim 2\times10^{-15}$ sec, a width $\Gamma \sim 10^{-15}$ to a life-time $\tau \sim 2 \times 10^{-6}$sec. The quantum effects are then very important in the case of a small black hole ($m_e M = 1 \rightarrow M \sim 10^{17}$g).

To obtain the eigenfunctions u(Eon,ℓ,m) we integrated the Klein-Gordon equation (26) numerically. Figure III illustrates some of the results.

When $\mu M \rightarrow \infty$ (i.e. when $\hbar/\mu c \ll \frac{GM}{c^2}$: classical limit), the discrete spectrum of resonances tends to the continuous spectrum of classical orbits: the classical positive and negative energy states are therefore explicitly checked to be the classical limit of the quantum particle and antiparticle states.

On the other hand when $\hbar/\mu c \gg GM/c^2$ (Compton wavelength large compared with the radius of the hole)$\Gamma \rightarrow 0$: this can easily be shown by studying the Klein-Gordon equation written in the $z = \int_r^\infty dr/\Delta$ coordinate. The particle is too "big" to

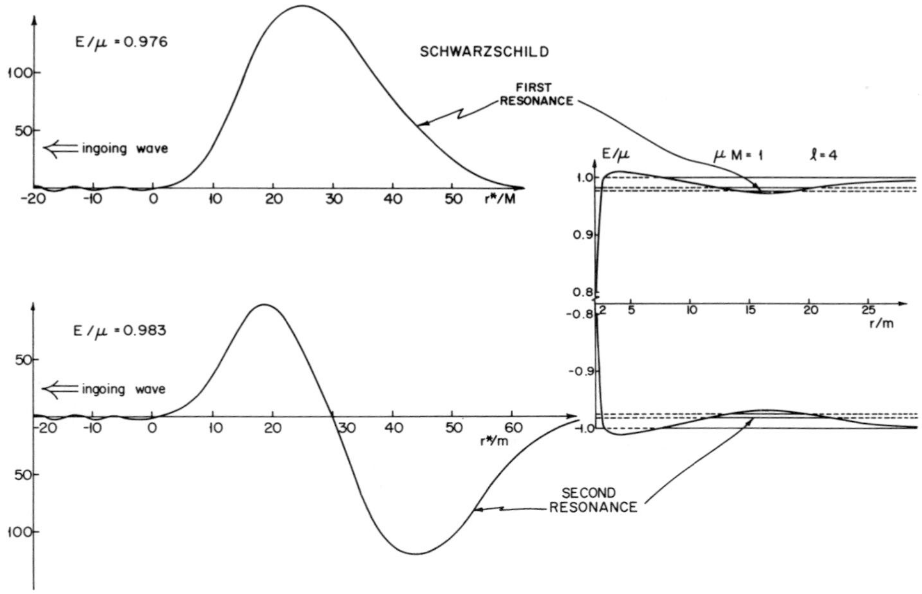

FIGURE III: The two first resonances obtained by numerical integration of the Klein-Gordon equation written in the Schwarzschild metric. The energy levels of the resonances are also represented n the classical diagrams $E_{0}^{\pm}(r)$. One then sees how the quantization modifies the classical results: the continuous spectrum of classical trajectories is replaced by a discrete set of resonances. A quantum particle initially placed in the potential well of the hole can tunnel through the potential barrier and is finally absorbed by the black hole.

Since $\int d^3x\, J^o$ diverges, the solutions can be interpreted by saying that at $t = -\infty$ an infinite number of particles was placed in the potential well and that they are progressively absorbed.

be absorbed by the black hole. This result agrees with that of Wheeler [34].

C. EXTREME KERR CASE

Let us now consider a quantum particle in the field of an extreme Kerr black hole (Q=o, a=M). The radial Klein-Gordon equation (19) becomes:

$$\frac{d^2u}{d\rho^{*2}} = \frac{1}{4}\left[-E^2\mu^2M^2\left(1+\frac{1}{\rho^2}+\frac{2}{\rho^3}\right)+\frac{4\mu MEm}{\rho^3}+\left(1-\frac{1}{\rho}\right)^2\left(\mu^2M^2+\frac{\lambda m\ell}{\rho^2}\right)-\frac{m^2}{\rho^4}+\right.$$
$$\left.+\,2\,\frac{(\rho-1)^3}{\rho^6}\right]u \qquad (34)$$

We used for $\lambda_{m\ell}$ the expansion [11] valid for $c = a^2\mu^2(E^2-1) \ll 1$:

$$\lambda_{m\ell} = \ell(\ell+1) - \tfrac{1}{2}[1-(4m^2-1)(2\ell-1)^{-1}(2\ell+3)^{-1}]\,c + O(c^2) \qquad (35)$$

Let us first consider the case where there is no level crossing i.e. where

$$\frac{am + e\Omega r_+}{r_+^2 + a^2} = \frac{m}{2M} < \mu E.$$

The quantization condition is then condition (23) and the problem is totally analogous to the Schwarzschild case: the continuous spectrum of stable orbits is replaced by a discrete spectrum of resonances [7]. See fig. IVb.

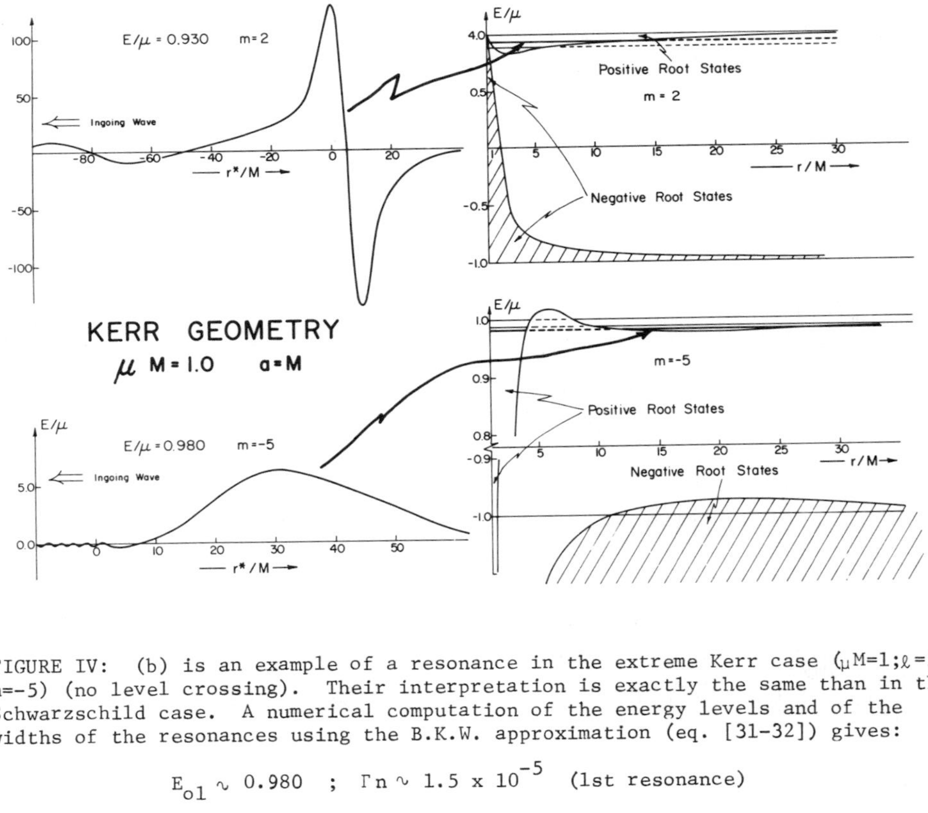

FIGURE IV: (b) is an example of a resonance in the extreme Kerr case ($\mu M=1; \ell=5; m=-5$) (no level crossing). Their interpretation is exactly the same than in the Schwarzschild case. A numerical computation of the energy levels and of the widths of the resonances using the B.K.W. approximation (eq. [31-32]) gives:

$$E_{01} \sim 0.980 \quad ; \quad \Gamma n \sim 1.5 \times 10^{-5} \quad \text{(1st resonance)}$$

$$E_{02} \sim 0.986 \quad ; \quad \Gamma n \sim 2.5 \times 10^{-5} \quad \text{(2nd resonance)}$$

(a): Example of a resonance with level crossing between the positive and the negative energy states in the field of an extreme Kerr black hole. The scalar field is assumed to have $\ell = m = 2$ and a mass μ such that $\mu M = 1$. The energy of the resonance $E = 0.930$ corresponds to the first excited state. The width of this growing resonance is $\Gamma \sim - 4 \times 10^{-5}$.

More interesting is the case where there is a level crossing between the positive and negative energy states i.e. where:

$$o < E < m/2\mu M \tag{36}$$

The quantization condition is then condition (25) (in-going wave near the horizon). The B.K.W. approximation then leads to (cf. eq. (29)):

$$+ i \sinh e^{-\varepsilon} + 4 \cosh e^{\varepsilon} = 0 \tag{37}$$

which has solutions $En = Eon - i\Gamma n/2$ where Eon is such that:

$$h(Eon) = \int_{b(Eon)}^{c(Eon)} \sqrt{-W(\rho^*, Eon)}\, d\rho^* = (n + \frac{1}{2})\pi$$

and where Γn is given by:

$$\Gamma n = - \left[e^{-2\varepsilon}/2(\partial h/\partial E)\right]_{E = Eon} ; \quad \varepsilon = \int_a^b \sqrt{W(\rho^*)}\, d\rho^* \tag{38}$$

Since Γn is negative and the solutions represent growing resonances [40] the radial function decreases exponentially near the horizon:

$$u \sim \exp - i(\mu Eon - m/2M)r^* \times \exp[-\mu\, \Gamma r^*/2]$$

and the norm $\int d^3x\, J^o$ converges and is identically zero [35, 18] instead of diverging as in the case $\Gamma n > o$ (no level crossing). One can interpret this result by saying that at $t = -\infty$ no particle was present in the potential well of the black hole and that an equal number of particles and antiparticles were created from the vacuum. See fig. IVa.

III. DIFFUSION STATES IN THE KERR GEOMETRY

We shall only consider here, in the extreme Kerr case, the diffusion states leading to the phenomenon of pair creation i.e. the states such that: [37]

$$1 \leq E \leq \frac{am + eQr_+}{\mu(r_+^2 + a^2)} \equiv \frac{m}{2\mu M} \qquad \text{(level crossing)} \tag{40}$$

(When there is no level crossing the problem is the usual problem of diffusion in the potential (19). See e.g. for the Schwarzschild case rf. [38].)

Since we are working in the framework of a first quantized theory we have to choose a boundary condition on the horizon. We shall require the solution to represent a flux of antiparticles going towards the horizon, i.e. an outgoing wave of the form:

$$u(r^*) \overset{r^* \to \infty}{\sim} \exp - i(\mu E - m/2M)r^* \tag{41}$$

corresponding to the physical situation of an observer at spatial infinity throwing particles towards the horizon.

In a second quantized theory in order to form a complete basis of solutions of the wave equation one would have to consider not only the solutions corresponding to antiparticles going towards the hole but also the solutions corresponding to antiparticles going outwards from the horizon (outgoing waves). But since, as we said before, one set of solutions is merely the time reversal of the other it is sufficient to consider the solutions satisfying condition (41).

The solution at spatial infinity has the form:

$$u(r^*) \sim A \exp - \frac{i\mu}{2} \sqrt{E^2 - 1} \quad r^* + B \exp + \frac{i\mu}{2} \sqrt{E^2 - 1} \quad r^* \tag{42}$$

the reflection coefficient at the barrier separating the positive from the negative energy states (see fig. II) is defined by $R^2 = |B/A|^2$ and the transmission coefficient by:

$$T^2 = \frac{m - 2\mu ME}{\mu M \sqrt{E^2-1}} \quad \left|\frac{C}{A}\right|^2$$

Since the probability density J^o has not a constant sign, the properties of the Wronskian imply:

$$T^2 + 1 = R^2 \quad (\rightarrow R^2 \geq 1) \tag{43}$$

Within the framework of a single particle theory T^2 is proportional to the probability that an incident particle will create a pair. In a second quantized theory T^2 is the basis for computing the number of pairs created from the vacuum [18]:

$$<N> = \sum_{\text{all possible channels}} T^2 \tag{44}$$

We have computed [37] the transmission coefficient T^2 by numerical integration of the radial Klein-Gordon equation (34). The results are summarized in fig. V and fig. VI.

The main conclusions are: 1. The probability of pair creation is negligible when the Compton wavelength, and therefore the de Broglie wavelength, of the incident particle is small compared with the radius of the black hole ($\hbar/\mu c << GM/c^2$: classical limit). Therefore this phenomenon is efficient in the Kerr case only for small black holes. However Damour and Ruffini [9] showed that, in the Kerr-Newman case, although T^2 is also very small when $\mu M \to\infty$ the total rate of created particles (dN/dt, see eq. (44)) may be non negligible even for astrophysical black holes because in this case, the number of channels can be huge. See

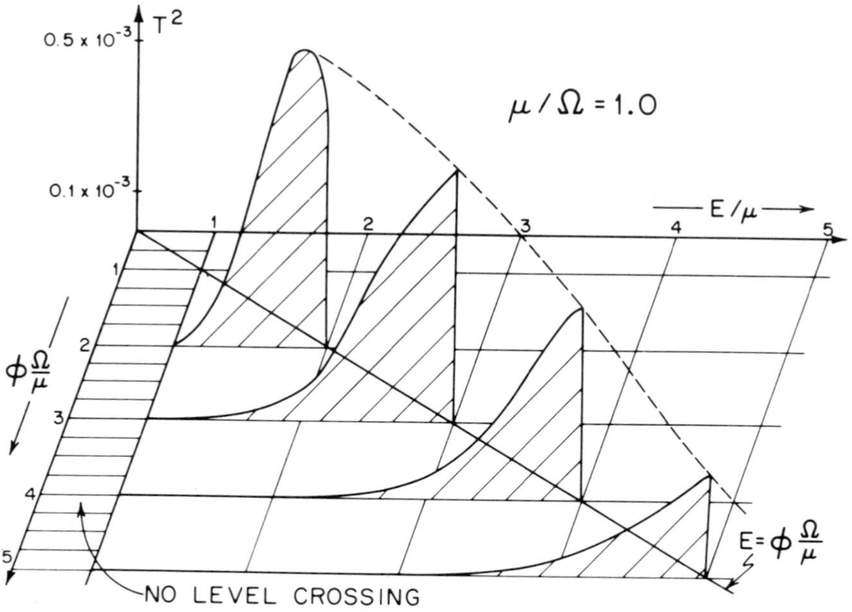

FIGURE V: The transmission coefficient has been computed by numerical integration of the Klein-Gordon equation in the field of an extreme Kerr black hole. We used the expansion of the eigenvalues $\lambda_{m\ell}$ up to the four order in $c=\mu^2M^2(E^2-1)$ and took $\ell=m$. The transmission coefficient is here represented as a function of $\phi\Omega/\mu \equiv m/2\mu M$ and the specific energy E/μ in the case $\mu/\Omega \equiv 2\mu M = 1.0$.

also rf. [20, 8, 26].

2. When $(GM/c^2)/(\hbar/\mu c)$ decreases, the probability of pair creation increases and tends asymptotically to the finite value $T^2 \sim 0.35 \times 10^{-2}$. This result is in agreement with rf. [39].

3. In the case of a small black hole ($\hbar/\mu c > GM/c^2$) the transmission coefficient reaches its maximum when:

$$\mu E \sim \hbar\Omega \tag{45}$$

i.e. when the de Broglie wavelength of the particle is comparable to the characteristic radius of the black hole. This result is in a nice agreement with rf. [40, 41].

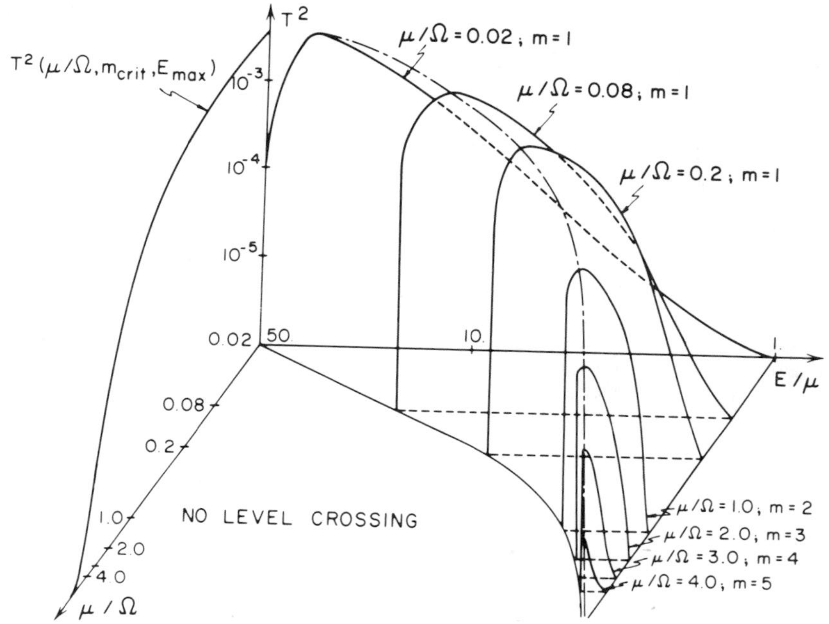

FIGURE VI: The transmission coefficient is here given as a function of E and μM. It is clear that $T^2 \to o$ when $\mu M \to \infty$ and that T^2 tends to a finite value when $\mu M \to o$. In this last limit the transmission coefficient reaches a maximum for an energy $E \sim \Omega$ ($E \sim \hbar\Omega$ in conventional units).

The law of conservation of the 4-momentum implies that this phenomenon of pair creation is a way of extracting rotational and Coulomb [40, 41] energy from the hole (a quantum analogue of the Floyd-Penrose process [5, 6]. A particle of energy $E>o$ and angular momentum $m>o$ may create an antiparticle of energy $-E<o$ and quantum number $-m<o$ which will be absorbed by the black hole thereby reducing its mass and its angular momentum, together with a particle of energy E and quantum number m which will flie to infinity with the stolen energy. However this process is irreversible as explicitely evaluated in ref. [9] and always increases the irreducible mass of the black hole in sharp contrast to the phenomenon of black hole evaporation [27].

ACKNOWLEDGEMENTS

I am grateful to Remo Ruffini for his guidance and help and to Thibaut Damour for enlightening discussions.

REFERENCES AND NOTES

[1] B. Carter, Phys. Rev. 174 (1968) 1559.
[2] D. Christodoulou and R. Ruffini, Phys. Rev. D4 (1971) 3552.
[3] G. Denardo and R. Ruffini, Phys. Lett. 45B (1973) 259.
[4] G. Denardo, L. Hively and R. Ruffini, Phys. Lett. 50B (1974) 270.
[5] R. Penrose and R. Floyd, Nature 229 (1971) 193.
[6] R. Penrose, Rev. Nuovo Cimento 1 (1969) 252.
[7] N. Deruelle and R. Ruffini, Phys. Lett. 52B (1974) 437.
[8] T. Nakamura and H. Sato, Phys. Lett. in press.
[9] T. Damour and R. Ruffini, Phys. Rev. Lett. 35 (1975) 463.
[10] See e.g. B. Carter, Commun. Math. Phys. 10 (1968) 280.
D. Brill, P. M. Chrzanowski, C. M. Pereira, E. D. Fackerell and J. R. Ipser, Phys. Rev. D5 (1972) 1913.
[11] Handbook of mathematical functions M. Abramowitz and I. A. Stegun, National Bureau of Standards (1964).
[12] T. Damour, Nuovo Cimento Letters 12 (1975) 315.
[13] P. Klein, Zeit. fur Phys. 53 (1929) 157.
[14] F. Sauter, Zeit. fur Phys. 69 (1931) 742.
[15] H. Euler and B. Kockel, Naturwissench. 23 (1935) 246.
[16] W. Heisenberg and H. Euler, Zeit. fur Phys. 98 (1935) 714.
[17] W. Pauli, The theory of positron and related topics, lectures notes by B. Hoffman unpublished, Princeton 1935.
[18] T. Damour: These proceedings.
[19] S. Fulling King's college preprints London 1975.
[20] G. Gibbons: These proceedings.
[21] H. Rumpf in these proceedings.
[22] H. Feshbach and F. Villars, Rev. Mod. Phys. 30 (1958) 24.
[23] E. Corinaldesi and F. Strocchi "Rel.Wave Mech." North-Holland,Amsterdam('63).
[24] R. Ruffini in "Les Houches Lectures" 1972 ed. by de Witt.
[25] D. Bohm "Quantum Theory" Prentice Hall N. Y. 1951.
[26] See e.g. G. Gibbons in these proceedings for a review as well as T. Damour in these proceedings.
[27] S. Hawking, Nature 248 (1974) 30.
[28] S. Hawking, Commun. Math. Phys. 43 (1975) 199.
[29] D. Christodoulou and R. Ruffini "On the electrodynamics of collapsed objects" in "Black Holes" B. and C. De Witt ed. Gordon and Breach 1973.
[30] B. Carter in Black Holes, B. and C. de Witt 1973.
[31] P. M. Morse and H. Feschbach "Methods of theoretical physics" McGraw Hill 1953.
[32] G. Gamow Zeit. Fur Phys. 51 (1928) 204.
G. Gamow "Structure of atomic nuclei and nuclear transformations" Clarendon Press Oxford (1937).
There is however a difference between the Gamow resonances and those considered here since the leakage occurs towards the horizon instead of towards infinity.
[33] W. T. Zaumen Nature 247 (1974) 530. See also Zaumen in these proceedings.
[34] J. A. Wheeler, "Transcending the law of conservation of leptons" Quaderno N. 157 Roma, Accademia Nazionale dei Lincei 1971.
[35] T. Damour, N. Deruelle and R. Ruffini submitted for publication.
[36] Even in the case l=0. See e.g. A. Messiah "Mecanique Quantique" Dunod 1962.
[37] N. Deruelle and R. Ruffini, Phys. Lett. 57B (1975) 248.
[38] R. A. Matzner, J. Math. Phys. 9 (1968) 163.
[39] S. Teukolsky and W. Press, Ap. J. 193 (1974) 443.
[40] Ya.B. Zel'dovich, Soviet Phys. JETP 35 N.5 (1972).
[41] A. A. Starobinsky and S. M. Churilov Soviet Phys. JETP 65 (1973) 3.

QUANTUM EFFECTS IN COSMOLOGY AND BLACK-AND-WHITE HOLE PHYSICS

A. A. Starobinsky

The Landau Institute for Theoretical Physics, Moscow, USSR

The paper presents some new results in the following questions: particle creation and vacuum polarization of quantum fields in strong varying background gravitational fields near the end of collapse and cosmological singularity, cosmological consequences of the particle creation effect, vacuum polarization due to complicated topology.

It is well known that electron-positron pairs can be created by strong background electromagnetic fields considered as classical ones. Recent theoretical investigations make clear that similar quantum effect is possible in classical gravitational fields too. Gravitational fields can create both charged and uncharged particles, among them ones with zero rest mass. It is also possible to speak about graviton production by a background gravitational field (i.e. a classical curved space-time) in case the wavelength of these gravitons is much less than the radius of curvature of the background space-time.

In cosmology this effect was first considered by Parker [1]. It appears (see papers [2,3]) that, firstly, the particle creation effect is usually much more significant than the effects due to quantization of the universe as a whole near cosmological singularity and, secondly, this effect is large near anisotropic singularity and small (or even absent at all) for all particles except gravitons near isotropic (Friedmann) one.

If space-time belongs to the Bianchi I type, i.e., its metric is of the form:

$$ds^2 = dt^2 - a^2(t)(dx^1)^2 - b^2(t)(dx^2)^2 - d^2(t)(dx^3)^2 \tag{1}$$

and $a \sim |t|^{q_1}$, $b \sim |t|^{q_2}$, $d \sim |t|^{q_3}$, when $t \to 0$ ($t < 0$ - the case of collapse is considered), where all $q_\alpha < 1$, then the vacuum expectation value of the energy-momentum tensor of a quantum field has the following form in the region

$$|t| \ll \frac{\hbar}{mc^2}$$

(with the logarithmic accuracy):

$$E \equiv \langle T_0^0 \rangle = \phi_0 \hbar\, c^{-3}\, t^{-4}\, \ln\left|\frac{t}{t_0}\right| , \tag{2}$$

$$P_1 \equiv -\langle T_1^1 \rangle = \phi_1 \hbar\, c^{-3}\, t^{-4} \ln\left|\frac{t}{t_0}\right|$$

and so on, with the additional conditions imposed on the coefficients ϕ_0, ϕ_α:

$$\phi_0 = \sum_{\alpha=1}^{3} \phi_\alpha, \tag{3}$$

$$(4 - \sum_{\alpha=1}^{3} q_\alpha)\,\phi_0 = \sum_{\alpha=1}^{3} q_\alpha \phi_\alpha \tag{3}$$

When $m \neq 0$, $t_0 \sim \hbar/mc^2$; in case $m = 0$, t_0 is determined by the moment of the change-over from isotropic collapse to anisotropic one. ϕ_0 and ϕ_α $(\alpha = 1,2,3)$ do not depend on m, so the conformal covariance of massless matter quantum fields can be used to represent ϕ_1 in the form:

$$\phi_1 = (1-q_1)^5\, \Phi\left(\frac{1-q_2}{1-q_1}, \frac{1-q_3}{1-q_1}\right). \tag{4}$$

The expressions for ϕ_2 and ϕ_3 are obtained from (4) by permutation of indices, the function Φ being the same, and $\phi_0 = \sum_{\alpha=1}^{3} \phi_\alpha$. Here $\Phi(y,z) = \Phi(z,y)$ is the function whose type does not depend on q_α. Besides, $\Phi(1,1) = \Phi(0,1) = 0$; this is the consequence of the absence of massless particle creation in the isotropic case and the degenerate Kasner one with exponents (1,0,0). By the use of the perturbation theory we can investigate the behavior of Φ in the vicinity of these two points. The results for the conformally-invariant scalar field are:

$$\left.\frac{\partial\Phi(y,z)}{\partial y}\right|_{y=z=1} = \left.\frac{\partial\Phi(y,z)}{\partial z}\right|_{y=z=1} = -\frac{1}{240\pi^2}, \tag{5}$$

$$\left.\frac{\partial\Phi(0,z)}{\partial z}\right|_{z=1} = 0, \quad \left.\frac{\partial\Phi(y,1)}{\partial y}\right|_{y=0} = \frac{1}{360\pi^2}$$

In case of neutrinos these expressions should be multiplied by 3, in case of photons, by 12.

The investigation of cosmological consequences of the particle creation effect (see papers [2,4,5]) gave the following results.

1) Particles are so intensively produced near anisotropic cosmological singularity that the amount of matter in the universe proves to be sufficient to make the subsequent expansion isotropic soon after the moment $t = t_{p\ell} \simeq 10^{-44}$ sec. (time is measured from the singularity).

2) The influence of particles produced near singularity on the properties of white holes results in the impossibility of long delay of cosmological white hole explosions. White holes either explode with practically no delay in the period $t < r_g/c$, where r_g is their gravitational radius, or do not explode at all.

3) If there exists a rotational perturbation in the universe near singularity and its characteristic scale is much more than the dimension of the horizon at the moment $t \sim t_{p\ell}$ (that corresponds to the scale $L \gg 10^{-1}$ cm in the present epoch) then this perturbation is significantly reduced soon after $t \sim t_{p\ell}$ due to particle creation. The main point is that the matter consisting of created particles does not move on the average and its energy density is

much more than the energy density of the initially present matter producing given rotational perturbation through the Einstein equations. The whole angular momentum per unit volume is approximately conserved when the initially present matter and the created one are intermixed, but the angular momentum per particle and the value of the rotational velocity decrease drastically during this process. The rotational velocity remained in the subsequent approximately quasi-isotropic stage of the universe expansion is in no way sufficient for the explanation of galaxies rotation within the limits of the vortex theory.

The nonzero vacuum polarization of matter quantum fields can be not only due to a curvature but also to a non-euclidean topology of a space. For example, in case of the closed Friedman model we obtain (R is the radius of curvature).

$$E = 3P = \frac{1}{480\pi^2 R^4} \tag{6}$$

for the conformally invariant massless scalar particles ($E = \langle T_0^0 \rangle$, $P = -\langle T_1^1 \rangle = -\langle T_2^2 \rangle = -\langle T_3^3 \rangle$). In case of photons the expression (6) should be multiplied by 22. This polarization is evidently due to the non-euclidean topology because wave equations of massless particles (except gravitons) are conformally covariant and the spherical Friedmann metric is conformally flat. The other example is the vacuum polarization of the scalar field in flat space-time with the 3-torus space topology with the scales a, b, c. As in the previous case, E and P_α are constant along the whole space. If a = b = c, then

$$E = 3P \simeq -\frac{8\pi}{a^4} \quad 3.33\text{x}10^{-2}. \tag{7}$$

Let us consider the more interesting example where E and P_α do depend on spatial coordinates: flat space-time with the identification of all points obtained from each initial point by rotation through the angle $\frac{2\pi}{n}$, where n is the integer, n > 2, about the Z axis. Then we find in case of the scalar massless particles that

$$E = -\frac{1}{\rho^4}\frac{1}{96\pi^2}\sum_{\alpha=1}^{n-1}\frac{3-2\sin^2(\pi\alpha/n)}{\sin^4(\pi\alpha/n)} \tag{8}$$

$$P_z = P_\rho = -E, \quad P_\phi = 3E,$$

where $\rho^2 = x^2+y^2$.

REFERENCES

[1] L. Parker, Phys. Rev. Lett. 21, 562, 1968; Phys. Rev. 183, 1057, 1969.
[2] Ya. B. Zeldovich. Pisma V Zh. E. T. F., 12, 443, 1970. [JETP Lett. 12, 307, 1970].
[3] Ya. B. Zeldovich, A. A. Starobinsky. Zh. E. T. F. 61, 2161, 1971 [Sov. Phys.-JETP 34, 1159, 1972].
[4] V. N. Lukash, A. A. Starobinsky. Zh. E.T.F. 66, 1515, 1974.
[5] Ya. B. Zeldovich, I. D. Novikov, A. A. Starobinsky. Zh.E.T.F. 66, 1897, 1974.

A REMARK ON RADIATION DAMPING*

P. C. Aichelburg

Institute for Theoretical Physics
University of Vienna, Austria

Professor Penrose in his talk touched upon the "arrow-of-time" problem and speculated about the relevance of non-time-symmetric physics in the early expanding universe. I would like to speak briefly on a subject somewhat related to his: The phenomenon of radiation damping.

It is often said that electrodynamics, because of its time symmetry, does not necessarily predict the irreversible features of radiation damping as observed in nature. The argument is that one has to impose boundary conditions, which do not follow from the theory, which lead to radiation damping. Therefore, people searched for a "source of time-asymmetry" and found it within thermodynamics [1] or non-static cosmological models [2,3,4]. However, I wish to stress here that a field theory in flat space-time, viewed as an initial value problem for particle and field, is sufficient to deduce the dissipative features of radiation phenomena.

Consider the following system of an oscillator $Q(t)$ coupled to a scalar field Φ:

$$\ddot{Q}(t) + \omega_0^2 Q(t) = \lambda \int d^3x \rho(|\vec{x}|)\ \Phi\ (\vec{x},t) \tag{1}$$

$$\Box\Phi(x,t) = \lambda\rho(|\vec{x}|)Q(t) \tag{2}$$

here λ is a coupling constant, $\rho(|\vec{x}|)$ the "charge" distribution and $\cdot \equiv d/dt$. This model parallels electrodynamics in the dipole approximation and was first considered by Schwabl and Thirring [5]. It can be solved exactly in terms of the initial values $Q(0)$, $\dot{Q}(0)$, $\Phi(\vec{x},0)$ and $\dot{\Phi}(\vec{x},0)$. For simplicity I give here the solution in the "point-limit," i.e. $\rho(\vec{x}) = \delta^3(\vec{x})$, for $t > 0$:

$$Q(t) = e^{-\Gamma t}\{[\cos\omega t - \frac{\Gamma}{\omega}\sin\omega t]Q(0) + \frac{\sin\omega t}{\omega}\dot{Q}(0)\} + \tag{3}$$

$$+ \lambda\int_0^t dr\{\dot{G}(t-r)\psi(r) + G(t-r)\ \chi\ (r)\}$$

$$\Phi(\vec{x},t) = \frac{\lambda}{4\pi}\Theta(t-r)\ \frac{Q(t-r)}{r} + \Phi_H(\vec{x},t) \tag{4}$$

where $G(t) = e^{-\Gamma t}\,\frac{\sin\omega t}{\omega}$, $\Gamma = \lambda^2/8\pi$, $|\vec{x}| \equiv r$, $\omega = (\bar{\omega}^2 - \Gamma^2)^{\frac{1}{2}}$, and $\bar{\omega}^2 = \omega_0^2 - $ "∞" is the renormalized spring constant; the s-wave amplitude of the field at $t = 0$ is given by

$$\psi(r) = \frac{r}{4\pi}\int d\Omega\ \Phi(\vec{x},0)\,,\ \chi\ (r) = \frac{r}{4\pi}\int d\Omega\ \dot{\Phi}(\vec{x},0)$$

and Φ_H is the homogenous solution to the initial values $\Phi(\vec{x},0)$ and $\dot{\Phi}(\vec{x},0)$.

---

*talk based on a paper by P. C. Aichelburg and R. Beig, "Radiation Damping as an Initial Value Problem," Univ. Vienna preprint R 1975/3.

If one requires the following asymptotic behaviour for the field at the initial surface

$$\Phi(\vec{x},0) = 0(\frac{1}{r}) \;, \quad \vec{\nabla}\Phi(\vec{x},0) = 0(\frac{1}{r^2}) \,, \quad \dot{\Phi}(\vec{x},0) = 0(\frac{1}{r^2}) \tag{5}$$

which is necessary for the field energy to be finite, then one can show for the solution (3) (but also in the general case) that

$$\lim_{t\to\infty} E_Q = 0 \;, \tag{6}$$

where $E_Q = \frac{1}{2}[\dot{Q}^2 + \bar{\omega}^2 Q^2]$ is the energy of the oscillator. Therefore, the dissipative behaviour is guaranteed by the physical requirement of finite energy (more precisely by conditions (5)) for the initial data.

I next discuss the role of boundary conditions and their connection with the initial values. Defining the asymptotic fields by

$$\Phi^{\substack{in \\ out}}(\vec{x},t) = \Phi(\vec{x},t) - \frac{\lambda}{4\pi}\frac{Q(t\mp r)}{r} \tag{7}$$

let us consider first the condition $\Phi^{out} = 0$. Then

$$\Phi(\vec{x},0) = \frac{\lambda}{4\pi}\frac{Q(r)}{r} \quad , \quad \dot{\Phi}(\vec{x},0) = \frac{\lambda}{4\pi}\frac{\dot{Q}(r)}{r} \tag{8}$$

Because $Q(r)$ can be expressed in terms of the initial values, Eq. (8) imposes constraints among them. These constraints then lead to a solution for the oscillator with an exponentially growing amplitude:

$$Q(t) = e^{+\Gamma t}\{(\cos\omega t - \frac{\Gamma}{\omega}\sin\omega t)Q(0) + \frac{\sin\omega t}{\omega}\dot{Q}(0)\} \tag{9}$$

Because $\Phi(\vec{x},0) \sim e^{+\Gamma r}$, this solution violates our conditions (5) and is unacceptable.

Consider now $\Phi^{in} = 0$. Because in this case the constraints read

$$\Phi(\vec{x},0) = \frac{\lambda}{4\pi}\frac{Q(-r)}{r} \quad , \quad \dot{\Phi}(\vec{x},0) = \frac{\lambda}{4\pi}\frac{\dot{Q}(-r)}{r} \;, \tag{10}$$

we have to know the dynamics of the system for times $t < 0$. This makes it necessary to distinguish between <u>closed</u> and <u>open</u> systems.

In a <u>closed system</u>, Eqs. (1) and (2) are valid for all t. Then, since the model is time symmetric, it immediately follows that, if the field satisfies conditions (5), then also

$$\lim_{t\to-\infty} E_Q = 0. \tag{11}$$

The boundary condition $\Phi^{in} = 0$ leads to damped oscillations for <u>all</u> t:

$$Q(t) = e^{-\Gamma t}\{[\cos\omega t + \frac{\Gamma}{\omega}\sin\omega t]Q(0) + \frac{\sin\omega t}{\omega}\dot{Q}(0)\} \tag{12}$$

Again, because $\Phi(\vec{x},0) \sim e^{\Gamma r}$ this solution violates conditions (5) and also Eq. (11).

In an <u>open system</u>, for $t > 0$ Eqs. (1) and (2) are satisfied, but for

$t < 0$ Eq. (2) is valid while $Q(t)$ is arbitrarily driven "by hand." The boundary condition $\Phi^{in} = 0$ determines $\Phi(\vec{x},0)$ and $\dot{\Phi}(\vec{x},0)$ via Eq. (10); these satisfy conditions (5) if $Q(t)$ is such that Eq. (11) is fulfilled. From this, damping follows for $t \to \infty$.

Thus we see that for the boundary condition $\Phi^{out} = 0$ no solutions with finite field energy exist. For $\Phi^{in} = 0$ physically acceptable solutions exist only if the system has been open in the past of some initial hypersurface.

REFERENCES

[1] J. A. Wheeler and R. P. Feynman, Rev. Mod. Phys. 17 (1945), 157.

[2] J. E. Hogarth, Proc. Roy. Soc. A 267 (1962), 365.

[3] F. Hoyle and J. V. Narlikar, "Action at a Distance in Physics and Cosmology," Freeman, San Francisco, 1974.

[4] D. W. Sciama, Proc. Roy. Soc. A 273 (1963), 484.

[5] F. Schwabl and W. Thirring, Erg. d. exakt. Naturw. 36 (1964), 219.

PARTICLE CREATION AND GEOMETRY

P. C. W. Davies

Department of Mathematics, King's College, London

The disturbance of the vacuum state of quantum fields by geometrical effects has assumed great importance in recent years by affording a description of phenomena such as particle production by (classical) gravitational fields, which may well be crucial for properly understanding black holes and various cosmological models.

So far, only one experiment [1] has been capable of detecting a geometrical disturbance of the vacuum--the famous electromagnetic Casimir force between two parallel reflecting plates (mirrors). This effect, which arises because the presence of the plates modifies the geometrical nature of the electromagnetic boundary conditions, may be calculated by evaluating the vacuum expectation value of the energy-momentum tensor $T_{\mu\nu}$ as a formally divergent mode sum [2].

$$<T_{\mu\nu}> = \sum_i T_{\mu\nu} (\phi_i , \phi_i^*) . \quad (1)$$

The physically meaningful part may be extracted from (1) by introducing a cut-off in the summation at some energy, and developing the answer as a power series in the cut-off parameter. The divergent parts are then discarded, the remaining finite part being identified with the Casimir effect. Mathematically this 'regularisation' procedure is equivalent to evaluating ϕ_i and ϕ_i^* on the right of (1) at slightly separated space-time points. It turns out that in flat space-time the finite results are independent of the direction of point-splitting and the coordinate system used provided that the splitting is carried out along geodesics, and the derivatives which occur in $T_{\mu\nu}$ are parallel transported to the displaced points.

If the plates are allowed to move, a new phenomenon arises. Off-diagonal terms appear in $<T_{\mu\nu}>$ corresponding to an energy flux attributable to creation of particle pairs out of the vacuum. In the case of a massless scalar field in two dimensions $<T_{\mu\nu}>$ may be evaluated exactly in the vacuum associated with an initially static system, by applying a conformal transformation which maps the space-time (t,x) between the moving plates onto the static slab $0 < x < 1$:

$$t \pm x \rightarrow R(t \pm x) \quad (2)$$

with an appropriate function R. It then follows that the finite part of $<T_{\mu\nu}>$ is given by

$$<T_{00}> = <T_{11}> = - [f(u) + f(v)] \quad (3)$$

$$<T_{10}> = <T_{01}> = f(u) - f(v) \quad (4)$$

where

$$f = (24\pi)^{-1}\left[\frac{R'''}{R'} - \frac{3}{2}\left(\frac{R''}{R'}\right)^2 + \frac{1}{2}\pi^2(R')^2\right]. \tag{5}$$

The first two terms in the square brackets of equation (5) represent the effect of the motion of the plates in producing a flux of radiation (4) between the plates, whilst the $(R')^2$ term reduces in the static case to the usual Casimir effect.

Similar results apply for the even simpler case of a single moving boundary or mirror. The conformal transformation used to map the mirror trajectory $x = z(t)$ onto a static boundary is

$$\begin{aligned} t + x &\to t + x \\ t - x &\to p(t-x) \end{aligned} \tag{6}$$

with p determined by the trajectory z(t) through the equations

$$\begin{aligned} \tau_u - z(\tau_u) &= u \equiv t - x \\ p(u) &= 2\tau_u - u \, . \end{aligned} \tag{7}$$

The resulting finite part of $\langle T_{\mu\nu}\rangle$ to the right of the moving mirror is given by

$$\langle T_{00}\rangle = \langle T_{11}\rangle = -\langle T_{01}\rangle = -\langle T_{10}\rangle$$

$$= -(24\pi)^{-1}\left[\frac{p'''}{p'} - \frac{3}{2}\left(\frac{p''}{p'}\right)^2\right] \tag{8}$$

and is a function of retarded null rays (t - x) alone. Note that $\langle T_{\mu\nu}\rangle$ in both (4) and (8) is traceless and satisfies the covariant conservation condition $\nabla_\mu T^{\mu\nu} = 0$.

The physical interpretation of (8) is very transparent. The motion of the mirror causes the production of radiation at its surface, which then propagates away along retarded null rays. The energy radiated is positive (negative) whenever the mirror acceleration is decreasing (increasing). The total energy radiated between two periods of uniform motion is, however, always positive.

One example of interest occurs when the mirror accelerates uniformly. In this case no radiation is produced. Uniform acceleration has been considered before in the context of a static universe (Rindler space) [2,3]. Surprisingly, the vacuum state in these considerations differs from the one being used here--there are <u>two</u> stable vacuum states above a uniformly accelerating mirror. Although both predict no radiation flux, the static vacuum contains a cloud of negative energy which falls off as the inverse square of the proper distance from the mirror.

Another example of interest is the asymptotically null trajectory $z(t) = -\log\cosh t$ for $t > 0$. Equation (8) predicts that a mirror moving along this trajectory will emit a flux of radiation to the right which settles down asymptotically to a constant rate. This is one of the few cases where an expression for the Bogolubov transformation between initial and final vacuum states may be evaluated directly. The spectrum of radiation may then be obtained in this steady-state region of constant flux; it turns out to be a Planck (thermal radiation) spectrum, meaning that the mirror shines with a characteristic temperature. This example provides a close analogue of the exploding black hole phenomenon discovered by Hawking [4]. There is a latest advanced time for which a null ray may reflect from the surface of the mirror out into the right hand region (see figure). This closely corresponds to the formation of a black hole horizon which traps all incoming null rays after a certain advanced time, preventing them from re-emerging from the black hole. This sudden 'switch-off' gives rise to the particle production. In fact, the calculation of the Bogolubov transformation is almost identical in both cases.

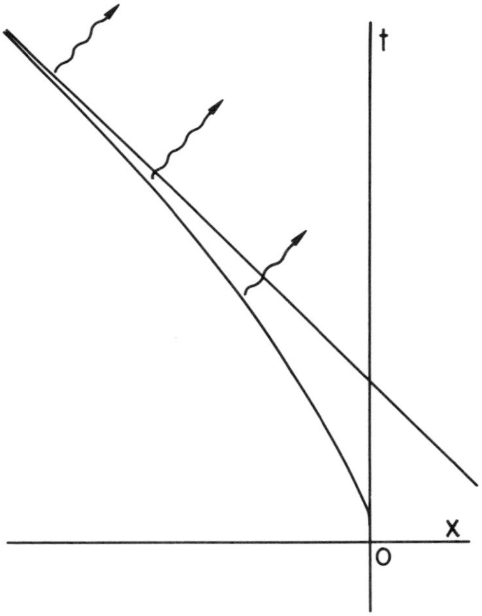

All the results described here have been generalised to arbitrary, curved two-dimensional space-times, both with and without mirrors present [5]. The utility of these investigations rests on the transparent nature of the results, and the fact that they represent the simplest models which still contain many non-trivial features that occur in more sophisticated systems. The fact that any radiation production occurs at all is of considerable interest, because the models considered are all conformally static and the massless scalar wave equation in two dimensions is invariant under conformal transformations. Thus, although the classical $T_{\mu\nu}$ is a conformally invariant quantity, the quantum $<T_{\mu\nu}>$ is not.

The work described here has been carried out in collaboration with Dr. Stephen Fulling, and full details will be published elsewhere [6].

REFERENCES

[1] See for example D. Tabor and R. H. S. Winterton, Proc. Roy. Soc. A. 312, 435 (1969).

[2] B. S. De Witt, Phys. Reports, 19, 295-357 (1975).

[3] S. A. Fulling, Phys. Rev. D, 7, 2850 (1973).

[4] S. W. Hawking, Comm. Math. Phys. 43, 199-220 (1975).

[5] P. C. W. Davies, S. A. Fulling and W. G. Unruh, 'Energy-Momentum Tensor Near An Evaporating Black Hole' submitted to Phys. Rev. (1975).

[6] S. A. Fulling and P. C. W. Davies, 'Radiation From a Moving Mirror in Two-Dimensional Space-Time: Conformal Anomoly,' submitted to Proc. Roy. Soc. (1975).

✓

Pair Production by Photons in a Constant Electromagnetic Field

Lincoln R. Davis*

Joseph Henry Physical Laboratories, Princeton University, Princeton, N. J. U.S.A.

The production of e^+e^- pairs by incident photons in an external field could polarize light passing by a charged and rotating black hole, tend to neutralize the charge, and extract energy from the black hole. This process is here treated as a problem in flatspace Quantum Electrodynamics, using the methods of Schwinger [1] and of Tsai and Erber [2], who considered the case of a pure magnetic field. Their results are extended to the general case of a field which is slowly varying in spacetime compared to a compton wavelength.

Vacuum polarization by stationary black holes was studied by Deruelle and Ruffini [3] and by Damour and Ruffini [4], who showed that a Kerr-Newman geometry creates pairs. To find the rate of production in the charged case, they applied the flat-space results of Heisenberg and Euler [5] and of Ref. [1] to the electromagnetic field in an orthonormal frame. This was equivalent to a WKB approximation in solving the wave equation in the curved background.

What Schwinger does in Ref. [1] is analogous to computing the sum of all one loop vacuum graphs with no external lines, no radiative corrections, and any number of external-field insertions. In fact, the amplitude for pair creation by a constant external field would vanish to all orders in perturbation theory, essentially because such a field can transfer no momentum to any finite order. But by using the exact Dirac propagator in a classical external field $F_{\mu\nu}$ Schwinger gets an effective Lagrangian for vacuum polarization, of which the imaginary part represents real pair creation.

The same approach is used in Ref. [2] to compute the sum of all one-loop graphs with two external photon lines. This is done by adding a radiation field $f_{\mu\nu}$ to $F_{\mu\nu}$, and expanding the action to second order in f, still treating F exactly. This is done in Ref. [2] for a pure magnetic field B ($*F_{\mu\nu}F^{\mu\nu}=0$, $F_{\mu\nu}F^{\mu\nu} = B^2 > 0$).

For a general field F, one proceeds as follows. In Ref. [1], the bare Lagrangian density $\mathcal{L}$ is given as

$$\mathcal{L}(x) = \frac{1}{2} i \int_0^\infty ds\, s^{-1} e^{-im^2 s}\ \mathrm{tr} \langle x | U(s) | x \rangle$$

*Present address: Graduate School, Cambridge University, England.

$$U(s) = e^{-iHs}, \quad H = \not{\pi}^2 = \pi^2 - \frac{1}{2}\sigma_{\mu\nu}F^{\mu\nu} \tag{1}$$

Here $\pi^\mu = P^\mu - eA^\mu$, $F_{\mu\nu} = A_{\mu,\nu} - A_{\nu,\mu}$, and the notation $\langle x|U(s)|x\rangle$ means that H is to be regarded as the "Hamiltonian" of a quantized system, s as the "time", and the $|x\rangle$ as an orthonormal basis for the space of states of this system. The states $|x\rangle$ are eigenstates of the four-dimensional "position operator" $x : x|x'\rangle = x'|x'\rangle$, $\langle x''|x'\rangle = \delta(x''-x')$. This purely formal analogy serves as an excuse for some equally formal manipulations of "operators" that are familiar in elementary quantum mechanics.

If $F\mu\nu$ is replaced by $F\mu\nu + f\mu\nu$, then $H = H_0 + H'$, where H_0 is the constant-field Hamiltonian, and

$$H' = e\left[\pi a(x) + a(x)\pi + \frac{1}{2}\sigma f(x)\right] + e^2 a^2(x)$$

$$(f\mu\nu = a_{\nu,\mu} - a_{\mu,\nu}) \tag{2}$$

Defining $U_0(s) = \exp(-iH_0 s)$, $U(s) = \exp(-iHs)$, and $V(s) = U_0^{-1}(s)U(s)$, we have an integral equation for V, [1]

$$V(s) = 1 - i\int_0^s ds' \; U_0^{-1}(s)H' \, U_0(s')V(s') \tag{3}$$

Left-multiplying by $U_0(s)$, and using the group properties of U_0, we have the lowest Born approximation

$$U(s) = U_0(s) - \frac{i}{2}s\int_{-1}^{+1} dv \; U_0(\tfrac{1}{2}(1-v)s) \; H' \; U_0(\tfrac{1}{2}(1+v)s) \tag{4}$$

with the substitution $s' = \frac{1}{2}(1+v)s$.

To get the effective action $W^{(2)}$ arising from the 2nd order term in a, we have to compute

$$\int dx \, \mathrm{tr}\langle x|U(s) - U_0(s)|x\rangle \equiv \mathrm{Tr}\left[U(s) - U_0(s)\right] \; ; \tag{5}$$

i.e., the trace of $U(s) - U_0(s)$ over both spacetime and spinor indices. This is done using the representation [1]

$$\mathrm{Tr}\left[U(s) - U_0(s)\right] = -is\int_0^1 d\lambda \; \mathrm{Tr}\, H' e^{-i(H_0+\lambda H')s} \tag{6}$$

by scaling $H' \rightarrow \lambda H'$ in Eq. (4), and inserting this in the right hand side of (6). For $W^{(2)}$ we thus have [2]

$$W^{(2)} = \frac{-1}{4} ie^2 \int_0^\infty ds \; se^{-im^2 s}\left[I_a + I_b\right] \tag{7}$$

where

$$I_a = (2i/s) \mathrm{Tr}\, U_0(s) a^2 \tag{8}$$

$$I_b = \int_{-1}^{+1} \frac{dv}{2} \mathrm{Tr}\, \{ U_0(\frac{s}{2}(1-v)) \left[\pi a + a\pi + \frac{1}{2}\sigma f\right] \times U_0(\frac{s}{2}(1+v)) \left[\pi a + a\pi + \frac{1}{2}\sigma f\right] \} . \tag{9}$$

In order to compute eq. (7) one takes the Fourier transform of a and works in a Lorentz frame where the external fields $\underset{\sim}{E}$, $\underset{\sim}{B}$ are parallel [4-6]. After renormalization of the charge [2] we have the finite result.

$$W^{(2)} = \int dk\, a^{\mu}(-k) a^{\nu}(k) K_{\mu\nu}(k) \tag{10}$$

where the explicit expression of $K_{\mu\nu}$ in function of the invariant strengths of the magnetic and electric fields can be found in Ref. [6]. One checks that $W^{(2)}$ is gauge-invariant in the radiation field a : $k^{\mu} K_{\mu\nu}(k) = 0$.

The quantity of physical interest is the absorption coefficient $\kappa(\omega)$ of the electron vacuum for photons of energy ω, due to pair creation. This is related to Im $W^{(2)}$, as follows: with the external fields $\underset{\sim}{E}$, $\underset{\sim}{B}$ in the z-direction, and the photon momentum

$$k^{\mu} = (\omega, k \sin\theta, 0, k\cos\theta), \tag{11}$$

we may take as a polarization basis,

$$\varepsilon^{\mu}_{\parallel} = (0, -\cos\theta, 0, \sin\theta)$$

$$\varepsilon^{\mu}_{\perp} = (0, 0, 1, 0) . \tag{12}$$

The dispersion relation for a plane wave in either of these modes, in view of the constitutive equations

$$D_i = \partial\mathcal{L}/\partial E_i$$

$$H_i = -\partial\mathcal{L}/\partial B_i$$

as applied to the radiation field, is [2, 7]

$$k^2 = M_{\parallel,\perp} \tag{13}$$

where $M_{\parallel,\perp} = -2\varepsilon_{\parallel,\perp} K(k) \varepsilon_{\parallel,\perp}$. To compute these matrix elements of K, we use the approximation $k^2 \simeq 0$ or $k \simeq \omega$ in eq. (11).

This yield explicit formulae for $M_{\parallel,\perp}$ to be found in ref. [6].

The dispersion relation (13) then gives, in Eq. (11),

$k \simeq \omega - \frac{1}{2\omega} M_{\parallel,\perp}$, which yields the index of refraction

$$n_{\parallel,\perp}(\omega) = 1 - \frac{1}{2\omega^2} \mathrm{Re} M_{\parallel,\perp}(\omega) \tag{14}$$

and the absorption coefficient

$$\kappa_{\parallel,\perp}(\omega) = \frac{-1}{\omega} \mathrm{Im}\, M_{\parallel,\perp} . \tag{15}$$

REFERENCES

[1] J. Schwinger, Phys. Rev. 82, 664 (1951).

[2] Wu-Yang Tsai and T. Erber, Phys. Rev. D10, 492 (1974).

[3] N. Deruelle and R. Ruffini, Phys. Lett. 52B, 437 (1974).

[4] T. Damour and R. Ruffini, Phys. Rev. Lett. 35, 463 (1975).

[5] W. Heisenberg and H. Euler, Z. f. Phys. 98, 714 (1936).

[6] L. Davis, Princeton A. B. Thesis (unpublished), 1975.

[7] S. Adler, Ann. Phys. (N. Y.) 67, 599 (1971).

EFFECT OF ZERO POINT FLUCTUATIONS ON THE FORMATION OF A BLACK HOLE

Ulrich H. Gerlach
Department of Mathematics, The Ohio State University, Columbus, Ohio, U. S. A.

The existence of a vacuum black hole formed from an already collapsed star is a sufficient condition for the existence of black body radiation emerging from the collapsed configuration [1]. That the existence of a vacuum black hole is a necessary condition for the emission of such black body radiation is however far from obvious. As a matter of fact, it is not true if the collapsing star is a spherically symmetric hollow shell star.

The purpose of my talk is to give an alternative picture of black body radiation emitted from a collapsing configuration, a picture, which is based on classical field theory applied to zero point radiation associated with a collapsing star [2].

Figure 1 gives the setting, It consists of standing waves evolving in a spherical cavity of very large radius a, in the center of which a hollow shell star undergoes spherically symmetric collapse. During the initial stages a typical standing wave $\Phi_{n\ell m}$ has a pattern roughly as depicted in the picture. Each standing wave consists of an ingoing and an outgoing radiation mode, $\Phi_{n\ell m} = \Phi^{outgoing}_{n\ell m} + \Phi^{ingoing}_{n\ell m}$. The finite size of the cavity allows us to treat the modes as discrete entities. We are focusing our attention on waves which are vacuum fluctuation modes. Each mode has therefore the energy of a quantum mechanical oscillator in its ground state, $\frac{1}{2}\hbar\omega_n$. Consequently, the ingoing and outgoing zero point fluctuation modes are

$$\Phi^{ingoing}_{n\ell m}{}^{outgoing} = \pm\sqrt{\frac{2\hbar}{a\omega_n}}\,\frac{1}{2r}\,\{\cos[\omega_n(t_{in}\pm r) + \beta_{n\ell m}]\}Y^m_\ell(\Theta,\phi). \tag{1}$$

Here t_{in} is the time associated with the flat geometry inside the star. It coincides with the exterior time during the early stages of collapse. The frequency of a mode is

$$\omega_n = \frac{\pi}{a}(n + \frac{\ell}{2}) . \tag{2}$$

The phase angle $\beta_{n\ell m} = \beta(n,\ell,m)$ is a random function of the integers n,ℓ, and m and thus expresses the lack of correlation between different modes.

The random phase angle $\beta_{n\ell m}$ together with Planck's constant $\hbar$ in the above expression for the vacuum fluctuation mode $\Phi_{n\ell m}$ constitute the only quantum mechanical input into our picture. The rest is merely classical field theory, i.e., solving the wave equation

$$\nabla^\mu\nabla_\mu\Psi = 0$$

in the setting of a collapsing hollow shell star during the late stages of collapse.

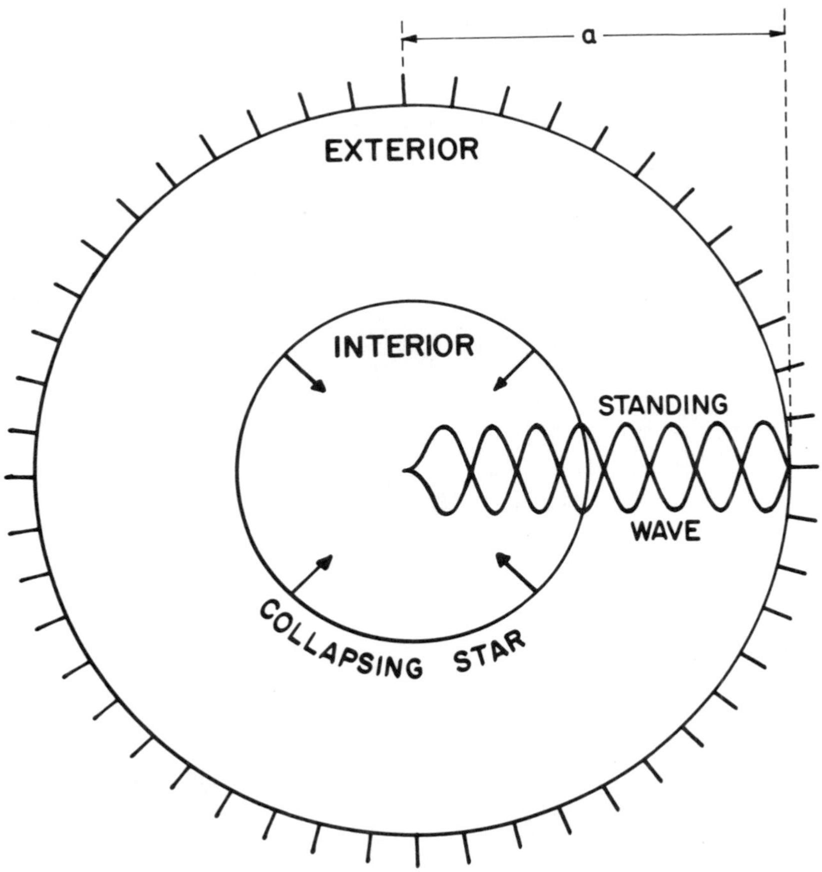

Figure 1. Setting for a standing wave evolving in a region occupied by a collapsing star. Inside a spherical cavity of a very large radius a, a hollow spherical shell star undergoes spherically symmetric gravitational collapse. Inside that star space is flat. Outside, during the early collapse stage, space is also flat. Thus a typical standing wave $\Phi_{n\ell m} = \Phi_{n\ell m}^{outgoing} + \Phi_{n\ell m}^{ingoing}$ has a pattern as indicated. The cavity radius a $\gg$ 2M; thus during the late stages of collapse the cavity wall will not reflect any manufactured radiation into the "incipient" black hole. The finite size of the cavity allows the modes $\Phi_{n\ell m}$ to be treated as discrete entities.

Figure 2 exhibits the salient features. The geometry in the interior is flat and in the exterior is the Schwarzschild geometry,

$$ds^2 = (1 - \frac{2M}{r})(-dt^2 + dr^{*2}) + r^2(d\Theta^2 + \sin^2\Theta d\phi^2), \tag{3}$$

given in terms of the usual Reege-Wheeler coordinates, better known as the tortoise coordinate,

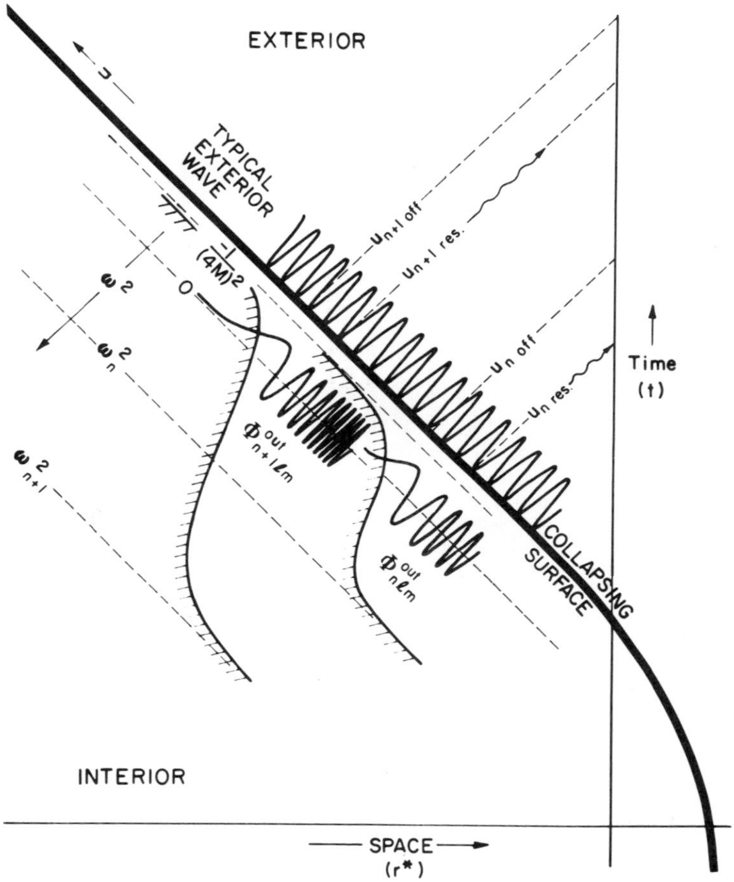

Figure 2. Production of black body radiation due to non-adiabatic red shifting of outgoing zero point radiation modes. The coordinates are the Schwarzschild time (t) and the tortoise coordinate (r*). The surface of the collapsing star asymptotically approaches, in these coordinates, an ingoing null-cone coordinatized by the retarded (Finkelstein) time $u = t - r^*$. Wave crest histories (null rays) are 45° lines in these coordinates. The outgoing part of two typical vacuum fluctuation modes, $\Phi^{out.}_{n\ell m}$ and $\Phi^{out.}_{n+1\ell m}$, undergo a red shift as indicated. The stiped curves are the respective instantaneous (squared) frequencies drawn as a function of retarded time u. As the mode $\Phi^{out.}_{n\ell m}$ crosses the star's surface it will cause the emission of photons into the exterior. Photons associated with a "typical exterior wave" will be produced by $\Phi^{out.}_{n\ell m}$ only in a neighboorhood $\Delta u = 8\pi M$ surrounding $u_{n\ res}(\omega)$, the time at which the instantaneous frequency ω of $\Phi^{out}_{n\ell m}$ agrees with that of the "typical exterior wave". An exterior wave with frequency say ω', lower (or higher) than the "typical" one in this picture will be in resonance with $\Phi^{out}_{n\ell m}$ in a neighborhood $\Delta u = 8\pi M$ surrounding a retarded time $u_{res}(\omega')$ later (or earlier) than the resonance time $u_{n\ res}(\omega)$. See Reference [2]. Thus the mode $\Phi^{out}_{n\ell m}$ gets amplified, high frequencies first, low frequencies later. Finally $\Phi_{n\ell m}$ gets effectively switched off at $u_{n\ off}$. Even before $\Phi^{out}_{n\ell m}$ has finished producing photons, the next zero point fluctuation mode $\Phi_{n+1\ell m}$ is getting amplified at a slightly later retarded time. The process for n+1 is however identical to the one for n.

$$r^* = r + 2M \quad (\frac{r}{2M} - 1). \tag{4}$$

These coordinates have a unique physical significance which is both local and global. This significance manifests itself in the scalar wave equation, which is

$$\{-\frac{\partial^2}{\partial t^2} + \frac{\partial^2}{\partial r^{*2}} - (\frac{\ell(\ell+1)}{r^2} + \frac{2M}{r^3})(1 - \frac{2M}{r})\}(r\ \Psi_{\ell m}) = 0 \tag{5}$$

The local significance lies in the fact that waves represented by $\Psi_{\ell m}$ propagate without amplification (i.e. parametric excitation) or attenuation. This follows from the constancy of the coefficients along the null rays. The waves are at most merely scattered by the time independent potential barrier. The global significance lies in the fact that these same coordinates go over into the usual Minkowskian coordinates (t,r) and thus are used by a distant observer to define the radiative energy flowing out of a black hole.

During the late stages of collapse the history of the star's surface is best parametrized by the retarded (Finkelstein) time

$$u = t - r^* \tag{6}$$

The wave field due to an outgoing vacuum fluctuation mode $\Phi^{\text{outgoing}}_{n\ell m}$ crosses the star's surface, where its field is

$$\Psi_{\ell m}(u) = \Phi^{\text{outgoing}}_{n\ell m} = \sqrt{\frac{2\hbar}{a\omega_n}}\ \frac{1}{2r}\ \{\cos[\exp \frac{u_{n\ \text{off}} - u}{4M} + \beta_{n\ell m}]\}Y^m_\ell(\Theta,\phi). \tag{7}$$

Figure 2 exhibits along the star's surface history the graphs of two typical fields due to two outgoing vacuum fluctuation modes, $\Phi^{\text{outgoing}}_{n\ell m}$ and $\Phi^{\text{outgoing}}_{n+1\ \ell m}$. Evidently these and all such fields undergo the well-known time dependent gravitational redshift [3]. The squared "instantaneous" frequency [2]

$$\begin{aligned}\omega^2(u) &= (\omega_n^2 - \frac{\ell(\ell+1)}{r^2})(1 - \frac{2M}{r})^2 - \frac{2M}{r^3}(1 - \frac{2M}{r}) - \frac{M^2}{r^4}\Bigg|_{\text{star's surface}} \\ &= (\frac{1}{4M})^2 \exp \frac{u_{n\ \text{off}} - u}{2M} - (\frac{1}{4M})^2 ,\end{aligned} \tag{8}$$

which is given by the two respective stiped graphs, refers to the frequency as seen by a distant observer. The retarded time $u_{n\ \text{off}}$, for which $\omega^2(u_{n\ \text{off}}) = 0$, separates the oscillatory from the exponential behavior of the field $\Phi^{\text{outgoing}}_{n\ell m}$ along the history of the star's surface. It evidently is

$$u_{n\ \text{off}} = 4M\ell n\ 4M\ \omega_n. \tag{9}$$

It constitutes the time at which the mode $\Phi^{\text{outgoing}}_{n\ell m}$ gets switched off from the view of a distant observer, and it evidently is, as it must be, later for higher frequencies ω_n. The corresponding Schwarzschild time is

$$t_{n\ \text{off}} = 2M\ell n\ 4M\ \omega_n + 2M \tag{10}$$

The most important feature of the red shifting process is that it is independent of how the respective mode $\Phi^{outgoing}_{n m}$ is being switched off:

$$\lim_{u\to\infty} \omega^2(u) = -\left(\frac{1}{4M}\right)^2 \tag{11a}$$

$$\frac{d\,\omega^2(u_{n\ off})}{du} = -\left(\frac{1}{2M}\right)^3 \tag{11b}$$

In other words, near the switch off time, the oscillatory behaviours of all modes $\Phi^{outgoing}_{n\ell m}$ to be switched off are identical. The behaviour is totally independent of the integers n,ℓ and m. The only difference in the behavior of the fields $\Phi_{n\ell m}$ is their dependence on the random phase $\beta_{n\ell m}$.

The outgoing energy spectrum due to the single zero point radiation mode $\Phi^{outgoing}_{n m}$ being amplified is

$$\frac{(\text{energy due to } \Phi_{n\ell m})}{(\text{unit area})(\text{unit frequency}}\ \frac{1}{M}\, d(M\omega)$$

$$= \frac{1}{4\pi r^2}\frac{\hbar/M}{a\omega_n}(M\omega)\left[\frac{1}{2} + \frac{1}{\exp 8\pi\omega M - 1} + \frac{1}{2}\ \frac{\cos\beta_{n\ell m}}{\sin h 4\pi\omega M}\right]d(M\omega). \tag{12}$$

The first term refers to the zero point energy associated with $\Phi^{outgoing}_{n\ell m}$. The second term is the black body energy spectrum. The third term is the thermal fluctuation in the energy; it goes to zero when averaged over several modes.

The rate at which modes get amplified and then switched off by the collapsing star is readily determined. From Eq. 2 and Eq. 10 we have

$$\frac{\Delta\omega_n}{\omega_n} = \frac{\pi}{a\omega_n}\Delta n \tag{13}$$

and

$$\Delta t = 2M\frac{\Delta\omega_n}{\omega_n}, \tag{14}$$

from which we have

$$\frac{(\#\ \text{amplified modes})}{(\text{unit time})} = \frac{(\#\ \text{switched off modes})}{(\text{unit time})} = \frac{dn}{dt} = \frac{a\omega_n}{2\pi}\frac{1}{M}. \tag{15}$$

Multiply this rate by the energy, Eq. 12, average over many modes, and obtain thereby the spectral power flux from an incipient black hole

$$\frac{(\text{energy})}{(\text{unit area})(\text{unit frequency})(\text{unit time})}\ \frac{1}{M}\, d(M\omega)$$

$$= \frac{\hbar/M}{2\pi M}(M\omega)\left[\frac{1}{2} + \frac{1}{\exp 8\pi\omega M - 1}\right]d(M\omega) \tag{16}$$

which is independent of the radius a of the spherical cavity.

There is a non-trivial technical remark that should be made about this expression. The total power flux as computed by integrating Eq. 16 over all frequencies gives evidently a divergent and hence contradictory (to the facts of reality) result. The reason, of course, is that we have omitted the zero point energy carried into the collapsing configuration by the ingoing modes $\Phi^{\text{ingoing}}_{n\ell m}$, Eq. 1. Their inclusion into the spectral power flux ledger will give a perfectly finite total power flux. In other words, the spectral power flux due to the ingoing zero point radiation modes $\Phi^{\text{ingoing}}_{n\ell m}$ cancels precisely, frequency by frequency [2] the first term in Eq. 16. The net spectral power flux from an incipient black hole is therefore

$$\frac{(\text{net energy})}{(\text{unit area})(\text{unit frequency})(\text{unit time})}\frac{1}{M}\,d(M\omega)$$
$$= \frac{\hbar/M}{2\pi M}(M\omega)\frac{1}{\exp 8\pi\omega M-1}\,d(M\omega). \tag{17}$$

Integrating this difference over all those modes that actually surmount the centrifugal barrier [2] in Eq. 5 gives the net rate of energy loss by an incipient black hole,

$$(\text{energy})/(\text{time}) = (27/2\pi . 15 . 8^5) . \hbar/M^2$$

In summary, let us list the most important features and conclusions that are based on classical field theory applied to zero point fluctuations in vacuum.

1. The outgoing vacuum fluctuation mode $\Phi^{\text{outgoing}}_{n\ell m}$ gives rise to a finite black body radiation packet at the star's surface--a packet--which is created in a retarded time interval preceding the switch off time $u_{n\ \text{off}}$.

Indeed, a simple Fourier analysis of a red shifted fluctuation mode $\Phi^{\text{outgoing}}_{n\ell m}$ in terms of orthonormal wave packets [2] shows that wave packets of frequency ω are produced predominantly in a retarded time interval $\Delta u = 8\pi M$ surrounding the time of "mutual resonance"

$$u_{n\ \text{res}}(\omega) = u_{n\ \text{off}} - 4M\ \ell n\ 4M\omega\ .$$

The width Δu during which these wave packets are produced depends only on the black hole descriptor M. Furthermore, the production process is present only to the extent that the W.K.B. approximation for waves propagating across the star's surface is inapplicable. An attempt to determine the wave field emerging from the star within the W.K.B. approximation only would yield the unamplified zero point radiation wave field only. The amplified spectral components arising from $\Phi^{\text{out}}_{\ell mn}$ have the relative intensities of the Planck spectrum (plus its fluctuation spectrum). They are produced before $\Phi^{\text{out}}_{\ell mn}$ gets switched off.

In view of the fact that a black body radiation packet due to $\Phi_{n\ell m}$ is confined to a retarded time interval preceding $u = u_{n\ \text{off}}$, and hence is bounded away from $u = \infty (r = 2M)$, it is unreasonable in our picture to associate with any emitted particle (photon) its anti-particle (again a photon) tunneling through the $r = 2M$ surface.

2. The radiation packets due to the various outgoing zero point radiation modes are statistically identical [2]; they are, in other words, independent of the integers n, ℓ, and m that characterize the vacuum fluctuation modes $\Phi_{n\ell m}$.

3. A vacuum black hole formed by a collapsed hollow shell star is not the proper setting for the production of black body radiation. Indeed if the star did collapse through its instantaneous r = 2M surface, an unlimited number of statistically identical radiation packets would have to be produced by the surface of the star. Energy conservation however require that the requisite (infinite) energy be furnished by the collapsing shell itself. It follows that the star will settle into an evolutionary pattern which is characterized by the collapsing configuration continually radiating away its irreducible mass, but the surface of the collapsing star never crossing its instantaneous r = 2M surface. It is most probable that such an evolving configuration is characterized by a Vaidya geometry [5] matched [6] onto a hollow flat space interior.

4. The statistical fluctuations in the emitted radiation are thermal fluctuations, i.e. a black hole is a black body [4].

5. Black hole entropy = k {#of emitted black body radiation packets} Indeed, the entropy of a black hole is known to be

$$S_{b.h.} = 4\pi k M^2/L_w^2. \tag{19}$$

Here k is Bolzman's constant and $L_w^2 = \hbar G/c^3$ is the squared Wheeler length. On the other hand, from Eqs. 2, 15, and 18 one finds [2] that if n_M is called the number of standing waves $\Phi_{n\ell m}$ (= number of black body radiation packets) responsible for causing the total evaporation of an incipient black hole with initial mass M, then

$$n_M = 3.53 \quad 10^3\ M^2/L_w^2.$$

In other words, modulo a factor ~ 280 we have Eq. 19.

6. The overall picture of an incipient black hole that the previous five conclusions point towards is that it constitutes a (macroscopic) system of many (microscopic) degrees of freedom [7,8] whose state of excitation is best described in terms of temperature, and which is making transitions towards states of lower irreducible mass. Alternatively or perhaps rather in addition, an incipient black hole constitutes a macroscopic system dissipating its energy against a viscuous medium: the surface of the collapsing star is "rubbing" against the vacuum fluctuations of space. The excitation of these fluctuations by the collapsing surface of the star reminds one of a bead dissipating its energy as it falls in a "wet" fluid. The energy dissipation in this analogy occurs in a layer at the surface of the bead. The dissipating mechanism itself is a microscopic process involving the fluctuating features (Brownian motion) of the fluid and is an example of the dissipation fluctuation theorem of Einstein [9], Nyquist, Callen and Welton [10]. The formulation and application of this theorem to the context of an incipient

black hole, if possible, would do much to illuminate on that part of its nature responsible for the emission of black body radiation.

Questions about space-time singularities are not applicable to an incipient black hole. However, by the time the irreducible black hole mass approaches the Wheeler mass (= $\sqrt{\hbar c/G}$) its statistical thermodynamical features will be replaced by features that have yet to be identified. In other words, the ultimate dilemma posed by a collapsing star is the question "What is the ground state (if any) of an incipient black hole?"

REFERENCES

[1] S. W. Hawking, "Particle Creation by Black Holes," preprint (1974).

[2] U. Gerlach, "The Mechanism of Black Body Radiation from an Incipient Black Hole," Phys. Rev. D (1976) (to appear).

[3] See, for example, W. L. Ames and K. S. Thorne, Ap. J. 151, 659 (1968), who consider the dynamical gravitational red shift together with the doppler shift from the surface of a collapsing star.

[4] This fact also follows from the nature of the density matrix found by R. M. Wald, "On Particle Creation by Black Holes" preprint (1975).

[5] See for example R. W. Lindquist, R. A. Schwartz, and C. W. Misner, Phys. Rev. 137, B1364 (1965).

[6] C. W. Misner, Phys. Rev. 137, B1360 (1965).

[7] U. Gerlach, "Why is a Black Hole Hot?" Talk delivered at the 1976 Washington Meeting of the American Physical Society; Bull. Am. Phys. Soc. (1976).

[8] J. D. Bekenstein, Lett. Nuovo Cimento 11, 467 (1974).

[9] A. Einstein, Phys. Z. 18, 121 (1917).

[10] H. Callen and T. A. Welton, Phys. Rev. 83, 34 (1951).

COVARIANT TREATMENT OF PARTICLE CREATION IN CURVED SPACE-TIME

H. Rumpf

Institut fur Theoretische Physik der Universitat Wien

ABSTRACT

We propose a particle definition in external electromagnetic and gravitational fields, using two Fock spaces ('in- and outgoing'), with the features: i) manifest gauge covariance, ii) coincidence of the vacuum persistence amplitude with that of Schwinger's formalism, iii) asymptotically quasiclassical modes.

It has been unknown, in general, how to define particles in curved space-time, even in a static one. For some very special situations [1], 'physical' modes and corresponding 'vacuum' states have been selected using the special features exhibited by the metrics considered. On the other hand, the physical relevance of the covariant particle definitions (i.e. definitions that work with 'generic' metrics) due to Lichnérowicz [2] and Nachtmann [3] is rather doubtful, since they exclude particle creation. (It can be shown that both coincide if the first one is unique). In this letter, we propose a definition which is covariant and strongly promises to be physical, too. Although it might be generalized to fields of other spin acted upon by external gauge fields of a different type, we present it here only for a scalar field Φ obeying the Klein-Gordon equation in presence of an external gravitational and/or electromagnetic field:

$$(\Box + \frac{R}{6} + m^2)\Phi = 0 \tag{1}$$

$$\Box := \frac{1}{\sqrt{-g}} (\partial_\mu + ie A_\mu)\sqrt{-g}\, g^{\mu\nu}(\partial_\nu + ie A_\nu). \tag{2}$$

(The inclusion of the term containing the curvature scalar R, assuring conformal covariance for $m^2 = 0$, is not essential in the following, but reflects our personal preference.) The operator $\Box + \frac{R}{6}$ is self-adjoint with respect to the scalar product

$$\langle f,g\rangle := \int d^4x\sqrt{-g}\, f^*(x)g(x); \tag{3}$$

hence the resolvent $(\Box + \frac{R}{6} + m^2)^{-1}$ exists for <u>complex</u> values of m^2, and we consider the boundary values of its integral kernel with respect to (3) when the real axis of the complex m^2-plane is approached from below and above. Denoted by $-K(x,y)$, $-\tilde{K}(x,y)$, they obey

$$(\Box_x + \frac{R(x)}{6} + m^2 - i0)K(x,y) = -\frac{1}{\sqrt{-g}}\delta^{(4)}(x-y) \tag{4}$$

$$K^A(x,y) = K^{-A}(y,x) \tag{5}$$

$$\tilde{K}(x,y) = K^*(y,x) \tag{6}$$

where in (5) the dependence on the potential $A_\mu(x)$ has been indicated. There are several reasons to consider K as the most natural generalization of the free Feynman propagator, so that we are led to define particle states in presence of the external field by using the propagation properties of K. We exhibit these by studying the action of K, $\Psi \to K*\Psi$, on the linear space H of solutions Ψ of the Klein-Gordon equation where, with Σ a spacelike hypersurface,

$$(K*\Psi)_\Sigma(x) := \int_\Sigma d\sigma'^\mu K(x,x')(\overleftrightarrow{\partial}'_\mu + 2ie\,A_\mu(x'))\Psi(x') \tag{7}$$

This action allows us to define two operators P↑, P↓ on H by

$$(P{\uparrow}\Psi)(x) := (K*\Psi)_\Sigma(x) \qquad x \text{ in the future of } \Sigma \tag{8}$$

$$(P{\downarrow}\Psi)(x) := -(K*\Psi)_\Sigma(x) \qquad x \text{ in the past of } \Sigma \tag{9}$$

In virtue of the conserved current, the choice of the spacelike hypersurface Σ is arbitrary except that x must be contained in its future or past, respectively. As a consequence of Stokes' theorem, one has

$$P{\uparrow} + P{\downarrow} = id_H \ , \tag{10}$$

and by an appropriate analytic continuation of the elements of H to the lower complex m^2 half plane one can see that P↑, P↓ are projection operators:

$$(P{\uparrow})^2 = P{\uparrow},\ (P{\downarrow})^2 = P{\downarrow},\ P{\uparrow}P{\downarrow} = 0 \tag{11}$$

Eqs. (11) state the desired propagation properties. Now the most direct application of them would be to define particle ('positive frequency') solutions as being propagated into the future by K, antiparticle ('negative frequency') solutions as being propagated to the past. However, such a definition excludes particle creation by the external field, just as it is excluded in refs. [2,3].

Indeed it was suggested by Menskii [4] that creation is possible only if P↑, P↓ are hypersurface-dependent ('vacuum instability') or not projectors ('vacuum degeneracy') due to some 'pathologies' tacitly suppressed in the constructions above. Of course, they might become of great importance in the gravitational case when global aspects like the appearance of horizons and singularities are to be taken into account, which will be the subject of further study. But neither phenomena occur in the by now familiar situations of a time-varying electric field or the Klein paradox, where creation is expected. Clearly, if one wants to allow for particle creation, one has not only to distinguish between particle and antiparticle modes, but also between in- and outgoing modes. We accomplish this by the additional introduction of the antipropagator $\tilde{K}$, defined in (6), together with projectors $\tilde{P}{\uparrow}$, $\tilde{P}{\downarrow}$ associated with it in analogy to (8), (9). Thus we define a space of outgoing (ingoing) particle/antiparticle modes $H^\pm(H_\pm)$ by

$$H^+ = P{\uparrow}H,\ H^- = \tilde{P}{\uparrow}H,\ H_+ = \tilde{P}{\downarrow}H,\ H_- = P{\downarrow}H \tag{12}$$

Particle creation occurs if

$$H^+ \neq H_+,\ H^- \neq H_- \tag{13}$$

(as a criterion we mention that in a pure gravitational field (A = 0) the real part of K then no longer vanishes for spacelike separations of the arguments). If the 'charge form' is defined by

$$(f,g) := \int d\sigma^{\mu} f^{*}(i\overleftrightarrow{\partial}_{\mu} - 2eA_{\mu})g, \tag{14}$$

the following relations may be derived ($\oplus$ denotes a direct (,)-orthogonal sum):

$$H = H^{+} \oplus H^{-} = H_{+} \oplus H_{-} \tag{15}$$

$$(\Psi,\Psi) \geq 0 \text{ for } \Psi \in H^{+} \text{ or } H_{+},\ (\Psi,\Psi) \leq 0 \text{ for } \Psi \in H^{-} \text{ or } H_{-} \tag{16}$$

These equations assure that particles and antiparticles get opposite signs of charge and that two Fock spaces can be built up with the in- and outgoing solutions in the standard manner. The ambiguity of the particle-antiparticle concept in external fields stems from the fact that there exist infinitely many decompositions of H obeying (15) and (16). The choice we have made depends on the employment of the additional scalar product (3), but in a way different from Nachtmann's [3].

We can now establish contact between our Fock space formulation of quantum field theory and the Schwinger-deWitt formalism [5,6,7] for the computation of the vacuum persistence amplitude. As shown in [7], the latter is completely determined by the Green's function

$$iK_{T} := \langle 0\ \text{out}|T(\Phi(x)\Phi^{\dagger}(y))|0\ \text{in}\rangle \;/\; \langle 0\ \text{out}|0\ \text{in}\rangle \tag{17}$$

In the Schwinger formalism, K_T is taken to be a solution of (4), which only a posteriori defines $|0\ \text{out}\rangle$, $|0\ \text{in}\rangle$ in an implicit way. It is possible to show that those states $|0\ \text{out}\rangle$ and $|0\ \text{in}\rangle$, which result from our formalism, make K_T equal to K. Thus our construction provides a clarification as to what the particles are that correspond to the vacua appearing in the Schwinger formalism.

We now briefly state some results obtained with our formalism. (i) In time dependent external fields, H^{+}, H^{-} (H_{+}, H_{-}) are separately spanned by wave functions whose phase approaches a solution of the Jacobi-Hamilton equation ('quasiclassical' behaviour) at late (early) times. (ii) In de Sitter space (maximally extended manifold) this leads to creation at a total rate $\sim\exp(-2\pi m R)$ in every mode. (iii) In electrostatic Klein paradox type situations the creation rate equals the transmission coefficient of the 'paradox' modes that become quasiclassical in one (space-like) direction off the transition zone. (iv) The 'resonance' states discovered by Schiff, Snyder and Weinberg [8] correspond to continuous pair creation and annihilation.

An interesting criterion for creation can be derived by using Nachtmann's [3] operator N which we now interpret (and normalize) differently. N is the self-adjoint operator defined by the charge form (14) with respect to the scalar product (3) as

$$\langle f', Nf''\rangle := 2\pi\delta(m'^{2} - m''^{2})(f', f'') \tag{18}$$

where f', f'' satisfy (1) with mass parameters m', m''. One can show

$$N\Psi = (K - \tilde{K}) * \Psi, \quad N = P\uparrow - \tilde{P}\uparrow = \tilde{P}\downarrow - P\downarrow. \tag{19}$$

Creation occurs if and only if the spectrum of N differs from {1,-1}. If the field equation is completely separable--the only case that can be dealt with in practice--the problem of decomposing H reduces to a two-dimensional one for every set of values of the separation constants. If n is an eigenvalue of the corresponding 2-dimensional restriction of N, $\frac{1}{n^2} - 1$ gives the total number of created particles in that mode.

In the case of separability, employment of the scalar product (3) for the computation of the relative probabilities for scattering and pair creation yields the same results as second quantization, if antiparticles are interpreted as particles propagating backward in time. This result underlines the physical relevance of the 'fifth parameter' formalism introduced by Nambu [9] and Feynman [10], since <,> is its natural scalar product. Indeed, Feynman's theory of antiparticles [11] can be interpreted as amounting to nothing else than the introduction of a "scalar product" (. , K*.), which is equivalent to <.,.>. That the fifth parameter formalism might be the more fundamental one and the role of the propagator just that to connect it with conventional theory, is also suggested by the fact that our basic definition (12) may be reformulated in such a way that it does no longer contain K and $\tilde{K}$ explicitly: Roughly speaking, $\genfrac{}{}{0pt}{}{\text{outgoing}}{\text{ingoing}}$ particle (antiparticle) modes are uniquely determined by requiring that they fulfill a natural regularity condition near $\genfrac{}{}{0pt}{}{\text{future}}{\text{past}}$ infinity after an appropriate analytic continuation in m^2 to the $\genfrac{}{}{0pt}{}{\text{lower}}{\text{upper}}$ $\genfrac{}{}{0pt}{}{\text{(upper)}}{\text{(lower)}}$ half plane. (Analytic continuation also provides the most convenient tool to prove the basic properties on which our formalism is built.)

Further results and the proofs will be published elsewhere.

REFERENCES

[1] E.g. S. A. Fulling, Phys. Rev. D7, 2850 (1973); W. G. Unruh, Phys. Rev. D10, 3194 (1975); D. G. Boulware, Phys. Rev. D11, 1404 (1975).

[2] A Lichnérowicz, in "Relativity, Groups and Topology" ed. by deWitt and deWitt (Gordon and Breach, 1964).

[3] O. Nachtmann, Zeitschr. f. Phys. 208, 113 (1968); Sitzungsber. d. Osterr. Akademie d. Wiss. II, 176, 363 (1968).

[4] M. B. Menskii, Teoreticheskaya i Matematicheskaya Fizika 18, 190 (1974).

[5] J. Schwinger, Phys. Rev. 82, 664 (1951).

[6] B. S. deWitt, Dynamical Theory of Groups and Fields (Gordon and Breach, 1965).

[7] B. S. deWitt, Phys. Rep. 19C, Nr.6 (1975).

[8] L. H. Schiff, H. Snyder, J. Weinberg, Phys. Rev. 57, 315 (1940).

[9] Y. Nambu, Progr. Theor. Phys. 5, 82 (1950).

[10] R. P. Feynman, Phys. Rev. 80, 440 (1950).

[11] R. P. Feynman, Phys. Rev. 76, 749 (1949).

PARTICLE DETECTORS AND BLACK HOLES

W. G. Unruh*

Department of Applied Math, McMaster University
Hamilton, Ontario, Canada

I would like to begin this talk by going back to the famous gedanken experiment which ultimately led Einstein to his formulation of General Relativity, namely the elevator experiment [1]. In addition to the classical falling weights, flashlights, etc. I would like the experimenter also to take along a particle detector - what kind of particles doesn't really matter, so lets assume its a detector of massless scalar particles, called Φ particles[2].

This detector will be a particularly simple detector - it consists of a Schroedinger particle of mass m in a box with walls impermeable to the detector particle.

From the equivalence principle, the detector particle in our accelerated elevator will be in an effective potential maz where z is the coordinate in the direction of the acceleration, a , measured from, lets say, the center of the detector.

The Schroedinger equation for this detector particle will be given by

$$i \frac{\partial\psi}{\partial\tau} = -\frac{1}{2m}\left(\frac{\partial^2}{\partial x^2} + \frac{\partial^2}{\partial y^2} + \frac{\partial^2}{\partial z^2}\right)\psi + maz\,\psi \tag{1}$$

where τ is the proper time measured at the center of the detector.

The simplest way to show this last statement, is by again going back to Einstein and noticing his statement that "the clock at point P for an observer anywhere in space runs $1 + \Phi/c^2$ times faster than the clock at the coordinate origin" [3]. One obtains that the observer is in a metric which can be written as

$$ds^2 = (1 + az)^2\, d\tau^2 - dx^2 - dy^2 - dz^2 \tag{2}$$

By reducing the scalar wave equation

$$\frac{1}{\sqrt{-g}} \frac{\partial}{\partial x^\nu}\left(\frac{\partial\phi}{\partial x^\mu} g^{\mu\nu}\sqrt{-g}\right) + m^2\phi = 0 \tag{3}$$

to a Schroedinger type equation, the above equation (1) results. The particle in the box-obeying (1) will have a series of eigenfunctions and energy eigenvalues which I will denote by ψ_j and E_j respectively.

This detector interacts with the Φ particles which I assume to be described by a massless scalar relativistic field Φ which obeys equation (3) (with m = 0).

This field is second quantized in the usual way, and let us assume that the field Φ is in its vacuum state $|0\rangle$. I will make this more definite in a moment.

The interaction between the Φ field and the detector is assumed to be a simple first order coupling with coupling constant ε, which gives the first order prob-

*Research partially supported by National Research Council of Canada. Preparation of manuscript partially supported by NSF Grant GP-30799X at Princeton University.

ability for the transition rate of the detector [4] initially in the ground ψ_o, to the excited state ψ_j as

$$\frac{dP}{d\tau} = \lim_{T\to\infty} \frac{\varepsilon^2}{T} \sum_{|p>} \left| \int_o^T d\tau \int_{box} d^3x \, \bar{\psi}_j \, \psi_o \, <p|\Phi|o> \right|^2 \quad (4)$$

where $|p>$ is one of a complete set of states of Φ.

This term is then directly proportional to the number of Φ particles of energy $E_j - E_o$ passing the detector per unit time.

The state ψ_j can be written as

$$\psi_j = e^{-i E_j t} h_j(x,y,z) \quad (5)$$

where h_j is the eigenfunction associated with the right side of eq. (1).

Substituting into eq. (4) gives

$$\frac{dP}{d\tau} = \lim_{T\to\infty} \frac{\varepsilon^2}{T} \sum_{|p>} \left| \int_o^T d\tau \, e^{i(E_j - E_o)\tau} \int_{box} d^3x \, \bar{h}_j \, h_o \, <p|\Phi|0> \right|^2 \quad (6)$$

The only term which survives the t integration will be the component of $<p|\Phi|0>$ which has a time dependence of $e^{-i(E_j - E_o)t}$.

The time has come to define the state $|0>$. The metric (eq. 2) is in fact just flat spacetime in a strange coordinate system [5]. Under the transformation

$$\tilde{t} = (1 + az) \sinh(a\tau)/a$$
$$\tilde{z} = (1 + az) \cosh(a\tau)/a \quad (7)$$

the metric becomes

$$ds^2 = d\tilde{t}^2 - dx^2 - dy^2 - d\tilde{z}^2$$

i.e. just flat spacetime. We know what $|0>$ is for a field in flat spacetime, and we know how to quantize it. Rather than expand Φ in the usual normal modes however, I will expand it in a set of modes adapted to the coordinate system of eq.(2).

The Rindler coordinates of eq. (2) are actually a double map on Minkowski space, i.e. each value of t, z, x, y corresponds to two points in flat spacetime. For $z > -1/a$, these two regions are R^+ and R^- of figure 1 whereas for $z < -1/a$, these are R_f and R_p.

I define the pre-superscript + or - on a function to mean a function which is zero in R^- or R^+ respectively, and non zero in the other region.

The scalar equation (eq. (3) with m = 0) can be separated in Rindler coordinates to give a set of modes.

$$\phi_{\omega k} = e^{-i\omega t} e^{i(k_1 x + k_2 y)} g_{\omega k}(z)/\sqrt{(2\pi)^3 \omega} \quad (8)$$

The function $g_{\omega k}(z)$ obeys the equation

$$\{(1 + az) \frac{\partial}{\partial z} (1 + az) \frac{\partial}{\partial z} + \omega^2 - (1 + az)^2 (k_1{}^2 + k_2{}^2)\} g_{\omega k} = 0 \quad (9)$$

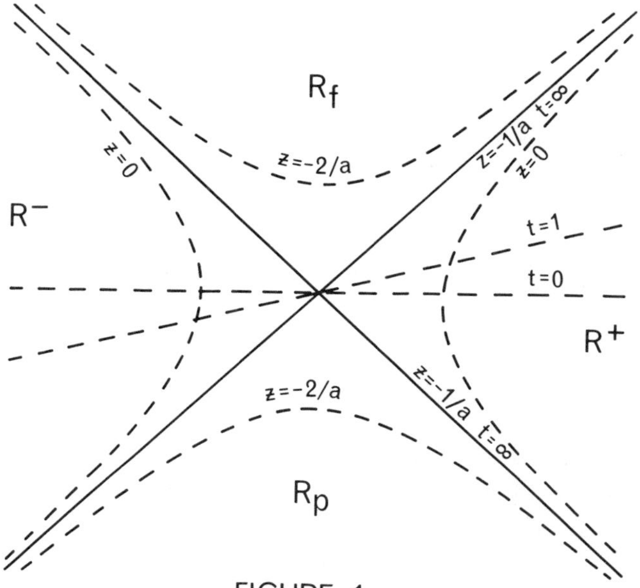

FIGURE 1

which has as its solutions

$$g_{\omega k}(z) = (\pi|\omega|/a)^{\frac{1}{2}} H^{(1)}_{i|\omega|} (i\ k(z + 1/a))\ e^{-\pi\omega/2a}/\sinh^{3/2}(\pi\ \omega/a) \tag{10}$$

where $H^{(1)}$ is the first Hankel function and $k = \sqrt{k_1^{\ 2} + k_2^{\ 2}}$.

Then it is possible to show that the function [6]

$$\hat{\phi}_{\omega k} = \left(e^{\pi\omega/2a}\ {}^{+}\phi_{\omega k} + e^{-\pi\omega/2a}\ {}^{-}\phi_{\omega k}\right)/(2\ \sinh(\pi\omega/a))^{\frac{1}{2}}. \tag{11}$$

is positive frequency (in the usual Minkowski sense) for all values of ω, positive or negative. Similarly $\bar{\hat{\phi}}_{\omega k}$ is a negative frequency function.

These functions form a complete set of functions for the scalar wave equation in flat spacetime. We therefore expand the quantum field operator Φ in terms of this set of modes

$$\Phi = \int d\omega d^2k \left(\hat{a}_{\omega k}\ \hat{\phi}_{\omega k} + \hat{a}^{\dagger}_{\omega k}\ \bar{\hat{\phi}}_{\omega k}\right) \tag{12}$$

with the vacuum now being given by the relation

$$\hat{a}_{\omega k}\ |0\rangle = 0 \tag{13}$$

This vacuum is the usual flat spacetime vacuum state.

Substituting into eq. (6), and using (12) therefore gives

$$\frac{dP}{d\tau} = \lim_{T\to\infty} \frac{\varepsilon^2}{T} \sum_{|p>} \Bigg| \int_0^T d\tau\, e^{i(E_j - E_o)\tau} \qquad (14)$$

$$\times \int_{box} d^3x\, \bar{h}_j\, h_o \int d^2k \int_{-\infty}^{\infty} d\omega\, <p|\, \hat{a}^{\dagger}_{\omega k}\, |0>\, \hat{\bar{\phi}}_{\omega k} \Bigg|^2$$

As the detector is in R^+, and as the inner product $<p|\hat{a}^{\dagger}_{\omega k}|0>$ will be non zero only for the set of states $|p> = \hat{a}^{\dagger}_{\tilde{\omega}\tilde{k}}|0>$ we obtain

$$\frac{dP}{d\tau} = \lim_{T\to\infty} \int d\tilde{\omega}\, d^2\tilde{k}\, \frac{\varepsilon^2}{T}\, \frac{e^{\pi\tilde{\omega}/a}}{2\sinh(\pi\tilde{\omega}/a)} \qquad (15)$$

$$\times \Bigg| \int_0^T dt\, e^{i(\tilde{\omega} + E_j - E_o)t} \int_{box} d^3x\, \frac{\bar{h}_j\, h_o\, e^{-i(k_1 x + k_2 y)}}{(2\pi)^{3/2}\, \tilde{\omega}^{\frac{1}{2}}}\, \bar{g}\, _{\tilde{\omega}\tilde{k}}(z) \Bigg|^2$$

For $\tilde{\omega} > \tilde{k}$, one can use the WKB approximation to solve for $g_{\tilde{\omega}\tilde{k}}(z)$ in eq. (9), and can then match to eq. (10) near $z = - 1/a$. One finds that

$$g_{\tilde{\omega}\tilde{k}}(z) \approx \sqrt{\frac{\tilde{\omega}}{\tilde{k}_3}} \left(e^{i(\tilde{k}_3 z + \phi(\tilde{\omega}))} + C.C \right) \qquad (16)$$

where $\tilde{k}_3 = \sqrt{\tilde{\omega}^2 - \tilde{k}^2}$ and $\phi(\tilde{\omega})$ is a rapidly varying phase. Note that $\tilde{k}_3/\tilde{\omega}$ is just equal to the velocity in the z direction, v_z.

Furthermore, the expression for $dP/d\tau$ can now be simplified [6] to

$$\frac{dP}{d\tau} \approx \int dk_1\, dk_2\, dk_3\, \frac{\sigma(k_1, k_2, k_3)\, e^{-\pi\tilde{\omega}/a}}{2v_z\, \sinh(\pi\tilde{\omega}/a)} \qquad (17)$$

where $\sigma\,(k_1,k_2,k_3)$ is the detection cross section of the detector for a mode with wave numbers k_1,k_2,k_3. The effective density of particles at $z = 0$ is therefore given by the expression

$$N(\tilde{\omega}) = [v_z\, (e^{2\pi\tilde{\omega}/a} - 1)]^{-1} \qquad (18)$$

The detector sees a thermal spectrum of particles (i.e. it responds exactly as if it were immersed in a thermal bath of temperature $a/2\pi$.).

(Before anyone rushes out and destroys his favorite detector trying to measure this effect, I'd better get the units back in:

$$\text{Temp} = \hbar\, a \Big/ \left(2\pi\, k_{Boltzmann}\right) \approx 10^{-24}\ {}^{\circ}\text{K sec}^2/\text{cm} \qquad (19)$$

i.e. $a \approx 10^{24}$ cm/sec^2 for $T = 1°K$. As with most quantum effect, this result is far too small to be macroscopically measured.)

We can now apply this analysis to the black hole situation. In the following I will replace the collapsing star by boundary conditions on the past horizon of a complete Schwarzschild metric. This may be regarded as purely a mathematical convenience, or may be regarded as an examination of the particle production process near an eternal black hole (i.e. one not produced by collapse). Although I have outlined reasons elsewhere [2] as to why I favor the latter interpretation, I will not push that viewpoint here.

The Schwarzschild metric is given by

$$ds^2 = (1 - 2M/r)dt^2 - (1 - 2M/r)^{-1} dr^2 - r^2(d\theta^2 + \sin^2\theta d\phi^2) \quad (20)$$

Let us assume that our accelerated observer is stationed at r = R, very near the horizon at 2M. If we let

$$z = \int_R^r \frac{dr}{\sqrt{1 - 2M/r}} = 2M\left(\sqrt{r/2M}\sqrt{r/2M - 1} + \ln(\sqrt{r/2M} + \sqrt{r/2M - 1})\right) \quad (21)$$

we find

$$(1 - 2M/r) \approx \left[\left(1 + \frac{2Mz}{R^2\sqrt{1 - 2M/R}}\right)^2 - \frac{2Mz^2}{R^3} + 0(z^3)\right](1 - 2M/R) \quad (22)$$

Letting $\tau = t\sqrt{1-2M/R}$ we obtain the metric.

$$ds^2 = \left(1 + \frac{Mz}{R^2\sqrt{1-2M/R}}^2 - \frac{2M}{R^3}z^2 + 0(z^3)\right)d\tau^2 - dz^2 - r(z)^2(d\theta^2 + \sin^2\theta d\phi^2) \quad (23)$$

Note that this is very similar to the metric of eq. (2) with

$$a = M/(R^2(1-2M/R)^{\frac{1}{2}}). \quad (24)$$

Furthermore, if $1-2M/R << 1$, then $z^2a^2 >> z^2 2M/R^3$, and the metric is dominated by the acceleration term. A detailed examination of the r = R curve shows that a is exactly the acceleration of the detector.

Placing our box detector at this point leads to a Schroedinger type equation for the particle in the box of the form

$$i\frac{\partial\psi}{\partial\tau} = -\left[\frac{1}{2m}\left(\frac{\partial}{\partial z^2} + \frac{1}{r^2}\left(\frac{1}{\sin\theta}\frac{\partial}{\partial\theta}\sin\theta\frac{\partial}{\partial\theta} + \frac{1}{\sin^2\theta}\frac{\partial}{\partial\phi^2}\right)\right) - maz\right]\psi$$

(note that by adding some additional external force on the detector particle [e.g. electromagnetic] the potential term could be cancelled out).

Again this detector particles will have energy levels E_j, and eigenfunctions going as

$$\psi_j = e^{-iE_j\tau}h_j(z,\theta,\phi) \quad (26)$$

We now turn our attention to the field Φ and the vacuum state $|0>$ of the field Φ. In this case we run into deeper problems than in flat spacetime in that we are not yet sure of what the correct quantization of the field is. However before proceeding let us solve the wave equation for a mode of the field Φ. The scalar wave equation can be separated in Schwarzschild coordinates to give

$$\phi_{\omega\ell m} = e^{-i\omega\tau} Y_{\ell m}(\theta\phi)\, g_{\omega\ell}(r)/\sqrt{2\pi\omega} \tag{27}$$

where $g_{\omega\ell}$ obeys

$$\left[\frac{(1 - 2M/r)}{r^2}\frac{d}{dr}r^2(1 - 2M/r)\frac{d}{dr} + (1 - 2M/R)\omega^2 - (1 - 2M/r)\frac{\ell(\ell + 1)}{r^2}\right] g_{\omega\ell}(r) = 0 \tag{28}$$

This equation has two solutions, one representing waves incident on the black hole from infinity, and the other representing waves emerging from the past horizon of the black hole. For the former set of waves we shall assume the obvious quantization procedure - namely that $\omega > 0$ represents positive frequencies. These waves are designated by $\phi^+_{\omega\ell m}$. The normalized modes coming out of the past horizon will be designated by $\phi^-_{\omega\ell m}$, but in this case I will not take $\omega > 0$ as my definition of positive frequency. Near the horizon, the Schwarzschild metric very closely resembles the Rindler metric which we previously studied.

It turns out that there exists an additional Killing vector on the past horizon [2], besides the usual one represented by $\partial/\partial t$. The relation of this additional Killing vector to the usual $\partial/\partial t$ Killing vector is very similar to the relation between the Killing vector $\partial/\partial\bar{t}$ associated with ordinary Minkowski time translations and the Killing vector $\partial/\partial t$ associated with Rindler time translations.

For those states coming out of the black hole, I therefore define positive frequency with respect to this new Killing vector on the past horizon of the black hole. (See Unruh (1975) [2] for more details).

This necessitates my saying something about how this definition of positive frequency is extended into the unobservable regions of spacetime behind the future horizon.

In my previous paper I extended the metric to the full Kruskal metric, and defined positive frequencies by means of this additional Killing vector along the full extension of the past horizon. It is, however, possible to show that the results in the exterior region are independent of how this definition of positive frequency is extended into the unobservable regions of space time behind the future horizon (see Appendix A).
The result is that a complete basis set of positive frequency modes originating from the past horizon is given by

$$\hat{\phi}_{\omega\ell m} = \frac{e^{-2\pi\omega M\sqrt{1-2M/R}}}{2\sinh(2\pi\omega M\sqrt{1-2M/R})}\phi^-_{\omega\ell m} + \begin{pmatrix}\text{terms which are zero}\\ \text{outside the horizon}\end{pmatrix} \tag{29}$$

Note that by using eq. (24), the factor in the exponent of e in (29) becomes

$$2\pi\omega\, M\sqrt{1 - 2M/R} \quad = \quad (\pi\,\omega/2a)\ (2M/R^2)^2$$

We can now expand Φ as

$$\Phi = \left\{ \int_0^\infty d\omega \sum_{\ell m} a^+_{\omega\ell m}\ \phi^+_{\omega\ell m} + \int_{-\infty}^{\infty} d\omega \sum_{\ell m} \hat{a}_{\omega\ell m}\ \hat{\phi}_{\omega\ell m} \right\} + \text{Herm. conj.} \qquad (30)$$

with the vacuum state $|0>$ defined by

$$a^+_{\omega\ell m}\ |0> = \hat{a}_{\tilde{\omega}\ell m}\ |0> = 0$$

for all ℓ, m, $\tilde{\omega}$ and for $\omega > 0$.

Examining the probability of transition of the detector, one finds that this is zero for all of the waves coming in from infinity (i.e. the detector sees no particles incident on the black hole from infinity). However for those modes coming out of the black hole, the analysis proceeds exactly as in the Rindler case to give a transition probability of

$$\frac{dP}{d\tau} = \int_{-\infty}^{\infty} d\tilde{\omega} \sum_{\tilde{\ell}\tilde{m}} \lim_{T\to\infty} \Bigg[\frac{\varepsilon^2}{T} \Bigg| \int_0^T d\tau\, e^{i(\tilde{\omega} + E_j - E_o)\tau} \int_{\text{box}} r^2\, dr\, d\cos\theta\, d\phi \qquad (31)$$

$$\times(\bar{h}_j\ h_o\ Y_{\tilde{\ell}\tilde{m}}(\theta, \phi)\ \bar{g}_{\tilde{\omega}\tilde{\ell}}\ (r)) \Bigg|^2 \left(\frac{e^{-(\pi\omega/a)(2M/R)^2}}{2\sinh((\pi\omega/a)(2M/R)^2)} \right) \Bigg]$$

This expression is applicable for a detector positioned at any radius R from the black hole. I shall examine the two cases when $R \approx 2M$ (detector near the horizon) and $R \to \infty$.

For $R \simeq 2M$, I shall-use a W.K.B. approximation for $g_{\bar{\omega}\ell}$ for r near R(i.e. for $z \simeq 0$). If $\omega^2 < \ell(\ell+1)/R^2$, one finds that $g_{\bar{\omega}\ell}$ becomes vanishingly small. For $\omega^2 > \ell(\ell+1)/R^2$ on the other hand, the solution for $g_{\bar{\omega}\ell}$ becomes

$$g_{\bar{\omega}\ell}\ (r(z)) \simeq v_r^{-\frac{1}{2}} \left\{ e^{i(\omega v_r z + \phi(\omega))} + A(\omega) e^{-i(\omega v_r z + \phi(\omega))} \right\} \qquad (32)$$

where $v_r = [1 - \ell(\ell+1)/(R^2\omega^2)]^{\frac{1}{2}}$ and $A(\omega)$ is approximately unity for $\omega^2 < \ell(\ell+1)/9M^2(1-2M/R))$ and drops rapidly to zero thereafter. From eq. 31 we obtain

$$\frac{dP}{d\tau} \approx \sum_{\tilde{\ell}\tilde{m}} \int_{\tilde{\ell}/r}^{\infty} d\tilde{\omega}\ \sigma(\tilde{\omega}\tilde{\ell}\tilde{m}) \left(\frac{1 + |A(\omega)|^2}{v_r(e^{-2\pi\omega/a} - 1)} \right) \qquad (33)$$

where I have assumed $\phi(\omega)$ varies more rapidly than $\sigma(\tilde{\omega},\tilde{\ell}\tilde{m})$, the detection cross section of the detector. Again the detector acts as though immersed in a thermal flux coming out of the black hole with temperature $a/2\pi$ when they reach the detector. For a high enough energy, some of these photons can escape the black hole entirely, ($|A(\omega)|\to 0$) and the detector does not see those drop back into the black hole.

Notice that as $R \to 2m$, $a \to \infty$, and thus the temperature of the horizon as seen by one of these stationary detectors is infinite.

The other case, when $R \to \infty$, follows exactly as the first case. However, now the factor in the exponent of (29) is not equal to $\pi\omega/2a$ but rather to $2\pi\omega M$. Furthermore, $g^-_{\omega\ell}$ becomes, for $\omega^2 > \ell(\ell+1)/R^2$,

$$g^-_{\omega\ell} \approx B(\omega)\ e^{i\omega v_r z}/\sqrt{v_r}$$

where $B(\omega)$ is related to the previously defined $A(\omega)$ by

$$|B(\omega)|^2 = 1 - |A(\lambda\omega)|^2$$

with λ being the ratio of the factor $\sqrt{1-2M/R}$ for the detector near the horizon to that at infinity. (This factor arises because ω is the frequency with respect to the proper times of the respective observers - i.e. ω is the energy of the mode as seen by the observer, not by someone at infinity.) Substituting into (31) we obtain

$$\frac{dP}{d\tau} \approx \sum_{\tilde{\ell}\tilde{m}} \int_{\tilde{\ell}/R}^{\infty} d\tilde{\omega} \left[\tilde{\sigma}(\omega\ \tilde{\ell}\ \tilde{m})\ \frac{|B(\tilde{\omega})|^2}{v_r(e^{-8\pi\omega M} - 1)}\right]$$

In this case the effective temperature is not equal to the acceleration of the detector, but is rather related to the mass of the black hole (i.e. $(8\pi M)^{-1}$), which is exactly the Hawking result [7].

Returning to the flat spacetime accelerated detector, I would like to point out one additional interesting feature.

By our derivation of the detection process, one sees that the Φ field ends up in a state $|p\rangle$ after the detection, a state which is a one particle state from the point of view of an inertial observer. The detection of a quanta by the accelerated detector is regarded as the emission of a quanta by an inertial observer.

This can also be applied to inertial observers near the black hole. They will not see a thermal spectrum of particles near the horizon, but will see nothing coming out of the black hole. On the other hand, they will regard what the stationary detector sees as a detection of a quanta as an emission by that accelerated detector.

The implication of these results to the back reaction of the Hawking process on the metric will be dealt with elsewhere.

ACKNOWLEDGEMENT

I wish to thank the Relativity Group at Princeton University for their hospitality during the preparation of this manuscript.

APPENDIX A

The Kruskal extension of the Schwarzschild metric is given by

$$ds^2 = 2Mr^{-1} e^{-r/2M} dUdV - r^2 (d\theta^2 + \sin^2 \theta d\phi^2)$$

where the relation

$$2M \ln\left(\frac{UV}{(2M)^2}\right) = r + 2M \ln(r - 2M)$$

defines r as a function of U and V. The past horizon is given by the surface V=0, $U < 0$. The vector field Killing on this surface is $\partial/\partial U$. In the text I have defined positive frequency as the space spanned by the positive eigenvalues of the operator $i\ \partial/\partial U$. However, the region $U > 0$, $V = 0$ is actually the future horizon of the other half Kruskal extension. Furthermore, the other half of the Kruskal extension could be replaced by some other metric, which wouldn't affect the metric outside our side of the black hole at all. In a non-rigorous fashion, I wish here to demonstrate that the method I use to extend my definition of positive frequency to the region $U > 0$ does not affect the quantization outside our side of the black hole.

The inner product on a scalar wave along the V = 0 surface is given by (having suppressed the angular coordinates)

$$\langle\phi_1, \phi_2\rangle = \frac{i}{2} \int_{V=0} \sqrt{-g}\ g^{UV}\ \phi_1 \overleftrightarrow{\frac{\partial}{\partial U}}\ \phi_2\ dU = -\ i \int_{V=0} \sqrt{-g}\ g^{VU} \left(\frac{\partial\phi_1}{\partial U}\right) \phi_2 dU$$

The second form arises by an integration by parts, since $g_{\mu\nu}$ is independent of U on V = 0.

Positive frequency is now defined as the eigenfunctions of an operator L which for $U < 0$ goes as $i\ \partial/\partial U$ and is unknown for $U > 0$. I assume that the eigenfunctions of this operator will be a complete orthonormal set for the scalar field along V = 0. But for $U < 0$, these eigenfunctions take the form $C(\omega)e^{-i\omega U}$ where $C(\omega)$ is some normalization constant. As this is, by assumption, a complete orthonormal set, for any U, U' both less than zero we have

$$i \int_{-\infty}^{\infty} |C(\omega)|^2\ e^{-i\omega U'} \frac{\partial}{\partial U}\ e^{i\omega U}\ d\omega = \delta(U - U')$$

or

$$\int_{-\infty}^{\infty} \omega|C(\omega)|^2\ e^{-i\omega(U' - U)}\ d\omega = \delta(U - U')$$

Although, U, U' are less than 0, $U - U' = \xi$ is unrestricted and the equation

$$\int_{-\infty}^{\infty} \omega|C(\omega)|^2\ e^{-i\omega\xi}\ d\omega = \delta(\xi)$$

has as solution

$$|C(\omega)| = 1/(2\pi\omega)$$

i.e. the normalization must be exactly the same, no matter what the behavior of the operator L for U > 0. However this immediatley implies that the restriction to U < 0 of the positive frequency portion of a function f(U) which is zero for U > 0, must be independent of the form of L for U > 0. This will immediately imply that the Feynman propagator between two points, both of which lie in our exterior region of the black hole is independent of the extension of the definition of positive frequency to the other side of the future horizon.

NOTES AND REFERENCES

[1] For one of the first references to the use of an "elevator" in the demonstration of the equivalence of inertial to gravitational mass see Einstein, A., Grossman, M., Zeit. Math. Phys. 62 (1913) 225. "Anschaulich läßt sich diese Hypothese so aussprechen: Ein in einem Kasten eingeschlossener Beobachter kann auf keine Weise entscheiden, ob der Kasten sich ruhend in einem statischen Gravitationsfeld befindet, oder ob sich der Kasten in einem von Gravitationsfeldern freien Raume in beschleunigter Bewegung befindet, die durch an dem Kasten angreifende Kräfte aufrecht erhalten wird (Äquivalenz-Hypothese).

[2] This talk is based on work reported in "Notes on Black Hole Evaporation" preprint McMaster U., Hamilton, Ont. Canada.

[3] See Einstein,A., Jarb. Radioakt. 4, (1907) 411. See p. 454 on and in particular p. 458. "Befindet sich in einem Punkte P vom Gravitationspotential ϕ eine Uhr, welche die Ortszeit angibt, so sind gemäß (30a) ihre Angaben $1+\frac{\phi}{c^2}$ mal größer als die Zeit τ, d.h. sie läuft $1+\frac{\phi}{c^2}$ mal schneller als eine gleich beschaffene, im Koordinatenanfangspunkt befindliche Urh.

[4] The detector described here is slightly inconsistant as it mixes both relativistic and non-relativistic fields. The interaction is assumed to be of the form $\varepsilon\delta(\chi-x')\Phi(x')$ in the Lagrangian density of the coupled system, where χ is the spacetime position operator for the non-relativistic field and x' a spacetime coordinate.

[5] The use of accelerated or Rindler coordinated for flat spacetime as a test case for quantum field theories near horizons was first done by Fulling, S. in his thesis, Princeton University (1972)(reported in Phys. Rev. D7 (1973) 2800). It is amazing that a crude application of Einstein's 1907 arguments actually leads to flat spacetime in an accelerated coordinate system. See also Rindler, W., Am. J. Phys. 34 (1966) 1174 where the relation between these accelerated coordinates and the Schwarzschile metric near the horizon is emphasized.

[6] The reduction of 15 to 17 comes about by comparing 15 with an equivalent unaccelerated detector (with the detector particle inside the box also placed in a potential equal to maz, possible by electromagnetic means) and recalling the definition of cross section as the transition probability rate over the incoming flux.

[7] Since the original work by Hawking (Hawking, S., Nature 248 (1974) 30; Comm. Math. Phys. (in press)) many people have studied the Black Hole evaporation problem. A partial list is Boulware, D., "Quantum Field Theory in Schwarzschild and Rindler Spaces", preprint (1974) U. of Washington, Seattle; DeWitt B., "Quantum Field Theory in Curved Spacetime", preprint (1974) U. of Texas, Austin; Davies, P., "Scalar Particle Production in Schwarzschild and Rindler Metrices" preprint (1974) Kings College, London, England.

SPONTANEOUS DISCHARGE OF KERR-NEWMANN BLACK HOLES

W. T. Zaumen

Lockheed Palo Alto Research Laboratory
Palo Alto, California 94304, U.S.A.

The physical motivation for the recent work on the vacuum polarization of black holes is based on the idea that pair production becomes feasible in a sufficiently high electric field. Regardless of whether one considers this problem in terms of the Klein Paradox approach or in terms of spontaneous creation in a field external to the event horizon via a pair production mechanism (i.e., a quantum field theoretic treatment), as long as we require that the change in the irreducible mass $\Delta m_{ir} > o$, the decrease of the black hole's electric charge e must decrease the black hole's total energy E enough to produce a free electron. The condition for this is

$$1 \leq \frac{E(m_{ir}, Q, 0)}{E(m_{ir}, Q, L)} \frac{e}{m_e} \left(\frac{Q}{2m_{ir}}\right)$$

where $E(m_{ir}, Q, L)$ is the Christodoulou-Ruffini formula for the total energy,

$$E^2 = \left(m_{ir} + \frac{Q^2}{4m_{ir}}\right)^2 + \frac{L^2}{4m_{ir}^2} ,$$

Q is the electric charge of the black hole, L is the angular momentum, m_e the electron's rest mass , and e, the electron charge.

For a Reissner-Nordstrom black hole, we can calculate the rate of pair production by using a fromula derived by Schwinger [1]. The rate of pair production per unit volume is given by the expression

$$\alpha\pi^{-2} \varepsilon^2 \sum_{n=1}^{\infty} n^{-2} \exp(-n\pi m_e^2/e\varepsilon)$$

where ε is the electric field, and α is the fine structure constant. Since $m_e^2/e\,\varepsilon \lesssim 1$ for a large rate, then $\alpha(\varepsilon^2 \lambda_e^3)/m_e c^2 \gtrsim 1$, where λ_e is the electron Compton wavelength. The physical interpretation of this is that the energy ($\sim \alpha\ \varepsilon^2$) within a region of space a few Compton wavelengths across must be large enough to account for the rest mass of the electron and positron that are produced if the rate of pair production is to be large. This condition can be written as

$$\frac{\lambda_e}{\pi R} \frac{e}{m_e} \frac{Q}{2M} \geq 1$$

with $\varepsilon \sim Q/R^2$ and $R = 2M$, where M is the mass of the black hole. For very large black holes ($\lambda_e/R << Q_e/m_e$), the rate of pair production becomes very small even when $Q = 2M$.

The expression for the rate of pair production can be generalized to the Kerr-Newman case by calculating twice the imaginary part of Schwinger's Lagrangian [1]

$$L = - F - \frac{1-}{8\pi^2} \int_0^\infty ds \ s^{-3} \exp(-m_e^2 s) \ [$$

$$(es)^2 \ G \ \frac{Re \cosh(esx)}{Im \cosh(esx)} - 1 - \frac{2}{3} (es)^2 F \ ]$$

where $F = \frac{1}{2}(b^2 - \varepsilon^2)$, $G = \varepsilon \cdot b$, b is the magnetic field and

$$x = [2(F + iG)]^{\frac{1}{2}} = \sqrt{2} \quad i \sqrt{-F} (1 - iG/F)^{\frac{1}{2}} = b_1 + i \varepsilon_1$$

The rate per unit volume for pair production is then given by the expression

$$\alpha\pi^{-1} G \sum_{n=1}^{\infty} n^{-1} \coth(n\pi \ b_1/\varepsilon_1) \exp[-n\pi m_e^2/e \varepsilon_1]$$

These results are identical to what one would obtain by calculating the rate with the level crossing methods. The difference between these methods is that in the level crossing methods, one uses a single particle model initially, and finds that one needs a many particle model. If one starts, conceptually at least, from a field theory point of view, one is simply calculating the rate of pair production in an external field. The many particle nature of the problem is built in to the field theory approach. The level crossing method, incidentally, is somewhat dependent on describing the external field in terms of a potential. If one should wish to calculate the rate of production of some other particles for which a potential is not convenient, then one must use field theory. It may seem surprising that such diverse methods should give the same result; however, the situation is the same as in the calculation of the Einstein A and B coefficients. The rate of spontaneous atomic transitions can be calculated by a thermodynamic argument and by a field theory approach. Both give the same answer, and yet the thermodynamic argument does not really depend on the fundamental nature of the interaction in an obvious way.

REFERENCES

[1] J. Schwinger, Phys. Rev. 82 (1951) 664.

THEORETICAL SIGNIFICANCE OF PRESENT-DAY GRAVITY EXPERIMENTS

K. Nordtvedt, Jr.

Montana State University
Bozeman, Montana U.S.A.

Einstein's General Relativity Theory includes three independent assertions about space, time, and gravity:

1. That the geometry and chronometry of matter are not a priori categories of experience or rational assumption, but rather are attributes of a dynamical field, the metric field $g_{\mu\nu}$, and its universal coupling to matter.

2. That physical law is generally covariant. (Though this principle was quickly criticized as having no physical content, I wish to reassert that this principle is of important physical significance. If all physical laws are tensor equations, and if we continue in this spirit by demanding that tensors in physical law may not have the numerical value of components designated a priori, but only a posteriori through the equations of nature themselves, then a world emerges with no "prior geometry." It is a world in which the onotological status of space-time metrical structure and the location of physical events is downgraded to an attribute of the world's distribution of stress-energy (matter). We then have truly a unified theory of nature, but one in which space-time geometry is subordinated to matter rather than the opposite and more usually stated goal of unified theories. Fulfillment of the general covariance principle as here described would be strong circumstantial evidence for the merely algorithmic and transient-historical role of space-time coordinates in physical law.)

3. And that the dynamical field equations for the gravitational metric field $g_{\mu\nu}$ and its universal coupling to all other forms of stress-energy are specifically

$$R_{\mu\nu} = - k(T_{\mu\nu} - \frac{1}{2} T g_{\mu\nu}) \tag{1}$$

where $R_{\mu\nu}$ is the Ricci tensor formed from $g_{\mu\nu}$ and its derivates, and $T_{\mu\nu}$ is the stress-energy tensor for all other forms of matter.

I want to stress the independence of these three parts of General Relativity. Empirical observation supports or refutes these principles individually. Present-day gravity experiments are now reviewed with focus on their significance as tests of these foundations and contents of General Relativity.

Theories of gravity which possess at least the metric field $g_{\mu\nu}$, but also possibly additional cosmological-gravitational fields (like scalar, vector, or tensor, etc.), but which have solely the metric field coupled to matter in a particular universal way, are called "metric theories of gravity" [1]. This is a wide and rich class of theories which includes General Relativity and such competitors as the Jordan-Brans-Dicke scalar-tensor theories [2], the vector-metric theories [3], and others. It has been found profitable to consider the class of metric theories as a whole when inventing and interpreting new empirical tests of General Relativity. Figure 1 outlines the logical structure of metric gravity and how the so-called PPN (parameterized post-Newtonian) [4] metric expansion allows us to analyze weak field gravitational experiments.

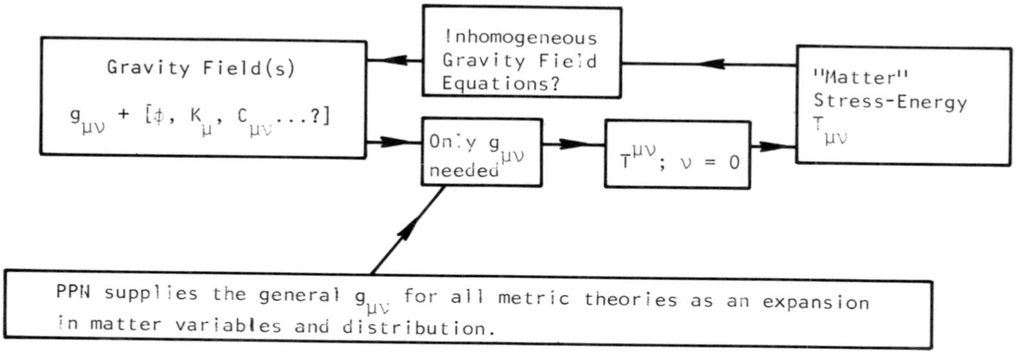

Figure 1. Logical Structure of Metric Gravity and PPN Theory.

For a complete theory of metric gravity and matter the gravitational fields and their dynamical field equations are specified including the inhomogeneous matter source terms. The matter equations of motion are given and in the presence of gravity the covariant divergence of the matter's stress-energy tensor vanishes; hence only $g_{\mu\nu}$ need be known to obtain matter's response to gravity. The gravity fields and matter dynamics are solved self-consistently (usually by a weak field expansion and iterative process). In the PPN theory a <u>general</u> weak field expansion for $g_{\mu\nu}$ in terms of the matter's variables and distribution is supplied. Dimensionless parameters appear in this general expansion which are numerically unique and calculable for any specific metric theory of gravity. Matter's response to gravity in any experimental situation and for any metric theory can now be determined and quoted in terms of the PPN parameters.

The strongest evidence that nature is described by a metric theory of gravity is the Eotvos-type experiment which shows the universality of free-fall rates to a part in 10^{11} or 10^{12} for various laboratory materials [5,6,7]. This together with the continued observance of Lorentz invariance in high energy elementary particle experiments supports the assumption that only one gravitational field $g_{\mu\nu}$ couples to matter and can always, by a coordinate system choice, be locally set equal to the Minkowski metric;

$$g_{\mu\nu} \rightarrow \eta_{\mu\nu} = \begin{matrix} 1 & 0 & 0 & 0 \\ 0 & -1 & 0 & 0 \\ 0 & 0 & -1 & 0 \\ 0 & 0 & 0 & -1 \end{matrix} \quad (2)$$

neglecting only corrections second order in coordinate separations, thus establishing free-falling inertial Lorentz frames without gravity (the equivalence principle). Terrestrial clock experiments confirm the gravitational frequency shift predicted by metric theories to 1% [8] while the near-future rocket launching of a hydrogen maser clock should check this effect to one or two magnitudes higher precision [9]. A solar orbitting clock experiment could do even better, perhaps to a part in 10^6.

For the future it seems important to continue Eotvos-type and clock-type experiments using matter of unusual quantum numbers and configurations such as non-zero spin or strangeness, stress-energy coupling to the laboratory, high charge to mass ratio, etc., in order to better eliminate the chance that "exotic" non-metric fields of a long range nature exist and couple to matter.

Elsewhere I have shown that oscillator clocks (O-clocks) like the hydrogen maser are unlikely to ever exceed Eotvos-type experiments in sensitivity as tests of the metrical foundations of gravitational theory [10]. But it is interesting that new possibilities for experiments now exist with ruler clocks (R-clocks) produced by a superconducting cavity stabilized oscillator (SCSO) [11]. In a changing gravitational field these clocks can be considered to be testing the adjustment of matter to space-like metrical field structure while O-clocks adjust to the time-like metrical field structure. Future experiments to compare these two types of clocks to high precision will be most significant in my judgment. The third type of clock, the decay clock (D-clock) such as an alpha or beta decay source, tests matter's coupling to the metrical field in yet a different way [10]. Such clocks can be particularly sensitive to specific matter interactions (like the weak interaction) and their coupling to gravity. Dyson's review of work on the constancy of the fundamental constants [12] leaves open the possibility of substantial non-metric coupling of gravity to the weak interaction.

The principle of general covariance is observationally supported by the absence of preferred inertial frames in gravitational physics, a situation not assured by the metric nature of gravity and the resulting Lorentz invariance of the local non-gravitational laws of physics. Metric theories with "prior geometry" most often possess PPN gravitational potentials dependent on the velocity of an experiment's inertial frame relative to some preferred inertial frame. Earth-like gravimeter experiments and Earth rotation rate observations [13], the orbital motion of the moon (see lunar laser ranging below), and observations of the binary pulsar PSR 1913+16 [14] and perihelion advance of the inner planets can put small limits on the preferred frame PPN parameters. These experiments also tend to rule out theories without "prior geometry" but which nevertheless have preferred inertial frames established in a Machian fashion by, for example, long range vector fields [3].

Fomalont and Sramek's recent interferometer measurement of the deflection of radio signals passing close by the Sun [15] is a quantitative culmination of years of many workers' observations of the effect of gravity on propagation of electromagnetic radiation, starting with the Eddington expedition of 1919. This latest experiment measures the quasi-static, linearized metric field of the Sun to about 2% precision. This metric field takes the form (in isotropic coordinates);

$$g_{00} = 1 - 2\left(\frac{GM}{c^2R}\right) + \ldots \tag{3a}$$

$$g_{xx} = g_{yy} = g_{zz} = = 1 - 2\gamma\left(\frac{GM}{c^2R}\right) + \ldots \tag{3b}$$

$$\text{other } g_{\mu\nu} = 0 \tag{3c}$$

GM is the mass parameter of the sun, c is the speed of light, R is the radial coordinate from the sun, and γ is the single PPN parameter in this metric which may vary from theory to theory. Their experiment yields

$$\gamma \simeq 1.03 \pm .02 \tag{4}$$

while Einstein's theory predicts a γ value of one.

Observations of the perihelion motion of the inner planets both optically and by radio ranging methods tests the quasi-static but non-linear metric field of the sun. Now we expand g_{00} in the classical form of Eddington and Schiff (Preferred frame or anisotropic PPN potentials are neglected here, such potentials make additional contributions to perihelion precession [16];

$$G_{00} = 1 - 2\left(\frac{GM}{c^2R}\right) + 2\beta\left(\frac{GM}{c^2R}\right)^2 + \ldots \tag{5}$$

where now a second PPN parameter β is needed. The interpretation of this experiment's data has been complicated by the report of an unusually large solar visual oblateness [17], but if we assume negligible solar oblateness based on Hill's recent work [18] then precession observations give a result of about

$$2\gamma - \beta \simeq 1.00 \pm .03 \tag{6}$$

General Relativity gives a β value of one.

The lunar laser ranging experiment very likely offers the most sensitive test of Einstein's field equations for the near future. Figure 2 outlines the essential features of this experiment. By measuring the time-dependent part of the earth-moon separation to 4 centimeter accuracy, a level now apparently being reached, the earth's gravitational to inertial mass ratio is determined to an accuracy of $2\ 10^{-12}$ [19]. Calculation within PPN theory yields a gravitational to inertial mass ratio for a massive body which differs from one by a term of order the body's gravitational self (internal) energy divided by the body's total mass energy [20]. This ratio is $4\ 10^{-10}$ for the earth. The PPN potentials measured in this effect not only include the γ and β potentials, but additional

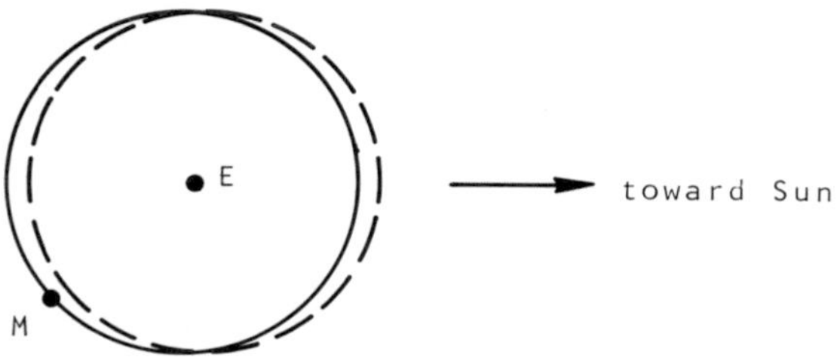

Figure 2. Outline of the Lunar Laser Experiment:

$a = [1 + \eta\, U_G/Mc^2]\, g$ Massive Body Equation of Motion from PPN

$U_G = -\frac{1}{2} G \int\int \frac{\rho(x)\rho(y)}{|x-y|}\, d^3x\, d^3y$ Body's Gravitational Self-Energy

$U_G/Mc^2 = 4 \cdot 10^{-10}$ for the earth. Lunar laser ranging limits $|\eta| \simeq 5 \cdot 10^{-3}$ by observing the polarization of the moon's orbit toward the sun to ± 4 cm. accuracy.

$\eta = 4\beta - 3 - \gamma - \alpha_1 + \frac{4}{3}\alpha_2$

$\eta = 0$ in General Relativity but not in most alternative theories.

$\eta = 3 \times 3$ anisotropic matrix for spinning massive body or in anisotropic PPN.

non-linear potentials and motional gravitational potentials. This experiment is particularly sensitive to the β parameter. A 4 centimeter accuracy in measuring the earth-moon range oscillation of lunar synodic frequency yields PPN parameters limits of

$$(4\beta - 3 - \gamma - \alpha_1 + \frac{4}{3}\alpha_2) \simeq 0 \pm 5 \times 10^{-3} \qquad (7)$$

General Relativity and scalar-tensor theories predict no preferred frame potentials; $\alpha_1 = \alpha_2 = 0$. However, the Jordan Brans-Dicke theory becomes compatible with (7) only for a coupling constant value of $\omega \simeq 200$ [23].

The future orbitting gyroscope experiment plans to measure a "pure" motional gravitational potential through the effect in which the spinning earth produces a magnetic-like gravitational vector potential which causes another spinning mass (the gyroscope) to precess, as well as possibly measure the γ parameter contribution to "geodetic precession" to a part in 10^3 or 10^4 precision.

Finally, I wish to remark on the use of the observations of the newly discovered binary pulsar PSR 1913+16 [21] to test General Relativity. The perihelion of this binary appears to be advancing at about 4° per year. If there are anisotropic gravitational potentials resulting perhaps from 2-tensor gravity, this 4° per year perihelion precession will include a contribution with cosine

$2(L-L_0)$ modulation where L is perihelion longitude [14] as a consequence of an anisotropic part of the pulsar's gravitational to inertial mass ratio. This can be said to be the "Hughes-Drever" experiment [22] of gravitational physics. The absence of this precession modulation would significantly rule out 2-tensor theories of gravity. Several years of observational data will be needed to search for this modulation.

In summary then, the observational situation confirms General Relativity and its several assumptions throughout a variety of experimental tests. There is no evidence that an alternative theory of relativistic gravity is needed as we enter a period of parts-in-a-thousand rather than parts-in-a-hundred check of Einstein's field equations in the weak field physical environment. These results permit us to more confidently apply Einstein's equations to the far-reaching environments and questions of cosmology, gravitational collapse, and gravitational radiation.

REFERENCES

[1] K. S. Thorne and C. M. Will, Astrophysical Journal 163, 595 (1971).
[2] C. H. Brans and R. H. Dicke, Phys. Rev. 124, 925 (1961).
[3] R. W. Hellings and K. Nordtvedt, Jr., Phys. Rev. D., Vol. 7, No. 12, 3593 (1973).
[4] C. M. Will and K. Nordtvedt, Jr., Astro. J. 177, 757 (1972).
[5] R. V. Eotvos, et al., Ann. der Phys. 68, 11 (1922).
[6] P. G. Roll, R. Krotkow, and R. H. Dicke, Ann. Phys. (N.Y.) 25, 442 (1964).
[7] V. B. Braginsky and V. I. Panov, Journ. Eksp. Teor. Fiz. 61, 875 (1971).
[8] R. V. Pound and J. L. Snider, Phys. Rev. 140, B788 (1965).
[9] R. F. C. Vessot, Proceedings of the International School of Physics "Enrico Fermi": Experimental Gravitation, B. Bertotti, ed., Academic Press (New York, 1974).
[10] K. Nordtvedt, Jr., Phys. Rev. D., Vol. 11, No. 2, 245 (1975).
[11] J. P. Turneaure, in Proc. 1972 Applied Superconductivity Conference (IEEE, New York, 1972), p. 621.
[12] F. J. Dyson, in Aspects of Quantum Theory, edited by A. Salam and E. P. Wigner (Cambridge University Press, 1972).
[13] K. Nordtvedt, Jr. and C. M. Will, Astro. J. 177, 775 (1972).
[14] K. Nordtvedt, Jr., Astro. J. 1975 (in press).
[15] E. B. Fomalont and R. A. Sramek, talk at VII Texas Symposium on Relativistic Astrophysics, Dallas, Texas, December 16-20, 1974.
[16] K. Nordtvedt, Jr. and C. M. Will, Astro. J. 177, 775 (1972).
[17] R. H. Dicke and H. M. Goldenberg, P.R.L. 18, 313 (1967).
[18] H. Hill, et al., P. R. L. 33, 1497 (1974).
[19] K. Nordtvedt, Jr., Phys. Rev. 170, 1186 (1968).
[20] K. Nordtvedt, Jr., Phys. Rev. 169, 1017 (1968).
[21] R. A. Hulse and J. H. Taylor, Ap. J. Letters 195, L107 (1975).
[22] V. W. Hughes, et al., P. R. L. 4, 342 (1958).
[23] J. G. Williams, et al. (submitted to P.R.L.). Analysis of 6 years of lunar laser ranging data yields 0 ± 30 cm for this effect with "an order of magnitude refinement .00 expected in the next few years. . . ."

GRAVITATION, RELATIVITY AND PRECISE EXPERIMENTATION*

C. W. F. Everitt
W. W. Hansen Laboratories of Physics
Stanford University, Stanford, California 94305

ABSTRACT

Experimentalists working in gravitation and relativity are heirs to a noble tradition of precise measurement which began in 1798, when Henry Cavendish, at the age of 67, carried out Michell's torsion balance experiment to measure the gravitational constant and "weigh the Earth". The tradition was carried forward in the Michelson-Morley experiment, the Eötvös experiment and in C. V. Boys's repetition of the Cavendish experiment, begun in 1889, which was perhaps the earliest attempt to analyse the influence of the size of an apparatus on its accuracy. Boys found that the accuracy of the Cavendish experiment was improved by making it smaller.

During the past fifteen years two new areas of research, space technology and large scale cryogenics technology have opened the way for an important class of modern gravitational experiments which depend on measuring small angular or linear displacements of suspended bodies just as the Cavendish and Eötvös experiments did. This article describes four such experiments (1) the Stanford Gyro Relativity experiment, (2) a new cryogenic test of the equivalence of gravitational and inertial mass to be performed both on Earth and in space, (3) the Stanford-LSU-Rome gravitational wave antennas, (4) an experiment to measure the Lense-Thirring nodal drag on a satellite and also obtain new geophysical data by means of precise Doppler ranging measurements between two counter-orbiting satellites in polar orbit around the Earth.

The fundamental limits to precision of these and similar experiments are reviewed.

CONTENTS

*Supported in part by NASA Grant 05-020-019.

1. THE CLASSICAL EXPERIMENTS 1798 - 1922

Just as the gravitational interaction occupies a special place in theoretical physics, so does experimental research on gravitation and relativity in physics history. Progress here is over a longer time span than elsewhere and experiments are fewer. Although space technology and cryogenic technology have brought enough new ideas in the last fifteen years to revolutionize our field, a single person can still read through the account of every experiment done in it. Those who experiment on gravitation are more likely to be written off as crackpots than to face the threat of instant competition by which so many scientists are perplexed.

The noble tradition of precise measurement to which we are heirs began in 1798 with Henry Cavendish (1731-1810). The aristocratic recluse who did so much and published so little was 67 years old when he "weighed the Earth". In performing this wonderful experiment Cavendish drew on a lifetime of careful observation from the accurate measurements of gas densities that underlay his discoveries of hydrogen, carbon dioxide, and even, unknown to himself, of argon, to the beautifully ingenious unpublished experiment of 1773 by which he verified the inverse square law of electrostatics to one part in 50, an accuracy far higher than that reached ten years later by Coulomb. The fundamental idea for measuring the gravitational constant, however, came not from Cavendish but from another 18th century Englishman too little known to physicists, the Reverend John Michell (1724-1793). Michell, who is best remembered as the discoverer of binary stars and the first man to make realistic estimates of stellar distances, had invented the torsion balance around 1750 independently of Coulomb. He conceived the principle of measuring the gravitational force between two large masses and two smaller ones attached to a torsion balance, and then determining the elastic constant of the suspension wire from the period of the torsional oscillations. Michell built part of the apparatus. On his death it passed to F. J. H. Wollaston of Cambridge (1762 -1823),brother of the better-known W. H. Wollaston, and from him to Cavendish. Cavendish saw the crucial need to isolate the balance-arm from disturbing forces, especially forces of convection; he had the simple but profound idea of placing the apparatus in a thermally isolated room and operating it by remote control. This difference in manner of observing, said Cavendish,"rendered it necessary to make some alterations in Mr. Michell's apparatus; and as there were some parts of it which I thought were not so convenient as could be wished, I chose to make the greater part of it anew" [1]. So the first modern physics experiment was born.

Figure 1, reproduced from Cavendish's paper, is a longitudinal section through the instrument and the room in which it was placed. Two lead balls, two inches in diameter, were hung from a six foot deal rod braced by a silver wire, a design for the balance arm fixed on by Cavendish as being strong and light, but at the

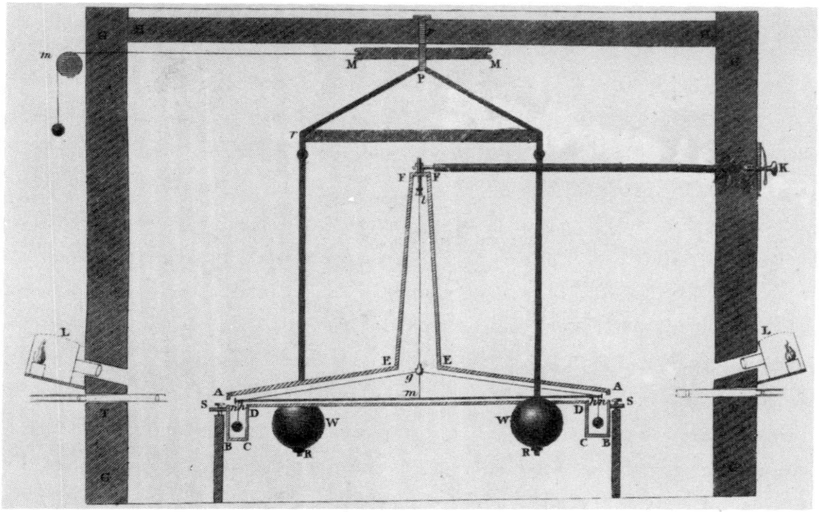

FIGURE 1

THE CAVENDISH EXPERIMENT

same time meeting with very little air resistance and having a form simple enough to compute easily its attraction to the external masses. The balance-arm was suspended inside a close mahogany box "to defend it from the wind", by means of a silvered copper wire which had a zero adjustment worked by remote control through the worm gear and rod FK. The box stood on piers and was levelled by four adjustment screws SS. The external masses were lead spheres, one foot in diameter, suspended first by iron rods, and afterwards by copper ones, from a beam which could be swung between two positions on either side of the mahogany case by a pulley and weights outside the building. The lead spheres were brought up within 0.2 inches of the case by coming against wooden stops fastened to the walls of the building: "I find" (wrote Cavendish) "that the weights may strike against [the stops] with considerable force without sensibly shaking the instrument". The final critical detail was the method of reading out the angular position of the balance arm. Slips of ivory engraved with scales to 1/20 inch were placed inside the instrument case at each end, very close to but not quite touching the balance arm, and corresponding vernier scales subdividing the main divisions into five parts were mounted on the arm. The scales were illuminated by collimating lanterns and observed through telescopes let into the end walls of the room. In this way the position of the arm was "observed with ease to 100 ths of an inch and . . . estimated to less" which corresponds to an angular resolution of about ±20 arc-seconds over the six foot balance arm.

Cavendish, at the accuracy he aimed for, encountered most of the disturbing forces that trouble modern experimenters engaged in observations of this kind. Two that he escaped were Brownian motion of the balance arm and seismic noise. For a system with 15 minute period (the longest he used) the Brownian motion is about 8×10^{-3} arc-seconds, three orders of magnitude below his limit of observation. Seismic noise might have contributed 0.1 arc-seconds of motion, also well below his limit. On the other hand Cavendish's study of magnetic disturbances is all too familiar. He decided in advance to make the balance of non-magnetic materials but cautiously checked for residual effects from the iron rods on which the large masses were hung. There was a displacement amounting to 3% of the gravitational deflection which disappeared when the iron was replaced by copper. Later Cavendish noticed a drift in the apparatus which he thought might come from the lead weights picking up magnetization in the earth's field. He hung the weights on pivots and reversed them each morning but saw no improvement. He then exercised the grand principles of experimental physics: if you are uncertain whether a disturbing effect is small try making it bigger and see how bad it is. He replaced the lead balls with ten inch magnets and reversed them instead. The displacement was negligible; a discovery which cast doubt on the earlier interpretation of the disturbance from the iron rods and left Cavendish very puzzled. He next investigated whether the drift in the torsion balance might be from elastic after-working in the suspension wire -- a phenomenon he accurately described though its discovery is usually credited to Kohlrausch in 1863. Having disposed of this possiblity (again by deliberately exaggerating the effect) Cavendish in a <u>tour de force</u> of experimental detective work traced the drifts to convection currents generated by the different thermal time constants in the system. The large lead masses did not cool overnight as much as the mahogany case surrounding the balance arm; their radiation induced temperature gradients in the case and hence convection in the air inside. The Bernoulli pressures then set up transverse forces driving the suspended masses towards the sides of the case. Cavendish was able to keep track of this effect by sealing a small thermometer in each of the large masses and hanging another thermometer next to the instrument case, reading the thermometers through the same telescope with which he observed the positions of the balance.

With these methods, having also made other admirable calculations, including a correction for the lengthening of the period of the torsion pendulum by air drag and an elaborate analysis of the attraction of the mahogany case on the balance arm, Cavendish achieved a measure of the transverse masses to 2×10^{-10} g and fixed the mean density of the earth as 5.48 ± 0.10 [2]. The modern value is 5.57. It is pleasing that this "heroic experiment" [3] should have had so brilliant a result, proving as it did that the Earth's core has a density much greater than the typical densities of 2.5 to 3.0 for surface rocks.

The Cavendish experiment was repeated three or four times up to 1880, but not until C. V. Boys (1855 - 1944) renewed the attack on it in 1889 did Cavendish find a worthy successor. Boys introduced two practical advances on Cavendish's apparatus, added one profoundly important new experimental idea, and made innumerable improvements in detail. His first innovation, the one that started him thinking about the experiment, was his invention in 1887 of a method of drawing fine quartz fibres for the torsion suspension. Quartz fibres are still unrivalled in strength and elastic qualities. The second advance on Cavendish was less original but equally important: this was to measure the angular displacement by an optical lever, by observing telescopically the reflection of a scale from a mirror mounted on the balance arm. The optical lever had been invented in 1826 by J. C. Poggendorf and been applied to many instruments by Gauss, Weber, William Thomson and others: Thomson's mirror galvanometer had even been the subject of a poem by Maxwell*: but it was this simple device in combination with the quartz fibre that made possible Boys's third and greatest innovation: his analysis of the influence of the size of the apparatus on its sensitivity. Boys was, I suspect, influenced by William Froude's researches during the 1850's on the scaling laws for ship models in towing tanks, as well as the wider extension of the dimensional analysis around that time: be that as it may he made the remarkable discovery that the accuracy of Cavendish's experiment was improved by making it smaller. Boys's balance arm was $\frac{5}{8}$ inch long rather than 6 feet. The advantages of a smaller apparatus is first that

---

* The lamplight falls on blackened walls
  And streams through narrow perforations
The long beam trails o'er pasteboard scales
  With slow decaying oscillations
Flow,current, flow, set the quick light-spot flying
  Flow current, answer light-spot, flashing, quivering, dying.

in parody of Tennyson's "Bugle Song".

whereas the angular sensitivity remains constant if all the parts are scaled down, the external masses can be relatively larger. The attacting spheres used by Boys were $4\frac{1}{4}$ inches in diameter. To get an equivalent effect with Cavendish's apparatus their diameter would have had to be 27 feet. In order to realize this advantage with a short balance arm Boys separated the masses vertically as shown in Figure 2. A second advantage of smallness was the reduction of thermal

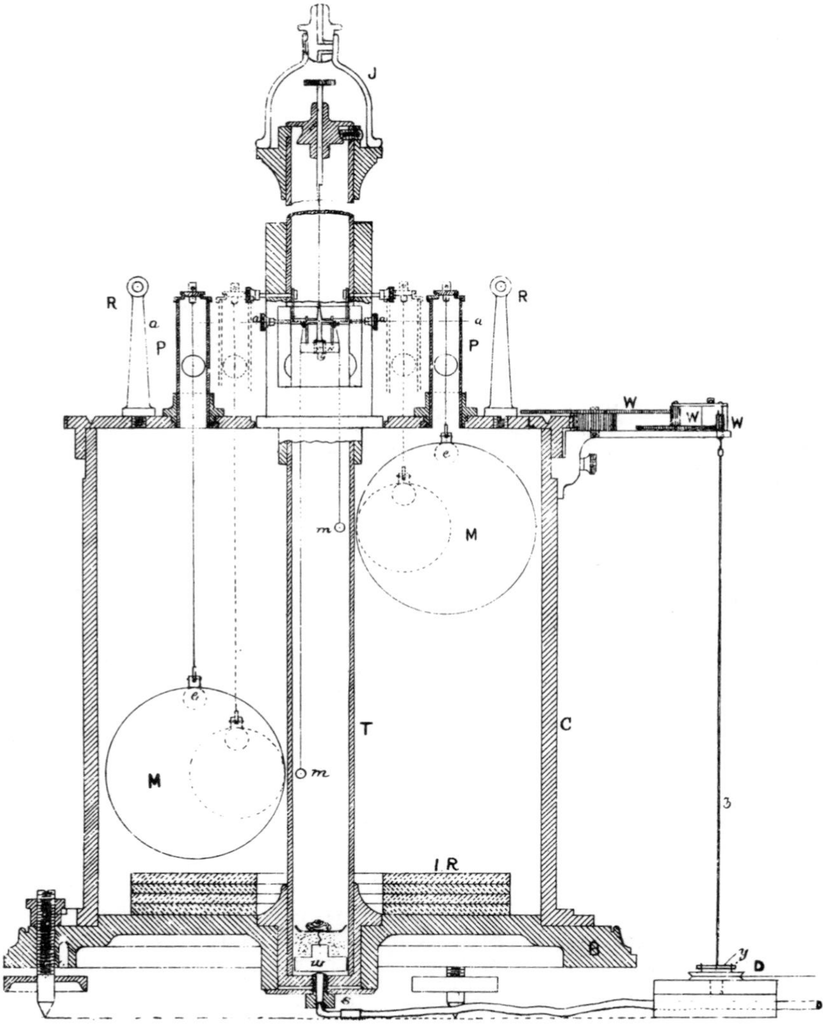

FIGURE 2: THE BOYS EXPERIMENT

time constants and temperature gradients: the convection currents which plagued Cavendish were eliminated. Boys considered evacuating his apparatus, but the lowest pressures then attainable were around 10^{-3} torr, a highly undesirable regime to work in because of the radiometer effect. The viscosity of air was a limiting factor in Boys' experiment. Unlike Cavendish Boys did reach the level of seismic disturbance. He worked at nights and weekends to avoid vibrations from passing trains; he saw the shock from an earthquake with epicenter in Romania , and the shaking from high winds in trees near his laboratory. All of these are problems painfully familiar to modern workers in our field.

Boys's optical lever was itself a triumph. Poggendorf had achieved a angular resolution of 5 arc-seconds, a little better than Cavendish's measurement. Boys could measure angular displacement of 0.7 arc-seconds (well below the 10 arc-second diffraction limit of the mirror) in accurately calibrated steps over a total scale width of 33,600 arc-seconds, that is very nearly to 16 bit accuracy --a precision few modern instruments reach. Consideration of this point leads to another view of Boys's achievement. He had separated the problems of optimizing his detector (the suspended bodies) and optimizing his position monitor (the optical lever). This is a recurrent theme in experimental gravitation.

Questions of mechanical stability were central in another classic investigation from the same era as Boys: the Michelson-Morley experiment. I have sketched the background elsewhere [4] . Michelson undertook the experiment in 1881 in response to a remark of Maxwell's about the change in velocity of light to be expected if the Earth is moving through a fixed aether of the kind advocated by Fresnel. Any earth-based experiment to see this effect requires a round-trip measurement and therefore depends on the second order quantity $(v/c)^2$ which Maxwell thought too small to detect. The alternative to Fresnel's theory was Stokes's theory of a convected aether. Michelson's interferometer compared the velocities of two light beams split and reunited in two orthogonal paths. Figure 3 illustrates the original apparatus with mirrors cantilevered on long steel arms. Although the structure was sensitive to vibrations and thermal distortions, the result was convincingly negative, and therefore in Michelson's view, decisive in support of Stokes's aether. Afterwards Lorentz pointed out an error in Michelson's analysis: he had not seen that a light ray moving transversely to the earth's motion would also suffer a second order time delay half that of the longitudinal ray, so the expected shift in the interference pattern would be half Michelson's prediction. Lorentz also discovered a difficulty in Stokes's aether. It is commonly said that Lorentz's observation made the error in Michelson's first experiment so large as to discredit it, but this is an

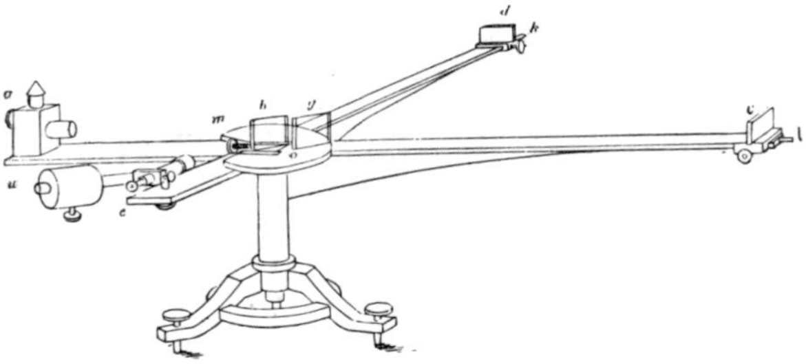

FIGURE 3 MICHELSON'S APPARATUS (1881)

exaggeration. Still a measurement so fundamental needed repeating. At the 1884 Baltimore Lectures Sir William Thomson (Lord Kelvin) used his magnetic persuasiveness to convice E. W. Morley that it must be done again. Figure 4 illustrates the famous apparatus of 1887 with mirrors mounted on a granite slab floating on mercury and path lengths increased by multiple reflection to eight times that of the original apparatus. The measurement was accurate to one-twentieth of an interference fringe or about 1% of the hypothetical displacement.

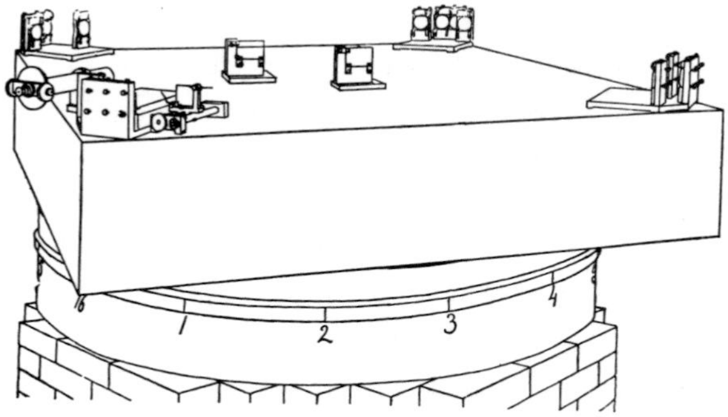

FIGURE 4

THE MICHELSON-MORELY EXPERIMENT (1887)

The many, still far from satisfactory, discussions of the historical and scientific significance of this experiment and its offshoots need not delay us here. The most searching theoretical analysis is still probably the one given by H. E. Ives in 1937. To the experimentalist an equal interest attaches to the change from Figure 3 to Figure 4, which speaks of one thing: mechanical stability. A measurement to a twentieth of a fringe corresponds to a 130 Å displacement of any mirror. A steel I-Beam 100 cm long and 4 cm deep deflects under its own weight by 1 arc-second, which means that a 2 cm mirror resting on it is tilted back at its upper edge by 2000 Å through beam flexure. Low frequency seismic vibrations of amplitude 10^{-3} g then cause motions of 2 Å and vibrations near the resonant frequency of the beam are much larger. Thermal distortions are even more serious. A temperature difference of 0.02° across the beam warps it by 1 arc-second. Such were the effects Michelson and Morley had to contend with. When, as in the Stanford Gyro Relativity experiment, one aims for measurements of 0.001 arc-second accuracy, questions of mechanical stability shape the whole experimental approach.

The famous experiment on the equivalence of gravitational and inertial mass, done first in 1890 by Baron Roland von Eötvös (1848-1919), and repeated over the years up to 1922 by Eötvös and his two colleagues Desiderius Pekár and Eugen Fekete, is notable more for the originality of its idea than for novelty in measurement technique. The apparatus was a gravity gradiometer invented by Eötvös for geophysical work, whose sensitivity to field gradients made it in some ways rather ill-adaped to an equivalence principle measurement. As a piece of instrumentation it fell far short of Boys's apparatus. But nothing can detract from the originality of the idea. Rarely does anyone think up an experiment based on a slight modification of a standard instrument that advances the accuracy of a fundamental measurement by four orders of magnitude. When H. H. Potter repeated Newton's pendulum measurement of equivalence in 1923 his accuracy was only one part in 10^5, three orders of magnitude short of Eötvös's. Figure 5 illustrates Eötvös's apparatus [5]. The experiment consisted in suspending two masses of different composition from a torsion balance, rotating the torsion head through 180° and checking whether the balance arm turns through exactly the same angle. Since both masses are simultaneously subject to gravity and to centrifugal acceleration from the Earth's rotation, any departure from equivalence causes a torque

$$\Gamma^E = \tfrac{1}{2}\eta \, MDf \sin\alpha \qquad (1)$$

where M is the sum of the masses, D the length of the balance arm, f the driving acceleration (the centrifugal acceleration of 1.4 cm/sec^2), α the angle between the balance arm and north, and η the Eötvös ratio

$$\eta = \frac{2\left[\left(\frac{M}{m}\right)_A - \left(\frac{M}{m}\right)_B\right]}{\left(\frac{M}{m}\right)_A + \left(\frac{M}{m}\right)_B} \tag{2}$$

where m and M are the inertial and passive gravitational masses of the two materials A and B. The quantity ηf may conveniently be called the Eötvös acceleration.

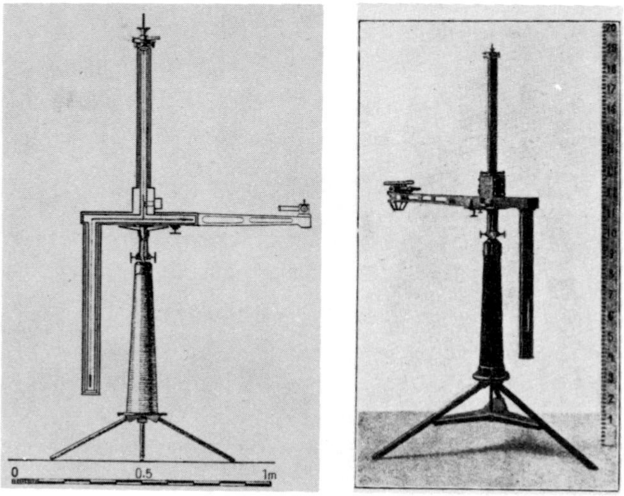

FIGURE 5 EÖTVÖS' APPARATUS

In appearance Eötvös' apparatus resembles Boys' in that the masses are hung at different levels, but this feature was conceived independently of Boys, and for a different reason, namely, to make the balance sensitive to vertical gravity gradients. The chief limitation of the apparatus aside from its sensitivity to gradients, was elastic hysteresis in the suspension. Hysteresis errors can be avoided, as Eötvös and his colleagues recognized, by keeping the apparatus fixed and looking for a daily torque from the sun's acceleration. This was the approach taken by Roll, Krotkov and Dicke [6] and by Braginsky and Panov [7] in their repetitions of the Eötvös experiment. Using the sun as source is not without its disadvantages. The driving acceleration is only 0.6 cm sec^{-2} , as compared with 1.4 cm sec^{-2} for the Earth's centrifugal acceleration, and the apparatus becomes susceptible to other diurnal disturbances -- for example Earth tides and temperature effects driven by the day-night cycle.

2. MORE RECENT EXPERIMENTS AT ROOM TEMPERATURE AND THE APPROACH TO FUNDAMENTAL LIMITS

In 1925 W. J. H. Moll and H. C. Burger invented a device to read an optical lever automatically from the differential heating of the light spot on a split thermopile [8] . The angular resolution of this "thermoelectric relay" was about 1/20 arc-second. Applying it to a very stable galvanometer they noticed a small jittering motion in the output, of amplitude about 0.5 arc second, which they attributed to microseismic disturbance, but which G. Ising [9] in "one of the most vital papers in the history of physical observations" [10] identified as the Brownian motion of the galvanometer. Ising had discussed this instrumental limit as early as 1917, following Smoluchowski and de Haas-Lorentz, but his paper of 1926 -- one hundred years after Poggendorf's invention of the optical lever -- was its first recognition in an actual instrument. The corresponding statistical noise limit in an electrical circuit, the Johnson noise, was discovered by J. B. Johnson and interpreted by H. Nyquist one year later, in 1927.

A fundamental measurement limit exists for the optical lever as well as for the suspended body. The detector measures the location of the diffraction pattern from the mirror by a finite number of photons; its output is disturbed by random fluctuations in intensity of the two halves of the beam. Optical levers and autocollimators that reach photon noise limits have been widely built; the best resolution so far being in R. V. Jones's [10] split-grating optical lever which can detect angular changes of a 2 mm x 2 mm mirror to about 10^{-5} arc-sec in a 10Hz bandwidth. Formally the limiting angular resolution with diffraction limited optics is

$$\delta\Theta \sim \frac{1}{D^2}\sqrt{\frac{hc\bar{\lambda}}{\phi\varepsilon}\Delta\nu} \qquad (3)$$

where h is Planck's constant, c the velocity of light, D the dimension of the mirror, $\Delta\nu$ the bandwidth, ϕ the light flux on the mirror, $\bar{\lambda}$ the colour temperature of the source, and ε the overall photoelectric efficiency, that is, the ratio of the number of photoelectrons excited in the detector to the number of photons impinging on the mirror, or, in other words, the product of the optical efficiency of the light system with the quantum efficiency of the photodectector.

For a body of inertia I suspended from a torsion fibre, and lightly damped, having a damping coefficient β and a natural period τ_0, the Brownian motion occurs as oscillations of period τ_0 coherent over times of order $Q\tau_0$, where $Q = \pi/\beta\tau_0$ is the quality factor of the system. The amplitude $\langle\Delta\Theta\rangle$ of the oscillations is

$$\langle\Delta\Theta\rangle \sim \frac{\tau_0}{2\pi}\sqrt{\frac{kT}{I}} \tag{4}$$

where k is Boltzmann's constant and T the absolute temperature. The magnitude of $\langle\Delta\Theta\rangle$ for a body of mass 10 gm and diameter 1 cm, hung at room temperature with a period of 10 sec, is 0.6 arc-sec. Defining a discrimination factor $\partial_B \equiv \frac{\delta\Theta}{\langle\Delta\Theta\rangle}\sqrt{\frac{\nu_0}{\Delta\nu}}$ which gives the precision in $\delta\Theta/\langle\Delta\Theta\rangle$ in a given fraction of the natural oscillation period, and writing the moment of inertia of the suspended body proportional to ρD^5, where ρ is the density and D the characteristic dimension we have

$$\partial_B = C\left(\frac{hc}{k}\right)^{\frac{1}{2}}\left[\frac{\rho D}{\tau_0{}^3 T}\right]^{\frac{1}{2}}\left[\frac{\overline{\lambda}}{\varepsilon\phi}\right]^{\frac{1}{2}} \tag{5}$$

where the terms in the first square bracket characterize the suspended body and those in the second square bracket characterize the optical lever. The quantity hc/k has the value 1.4 cm^{-1} $°K^{-1}$. The quantity C depends on the shapes and relative size of the mirror and suspended body: for a sphere surmounted by a weightless circular mirror of equal diameter C is 0.063. A criterion for resolving the Brownian motion is that ∂_B be less than unity, though greater resolution may be useful. R. V. Jones and C. W. McCombie [11] used a discrimination factor of about 5×10^{-4} in their experiments on the statistical mechanical limits on galvanometers. Equation (5) tells the light flux needed for a given discrimination; alternatively it tells the dimensions of an apparatus that can usefully be operated with a given optical lever. The discrimination requirement becomes more severe as the size and density of the body are increased, the period shortened and the temperature lowered.

The first measurement in our field that approached a Brownian limit was the negative experiment on magnetism and the Earth's rotation started in 1949 by P. M. S. Blackett. Blackett had become interested in the hypothesis due originally to Arthur Schuster (1851 - 1934) that the primary magnetic fields of the Earth, the sun and other massive rotating bodies are a fundamental property of their rotation to be understood eventually in terms of some unified field theory. As formulated by Blackett the hypothesis may be written as an equation between magnetic field and a virtual current density vector

$$\text{curl}\ \underline{H} = 4\pi\beta\frac{G^{\frac{1}{2}}}{c}\ \rho\ a\ \underline{\omega} \tag{6}$$

where G is the gravitational constant, c the velocity of light, $\underline{\omega}$ the absolute angular velocity, ρ the matter density at distance a from the rotation axis, and β is a numerical constant with value estimated from the known fields of the Earth, the sun and certain stars as about 0.3. A gold cylinder of dimension 10 cm X 10 cm rotating with the Earth generates on this hypothesis a magnetic field of 3.4×10^{-8} gauss at 5 cm above its surface. Blackett [12] developed an extremely sensitive astatic magnetometer, applying design principles that has been worked out in 1926 for the Paschen galvanometer by H. A. Daynes [13] and A. V. Hill [14] . Hill, who later achieved fame as a physiologist, had extended Ising's work to show that with a given optical lever there is an upper limit to the useful suspension period τ_0 of the galvanometer and hence a lower limit on current sensitivity. Blackett found that the astatic magnetometer, when optimized with respect to the choice and shape of magnet, had a minimum detectable field H_0 proportional to $1/\tau_0 D^{\frac{1}{2}}$ set by the Brownian motion; but there was a practical upper limit to τ_0 of about one minute set by the second order changes d^2H/dt^2 in the ambient magnetic field, and practical upper limits on the dimension D of the magnets set by the optical design and the reaction field of the magnets on the specimen to be measured. Blackett's most sensitive instrument was capable of detecting a field of 3×10^{-10} gauss in a single observation of 30 seconds duration, an improvement of three orders of magnitude over previous magnetometers. Its moment of inertia was 0.165 gm cm^2 which made the amplitude of Brownian motion with a 30 sec. period 0.7 arc-sec. The experiment showed that no rotational magnetic field existed greater than $[4.2 \pm 2.6]\%$ of the prediction of equation (6). As an explanation of the Earth's magnetic field the hypothesis was disproved. Interest in the magnetic fields of massive rotating bodies continues, however. The extreme Kerr-Newman solution for an isolated rotating black hole yields of gyromagnetic ratio in physical units of $\sqrt{G/c}$, corresponding to Blackett's formula with a β of unity, while Ruffini and Treves [15] have shown that even in the classical limit a magnetized sphere spinning in vacuo acquires a surface charge and has a gyromagnetic ratio proportional to the same quantities.

The equivalence principle experiment of Roll, Krotkov and Dicke [6] also called for a study of fundamental limits with results that are in instructive contrast to Blackett's. In Blackett's magnetometer, with its relatively small suspended body and short observation time, the dominant limitation was Brownian motion, and since the amplitude of the motion was only one order of magnitude less than the diffraction width of the mirror Blackett was able to ignore photon noise and made do with an old-fashioned visual optical lever simpler even than the one used by Boys. Very different was the situation of Roll, Krotkov and Dicke. Their balance had a moment of inertia of 270 gm cm^2 , three orders of magnitude larger than Blackett's, a natural period of four minutes and a Brownian motion of 0.09 arc-sec amplitude. With the design goal of measuring the Eötvös ratio to 1 part in 10^{11}, the detector had to be capable of resolving a signal of 4.5×10^{-4} arc-sec with 24 hour period. This, though a factor of 200 less than the Brownian motion was relatively easy to separate from it because the period was so long. On the other hand, the resolution, being four orders of magnitude less than the diffraction width of the mirror, called for a sophisticated optical lever. Not only did the resolution have to approach photon noise limits: the apparatus had to be stable, optically and mechanically, to 10^{-3} arc-sec a day. For mechanical stability Roll, Krotkov and Dicke established close temperature equilibrium throughout the apparatus. Drifts in the optical lever were minimized by using a single photomultiplier with a vibrating wire at the image plane, oscillating across the diffraction image with frequency ν (3kHz). With the image exactly centered on a detector of this kind the photomultiplier sees only even harmonics of ν, but as the balance turns and shifts the pattern off-centre the fundamental frequency begins to appear in the output. The phase relative to the driving oscillator tells the sign of the displacement; for small displacements the amplitude is proportional to the rotation angle. Vibrating wire or vibrating slit detectors have been used in many instruments. They have two merits. Changes in sensitivity of the photomultiplier do not affect the null point, and the angular sensitivity can be calibrated internally from the ratio of the fundamental frequency to the first even harmonic, independent of the light intensity and photomultiplier gain. The chief question is the stability of the centre of vibration of the wire.

The principal limitations on Roll, Krotkov and Dicke's experiment were (1) seismic disturbances (2) temperature effects, (3) gravity gradient torques. Seismic noise coupled from pendulum to torsional modes of the balance through non-linearities in the system. Large shifts were observed from construction activity near the instrument site as well as quarry blasting five miles away; like Boys the experimenters were driven to working mostly at weekends. Temperature changes caused distortions of the optical lever, torques on the balance and drifts in the electronics. Roll, Krotkov and Dicke considered it hopeless to account for all the effects individually; instead they controlled the temperatures as best they could and correlated the results with the measured temperature coefficients -- exactly the procedure Cavendish had followed 170 years earlier.

The gravity gradient torque deserves special comment because of its importance in later experiments. Consider a balance with unequal moments of inertia I_1, I_3 situated at a distance R from a point mass M. It will experience a torque

$$\Gamma^{g} = \tfrac{3}{2}\,(I_3 - I_1)\,\frac{GM}{R^3}\,\sin 2\alpha \tag{7}$$

where α is the angle of the principal axis of the balance to the vector $\underline{R}$. Since Γ^g depends on $\sin 2\alpha$ it is a maximum when the principal axis is at 45° to $\underline{R}$; furthermore the sun's gravity gradient exerts a torque with a period of 12 hours in contrast to the 24 hour period of the Eötvös torque.

The quantity $(I_3 - I_1)$ may be written $J_2 m l^2$, where m is the mass of the balance, l the radius of gyration and J_2 the quadrupole coefficient. For a balance with two masses made in half rings of diameter D the ratio of amplitudes of the gravity gradient and Eötvös torques is

$$\frac{\Gamma^g}{\Gamma^E} = \frac{3\pi}{8}\,\frac{J_2 D}{\eta f}\,\frac{GM}{R^3} \tag{8}$$

or in the particular case when M is the source of Eötvös acceleration

$$\left(\frac{\Gamma^g}{\Gamma^E}\right)_{\text{source}} = \frac{3\pi}{8}\,\frac{J_2 D}{\eta\, R} \tag{9}$$

where R is the distance to the source. For the sun R is 1.5×10^{13} cm, so to make the gravity gradient torque less than the Eötvös torque $J_2 D$ has to be less than 140: a modest requirement. It is a different story, however, if one tries to improve the measurement by using a torsion balance in a satellite with the earth as source. Then R is 7×10^8 cm and to make Γ^g/Γ^E unity $J_2 D$ must be less than 8×10^{-3} cm for an experiment to measure η at the 10^{-11} level and less than 8×10^{-9} cm for an experiment at the 10^{-17} level. Roll, Krotkov and Dicke had a $J_2 D$ of 3×10^{-2} cm; a practical manufacturing limit in a balance of reasonable size is 3×10^{-4} cm, five orders of magnitude larger than one would like in a 10^{-17} experiment. The torsion balance in a satellite is not a good idea.

It is useful to transpose (8) into a condition on the distance r which a body of density ρ and diameter d may be allowed to approach the balance before causing a disturbance comparable to the Eötvös torque

$$\frac{r}{d} > 0.85\left(\frac{G J_2 D}{\eta f}\right)^{1/3} \rho^{1/3} \tag{10}$$

For Roll, Krotkov and Dicke's apparatus with $J_2 D \sim 3 \times 10^{-2}$ and $\eta f \sim 6 \times 10^{-12}$ r/d had to be less than $6.6\,\rho^{1/3}$. A 100 kg man could not approach closer than

4m or a 10-ton truck closer than 18 m without disturbing the balance. With more ambitious design goals the restrictions become still more severe.

The foregoing arguments, due originally to Roll, Krotkov and Dicke, and in their present form to P. W. Worden, Jr., lead to a discussion of the size of the apparatus that would have appealed to Boys. From equation (1) the Eötvös torque increases as D^4 as the apparatus is scaled up. This being so it is tempting to argue, as J. Faller has done, that Earth-based equivalence principle experiments can be made enormously more sensitive by using a bigger apparatus. The catch is that Γ^g/Γ^E gets worse. Consider an attempt to improve the measurement of η by four orders of magnitude by means of a factor of ten increase in the size of the apparatus. The lowest plausible J_2 in the larger size is about 10^{-3}. Then from equation (10) no man can be allowed within 90 m of the apparatus; no truck within 400 m. Or consider atmospheric pressure effects. Taking ρ for air as 10^{-3} gm cm^{-3}, one finds that pressure differences in the nearby atmosphere as small as 0.2 mm can upset the balance, while the minimum r/d for a hurricane 200 km in diameter with a 20% pressure drop is 10. No hurricane can be allowed within 2000 km of the apparatus.

Although the Brownian motion in Roll, Krotkov and Dicke's experiment was 200 times the designed sensitivity, it was, as we have seen, characterized by the four-minute period of the balance and could be averaged over the 24 hour period of the measurement. It is important to grasp how far averaging can go. According to the mechanical counterpart of Nyquist's formula the mean square fluctuation torque on a body of inertia I during an observation time S is

$$\langle\Gamma^2\rangle = 2I\beta kT/S \tag{11}$$

where β is the damping coefficient. For a balance of total mass M with the two bodies formed in half-rings of diameter D, the mean value of the Eötvös torque through each half cycle is $2MD\eta f/\pi^2$. The moment of inertia is $MD^2/4$. The root mean square error in determining η in a time long compared with 12 hours is

$$\langle\eta\rangle \sim \frac{3.4}{f}\sqrt{\frac{\beta kT}{MS}} \tag{12}$$

where the numerical factor 3.4 is $\pi^2/2\sqrt{2}$. The damping coefficient β may be replaced by $\pi/\tau_0 Q$, where Q is the quality factor and τ_0 the period of the balance. Roll, Krotkov and Dicke had $f = 0.6$ cm sec^{-2}, $M = 90$gm, $\tau_0 \simeq 230$sec; according to the discussion on page 458 of their paper the natural Q of the balance was 10^5. The limit from (12) on determining η in one daily cycle is 1.5×10^{-13}: two orders of magnitude below the observed experimental limit of

3×10^{-11}.

The Nyquist limit on equivalence principle experiments was first discussed by V. B. Braginsky [16]. Equation (12) is due to P. W. Worden, Jr. and C. W. F. Everitt [17] who based their analysis on the discussion by C. W. McCombie [18]. It is worth emphasizing how fundamental equation (12) is. Except for slight differences in numerical factor every equivalence principle experiment is subject to it. No elaborations to the apparatus, however useful they may be on other grounds can change the fluctuations associated with natural damping. The addition of a controller with negative damping, for example, may give an effective Q higher than the natural damping: it does not reduce thermal noise. Reference to this point brings out one of the few mistakes in Roll, Krotkov and Dicke's paper. They added a controller with positive damping to kill off the thermal oscillations of the balance, making it in effect critically damped, and said that the noise was thereby reduced through a reduction in the apparent temperature of the system. This statement is plainly wrong. An increase in natural damping would increase the noise; an addition of feedback damping should ideally leave the noise unchanged. The only sense in which feedback might be said to reduce the noise is that for a given Q the noise is less if damping is provided dynamically rather than dissipatively, in which case the feedback could be thought of artificially as lowering the temperature. Possibly this was the meaning intended by Roll, Krotkov and Dicke.

Although Roll, Krotkov and Dicke's result was two orders of magnitude from the limit set by equation (12), Braginsky [7] seems to have thought it was limited by Nyquist fluctuations. Perhaps he assumed the feedback generated noise. For critical damping, with a Q of unity instead of 10^5, equation (12) yields a limit of 3×10^{-11} on η, very near the observed limit. Of course if the feedback servo has significant internal losses it can add noise, just as the external circuit of a galvanometer does. However, since Roll, Krotkov and Dicke adjusted the feedback gain to minimize the observed noise [19] the electronic damping must have been primarily dynamic not dissipative, so the experiment was most unlikely to have been limited by Nyquist fluctuations. Be that as it may in 1971 Braginsky and Panov repeated the experiment with an undamped system. Their apparatus had several differences in design from Roll, Krotkov and Dicke's. The balance was somewhat larger; it was made with eight bodies rather than three to reduce gravity gradient torques; and the period was much longer -- six hours instead of four minutes. The longer period increased the angular deflection for a given Eötvös signal, the predicted deflection for an η of 10^{-12} being 2×10^{-2} arc-sec as compared with 5×10^{-5} arc-sec in the Roll-Krotkov-Dicke apparatus. The Brownian motion was also larger, the characteristic amplitude, as calculated from (4), being 15 arc-sec instead of 0.09 arc-sec. The damping coefficient β was not much different from Roll, Krotkov and Dicke's: the natural damping time for a balance of moment of inertia 400 gm cm^2 was about two years as compared with one year for a 270 gm cm^2 body. Again the fundamental limit on η from equation (12) was 1.5×10^{-13}. The estimated limit from measurement was 0.9×10^{-12}.

Null experiments are always difficult to assess. The published accounts of Braginsky and Panov's experiment are much less complete than Roll, Krotkov and Dicke's. The chief advantage seems to have been a quieter seismic environment; the chief shortcoming the method of reading the orientation of the balance. A laser beam was reflected over a distance of 12 m to a photographic film on a rotating drum, the drum being inclined to the light path at an angle of 0.2 rad to increase the effective path length to 50m. The spot was 0.3 cm wide; its position was estimated microscopically to 5×10^{-4} cm. Such precision seems surprising in visual examination of so wide a diffraction image, especially since the spot was swinging with 6 hour period over a range of ±0.7 cm (the observed displacement, which corresponds to a balance swing of ±15 arc-sec, quite close to the calculated Brownian motion). Qualms on that point were partly offset, however, by averaging data from three observers. Still more surprising to anyone who has strugged with long optical paths is the absence of errors from atmospheric refraction. According to the mirage formula a beam of light passing a distance ℓ through a gas of refractive index n is deflected under a transverse temperature gradient dT/dx through a distance x_n

$$x_n = \frac{n-1}{2n} \frac{\ell^2}{T} \frac{dT}{dx} \tag{13}$$

If the light source stands next to the detector x_n has to be doubled. With ℓ = 12 m, $x_n/\ell < 10^{-7}$ rad, and n for air at STP = 1.00029 the transverse temperature gradient cannot exceed 7×10^{-5} °C cm^{-1}. The daily temperature cycle of the room was 2×10^{-2} °C [7] . The stability seems hard to achieve. A comparable problem is thermal distortion of the apparatus, including the building in which it was housed. A body of length ℓ and expansion coefficient α under a transverse temperature gradient dT/dx is deflected through a distance x_α equal to $\frac{1}{2}\alpha\ell^2\, dT/dx$, which also has to be doubled if the source stands next to the detector. For masonry $\alpha = 7 \times 10^{-6}$ °C^{-1}, so to make $x_\alpha/\ell < 10^{-7}$ the transverse temperature gradient must not exceed 1.1×10^{-5} °C cm^{-1}. Again this is an extraordinarily small gradient.

The thermal distortion formula is sometimes conveniently written in terms of the thermal conductivity K and transverse heat flux Φ for the apparatus, where $\Phi = KdT/dx$. Then

$$x_\alpha = \frac{1}{2}\left(\frac{\alpha}{K}\right) \ell^2 \Phi \tag{14}$$

and any apparatus used in precise angular measurements should be constructed of materials with low α/K.

Irrespective of the final assessment of Braginsky and Panov's experiment equations (13) and (14) show how undesirable long optical paths are. The angular errors x_n/ℓ and x_α/ℓ both increase with ℓ . Whatever success Braginsky and Panov had depended on the enormous period of the balance, which, though it did not alter the noise limit from equation (12) did increase the signal amplitude for a given Eötvös ratio.

It is interesting to return to Hill's argument that there is an upper limit to the useful period of a torsion balance set by competition between the Brownian motion and the resolution of the optical lever. In its original form the argument is not rigorous but it may be modernized using equation (3) and (11). Equation (11) gives the fundamental limit on torque measurement from the Nyquist criterion. The question is whether the optical lever has enough resolution to reach that limit. Assume the torque is either constant or sinusoidal with a period much longer than the natural period τ_0 of the balance. The deflection Θ due to a torque Γ is Γ/κ, and κ , the elastic constant of the suspension, equals $4\pi^2 I/\tau_0{}^2$. The minimum torque the optical lever can resolve is therefore

$$\delta\Gamma = \frac{4\pi^2 I}{\tau_0{}^2} \delta\Theta \tag{16}$$

where $\delta\Theta$ is the quantity from equation (3). To reach the fundamental limit, $\delta\Gamma$ must be less than $\langle\Gamma^2\rangle^{\frac{1}{2}}$ from equation (11). The ratio $\delta\Gamma/\langle\Gamma^2\rangle^{\frac{1}{2}}$ is identical with the discrimination equation (5) except that the term in the first square bracket of (5) is multiplied by $2\pi\,\tau_d/\tau_0$, where $\tau_d = Q\tau_0$ is the characteristic damping time of the system. Transposing to a limit on τ_0

$$\frac{\tau_0{}^4}{\tau_d} > C'\left(\frac{hc}{k}\right)\left[\frac{\rho D}{T}\right]\left[\frac{\overline{\lambda}}{\varepsilon\phi\partial^2{}_N}\right] \tag{17}$$

where C' is $2\pi C^2$ from (5) and ∂_N is a factor similar to ∂_B defined from the ratio of sensor noise to Nyquist limit rather than sensor noise to Brownian amplitude. Again ∂_N less than unity is the criterion for a good measurement. For both Dicke's and Braginsky's experiments, the combination of quantities $C'\rho D$ was about 10^{-5}, τ_d about 10^8, $\varepsilon\phi$ about 10^2 erg/cm^2, $\bar{\lambda}$ about 6×10^{-5} cm, making the shortest allowable suspension period τ_0 in each case about 10 sec.

Further discussion hinges on practical considerations. If one is interested as Hill was, in making multiple observations with a galvanometer, any suspension period longer than that from equation (17) is undesirable in theory and probably pointless in practice. For equivalence principle experiments on the other hand, with a fixed driving frequency, there is no intrinsic reason for sticking to short periods. Provided the optical lever reaches photon noise limits the practical bounds are set by null shifts and non-linearities. Null shifts in the readout become progressively less troublesome as the signal amplitude is increased by increasing τ_0, but if τ_0 is too long Brownian motion, or the signal itself, may cause errors through optical or mechanical non-linearities in the system. Alternatively, as in Blackett's magnetometer, there may be an upper limit on τ_0 from non-linearities in the time rate of change of an external disturbing torque. Such questions can only be resolved by detailed study of the performance limitations on each individual experiment.

An apparatus of a quite different kind where Brownian motion is critical is the much discussed gravitational wave antenna developed by J. Weber from 1959 onwards, and copied by several other groups since 1970. An aluminum bar of length 1.55 m and diameter 0.66 m was hung from steel cables inside a vacuum chamber, piezoelectric strain gauges cemented to the bar sensed displacements down to 10^{-15} cm in 150 mS sampling time. The resonant frequency was 1660 Hz and the mechanical Q about 10^5. The bar responds to gravitational waves much as a Hertzian dipole responds to radio waves. For a favourably polarized source whose line of sight is perpendicular to the long axis of the bar the energy absorbed in the $n^{\underline{th}}$ longitudinal mode from an incoming gravitational wave pulse of energy spectral density $F(\nu)$ is

$$E_s = \frac{8}{\pi} \left(\frac{G}{c}\right) \frac{M}{n^2} \frac{v_n^2}{c^2} F(\nu_n) \tag{18}$$

where M is the mass of the bar, ν_n the resonant frequency of the mode, and v_n the velocity of sound at ν_n. In 1967 Weber began to detect pulses not characteristically seismic in origin, and in 1970, after coincidence checks between bars 1600 km apart in Maryland and Argonne, announced statistically significant events which he identified as gravity waves from a source near the centre of the galaxy.

It is convenient to defer the main discussion of noise limits on gravitational wave detectors until Section 5. Controversy over the results apart, Weber's apparatus is interesting in an instrumental view for its analogy to those just described on torsionally suspended bodies. One is looking for a gravitational impulse on a massive detector (the bar) subject to Brownian motion using a sensor (the piezoelectric crystal) which is also subject to noise. With some assumptions about noise in the bar and the crystal, Gibbons and Hawking [20] found an optimum sampling time τ_s for Weber's apparatus. In times shorter than τ_s the sensitivity is limited by Johnson noise in the crystal; in longer times it is limited by the task of resolving the ringing signal in the bar from Brownian motion. Good coupling is critical. Under optimum conditions gravitational pulses can be detected if they impart to the bar an energy

$$E_{min} > 0.8\ kT_B \sqrt{\frac{\tan\ \delta}{\beta Q_B}} \tag{19}$$

where T_B and Q_B are the temperature and quality factor of the bar, $\tan\ \delta$ the loss tangent of the crystal, and β the coupling coefficient, that is, the proportional of elastic energy in the bar that can be extracted electrically from the crystal in one cycle. For Weber's bar with a β of 5×10^{-6} the optimum τ_s was 0.4 sec, and a single measurement could detect a gravitational pulse with energy approaching $kT_B/12$.

Experiments on gravitation may be divided into two classes depending whether the test object is a massive body, as in the Cavendish experiment or an electromagnetic wave, as in the Michelson-Morley experiment. Apart from the experiments just described, some improved Cavendish experiments, and laser ranging measurements on the Earth-moon system, most of the new experiments completed since 1960, have been on electromagnetic radiation. Examples are the Mössbauer measurements on gravitational redshift by Pound, Rebka and Snider [21] ; the radar-ranging measurements of the time delay of signals passing the sun suggested by Shapiro in 1963; the measurements by Dicke and Goldenberg [22] and by Hill [23] and his colleagues on the sun's visual oblateness; and Hill's daylight star detector for the Arizona solar telescope. The Dicke-Goldenberg experiment is based on a rotating slit detector; the daylight star detector on a vibrating slit counterpart to the vibrating wire for Roll, Krotkov and Dicke's optical lever. Ultimately all experiments on electromagnetic waves are limited by photon noise; the practical limits are complex. In optical telescopes the limitations are atmospheric seeing and mechanical stability. In radar and laser ranging the disturbances are many and large; they have to be elaborately modelled and taken out in data analysis.

The next sections describe four gravitational experiments with massive bodies currently being developed at Stanford University and elsewhere. Three -- the Gyro Relativity experiment, the Stanford-LSU-Rome gravitational wave experiment and a new equivalence principle experiment -- depend on measuring small angular and linear displacements of laboratory-sized objects, just as the Cavendish and Eötvös experiments did. Ultimately they too are limited by Brownian motion and measurement noise. Since these are lowered by lowering the temperature a prima facie case exists for applying cryogenic techniques to the experiments, and in fact all three are done at low temperatures. However the immediate reasons for low temperature operation are not noise, but other more technical considerations: improved mechanical properties of materials, reduction of disturbances from residual gas and black body radiation, and the use of superconducting magnetic shields to stabilize the field or reduce it to very low values. Critical to all three experiments is the use of Josephson junction devices as position detectors. With so much in common one might think the experiments could be reduced to common design principles. This is not so. Each is subject to its own constraints which make for variety; plenty of room exists for specific ingenuity.

In the fourth experiment the test bodies are a pair of drag-free satellites in polar orbit around the Earth.

3. THE GYRO RELATIVITY EXPERIMENT

The idea of testing General Relativity by gyroscopes was discussed soon after Einstein's theory appeared by deSitter [24] , Schouten [25] , Fokker [26] and Eddington [27] . Several of these authors suggested looking for the small geodetic precession of the Earth due to its motion about the sun, but since the change of the Earth's polar axis from this cause is only 8×10^{-3} arc-sec/year, it is a factor of five smaller than the uncertainty in the Chandler wobble. During the 1930's Blackett [28] examined the prospects for a laboratory experiment but concluded that with then-existing technology the task was hopeless. In 1959, two years after the launch of Sputnik, and following also the improvement in gyroscope technology since World War II, L. I. Schiff [29] and G. E. Pugh [30] independently proposed experiments with gyroscopes in space.

3.1 Description of the Experiment

Schiff's formula for the relativistic precession-rate of an ideal torque-free gyroscope in free fall about a rotating massive sphere is

$$\underline{\Omega} = \frac{3GM}{2c^2R^3} (\underline{R} \wedge \underline{v}) + \frac{GI}{c^2R^3} \left[\frac{3\underline{R}}{R^2} (\underline{\omega}_e \cdot \underline{R}) - \underline{\omega}_e \right] \tag{20}$$

where $\underline{R}$ and $\underline{v}$ are the coordinate and velocity of the gyroscope, and M, I and ω_e are the mass, moment of inertia and angular velocity of the central body. The first term gives the spin-orbit, or geodetic, precession Ω^G due to motion of the gyroscope through the gravitational field; the second gives the spin-spin, or motional, precession Ω^M due to rotation of the central body. Integrated around an elliptic orbit and expressed as the rate of change $\underline{n}_s$ of the unit vector $\underline{n}_s$ defining the gyro spin axis ($\dot{\underline{n}}_s = \Omega \wedge \underline{n}_s$) the geodetic term is

$$\dot{\underline{n}}_s^G = \frac{3(GM)^{3/2}}{2c^2a^{5/2}(1 - e^2)^{3/2}} \left[\underline{n}_0 \wedge \underline{n}_s \right] \tag{21}$$

where a and e are the semi-major axis and eccentricity and $\underline{n}$ is the unit orbit-normal. The motional term is

$$\dot{\underline{n}}_s^M = \frac{I\omega_e}{2c^2a^3(1 - e^2)^{3/2}} \left[\underline{n}_e \wedge \underline{n}_s + 3(\underline{n}_0 \wedge \underline{n}_e)(\underline{n}_0 \wedge \underline{n}_s) \right] \tag{22}$$

where $\underline{n}_e$ is the unit vector defining the Earth's axis. The geodetic term therefor scales as $a^{-5/2}$ and the motional term as a^{-3}. The geodetic precession lies in the orbit plane and in a 500 km circular orbit around the Earth amounts to 6.9 arc-sec/year. The motional precession depends on the orbit inclination c'. For a gyro perpendicular to the Earth's axis its magnitude in a 500 km polar orbit is + 0.05 arc-sec/year where the positive sign implies a precession in the same sense as the Earth's rotation; in an equatorial orbit it is -0.10 arc-sec/year; in an orbit where $\cos 2i = - {}^1/_3$ ($i = 54°16'$) it vanishes. The goal of the experiment is to make a gyroscope with residual drift-rate 10^{-16} rad/sec (6×10^{-4} arc-sec/year). If achieved this would measure Ω^G to 1 part in 10,000 and Ω^M in polar orbit to 1 part in 70.

The Schiff motional effect is sometimes miscalled the Lense-Thirring effect. The Lense-Thirring effect is a nodal drag of the orbit plane of a moon in orbit around a massive rotating body. It is measured by the twin-satellite experiment described in Section 6 and has different magnitude and different vector form from the Schiff effect.

The gyroscope precessions are measured in the framework of the fixed stars. An experiment to measure them requires one or more gyros and a reference telescope pointed at an appropriate star. In addition to the principal terms there are three smaller relativistic effects measurable by a gyro with 6×10^{-4} arc-sec/

drift-rate: (1) the geodetic precession the Earth's motion about the sun (0.021 arc-sec/year), (2) the higher order geodetic term calculated by Barker and O'Connell [31] and by Wilkins [32] from the Earth's quadrupole mass-moment (0.010 arc-sec/year in a 50 km polar orbit), (3) deflection by the sun of the starlight signal for the reference telescope. During the time of year when the line of sight approaches the sun the starlight deflection superimposes on the gyro drifts an apparent motion away from the sun which reaches a maximum at closest approach. It can be extracted from the data by in effect turning the experiment around and using the gyros as reference for the telescope. For Rigel, which is 30° from the ecliptic plane the maximum deflection is 0.016 arc-sec. There are also much larger non-relativistic deflections of the apparent position of the star due to aberration: ±20 arc-sec in the ecliptic plane from the Earth's motion about the sun, and ±5 arc-sec in the orbit plane from the satellite's motion about the Earth. The aberrations are calculated very exactly from the known ratios of the orbit-velocities to the velocity of light. They supply handy scaling checks of the gyro and telescope readouts, since the relativistic effects can be expressed as ratios of the known aberrations.

Various possibilites exist for the choice of orbit and the configurations of the gyroscopes. The simplest is an ideal polar orbit with two gyro pairs, one parallel and antiparallel to the Earth's axis and sensitive only to Ω^G, the other parallel and antiparallel to the orbit-normal and sensitive only to Ω^M. The telescope is then pointed at a bright star on the celestial equator orthogonal to both gyro axes. In reality no star is in the right place and no orbit is exactly polar. The Newtonian regression of the orbit-plane from the Earth's quadrupole mass-moment causes a mixing of terms, as a result of which some people have argued that the experiment cannot distinguish Ω^G and Ω^M unless the orbit is within a few arc-minutes of the poles. This opinion is mistaken; the nodal regression actually makes inclined orbits richer in relativity information than polar orbits. The information that can be extracted from different orbits depends on practical considerations reviewed elsewhere [33].

A better configuration is a spacecraft that rolls slowly around the line of sight to the star, containing two gyroscopes with axes parallel to the boresight or the telescope and two at right angles to the telescope and approximately parallel and perpendicular to the Earth's axis. As before one of the perpendicular gyros primarily sees Ω^G and the other Ω^M; both serve also as accurate roll references. With a star lying on the celestial equator and an ideal polar orbit the two gyros parallel to the boresight see a signal periodic in the roll-rate, of amplitude $t\sqrt{\Omega^{G^2} + \Omega^{M^2}}$ and phase $\tan^{-1} \Omega^M/\Omega^G$. I have shown elsewhere [33] how the separation of terms is carried through in inclined orbits with real stars. The advantage is that torques on the gyros, and drifts in the gyro and telescope readouts are strongly averaged by roll. A typical roll-period is 25 minutes.

Figure 6 is a general view of the experiment. The telescope, four gyroscopes and the proof mass for a drag-free control system from a single package mounted inside a superinsulated dewar vessel containing 800 liters of liquid helium and designed to maintain cryogenic temperatures for two years. Boil-off of helium

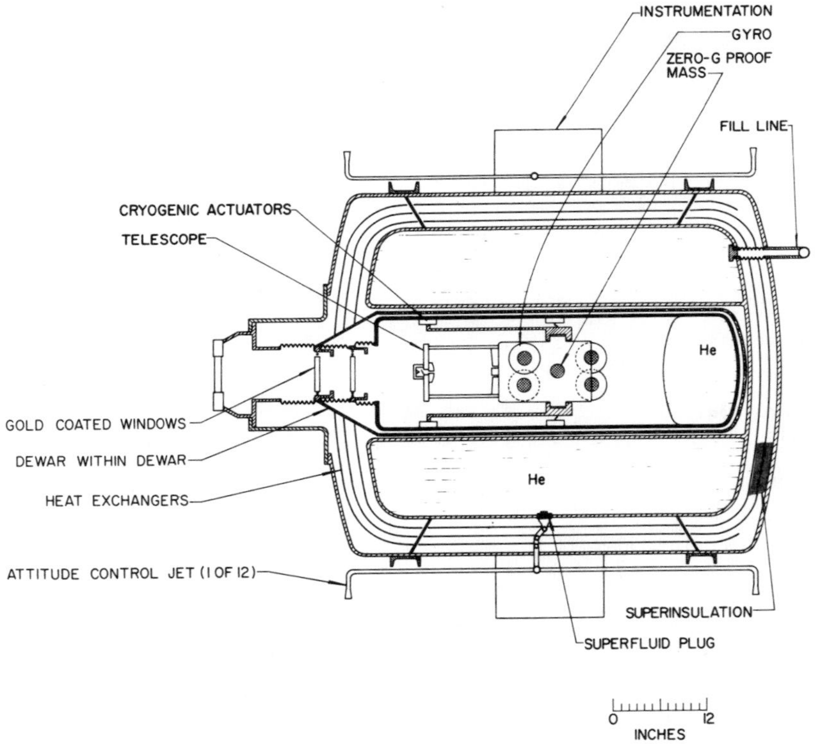

FIGURE 6: GENERAL VIEW OF GYRO RELATIVITY SATELLITE

in the zero-g environment is controlled by a porous plug device invented by Selzer, Fairbank and Everitt [34] . The gyroscopes are quartz spheres, coated with superconductor and suspended electrically in a quartz housing attached to the telescope. Each gyro is surrounded by a spherical superconducting magnetic shield. The telescope is a folded Schmidt-Cassegrainian system of $5\frac{1}{2}$ inch aperture and 150 inch focal length, also made entirely of fused quartz.

For mechanical stability no cements are used in the structure: the gyro and telescope parts are held together in molecular adhesion by optical contacting.

Pointing control of the spacecraft is based on signals from the telescope, switched automatically to the gyroscopes during the portion of each orbit when the star is occulted. Thrust is obtained from the helium boil off from the dewar, which is copious enough to mechanize in a very smooth proportional control system. Drag-free control is mechanized through the same thrustors referenced to the internal proof-mass. Making the satellite drag-free helps in two ways: it improves averaging of residual accelerations on the gyroscopes and it reduces errors in the orbit determinations needed in analysing relativity data. The first drag-free satellite, with the DISCOS control system developed by members of the Stanford Department of Aeronautics and Astronautics, was successfully launched by the U. S. Navy in July 1972 [35] . The performance level was 5×10^{-12} g.

The superconducting gyroscope has to operate in a magnetic field below 10^{-7} gauss. To obtain this the experimental package is held in a separate cylindrical dewar vessel within the main dewar, having a diameter between 12 and 16 inches, built entirely from non-magnetic materials. The inner dewar is first cooled in an independent ultra-low field facility developed at Stanford by B. Cabrera [36] ; after trapping the low field in the gyro shields it is removed from the facility and slid into place in the main dewar.

3.2 Principles of Gyro Design.

A gyroscope may be a spinning body, a nucleus, a superconducting current or a circulating light beam. Present laser gyros are orders of magnitude from the performance needed. Nuclei and currents have the shortcoming of being highly susceptible to magnetic torques; their gyromagnetic ratios are up to 10^{14} times those of ordinary bodies; a free precession He^3 gyroscope would have to be in a field below 10^{-20} gauss to do the experiment [37] . The only horse in the race is a supported spinning body, and no elaborate thought is needed to see that the most torque-free body is a very round, very homogeneous sphere. The problems come down to four: the size of the sphere, and how it is to be supported, spun up and read out.

Size turns out not to be critical. The drift-rate $\dot{\underline{n}}_s^r$ of the gyro spin vector $\underline{n}_s$ due to some extraneous non-relativistic torque $\underline{\Gamma}^r$ is $\underline{\Gamma}^r \wedge \underline{n}_s / I\omega_s$ where I is the moment of inertia and ω_s the spin angular velocity. Substituting

$(8\pi/15)\ \rho r^5$ for I and replacing ω_s by v_s/r, where v_s is the peripheral velocity, we have

$$\dot{\underline{n}}_s^r = \left(\frac{15}{8\pi}\right) \frac{\underline{\Gamma}^r \wedge \underline{n}_s}{\rho r^4 v_s} \tag{23}$$

with a limit on v_s from elastic distortion under centrifugal forces given by

$$(v_s)_{max} = 1.88 \left(\frac{\Delta r}{r}\right)_{max} \left[\left(\frac{E}{\rho}\right)^{\frac{1}{2}} (1 - 11\sigma/28)\right] \tag{24}$$

where E is the Young's modulus and σ the Poisson's ratio for the ball, and $\left(\frac{\Delta r}{r}\right)_{max}$ the maximum allowed difference between the polar and equatorial radii. The torques may be divided into two categories: those related to the surface area of the ball and those related to its volume. Each surface dependent torque Γ^σ is proportional to (area) x (radius) x $\sigma(r)$, where $\sigma(r)$ is a function which in some instances is constant and in others depends on deviations of the ball from perfect sphericity. Each volume dependent torque is proportional to (density) x (volume) x (radius) x $\phi(r)$, where $\phi(r)$ is a function measuring deviations from perfect homogeneity. Over a fair range of radii $\sigma(r)$ may be taken proportional to r^s and $\phi(r)$ proportional to r^v where s and v each lie between 0 and 1. Thus Γ^σ varies as $r^{(3+s)}$ and Γ^ϕ as $r^{(4+v)}$ and from (23) the drift rates $\dot{\underline{n}}_s^\sigma$ and $\dot{\underline{n}}_s^\phi$ from all torques in these categories vary as $r^{(s-1)}$ and r^v. Thus some errors increase and some decrease with increasing rotor size; in neither case is there much advantage to a change of diameter.

The actual rotor is a ball 4 cm in diameter made from optically selected fused quartz, homogeneous in density to 1 part in 10^6 and spherical to a few parts in 10^7. It is coated with superconducting niobium and supported inside a quartz or ceramic housing by voltages applied to three mutually perpendicular pairs of electrodes. Figure 7 illustrates piece parts of a ceramic housing manufactured by Honeywell for experiments at Stanford. The electrodes are 2 cm diameter circular pads 4×10^{-3} cm from the ball. The suspension system used in most of the work was designed by the late J. R. Nikirk [38] . It holds the ball against accelerations up to 5 g by 20 kHz signals of amplitude between 2 and 3 kV rms. The ball position is sensed by a 1 MHz 2 V signal. The suspension servo has a bandwidth of 600 Hz and long term centering stability good to 10^{-5} cm. In space the support voltage is about 0.5 V.

The gyro is spun to 200 Hz in 30 minutes by passing gas at approximately 16 torr pressure through circumferential channels around the ball, after which the gas is pumped out and the ball runs freely in a 10^{-9} torr vacuum. The configuration of the channels is seen in Figure 7. To prevent electrical breakdown and reduce

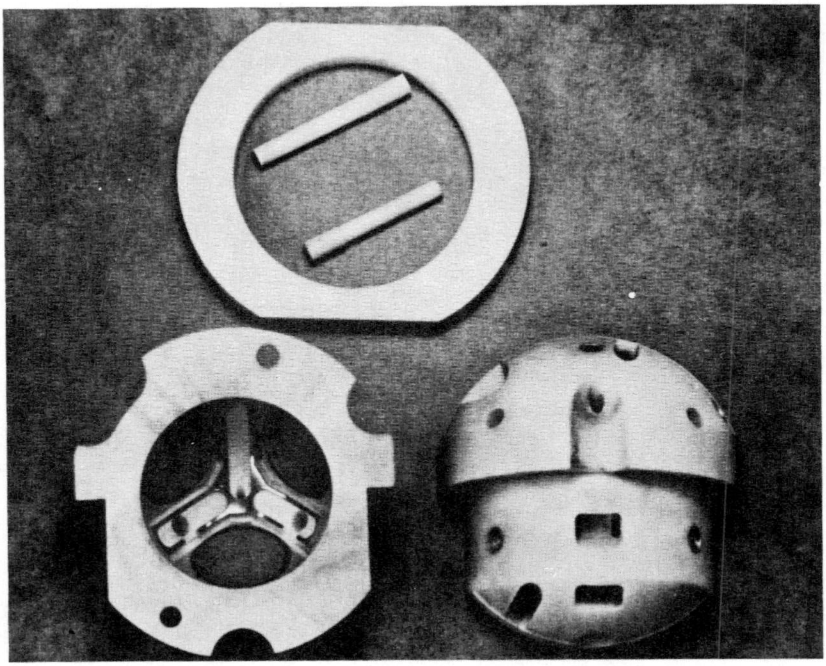

FIGURE 7 : CERAMIC GYRO HOUSING

gas drag in the cavity the channels are surrounded by raised lands 5×10^{-4} cm from the ball, and differential pumping is provided by ports in the housing. About 95% of the gas exits at high pressure through the main channels; the rest seeps over the lands and is exhausted through a separate low pressure system. Optimization of the design is due to Bracken and Everitt [39] . In space, where the support voltages are low, differential pumping might seem unnecessary, but the pressure needed to spin without it is much higher (about 0.5 atmosphere) and the ball is therefore subject to large gas forces which cannot readily be overcome except by increasing the voltages. So it is best to stick to the same design.

The gyro is read out by magnetic observations of the London moment in the spinning superconductor. According to the London equations of superconductivity there is in a rotating superconductor a magnetic moment aligned with the instantaneous spin axis which reduces in spherical geometry to a dipole of magnitude $M_L = \frac{mc}{2e} r^3\omega_s = 5 \times 10^{-8} r^3\omega_s$ gauss cm^3. Figure 8 illustrates the principles of the readout. The ball is surrounded by a superconducting loop connected to a sensitive Josephson junction magnetometer. Any change in

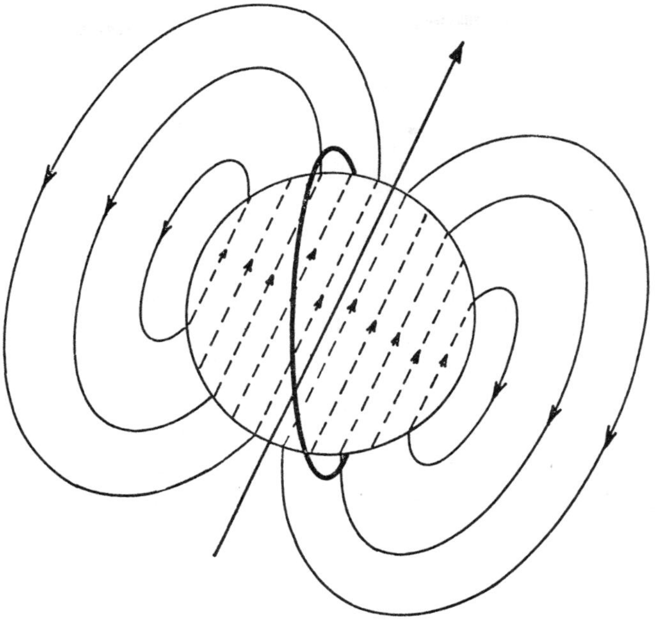

FIGURE 8
PRINCIPLES OF THE LONDON-MOMENT READOUT

orientation of M_L changes the flux through the loop and can therefore be measured by the magnetometer. Three mutually perpendicular loops give a three axis readout.

The following are some of the advantages of the London moment readout: (1) The angular resolution in a 100 radian bandwidth with realistic coupling to commercially available shake-tested SQUIDs is 1 arc-sec, or 0.001 arc-sec after 10^4 sec integration. Better performance is likely with more advanced magnetometers. Existing optical readouts give 15 arc-sec resolution in a 100 radian bandwidth. (2) application to homogeneous rotors. Other readouts for mechanical gyroscopes depend on observations tied to the body axes of the rotor and require a rotor with unequal moments of inertia. The London moment is aligned with the instantaneous spin axis, which cones about the angular momentum vector at a rate ω_p in inertial space nearly equal to ω_s ($\omega_p < \omega_s \pm \Delta I/I$) and with an amplitude Θ_s moving slowly between limits of order $\Theta_I \ \Delta I_{13}/I$ and $\Theta_I \ \Delta I_{12}/I$, where Θ_I's are maximum and minimum values of the body cone, and ΔI_{12}, ΔI_{13} are the residual inertia differences. With $\Delta I/I \sim 10^{-6}$ the coning of the spin axis is never more than a fraction of an arc-second and averages to much less: the London moment readout works regardless of polhoding. Other readouts do not. This is a point of crucial significance since a homogeneous rotor is essential to the experiment. (3) The gyro drift from reaction currents in the readout is completely negligible -- below 10^{-5} arc-sec/year. (4) The centering errors due to a displacement t of the gyro are negligible with the London moment readout since they vary as $(t/r)^2$, whereas in standard optical readouts they vary as (t/r) and may cause null drifts up to 2-3 arc-sec. (5) The readout has the high intrinsic linearity and null stability of superconducting circuits and shields. (6) mechanization as an all angle readout. A Josephson junction magnetometer has a saw-tooth response quantized in fixed steps of hc/2e, each of which can be resolved with great precision. By combining the normal feedback measurement with flux counting techniques, an all angle readout may be developed having a resolution of twenty or more bits per quadrant, or 0.1 arc-sec throught the range [40]. The best conventional angle encoders have resolutions of 17 bits per quadrant.

The all angle readout is not needed in the Gyro Relativity experiment, but would be in applying the gyroscope in inertial references for an astronomical telescope.

The choice of magnetic readout, electrical suspension and gas spin up neatly separates problems in the three areas. The separation is less complete on Earth than in space because the large 20 kHz suspension signals generate pickup in the readout. In fact suspension interference was an awkward problem in the first stages of the laboratory experiment. It may be reduced by reducing the support frequency. J. A. Lipa has shown that a lower support frequency also eases other problems. One cannot on Earth conveniently go below 2 kHz since the ball drops too far during each half-cycle. A d.c. suspension is possible, but hard to mechanize.

Assuming a London moment readout one might ask if other alternatives are worth considering for suspension and spin up. The best claim another suspension might have would be that it exerts a smaller torque on the ball, particularly if the torque were low enough to do a relativity experiment on Earth. Various support schemes have been proposed for spherical gyro rotors -- gas bearings, superconducting magnets, flotation in superfluid helium, and so on -- and amazing claims are sometimes heard about torque levels. No universal judgement can be offered; the following argument shows where the heart of the problem lies.

Any scheme for supporting a massive body against gravity depends on creating pressure differences across the surface. If the body is nearly but not quite spherical a change in orientation results in work being done against these pressures; conversely the pressures exert a torque that drives the body towards an energy mimimum. The torque evidently vanishes for a true sphere; it depends on the size and shape of deviations from perfect sphericity. Discounting extraneous effects, such as the interaction of a magnetic suspension with the London moment or turbine torques in a gas bearing, we may expect different suspensions to exert similar torques on bodies of the same form, for the pressure that has to be applied over a given area to balance a given acceleration is always the same. More accurately we may think of two extreme suspension mechanizations between which all others fall. One which may be called the plain man's suspension generates pressure simply to counteract gravity. To fix ideas think of a light sphere floating on mercury. If the total deviation from sphericity is Δr the difference between maximum and minimum energies is $Mf\Delta r$ where f is the residual acceleration. Upper bounds on the torques are found by expanding the shape of the spinning body in Legendre polynomials and identifying Δr with each polynomial in turn. The drift-rates are

$$\dot{\underline{n}}_r^m = \frac{5m\chi}{2v_s}\left(\frac{\Delta r}{r}\right)\underline{f} \wedge \underline{n}_s \qquad (25)$$

where m is the order of the polynomial and χ is a quantity between 0 and 1 depending on m and the size of the support pads. If the diameter of the pad is d then $m_\chi \rightarrow 0$ as $m \gg \pi r/d$, and satisfactory limits are got by considering the first few even and odd harmonics. Taking $\frac{\Delta r}{r}$ as 3×10^{-7} and f in space as 10^{-9} g the expected drift rate with a plain man's suspension and $\underline{f}$ perpendicular to $\underline{n}_s$ is around 10^{-15} rad/sec, a factor of ten higher than the design goal for the experiment. Further improvements depend on the extent to which f averages through the orbit or the plain man's suspension can be improved on.

To do better the energy put in at one point must be taken out elsewhere. The simple picture is not a sphere on mercury but a neutral density body immersed in an incompressible fluid. The pressure extends over the whole surface and the torque vanishes. Remembering Archimedes we may call this the eureka suspension. An electrical suspension can be arranged to mimic an eureka suspension by applying voltages to all six electrodes at once. It is then said to be preloaded. In the most common mechanization the voltages V_2, V_4 on opposite electrodes are adjusted to keep $(V_2 + V_4)$ constant and the preload acceleration h is the acceleration along a support axis required to send the voltage on one plate to zero. The most critical shape is the oblate spheroid, since it determines the gyro drift due to centrifugal distortion of the ball, and for a ball spinning at 200 Hz the centrifugal $\Delta r'/r$ is 3×10^{-6}, a factor of ten larger than the polishing errors. Defining a preload compensation factor ζ = $(hx - hy)/hz$ etc., the torque on an oblate spheroid inclined to the electrode axes turns out to be proportional not to f as in the plain man's suspension but to $\left[\frac{f}{h} + \zeta\frac{h}{f}\right] f$. There is therefore an optimum preload $h = f/\sqrt{\xi}$; and if the preloads are matched to 1%, as is reasonable, this torque is an order of magnitude less than with the plain man's suspension (actually nearer a factor of 40 less since the numerical coefficient is smaller).

Two other, more elaborate, mechanizations deserve mention. One known as "sum of the energies" control has voltages continually adjusted to hold $\sum C_i V_i^2$ constant where C_i and V_i are the capacitance and voltage of the ith electrode. The energy is independent of orientation; this is a true eureka suspension. The second is "sum of the squares" control, for which the voltages on the three axes fulfil the condition $V_2^2 + V_4^2 = V_1^2 + V_6^2 = V_3^2 + V_5^2$. This leaves the higher order terms but makes the torque on an ellipsoid vanish. Defining a sum of squares control factor ξ analogous to the preload compensation factor , the ellipsoid torque is proportional to $\left[\frac{h}{f} + \zeta \frac{f}{h}\right] f\xi$, still further down from the plain man's suspension.

Other suspension techniques such as a superconducting bearing might also outperform the plain man's suspension. The trouble always is that eventually the analytic arguments break down through secondary effects like polishing errors in the housing. The eureka suspension is a will o' the wisp. In fact in Honeywell's studies of electrical suspensions sum-of-the-squares control really helps, sum-of-the-energies does not. One may conjecture that with comparable work all suspension techniques will beat the plain man's suspension in about the same degree.

Consider now an attempt to do a Gyro Relativity experiment on Earth, say in an observatory near the equator. The combined relativistic precession $(\dot{\underline{n}}_s^G + \dot{\underline{n}}_s^M)$ is 0.4 arc-sec/year. Suspension errors may be reduced either by approximating a eureka suspension or by averaging. If the spin axis $\underline{n}_s$ lies in the equatorial plane the quantity $\underline{f} \wedge \underline{n}_s$ in equation (24) averages to $g\bar{\lambda}$, where $\bar{\lambda}$ is the average uncertainty in $\underline{f} \wedge \underline{n}_s$ from fluctuations in the local vertical, say 10^{-5}

rad . Then with $\frac{\Delta r}{r}$ for a spinning ball equal to 3×10^{-6}, the uncertainty in gyro drift with a plain man's suspension is about 100 arc-sec/year. Better things might be hoped from a preloaded suspension. Alas not! The residual torques, instead of being parallel to the local vertical are in an unknown direction in the housing: experience with live gyros suggests a limiting drift-rate nearer 1000 arc-sec/year -- worse than straight averaging with the plain man's suspension.

Conside another torque: mass-unbalance from inhomogeneities in the rotor. If u is the distance from center of mass to center of geometry the torque is Mfu and the drift-rate

$$\dot{\underline{n}}_s^u = \frac{5}{2}\left(\frac{u}{r}\right) \frac{\underline{f} \wedge \underline{n}_s}{v_s} \qquad (25)$$

which with $(u/r) \sim 10^{-6}$ and $\overline{\underline{f} \wedge \underline{n}_s} - 10^{-5}g$ averages to 60 arc-sec/year, comparable to the suspension torques. In space with $f \sim 10^{-9}$ g, the drift-rate is below 0.001 arc-sec/year. The mass-balance might be improved by evaporating material on the surface of the ball and checking its pendulum period before spin up. If Ω_0 is the design goal and $_s$ the gyro spin-rate the pendulum period T must exceed $2\pi \sqrt{\sin\lambda/\Omega_0\, \omega_s}$, and with $\lambda \sim 10^{-5}$ the period for an 0.1 arc-sec experiment is four hours -- perhaps a factor of 20 beyond the limits of feasibility. An Earth based experiment is hopeless.

In space with a rolling drag-free spacecraft $\underline{f} \wedge \underline{n}_s$ and higher terms average extremely well. The estimates in my earlier papers were unnecessarily conservative; I am in process of revising them. In the final experiment support dependend drifts should go below 10^{-4} arc-sec/year. Drifts from other sources, calculated in earlier papers, should be below 10^{-3} arc-sec/year.

The choice of spin system turns on two questions. To spin a gyro a torque Γ^s has to be applied for time t_s, after which Γ^s is switched to its residual value Γ^r. A simple calculation establishes that Γ^r/Γ^s must be below $\Omega_0 t_s$, which for an 0.001 arc-sec gyro spun up in 30 minutes is 10^{-12}. Gas spin up, with its large change in operating pressure and favourable averaging provides this enormous torque-switching ratio; other methods do not. But there is a further consideration. A problem with gas is pumping down afterwards, especially through the pressure range 10^{-7} to 10^{-9} torr. Suppose we try another spin system. It will have losses which generate heat; the heat has to be removed to stop the ball warming above its superconducting transition; at low temperatures the only way is by exchange gas, and with any realistic efficiency the pressure has to be well above 10^{-7} torr [41] . One might as well stick to gas spin up.

The arguments for choosing a space experiment with electrical suspension, London moment readout and gas spin up, are persuasive. Ingenuity should never be discounted but in the Gyro Relativity experiment the issues ingenuity has to address are severely constrained.

3.3 Present Status

Research on the Gyro Relativity experiment has been going on at Stanford jointly between the Hansen Laboratories of Physics and the Department of Aeronautics and Astronautics with NASA support since 1963.* Areas of progress include: (1) the gyroscope (2) the telescope (3) ultra-low magnetic field technology (4) long hold-time dewars for space (5) the porous plug for controlling helium in space (6) attitude and translational control with helium thrustors (7) the relativity data instrumentation system (8) analysis. Two Mission Definition Studies and a Phase A study of the flight program have been performed for NASA by Ball Brothers Research Corporation. An opportunity for the first flight has been identified on Shuttle 10 in 1980. Complementing the work at Stanford has been research at NASA Marshall Center on precision manufacturing of quartz gyro rotors and piece parts, engineering development of the porous plug device, and low noise readout work. A rocket flight of the porous plug by NASA Marshall Center in cooperation with JPL and the Los Alamos Scientific Laboratory was successfully performed in June 1976.

The main thrust at Stanford has been to develop a laboratory model of the flight experiment with a gyro telescope package mounted in a long hold time helium dewar, tilted and aligned parallel to the Earth's axis with an artificial star reference. The dewar is a 100 litre superinsulated vessel made entirely from non-magnetic materials: aluminum, copper, titanium and low thermal conductivity plastic . It was fabricated to Stanford design by AGS Incorporated and delivered in 1969, since when it has been greatly modified and improved. Gyro designs were completed in 1967 in cooperation with Honeywell Incorporated and after experiments with quartz ang ceramic parts the ceramic housing (Figure 7) was delivered in 1971. An electrical suspension system had been purchased from Honeywell earlier. Besides gas spin up the gyro had several novel features for cryogenic operation, one of which was electrodes of relatively thin sputtered titanium with a gold overcoating rather than the electroless nickel electrodes used in standard Honeywell gyros. Although earlier test pieces had shown good electrical breakdown characteristics the housing itself suffered at first from severe arcing

* Present members of the research team are J. T. Anderson, B. Cabrera, R. R. Clappier, D. B. DeBra, C. W. F. Everitt, W. M. Fairbank, J. A. Lipa, B. Nesbit, F. J. van Kann, R. A. Van Patten.

under high voltage. Experiments by J. A. Lipa and J. R. Nikirk progressively overcome this and other difficulties and in January 1973 the gyro was spun for the first time at room temperature. Low temperature spin with a superconducting ball was achieved in June 1973. Between then and August 1975 about 1000 hours of cryogenic operations were logged at spin speeds up to 40 Hz. We have not tried to go faster, except once by mistake, because residual problems in the ceramic housing impart an element of risk to high speed spin. Higher speeds are expected in a new apparatus.

Readout magnetometers were developed by J. T. Anderson and R. R. Clappier. Elaborate shielding and filtering against suspension interference was needed in applying them to the gyro. Simultaneously with the readout work J. R. Nikirk developed a new suspension system, having better servo response and centering capability, with sensing bridges operating at a single 1 MHz frequency of 2V rms amplitude rather than the three 30 V signals at independent frequencies used by Honeywell. This reduced suspension interference. The hardest problem was trapped magnetic flux in the gyro rotor. The apparatus has two Mu-metal shields to bring the ambient field to 10^{-5} gauss, but initial trapped fields were commonly nearer 3×10^{-4} gauss, because of residual magnetism and thermoelectric currents in the experimental chamber. The London moment as 35 Hz corresponds to 2.1×10^{-5} gauss. Laborious procedures with field coils around the gyro brought the trapped fields down to 10^{-5} gauss, after which London moment data was obtained by processing signals from a three-axis readout. Figure 9 from a paper by Lipa, Nikirk, Anderson and Clappier [42] compares theory and experiment for the London moment as a function of spin speed.

A precise readout needs a much lower field: around 10^{-7} gauss. During the past six years B. Cabrera of Stanford has developed methods of creating superconducting shields up to 8 inches in diameter with fields below 10^{-7} gauss over a 30 inch length. The technique depends on two special properties of superconductors. One is that the magnetic flux through a closed superconducting surface is conserved. A tightly compressed lead shield is cooled in a low field and then expanded. Since flux is equal to field times area, the increase in area causes a corresponding reduction in field. The second method of field reduction is by heat flushing. If a temperature gradient is applied along a superconductor as it is cooled one end will be superconducting and the other normal, with a transition zone somewhere in between. Further cooling makes the boundary move steadily forward and in suitable circumstances the magnetic field is progressively pushed out of the enclosed volume. Flushing is complicated by the magnetic fields from thermoelectric currents, but by flushing and expanding a series

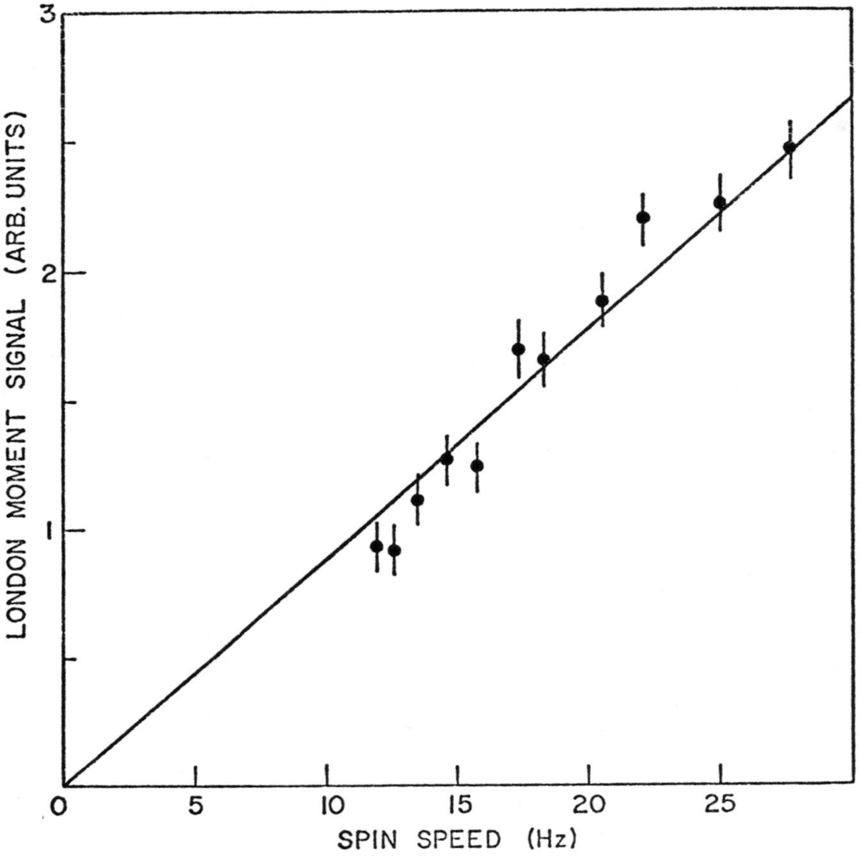

FIGURE 9
LONDON MOMENT DATA FROM A SPINNING GYROSCOPE

of balloons one inside the next, a 10^{-7} gauss region can be produced.

F. J. van Kann and B. Cabrera have developed a new test apparatus, shown in Figure 10, in which the gyro is held in an 8 inch ultra-low field shield. The gyro probe assembly is suspended from a rigid frame which stands on a concrete pad and the dewar containing the shield is slowly raised around it on a hydraulic piston, using an airlock to prevent condensation of solid air. The gyro probe can be transferred to a laminar flow clean bench nearby for assembly and disassembly under ideal conditions. Gyro operations in a low field should begin in August 1976.

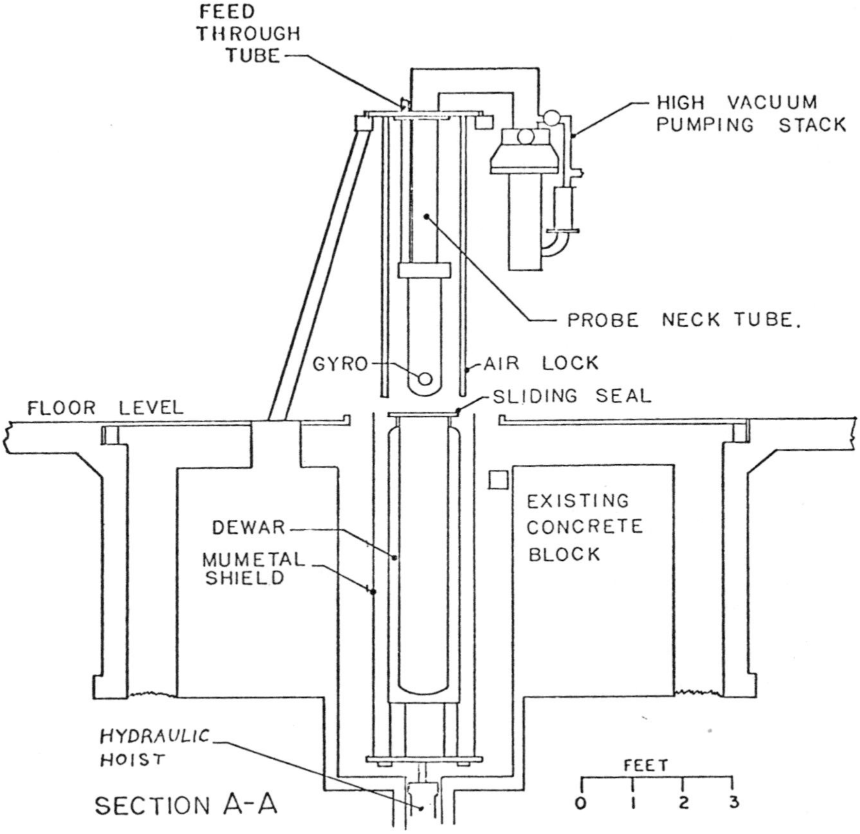

FIGURE 10

APPARATUS FOR OPERATING GYROSCOPE IN ULTRA-LOW MAGNETIC FIELDS

Work on the telescope has gone on in parallel. Figure 11 is a view of the telescope designed in conjunction with D. E. Davidson, which is a folded Schmidt-Cassegrainian system of 150 inch focal length and 5.5 inch aperture, fabricated entirely of fused quartz. The light is divided by a beam splitter near the focal plane to give two star images, one for each readout axis. Each image falls on the sharp edge of a roof prism where it is again subdivided and passed to a light-chopper and photodetector at room temperature. The limitations are the sharpness of the prism edge and photon noise. A simple analysis [43] shows that the prism edge should have no nicks greater than 3500 Å. The prisms were made by polishing two optical flats, contacting them together, lapping a second

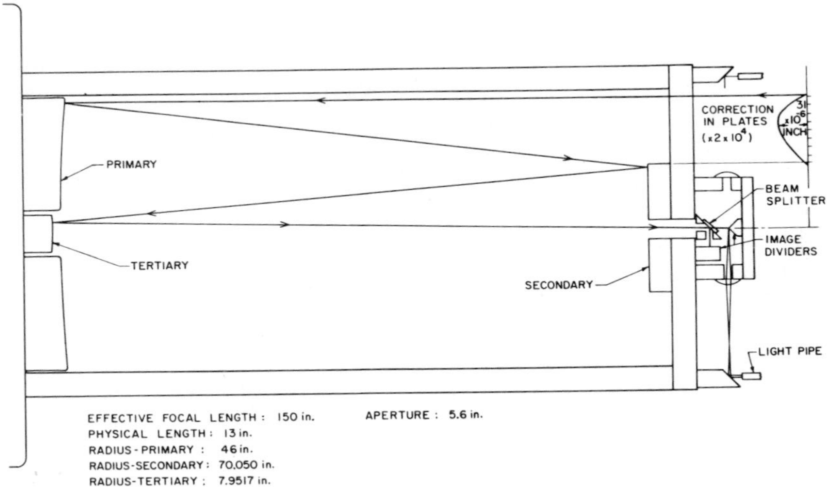

FIGURE 11: DAVIDSON TELESCOPE

surface at right angles, and separating the two halves. Each edge protected the other. The final prisms have no nicks greater than 1500 Å.

Much attention has been paid to mechanical and optical stability. The space telescope, being in zero g and looking through the vacuum of space, is free from effects of sag or atmospheric turbulence. Low temperature operation is equally important. As equation (14) shows, the thermal distortion depends on the transverse heat flux and the ratio (α/K) of the expansion coefficient to thermal conductivity. At ambient temperature the radiation falling on the spacecraft would have to be cancelled to 1 part in 10^5 to avoid 10^{-3} arc-sec distortions. At low temperatures (α/K) is orders of magnitude smaller; the heat flux can be a factor of 100 higher than the total heat input into the dewar. Problems of stress-relaxation in the quartz, ultraviolet darkening, tarnishing of the mirrors and aging of the light pipes are discussed elsewhere [44].

The photon noise is given by an equation similar to (3). Experiments by R. A. Van Patten, H. Gey and myself on a single axis model telescope and artifical star give results for slightly defocused images (Figure 12) in good agreement with theory. A star equal to Procyon was resolved to 0.02 arc-sec in 0.3 sec. With diffraction limited optics, Rigel would be resolved to 0.01 arc-sec in 0.1 sec.

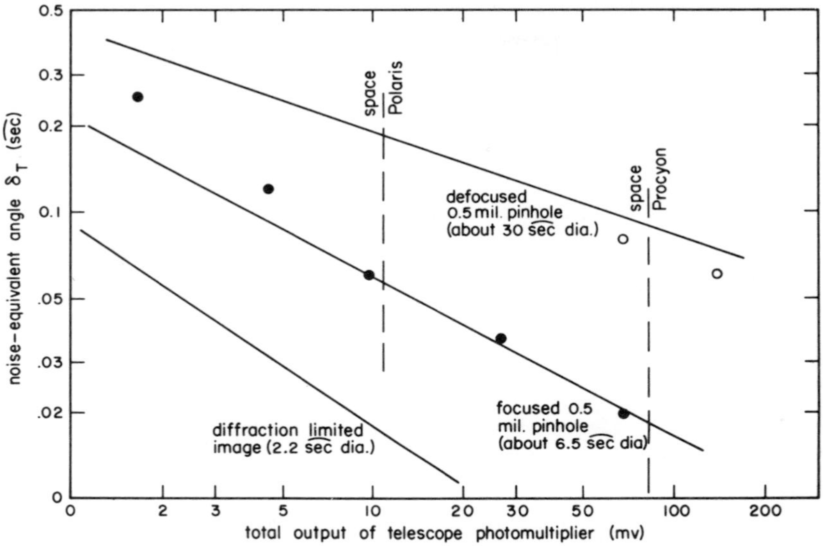

FIGURE 12: COMPARISON OF OBSERVED DATA WITH CALCULATED PHOTON NOISE FOR A 2.5 INCH APERTURE MODEL TELESCOPE WITH ROOF PRISM IMAGE-DIVIDERS

Recently we took delivery of the North Star simulator designed by D. E. Davidson that will be used in testing the real telescope. The device has a 200 inch focal length 8 inch aperture off-axis parabola mirror, support optics and tipping and dither plates for exploring the star image, all enclosed in an evacuated chamber. The telescope is expected to be linear to 0.001 arc-sec over a range of ±0.05 arc-sec.

Other laboratory research at Stanford is described in papers by T. Dan Bracken [39] (gas spin up), J. Bull [45] (attitude control), D. Klinger [46] (fixed base simulation), J. R. Nikirk [38] (suspension system), P. M. Selzer [34] (porous plug for controlling helium in space), R. A.Van Patten [47] (relativity data instrumentation system), D. C. Wilkins [32] (relativity effects in perturbed orbits). More about the dewar appears in articles by Lipa, Fairbank and Everitt [48] .

4. TESTS OF THE EQUIVALENCE OF GRAVITATIONAL AND INERTIAL MASS BASED ON CRYOGENIC TECHNIQUES

The possibility of testing the equivalence of gravitational and inertial mass in Earth orbit has been discussed several times. An orbital experiment has two advantages. The disturbances are smaller than on Earth and the driving acceleration, instead of being derived from the sun or the Earth's rotation, may be the full gravitational attraction of the Earth, which at 300 nautical miles is 950 cm sec^{-2}, three orders of magnitude larger than either of the sources available in ground-based experiments. During the past four years P. W. Worden, Jr. and I, with other colleagues at Stanford, have been designing an orbital experiment using cryogenic techniques to measure the Eötvös ratio η to 1 part in 10^{17} . A preliminary laboratory experiment now in progress should reach an accuracy approaching 1 part in 10^{13}.

The difficulty in space is gravity gradients. For a torsion balance the ratio Γ^g/Γ^E of gravity gradient to Eötvös torques from a common source is proportional (equation (9)) to $J_2D/\eta R$ where D is the diameter of the balance and R the distance to the source. The ratio is five orders of magnitude higher with the Earth as source than with the sun. For an orbital experiment with J_2D around 10^{-3} , Γ^g exceeds Γ^E when η is less than 2×10^{-12} . The gravity gradient disturbance, being doubly periodic with the orbit, can be partially suppressed by resonating the balance with the orbit-period, but even if the system were given three months to come to equilibrium (a startling thought!) the maximum allowable Q would be 1000. At that level the attenuation of the second harmonic is only 0.0016, and the gravity gradient signal in a 10^{-17} experiment is still 400 times the Eötvös signal. Resolution is difficult in the presence of such a large disturbance and non-linearities may generate suharmonics of Γ^g that masquerade as Eötvös signals.

The difficulties are removed by abandoning the time-honoured torsion balance, for which the measured quantity is an angular displacement, in favour of an experiment to measure the relative linear displacement of two nearly coincident freely falling masses, for example two coaxial cylinders. The ratio a^g/a^E of gravity gradient to Eotvos accelerations is then $2\ \Delta R/R\eta$ where ΔR is the displacement between the mass-centres of the cylinders. Again a^g is doubly periodic with the orbit; however in an elliptic orbit of eccentricity e it has a singly periodic subharmonic $eg'\ \Delta R/R$. No such term occurs in the gravity gradient torque Γ^g on a torsion balance. If the cylinder axes are aligned with the semi-major axis of the orbit the subharmonic masquerades as an Eötvös signal. The error is eliminated by centering the masses, driving the twice orbital signal a^g to null with a centering servo. To make a^g/a^E unity $\Delta R \sim R\eta/2 \sim 3.5 \times 10^{-8}$ cm., which for an η of 10^{-17} requires coincidence in the average position of the mass-centers to about 0.35 Å. This is an unnecessarily restrictive centering criterion, but one that can be achieved.

Recently P. W. Worden, Jr. has extended the comparison of the free-fall and torsion balance experiments by forming expressions for the higher order gravity gradient terms on the free-fall cylinders and comparing the ratios (S/N) of a given Eötvös signal S to gravity gradient disturbances N for the two experiments. Provided the cylinders are centered enough to make the first order term negligible, the ratio of the two (S/N)s is

$$\frac{(S/N)_{\text{free fall}}}{(S/N)_{\text{torsion balance}}} \sim \frac{\varepsilon\xi^2}{J_2 DR} \tag{26}$$

where R is the distance to the disturbing mass, ξ the radius of gyration of the outer cylinder in the free fall experiment, and ε the quantity which when multiplied by the principal moment of the cylinder gives the difference between its largest and smallest principal moments. For given manufacturing procedures the quantities J_2 and ε are comparable, as are the characteristic dimensions ξ and D for the two apparatuses. Hence the ratio of the two (S/N)s is of order ξ/R. Take R as the radius of the Earth: the free fall experiment is eight orders of magnitude less sensitive to the Earth's gradient than a torsion balance. Next consider gradient disturbance within the spacecraft, say from sloshing motions of the liquid helium. With R about 100 cm the free fall experiment is in this respect one order of magnitude quieter than the torsion balance. The centering requirement to make the first order term in (26) negligible is $\Delta R \ll \varepsilon\xi^2/R$. With $\varepsilon \sim 10^{-3}$ and $\xi \sim 3$ cm, R has to be below 10,000 Å to remove the term for the helium tides, but nearer 0.001 Å to remove it for the Earth's gradient. In reality R is more likely to be about 0.1 Å, so the free fall experiment is six rather than eight orders of magnitude better than the torsion balance. That is enough.

The problem in any improved determination of η is the minuteness of the displacements. Consider the free fall experiment. If the masses are essentially free floating, a differential acceleration ηg with period T will cause a relative displacement $T^2\,\eta g/4\pi^2$ which implies a displacement of 0.6 Å for a satellite period of 90 minutes and an η of 10^{-17}. If, as is likely, the masses are restrained in a harmonic oscillator with natural period shorter than the orbital period the amplitude is correspondingly smaller. Large advances in sensitivity and stability over existing apparatus are needed.

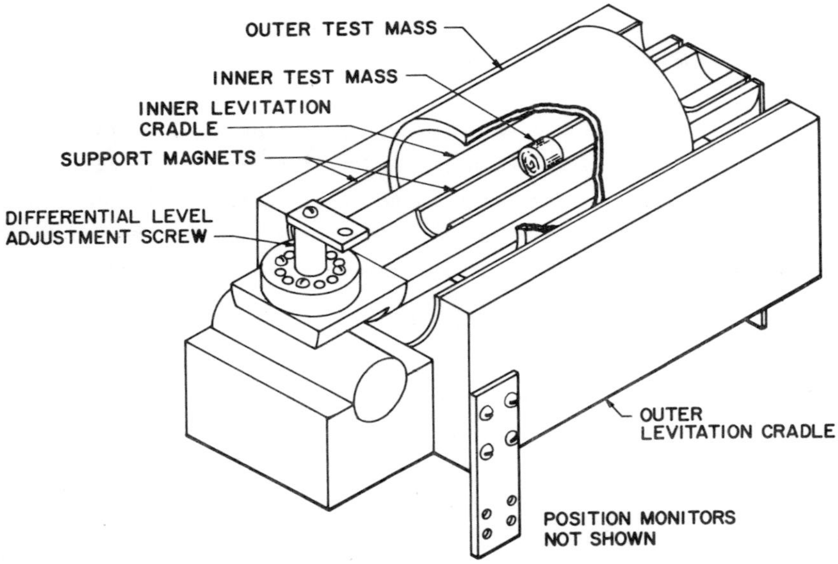

EQUIVALENCE PRINCIPLE ACCELEROMETER

FIGURE 13

TEST MASSES AND ADJUSTABLE SUPERCONDUCTING SUPPORT CRADLES FOR THE GROUND-BASED VERSION OF THE CRYOGENIC EQUIVALENCE EXPERIMENT

(In the present apparatus the inner test mass is made of niobium and the outer test mass is of niobium coated aluminum.)

Figure 13 illustrates the arrangement of test masses for the laboratory experiment now under development at Stanford. They consist of a rod of superconducting niobium approximately 1 cm x 1 cm and a concentric hollow cylinder of niobium coated aluminum, approximately 5 cm x 5 cm. Later other materials will be tried. The two bodies are independently supported by two superconducting magnetic bearings and are free to move along their common axis. The experiment consists in measuring the relative accelerations of the masses throughout the day with the sun as source. The position of each mass is controlled by a servo-mechanism which in effect tilts the support plane; the difference in control efforts to centre the two masses contains the signal. Data is recorded digitally and computer analysed to detect any 24 hour component in the difference output.

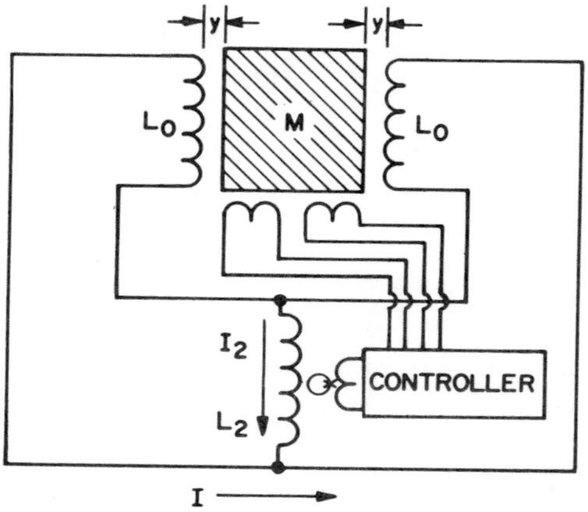

FIGURE 14: POSITION DETECTOR FOR EQUIVALENCE PRINCIPLE EXPERIMENT

Figure 14 illustrates the principle of the position readout. Each superconducting test body is placed between two coils of inductance L_0 joined in a continuous superconducting loop in which a magnetic field is trapped. A third inductance L_2 is connected in parallel. The test body being a perfect diamagnet, its motion causes a redistribution of field in the coils. If the current in L_2 is initially zero and the current trapped in the main loop is I, a motion Δx of the test body generates a current in L_2

$$\Delta I_2 = 2I \frac{L_0}{L_0 + 2L_2} \frac{\Delta x}{y} \tag{27}$$

where y is the distance from the coils L to the test mass. The current ΔI_2 is read by a Josephson junction magnetometer, which then drives the centering servo.

For a quantum interference detector capable of resolving in τ sec an amount of magnetic flux equal to a small fraction $\varepsilon(\tau)$ of the flux quantum ϕ_0 the minimum detectable current change is $\delta I_2 = \varepsilon\phi_0/L_2$ and the minimum detectable position change is

$$\delta x = \frac{y\phi_0}{I}\left(\frac{L_0 + 2L_2}{2L_0L_2}\right) \varepsilon\ (\tau) \tag{28}$$

and in the ideal case where resolution is limited solely by measurement noise $\varepsilon = \gamma^{\frac{1}{2}}\tau^{-\frac{1}{2}}$, with $\gamma^{\frac{1}{2}}$ being typically of order 10^{-4} Hz$^{-\frac{1}{2}}$. Following the approach of equation (5) we may define a discrimination factor ∂_B giving the resolution of the Brownian motion $<\Delta x>$, where $<\Delta x>$ for an oscillator of natural period is again $\frac{\tau_0}{2\pi}\sqrt{\frac{kT}{M}}$. Putting ϕ_0 = hc/2e

$$\partial_B = C\left(\frac{hc}{ek^{\frac{1}{2}}}\right)\left[\frac{1}{\tau_0 T}\right]^{\frac{1}{2}}\left[\frac{\gamma(1 + 2L_2/L_0)}{L_0}\right]^{\frac{1}{2}} \tag{29}$$

where in analogy with (5) the terms in the first square bracket characterize the suspended body and those in the second square bracket characterize the readout.

The quantities determining τ_0 are different in space and on Earth. On Earth the masses tend to be pendulums, rocking back and forth in local depressions of their support cradles and τ_0 is set by machining tolerances in the cradles. In space the support is small and the largest restoring force is the reaction of the readout. In the ideal case of a single body the system becomes a simple harmonic oscillator whose period is $\tau_0 = \pi y \sqrt{2M(L_0 + 2L_2)} / IL_0$ and the characteristics of the readout and suspension are no longer separable. An interesting consequence is that the sensitivity (i. e. the ratio of the Eötvös signal $|\Delta x|$ to δx) is increased by decreasing the readout current I. In fact to make τ_0 resonant with the orbit, which gives maximum sensitivity, I may be only a few μA.

In all circumstances the discrimination is excellent; the readout noise is typically some four orders of magnitude less than the Brownian motion.

The fundamental limit is given by an equation identical with (12) except for a smaller numerical factor

$$<\eta> \sim \frac{1.4}{f}\sqrt{\frac{\beta kT}{MS}} \tag{30}$$

With M ~100g, $\beta \sim 10^{-5}$ sec^{-1}, T ~ 2K, f ~ 950 cm sec^{-2}, η can be resolved in a space experiment to 10^{-17} in 4 hours and 10^{-18} in 16 days.

The practical limit in a space experiment is set by tidal slosh in the liquid helium. Taking ℓ as the length of an optimized outer cylinder and z as the mean distance to the helium, the mass amplitude M_H of the tide must satisfy

$$M_H < \frac{fz^6}{G\,\ell^4}\,\eta_0 \tag{31}$$

to avoid gravity gradient disturbances greater than η_0. With ℓ around 10 cm, M_H must be less than 20 gm for a 10^{-17}experiment. The corresponding tidal amplitude is 1 mm. Several techniques are available to control slosh. Since M_H scales as ℓ^{-4} there is great advantage in making the apparatus smaller.

Other limits to the experiment, including effects of residual gas, residual electric charge, trapped magnetic fields, and black body radiation in the experimental chamber, are discussed elsewhere [17] .

The copper support block for the laboratory test masses (Figure 13) comprises two hemicylindrical cradles joined by a flat spring hinge with a differential screw adjustment to tilt the inner cradle with respect to the outer one. The levitation coils are 5 mil niobium-titanium wires wound back and forth parallel to the axes of the cradles. They were made by winding the wires on a precisely machined aluminum mandrel, epoxying the structure into the cradle and then etching out the mandrel with sodium hydroxide. The inner coil has no deviations from an ideal cylinder greater than 2.5×10^{-4} cm. The positions of the test masses are read out by two circuits of the kind already described, with the outputs being fed back to subsidiary tilt coils on each cradle. Any differential acceleration between the two bodies is found by comparing and subtracting the control efforts needed to keep them centered. The control gains have to be very closely matched; otherwise spurious signals appear in the output if the apparatus is tilted back and forth through a small angle once per day, as it may be for example through ground motions. The gain match is done in two stages: (1) by matching superconducting control transformers inside the dewar to 1 part in 10^3 or better, (2) by sending the control currents in opposite directions through a common external resistance R_{diff} and adjusting a precision decade box in one half of the system to remove the residual mismatch. The electronic components all have precision of 1 part in 10^6. The combined matching should be better than 1 part in 10^8. Data from the experiment appears in the form of three 16-bit binary words once per second, corresponding to the three signals A,B, (A-B) where A and B are the two control efforts. It is fed to a computer for analysis and gain matching.

Low temperature operations began in late 1974. The inner test mass has been tried in several runs. The free period varies with the location of the mass in the cradle: The maximum useful working sensitivity of the position readout is at present 5 Å , although 5×10^{-3} Å resolution has been reached in special circumstances. The 5 Å limit is set by the need to handle seismic noise with the controller off. Figure 15 shows a sequence of events during levitation of the inner mass. In the top record the mass is floating in the bearing but rests initially at one end because the cradle is tilted. Seismic noise kicks the mass away from the end, after which it follows a parabolic trajectory, strikes and bounces away again. The middle chart record shows the mass floating freely after levelling. The third record shows excitation of the test mass by seismic noise.

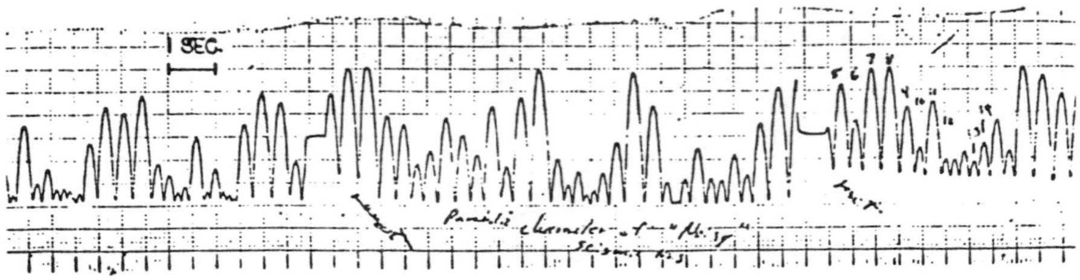

TEST MASS EXCITED BY SEISMIC NOISE
AND BOUNCING AGAINST STOP
$1' < \theta < 9'$ $1 < X < 10$ MICRONS

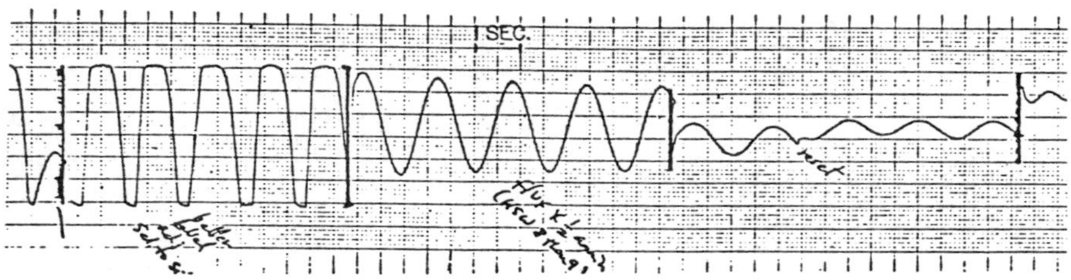

FREE OSCILLATIONS OF TEST MASS AFTER LEVEL ADJUSTMENT, SHOWING EFFECT OF RFDUCING FLUX IN POSITION MONITOR

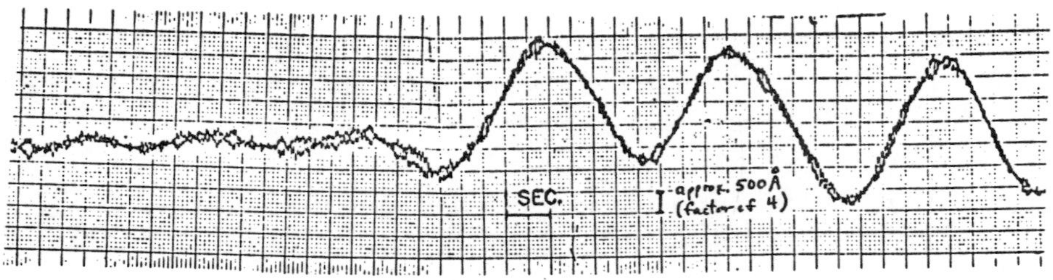

TYPICAL SEISMIC NOISE EXCITING THE SYSTEM

FIGURE 15

The outer test mass has not been levitated. Various levitation attempts early in 1975 failed because it was jammed in the cradle, which had been damaged. The cradle has been repaired. Although the failure has delayed the main experiment, Dr. Worden has tried an interesting subsidiary experiment with the Earth and the inner test mass as proof bodies, by searching for evidence of a differential acceleration between the test mass and the Earth when the apparatus was run overnight. The best of several such runs verified the equivalence principle for niobium versus the Earth to 1 part in 10^5. With the cradle repaired and the controller operating,the main experiment should reach 1 part in 10^{12} or 10^{13}. The limiting factors are gravity gradient disturbances and seismic noise.

Analysis of the flight experiment is continuing.

5. THE STANFORD-LSU-ROME GRAVITATIONAL WAVE ANTENNAS

The observation by Joseph Weber in 1969 of coincident pulses in widely separated gravitational wave antennas started a flood of theoretical and experimental research. Six groups at least have operated gravity antennas, and more are on the way but none has confirmed the original observations. Theorists have balked at finding sources for Weber's pulses, which require the annihilation of 100 solar masses per year in our galaxy. However substantial quantities of energy should be released through the collapse of stars and the capture of stellar material by black holes or neutron stars. The ideal would be to correlate gravitational wave observations with the visible supernovae following stellar collapse. For 408 galaxies (including the Virgo cluster) within 22 Mpc of the Earth the estimated rate of visible supernovae over a sample period corresponding to 32 years is 3.2 events per year. The predicted energies range up to 10^{53} ergs, or 0.04 solar masses, released in a few millisec. To see this at 22 Mpc the detector needs six orders of magnitude more energy sensitivity than Weber's bar.

Various ideas, more or less exotic, have been mooted for better detectors. For Weber's detector equation (18) shows that the bar should be massive and made from a material with high velocity of sound: in practice, that is, the largest possible hunk of aluminum. An improvement discussed by Aplin, Gibbons and Hawking [20] and implemented in a room temperature bar by Drever [49] is to increase the coupling coefficient β by sandwiching piezoelectric crystals between the two halves of the bar. Drever and his colleagues reached a sensitivity of $kT_B/100$ in a bar smaller than Weber's. Other possibilities include operating at cryogenic temperatures and increasing the mechanical Q of the bar. The Gibbons-Hawking equation (19) suggests that the sensitivity might be increased almost indefinitely by increasing Q_B, and Braginsky [50] has shown that Q's as high as 10^9 -- four orders of magnitude higher than Weber's -- are attainable in large single crystals.

To gain the advantage of high Q one must have the right detector. Recently R. P. Giffard [51] in a very penetrating paper has shown that the Gibbons-Hawking argument is incomplete. In general the input noise of the amplifier for the electrical signal is more significant than the transducer noise. The critical problem is matching the bar to this amplifier, in which process current noise must be taken into account as well as the Johnson or voltage noise. With perfect matching by an ideal transducer to a noiseless parametric amplifier of gain G feeding an amplifier of noise temperature T_N the smallest excitation detectable in the $n^{\underline{th}}$ mode of the bar has equivalent energy

$$E_{min} = 2k \left[\frac{\pi(\nu_n \tau_s) T_B}{Q_B} + \frac{T_N}{G} \right] \ln (1/\dot{N}\tau_s) \qquad (32)$$

where $\dot{N}$ is the "accidentals rate", that is the maximum rate of occurence of spurious pulses acceptable to the experimenter. Equation (32) fixes $\tilde{Q}_B$, the maximum useful Q of the bar, for a particular amplifier and sampling time: $\tilde{Q}_B \sim \pi (\nu_n \tau_s)\ G\ T_B/T_N$. With the best available FET amplifier ($T_N \sim 0.2$ K) and a sampling time of 10 mS, $\tilde{Q}_B$ is 1.5×10^5 for a room temperature bar and 10^4 for a 2 K bar. Further progress therefore requires at least as much attention to developing an adequate parametric amplifier as it does to increasing Q_B or lowering the temperature of the bar.

Giffard's formula, unlike that of Gibbons and Hawking, does not yield an optimum sampling time. This is because the amplifier, having current as well as voltage noise, can be characterized by a noise temperature T_N and in measuring an energy pulse (in contrast to a continuous signal from a power source) there is no improvement in resolution with time. Theoretically τ_s should be as short as possible, subject to the constraints that it should be longer than the time B_t^{-1} reciprocal to the bandwidth of the transducer and that both times should be longer than τ_p the duration of the gravitational pulse. In practice good matching to the amplifier may require a transducer of narrow bandwidth and a compromise may have to be struck between good matching and the condition $B_t^{-1} > \tau_p$. The conclusion should be contrasted with the earlier discussion comparing Brownian motion and photon noise in a torsion balance with optical readout. Equation (17), modernizing Hill's argument described above, shows that readout noise is dominant in a torque measurement unless the suspension period of the balance, and hence the observation time, exceeds a certain minimum value. Equation (32) shows that the sampling time in a pulse measurement should be short. The difference is attributable partly to the difference between a steady torque and an energy pulse and partly to the different characters of readout noise.

During the past five years research teams at Stanford, Louisiana State University and the University of Rome have cooperated in applying cryogenic techniques to gravitational wave antennae. The obvious advantage of low temperature operation is the reduction in thermal noise -- by a factor of 10^5 if the bar is cooled to 3 mK -- but this, as just remarked, has to be qualified in the light of equation (32). The real improvement comes through applying superconducting techniques to develop low noise high gain parametric amplifiers and match them to the cold bar. Another advantage of low temperatures is the improved mechanical and electromagnetic isolation obtained with superconducting magnetic shields and superconducting supports.

Similar large antennas are being built at Stanford, LSU and Rome, each 3 m long and 0.9 m in diameter. The Stanford team* has also built the small prototype illustrated in Figure 16, which has been run at liquid helium temperatures since early 1975. The dewar is a modification of the unit for the Stanford superconducting accelerator, with the helium tank enclosed by two helium vapour cooled shields. The boil-off rate is about 1 ℓ/hour, which necessitates transfer from a storage dewar every two weeks. The antenna is an aluminum cylinder 2 m long and 0.4 m in diameter, clad with a layer of niobium-titanium foil 0.38 mm thick. Its resonant frequency is 1312 Hz. The bar is levitated magnetically with coils on the helium tank, wound from 0.6 x 1.3 mm rectangular superconducting wire. The support field is 2400 gauss. The apparatus has been run with the bar up for as long as 900 hours. The resonant frequency of the magnetic support is 5 Hz; its attenuation at bar frequency is 57 db. The helium tank is suspended by acoustic isolation stacks constructed from alternate layers of rubber and iron, mechanically independent of the rest of the cryostat. With the bar levitated in exchange gas at 10^{-4} torr, the longest decay time for the 1312 Hz mode was 40 sec, which corresponds to a Q of 3.3×10^5.

The position detector conceived by Paik, Opfer and Fairbank [52] for the Stanford bar has elements in common with the detector for the free fall equivalence principle experiment. The two systems, developed at the same time in the same institution, were independently conceived. The principle is illustrated in Figure 17. Attached to the end of the bar is a superconducting diaphragm

---

* Present members include S. P. Boughn, W. M. Fairbank, R. P. Giffard, M. W. McAshan, H. J. Paik and R. C. Taber.

FIGURE 16: ASSEMBLY OF THE 2 M LONG STANFORD CRYOGENIC GRAVITY WAVE ANTENNA resonant at antenna frequency. Two flat coils, wound with 2 mil niobium-titanium wire, are placed 0.1 mm from the opposite faces of the diaphragm. They and the circular edge of the diaphragm are rigidly clamped to the bar. A large

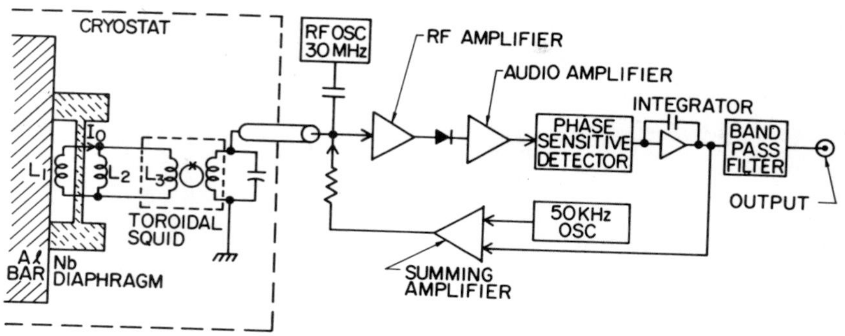

FIGURE 17: SUPERCONDUCTING TRANSDUCER AND READOUT FOR CRYOGENIC GRAVITY WAVE ANTENNA

persistent current of about 5 A is stored in the double loop formed by L_1, L_2. As the diaphragm oscillates it modulates the inductance L_1 , L_2 and a net alternating current appears in L_3 , where it is measured by a SQUID magnetometer as in the similar three loop circuit for the equivalence principle experiment. The large persistent current in L_1, L_2 acts as a magnetic spring which centers the diaphragm between the coils and also changes its resonant frequency. The frequency is raised in one version from about 300 Hz to 900 Hz or 1312 Hz and the diaphragm is tuned to the bar by adjusting the current.

In both the orbiting equivalence principle experiment and the gravity wave apparatus the trapped flux in the position detector serves as a spring and maximum sensitivity is reached when the oscillator resonates with the driving signal. However since the Eötvös acceleration is at 1.8×10^{-4} Hz (90 minute period) while the antenna frequency is 900 Hz or 1312 Hz the optimization conditions are very different. The trapped flux needed to resonate the equivalence principle masses is a few μA as compared with 5 A for the resonant diaphragm. The equivalence principle detector has, as already indicated, ample sensitivity even when used off-resonance.

One way of looking at the resonant diaphragm detector is as a mechanical transformer, which increases the amplitude of the bar motion enough to be seen by the loop detector. The amplitudes of diaphragm motion to bar motion are in the ratio $\sqrt{M/m}$ where M and m are the effective masses of the bar and the diaphragm. More formally, the diaphragm serves as a linear network connecting the bar to the magnetometer, whose operation can be described by a complex transformation matrix. The transformation coefficients are varied by changing the mass of the diaphragm, the spacing of the coils, and the trapped magnetic field. Ideal conditions are when the input reactance of the magnetometer is tuned out by the diaphragm, and the mechanical impedance of the antenna is transformed into the optimum noise matched source impedance for the magnetometer. The diaphragm used at present is 10 cm in diameter. A larger diaphragm and coils would give better matching, but is hard to make.

The transducer and magnetometer have been operated on the antenna. The output signal was processed in the following way. The signal was fed to a two-phase lock-in amplifier whose reference was driven at antenna frequency by means of a frequency synthesizer. The two output channels were then smoothed by R. C. filters, digitized and stored and processed in a PDPH-45 minicomputer. The output noise was Gaussian. To investigate its properties the two channels of recorded data were squared and summed to yield an output proportional to the signal power from the magnetometer. The upper curve in Figure 18 is a sample pulse height analysis with $\ln P(E)$ plotted against energy. The slope implies a noise temperature of 5.9 ± 1.5 K. Since the observed noise was primarily narrow band a differencing technique reduces the effective noise temperature for pulse detection. The lower curve illustrates the effect of differencing the signal in increments of 0.2 sec and integrating via two R-C filters each with 0.1 sec time constant. The observed noise temperature was 0.22 K. Of this 80% was from wideband transducer noise and only 20% from narrowband noise in the bar. After various corrections the final effective noise corresponds to 0.7 K energy in the bar.

Like all gravitational wave antennas the bar is sensitive to acoustic noise. One source was the violent boiling of nitrogen in the liquid nitrogen shield originally used in the dewar. It was replaced by a helium vapour cooled shield. A sound insulating room has also been put around the cryostat. External noise has been much less in recent runs, but occasional bad pulses still occur, probably from motions between the outer heat shield and the dewar.

Giffard's analysis shows the need for using a high gain parametric amplifier with the bar. If the amplifier receives its input signal at the bar frequency ω_b and delivers its output in the form of sidebands on a higher pump frequency ω_p the input noise temperature T_N cannot be less than $T_A\ (\omega_b/\omega_p)$, where T_A is the noise temperature of the amplifier detecting the sideband power. For best performance the amplifier should be cooled to reduce dissipation in the parametric element itself; with superconducting elements the dissipation can be minute. Recently Josephson junction devices have been demonstrated with self-pumping frequencies of order 500 MHz and loud noise at 4.2 K. If they function

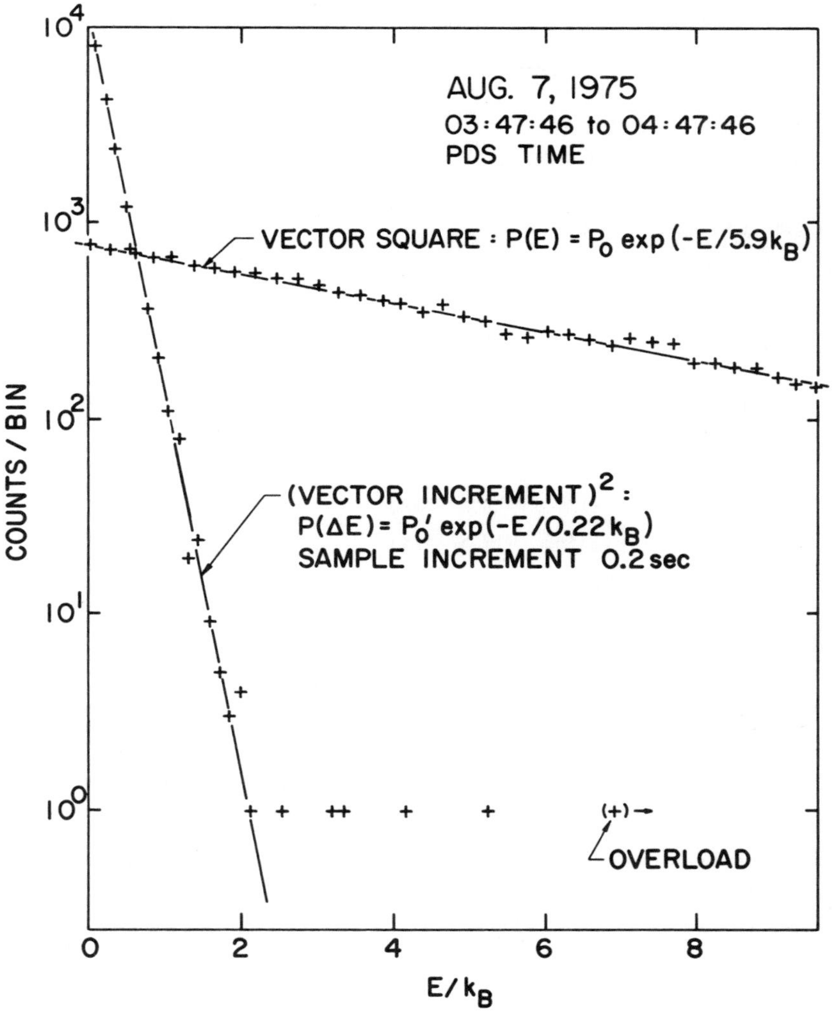

FIGURE 18: OUTPUT STATISTICS OBTAINED IN 8.64×10^6 SEC FROM PROTOTYPE GRAVITATIONAL WAVE ANTENNA

as parametric amplifiers the effective noise temperature could be as low as 10 μK. All such devices have electrical inputs and require transducers such as the resonant diaphragm. Until the optimum source impedance of the following amplifier is determined it is impossible to say how close the resonant diaphragm is to an ideal transformer.

An alternative to the detector of Figure 17 is the one being developed by the LSU team*, which is a resonant microwave cavity modulated by a diaphragm attached to the bar. It too is a parametric amplifier. A calculated minimum noise temperature is 0.1 μK. For a 5×10^6 kg bar the antenna noise becomes negligible when T_B/Q_B is below 3×10^{-8} K. With a Q of 4×10^6 the bar should be cooled to 0.1 K. The minimum detectable energy flux is 80 erg cm^{-2}, which corresponds to detecting an isotropic mass conversion of 4×10^{-7} solar masses at the centre of the galaxy or 1.6 solar masses at 20 Mpc. Since the energy converted to gravitational radiation in a typical supernova explosion is calculated to be less than 0.04 solar masses there is small chance of detecting pulses from visible events in the Virgo cluster unless some device is made not subject to the limitations of conventional amplifiers. The gravitational wave astronomer will not lack challenges for many years to come.

6. THE TWIN SATELLITE EXPERIMENT

A spacecraft orbiting a massive body like the Earth may be regarded as a gyroscope. The angular momentum of such a global gyroscope, being determined by the orbit radius and period and the mass of the spacecraft, is about fourteen orders of magnitude larger than that of the spinning spheres in the Gyro Relativity experiment. If the satellite is constrained to follow an internal shielded proof mass by means of drag-free controllers and thrustors its orbit is vey little disturbed by air drag, solar radiation pressure or other non-gravitational forces. The question arises whether it too might detect small relativistic precessions like those investigated by Schiff for an ordinary gyroscope.

Two principal relativistic effects influence a satellite's motion about the Earth. In track there is a perigee precession similar to the perihelion advance of Mercury about the sun, equal in an 800 km orbit to 12.9 arc-sec/year or 360 m displacement along the path. Cross-track the Earth's rotation causes a nodal dragging of the orbit plane, first calculated by Lense and Thirring [53] in 1918, giving a rate of advance of the right ascension of the ascending node

$$\underline{\dot{\Omega}}_{LT} = \frac{2I\omega_e}{c^2a^3(1 - e^2)^{3/2}}\,\underline{n}_e \tag{33}$$

equal to 0.18 arc-sec/year in an 800 km orbit.

* T. Bernat, W. O. Hamilton, W. Oelske, J. M. Reynolds

The DISCOS controller of the TRIAD satellite was good to about 5×10^{-12}g. With any drag free satellite the residual error, due chiefly to self-gravitation, tends to stay fixed in spacecraft coordinates. Consider an Earth-oriented satellite. In track the displacement builds up continuously according to Hill's equation $S = (3/2)ft^2$. After one year the error from a constant 5×10^{-12} g acceleration is 60 km, a factor of 200 times the relativistic perigee advance, but this could be much reduced by spinning the satellite or periodically turning it through 180° about the local vertical. Cross-track a constant 5×10^{-12}g acceleration simply displaces the orbit plane sideways through a fraction of a mm, but with DISCOS residual errors may have caused a cumulative orbit plane drift of about 2cm/year, or in angular measure 10^{-16} rad sec^{-1}. Thus the drift-rate of the global gyroscope from non-gravitational sources is comparable with that expected from the 4 cm diameter electrically suspended gyroscope and only about 0.3% of the Lense-Thirring nodal drag.

Equation (33) should be compared with Schiff's equation (23) for the motional precession $\dot{\underline{n}}_s^M$ of the spin axis of an orbiting gyroscope. The two effects have the same functional dependence on I, ω_e, a and e, but whereas the gyro precession $\dot{\underline{n}}_s^M$ depends on the orbit inclination i and reverse sign when i is 54° 16', the nodal advance $\underline{\dot{\Omega}}_{LT}$ is independent of i. This surprising result occurs because the perturbing function and the transverse component of orbital angular momentum both depend on sin i, which cancels from the final expression. In identical polar orbits $\underline{\Omega}_{LT} = 4\dot{\underline{n}}_s^M$. D. C. Wilkins has pointed out that both effects originate in the off-diagonal terms g_{14}, g_{24} of the metric for the rotating Earth, but the centre of mass of the satellite follows a geodesic and has time-like four velocity, while the gyro spin vector obeys Fermi-Walker transport and is space-like.

In addition to the Lense-Thirring drag the global gyroscope undergoes a geodetic precession due to the Earth's motion about the sun, just as the electrically suspended gyro does. The motion in the plane of the Earth's equator is 0.020 arc-sec/year; there is also a change of 0.0084 arc-sec/year in orbit inclination. The total relativistic displacement with respect to inertial space is $(\Omega_{LT} + \Omega_G)$ which amounts after 2.5 years to a lateral shift of (13.9 + 1.51) m referred to the Earth's surface. Since the Earth and moon also undergo geodetic precessions in their motion about the sun, the nodal advance of the orbit-plane with respect to the Earth-moon system is Ω_{LT} rather than $(\Omega_{LT} + \Omega_G)$.

In April 1973 R. A. Van Patten conceived the idea of looking for the Lense-Thirring drag in tracking data from TRIAD. The U. S. Navy's TRANET Döppler tracking network can maintain a running average measurement of the cross-track position of a satellite from each ground station to about 1 m over any 15 day arc, with integration to lower values over longer times. Thus both the non-gravitational drift of the global gyroscope and the tracking errors in determining its nodal position with respect to a ground station are well below the cumulative value of $(\Omega_{LT} + \Omega_G)$ after 2.5 years. The idea of performing a relativity experiment with a polar-orbiting drag-free satellite seems promising.

The limitation to a measurement of $(\Omega_{LT} + \Omega_G)$ with a single satellite comes in estimating the non-relativistic nodal regression Ω_Q due principally to the Earth's oblateness

$$\dot{\underline{\Omega}}_Q = -\frac{3}{2}\,\overline{\omega}_0\,J_2'\left[\frac{R_e}{a(1-e^2)}\right]^2 \cos i\,\underline{n}_e \tag{34}$$

where $\overline{\omega}_0$ is the mean motion, R_e the Earth's equatorial radius and J_2' is a regression coefficient formed from the Earth's quadrupole mass-moment J_2 with corrections for higher terms. The quantities J_2' and R_e are well-known, and $\overline{\omega}_0$, a, and e can be accurately determined, but appreciable uncertainties remain in the time history of the inclination angle i from both tracking errors and uncertainties in the Earth's polar position, which is known only to about 0.3 m or 10^{-2} arc-sec. Small as the uncertainties are, they are enough to cause errors in determining the time history of $\dot{\Omega}_Q$ with a single satellite about six times the relativistic modal shift $(\Omega_{LT} + \Omega_G)$.

In January 1974 R. A. Van Patten and I conceived an experiment which has the remarkable property of measuring $(\Omega_{LT} + \Omega_G)$ and simultaneously yielding information about perturbations of the orbit plane and altitude some three orders of magnitude more accurate than any previously obtained. The experiment depends on two counter-orbiting drag-free satellites in polar orbits, initially adjusted by in-flight corrections to be very nearly equal and opposite but planned to avoid collisions. The orbits are chosen so that encounters occur close to the north and south poles. At each polar passing the distance between the spacecraft is measured to about 1 cm. by satellite to satellite Döppler ranging, while at lower latitudes each satellite is tracked independently from existing ground stations. The satellite to satellite data yields a measurement of the angle 2α which is the sum of the coinclinations $(i_1' + i_2')$ for the two orbits. Since orbit inclinations are defined from the ascending node, counterorbiting satellites in the same plane have opposite coinclinations and the nodal rates are

$$\dot{\Omega}_{Q_1} + \Delta\dot{\Omega}_Q = KJ_2'\,(i_1' + \Delta i') \tag{35}$$

$$\dot{\Omega}_{Q_2} - \Delta\dot{\Omega}_Q = KJ_2' (i_2' - \Delta i') \tag{36}$$

where $\Delta i'$ is the angle from the true pole to an estimated pole location and $\Delta\dot{\Omega}_Q$ is the error in nodal rate associated with $\Delta i'$. When (35) and (36) are combined with the definition of α and integrated over the life of the experiment the calculated sum of the two nodal regressions is

$$\left[\Omega_{Q_1} + \Omega_{Q_2} \right]_{calc} = 2KJ_2' \int_{t_1}^{t_2} \alpha \, dt \tag{37}$$

independent of the uncertainty in orbit inclinations. The regression sum calculated from (37) contains the Newtonian perturbations from J_2', but the measured regressions $\Omega_{1_{meas}}$ and $\Omega_{2_{meas}}$ of the two satellites, individually tracked from ground stations at lower latitudes, each also contain the relativistic term $(\Omega_{LT} + \Omega_G)$

$$2(\Omega_{LT} + \Omega_G) = \Omega_{1_{meas}} + \Omega_{2_{meas}} - \left[\Omega_{Q_1} + \Omega_{Q_2} \right]_{calc} \tag{38}$$

Thus the combination of polar ranging and ground-tracking data yields $(\Omega_{LT} + \Omega_G)$.

The experiment dictates the use of closely matched very nearly polar orbits obtained by in-flight corrections during the first phase of the mission. The mission lifetime is 2.5 years. Occasional (i.e. every 3 month) orbit adjustments are needed to assure collision avoidance. The resultant errors may be kept small by restricting the adjustments to impulses applied at equatorial crossings.

The point that the Lense-Thirring effect drags counter-orbiting satellites in the same sense, while Newtonian perturbations are opposite, was recognized by Shapiro, Miller and Jaffé [54] for a suggested experiment in polar orbit around the sun, and by R. W. Davies [55] for a suggested experiment in equatorial orbit around the Earth. However the idea of eliminating the error in the calculated nodal regression by satellite to satellite ranging between polar orbiting spacecraft is new, and crucial to this experiment.

The satellite to satellite Döppler ranging measurement picks up more than just a mean plane separation. The orbits are perturbed by solar and lunar gravity gradients and by the higher order mass-moments of the Earth. The gravity perturbations may be calculated in various ways: one nice method is to utilize the concept of the satellite as a global gyroscope and calculate the drift rate $\dot{\underline{n}}_s^g$ of the gyroscope. This leads to an interesting comparison given below between gravity gradient effects in the twin satellite and Gyro Relativity experiments. The formula is the same as that derived by Laplace in 1789 for the precession of the equinoxes; for a gyroscope of inertia ratio $\Delta I/I$ spinning with angular velocity ω_s at a distance R from a point mass M

$$\dot{\underline{n}}_s^g = \frac{3}{2}\frac{\Delta I}{I}\frac{GM}{\omega_s R^3}(\underline{n}_s \cdot \underline{n}_g)(\underline{n}_s \wedge \underline{n}_g) \tag{39}$$

where $\underline{n}_g$ is the unit vector defining the gradient direction. Taken around an elliptic orbit (39) yields periodic terms and a secular drift $\bar{\dot{\underline{n}}}_s^g$ towards or away from the orbit normal

$$\bar{\dot{\underline{n}}}_s^g = \frac{3}{4}\frac{\Delta I}{I}\frac{GM}{\omega_s a^3(1 - e^2)^2}\sin 2i \tag{40}$$

where i is the angle between $\underline{n}_s$ and the orbit normal [56] .

For a satellite ω_s is just the mean motion $\bar{\omega}_0$ and $\Delta I/I$ is equal to 0.5, the inertia ratio for a ring of matter about axes in and normal to its orbit-plane. With matched counter-orbiting satellites the secular and periodic disturbances from lunar and solar gradients are equal in magnitude but opposite in direction. Since the orbit-normal is tilted with respect to the ecliptic plane and the plane of the moon's orbit there are two secular terms. The lunar effect also has a long term variation since the moon's orbit vector moves along the eliptic vector in a circle of radius 5° with nineteen-year period. The combined secular and nineteen year terms can be very nearly balanced out for any launch date by choosing a suitable orbit node. After removal of the long term effects the remaining direct variations in $2a\alpha$ are a twice yearly term of amplitude ± 1100 m and a twice monthly term ± 180 m. Tides raised on the Earth's surface add components nearly but not quite in phase with the main solar and lunar perturbations, having amplitudes about 15% of the direct terms. All ten effects (secular and periodic) propagate into the nodal motions through the regression equation (34). There are in addition direct perturbations of the nodes from the lunar and solar gradients, having amplitudes 30% to 50% of the 1100 m and 180 m perturbations in $2a\alpha$. None of the nodal perturbations cause significant errors in $(\Omega_{LT} + \Omega_Q)$ since all have equal and opposite effects on the two satellite orbits.

The effects of the Earth's higher order mass-moments have been investigated by D. Schaechter and J. V. Breakwell [57] . Tesseral harmonics ($J_{\ell m}$) of the field with even ℓ and m produce fluctuations in the lateral separation at the poles in times per day, the amplitude of the largest term J_{22} being ± 550 m in a nearly circular 800 km orbit. Tesseral harmonics with ℓ and m odd produce vertical perturbations in times per day at the poles, the largest amplitude (J_{31}) being ± 174 m. Harmonics like J_{41} with even ℓ and odd m have relatively minor effect on the vertical separation but introduce a small difference between the daily histories of the vertical separations at the two poles. Harmonics like J_{32} with odd ℓ and even m have a similar small effect on the lateral separations. Thus the m times daily fluctuations (both lateral and vertical) are not quite the same at the two poles.

In addition to the foregoing effects slight orbital eccentricity differences cause altitude fluctuations from perigee regression, having a period of 15 weeks. The direct effect of the gradient of the lunar gravity gradient cause monthly relative altitude fluctuations of about 10 m. Finally Graziani and Breakwell [58] have shown that longitude differences in the elastic tidal response of the Earth will appear as lateral fluctuations at integral numbers of times per day, modulated by the solar and lunar periods, in other words, apparent seasonal fluctuations in tesseral harmonics. North-south elastic asymmetry of the Earth will cause twice yearly and twice monthly vertical perturbations.

The slant range distance r between the two spacecraft is determined by Döppler ranging with a standard deviation σ_r given theoretically by a formula supplied independently by J. D. Anderson and J. V. Breakwell

$$\frac{\sigma_r}{r} = \left(\frac{8}{\pi}\right)^{\frac{1}{2}} \frac{c}{V} \frac{\sigma_f}{f} \left(\frac{Vt_c}{C}\right)^{\frac{1}{2}} \tag{41}$$

where V is the relative velocity, f the nominal frequency, σ_f the standard deviation of f, and t_c the correlation time for σ_f. Equation (41) assumes the spacecraft pass on parallel courses with data taken over all time (sight angles ϕ between $\pm\pi/2$). R. A. Van Patten has made preliminary studies of a practical system based on a transmitted frequency of 4 GHz (chosen to give a reasonable antenna size) in which data is taken for a few seconds between ϕ's of $\pm 11\pi/24$, i.e. $\pm 82\frac{1}{2}°$. This yields an amount of data reasonable to process with an on-board computer between passes: from 170 to 3400 15 bit words, depending on range. The σ_V is within $\sqrt{2}$ of the theoretical value from equation (41). At 1000 m r can be determined to 1 or 2 cm.

Figure 19 a illustrates the slant range data to be expected from the experiment during a typical six month period, with the lunisolar perturbations at the poles and the envelope of the geophysical perturbations. Each 90 day period yields about 2600 data points. A good mathematical model must include all parameters that perturb the orbit down to the Döppler limit of 1 cm. Based on theoretical

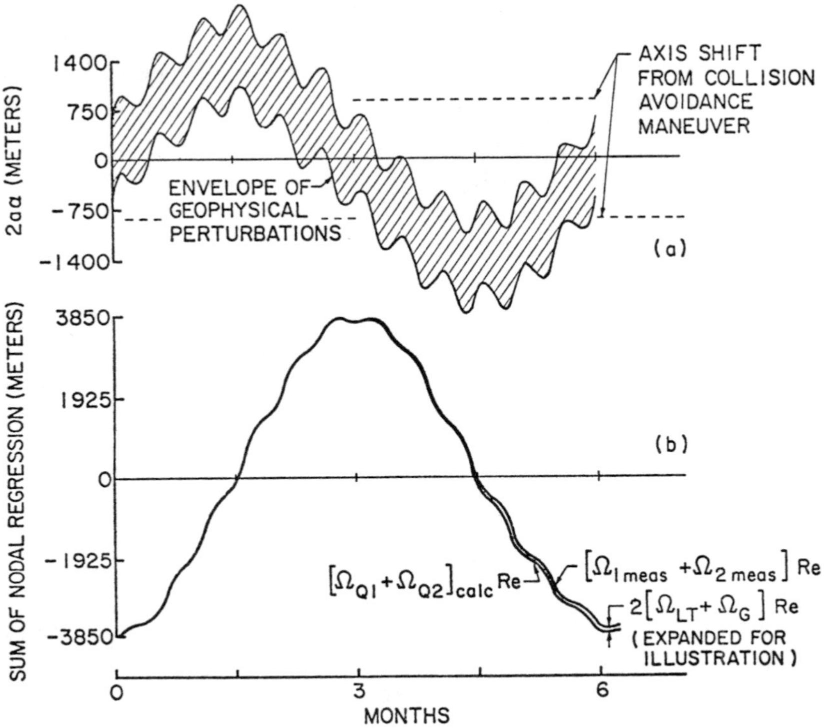

FIGURE 19: FORM OF DATA FROM TWIN SATELLITE EXPERIMENT
(a) lateral polar data (b) extraction of relativity information from nodal experiment

estimates by Allan of the magnitudes of $J_{\ell m}$, Breakwell and Schaechter have developed a model containing 202 parameters including the magnitudes and phases of tesseral harmonics up to m = 60 in the lateral perturbations and m = 59 in the vertical perturbations. Solar and lunar gradient and tidal effects were included in the lateral direction as well as the gradient of the lunar gravity gradient in the vertical direction, and frequencies corresponding to seasonally varying tesseral coefficients were included in the lateral direction through m = 8. Two plane separations and two eccentricities were also included. Although the slant-range data contains the odd-frequency differential altitude functions in quadratic combination with the lateral fluctuations, so that altitude perturbations cause twice daily fluctuations in slant range, the lateral and vertical perturbations can be separated in data processing because the slant-range sensitivity to vertical perturbations changes appreciably during the lunisolar cycle.

A covariance analysis of model data for a typical six month period at one pole has been performed by Schaechter and Breakwell. The covariances were very insensitive to different recent nominal descriptions of the Earth's gravity field. The accuracy of the Döppler determination of slant range was assumed to vary with the range to be measured in accordance with equation (41). A computer program employing a Householder transformation technique was used to sequentially update the covariance matrix, starting with a pessimistic initial covariance. The following results were obtained for a six month experiment:

1. All periodic lateral fluctuations (m even) are measurable to a statistical uncertainty of 0.5 mm.
2. All periodic vertical fluctuations (m odd) are measurable to 2.5 cm.
3. Mean plane separations are measureable to 0.2 cm.

Figure 19 b illustrates extraction of the relativity data when α is integrated over the experimental period. The high frequency geophysical terms are small, and the calculated regression $\left[\Omega_{Q_1} + \Omega_{Q_2}\right]_{calc}$ has the form illustrated in the lower curve of 19 b, dominated by solar and lunar terms. The quantity $\left[\Omega_{1_{meas}} + \Omega_{2_{meas}}\right]$ from tracking data is the upper curve of 19 b. The relativistic quantity $2(\Omega_{LT} + \Omega_G)$ is the secularly increasing difference between the two curves. Error sources in the measurement include the 0.2 cm uncertainty in mean plane separation of the orbits at the poles, uncertainties in the J_2 used to calculate the nodal regression, fluctuations in the self-gravitation of each satellite on its proof mass, nodal shifts from the collision avoidance manoeuvers, and errors in $\Omega_{1_{meas}}$ and $\Omega_{2_{meas}}$ from ground tracking. They are analysed in detail elsewhere [59] . The greatest present limitation is the uncertainty in time base for ground tracking. The quantities $\Omega_{1_{meas}}$ + $\Omega_{2_{meas}}$ have to be referred to inertial space, and the uncertainty in UT1(BIH) may be 0.5 to 1.0 m over the 2.5 year duration of the experiment, giving an error in $(\Omega_{LT} + \Omega_G)$ of 3 to 4 %. However there are good prospects for improving UT1 over the next few years by corrections from lunar laser ranging. With reasonable assumptions about the improvement the error should come down to 0.1 m, making the net error in $(\Omega_{LT} + \Omega_G)$ from all sources 1.1% for a 2.5 year experiment.

The satellite-to-satellite ranging data determines the orbit perturbations due to combinations of Earth harmonics $J_{\ell m}$ with the same m to accuracies some three orders of magnitude higher than they are now known. Thus the differences between vertical perturbations calculated [57] from two recent gravity models (SAO 1973 and GEM 1974) are as much as 50 m whereas the experiment determines the quantities to a few cm. New geophysical data is also obtained from the tidal perturbations. K_2, the Love number associated with the Earth's elastic response to a second harmonic disturbing function, is measurable to 1 part in 10^5, two orders of magnitude better than it is now known. K_3, the Love number associated with the Earth's elastic response to a third harmonic disturbing function, is measuable to 1 part in 10. No determination of K_3 has been made.

The relativistic perigee advance, large as it is, is not measurable in the twin satellite experiment. The difficulty is the 15 week non-relativistic perigee regression. Unfortunately both relativistic and non-relativistic terms reverse sign in a reversed orbit. The greatly improved geophysical data offers some help in calculating out the non-relativistic term, but not enough. The best hope would be to put another satellite in an orbit with an inclination near the value 64° 16' at which the non-relativistic perigee regression changes sign. If the regression were solely due to the Earth's oblateness, the combination of an orbit near 64° 16' and accurate calculations based on tracking data and the improved knowledge of J_2 from the twin satellite experiment, might succeed. However the perigee regression also depends on high harmonics in different combinations from those measured by the twin-satellite experiment. A measure of perigee advance does not seem to be feasible.

7. CONCLUSION

From the Cavendish experiment to the four new experiments on massive bodies described here runs a smooth arc , on which Poggendorf's invention of the optical lever, Boys's investigation of the optimum size of an experiment, Ising's study of the Brownian limit on galvanometers, and the more recent developments in cryogenic technology, serve as defining points. Comparisons among the four new experiments are useful, both to understand the experiments themselves and as background for other possible gravitational experiments.

Consider first experimental size. The equivalence principle experiment needs to be small in order not to be disturbed by gravity gradients. The gravitational wave antenna needs to be large to increase its cross-section to incoming waves. The cryogenic gyroscope is little affected by changes in size. It is indeed striking, and not wholly fortuitous, that the global gyroscope, with fourteen orders of magnitude higher angular momentum, has approximately the same residual drift-rate from non-gravitational forces as the 4 cm diameter spinning sphere.

Consider next the frequency range of the effects to be measured. The Gyro Relativity and twin satellite experiments measure quantities whose periods for a complete 360° rotation are from 2×10^5 to 3×10^7 years. The equivalence principle effect has characteristic periods of 90 minutes or 24 hours. Gravitational wave antennas have resonant frequencies around 1000 Hz, and look for pulses with charactersitic times of order 0.01 sec. These differences in frequency domain affect the experiment plan, even to the extent of determining whether a particular experiment should be done on Earth or in space. The gyroscope drift, being essentially a d. c. effect, cannot on Earth be separated from the d. c. component of drift due to suspension forces; the experiment must be done in space. An equivalence principle experiment is subject on Earth to tides and other diurnal disturbances in the same frequency domain as the effect to be measured; it too is better performed in space, even apart from the increase in driving acceleration. For a gravity wave experiment, however, there is little advantage in space operation, since seismic disturbances are easy to filter at 1000 Hz.

The arguments for cryogenic operation require thought. One line, dear to the heart of a certain kind of theoretical physicist, is the reduction in thermal noise. For gravity wave antennas this has some truth, but the relevant comparison is between the two ratios T_B/Q_B for the bar and T_N/G for the amplifier; the real advantage of low temperatures is in the combination of a high gain parametric amplifier cryogenically coupled to the bar. For the gyroscope thermal noise in the rotor has three principal effects (1) a jitter of the apparent spin axis due to exchange of energy between the lattice angular momentum and the molecular degrees of freedom, (2) a much higher frequency jitter due to thermal fluctuation currents in the superconductor, (3) random walk drifts of the gyro spin axis from the transfer of angular momentum to the ball through impacts of photons or gas molecules. I have shown elsewhere [60] that the random walk drifts are completely negligible even at room temperature. The current fluctuations correspond to displacements of several arc-sec in the readout but being at a frequency of about 1 GHz they are unobservable. The lattice exchange jitter is about 10^{-6} arc sec. Readout noise is therefore wholly dominant, or to put the same point in other words, several order of magnitude improvement in readout sensitivity are needed to achieve a discrimination factor ∂_B of unity. Thermal noise in the rotor has no bearing on the cryogenic operation of the Gyro Relativity experiment.

Nor is thermal noise a determining factor yet in equivalence principle experiments. The limiting $\langle\eta\rangle$ for both Dicke's and Braginsky's experiments was 10^{-13}

after a single day, well below the practical limit, while in the space experiment $\langle\eta\rangle$ is around 10^{-19}, again well below other limits. It is indeed a nice point, and one whose explanation lies but a short way below the surface, that Brownian motion has so far only been a limit for very small bodies like Blackett's magnetometer or very large bodies like the gravity wave antennas. For objects of intermediate size like the Cavendish, Eötvös and gyroscope experiments it does not yet determine a limit.

An Eötvös or Cavendish experiment is ultimately a measurement of an acceleration f on a suspended body. The limit from (11) on determining f in time S is

$$\langle f\rangle \sim \sqrt{\frac{2\beta kT}{MS}}, \tag{42}$$

where β is the damping coefficient and M the mass of the suspended body. In nearly all experiments, particularly those done at low temperatures, the damping is predominantly that from the residual gas surrounding the balance which for a body of characteristic dimension D and gas at pressure p is given by

$$\beta = CD^{-1}\sqrt{\frac{2\pi m}{kT}}\; p, \tag{43}$$

where for a spherical body C is $5/32\pi$. Substituting (43) in (42) and putting M proportional to ρD^3 we have

$$\langle f\rangle \sim \frac{A}{D^2}\left(\frac{1}{\rho S}\right)^{\frac{1}{2}} p^{\frac{1}{2}}\; T^{\frac{1}{4}} \tag{44}$$

so at constant pressure, lowering the temperature yields only an improvement proportional to $T^{\frac{1}{4}}$: a meagre gain.

The reduction of noise with temperature becomes still more questionable if the experiment depends on an optical readout. As the temperature is lowered an optical readout faces new problems of three kinds, each of which limits the useful reduction of T. The first, of academic interest, applies even in a perfect vacuum. Since in lowering the temperature the Brownian motion decreases while the number of photons needed to discriminate the signal increases, there must be some temperature T_ϕ at which disturbances from the random impacts of photons exceed the ordinary Brownian motion. For a balance of period τ_0 observed by means of an optical lever of efficiency ε and discrimination factor ∂_B the temperature is

$$T_\phi = 6.9\,\frac{h}{k}\,\frac{1}{\varepsilon^{\frac{1}{2}}\,\tau_0\,\partial_B} \tag{45}$$

from which for a ∂_B of 10^{-3}, τ_0 of 10 sec and ε of 1% the operating temperature has to be above 0.3 μK. The second limit is that the light beam, being a photon gas, introduces damping which again goes up as the temperature is reduced. Further related limits have been suggested by Braginsky [61] from effects of an imperfectly centered light beam; his assumptions appear unnecessary since the light can be centered on the mirror very accurately by means of servo driven tipping plates controlled by signals derived from a chopped light source.

The real limit to cryogenic operation with an optical lever is that when the light beam falls on the mirror some radiation is absorbed and must be got rid of. At room temperature heat can be removed either by radiation or exchange gas, but at low temperatures radiative transfer becomes ineffective. The addition of exchange gas increases the damping and hence the Nyquist limit. Since the amount of gas needed depends on the intensity of the photon beam, there is an optimum working pressure which is found by minimizing the sum of the squares of the Nyquist limit, expressed as the angular displacement corresponding to the limiting torque from equation (11), and the measurement noise (equation (3)). The result depends on the absorption coefficient Λ of the mirror and ΔT the maximum allowed temperature difference between the suspended body and its surroundings. For a body of diameter D surmounted by a mirror of diameter D'

$$< \Theta_N^2 > + \delta\Theta^2 = \frac{1}{D^2}\left[\frac{M}{P} + Np\right]\frac{T^{\frac{1}{2}}}{S} , \qquad (46)$$

where for a spherical body

$$M = 1.24\ hc \left[\frac{m}{k}\right]^{\frac{1}{2}} \frac{\overline{\lambda}\ \Lambda}{\varepsilon\ \Delta T\ D'^2} \qquad \text{and} \qquad N = 4 \times 10^{-5}\ (mk)^{\frac{1}{2}}\ \frac{\tau_0^4}{\rho D^4} .$$

Thus in any experiment where exchange gas provides the only method of removing heat from the suspended body there is an optimum pressure equal to $\sqrt{M/N}$ independent of the working temperature T. For a reasonable Λ and ΔT (say 1% and 10 mK) the operating pressure has to be $1.9 \times 10^{-2} \sqrt{\rho}\, D^2/D'\tau_0^2$. For Dicke's balance with period 230 sec. the optimum pressure is 10^{-5} torr. From equation (44) the minimum detectable acceleration $<f>$ scales as $p^{\frac{1}{2}}\, T^{\frac{1}{4}}$. Suppose one starts cooling the balance from room temperature at initial pressure 10^{-7} torr. First $<f>$ goes down, but near 90 K radiation cannot provide adequate cooling and the pressure must be increased two orders of magnitude, so $<f>$ increases a factor of 10. To recover the lost ground the balance must be cooled below 9 mK. The difficulty can only be avoided by using a balance whose suspension period τ_0 exceeds $0.14\ \rho^{\frac{1}{4}}\ D/D'\ p_{min}^{\frac{1}{2}}$, where p_{min} is the lowest operating pressure. In our example with p_{min} of 10^{-7} torr, τ_0 has to exceed one hour. Should one wish to work at lower pressures, the natural period of the balance has to be correspondingly longer.

Such difficulties are avoided with the linear and angular position readouts based on Josephson junction devices for which heat transfer to the suspended body is negligible. However the London moment readout for the gyroscope, superior as it is to standard gyro readouts, does not yet attain the resolution of the Jones optical lever. The optimization conditions of each experiment have to be found individually.

Besides specific ideas like the London moment readout, the chief assets of cryogenic technology are (1) the extraordinary improvements in mechanical stability, thermal stability and magnetic shielding at liquid helium temperatures, (2) the many applications of superconducting devices, (3) the reduction in disturbing forces on a suspended body, especially radiation pressure. Consider the equivalence principle experiment. At room temperature the unbalanced pressure from black body radiation at one end of a cylinder of density 10 causes an acceleration of 1.3×10^{-9} g, five orders of magnitude larger than the Eötvös acceleration for an η of 10^{-17}. Cyclically varying temperature differences of 0.001 C across the experimental chamber could masquerade as an Eötvös signal of 2×10^{-17}. Such effects, and corresponding effects of residual gas, become almost negligible at low temperatures.

One last instructive comparison is in the influence of gravity gradient torques on the Gyro Relativity, twin satellite and orbiting equivalence principle experiments. Both the 4 cm spinning sphere for the Gyro Relativity experiment and the global gyroscope of the twin satellite experiment are subject to periodic and secular drift terms derived from equations (39) and (40), but with quite different effects on experiment planning in the two cases. For the twin satellites there are two important secular terms, since the orbit vectors are inclined to both the ecliptic plane and the plane of the moon's orbit, but by choosing a suitable orbit node these can be balanced out. The critical terms for experiment planning are the periodic ones which cause excursion of the ascending node up to ± 3800 m (Figure 19b) or ± 120 arc-sec, two orders of magnitude larger than the expected relativity drift. The terms are separated in data analysis, in which, as already explained, the large lunisolar effects play an important part in separating the lateral and vertical geophysical perturbations. For the Gyro Relativity experiment solar and lunar gradients are negligible, but the gradient of the Earth's field is all important. As was remarked in the discussion of the orbiting equivalence principle experiment, the gravity gradient from the Earth on a body a few hundred km above its surface is five orders of magnitude larger than the sun's gradient at the same point. Hence, the secular drift term from equation (40) is critical and just as the equivalence principle masses must be well-centered, so the $\Delta I/I$ for the spinning sphere of the Gyro Relativity experiment must be kept small. A criterion derived from (40) is that density inhomogeneities should be below one part in 10^6, which is comparable to the criterion obtained from the mass-unbalance equation (26). The periodic gravity gradient terms, on the other hand, have negligible significance in the Gyro Relativity experiment. Their amplitude depends inversely on the frequency of the torque, and for a gyroscope orbiting the Earth the period is 45 minutes rather than the twice monthly and twice yearly periods characterizing the lunisolar torques on the counter-orbiting satellites. The maximum amplitude of the sine wave for a gyrosocpe with $\Delta I/I$ of 10^{-6} is 2×10^{-4} arc-sec. Procedures separating the gravity gradient torques from the relativity terms in inclined orbits are discussed elsewhere [33] .

To find so much variety among experiments having so much in common is one of the pleasures of working in experimental gravitation. The opportunities for ingenuity will not soon be exhausted.

ACKNOWLEDGEMENTS

I thank W. M. Fairbank and other colleagues named in the text for innumerable discussions, and R. P. Giffard, R. A. Van Patten and P. W. Worden, Jr. for criticizing portions of the manuscript.

REFERENCES

[1] H. Cavendish, Phil. Trans. Roy. Soc. 83 (1798) 470. Reprinted in the 1809 abridgement of the Transactions by C. Hutton, G. Shaw and R. Pearson (London 1809) vol. 18, p. 389.

[2] The quoted error is the 95% confidence limit on the mean of Cavendish's 23 measurements.

[3] Maxwell to Joule, quoted in History of Cavendish Laboratory (London 1910) 31.

[4] C. W. F. Everitt, James Clerk Maxwell: Physicist and Natural Philospher (New York, Scribners 1975) 118-123.

[5] R. von Eötvös, D. Pekar and E. Fekete, Ann. Phys. 68 (Leipsig, 1922)11.

[6] P. H. Roll, R. Krotkov and R. H. Dicke, Ann. Phys. 26 (New York, 1964) 442.

[7] V. B. Braginsky "Verification of the Equivalence of Gravitational and Inertial Mass" B. Bertotti, ed. in Experimental Gravitation (New York, Academic Press, 1974) 252-258.

[8] W. J. H. Moll and H. C. Burger, Phil. Mag. 50 (6th Series, 1925) 624, 626.

[9] G. Ising, Phil. Mag. 1 (7th Series, 1926) 827.

[10] R. V. Jones, J. Sci. Instrum.,38 (1961) 37.

[11] R. V. Jones and C. W. McCombie, Phil. Trans. Roy. Soc. 244 (1951) 205.

[12] P. M. S. Blackett, Phil. Trans. Roy. Soc. 245 (1952) 309.

[13] H. A. Daynes, J. Sci. Instrum. 3 (1926) 7.

[14] A. V. Hill, J. Sci. Instrum. 4 (1926) 72.

[15] R. Ruffini and A. Treves, Astrophysical Letters 13 (1973) 109.

[16] V. B. Braginsky, Physical Experiments with Test Bodies (Moscow 1970, Washington D. C. NASA translation 1972) 53f.

[17] P. W. Worden, Jr. and C. W. F. Everitt, "Tests of the Equivalence of Gravitational and Inertial Mass Based on Cryogenic Techniques" in B. Bertotti (ed) Experimental Gravitation (New York, Academic Press 1974) 393.

[18] C. W. McCombie, Rep. Prog. Phys. 16 (1953) 266.

[19] Reference [6] p. 474.

[20] G. W. Gibbons and S. Hawkins, Phys. Rev. D. 4 (1971) 2191.

[21] R. V. Pound and G. A. Rebka, Phys. Rev. Letters 3 (1959) 439; R. V. Pound and J. L. Snider, Phys. Rev. B. 140 (1965).

[22] R. H. Dicke and H. M. Goldenberg, Phys. Rev. Letters. 18 (1967) 313.

[23] H. A. Hill "Light Deflection" in R. W. Davies (ed) Proceedings of the Conference on Experimental Tests of Gravitation Theories" JPL Technical

Memorandum (Pasadena, 1970) 33-499, p. 89.

[24] W. de Sitter, M. N. Roy. Astr. Soc. 76 (1916) 699; 77(1916) 155, 481.

[25] J. A. Schouten, Proc. Amst. Acad. 21 (1919) 533.

[26] A. D. Fokker, Proc. Kon. Akad. Weten 23 (Amsterdam, 1920) 729.

[27] A. S. Eddington, Mathematical Theory of Relativity (Cambridge, 1926) 99.

[28] Letter from P. M. S. Blackett to C. W. F. Everitt, 20 November 1962 and personal conversations.

[29] L. I. Schiff, Proc. Nat. Acad. Sci. 46 (1960) 871.

[30] G. E. Pugh, WSEG Research Memorandum, No. 11, Weapons System Evaluation Group, the Pentagon, Washington 25 D. C. (12 November 1959).

[31] B. M. Barker and R. F. O'Connell, Phys. Rev. D. 2 (1970) 1428.

[32] D. C. Wilkins, Ann. Phys. 61 (New York, 1970) 277: "Precession of a Gyroscope in a Perturbed Orbit about the Earth", Stanford University Memorandum, May 1970.

[33] C. W. F. Everitt, "Review of the Significance of Gravity Gradient Torques in Different Versions of the Gyro Relativity Experiment" (Stanford University, W. W. Hansen Laboratories of Physics 1974).

[34] P. M. Selzer, W. M. Fairbank and C. W. F. Everitt, Adv. Cry. Eng. 16 (1971) 277.

[35] Staff of the Space Department of the Johns Hopkin's University Applied Physics Laboratory and Staff of the Guidance and Control Laboratory, Stanford University, Journal of Spacecraft and Rockets 2 (1974) 631.

[36] B. Cabrera, "Generating Ultra Low Magnetic Field Regions with Superconducting Shields and Their Use with a Sensitive Magnetic Charge Detector" in M. Crusius and M. Vuorio, Low Temperature Physics LT 14 (Amsterdan, North Holland Publishing Co. 1975) vol. 4 p. 270.

[37] C. W. F. Everitt, W. M. Fairbank, W. O. Hamilton "From Quantized Flux in Superconductors to Experimens on Gravitation and Time Reversal Invariance" in John R. Klauder ed. Magic without Magic: John Archibald Wheeler (San Francisco, W. H. Freeman 1972) p. 217.

[38] J. R. Nikirk, "Fabrication of an Electronic Suspension Subsystem for a Cryogenic Electrostatically Suspended Gyroscope for the Relativity Experiment" Final Report on NASA Contract NAS8-27333 (Stanford University Center of Systems Research, January 1973).

[39] T. D. Bracken and C. W. F. Everitt, Adv. Cry. Eng. 13 (1968) 168.

[40] J. T. Anderson and C. W. F. Everitt, "A High Accuracy All-angle Gyroscope Readout Using Quantized Flux" Paper submitted to the Applied Superconductivity Conference, Stanford 1976.

[41] C. W. F. Everitt "The Gyroscope Experiment I: General Description and Analysis of Gyroscope Performance" in B. Bertotti (ed) Experimental Gravitation (New York, Academic Press 1974) p. 357, equation(29).

[42] J. A. Lipa, J. R. Nikirk, J. T. Anderson and R. R. Clappier, "A Superconducting Gyroscope for Testing General Relativity" in M. Crusius and M. Vuorio, Low Temperature Physics LT 14 14(Amsterdam, North Holland Publishing Co., 1975) 250.

[43] Reference[41], p. 341.

[44] C. W. F. Everitt, Final Report on Contract NAS 8-25705 to build and test a Precise Star Tracking Telescope (Stanford, July 1972).

[45] J. Bull and D. B. DeBra "Precise Attitude Control of the Stanford Relativity Satellite, Joint Automatic Control Conference, Ohio State University (June, 1973).

[46] D. Klinger,"Error Modeling of Precision Orientation Sensors in a Fixed Base Simulation" SUDAAR No. 481 (Ph.D. Thesis, Stanford University, 1974).

[47] R. A. Van Patten in Annual Report for the Program a Gyro Test of General Relativity and Develop Associated Control Technology" (June, 1973).

[48] J. A. Lipa, W. M. Fairbank and C. W. F. Everitt,"Gyroscope Experiment II: Development of the London Moment Gyroscope and of Cryogenic Technology for Space" in B. Bertotti (ed.) Experimental Gravitation (New York, Academic Press, 1974) 361; iden Proceedings of the Cryogenic Workshop (NASA George C. Marshall Space Flight Center, 1972) 169.

[49] R. W. P. Drever, J. Hough, R. Bland and G. W. Lessnoff, Nature 246 (1973) 340.

[50] Reference [7] page 245.

[51] R. P. Giffard "Sensitivity Limit of Resonant Gravitational Wave Detectors Using Conventional Amplifiers", Stanford University Memorandum,(December, 1975).

[52] H. J. Paik, "Low Temperature Gravitational Wave Detector" in B. Bertotti (ed.) Experimental Gravitation (New York, Academic Press, 1974) 515.

[53] J. Lense and H. Thirring, Phys. Zeits. 19 (1918) 156.

[54] Lewis F. Miller, "The Lense-Thirring Effect: A Theoretical Investigation of Possible Experiments to Measure It", Thesis submitted in Partial Fulfillment of the Requirements for the Degree of Bachelor of Science at the Massachusetts Institute of Technology, (January 1971) 18.

[55] R. W. Davies, "A Suggested Space Mission for Measuring the Angular Momentum of the Sun" in B. Bertotti (ed.) Experimental Gravitation (New York, Academic Press, 1974) 412, Appendix II to paper discussing an experiment circulating the Earth.

[56] Reference [43] contained an error of a factor of 2 in the equation (11) corresponding to (40) here, which has been pointed out by B. M. Barker and R. F. O'Connell, Phys. Rev. D. (1976) in press.

[57] D. Schaechter, J. V. Breakwell, R. A. Van Patten and C. W. F. Everitt "A Covariance Analysis for Parameters Determined in the Relativity Mission with Two Counter-Orbiting Satellites" Astrodynamics Conference of the AIAA/AAS, San Diego, California, August, 1976) to be published.

[58] F. Graziani, J. V. Breakwell, R. A. Van Patten, C. W. F. Everitt, "Earth Tide Information from Two Counter Orbiting Polar Satellites"

International Astronautical Federation (IAF), XXVI$^{\underline{th}}$ Congress, Lisbon, Portugal, (September, 1975).

[59] R. A. Van Patten and C. W. F. Everitt, "A Possible Experiment with Two Counter-Orbiting Drag-Free Satellites to Obtain a New Test of Einstein's General Theory of Relativity, and Improved Measurements in Geodesy", Celestial Mechanics (1976) in press.

[60] Reference [43] p. 358.

[61] Reference [7] pp. 238 f.

OBSERVATIONAL COSMOLOGY

R. B. Partridge
Haverford College
Haverford, Pennsylvania

ABSTRACT

Some sixty years after the development of relativistic cosmology by Einstein and his colleagues, observations are finally beginning to have an important impact on our views of the Universe. The available evidence seems to support one of the simplest cosmological models, the hot Big Bang model.

The aim of this paper is to assess the observational support for certain assumptions underlying the hot Big Bang model. These are that the Universe is isotropic and homogeneous on a large scale; that it is expanding from an initial state of high density and temperature; and that the proper theory to describe the dynamics of the Universe is unmodified General Relativity. The properties of the cosmic microwave background radiation and recent observations of the abundance of light elements, in particular, support these assumptions. Also examined here are the data bearing on the related questions of the geometry and the future of the Universe (is it ever-expanding, or fated to recollapse?).

Finally, some difficulties and faults of the standard model are discussed, particularly various aspects of the "initial condition" problem. It appears that the simplest Big Bang cosmological model calls for a highly specific set of initial conditions to produce the presently observed properties of the Universe.

For millenia cosmology was a branch of the natural sciences in which observations lagged far behind theory. The gap between theory and observation remained large until about ten years ago, five decades after Einstein and others brought the theory of General Relativity to bear on the Universe as a whole. It is a good measure of the genius of Einstein, Lemaître, Friedmann, and their collaborators, that they were able so long ago to predict so many properties of the Universe for which we are just now finding observational support.

Now, however, the gap between theory and observation is slowly narrowing--it is a time of great excitement for cosmologists. We now have evidence which supports many of Einstein's prescient conjectures about the nature of the Universe. We stand on the verge of being able to decide which of the several models for the Universe permitted by General Relativity is the most accurate description of the Universe we inhabit. We are not yet, however, at that fruitful stage of interaction between theory and observation where observations lead to refinements and revisions of theory. So far, cosmological data are too "tolerant".

My aim in this review is to present a brief survey of the current state of observations in cosmology. I shall deal almost exclusively with observational results, assuming that the reader is familiar both with the fundamentals of relativistic cosmology and with some of the recent theoretical developments discussed elsewhere in this volume.

I shall begin by discussing the observational evidence supporting the Big Bang model for the Universe, a model which is now generally accepted by most workers in the field. Next, I shall deal with some of the problems lurking in this model of the Universe. Some of these problems are not entirely relativistic in nature, but may nevertheless prove to be the kind of grit in the mechanism which forces us to re-examine our theories, including (possibly) General Relativity.

I. THE STANDARD COSMOLOGICAL MODEL

Twenty years ago there was no generally accepted description, or <u>model</u>, for the properties of the Universe as a whole. There is now. The Big Bang model* is "standard" not in the sense that all workers in the field "believe" it, but in the sense that it is normally adopted as a basis for work on cosmological problems (often almost casually, so deeply ingrained is the model by now). The standard Big Bang model Universe is isotropic and homogeneous on a large scale. General Relativity in its unmodified form (that is, with no cosmological constant and with no Brans-Dicke-Jordan scalar component) is taken as the proper theory for gravity. The Universe is assumed to be expanding from an initial state of high density and temperature, hence the nomenclature "hot Big Bang". Although there is substantial agreement on the <u>past</u> properties of the Universe, there is considerably less agreement on the probable <u>future</u> properties of the Universe. As is well known, General Relativity admits two general types of solution for an expanding, isotropic, homogeneous model Universe: one corresponds to a Universe which continues to expand forever; the other to a Universe which eventually reaches

*For a general review of the properties of this cosmological model, see Peebles [4] or Weinberg [6].

a point of maximum expansion and then contracts back to a state of high density. In 1975 there is some sentiment in favor of the former view, but as we shall see the observational evidence in support of this opinion is far more tenuous than the observational support for other aspects of the standard cosmological model.

If the Universe is strictly isotropic and homogeneous, as assumed in the standard model, then the geometry of the Universe is specified by the Robertson-Walker metric (see, for instance, Weinberg, [6]:

$$ds^2 = c^2dt^2 - \frac{R^2(t)dr^2}{1 - kr^2} - R^2(t)r^2(d\theta^2 + \sin^2\theta d\phi^2) \quad , \quad . \quad (1)$$

The metric has spatial hypersurfaces of constant curvature, which may be positive ($k > 0$), zero ($k = 0$), or negative ($k < 1$). The evolution of the Universe is contained in the function $R(t)$, the <u>scale factor</u>. In an expanding Universe, $\dot{R}(t) > 0$. For future reference, we introduce the standard notion

$$H_o \equiv \frac{\dot{R}(t_o)}{R(t_o)} \quad , \quad . \quad . \quad . \quad . \quad . \quad . \quad . \quad . \quad . \quad . \quad . \quad . \quad (2)$$

where the subscript zeros here, as elsewhere, indicate the present epoch. H_o is the present value of the Hubble constant, a measure of the rate of expansion of the Universe*.

Each different cosmological model corresponds to a different time-dependence of R. To determine R(t), we need a theory of gravitation (and, unless the Universe is pressure free, an equation of state). If General Relativity is taken, then Einstein's field equations yield the differential equations for R(t). The field equations are:

$$\mathcal{R}_{\mu\nu} - \tfrac{1}{2}\mathcal{R}\, g_{\mu\nu} + \lambda g_{\mu\nu} = - 8\pi G T_{\mu\nu} \quad , \quad . \quad . \quad . \quad . \quad . \quad . \quad (3)$$

where $\mathcal{R}_{\mu\nu}$ is the Ricci tensor. There are two physical quantities

*The most recent observational value for H_o is 55 km/sec per Mpc in the usual units [19] or 1.8×10^{-18} sec^{-1}, since one Mpc = 3×10^{24} cm.

which enter (3)--and hence eventually determine $R(t)$. The first is the cosmological constant λ [carrying units of $(\text{time})^{-2}$ in the notation employed here]. It is generally assumed (or argued on formal or aesthetic grounds; see [17]) that $\lambda \equiv 0$. The second physical quantity is the stress-energy tensor $T_{\mu\nu}$, which involves both the pressure and the mean mass density ρ of the Universe. It is easy to show that known particles and fields now contribute a negligibly small pressure in the Universe. In this simple case, $T_{\mu\nu}$ reduces to

$$T_{oo} = \rho c^2 \text{ , ... (4)}$$

Substituting (4) into (3) leads to

$$\dot{R}^2(t) = \frac{8\pi}{3} G\rho_o R^3(t_o) R^{-1}(t) - kc^2 \text{ , ... (5)}$$

This equation for $R(t)$ may be taken as the mathematical definition of the standard model.

A. The Friedmann Solutions

As is well known, there are three solutions to (5), depending on the value for k. These will be referred to as the Friedmann solutions*. For isotropic, homogeneous, pressure-free models with $\lambda = 0$ (i.e., for the standard models), the present value of the mass density ρ_o and k are directly related: $k \lesseqgtr 0 \leftrightarrow \rho_o \lesseqgtr \rho_c$ where ρ_c is a critical density given by

$$\rho_c \equiv \frac{3H_o^2}{8\pi G} \text{ , ... (6)}$$

A standard model with $\rho_o > \rho_c$ thus necessarily has positive space curvature. In addition, from a physical point of view [13], a model with $\rho_o > \rho_c$ is one in which the gravitational forces between particles in the Universe are sufficient to reverse the expansion and cause recontraction (see figure 1). On the other hand, models with $\rho_o \leq \rho_c$ continue to expand forever (figure 1).

*This usage is general, despite the fact that Friedmann actually considered a wider class of models.

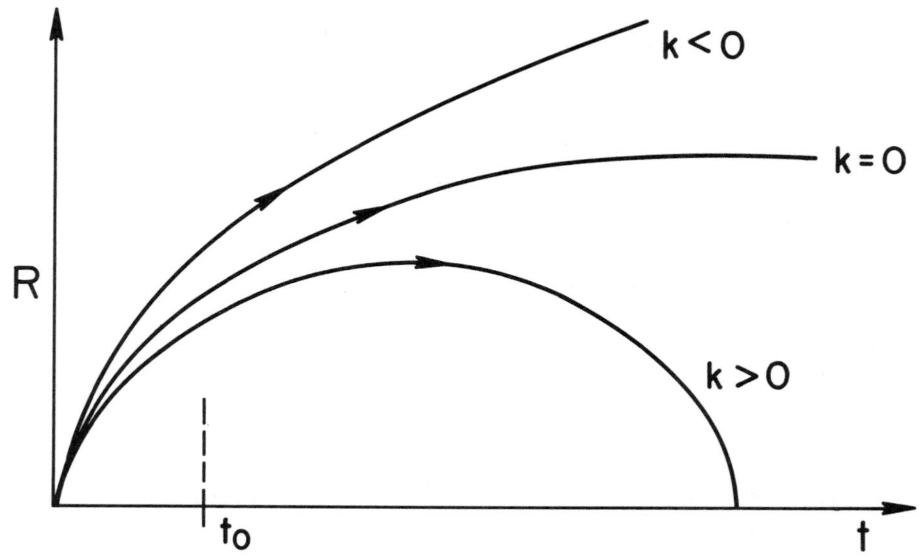

1. Scale factor R as a function of epoch for the three Friedmann models, solutions of eqn. 5. If $k > 0$ (a high density Universe, with $\rho_o > \rho_c$), the Universe recollapses. The present epoch is indicated schematically by t_o.

Hence, the magnitude of the deceleration can also be used to distinguish the Friedmann solutions. A convention we will follow is the use of the <u>deceleration parameter</u>,

$$q_o \equiv \frac{-R(t_o)\ddot{R}(t_o)}{[\dot{R}(t_o)]^2}, \quad . \quad . \quad . \quad . \quad . \quad . \quad . \quad . \quad . \quad . \quad . \quad (7)$$

to distinguish between the various standard models. For the standard models, $q_o \lesseqgtr \frac{1}{2} \leftrightarrow k \lesseqgtr 0$.

Clearly, to find a proper description of the properties of the Universe as a whole, a cosmologist must not only justify the use of the standard model (5); he or she must determine which of the Friedmann models best fits the Universe. There are three ways to proceed: to measure ρ_o/ρ_c*, to determine the deceleration directly,

*Which, from (6), requires a knowledge of H_o.

or to determine the curvature of space (k). Measurements along these lines have been attempted, and will be discussed in section II F.

Finally, as is clear either from (5) or figure 1, all variants of the standard model possess a singularity with R = 0*. The value of t for which R = 0 is usually taken as the initial epoch, t = 0, in the standard Big Bang models. For non-standard models (e.g., anisotropic cosmological models), the nature and even the existence of a real physical singularity has been frequently debated (see, for instance, the contribution of Lifschitz to this volume).

B. Non-standard Models

Before turning to the observations, we wish to catalog briefly some properties of the major non-standard cosmological models, some of which have (or had) long competed with the hot Big Bang model.

Consider first those models which accept General Relativity as the correct theory of gravitation. If we permit non-zero values of the cosmological constant, a range of new models is possible. R(t) for some of these is shown in figure 2. The Lemaître model in particular has attracted some attention [18].

Two major modifications of General Relativity have been proposed over the past six decades--or at least two which give rise to new cosmological models. In the first class are theories such as that due to Jordan [16] or to Brans and Dicke [12] which add a scalar field to the usual tensor formulation of General Relativity. Given the observational limits on the magnitude of the scalar component, it is safe to say that the present cosmological effect of the scalar field is small. On the other hand, the scalar field, if present, would have played a major role at earlier epochs (see section II). The second major modification to General Relativity was made for essentially cosmological reasons by Bondi and Gold [11] and Hoyle [14], [15]. Their Steady State Theory is based on the notion that the large-scale properties of the Universe are the same at all times as well as at all places (the "Perfect Cosmolog-

*This remains true for models with non-zero values of pressure and λ, provided $P > 0$ and $\lambda \leq 0$ (see [1] for details).

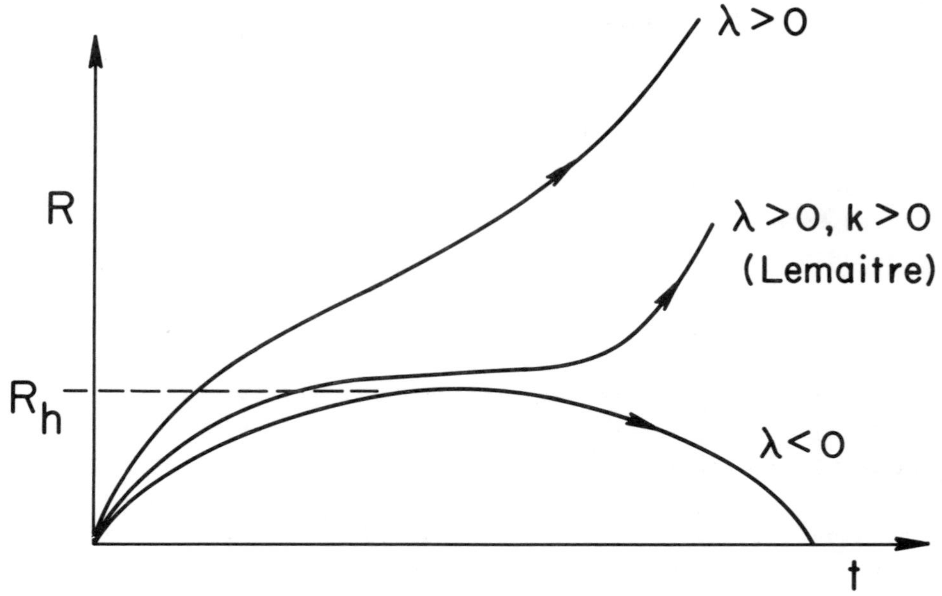

2. Some non-standard models with $\lambda = 0$. No new features are present if $\lambda < 0$. If $\lambda > 0$, however, $R(t)$ is more complicated. Lemaître's model includes an interval where $R(t)$ changes very slowly. Associated with this interval is a characteristic redshift, $z_h = R_o/R_h$ (see [1] and [18]).

ical Principle"). To keep ρ_o constant as the Universe expands, new matter is created ex nihilo; the creation field responsible appears in the field equations like a negative pressure of magnitude $\rho_o c^2$; equation (3) reduces to

$$\ddot{R}(t)R(t) - \dot{R}^2(t) = 0 , \qquad (8)$$

so that $q_o = -1$ for the Steady State Theory. In addition, the Perfect Cosmological Principle prohibits any evolution of the large scale properties of the Universe: individual objects--stars or galaxies--may evolve, but no temporal change in the statistical properties of a large sample of objects is permitted. The Universe itself can have no history.

II. THE OBSERVATIONAL EVIDENCE

The standard Big Bang model is one of the simplest of the modern cosmological theories; it assumes unmodified General Relativity (with no cosmological constant) and the simplest distribution of matter. No doubt part of its appeal lies in its simplicity. We do not yet know whether Nature is as fond of simplicity as we are. As we turn to a consideration of the observational data, it is appropriate to keep before us the central question: "Are there observations which will force us to abandon the standard model?"

Let us begin, however, with the data we can offer to support the assumptions made in the standard model. We must admit at the outset that the evidence is nowhere absolutely and indisputably convincing. We must also admit that the data we do possess have been (and no doubt will continue to be) interpreted in different ways. With these confessions made, let us proceed to the evidence. *En passant*, we will also list observations which tend to falsify rival theories.

A. *Evidence that Einstein's General Relativity is the correct theory of gravity.*

This topic has been discussed by other, more knowledgeable contributors to this volume. Perhaps the most convenient approach to the question is that of Thorne and his collaborators and colleagues [28]. They show that all *metric* theories of gravitation may be conveniently classified using 9 parameters; they then proceed to consider observational limits on each of the parameters. Interestingly, most of the data are locally derived, either from laboratory experiments, or observations of the earth and the solar system. They conclude that only a very restricted set of gravity theories (including the Jordan-Brans-Dicke scalar-tensor theories and of course General Relativity) meet the observational tests.

One local observation is of particular interest to cosmology: a measurement of the deflection of light by the sun, using radio interferometric methods (see *Physics Today*, April, 1975, p 17). The close agreement between the radio results and the value predicted by Einstein increases our confidence in General Relativity.

It also sets an upper limit of $\sim$ 2% on the present value of the scalar component of gravitational forces in the framework of the Brans-Dicke theory, negligibly small from the point of view of cosmology.

Two new developments which bear on cosmology may now be mentioned. First, on the observational front, van Flandern [22] has reported evidence for a secular change in the gravitational "constant", $\frac{\dot{G}}{G} = (-8 \pm 5) \times 10^{-11}\ \mathrm{yr}^{-1}$. However, his value lies just above an upper limit derived by Dearborn and Schramm [21] from a study of clusters of galaxies. Better evidence is clearly needed before we reject the naive view that G is constant. The need for more direct and more precise evidence on this question has been sharpened by the second new development, Hoyle and Narlikar's new cosmological theory [24]. In one of its variants, this theory requires a changing value of G, $\frac{\dot{G}}{G} \approx -1.1 \times 10^{-10}\ \mathrm{yr}^{-1}$, for $H_o = 55$.

We return now to the oft-debated question of the cosmological constant, λ. In general the debate has centered on whether it should be "allowed" in the field equations at all [17]. On the strictly observational side, no useful limits on λ are available from astronomical measurements of subsystems within the Universe (such as the solar system or clusters of galaxies). That is to say, a value of $|\lambda|$ large enough to have pronounced effects on cosmology cannot be so excluded. Indeed, the best limits on λ are arrived at indirectly by comparing measurements of other cosmological parameters. It may be shown that

$$\lambda = 3H_o^2 (\rho_o/2\rho_c - q_o) \quad , \quad . \; . \; . \; . \; . \; . \; . \; . \; . \; . \quad (8)$$

Limits on ρ_o and q_o to be discussed below imply conservatively that $|\lambda| \lesssim 5H_o^2$. Of course, to exclude non-zero values of $|\lambda|$ rigorously would require precision measurements of both ρ_o and q_o: it seems likely that the best we can hope for is to show that λ is too small to have any appreciable effect on the dynamics of the Universe.

Now let us look at the cosmological constant in a different light: is there any evidence <u>for</u> a non-zero value? Five to ten years ago, there was a flurry of interest in the Lemaître model [27], because its protracted period with $\dot{R} \stackrel{\sim}{\sim} 0$ offered a way of explaining the concentration of quasar redshifts at $z \sim 2$. In the past eight years, the observational evidence supporting the Lemaître model has grown weaker, and the recent observations of bright QSO's with redshifts above three cast grave doubt on it. As one of the original proponents of the Lemaître model recently put it [18], we must put the Lemaître models "back on the shelf"*. Nor is there yet any other evidence which compels us to accept a non-zero value of λ. (Reliable evidence that $q_o < 0$ would raise the possibility of a positive value for λ.)

B. Evidence for the expansion of the Universe.

One of the great strides in modern science was Hubble's demonstration that the redshifts observed in the spectra of distant galaxies were proportional to the apparent distances to the galaxies. That is, $z \propto D$, where $z \equiv \Delta\lambda/\lambda_e$, λ_e being the emission wavelength. Hubble immediately interpreted the redshift as a Doppler shift arising from the expansion of the Universe. In the usual notation, $z = \frac{v}{c} = \frac{H_o D}{c}$ for recession velocities $v << c$. This follows directly from the relation

$$z + 1 = \frac{R(t_o)}{R(t)}, \quad \ldots \ldots \ldots \ldots \quad (9)$$

which in turn comes from (1).

It is important to remember, however, that the observed quantity is redshift, not velocity. As a consequence, Hubble's interpretation has not been uniformly accepted. Some have argued that electromagnetic radiation becomes "tired" (and hence redshifted) as it travels large distances through intergalactic space (see, for example, [25] and [26]). Geller and Peebles [23] have shown that straightforward "tired light" theories are not supported by the observational evidence. Recently, Hoyle and Narlikar [24] have put forward a static cosmological model in which the apparently distance-dependent redshift actually arises from the time

*But see Occhionero's contribution to this volume for a different viewpoint.

dependence of the mass of elementary particles. In this theory, elementary particles were less massive in the past, hence the wavelength of spectral lines was greater. At time $t = 0$, there was no singularity in the sense that $R \to 0$, since R is constant in their theory; rather there was a singularity as <u>masses</u> $\to 0$. Now Hoyle and Narlikar's theory explains the standard observations of cosmology (e.g., Hubble's observations, and the microwave background radiation), and the rate of change of mass it requires lies below our laboratory capabilities (I think). It is not clear, however, that it meets the Geller-Peebles test; nor is it clear that the $\dot{G}/G$ it requires is in accord with the observations (see section II A above)*.

Finally, Hubble's interpretation, unlike alternative ones, offers a natural explanation for the observation that the ages of the oldest systems in the Universe roughly equal H_o^{-1} [19].

C. <u>Evidence for the primeval fireball (a hot, dense state)</u>.

If the Universe is expanding, what kind of state is it expanding from? The detailed answer was worked out by Gamow and his collaborators thirty years ago (see [34] and references therein): the Universe expands from a state of high density and temperature, variously known as the primeval fireball or the hot Big Bang**. How strong is the evidence for this aspect of the standard model?

Of course the expansion of the Universe in itself provides evidence that the Universe was denser (and therefore hotter) in the past than it is now (provided that no new matter is being created as in the Steady State theory). However, the crucial physical question is whether the Universe has evolved from a past state so dense that its properties were completely different from those we now encounter.

The discovery of the cosmic microwave background (Penzias and Wilson [38]) answered this question affirmatively. The observa-

*Another objection to the Hoyle-Narlikar theory is raised in section II D.

**I prefer to duck here the issue of whether or not there was a real physical ($R = 0$) singularity at $t = 0$.

tions to date suggest that the cosmic microwave background (henceforward abbreviated cmb) has a blackbody spectrum. Thus the radiation must have been generated at an earlier epoch of thermal equilibrium. It is easy to show that microwave radiation cannot be thermalized in the Universe now (the mean free path of a photon exceeds cH_o^{-1}). To ensure thermalization, we require an epoch of higher density and temperature. Indeed, the requirements are that $\rho(t) > 10^9 \rho_o$, and that the matter was largely ionized. Under such conditions, the formation of stars or other bound systems would have been impossible. The properties of the Universe would have been radically different from those obtaining today.

Since the acceptance of the hot Big Bang models depends so strongly on the interpretation of the cmb we have given, it is proper to examine the data with care. Evidence for a Rayleigh-Jeans ($i_\nu \propto \nu^2$) spectrum is very good for measurements at wavelengths greater than one cm. These observations establish a value for the present temperature of the cmb, T = 2.7 K, which is accurate to within $\sim$ 10% (see [4] or [37] and references therein). The situation at shorter wavelengths is both observationally more difficult (the atmosphere becomes opaque and emissive) and less certain [30]. Measurements of two sorts have been published: those based on excited states of interstellar molecules [42], and direct measurements from balloons and rockets (for the most recent results, see [45]). Initially, the two methods appeared to produce discordant results, but experimental difficulties with the latter type of measurements have now been overcome, and the most recent balloon measurement [45] is in good agreement with both the molecular line values and the 2.7 K value obtained by those observing at longer wavelengths. The situation is summarized in figure 3.

Thus the spectral measurements agree well with the hot Big Bang picture. Are other interpretations of the data possible, however? In particular, is there one which is consistent with the Steady State theory, which has no hot initial phase? One possibility is that the cmb is the sum of emissions from numerous discrete sources ([40], [44]). Such suggestions have run into trouble on two fronts, (1) the spectrum of the cmb (especially at short wavelengths: compare the predicted spectra of Wolfe and Burbidge with the observational results of Woody, et al), and (2) the small scale

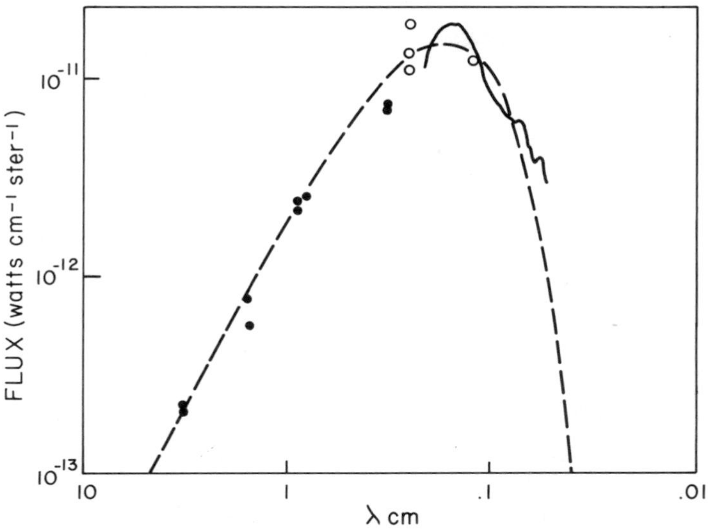

3. The cosmic microwave background spectrum. The theoretical curve for a blackbody is fit well by radio frequency measurements (dots), observations of interstellar molecules (open circles) and the new results of Woody, et al [45]. (Adopted from Physics Today, July, 1975.)

angular isotropy of the radiation ([31] and [41]). The upshot is that discrete source models appear to require a new population of sources, more numerous than galaxies, none of which are observed locally, and which must have quite precisely defined spectral properties and evolution*. Even allowing this freedom, none of the published discrete source models agrees with all the observations now available.

A second alternative explanation of the cmb is that of Layzer and Hively [36], who argue for a cold Big Bang, the cmb being in their view the thermalized relic of emission from an early generation of massive stars. In this theory, thermalization is accomplished by the interaction of ultraviolet and visible photons with intergalactic grains. Three problems with this suggestion exist: the grains it calls for must consist of heavy elements which in turn must have been made during or prior to the epoch of emission by the massive stars (how?); the grains must have the right optical

*Not permitted, of course, in the Steady State theory.

properties to radiate as blackbodies over a wavelength range of 10^3; and the grains must have vanished by now or else they would produce unacceptably high absorption in the spectra of high-redshift quasars. The last two problems arise in any other "grain theory."

The third explanation is Hoyle's [35], in which the initial singularity occurs when the masses of elementary particles go to zero. Here "thermalization" occurs not because of an increase in density and temperature but, Hoyle argues, because of an increase in the Thompson scattering cross section as masses $\rightarrow 0$.

In my view, the most satisfactory explanation to date for both the spectrum and the isotropy of the cmb is the original Big Bang (or primeval fireball) hypothesis of Gamow. It follows, as we have seen, that the Universe must have been a factor of 10^9 times denser (and 10^3 times hotter) than at present. Since the density at that epoch was not more than about 10^{-20} gm/cm^3, it is a far cry from the infinite density singularity predicted by the standard model. However, we do have indirect evidence for a far denser state, with $\rho \gtrsim 10^{-5}$ gm/cm^3 and $T \sim 10^9$ K. This evidence is the universal, and apparently primordial, helium abundance of $\sim$ 27% by mass [39].

The possibility of element formation in the early phase of a hot Big Bang Universe has long been recognized (see [34] and [29]). More recent calculations [43] reach the following important conclusions:--

1.) In Big Bang nucleosynthesis, essentially no elements heavier than helium were produced unless the Universe departed wildly from the standard model (e.g., gross inhomogeneity).

2.) For all Friedmann models, the amount of helium depends <u>very</u> weakly on the density parameter ρ_o/ρ_c: the helium abundance is restricted to the range 20-30% by mass.

3.) On the other hand, even rather minor departures from the standard model (e.g., anisotropic or inhomogeneous models, or a scalar component of the magnitude suggested by Dicke [33]) change the helium production, generally giving essentially 0%

or 100% by mass.

4.) For the standard Friedmann models, the ^{3}He abundance, and particularly the ^{2}H abundance, depend sensitively on ρ_o/ρ_c (or alternatively on q_o). The dependence is shown in figure 4.

5.) Nucleosynthesis is essentially complete after the first 15 minutes of the history of the Universe, at which epoch the density of the Universe was approximately twenty-five orders of magnitude greater than its present value.

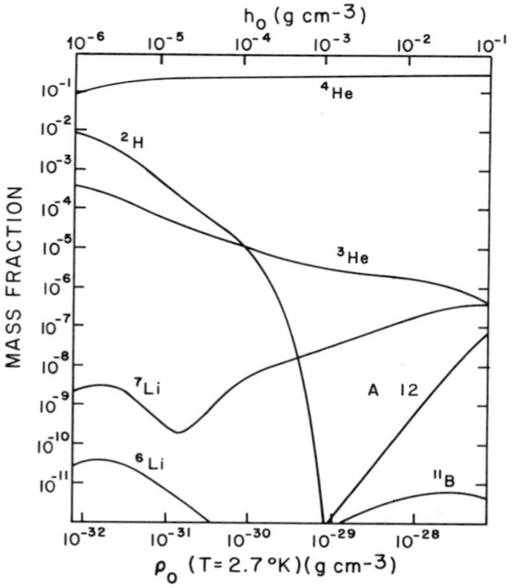

4. Abundances of light elements produced by a standard Big Bang model, calculated as a function of the present matter density ρ_o. Note the strong dependence of ^{2}H abundance on ρ_o. (Adopted from [43]).

If the present data on primordial helium hold up, we have a strong argument not only for a Big Bang with $\rho \gtrsim 10^{25}\ \rho_o$, but also for the suitability of the standard model. This last point cannot be emphasized too strongly--the close agreement between the observed and predicted ^{4}He abundance tells us that we understand nuclear physics, gravitation, and the present properties of the Universe well enough to project the properties of the Universe back to the first few minutes of its existence. To me, this is an achievement of which astrophysicists may justly be proud--at least in part because our observations so neatly vindicate conjectures of Einstein and his collaborators extending back six decades.

D. Evidence for the isotropy of the Universe.

Equation (1) represents a Universe which is isotropic and non-rotating. It is easy to see from (1) that any expansion of the Universe must also be isotropic. On the other hand, it is well known that Einstein's equations admit a number of classes of anisotropic cosmological models. Hence the use of the Robertson-Walker metric as a description of our Universe must be observationally justified. What data support the assumption of isotropy? We first ask how a possible anisotropy would be manifest. We could look either for an anisotropic distribution of matter or an anisotropy in the Hubble flow (the map of apparent recession velocities as a function of position). The latter may appear as shear or vorticity or a combination of the two [49].

A search for an anisotropic distribution of matter on a cosmological scale is impeded by the known local anisotropy in the distribution of galaxies. Although there are now searches underway for an anisotropy in the distribution of galaxies, it is not yet clear whether there is any evidence for an anisotropy in the distribution of matter on a cosmic scale. In any case, galaxy counts are unlikely to equal in precision other measurements to be discussed below.

As Kristian and Sachs pointed out [50], the Hubble flow in an anisotropic Universe would necessarily be anisotropic. That is, the Hubble "constant" would have different values in different directions. The detection of just such an effect, with $\Delta H/H \sim 0.25$, has been claimed by Rubin, Ford and Rubin [55], based on a sample of galaxies nominally lying on a spherical shell centered on the Galaxy. This result has been reexamined by Sandage and Tammann [19] who conclude that whatever was responsible for the observations of Rubin *et al*, it was *not* an anisotropy in the Hubble flow. In addition, the result of Rubin *et al* implies a motion for the solar system which disagrees strongly with measurements of the isotropy of the cmb (to be considered next).

The most sensitive direct limits on the present anisotropy of the Hubble flow are those provided by measurements of the microwave background. The shear component would produce a "quadrupole" or 180° anisotropy in the cmb. The observational upper limit on such an anisotropy, $\Delta I/I \lesssim 10^{-3}$ (see [52] and references therein), puts an upper limit of about 10^{-4} on the present value of the shear anisotropy of the Universe*. This limit, taken together with the work of Collins and Hawking [48] substantially restricts the range of permissible anisotropic cosmological models**.

A word of caution is in order on the subject of limits on the shear anisotropy. As Zel'dovich and Novikov have frequently pointed out (see, for example, [51]), a simple quadrupole anisotropy in the cmb is present only for closed anisotropic models. In open ($k < 0$) anisotropic models, the angular variation in the intensity of the microwave background is more complicated. In particular for models with $\rho_o \ll \rho_c$, the microwave background will in fact appear isotropic over most of the sky; the anisotropy is restricted to an angular region of $\sim (\rho_o/\rho_c)$ radians. Since observations of the cmb do not cover the entire sky, the possibility that we live in an open but anisotropic Universe itself remains open.

Limits on the vorticity of anisotropic or rotating models may be set by observations of the "dipole" or 360° anisotropy of the cmb [49]. Again, the limit depends on the epoch at which the microwave photons were last scattered. If the last scattering occurred at $z \sim 7$, the limit on the vorticity for the $k = 0$ model is

*The exact limit depends on the epoch at which the microwave photons were last scattered by matter (see, e.g., [57]). In the absence of intergalactic plasma, the photons propagate freely from an epoch corresponding to $z \sim 1000$; if an intergalactic plasma with $\rho_{HII} \sim \rho_c$ is present, the epoch is later, $z \sim 7$.

**In addition, this observation places constraints on the possible position of the sun within the four-dimensional plus and minus mass aggregates of the Hoyle-Narlikar theory (see [35] for details). Not only must we be near one of the space-like hypersurfaces where masses tend to zero; we must also be far from any time-like hypersurface where $m \to 0$, otherwise there would be a detectable anisotropy in the Hubble flow. In other words, because the Hoyle-Narlikar theory is not cosmologically homogeneous, special conditions are required to prevent the inhomogeneity from becoming apparent.

$\omega \lesssim 10^{-13}$ rad/yr: if the redshift of last scattering is ~ 1000, the limit for the same model is $\omega \lesssim 10^{-15}$ rad/yr. For an open model with $\rho_o << \rho_c$, the limit becomes $\omega \lesssim 10^{-14}$ rad/yr. These limits are based on a value of 10^{-3} for the dipole anisotropy of the cmb (see [52])*.

Finally, for completeness, we mention one further piece of indirect evidence: the abundance of helium and other light elements in the Universe. Substantial shear anisotropy during the epoch of element formation (roughly 1-1000 sec after the initial singularity) would have produced unacceptable abundances for the light elements [57]. Our present knowledge of the primordial helium abundance allows us to set an approximate limit of 50% on the amplitude of shear anisotropy during this early epoch [46]. For some classes of anisotropic but homogeneous models, the amplitude of the anisotropy decreases as the Universe expands**. Hence, as Barrow [46] shows, for such models the abundance argument places stronger constraints on the present value of the shear anisotropy than direct observations of the cmb. Barrow's argument cannot be used to set limits on the vorticity, however.

E. Evidence for the homogeneity of the Universe.

Evidence for the isotropy of the Universe may also be used to support the view that it is homogeneous on a large scale, provided we make the additional modest assumption that we are not in a favored location. We need this assumption to eliminate (somewhat fanciful) models with a spherically symmetric but non-constant density distribution centered on the solar system.

As we pointed out earlier, the Robertson-Walker metric gives a true description of the spacetime geometry of the Universe only if the matter distribution is homogeneous on all scales. We do not expect to find geometry of constant curvature near density

*This limit also sets an upper limit of ~ 300 km/sec on the motion of the solar system, in conflict with the motion of ~ 700 km/sec claimed by Rubin et al [55].

**Anisotropic models which may tend towards isotropy are those Bianchi types which include the isotropic Friedmann models as special cases. These are Bianchi types I, V, VII_o, VII_h, and IX. For further details, see Collins and Hawking [48], and section III below.

inhomogeneities such as Black Holes, or even the sun. Known density inhomogeneities in the Universe range upwards in size to a few Mpc (clusters of galaxies). For all known structures save Black Holes, the perturbations of the geometry are small. Consequently, a geometrical description of the Universe based on the assumption of homogeneity can serve as a first approximation, and we can in principle take account of local perturbations in the geometry produced by structures we recognize.

Our real concern in cosmology is with the possible existence of larger density inhomogeneities, of characteristic scale 3-3000 Mpc. The existence of clumps of matter on a scale larger than a few Mpc (often called superclusters, [58]) has been a lively topic for debate for decades. To make a sweeping generalization, the existence of large structures has been more favored by theoretical than by observational cosmologists. Two sorts of observations may settle this question: counts of galaxies, and measurements of the fine scale isotropy of the microwave background radiation.

Recently, Peebles and his colleagues have made an extended study of the distribution of galactic images on the plane of the sky [53] and [54]). A tendency for matter to cluster is found on scales up to $\sim$ 10 Mpc. This tendency is not strong ($\Delta\rho/\bar{\rho} \lesssim 1$), especially for larger systems, but it does provide some support for the gravitational instability picture for the formation of bound systems in the Universe. Neither Peebles' investigation nor any other has turned up evidence of matter inhomogeneities large enough to invalidate the use of equation (1).

Further evidence for the (approximate) homogeneity of the Universe on a large scale is provided by observations of the microwave background radiation. Gross inhomogeneities would produce fluctuations in the observed intensity of the radiation. These have been looked for on angular scales from one arcminute to tens of degrees [31]. The data are summarized in figure 5: on angular scales of an arcminute and above, the radiation appears uniform to less than a part in a thousand. This range of angular scales corresponds very roughly to aggregates of mass in the range 10^{14}-10^{21} gm*.

*The relation between angular size and mass given here depends strongly on the choice of cosmological model.

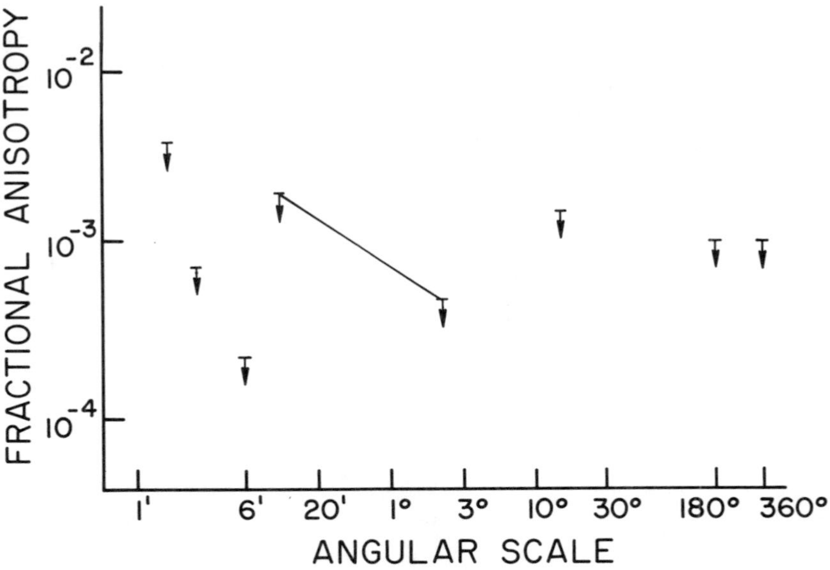

5. Observational upper limits on the anisotropy of the cmb on various angular scales. Unless otherwise indicated, the measurements were made at a wavelength of 3-4 cm. For details and references, see [31].

Unfortunately, calculations by Sunyaev and Zel'dovich [56] show that the magnitude of nonuniformities in the cmb expected for known density inhomogeneities (those that produce galaxies and clusters of galaxies) are of order 10^{-4}-10^{-5}, below our present detection threshold. At least the cmb measurements tell us that there are no unexpected large scale inhomogeneities, so we are free to continue to use a metric of constant curvature (1). In addition, as we shall see, these observations may establish rather interesting constraints on the initial conditions of the standard models.

F. Evidence that the Universe Will Continue to Expand Forever.

We come now to the single aspect of Einstein's cosmological vision which the data do not appear to favor. Although General Relativity includes both open ($k < 0$) and closed ($k > 0$) cosmological models, Einstein clearly favored the latter, a view now vigorously championed by Wheeler [65]. We recall from section I that the closed Friedmann model is a recontracting one.

At this time, the evidence seems to point in the other direction, towards an open, low density Universe which will continue to expand forever. But we must keep in mind that these observations are both more difficult and more subject to conflicting interpretation than the observations discussed earlier. It follows that the conflict with Einstein's vision of a closed Universe may pass away.

Let us begin with attempts to determine the deceleration parameter, q_o, which, for the standard models, fixes the geometry (section I). In a classic paper, Sandage [68] described three tests which in principle could produce a value of q_o. In one way or another, all three have been performed.

The first of these tests is the <u>magnitude</u>-<u>redshift</u> test: the functional dependence on the redshift of the observed magnitudes of sources of the same intrinsic luminosity is different for different values of q_o. For redshifts $z < 1$, one may use the approximation

$$m(z) = M + \text{const} + 5 \log z + 1.086(1 - q_o)z \quad . \; . \; . \; . \quad (10)$$

to give an indication of the nature of the variation with q_o. Here, M is the absolute magnitude of the set of sources used in the test, and $m(z)$ is the observed magnitude for a source at redshift z. In figure 6, taken from Sandage [69], exact expressions, rather than this approximate one, are used to generate the theoretical curves.

The major question facing astronomers who seek to apply this test is the obvious one: are there sources of the same intrinsic luminosity present in the Universe? Sandage and Hardy [70] have argued that the brightest galaxy in a cluster of galaxies may be used as a standard source. Although this view has been challenged, we shall adopt it as a working hypothesis. With this assumption, Sandage [69] has derived a value of

$$q_o = 0.96 \pm 0.40 \; , \qquad (11)$$

from a sample of galaxies reaching out to a redshift $z = 0.46$ (fig. 6).

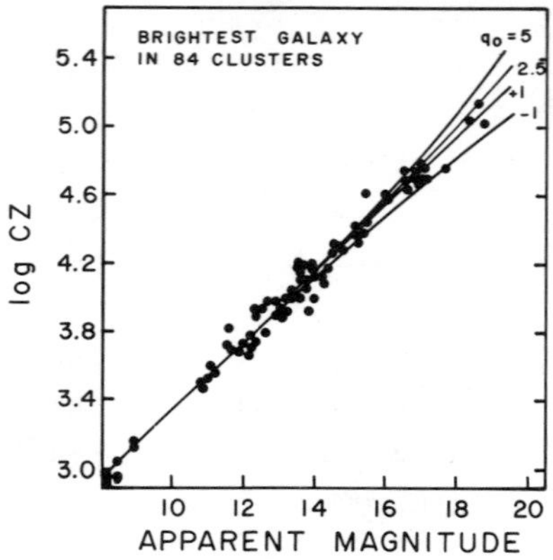

6. The magnitude-redshift relation. Theoretical curves for the standard model are shown for several values of q_o. The points are observations discussed by Sandage [69], the source of this figure. Note that $q_o = -1$ (the value for the Steady State model) is excluded.

The error cited is the formal value for the statistical probable error. Unfortunately, the result has little value unless we consider some of the many sources of systematic error which may arise in measurements of this sort. Many of these are discussed by Peach [66]. The single largest source of systematic error of which we are now aware is the evolution of the sources used in the test. If the luminosity of the sources changes with time, then M in equation (10) is no longer a constant, but becomes a function of the redshift, z. To use (10) to find q_o from measurements of m and z, we must correct for the evolution of the sources; and we do not know how to do so with any precision. It is probable that the galaxies used in the test were more luminous in the past. If so, the true value for q_o lies below that given in (11) above. Rough estimates ([62], [71]) of the correction for evolution suggest $-1 \lesssim q_o \lesssim \frac{1}{2}$. Fortunately, this question is under active investigation by several workers, so we may soon have an improved estimate of q_o from this test.

The second classical test is the angular diameter test. For a set of objects of uniform proper diameter D located at a range of redshifts from us, one finds [68] for the measured angular diameters

$$\alpha(z) = \frac{DH_o}{c} q_o^2 (1 + z)^2 \left\{ q_o z + (q_o - 1)(\sqrt{1 + 2q_o z} - 1) \right\}^{-1} , \quad (12)$$

Unfortunately, the proper diameter of galaxies is by no means easy to define, hence this test has received less attention from optical astronomers than the magnitude-redshift test. Recently, however, Baum [60] has devised an ingenious way to determine the angular diameters of photographic images of galaxies. His results, shown in figure 7, indicate $q_o \sim 0.3 \pm 0.2$. Other workers have abandoned galaxies as their test objects, employing instead clusters of galaxies [59], double radio sources [64], or very small single radio sources [63]. Although no final value of q_o has emerged from these studies, Austin and Peach give a preliminary value of $q_o \sim 1 \pm 1$, and Hewish et al favor a value above ½. What does clearly emerge from the radio studies is that the properties of radio sources must undergo strong evolution.

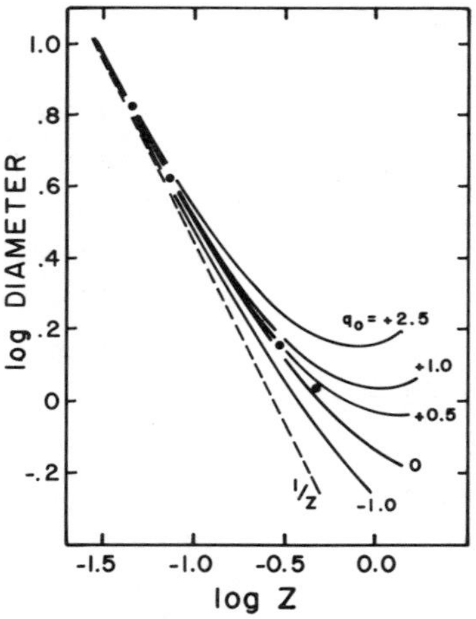

7. The angular diameter-redshift relation. Theoretical curves for the standard model are shown for several values of q_o. The observations are from [60]. Again, $q_o = -1$ is not favored.

Finally, there is the number-luminosity test; the number of sources above a certain luminosity limit depends on q_o. This test has been applied mainly by radio astronomers [67]. What they discover, sad to say, is not a value for q_o, but rather convincing evidence that the properties of radio sources evolve in time. Although this evidence helped defeat the Steady State theory--which of course does not permit secular changes of this sort--it does not help to distinguish between the Friedmann models.

The classical cosmological tests evidently fail to provide us with a good value for q_o and thus are unable to distinguish between the possible solutions to equation (5). To the extent that we can cope with the sources of systematic error like the one discussed above, we are probably safe in concluding that

$$-1 \lesssim q_o \lesssim +1^*, \qquad (13)$$

The interval $0 \lesssim q_o \lesssim \frac{1}{2}$ is rather weakly favored.

Much crisper (but also much more controversial) support for the open, ever-expanding model is provided by estimates of the present mass density of the Universe (see eqn. 6). The situation has recently been reviewed by Gott, Gunn, Schramm and Tinsley [61], who conclude that $\rho_o \sim 0.1\ \rho_c$. It should be noted, however, that their conclusion that $\rho_o \ll \rho_c$ is based very largely on a single argument--the high abundance of deuterium in the Universe. Recall from section IIC that the abundance of ^2H produced by the Big Bang depends sensitively on the ratio ρ_o/ρ_c (see fig. 4). Observations summarized by Gott <u>et al</u> lead to a ^2H abundance of $\sim 2 \times 10^{-5}$ by mass (see also [39]). Since it is easier to destroy than to form deuterium in known astrophysical processes, we may adopt the figure given above as a lower limit on the primordial ^2H abundance. It follows from figure 4 that $\rho_o \lesssim 0.05\ \rho_c$. This elegant argument has been challenged in the past few months, and it is simply too soon to accept the result as final. If the argument survives, it may turn out that nuclear astrophysics will provide us with the crucial cosmological answers we have been so long seeking by other means.

*A negative value of q_o of course is incompatible with the standard model: if $q_o < 0$, a small positive value for the cosmological constant may be required.

G. Conclusions.

Put briefly, observational cosmology has reached two important goals in the past decade. First, it has eliminated, either completely or very probably, a vast range of non-standard cosmological models (including most anisotropic models and also, I would claim, the Steady State model). In so doing, observational cosmology has strengthened our faith in the major assumptions underlying the standard model, including the validity of General Relativity. Second, it appears that we have the evidence to predict the future of the Universe--unending expansion.

III. SOME PROBLEMS WITH THE STANDARD MODEL

In the preceding pages, I have played the role (perhaps too vigorously) of advocate for the standard hot Big Bang model. Many of the data support the model; none appears flatly to contradict it. But the model is by no means free of problems. Whether these are minor and temporary difficulties which we shall be able to put right--or whether they are major flaws which will call forth entirely new theoretical advances--we do not yet know.

Let us begin by reviewing some of the uncertainties we have encountered above. We have no firm information about the cosmological constant. Our knowledge of the homogeneity of the Universe is shaky. Alternatively, we may say that we do not yet know to what range of masses the observed inhomogeneities extend. Are there systems larger than clusters of galaxies? Finally, the evidence that the Universe is low-density and ever-expanding is very fragile. Better observations may help resolve these questions.

There are, however, far more fundamental problems with the standard model. These may be gathered under the heading: The problem of initial conditions. We may outline the problem as follows. The familiar laws of physics appear to provide a good description of the evolution of the Universe back to the first few minutes of its existence. They can reasonably be extrapolated back to the first few microseconds before we reach regimes of temperature and density beyond the grasp of present theories. Thus we can in principle apply the laws of physics retrodictively to ask what "initial" conditions (at some chosen epoch, say t = 1 yr) were necessary to produce the present properties of the Universe. It appears that quite special--and quite a few--"initial" conditions were necessary. The need to add a long list of "initial" conditions detracts from the appealing simplicity of the standard model.

Let us begin by asking why the Universe is (or appears to be) so isotropic, since anisotropic solutions are perfectly consistent with General Relativity. There are two general answers to this question. The first is that there exist physical processes which rapidly and generally lead to isotropic expansion, for any "initial" anisotropy. Thus no special "initial" conditions are required to explain the present isotropy of the Universe. Neutrino viscosity [77] and pair production in a rapidly changing gravitational field [83] have been suggested as possible physical mechanisms which would lead to isotropization. Both deserve further study.

A quite different approach is that of Collins and Hawking [48] (see [76] and, for a different view, [51]). They argue on quite general grounds that the set of anisotropic cosmological models that tend eventually to isotropic expansion is of measure zero*. In other words, the observed isotropy of the Universe requires rather special "initial" conditions. In particular, they show that the present isotropy of the Universe implies limits on both the "initial" anisotropy and on the mean mass density ρ_o (see [76] for a provocative discussion of this point--and an interesting explanation of why the Universe should be nearly isotropic). If Collins and Hawking are right, then an important present property of the Universe--its isotropy--can be explained only by "initial" conditions.

The problem presented by isotropy becomes sharper when we consider an interesting paradox involving causality. The argument--which I first heard from Ya. B. Zel'dovich--is the following. Our measurements of the isotropy of the Universe are based on observations of the cosmic microwave background, and thus include most of the Universe within the particle horizon, ct_o. Regions we now observe, however, were not causally linked at earlier epochs--particularly the epoch when the cmb radiation was last scattered

*They consider only <u>homogeneous</u> anisotropic models.

(see figure 8)

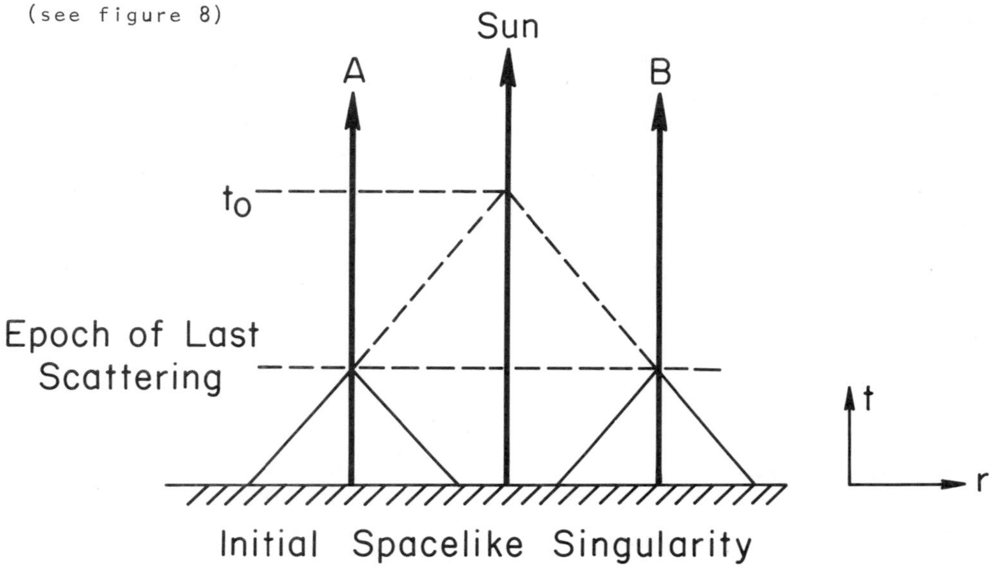

8. Cosmic microwave radiation from two regions, A and B now reaches the Solar System. At the epoch of last scattering, however, A and B were not causally connected (A was not in B's past light cone, and vice versa).

To pose the paradox somewhat anthropomorphically, how did regions A and B of figure 8 "know" they should both participate in an isotropic expansion? Our only response within the confines of the standard model is: "initial conditions."

Naturally, this same question applies with equal force to the observed homogeneity of the Universe: how did two regions like A and B in figure 8, not causally connected until recently, "decide" to have the same average density ρ_o?

Furthermore, there is another paradox hidden in the question of homogeneity. The data discussed in II E above suggest that the Universe is homogeneous on the large scale, despite its obvious inhomogeneity on a small scale. Why? One could argue that the growth of small scale inhomogeneities is favored and that of larger ones retarded. However, on the basis of our present understanding of the growth of density perturbations in the standard model, the opposite appears to be true (for reviews, see [74] and [81]). Even if this question is satisfactorily resolved, there remains the difficulty of explaining the <u>amplitude</u> of the density perturbations which eventually led to galaxies and clusters. The problem assumes a clear form if we consider the amplitude of proto-galactic and proto-cluster perturbations at the epoch when the cmb radiation last interacted with matter (nominally at $z \sim 1000$). Peebles [80] has shown that the growth of density perturbations is impeded before this epoch; hence we may reasonably select it as the "initial" epoch for the formation of gravitationally bound systems. Had the density perturbations at this "initial" epoch been too small in amplitude, bound systems would not have formed (this holds especially for low density models). On the other hand, if the perturbations had been too large, galaxies would have formed too rapidly, and would thus have higher average densities than those observed. The required density perturbations (should we not call them "initial" conditions?) are of the order of $\Delta\rho/\bar{\rho} \sim 10^{-4}$ [56]. Here, incidentally, is an area where observational cosmology may soon play a deciding role. Observations of the fine scale anisotropy of the cmb already place limits* of $\sim 10^{-3}$ on density fluctuations corresponding to masses in the range $\sim 10^{14}$ to 10^{21} $M_{\odot}$.

Another thorny problem is the baryon number of the Universe. In the laboratory, since baryon number is conserved, particles and antiparticles are created in equal number. Why then should the baryon number of the Universe as a whole be non-zero? In the standard model, one has to accept this as yet another "initial" condition. Cosmological models having a zero net baryon number have been constructed by Alfvén and his collaborators (see [72] and [73]) and by Omnès and his colleagues ([78], [79]). In these models, regions of matter are balanced by regions of antimatter. Direct and indirect searches for antimatter (or annihilation products) tell us only that these regions cannot be smaller than galaxies.

This result in turn implies the need for a mechanism to separate matter and antimatter early in the history of the Universe--and to keep it separated. Various mechanisms have been suggested to ensure this separation (the Leidenfrost layer by Alfvén, a phase separation by Omnès). The need for a separation mechanism becomes more emphatic when we consider a (rough) observational limit on

*I.e., $\Delta T/T \lesssim 3 \times 10^{-4}$; see II E.

the amount of matter-antimatter annihilation which can have taken place early in the history of the Universe. Consider a matter-antimatter symmetric Big Bang model as it cools below $T \sim 10^9$ K. At lower temperatures, electron pair production is no longer possible [75]. Consequently, any further annihilation will increase the number of photons relative to the number of electrons. But the ratio n_γ/n_e is known to be $\lesssim 10^9$ (n_γ may be calculated from the known temperature of the cmb)*. This limit immediately rules out the possibility of a merely statistical separation. It also sets tight constraints on the models proposed by Omnès and others (see [82] for a recent critical review). Only if we can overcome the theoretical difficulties posed by the need to separate matter and antimatter can we entertain these symmetric cosmological models. Until they are further tested, we must accept the non-zero baryon number of the Universe as yet another "initial" condition.

IV. SOME CONCLUSIONS

How much have cosmological observations contributed to the progress of our understanding of General Relativity? I suspect that observations of the large scale properties of the Universe are just beginning to play a role. The most exacting tests of Einstein's theory are still the local tests discussed in this volume by Everitt and Nordvedt. It has been my thesis in this paper, however, that the available observations are consistent with a straightforward and simple cosmological model based on General Relativity. Indirectly, observational support for the standard model also helps confirm our faith in General Relativity. The problems I see with the standard model do not (yet!) appear to involve General Relativity itself. Whether or not one looks forward to a time when observational cosmology can fruitfully challenge our theories of gravity depends on one's philosophy; I do.

Needless to say, a general review of this nature owes much to many other workers in cosmology. In particular, I would like to thank Martin Rees not only for his hospitality at the Institute

*Incidentally, this ratio itself has no ready explanation in the standard model: it is yet another "initial" condition.

of Astronomy, Cambridge, where this paper was prepared, but also for his many helpful comments. Finally, the references I have listed represent only a tiny fraction of the current research in the field. I apologize to colleagues whose work I have omitted, and to readers who may find my choices arbitrary.

REFERENCES

This list is by no means complete. In keeping with the spirit of this paper, I have emphasized observational work. In general, three sorts of papers are listed:

1.) classical papers in the field, (e.g., Sandage's 1961 article)
2.) good review or survey articles (e.g. [39])
and 3.) papers presenting the most recent observational results, (e.g., Woody *et al* [45])

Finally, I must apologize for depending heavily on journals and books in English.

References are given for each section separately.

I

For general surveys of the mathematical structure of the standard model, see

[1] H. Bondi, *Cosmology*, Cambridge Univ. Press (1961)
[2] G. C. McVittie, *General Relativity and Cosmology*, Chapman and Hall, London (1965)

Recent reviews of both the canonical model and the observational data supporting it are provided by

[3] D. W. Sciama, *Modern Cosmology*, Cambridge Univ. Press (1971)
[4] P. J. E. Peebles, *Physical Cosmology*, Princeton Univ. Press (1971)
[5] S. Mavridès, *L'Univers Relativiste*, Masson et Cie., Paris (1973)
[6] S. Weinberg, *Gravitation and Cosmology*, J. Wiley, N.Y. (1972)

These works are listed in rough order of increasing difficulty. Also very helpful are:

[7] M. J. Rees, R. Ruffini, and J. A. Wheeler, *Black Holes, Gravitational Waves and Cosmology*, Gordon and Breach, N.Y. (1974)--see chapters 11-19.
[8] M. S. Longair, *Rep. Prog. Phys.*, *34* (1971) 1125
[9] M. S. Longair (editor), *I.A.U. Symposium 63, Confrontation of Cosmological Theories with Observations*, Reidel, Holland (1974)
[10] I. D. Novikov, and Ya. B. Zel'dovich, *An. Rev. Astron. and Astrophys.*, *5* (1967) 627

Other papers cited in this article:--

[11] H. Bondi, and T. Gold, *Mon. Not. Roy. Ast. Soc.*, *108* (1948) 252

[12] C. Brans, and R. H. Dicke, Phys. Rev., 124 (1961) 925
[13] C. G. Callan, R. H. Dicke, and P. J. E. Peebles, Am. J. Phys., 33 (1965) 105
[14] F. Hoyle, Mon. Not. Roy. Ast. Soc., 108 (1948) 372
[15] F. Hoyle, ibid 109 (1949) 365
[16] P. Jordan, Z. Physik, 157 (1959) 112
[17] W. H. McCrea, Quart. J. Roy. Ast. Soc., 12 (1971) 140
[18] V. Petrosian in [9] above
[19] A. Sandage, and G. A. Tammann, Astrophys. J., 196 (1975) 313

II

For general surveys of the data, see the works by Longair, Peebles and Weinberg given above ([8], [4], [6]). Some useful papers also appear in

[20] D. S. Evans (editor), I.A.U. Symposium 44, External Galaxies and Quasi-Stellar Objects, Reidel, Holland (1972)

A and B

[21] D. S. Dearborn, and D. N. Schramm, Nature, 247 (1974) 441
[22] T. C. van Flandern, Mon. Not. Roy. Ast. Soc., 170 (1975) 333
[23] M. J. Geller, and P. J. E. Peebles, Astrophys. J., 174 (1972) 1
[24] F. Hoyle, and J. V. Narlikar, Action-at-a-Distance in Physics and Cosmology, W. M. Freeman, San Francisco (1974)
[25] E. Hubble, and R. C. Tolman, Astrophys. J., 82 (1935) 302
[26] J. C. Pecker, A. P. Roberts, and J. P. Vigier, Nature, 237 (1972) 227
[27] V. Petrosian, E. E. Salpeter, and P. Szekeres, Astrophys. J., 147 (1967) 1222
[28] K. S. Thorne, C. Will, and W. T. Ni, proceedings of a conference on The Experimental Verification of Theories of Gravity, Jet Propulsion Laboratory, California (1971)

C

[29] R. A. Alpher, and R. C. Herman, Phys. Rev., 75 (1949) 1089
[30] G. E. Blair in [9] above
[31] P. E. Boynton in [9] above
[32] C. Brans, and R. H. Dicke, Phys. Rev., 124 (1961) 925
[33] R. H. Dicke, Science, 184 (1974) 419
[34] G. Gamow, Rev. Mod. Phys., 21 (1949) 367
[35] F. Hoyle, Astrophys. J., 196 (1975) 661
[36] D. Layzer, and R. Hively, Astrophys. J., 179 (1973) 361
[37] R. B. Partridge, American Scientist, 57 (1969) 37
[38] A. A. Penzias, and R. W. Wilson, Astrophys. J., 142 (1965) 419
[39] H. Reeves, Ann. Rev. Astron. and Astrophys., 12 (1974) 437 (see also an article to appear in Rev. Mod. Phys. by V. I. Trimble)
[40] M. Rowan-Robinson, Mon. Not. Roy. Astr. Soc., 168 (1974) 45p
[41] M. G. Smith, and R. B. Partridge, Astrophys. J., 159 (1970) 737
[42] P. Thaddeus, Ann. Rev. Astron. and Astrophys., 10 (1972) 305
[43] R. V. Wagoner, Astrophys. J., 179 (1973) 343
[44] A. M. Wolfe, and G. R. Burbidge, Astrophys. J., 156 (1969) 345
[45] D. P. Woody, J. C. Mather, N. S. Nishioka, and P. L. Richards, Phys. Rev. Letters, 34 (1975) 1036

D and E

[46] J. Barrow, preprint, "Light Elements and the Isotopy of the Universe", Oxford University (1975)
[47] C. B. Collins, and S. W. Hawking, Astrophys. J., 180 (1973) 317
[48] C. B. Collins and S. W. Hawking, Mon. Not. Roy. Astr. Soc., 162 (1973) 307
[49] S. W. Hawking, Mon. Not. Roy. Astr. Soc., 142 (1969) 129
[50] J. Kristian, and R. K. Sachs, Astrophys. J., 143 (1966) 379
[51] I. D. Novikov in [9] above
[52] R. B. Partridge in [9] above
[53] P. J. E. Peebles, Astrophys. J. Letters, 189 (1974) L51
[54] P. J. E. Peebles, and E. J. Groth, Astrophys. J., 196 (1975) 1
[55] V. C. Rubin, W. F. Ford, and J. S. Rubin, Astrophys. J. Letters, 183 (1973) L111
[56] R. A. Sunyaev, and Ya. B. Zel'dovich, Astrophys. and Space Sci., 6 (1970) 358
[57] K. S. Thorne, Astrophys. J., 148 (1967) 51
[58] G. de Vaucouleurs, Science, 167 (1970) 1203

F

[59] T. B. Austin, and J. V. Peach, Mon. Not. Roy. Astr. Soc., 167 (1974) 437
[60] W. A. Baum in [20] above
[61] J. R. Gott, J. E. Gunn, D. N. Schramm, and B. M. Tinsley, Astrophys. J., 194 (1974) 543
[62] J. E. Gunn, and J. B. Oke, Astrophys. J., 195 (1975) 255
[63] A. Hewish, A. C. S. Readhead, and P. J. Duffett-Smith, Nature, 252 (1974) 657
[64] G. K. Miley, Mon. Not. Roy. Astr. Soc., 152 (1971) 477
[65] C. W. Misner, K. S. Thorne, and J. A. Wheeler, Gravitation, W. H. Freeman, San Francisco (1973)
[66] J. V. Peach in [20] above
[67] M. Ryle, ibid, 6 (1968) 249
[68] A. Sandage, Astrophys. J., 133 (1961) 355
[69] A. Sandage, ibid, 178 (1972) 1
[70] A. Sandage, and E. Hardy, ibid, 183 (1973) 743
[71] B. M. Tinsley, Astrophys. J. Letters, 173 (1972) L93
[72] H. Alfvén, Rev. Mod. Phys., 37 (1965) 652
[73] A. Elvius, E. T. Karlson, and B. E. Laurent in [9] above.
[74] G. B. Field, in Stars and Stellar Systems, Vol. 9, Univ. Chicago Press (to be published).
[75] E. R. Harrison, An. Rev. Astron. and Astrophys., 11 (1973) 155
[76] S. W. Hawking in [9] above
[77] C. W. Misner, Astrophys. J., 151 (1968) 431
[78] R. Omnès, Astron. and Astrophys., 10 (1971) 228
[79] R. Omnès, and J. L. Puget in [9] above
[80] P. J. E. Peebles, Astrophys. J., 142 (1965) 1317
[81] M. J. Rees, in General Relativity and Cosmology, Corso 47, Scuola Enrico Fermi, Italian Physical Society (1971)
[82] G. Steigman in [9] above
[83] Ya. B. Zel'dovich, Zh. Eksper. Teor. Fiz. Pizma, 12 (1970) 443, English Version in J.E.T.P. Letters, 12 (1970) 307

A NEW RELATIVISTIC THEORY OF GRAVITATION*

F.I. Cooperstock and G.J.G. Junevicus†
Department of Physics, University of Victoria
Victoria, B.C., Canada V8W 2Y2

ABSTRACT: The operational method of Schiff to obtain an approximate spherically symmetric metric is generalized as a phenomenological gravitational model. This leads to the introduction of a new, physically motivated relativistic theory of gravitation which derives from an action principle and which is simpler than general relativity. Both the model and theory yield the Schwarzschild solution in the spherically-symmetric case and hence satisfy the standard tests.

In recent years, many new relativistic theories of gravitation have been proposed to challenge the excellent but difficult theory of general relativity due to Einstein. Most of these have little, if any, physical motivation. In this paper, we develop a new, physically motivated theory of gravitation which has its foundation in the work of Schiff [1].

Schiff demonstrated that the gravitational red shift and the deflection of light could be derived from the principle of equivalence and special relativity. As expressed by Adler, Bazin, and Schiffer [2], the method consists in developing the approximate spherically-symmetric metric by the calibration of the radial (dr) and time (dt) increments of the non-inertial observer at rest in the field via the Lorentz time dilation and length contraction factors communicated by an adjacent inertial observer. The approximate kinematic relation $v^2 = 2gh$ is used throughout, gh is recognized as the change in gravitational potential $\Delta\phi$ over the small interval, and an intercomparison is finally invoked to bring in the Newtonian potential Gm/r in the metric expression.

We first recognized that the velocity of the adjacent inertial observer could be described more easily and more directly by the Newtonian energy conservation law

$$\tfrac{1}{2}v^2 + \phi = \text{constant}.$$

The constant is chosen to be zero in order to achieve a coincidence of the inertial and non-inertial observers at infinity. This implies

*Supported by National Research Council of Canada Grants A5340, T0624.

†N.R.C. Postdoctoral Fellow, present address: Department of Mathematics, Queen Elizabeth College, London. W8 7AH.

that there is no Lorentz calibration to be effected at spatial infinity which results in a Lorentz line element in the usual form asymptotically. It is interesting to note that what is achieved thereby is the exact Schwarzschild solution.

We were encouraged by this result to ask whether or not it is possible to step from this hybrid of Newtonian and relativistic concepts to a fully relativistic theory of gravitation. Analogous syntheses have led to great strides in the development of physics. A prime example is the Bohr theory which melded classical and quantum ideas and achieved full fruition as a purely quantum mechanical structure. We reasoned that the natural generalization of the Schiff method to arbitrary fields requires the calibration, at each field point, of the projection of the spatial element along the local field line direction, i.e. along the acceleration direction of a freely falling test particle at the point. As in special relativity, the time element is to be calibrated by the reciprocal factor and the spatial element tranverse to the free-fall or field-line direction is uncalibrated. The calibration function is not to be determined by a Newtonian energy conservation law but rather by the action principle formalism.

We take the flat space metric [3]

$$ds^2 = dt^2 - \eta_{\alpha\beta}\, dx^\alpha dx^\beta \tag{1}$$

where $\eta_{\alpha\beta}$ is the Euclidean metric tensor and separate the projection $dx^\alpha \xi_\alpha$ parallel to the field line direction characterized by the unit vector ξ_α .

The time element is calibrated by Φ and the field line element is calibrated by the reciprocal function Φ^{-1} leading to the metric form

$$ds^2 = \Phi^2 dt^2 - \Phi^{-2}(dx^\alpha \xi_\alpha)^2 - \{\eta_{\alpha\beta} dx^\alpha dx^\beta - (dx^\alpha \xi_\alpha)^2\} . \tag{2}$$

We now postulate that ξ_α is the unit normal to the equipotential Φ surfaces:

$$\xi_\alpha \equiv \frac{-\Phi_\alpha}{\sqrt{\eta^{\beta\gamma}\Phi_\beta \Phi_\gamma}} , \quad \Phi_\alpha \equiv \frac{\partial\Phi}{\partial x^\alpha} \quad \ldots \tag{3}$$

which leads to the metric form

$$ds^2 = \Phi^2 dt^2 - \frac{\Phi^{-2}(\Phi_\alpha dx^\alpha)^2}{\eta^{\beta\gamma}\Phi_\beta\Phi_\gamma} - \eta_{\alpha\beta}dx^\alpha dx^\beta + \left[\frac{(dx^\alpha\Phi_\alpha)^2}{\eta^{\beta\gamma}\Phi_\beta\Phi_\gamma}\right] . \quad (4)$$

However, in our metric formalism, the free-fall direction is determined by the geodesic equations. A straightforward calculation shows that the geodesic equations in conjunction with the metric form of Eq. (2) lead to an acceleration direction given by Eq. (3). Thus, the postulation of Eq. (3) as a natural extension of the Newtonian concept is completely consistent with the metric formalism.

We choose the total action function

$$S = S_G + S_I \quad (5)$$

where S_G, the action integral for the gravitational field is

$$S_G = K \int \Phi^i \Phi_i \sqrt{-g}\, d\Omega \quad (6)$$

with $\Phi^i = g^{ik}\Phi_k$ and g^{ik} the contravariant form of the metric components of Eq. (4). S_I is the action integral for all matter and non-gravitational fields and their mutual coupling with the gravitational field.

$$S_I = \int \Lambda \sqrt{-g}\, d\Omega \quad (7)$$

where

$$\tfrac{1}{2}\sqrt{-g}\, T_{ik} = \frac{\partial}{\partial x^\ell}\left[\frac{\partial\sqrt{-g}\,\Lambda}{\partial g^{ik}{}_{,\ell}}\right] - \frac{\partial(\sqrt{-g}\,\Lambda)}{\partial g^{ik}} . \quad (8)$$

The action principle $\delta S = 0$ leads to the field equation

$$2K\left\{\frac{\partial L_G}{\partial\Phi} - \left(\frac{\partial L_G}{\partial\Phi_\ell}\right)_{,\ell}\right\} +$$

$$+ \left\{T^{ik}\frac{\partial g_{ik}}{\partial\Phi}\sqrt{-g} - \left[T^{ik}\frac{\partial g_{ik}}{\partial\Phi_\ell}\sqrt{-g}\right]_{,\ell}\right\} = 0 \quad (9)$$

where $L_G \equiv \Phi^i\Phi_i\sqrt{-g}$ is the Lagrangian density for the gravitational field. In conjunction with the metric of Eq. (4), Eq. (9) yields the field equation for Φ in our theory.

For the spherically symmetric vacuum case, the solution is precisely the Schwarzschild metric. Thus, the new theory satisfies the usual tests as well as does the theory of general relativity. The theory leads to different solutions from that of Einstein for other symmetries, for example axial symmetry. Moreover, the theory is simpler, and can be regarded as a natural extension of both the Schiff method and the non-relativistic Newtonian theory with the scalar potential playing a somewhat analogous role. The new theory has the correct Newtonian and special relativistic limits. Moreover, for weak fields the field equation yields a d'Alembert equation for the first order approximation of Φ.

REFERENCES

[1]. L. I. Schiff, Am. J. Phys. 28,340 (1960).

[2]. R. Adler, M. Bazin, and M. Schiffer, Introduction to General Relativity (McGraw-Hill, New York) 1965.

[3]. Greek indices range from 1-3, Latin indices from 1-4. We set $G = c = 1$.

JEANS INSTABILITY IN LEMAÎTRE UNIVERSES

Franco Occhionero

Laboratorio di Astrofisica Spaziale, C.N.R.
C.P. 67, Frascati 00044, Italy

ABSTRACT

The gravitational condensation of protoclouds in a low density Universe becomes easier to understand if it is assumed that the Universe has undergone a long standstill: The cosmological model obtained by assuming the Lemaître type and the presently accepted value for the density parameter, $\sigma_o \cong 0.05$, coincides with the model proposed by Kardashev (1967) to explain the accumulation of quasar redshifts around 2; for $\sigma_o \cong 0.01$ this redshift is pushed upwards around 4. In either case the e-folding time of the gravitational instability during the quasi-static phase is of the order of 10 billion years, while this phase itself may contain up to 20 e-folding times without any risk of collapse for the Universe. The latter may then be up to 200 billion years old thus allowing the exponential gravitational instability operate over most of its lifetime.

Observational evidence [1], [2], indicates that the Universe is far from closure (if $p = 0$, $\Lambda = 0$); with the present value of the Hubble constant [3] the critical density is of the order of 5×10^{-30} g/cm^3 while the estimated present density is lower by a factor 10. This conclusion may be derived solely from the deuterium abundance (assumed of cosmological origin) since the canonical nucleosynthesis [4] yields the required abundances of both deuterium and helium only for very low present baryon densities (i.e. less than 6×10^{-31} g/cm^3 [5]). Consequently we will take $\sigma_o = 1/20$ as an upper value for the density parameter (see the definition below).

Then, as the total density approaches Oort's value for the density in Galaxies, $3\times10^{-31}(H_o/75)^2 = 1.4\times10^{-31}$g/cm^3 [6], the explanation of the high efficiency of the process of Galaxy formation becomes difficult: The work of Lifshitz [7] has shown that in an expanding Universe the gravitational instability does not produce the exponential growth of small scale, long wavelength density perturbations familiar from Jeans instability, but produces instead a much more modest power-law growth (e.g. linear in the scale-factor in an Einstein-de Sitter model). Furthermore, in a low density Universe on one hand the process of condensation of proto-clouds has ended well in the past, $z \gtrsim 1/(2\sigma_o) \gtrsim 10$ ([8] since later on $\delta\rho/\rho$ remains nearly constant), on the other hand the same process has occurred only recently, for $z < 10$, if clusters of Galaxies condensed first ([9] since at earlier times the clusters overlapped).

The resolution of this specific contradiction as well as of the more general difficulties connected with the process of condensation in an expanding Universe (see [10] for a review) may well lie within the properties of the seed fluctuations that must be postulated anyhow, since a spontaneous statistical fluctuation does not have a chance to be amplified to the non-linear regime. In reality, the ad hoc assumptions will be tested when these embryonic Galaxies or clusters will be detected either as temperature anisotropies of small angular scale in the background radiation, [11], [12], [13], or as redshifted 21-cm lines, [9]. Nevertheless, it is clear that gravitational instability may be quite efficient in carrying on a condensation--however initiated--if simply the Universe has undergone in the recent past some hibernated period of nearly vanishing expansion as it occurs in the Lemaître models. Attention to these models in connection with the formation of Galaxies has been called by Bonnor [14] and Rawson-Harris [15] who observed that, during the quasi-static phases, gravitational instability of density perturbations of long wavelength is restored to the Jeans-like exponential growth. Although not even in this case can a spontaneous statistical fluctuation of galactic mass be amplified to the non-linear regime [16], still this is the situation most favorable to Galaxy formation since the onset of the exponential instability is allowed over a long period which is a large fraction of the lifetime of the Universe, even in the low density case. The parameters of reasonable Lemaître models and the largest possible ratios between the duration of the quasi-static period and the gravitational e-folding time, are evaluated below, with the purpose of showing that such models are entirely plausible.

With the usual meaning of the symbols, the canonical equations governing the evolution of a cosmological model are

$$\left(\frac{\dot{R}}{R}\right)^2 + \frac{kc^2}{R^2} = \frac{8\pi G}{3c^2}\,\varepsilon + \frac{1}{3}\,\Lambda c^2 \,, \qquad (1)$$

$$\frac{\ddot{R}}{R} = - \frac{4\pi G}{3c^2} (\varepsilon + 3p) + \frac{1}{3} \Lambda c^2 \ . \qquad (2)$$

A Lemaître model (k = +1) may be defined by saying that when the gravitational deceleration vanishes,

$$4\pi G\rho_1 = 4\pi G\rho_o (1 + z_1)^3 = \Lambda c^2 \ , \qquad (3)$$

The residual expansion velocity is infinitesimally small,

$$\dot{R}_1 = \beta c \quad , \quad \beta << 1 \ . \qquad (4)$$

(A subscript "1" denotes the values of all quantities at this point while a subscript "o" denotes present values; here the fluid is assumed pressureless and $\varepsilon = \rho c^2$.) Then a convenient expression for the half-life T of the quasi-static phase,

$$|R-R_1| < R_1/2 \quad , \quad |t-t_1| < T,$$

is [6]

$$T = \ell n(1/\beta)/(c\sqrt{\Lambda}) \ . \qquad (5)$$

The model is now completely specified (aside from β) by the assumption that it is of the Lemaître type and by the value of the present density: The observational parameters,

$$\sigma_o = 4\pi G\rho_o/(3H_o^2) \quad , \quad q_o = - (\ddot{R}R/\dot{R}^2)_o \ ,$$

are related by

$$\sigma_o(z_1^3 + 3z_1^2) = 1 \quad , \quad -q_o = 1 + 3\sigma_o z_1 \ ; \qquad (6)$$

then the choice $\sigma_o = 0.05$ yields the solution

$$z_1 = 2 \quad , \quad q_o = - 1.3 \ , \qquad (7)$$

found by Kardashev [17] in connection with the suggestion (see also Petrosian et al [18]; Skhlovsky [19]) that a Lemaître model with a quasi-static phase around redshift 2 explains most naturally the accumulation of quasar absorption redshifts. Unfortunately the recent appearance of larger redshifts speaks against the model (Petrosian [18]), although one may expect that even a line of cosmological redshifts would be considerably broadened by local Doppler effects;

theoretical considerations (Saslaw [19]) suggest however that only non-relativistic velocities may be explained. Alternatively, it may be observed that higher redshifts for the quasi-static phase are implied in (6) by lower values of the present density: For instance, for $\sigma_o = 0.013$,

$$z_1 = 3.5 \quad , \quad q_o = -1.13 \quad ; \tag{8}$$

see also Fig. 1 and notice that for $\rho_o = \rho_{Gal}$, $\sigma_o = 0.014$. Now, Eq. (3) yields

$$\Lambda^{-1/2} \cong 10^{10} \text{ l.y.} \quad , \quad \Lambda \cong 10^{-56} \text{ cm}^{-2} \quad , \tag{9}$$

approximately both for (7) and for (8).

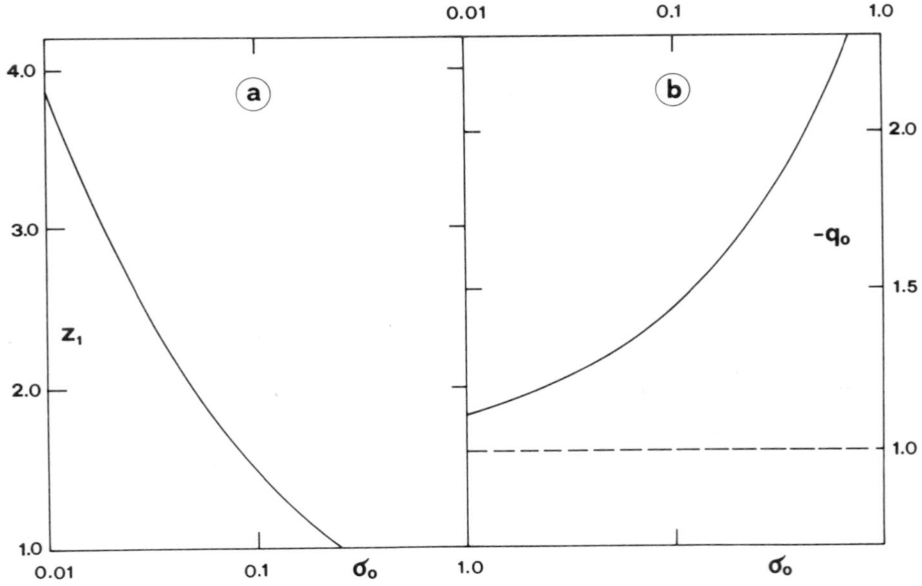

FIG. 1 - Plot of the central redshift z_1 defined by Eq. (3) as a function of σ_o, (a), and of the present value of the deceleration parameter q_o as a function of σ_o, (b). High redshifts are accommodated by the low density models, which approximate the empty de Sitter model.

The linear adiabatic perturbations in the limit of long wavelength and for the matter dominated case, are governed by the standard equation (Peebles [20]).

$$\left(\frac{1}{R^2} \frac{d}{dt} R^2 \frac{d}{dt} - 4\pi G\rho \right) \left(\frac{\delta\rho}{\rho} \right) = 0 \quad , \tag{10}$$

which incidentally is valid in general and not only for the cosmologically flat case, as it can be immediately seen by performing the algebra in the neighborhood of the origin (see Harrison [21], where the wave functions and eigenvalues for the spaces of positive and negative curvature are given). The solution of (10) exhibits the required exponential growth,

$$\delta\rho/\rho \sim \exp(t/\tau) \quad ,$$

around t_1 where $R \cong$ constant $= R_1$ with an e-folding time given by

$$\tau = (4\pi G\rho_1)^{-1/2} = (c\sqrt{\Lambda})^{-1} = 10^{10} y \quad . \tag{11}$$

The numerical estimate, obtained by making use of (9), is approximately valid for $\sigma_o = 0.05$ as well as for $\sigma_o = 0.01$.

From (5) and (11) follows now the convenient expression

$$T/\tau = \ell n(1/\beta) \quad , \tag{12}$$

which shows that the Universe is best engineered when β attains its smallest possible value.

Brecher and Silk [22] observed that the quasi-static phase of a Lemaître model cannot last arbitrarily long because of the destabilizing effect of the formation of Galaxies: In fact, the pressure of the electromagnetic radiation generated by condensing Galaxies yields a negative contribution to $\ddot{R}$, (2), which makes it again negative just after t_1 and changes the weak residual expansion of the Universe into a collapse unless the following obvious condition is satisfied:

$$|\Delta\dot{R}| = \int_{t_1}^{t_1+T} |\ddot{R}| dt \leq \dot{R}_1 = \beta c \quad . \tag{13}$$

From (2) just after t_1 one has

$$\ddot{R} \cong - \frac{4\pi G}{3c^2} (3p_{rad}) R_1 = - \frac{4\pi G}{3c^2} \varepsilon_{rad} R_1 \quad ,$$

where ε_{rad} keeps accumulating in the static Universe; in particular, ε_{rad} may be thought to be generated at a rate τ^{-1} with efficiency η (<<1) from rest-mass energy,

$$\varepsilon_{rad}(t) = \frac{1}{\tau} \eta \rho_1 c^2 (t-t_1) \quad .$$

Putting all together, (13) takes the compact form

$$\eta \, (\ell n \beta)^2 \leq \beta \ . \tag{14}$$

A conservative estimate for β_{min}, the smallest value of β satisfying (14), is obtained by choosing for η the ratio of the gravitational to the rest-mass energy of a Galaxy,

$$\eta \cong (GM^2/L)/(Mc^2) \cong 10^{-6} \quad ;$$

then (14) yields

$$\beta > \beta_{min} \cong 10^{-4} \tag{15}$$

Clearly in this way both η and τ^{-1} have been overestimated since i) not all the pregalactic gas will be condensing at the same time, and ii) the outflow of radiation from a condensing and hot cloud of gas will be slowed by Thomson scattering.

In conclusion, it is not impossible that our Universe spent a long time in a dormant state,

$$2T \cong 2\tau \ \ell n \ (1/\beta_{min}) \cong 20 \ \tau \cong 2x10^{11} \ y \quad ,$$

thus becoming unconventionally old. The advantage of this picture is not as much that of amplifying any initial density condensation by large factors such as $\exp(T/\tau) \cong 1/\beta \stackrel{\sim}{<} 1/\beta_{min}$, but that of allowing the exponential amplification to start nearly at any time over the Universe life. This may be important if (Doroshkevich et al. [23]) Galaxy formation requires the consecutive condensations of different generations of objects and the intermediate heating of the pregalactic gas.

REFERENCES

[1] Sunyaev, R. A. and Zel'dovich, Ya. B., 1969, Comm. Astrophys. Sp. Sci. 1, 159.

[2] Gott, J. R., Gunn, J. E., Schramm, D. N. and Tinsley, B. M., 1974, Astrophys. J., 194, 543.

[3] Sandage, A. and Tammann, G. A., 1974, Astrophys. J., 194, 223 and 559.

[4] Wagoner, R. V., 1973, Astrophys. J., 179, 343.

[5] Wagoner, R. V., 1974, in Confrontation of Cosmological Theories with Observational Data, I.A.U. Symposium No. 63, M. S. Longair, ed., Reidel Publ. Co., Boston, MA.

[6] Weinberg, S., 1972, Gravitation and Cosmology, J. Wiley, New York, N.Y.

[7] Lifshitz, E.M., 1946, J. Phys. USSR, 10, 116.

[8] Rees, M. J. and Sciama, D. W., 1969, Comm. Astrophys. Sp. Sci., 1, 153.

[9] Sunyaev, R. A. and Zel'dovich, Ya. B., 1975, MNRAS, 171, 375.

[10] Harrison, E. R., 1973, in Cargese Lectures in Physics, E. Schatzman, ed., Gordon and Breach, New York, N.Y.

[11] Silk, J., 1968, Astrophys. J., 151, 459.

[12] Longair, M.S. and Rees, M.J., 1973, in Cargese Lectures in Physics, E. Schatzman, ed., Gordon and Breach, New York, N.Y.

[13] Boynton, P.E., 1974, in Confrontation of Cosmological Theories with Observational Data, I.A.U. Symposium No. 63, M. S. Longair, ed., Reidel Publ. Co., Boston, MA.

[14] Bonnor, W. B., 1967, in Relativity Theory and Astrophysics, J. Ehlers, ed., American Mathematical Society, Providence, R.I.

[15] Rawson-Harris, D., 1969, MNRAS, 143, 49.

[16] Rees, M.J., 1971, in General Relativity and Cosmology, R. K. Sachs, ed., Academic Press, New York, N.Y.

[17] Kardashev, N., 1967, Astrophys. J. (Lett.), 150, L 135.

[18] Petrosian, V., 1974, in Confrontation of Cosmological Theories with Observational Data, I.A.U. Symposium No. 63, M. S. Longair, eds., Reidel Publ. Co., Boston, MA.

Petrosian, V. and Salpeter, E. E., 1968, Astrophys. J., 151, 411.

Petrosian, V. Salpeter E., Szekeres, P., 1967, Astrophys. J., 147, 1222.

[19] Saslaw, W. E., 1974, in The Formation and Dynamics of Galaxies, I.A.U. Symposium No. 58, J. R. Shakeshaft, ed., Reidel Publ. Co., Boston, MA.

Shklovsky, J., 1967, Astrophys. J. (Lett.), 150, L 1.

[20] Peebles, P. J. E., 1971, Physical Cosmology, Princeton University Press, Princeton, N.J.

[21] Harrison, E. R., 1967, Rev. Mod. Phys., 39, 862.

[22] Brecher, K. and Silk, J., 1969, Astrophys. J., 158, 91.

[23] Doroshkevich, A. G., Zel'dovich, Ya. B. and Novikov, I.D., 1967, Soviet Astronomy-AJ, 11, 233.

AUTHOR INDEX

SUBJECT INDEX